Handbook of Computational Quantum Chemistry

David B. Cook
Dept. of Chemistry
University of Sheffield

DOVER PUBLICATIONS, INC.
Mineola, New York

To Irene, Steve and Laura,
whose enthusiasm for numerical computing
cannot be underestimated,
and to Ellie; may she continue this tradition.

Copyright

Copyright © 1998, 2005 by David B. Cook
All rights reserved.

Bibliographical Note

This Dover edition, first published in 2005, is an unabridged and slightly revised republication of the work originally published by Oxford University Press in 1998. A new Preface is included in the present edition.

Library of Congress Cataloging-in-Publication Data

Cook, David B.
    Handbook of computational quantum chemistry / David B. Cook.
      p. cm.
    Originally published: Oxford ; New York : Oxford University Press, 1998.
    Includes bibliographical references and index.
    ISBN 0-486-44307-8 (pbk.)
    1. Quantum chemistry—Data processing. I. Title.

QD462.6.D38C66 2005
541'.28'0285—dc22

2005043296

Manufactured in the United States of America
Dover Publications, Inc., 31 East 2nd Street, Mineola, N.Y. 11501

Preface to Dover Edition

It must be the ambition—admitted or not—of any scientist who writes a monograph to have it reprinted by Dover; in my case it has been only moderately secret and I am delighted to have this work reprinted with Dover's excellent quality of production. I bought my first Dover book, Gerhard Herzberg on atomic spectra, as an undergraduate student in 1959 and I am pleased to see that it is still in print; I have been buying them ever since.

For this edition, the text is essential the same as the original; I have removed a few misprints and made slight, basically cosmetic, changes to the section numbering system in those cases where there are multiple appendices to a single chapter. I have, however, changed the format of the whole book from the rather Spartan style of the original. The new format includes, as well as a table of contents in the usual place, a contents list at the start of each chapter which, I think, helps one to use such a large volume of material more smoothly. The ease with which these changes were made is a credit to all the contributors to the LaTeX project.

There has been a great deal of progress in the technical implementation of the method of calculation of the energetics and distributions of electrons in molecules since this work was first written, much of it concentrated on improvements in density functional theory; however, I have not changed the text since it was not my intention to concentrate on particular implementations but to outline the underlying theories of molecular electronic structure, the software tools to implement those theories and the interpretation of the results of the calculations. I hope that this new edition will be useful to research workers and students in these fields.

D. B. Cook, Autumn 2004

Preface

One of the central problems to address in the increasingly mathematical and technical area of computational quantum chemistry is:

> To what extent is it desirable and even possible for a worker in the field to understand the mathematical structure and physical content of the theories being used *and* to be able to understand, maintain and modify the computer-coded implementations of these theories?

At first sight, it may seem reasonable to treat the program systems in much the same way that complex technical instruments are treated by many working scientists; the machines are simply "black boxes" which generate measurements. The key *scientific* point here being that the *science* underlying the measurements (NMR spectrum, Mass spectrum etc.) is understood and it is simply not necessary to be able to actually *build* an NMR machine to perform meaningful NMR measurements.

I am taking the view here that it is essential that there is a theoretically informed and computationally competent community which is as widespread as possible. There are at least three problems associated with this decision:

1. Familiarity with the basic theories, models and approximations involved and, if possible, the establishment of a general and flexible notation and nomenclature.

2. Understanding the basic structures and software tools involved in computer implementations.

3. Dealing with the existing large and complex "production" programs.

This book is mainly concerned with the first two of these problems although the programming techniques used should prove of general utility. Fortunately, there are now available very powerful software tools which help considerably with the enormous organisational tasks associated with maintaining complex software systems:

- The WEB system of programming.[1]
- Revision control.
- Hyper-text documentation (the World-Wide-Web html system); perhaps soon hyper-text WEB programming.

We shall provide introductions to these systems as the complexity of the codes generated grow. Equally important, the question of turning *existing* software into comprehensible systems will be addressed, however tentatively.

One might take the view that the practice of designing and coding the algorithms of computational quantum chemistry is to "reinvent the wheel"; there are already commercially available implementations of some of these models in existence. But I am not *inventing* the wheel, merely trying to help ensure that there are enough wheelwrights around to keep up a supply of novel and useful craftsman-made wheels for the use and enjoyment of everyone who is interested.

I take the view that it is a good thing *in itself* to be able to *build and do* things not simply to be able to use or experience ready-made things. In fact, it is the purpose of education both to be able to understand and therefore *do* things and to enjoy doing them. Obstinately, we still teach people to be able to solve differential equations, to be able to perform chemical experiments, to play music and to write good prose *in spite of the fact that these things have all been done before.*

Technology cannot replace understanding any more than photography makes drawing obsolete. One learns to draw not in order to be able to surpass or even approach photographic reproduction, but in order to *learn to see*. Describing the carvings on the outer archivolt of the facade of St Mark's in Venice, John Ruskin writes:

> Nor can you... know how good it is, *unless you learn to draw*: but some good things concerning it may be seen, by attentive eyes, ...

(my emphasis).[2] He wishes to emphasise that only someone who studies the carving sufficiently intently to *draw* it can possibly give it the attention required to appreciate its quality fully. One learns to implement theories not in order to surpass existing codes, but in order to *understand*. To risk another even earlier quote, this time from Giovanni Vico:

> One truly understands only what one creates

---

[1] Not the World-Wide-WEB (WWW) which is something quite different!
[2] *St Mark's Rest* (George Allen, 1884) p. 100.

PREFACE

If parts of science are split off as the special interest of technical groups both the scientist and the technician lose. The scientist loses "control" of his work; he can only do what the equipment will let him do. The technician becomes fascinated by mere technique and loses sight of the purpose of this technique.

There have been two revolutions in the computer software world; both generated by individuals or small groups; the "Software Tools/Unix" revolution in the 1970s and the Free Software Foundation in the 1980s. It is not now necessary to have to *buy* any software to have the best program development environment there is. All the material in this book has been prepared using Public Domain Software (emacs, FWEB, LaTeX, f2c, g77, RCS and all the Linux packages) and the products of this work are offered in the same spirit.

Computational quantum chemistry is now a very inexpensive area of study and research, and it is absolutely essential that this accessibility is taken advantage of to provide as broad a base of theoretically informed, reflective and computationally enthusiastic scientists as possible, since, notwithstanding the enormous *technical* advances of the past decade, some of the most challenging conceptual and scientific problems of molecular electronic structure theory still remain.

It is a pleasure to thank Patrick Fowler for reading most of the "manuscript" at a point when I thought it was close to completion. His comments both of detail and of principle have been most helpful in making the work much more comprehensible than it might have been. However, quite a lot of work was done since his reading so, even more than usual, I must admit to responsibility for remaining faults.

I am acutely conscious of the shortcomings of the work most of all because it simply did not prove possible to present such a large amount of theory and technique in book form without the whole project becoming too large to handle; the book is just a part of what I had planned. I have tried to make up for some of these deficiencies by keeping the most unwieldy material separate and making it available from the network via anonymous ftp, but I am most uncomfortable with those areas where I have not attempted even a sketchy implementation (STO integrals and VB theory) I apologise to *aficionados* of these areas of work.

Sheffield
October 1997

D.B.C

Contents

1	Mechanics and molecules	1
	1.1 Introduction	1
	1.2 Time-independent Schrödinger equation	5
	1.3 The Born–Oppenheimer model	7
	1.4 The Pauli principle	9
	1.5 The orbital model	11
	1.6 The determinantal method	14
	1.7 Physical interpretation	16
	1.8 Non-determinantal forms	18
	1.9 The variation principle	19
	1.10 Summary	23
1.A	Atomic units	24
1.B	Standard Notation for Quantum Chemistry	27
	1.B.1 Introduction	27
	1.B.2 The Hamiltonian	28
	1.B.3 Many-electron wavefunctions	29

1.B.4	Spin-orbitals	29
1.B.5	Linear expansions for the spatial orbitals	30
1.B.6	Primitive Gaussians	31
1.B.7	Single determinant energy expression	32
1.B.8	Notation for repulsion integrals	33
1.B.9	Spatial orbital repulsion integrals	34
1.B.10	Basis function repulsion integrals	35

2 The Hartree–Fock Method 37

2.1	Introduction	37
2.2	The variational method	38
2.3	The differential Hartree–Fock equation	40
2.4	Canonical form	47
2.5	Orbital energies	49
2.6	Physical interpretation	50
2.7	Direct parametric minimisation	51
2.8	Summary	53

2.A Single-determinant energy expression 54

2.A.1	Introduction	54
2.A.2	The normalisation integral	56
2.A.3	One-electron terms	61
2.A.4	Two-electron terms	65
2.A.5	Summary	70

CONTENTS

3	The matrix SCF equations	75
3.1	Introduction	75
3.2	Notation	77
3.3	The expansion	78
3.4	The energy expression	80
3.5	The numerator: Hamiltonian mean value	81
3.6	The denominator: normalisation condition	84
3.7	The Hartree–Fock equation	86
3.8	"Normalisation": the Lagrangian	87
3.9	Preliminary summary	88
3.10	Some technical manipulations	89
3.11	Canonical orbitals	92
3.12	The total energy	95
3.13	Summary	96
3.A	Atomic orbitals	97
3.B	Charge density	100
3.C	Properties of the J and K matrices	103
3.C.1	Mathematical properties	103
3.C.2	Physical interpretation	106
3.C.3	Supermatrices	107
3.D	An artifact of expansion	108
3.D.1	Lowest state of a given symmetry	108

CONTENTS

3.E Single determinant: choice of orbitals — 110
- 3.E.1 Orthogonal invariance — 110
- 3.E.2 Koopmans' theorem — 112
- 3.E.3 Localised orbitals — 113
- 3.E.4 "Zeroth-order" perturbed orbitals — 113

4 A special case: closed shells — 115
- 4.1 Introduction — 115
- 4.2 Notation for the closed-shell case — 116
- 4.3 Closed-shell expansion — 116
- 4.4 The closed-shell "HF" equation — 117
- 4.5 Closed-shell summary — 121

5 Implementation of the closed-shell case — 123
- 5.1 Preview — 124
- 5.2 Vectors, matrices and arrays — 125
- 5.3 The implementation: getting started — 131
- 5.4 The implementation: repulsion integral access — 148
- 5.5 Building a testbench: conventional SCF — 158
- 5.6 Another testbench: direct SCF — 165
- 5.7 Summary — 173
- 5.8 What next? — 174

5.A Jacobi diagonalisation — 175
- 5.A.1 Introduction — 175
- 5.A.2 The problem — 176
- 5.A.3 The solution — 177
- 5.A.4 Implementation — 179
- 5.A.5 Other diagonalisation methods — 182

CONTENTS

5.B	Orthogonalisation	184
	5.B.1 Introduction	184
	5.B.2 Functions of a matrix	186
	5.B.3 Implementation	187
5.C	getint and data for H_2O	190
5.D	Coding the standard index loops	194
6	Improvements: tools and methods	199
	6.1 Introduction	199
	6.2 Versions: conditional compilation	200
	6.3 Improved diagonalisation	207
	6.4 Simple interpolation	210
	6.5 Improving the formation of G(R)	212
	6.6 Summary	215
7	Molecular integrals: an introduction	217
	7.1 Introduction	218
	7.2 Basis functions	219
	7.3 AOs and atom-centred-functions	221
	7.4 Multi-dimensional integral evaluation	222
	7.5 Molecular integrals over STOs	223
	7.6 Basis functions of convenience	232
	7.7 Gaussian basis functions	234
	7.8 The contraction technique	251

CONTENTS

8 Molecular integrals: implementation — 255
- 8.1 Introduction ... 256
- 8.2 Data structures ... 256
- 8.3 Normalisation ... 260
- 8.4 Overview; the general structure ... 263
- 8.5 Complex code management: the WEB system ... 269
- 8.6 A working WEB ... 276
- 8.7 Some comments on the WEB ... 287
- 8.8 The full integral codes ... 288

8.A Source for the WEB of fmch — 289

9 Repulsion integral storage — 297
- 9.1 Introduction ... 297
- 9.2 A storage algorithm ... 298
- 9.3 Implementation: putint ... 300
- 9.4 A partner for putint; getint ... 307
- 9.5 Conclusion ... 309

10 "Virtual orbitals" — 311
- 10.1 Introduction ... 311
- 10.2 Virtual orbitals in practice ... 312
- 10.3 The virtual space in LCAO ... 317
- 10.4 Conclusions ... 322

10.A Perturbation theory — 323
- 10.A.1 Introduction ... 323
- 10.A.2 Perturbation theory ... 324
- 10.A.3 Perturbation theory for matrix equations ... 329

CONTENTS

11 Choice of tools — 331

- 11.1 Existing software 331
- 11.2 Why ratfor? 335
- 11.3 The Revision Control System: RCS 337

11.A RCS: version control — 339

- 11.A.1 Motivation 339
- 11.A.2 Introduction 340
- 11.A.3 Getting started with RCS 341

12 Open shells: implementing UHF — 343

- 12.1 Introduction 344
- 12.2 Choice of constraints 344
- 12.3 Organising the basis 347
- 12.4 Integrals over the spin-basis 348
- 12.5 Implementation 350
- 12.6 J and K for GUHF 351
- 12.7 The GUHF testbench 357
- 12.8 Interpreting the MO coefficients 360
- 12.9 DODS or GUHF? 363
- 12.10 Version 1 of the SCF code 363
- 12.11 WEB output for function scf 368
- 12.12 Comments 376

12.A WEB Source for the scf code — 377

12.B Blocking the Hartree–Fock matrix — 382

12.B.1 The block form of the HF matrix — 382

12.B.2 Implementation — 384

12.C The Aufbau principle — 394

12.C.3 Introduction — 394

12.C.4 The second variation — 395

12.C.5 Special case: a single excitation — 396

13 Population analysis — 399

13.1 Introduction — 399

13.2 Densities and spin-densities — 400

13.3 Basis representations: charges — 401

13.4 Basis-function analysis — 404

13.5 A cautionary note — 406

13.6 Multi-determinant forms — 407

13.7 Implementation — 408

14 The general MO functional — 409

14.1 A generalisation — 409

14.2 Shells of orbitals — 410

14.3 The variational method — 412

14.4 A single "Hartree–Fock" operator — 416

14.5 Non-orthogonal basis — 419

14.6 Choice of the arbitrary matrices — 420

14.7 Implementation: stacks of matrices — 422

CONTENTS

14.A	Projection operators and SCF	433
	14.A.1 Introduction: optimum single determinant	433
	14.A.2 Alternative SCF conditions	435
	14.A.3 R matrices as projection operators	436
15	Spin-restricted open shell	441
	15.1 Introduction	441
	15.2 The ROHF model	442
	15.3 Implementation	444
	15.4 A WEB for spin-restricted open shell	445
16	Banana skins: unexpected disasters	473
	16.1 Symmetry restrictions	474
	16.2 Anions	475
	16.3 Aufbau exceptions	477
	16.4 Summary	478
17	Molecular symmetry	479
	17.1 Introduction	479
	17.2 Symmetry and the HF method	480
	17.3 Permutational symmetry of the basis	483
	17.4 Implementation	488
	17.5 Permutation symmetry: summary	504
18	Symmetry orbital transformations	505
	18.1 Introduction	505
	18.2 Symmetry-adapted basis	509
	18.3 Generation of symmetry orbitals	512
	18.4 Conclusions	514

CONTENTS

19 A symmetry-adapted SCF method — 515
19.1 Introduction — 515
19.2 Permutations only — 518
19.3 Full implementation; linear combinations — 528
19.4 Summary — 533

19.A Kronecker product notation — 534
19.A.1 Basis transformations — 534
19.A.2 Basis-product transformations — 535
19.A.3 Density matrix transformations — 536
19.A.4 Transformations in the HF matrix — 537
19.A.5 Practice — 539

20 Linear multi-determinant methods — 541
20.1 Correlation and the Hartree–Fock model — 541
20.2 The configuration interaction method — 543
20.3 The valence bond method — 544
20.4 Restricted CI — 545
20.5 Symmetry-restricted CI — 550
20.6 More general CI — 553
20.7 Nesbet's method for large matrices — 554
20.8 "Direct" CI — 560
20.9 Conclusions — 565

20.A The "orthogonal VB" model — 566

20.B DCI matrix elements — 568

CONTENTS

21 The valence bond model — 571
- 21.1 Non-orthogonality in expansions — 571
- 21.2 Spins and spin functions — 573
- 21.3 Spin eigenfunctions and permutations — 576
- 21.4 Spin-free VB theory — 581
- 21.5 Summary — 585

22 Doubly-occupied MCSCF — 587
- 22.1 Introduction: natural orbitals — 588
- 22.2 Paired-excitation MCSCF — 590
- 22.3 Implementation — 595
- 22.4 Partial Paired-Excitations; GVB — 596
- 22.5 Details of GVB — 599
- 22.6 Implementation — 604

23 Interpreting the McWeenyan — 605
- 23.1 Introduction — 605
- 23.2 Stationary points — 607
- 23.3 Many shells — 608
- 23.4 Summary — 610

24 Core potentials — 611
- 24.1 Introduction — 612
- 24.2 Simple orthogonalization — 614
- 24.3 Transforming the Hartree–Fock equation — 615
- 24.4 The pseudopotential — 618
- 24.5 Arbitrariness in the pseudo-orbital — 621
- 24.6 Modelling atomic pseudopotentials — 624
- 24.7 Modelling atomic core potentials — 626
- 24.8 Several valence electrons — 629
- 24.9 Atomic cores in molecules — 633
- 24.10 Summary — 634

CONTENTS

25 Practical core potentials — 637
- 25.1 Introduction — 637
- 25.2 Forms for the core potentials — 638
- 25.3 Core potential integrals — 641
- 25.4 Implementation — 651

26 SCF perturbation theory — 653
- 26.1 Introduction — 654
- 26.2 Two forms for the HF equations — 654
- 26.3 Self-consistent perturbation theory — 657
- 26.4 The method — 659
- 26.5 Conclusions — 667

27 Time-dependent perturbations: RPA — 669
- 27.1 Introduction — 669
- 27.2 Time-dependent Hartree–Fock theory — 670
- 27.3 Oscillatory time-dependent perturbations — 671
- 27.4 Self consistency — 675
- 27.5 Implementation — 676

27.A "Random phase approximation" — 678

27.B Time-dependent variation principle — 680

28 Transitions and stability — 683
- 28.1 Introduction — 683
- 28.2 Transitions — 684
- 28.3 The transition frequencies — 685
- 28.4 Finite perturbations; oscillations — 686
- 28.5 Stability; the time-independent case — 688
- 28.6 Implementation — 689

CONTENTS

29	**Two-electron transformations**	**691**
29.1	Orbital transformations	691
29.2	Strategy	693
29.3	Transformation without sorting	695
29.4	Transformations with sorting	705
29.5	Summary	708
29.A	**A bit of fun: MP2**	**709**
29.A.1	Derivation	709
29.A.2	Implementation	712
30	**Geometry optimisation: derivatives**	**723**
30.1	Introduction	724
30.2	Derivatives and perturbation theory	725
30.3	Derivatives of variational solutions	727
30.4	Parameter-dependent basis functions	729
30.5	The derivative of the SCF energy	730
30.6	Derivatives of molecular integrals	734
30.7	Derivatives of non-variational energies	735
30.8	Higher derivatives	737
30.9	Summary	737
31	**The Semi-empirical approach**	**739**
31.1	Introduction	739
31.2	Use of Coulomb's law	741
31.3	Atomic data	742
31.4	Simulation or calibration?	744
31.5	General conclusions	745

32.1	Introduction	748
32.2	Hohenberg and Kohn's proofs	750
32.3	Kohn–Sham equations: introduction	755
32.4	Kohn–Sham equations	757
32.5	Non-local operators in orbital theories	760

33 Implementing the Kohn–Sham equations 763

33.1	A precursor: The Hartree–Fock–Slater model	764
33.2	Implementation of the Kohn–Sham method	766
33.3	The kinetic energy density	771
33.4	Gradients in the exchange-correlation energy	772
33.5	Numerical integration of densities	773
33.6	Summary	776

34 Semi-numerical methods 779

34.1	Non-variational expansions	779
34.2	The pseudospectral method	782
34.3	The discrete variational method	786

35 Additional reading and other material 791

35.1	Additional reading	791
35.2	Additional material by ftp	794

Handbook of Computational Quantum Chemistry

Chapter 1

Mechanics and molecules

This chapter contains the minimum of quantum-mechanical theory necessary to follow the main thrust of the rest of the work. No attempt is made to develop the theoretical techniques of quantum mechanics, the method used here is the one allegedly used by the ancient Egyptians who, in contrast to the ancient Greeks, knew a lot of rules but did not develop much in the way of general theories.

Contents

1.1	Introduction	1
1.2	Time-independent Schrödinger equation	5
1.3	The Born–Oppenheimer model	7
1.4	The Pauli principle	9
1.5	The orbital model	11
1.6	The determinantal method	14
1.7	Physical interpretation	16
1.8	Non-determinantal forms	18
1.9	The variation principle	19
1.10	Summary	23

1.1 Introduction

Chemistry is *essentially* a quantum-mechanical subject; there can be no satisfactory explanation of the existence, properties and transformations of matter in and between *preferred energy states* in terms of a mechanics which does not have the existence of discrete states at its very centre. We may wish to *visualise* the structure and transformations of matter in terms of properties and processes familiar to us at the macroscopic level, but we must always be aware that these qualitative pictures cannot be given quantitative expression *unless* we use a mechanics which is appropriate to the nature of the materials involved.

This mechanics is, of course, Schrödinger's mechanics and, insofar as chemistry is about electronic processes, Schrödinger's mechanics is capable of giving descriptions of the energies and distributions of the electrons which determine chemical structures and processes of *arbitrary accuracy*; the limitations are, as far as we know, technological rather than theoretical.

In common with other areas of science where quantum mechanics is the main theory, the evolution of a conceptual structure to describe electronic structures at the molecular level has lagged behind the quantitative applications of the theory. In this work we shall be almost solely concerned with the quantitative application of Schrödinger's Mechanics to the problem of the electronic structure of molecules.[1]

We start from Schrödinger's equation for the time development of a probability distribution amplitude for N particles:

$$\hat{H}\Psi(\vec{x}_1, \vec{x}_2, \ldots, \vec{x}_N; t) = i\hbar \frac{\partial}{\partial t} \Psi(\vec{x}_1, \vec{x}_2, \ldots, \vec{x}_N; t) \qquad (1.1)$$

where the *molecular Hamiltonian*, $\hat{H}$, includes all the energy terms which we wish to include in our description of the molecule. In the vast majority of cases, the only interactions between the particles of which the molecule is composed (nuclei and electrons) are those due to Coulomb's law; the electrostatic attractions and repulsions between *point-charges*; our nuclei and electrons have no internal structure.

The physical interpretation of the solutions (Ψ) is that

$$|\Psi|^2 = \Psi^*\Psi$$

is a probability distribution function in the $3N$-dimensional configuration space of the system in the sense that

$$\int_R \Psi^*\Psi d\tau$$

is the probability that the systems coordinates $(\vec{x}_1, \ldots \vec{x}_N)$ be in the region R at time t, where R is contained in the whole coordinate space spanned by $(\vec{x}_1, \ldots \vec{x}_N)$.

---

[1] Here, as elsewhere, "molecule" subsumes atom, ion, radical, a collection of these etc. and simply means a set of positively-charged nuclei and some captive electrons.

1.1. INTRODUCTION

Roughly speaking, for given values of $\vec{x}_i$ and t, $|\Psi(\vec{x}_1, \vec{x}_2, \ldots, \vec{x}_N; t)|^2$ is the probability that, at time t, particle 1 is at $\vec{x}_1$, particle 2 is at $\vec{x}_1$, etc. Clearly, this information (and selected integrals of it) are of the utmost chemical importance. When we are able to examine approximate solutions of the Schrödinger equation we shall extract this all-important information.

For well-understood historical reasons, solutions (Ψ) of the Schrödinger equation are known as *wavefunctions*.

The molecular Hamiltonian will contain operators representing

- Kinetic energy: the kinetic energy operator for a particle of mass m is, in SI units,
$$-\frac{1}{2\hbar^2 m}\nabla^2$$
where
$$\nabla^2 = \sum_{j,l=1}^{3} \frac{1}{\sqrt{g}} \frac{\partial}{\partial q^j}\left(\sqrt{g} g^{jl} \frac{\partial \psi}{\partial q^l}\right)$$
in general coordinates. In Cartesian coordinates, where the metric is trivial, ∇^2 takes a more familiar form since it actually looks like "the square of ∇" :
$$\nabla^2 = \frac{\partial^2}{\partial x^2} + \frac{\partial^2}{\partial y^2} + \frac{\partial^2}{\partial z^2}$$
The molecular Hamiltonian contains a kinetic energy term for each nucleus and electron.
$$\sum_{i=1}^{N} -\frac{1}{2\hbar^2 m_i}\nabla^2(i)$$

- Electrostatic interactions: all the particles in a molecule are charged and Coulomb's law applies to the interactions between all of them so that the total energy of interaction amongst all of them is given by the expression
$$\sum_{i=1}^{N}\sum_{j=1,j>i}^{N} \frac{q_i \times q_j}{4\pi\epsilon_0 |\vec{r}_i - \vec{r}_j|}$$
where $\vec{r}_i$ is the position vector of a typical particle and q_i is its charge (a signed integral multiple of the electronic charge).

- All other interactions between the particles comprising the molecule are, at least for the present, *neglected*. In particular there is no mention in the Hamiltonian of any spin-dependent (magnetic) interactions; both the phenomenological Pauli (two-component) and the relativistic Dirac (four-component) interaction terms amongst electron and nuclei are replaced by the simple electrostatic model.

With this information nothing could be simpler than to write out the molecular Hamiltonian and the associated Schrödinger equation. What is harder is to *solve* this equation. The rest of this work illustrates the concepts and technology involved in coming to grips with the approximate solution of some special forms of this equation in particular cases.

Since we shall always be working with electrons and nuclei it is appropriate and convenient to use a system of units which is adapted to the *size* of our quantities. In SI units, for example, Planck's constant $(2\pi\hbar)$ is tiny (6.626×10^{-34} Js) as is the charge on the proton (1.602×10^{-19} C).

In fact we shall always work in a system of *atomic units* which is described in Appendix 1.A to this chapter. Roughly speaking, choice of Planck's constant divided by 2π ($\hbar$), the charge on the *proton* (to avoid sign absurdities), and the mass of the *electron* as units fixes all mechanical quantities. More controversially, perhaps, choosing $4\pi\epsilon_0$ (the permittivity of the vacuum) as unity simplifies our Hamiltonian since we are always concerned with electrostatic interactions.

These choices give us a system of units in which the molecular electronic quantities are manageable numbers. The (derived) unit of length, for example is a bohr which is about half an Angstrom (0.529×10^{-10} m) and the unit of energy (the hartree) is twice the ionisation energy of a hydrogen atom (13.6 eV or 4.36×10^{-18} J). Not all the units are small on our (macroscopic) scale; the atomic unit of velocity is very large ($c/137$) but that is also realistic for molecular electrons. The details and accurate numerical values are given in Appendix 1.A. With this choice of units the molecular Hamiltonian and associated Schrödinger equation becomes:

$$\left(\sum_{i=1}^{N} -\frac{1}{2m_i}\nabla^2(i) + \sum_{i=1}^{N}\sum_{j=1, j>i}^{N} \frac{q_i \times q_j}{|\vec{r}_i - \vec{r}_j|} \right) \Psi(\vec{x}_1, \vec{x}_2, \ldots, \vec{x}_N; t) = \sqrt{-1}\frac{\partial}{\partial t}\Psi(\vec{x}_1, \vec{x}_2, \ldots, \vec{x}_N; t)$$

where:

- The masses m_i are unity for electrons.

- The charges q_i are -1 for electrons and $+Z_i$ (atomic numbers) for nuclei.

- $\sqrt{-1}$ has been written in explicitly because of the appearance of i as a summation index elsewhere.

This equation is a partial differential equation in $3N+1$ variables[2] which, because of the appearance of the $1/|\vec{r}_i - \vec{r}_j|$ terms in *not separable* into *any* equations of smaller dimension as it stands. In order to get to grips with it we shall

---

[2] "Spin" is considered later.

1.2. TIME-INDEPENDENT SCHRÖDINGER EQUATION

have to make some considerable approximations; these "approximations" will be:

- *Models* of the electronic structure; made in order to transform the equation into a form in which we recognise some of our physical and chemical *concepts* or

- *Numerical* approximations; made to facilitate the technical methods used.

Sometimes, however, a particular technique may well be regarded as either a model or a numerical approximation.

Let us start with the most important restriction of our area of study. For most of our studies we shall be dealing with *isolated* molecules; molecules which are not in interaction with their environment in any way. That is these systems will have their energy conserved; their energies and electron distributions are *independent of time*. This important restriction enables us to remove all time dependence from the Schrödinger equation leaving the wavefunction with only a trivial time dependence which is usually ignored.

1.2 Time-independent Schrödinger equation

The molecular Hamiltonian as we have given it in the last section does not depend explicitly on time and, if we exclude the possibility of interactions with the environment, the Schrödinger equation may be written

$$(\hat{H} - i\frac{\partial}{\partial t})\Psi = 0$$

where the operator on the left is a sum of two terms; one depending only on the spatial coordinates ($\hat{H}$) and one depending only on time ($i(\partial/\partial t) = \hat{T}$, say). In these circumstances the solution must be a *product* of a function of space and a function of time:

$$\Psi(\vec{x}_1, \vec{x}_2, \ldots, \vec{x}_N; t) = \Psi_x(\vec{x}_1, \vec{x}_2, \ldots, \vec{x}_N) \times \Psi_t(t)$$

Substitution of this product into the Schrödinger equation and division by the product gives:

$$\frac{\Psi_t \hat{H} \Psi_x}{\Psi_t \Psi_x} - \frac{\Psi_x \hat{T} \Psi_t}{\Psi_t \Psi_x} = 0$$

since $\hat{H}$ does not act on Ψ_t and $\hat{T}$ has no effect on Ψ_x. Cancellation shows that

$$\frac{\hat{H}\Psi_x}{\Psi_x} - \frac{\hat{T}\Psi_t}{\Psi_t} = 0$$

That is, the sum of a function of *space only* and a function of *time only* must be zero. These two functions must be separately equal to constants which sum to zero:

$$\frac{\hat{H}\Psi_x}{\Psi_x} = E$$

$$-\frac{\hat{T}\Psi_t}{\Psi_t} = -E$$

(say) where E is a constant with the dimensions of energy.

The two equations are to be solved *separately*:

$$\hat{H}\Psi_x(\vec{x}_1, \vec{x}_2, \ldots, \vec{x}_N) = E\Psi_x(\vec{x}_1, \vec{x}_2, \ldots, \vec{x}_N) \tag{1.2}$$

and (reverting to the full form for $\hat{T}$):

$$i\frac{\partial \Psi_t(t)}{\partial t} = E\Psi_t(t)$$

This latter equation has the trivial solution:

$$\Psi_t(t) = A\exp(-iEt)$$

for any E and amplitude A. Thus the full solution of the Schrödinger equation for a time-independent Hamiltonian is

$$\Psi(\vec{x}_1, \vec{x}_2, \ldots, \vec{x}_N; t) = \Psi_x(\vec{x}_1, \vec{x}_2, \ldots, \vec{x}_N) \times A\exp(-iEt)$$

Now, since the *physical interpretation* of the solutions of the Schrödinger equation is of

$$|\Psi|^2 = \Psi^*\Psi = A^2 \Psi_x^* \Psi_x$$

all the physics and chemistry lie in Ψ_x, as the squared modulus of Ψ_t is independent of t. Any solution of the equation for Ψ_x may be multiplied by a constant and still remain a solution so we absorb A into Ψ_x.

In these circumstances we may drop the x subscript on the spatial wavefunction Ψ_x and say that the solutions of the *time-independent Schrödinger equation* (values of E and the associated functions Ψ) generate all the physics and chemistry of the *isolated system*[3] with our assumed molecular (electrostatic) Hamiltonian. We therefore lower our sights a little and seek solutions of

$$\hat{H}\Psi(\vec{x}_1, \vec{x}_2, \ldots, \vec{x}_N) = E\Psi(\vec{x}_1, \vec{x}_2, \ldots, \vec{x}_N)$$

Recall that this is a non-separable partial differential equation in $3N$ spatial variables; of these $3N$ variables the majority are the spatial coordinates of electrons and the remaining minority are the position vector components of nuclei.

---

[3] Of course, if we want to deal with several molecules in mutual interaction or with molecules in interaction with fields, we simply extend out boundaries and include these in the larger isolated system.

1.3. THE BORN–OPPENHEIMER MODEL

It is clear that the motions of atomic nuclei should be capable of a different treatment from that used to describe electrons. Also, it should be possible to separate (at least approximately) the motions of the molecule *as a whole* (translation and rotation) from its *internal* motions (nuclear vibrations and electronic motion). So much is, perhaps, obvious conceptually but the Schrödinger equation is a *differential equation* it does not know any concepts; we must impose our concepts on the equation.

1.3 The Born–Oppenheimer model

Of the $3N$ spatial variables describing the motions of the N electrons and nuclei, it will always be possible (for a non-linear molecule) to regard six as describing the motions of the molecule *as a whole*. These six, if written out explicitly, will represent the translations of the molecular centre of mass and rotations of the molecule about axes containing the centre of mass. This may be difficult in practice but the concept is clear. The remaining variables describe the *internal* motions of the molecule; those normally pictured as the nuclear vibrations and the motions of the electrons "around" the nuclei. Conventionally, we regard the combined effect of the nuclei as holding the electrons captive in a more-or-less localised region of space, and the effect of the electrons as binding the nuclei to each other in an effective field of force. This model sees the electrons interacting with the nuclei and with each other via Coulomb's law to produce an *effective* law of force between the nuclei which is approximately given by Hooke's law, while the electrons "see" the nuclei and each other still interacting via Coulomb's law unaffected. The two sets of particles affect each other's behaviour in *qualitatively* different ways.

These very familiar concepts may be made more precise by examining the formal properties of the Schrödinger equation for the N electrons and nuclei. The full treatment was first given by Born and Oppenheimer very early in the application of quantum mechanics to molecules, and hinges on writing the total wavefunction for the combined system as a sum of products of wavefunctions for the electrons and for the nuclei. They showed that, in fact, to a high degree of approximation the total wavefunction could be written as a *single product* of a nuclear wavefunction Ψ_{nuc}, which depends only on the nuclear coordinates (generically $\vec{r}_{nuc}$[4] say), and an electronic wavefunction, Ψ_{elec} which depends on the electronic coordinates (generically $\vec{x}_{elec}$) and parametrically on the nuclear coordinates:

$$\Psi(\vec{x}_1, \vec{x}_2, \ldots, \vec{x}_N) = \Psi_{elec}(\vec{x}_{elec}; \vec{r}_{nuc}) \times \Psi_{nuc}(\vec{r}_{nuc})$$

Notice that the nuclear wavefunction has no parametric dependence on the electronic coordinates. Clearly the probability distribution of the nuclei must

---
[4]Using $\vec{r}$ rather than $\vec{x}$ for nuclei since nuclear spin will not feature in our deliberations.

involve the electrons since it is the electrons which hold the nuclei together. This basic asymmetry in the two sets of particles shows itself in the equations which the two wavefunctions satisfy:

$$\left(\hat{H}_{nuc}(\vec{r}_{nuc}) + E_{elec}\right)\Psi_{nuc}(\vec{r}_{nuc}) = E_{nuc}\Psi_{nuc}(\vec{r}_{nuc})$$

and

$$\hat{H}_{elec}\Psi_{elec}(\vec{x}_{elec};\vec{x}_{nuc}) = E_{elec}\Psi_{elec}(\vec{x}_{elec};\vec{x}_{nuc})$$

The *Hamiltonian* for the nuclei depends on the *electronic energy* as a function of nuclei positions, while the *wavefunction* for the electrons depends directly on the nuclear coordinates; the electrons provide an *effective field* for the nuclei but the nuclei are seen *individually* by the electrons.

This difference between the dynamics of the electrons and nuclei is colloquially "derived" in the following way:

1. The nuclei are typically tens of thousands of times heavier than the electrons.
2. The particles comprising the molecule are in equilibrium so that, in the mean, they have similar kinetic energies (equipartition).
3. Thus the ratios of their velocities will be roughly the inverse ratio of their masses ($\approx \sqrt{10000}$).
4. So the electrons see stationary nuclei and the nuclei see electrons in the mean.[5]

The full derivation of the Born–Oppenheimer model is rather more involved and can be found in *Dynamical Theory of Crystal Lattices* by M. Born and K. Huang (OUP, 1954).

In detail the electronic Hamiltonian is precisely what one might have anticipated, containing terms for the kinetic energy of each electron, its attraction to each nucleus and the mutual repulsions between the electrons:

$$\hat{H}_{elec} = \sum_{i=1}^{n} -\frac{1}{2}\nabla_i^2 + \sum_{i=1}^{n}\sum_{K=n+1}^{N} \frac{-Z_K}{|\vec{r}_i - \vec{r}_K|} + \sum_{i,j<i=1}^{n} \frac{1}{|\vec{r}_i - \vec{r}_j|}$$

where it has been assumed that the first n of the coordinates are those of electrons and the remaining $(N - n)$ are nuclear coordinates.

From now on we shall only be concerned with the energies and distributions of the n electrons and therefore drop the superfluous subscript on Ψ.

---

[5]In fact, a method completely analogous to the Hartree–Fock method developed in the next chapter may be used for the Hartree product of the two wavefunctions but this would be purely formal and take no account of the *qualitative* differences between the two sets of particles.

1.4 The Pauli principle

Now that we have an equation for the energies and distributions of the n *electrons* in a molecule we may investigate some of the properties of the solutions of this equation. Perhaps the most important constraint that we must place on the solutions of the equation is that due to the consequences of the fact that electrons (unlike nuclei) are *always all the same*; the wavefunction of a many-electron system must reflect the fact that electrons are indistinguishable.

One thing is elementary; the electrons do not know that we have labelled them. If we use $\vec{x}_1$ to describe the coordinates of one electron and $\vec{x}_2$ for another then it must be the case that

$$|\Psi(\vec{x}_1, \vec{x}_2, \ldots, \vec{x}_n)|^2 = |\Psi(\vec{x}_2, \vec{x}_1, \ldots, \vec{x}_n)|^2$$

The probability distribution of the electrons cannot possibly depend on the way that we have labelled them; of course the same must be true for any possible permutation of electronic coordinates. In fact, if $\hat{P}$ is any one of the $n!$ permutations of the n electronic coordinates then

$$\hat{P}|\Psi|^2 = |\Psi|^2$$

However, we are seeking solutions Ψ of the Schrödinger equation not $|\Psi|^2$ and there are just two possibilities:

$$\hat{P}\Psi = \Psi$$

and

$$\hat{P}\Psi = -\Psi$$

for any particular $\hat{P}$.

Since any permutation may be broken down into a sequence of *transpositions* the key question is the value (sign) of a (say) in

$$\hat{P}\Psi(\vec{x}_1, \vec{x}_2, \ldots, \vec{x}_n) = a\Psi(\vec{x}_2, \vec{x}_1, \ldots, \vec{x}_n)$$

There is no way of determining the value of a except by its consequences; we simply assert that for electrons

$$a = -1$$

for a simple exchange of electronic coordinates.

In general, therefore for an arbitrary permutation $\hat{P}$

$$\hat{P}\Psi = (-1)^p \Psi$$

where p is the *parity* of the permutation; the number of simple interchanges to which the permutation may be reduced (this *number* is unique).

We may now state the general result in words:

An acceptable wavefunction for many electrons must be *antisymmetric* with respect to the exchange of the coordinates of any two electrons.

This is, perhaps, the most general statement of the *Pauli Principle* for electrons.

The relevance of the Pauli principle to the computation of molecular electronic structure is enormous. The complications brought about by the antisymmetry requirement itself look rather trivial at first sight, for example for two *non-interacting* electrons (if such a system were possible) the fact that the Schrödinger equation is completely separable would generate a product wavefunction for the system:

$$\Psi(\vec{x}_1, \vec{x}_2) = \psi_1(\vec{x}_1)\psi_2(\vec{x}_2)$$

but this product is not antisymmetric with respect to exchange of the two coordinates. This is simply rectified, the function

$$\Psi(\vec{x}_1, \vec{x}_2) = (\psi_1(\vec{x}_1)\psi_2(\vec{x}_2) - \psi_1(\vec{x}_2)\psi_2(\vec{x}_1))$$

is antisymmetric and also satisfies the relevant Schrödinger equation.

This generates some technical difficulties by increasing the number of terms in the wavefunction.

The real complications arise when we consider what is meant by the "coordinates" of the electrons. We have, so far, not included anything in our description of the electrons to account for electron "spin" on the entirely reasonable grounds that, since there is no mention of electron spin in the Hamiltonian the spin should have as little relevance to the energy of the system as any other coordinate which does not occur in the Hamiltonian. However, the Pauli principle applies to exchange of *all* the coordinates of an electron; space *and* spin. So, in order to comply with the Pauli principle in full we have to bring electron spin into our theory, however formally.

To see the relevance of the inclusion of spin it is only necessary to recall the two-electron case. Since, in this simple case separation of the spatial and spin factors in the wavefunction into a simple product is possible, it is quite possible to have a wavefunction which is *symmetric* with respect to exchange of the spatial coordinates *provided* that it is multiplied by a spin factor which is antisymmetric with respect to exchange. Unless we explicitly include a coordinate for electron spin in our wavefunctions we cannot satisfy the Pauli principle. This leads us into the strange position that, even though the Hamiltonian contains no spin-dependent terms, the total energy of the system will, in general, depend on the spin state of the molecule.

For the moment, since we are unencumbered by practical considerations, we can give a formal solution to the problem by simply extending the scope of the

1.5. THE ORBITAL MODEL

definition of our electronic coordinates $\vec{x}_i$. Originally, each of them was a triple of numbers; the values of the three *spatial* coordinates of each electron (*e.g.* $\vec{x}_i = (x_i, y_i, z_i)$). We now use s_i to be the "spin coordinate" of particle i (deferring, for the time being the question of the spin-dependence of the wavefunction) so that our new definition for $\vec{x}_i$ is

$$\vec{x}_i = (x_i, y_i, z_i; s_i) = (\vec{r}_i; s_i)$$

where the (genuine) vector $\vec{r}_i$ is the essential spatial variable. With this change in the meaning of our notation nothing else need be changed. However, this brief formal introduction is at an end we must now begin to think about the practical tasks involved in the approximate solution of the Schrödinger equation. We can start with the simplest case, the one-electron system.

1.5 The orbital model

There are a few one-particle systems for which the Schrödinger equation is completely soluble. The ones of greatest chemical interest are:

- The (fixed-nucleus) hydrogen-like atom
- The harmonic oscillator

and full solutions for these systems may be found in standard texts.[6] For our purposes it is only necessary to note that the solutions of these one-electron equations (*orbitals*) are available.

Let us assume that we have access to the set of all solutions of the one-electron Schrödinger equation

$$\hat{h}\phi_i = \epsilon_i \phi_i$$

Mathematically, these solutions are the eigenvalues (ϵ_i) and eigenfunctions (ϕ_i) of a Hermitian operator ($\hat{h}$) and, as such, they have several important properties. One of these properties is that they are *complete*; any function of ordinary three-space (the coordinates of a single electron) with sufficiently similar boundary conditions can be expanded as a *linear* combination of these functions. That is, any function $f(\vec{r})$ can be written *exactly* as

$$f(\vec{r}) = \sum_{i=1}^{\infty} c_i \phi_i(\vec{r})$$

---

[6]The Schrödinger equation for the one-electron diatomic (homo- and hetero-nuclear) is also soluble but the solutions are not of wide applicability essentially because the solutions are numerical in one of the three dimensions.

This is a very powerful result since it means, in principle, that any one-electron Schrödinger equation may be solved by evaluating a set of *numerical coefficients*; that is *a partial differential equation may be replaced by an ordinary, algebraic, equation.*

There is no question of an antisymmetry requirement here since we are only dealing with a single electron but, in anticipation of future generalisations, let us extend this result by introducing an (algebraic) "spin space" of two "functions" $\alpha(s)$ and $\beta(s)$ which represent the two possible directions of electron spin:

$$\hat{S}^2 \alpha(s) = \frac{1}{2}(\frac{1}{2}+1)\alpha(s)$$
$$\hat{S}_z \alpha(s) = \frac{1}{2}\alpha(s)$$
$$\hat{S}^2 \beta(s) = \frac{1}{2}(\frac{1}{2}+1)\beta(s)$$
$$\hat{S}_z \beta(s) = -\frac{1}{2}\beta(s)$$

where the formal "spin operators" $\hat{S}^2$ and $\hat{S}_z$ have the mathematical properties of angular momentum operators.

Since we are assuming no spin-space interactions in the Hamiltonians the total wavefunction for a single spinning electron, the *spin-orbital*, is a simple product of the spatial part and a spin factor α or β. Both of these functions have the same energy of course because of the form of the Hamiltonian.

From now on we will have occasion to use these spin-orbitals which we denote by the generic symbol φ_k to distinguish them from the spatial orbitals ϕ_i. The two sets are related by the product rule:

$$\varphi_k(\vec{x}) = \phi_i(\vec{r}) \times \begin{cases} \alpha(s) \\ \text{or } \beta(s) \end{cases}$$

Now suppose that we wish to solve a *two-electron* Schrödinger equation. The required function is, say, $\Psi(\vec{x}_1, \vec{x}_2)$. We can get an idea of how to approach the problem by *fixing* the value of one of the coordinate sets ($\vec{x}_2 = \vec{a}$, say) and expanding the resulting function of the single variable $\vec{x}_1$ in terms of a complete set of *orbitals*:

$$\Psi(\vec{x}_1, \vec{a}) = \sum_{i=1}^{\infty} c_i(\vec{a}) \varphi_i(\vec{x}_1)$$

where the expansion coefficients clearly must depend on the actual value ($\vec{a}$) which the second coordinate ($\vec{x}_2$) takes. This has been emphasised in the notation by showing the coefficients as functions of $\vec{a}$. Obviously, since the two $\vec{x}_i$ are both electronic coordinates we could make a similar expansion if $\vec{x}_1$ were held fixed.

1.5. THE ORBITAL MODEL

The *coefficients* $c_i(\vec{a})$ are themselves functions of an electronic coordinate by assumption; different values of $\vec{a}$ will give rise to different functions $\Psi(\vec{x}_1, \vec{a})$. So that we may expand these coefficients too, which are functions of $\vec{x}_2$, in terms of our complete set of spin-orbitals:

$$c_i(\vec{x}_2) = \sum_{j=1}^{\infty} d_j^i \varphi(\vec{x}_2)$$

The coefficients d_j^i in this expansion comprise a potentially infinite set for *each* of the infinite set of c_i but they are now *numerical* coefficients, independent of the value of $\vec{x}_2$ and, by construction, they are independent of the value of $\vec{x}_1$.

These two expansions may be combined to give an expansion for the function of *two* electronic variables:

$$\Psi(\vec{x}_1, \vec{x}_2) = \sum_{i=1}^{\infty} \sum_{j=1}^{\infty} c_{ij} \varphi_i(\vec{x}_1) \varphi_j(\vec{x}_2)$$

where the coefficients have been given a more symmetrical notation:

$$c_{ij} = d_j^i$$

It is clear that this sketch derivation can be extended to more than two electronic coordinates.

It must be emphasised that this is not a *proof*; it is, for example, not obvious that the expansion is *unique* but it is strongly suggestive of a result which can be proved and generalised to functions of any number of electronic coordinates:

> Any well-behaved[7] function of the coordinates of n electrons can be expanded as a linear combination of *products* of the members of a complete set of spin-orbitals (functions of the coordinates of a single electron) taken n at a time.

That is

$$\Psi(\vec{x}_1, \vec{x}_2, \ldots, \vec{x}_n,) = \sum_{i_1, i_2, \ldots, i_n = 1}^{\infty} c_{i_1 i_2 \ldots, i_n} \varphi_{i_1}(\vec{x}_1) \varphi_{i_2}(\vec{x}_2) \ldots \varphi_{i_n}(\vec{x}_n) \qquad (1.3)$$

---

[7]This is deliberately vague; it means "continuous, square integrable, smooth and differentiable at least twice almost everywhere and having all the desirable properties which a many-electron function should have and which the spin-orbitals themselves are required to have".

> The importance of this result for the computation of molecular electronic structures and energetics cannot be over-emphasised; this expansion technique, by concentrating the *spatial* dependence of Ψ in the *spin-orbitals* φ_i, replaces a multi-dimensional partial *differential* equation by a set of ordinary *algebraic* equations for a set of *numbers* (the coefficients $c_{i_1,i_2...,i_n}$).

We have taken no account of the Pauli principle in this expansion; we must now investigate the consequences of the antisymmetry requirement.

1.6 The determinantal method

The Pauli principle states that every many-electron wavefunction must be antisymmetric with respect to exchange of the coordinates of any two electrons, and so in the case of the expansion of a many-electron wavefunction as a sum of products of spin-orbitals we must have:

$$\sum_{i_1,i_2,...,i_n=1}^{\infty} c_{i_1 i_2...,i_n} \varphi_{i_1}(\vec{x}_2) \varphi_{i_2}(\vec{x}_1) \ldots \varphi_{i_n}(\vec{x}_n)$$
$$= -\sum_{i_1,i_2,...,i_n=1}^{\infty} c_{i_2 i_1...,i_n} \varphi_{i_1}(\vec{x}_1) \varphi_{i_2}(\vec{x}_2) \ldots \varphi_{i_n}(\vec{x}_n)$$

where the coordinates of electrons 1 and 2 have been interchanged. Obviously, the spin-orbital *products* do not care about this interchange; they cannot change sign. The *coefficients* $c_{i_1 i_2...,i_n}$ must change sign when any pair of them is interchanged in order that the orbital-product-sum wavefunction satisfy the Pauli principle. By extension, therefore the coefficients must obey a permutation rule in their *subscripts* similar to the permutation rule for electronic coordinates:

$$\hat{P} c_{i_1 i_2...,i_n} = (-1)^p c_{j_1 j_2...,j_n}$$

where $\hat{P}$ permutes $(i_1, i_2, \ldots, i_n)$ to give $(j_1, j_2, \ldots, j_n)$ which is a permutation equivalent to p interchanges.

That is to say *the expansion of Ψ as a sum of products is highly redundant*; many of the "expansion coefficients" are not independent, they are related to each other by (at most) a sign change. The *same* orbital product is multiplied by $n!$ unknown coefficients all of which have to have the same absolute value. A more practical approach is to group together all the *known* quantities and concentrate attention on the *unknown* coefficients which have to be determined.

This can be done, for example, by:

1. Form a spin-orbital product with the spin-orbitals in some standard order (that of increasing subscript, say).

1.6. THE DETERMINANTAL METHOD

2. Collect together those sets of spin-orbital products involving the *same set* of spin-orbitals and that only differ by which spin-orbital is a function of which electron's coordinates.
3. Multiply each of these spin-orbital products by a sign determined by the parity of the permutation that relates each spin-orbital product to the standard one.
4. Form the sum of all these signed products (Φ_K, say).

When this is done these sums of signed products form a non-redundant set for the expansion of any function of the coordinates of n electrons which satisfies the Pauli (antisymmetry) principle.

Thus, in place of the highly redundant expansion in terms of spin-orbital products we have a non-redundant expansion in terms of *antisymmetrised* "products" in which the linear coefficients (D_K, say) are all independent.

$$\Psi(\vec{x}_1, \vec{x}_2, \ldots, \vec{x}_n,) = \sum_{K=1}^{\infty} D_K \Phi_K(\vec{x}_1, \vec{x}_2, \ldots, \vec{x}_n,) \tag{1.4}$$

Of course, there is, in principle, still an infinity of the Φ_K but since they are *independent*, there are considerable technical advantages in their use.

It only remains for us to recognise that the rather laborious procedure outlined above for the construction of the Φ_K from the spin-orbital products is nothing more or less than the definition of a *determinant* of order n whose rows are enumerated by the spin-orbitals φ_i and columns by the electronic coordinates $\vec{x}_j$:

$$\Phi_K(\vec{x}_1, \vec{x}_2, \ldots, \vec{x}_n) = \begin{vmatrix} \varphi_{i_1}(\vec{x}_1) & \varphi_{i_1}(\vec{x}_2) & \ldots & \varphi_{i_1}(\vec{x}_n) \\ \varphi_{i_2}(\vec{x}_1) & \varphi_{i_2}(\vec{x}_2) & \ldots & \varphi_{i_2}(\vec{x}_n) \\ \varphi_{i_3}(\vec{x}_1) & \varphi_{i_3}(\vec{x}_2) & \ldots & \varphi_{i_3}(\vec{x}_n) \\ \ldots & \ldots & \ldots & \ldots \\ \ldots & \ldots & \ldots & \ldots \\ \ldots & \ldots & \ldots & \ldots \\ \varphi_{i_{n-2}}(\vec{x}_1) & \varphi_{i_{n-2}}(\vec{x}_2) & \ldots & \varphi_{i_{n-2}}(\vec{x}_n) \\ \varphi_{i_{n-1}}(\vec{x}_1) & \varphi_{i_{n-1}}(\vec{x}_2) & \ldots & \varphi_{i_{n-1}}(\vec{x}_n) \\ \varphi_{i_n}(\vec{x}_1) & \varphi_{i_n}(\vec{x}_2) & \ldots & \varphi_{i_n}(\vec{x}_n) \end{vmatrix} \tag{1.5}$$

where the subscript K which enumerates the determinants is generated in some systematic way from the subscripts i_k of the orbitals which compose the determinant.

Expansion of the determinant by conventional means quickly shows that this expansion coincides with our recipe above for the generation of antisymmetrised spin-orbital products.

In fact, it is useful to think of the process of obtaining a determinant from a simple spin-orbital product as the result of an *operator* acting on that product. If we define the antisymmetrising operator $\hat{A}$, say, by

$$\hat{A} = \sum_{\hat{P}} (-1)^p \hat{P}$$

where the summation goes over all the $n!$ possible permutations $\hat{P}$ of the coordinates of n particles and p is the parity of each permutation $\hat{P}$, then the determinant Φ may be written in a more compact form as

$$\Phi(\vec{x}_1, \vec{x}_2, \ldots, \vec{x}_n) = \hat{A}\varphi_1(\vec{x}_1)\varphi_2(\vec{x}_2)\varphi_3(\vec{x}_3)\ldots\varphi_n(\vec{x}_n)$$

1.7 Physical interpretation

The justification given for the choice of antisymmetrised products of spin-orbitals as building blocks for molecular wavefunctions has been rather formal; it was simply shown that, if a set of spin-orbitals is available for the expansion of functions of the coordinates of one electron, then determinants of these orbitals form a basis for the expansion of functions of many electrons' coordinates. Now we shall *never* have access to a *complete* set of one-electron functions or, at least, never be able to *use* such a set in practice. We shall always be using some (truncated) incomplete set chosen on the basis of the physics and chemistry of the problem in hand.

With this in mind it is worth asking how *practical* the expansion of a wavefunction in terms of a large number of determinants actually is. This question can only be answered by carrying out detailed calculations but if we try to *interpret* the mathematics of the determinantal method we may get a clue.

It was shown above in obtaining the time-independent Schrödinger equation that, if a Hamiltonian operator is a *sum* of terms which each depend on a coordinate or group of coordinates, then the solutions of the associated Schrödinger equation are simply *products* of the solutions of the Schrödinger equation associated with each term in the Hamiltonian, i.e. if

$$\hat{H} = \sum_{i=1}^{n} \hat{h}(\vec{r}_i)$$

and

$$\hat{h}\varphi_k = \epsilon_k \varphi_k$$

then the solutions of

$$\hat{H}\Psi = E\Psi$$

1.7. PHYSICAL INTERPRETATION

are just products of the φ_k, taken n at a time, with energy E given by the sum of the n associated ϵ_k. Of course, the imposition of the Pauli principle means that the solutions would be *determinants* rather than products but the general idea is still the same.

The physical interpretation of a wavefunction whose square is a simple product of one-electron functions (squares of spin-orbitals) is that the probability distributions of the electrons are just the products of the probability distributions of the separate electrons; there are no *correlations* (conditional probabilities) involved here.[8] But electrons are *charged*, their motions *must* be interdependent, their distributions must be correlated!

Now, the key point here is that this method of solution depends on there being *no interaction terms* in the Hamiltonian. Now the true molecular Hamiltonian looks, not like the above, but like

$$\hat{H} = \sum_{i=1}^{n} \hat{h}(\vec{r}_i) + \sum_{i,j<i=1}^{n} \hat{g}(|\vec{r}_i - \vec{r}_j|)$$

where the $\hat{g}$s are the electron repulsion terms.

Thus, the physical interpretation gives a rather pessimistic look to our scheme; each determinant is attempting a description of a system of strongly interacting electrons by using a product of the distributions of non-interacting electrons. It is therefore *a priori* quite likely that an acceptable expansion of the true wavefunction might contain a rather large number of determinants.

It might be desirable, therefore, to use physical and chemical knowledge to set up a trial form for an approximate wavefunction which is more appropriate to the problem. If we set up such a *model* wavefunction we can ensure that it satisfies the Pauli principle by applying the antisymmetrising operator $\hat{A}$. Notice that, if a wavefunction is *already* antisymmetric, application of $\hat{A}$ does not change it:

$$\hat{A}\Phi = \hat{A}\left(\hat{A}\varphi_1(\vec{x}_1)\varphi_2(\vec{x}_2)\varphi_3(\vec{x}_3)\dots\varphi_n(\vec{x}_n)\right) = \Phi$$

or

$$\hat{A}^2 = \hat{A}$$

which is general and independent of Φ.

Notice also that any antisymmetric function whatsoever of n electronic coordinates may be expanded as a linear combination of determinants, so that, formally speaking, the use of physical or chemical models of molecular electronic

---
[8] When determinants are used, the antisymmetry requirement does actually, surprisingly, impose a degree of conditional probability on the distribution of electrons with like spin, a point which will be relevant later on.

structure which do not generate determinants of orbitals directly may be viewed mathematically as simply a *regrouping* of some of the terms in a determinantal expansion.

We now have a method of attempting to solve the Schrödinger equation for a system of n electrons. We must use the Schrödinger equation somehow to determine the only unknowns in the expansion

$$\Psi = \sum_{K=1}^{\infty} D_K \Phi_K$$

the D_K, since we assume that the spin-orbitals φ_i are *known functions*.

1.8 Non-determinantal forms

The use of spin-orbitals which puts the electronic spin variables on the same formal footing as the spatial variables has enabled a rather simple solution to be given to the problem of satisfying the Pauli principle. But the facts are, nevertheless, that:

- The spatial and spin variables are *very* different from each other.
- The usual molecular Hamiltonian does not depend on the spin "variable" at all.

The determinantal method generates acceptable trial functions which, however, have the spatial and spin variables inextricably entangled; a determinant of n spin-orbitals cannot be written as a product of a function of space and a function of spin, notwithstanding the fact that the wavefunction of each individual electron is such a product.[9] The many-electron analogy of the product form of the simple spin-orbital for a single electron would be to generate wavefunctions in which the spatial and spin factors are developed *separately* and then combined to form a total wavefunction which satisfies the Pauli principle. Experiments with trial forms for three electrons quickly convinces one that a *single-term* is not eligible and more powerful mathematical tools than we have yet to hand are required; we need methods of dealing with the theory of permutations.

The molecular Hamiltonian is unchanged by permutations of the spatial variables and the spin operators are unchanged by permutations of the spin variables. These facts have some consequences for the generation of chemically attractive forms for approximate model wavefunctions which we shall return to in Chapter 21.

---

[9]In fact, antisymmetric functions of *two* electrons *can* be written as a product but that is the largest case.

1.9 The variation principle

Mathematical variation principles have two very general kinds of application in physical theories:

1. Over-arching *theoretical* statements of physical laws from which the differential equations embodying the details of those physical laws may be derived. The Schrödinger equation itself and the boundary conditions which the solutions must satisfy are derived from a variation principle. In this application the variation principle is "above" the differential equations in the theoretical hierarchy.

2. Methods of modelling or approximating the solutions of a given differential equation. Here, the variation principle is "below" the differential equation in the hierarchy.

We shall be concerned with the latter type of application.

So far, we have been concerned to find *a* solution of the many-electron Schrödinger equation, the unstated assumption being that it would be the unique lowest-energy (ground-state) solution which would be sought. In fact any Schrödinger equation has an infinity of solutions

$$\hat{H}\Psi_i = E_i\Psi_i$$

which have the usual properties of the differential equation associated with a self-adjoint operator. In particular, they form an orthonormal set:

$$\int \Psi_i^* \Psi_j d\tau = \delta_{ij}$$

provided each E_i is associated with only one Ψ_i. If any E_i is *degenerate* (has more than one Ψ_i associated with it), then the Ψ_i may not have the orthonormality property "naturally" but orthonormality may be imposed by suitable linear combinations of these Ψ_i.

Although the Ψ_i are all unknown their formal properties ensure that they form a complete set for the expansion of any function of the variables of n electrons in a way completely analogous to the completeness of the orbital solutions of a one-electron Schrödinger equation. We may reverse the procedure used so far and expand a function of the coordinates of n electrons ($\tilde{\Psi}$, say) in terms of these solutions:

$$\tilde{\Psi} = \sum_{j=1}^{\infty} B_j \Psi_j$$

If this general function is regarded as an *approximate* solution of the Schrödinger equation it is of interest to calculate the mean value of the energy associated with $\tilde{\Psi}$:

$$\tilde{E} = \frac{\int \tilde{\Psi}^* \hat{H} \tilde{\Psi} d\tau}{\int \tilde{\Psi}^* \tilde{\Psi} d\tau}$$

where the denominator ensures that the result is independent of the normalisation of $\tilde{\Psi}$. Replacing $\tilde{\Psi}$ by its expansion in terms of the Ψ_j and using the fact that these Ψ_j solve the Schrödinger equation the expression for $\tilde{E}$ becomes:

$$\tilde{E} = \frac{\sum_{j=1}^{\infty} |B_j|^2 E_j}{\sum_{j=1}^{\infty} |B_j|^2}$$

Now subtracting the *lowest* (ground-state) value of the energy (E_1) from $\tilde{E}$ gives

$$(\tilde{E} - E_1) = \frac{1}{\sum_{j=1}^{\infty} |B_j|^2} \times \sum_{j=1}^{\infty} |B_j|^2 (E_j - E_1)$$

Each individual modulus squared is positive and the sum of them is positive. Since E_1 is, by definition, the lowest of the E_j, $(E_j - E_1)$ is also never negative, thus the whole of the right-hand-side of the equation is never negative. We therefore have the result that, *for any function $\tilde{\Psi}$ of the coordinates of n electrons*[10]

$$\tilde{E} = \frac{\int \tilde{\Psi}^* \hat{H} \tilde{\Psi} d\tau}{\int \tilde{\Psi}^* \tilde{\Psi} d\tau} \geq E_1$$

with equality obviously holding if $\tilde{\Psi} = \Psi_1$.

This is a rather surprising result. It means that, whatever function of n electronic coordinates one chooses, the mean value of the Hamiltonian operator is *always greater* than the lowest true energy of the associated Schrödinger equation. Clearly, this result can be made the basis of a very general and very powerful method of finding approximate solutions of any Schrödinger equation for which we can set up the Hamiltonian. For, if $\tilde{E}$ is always *above* the required energy, then if we *minimise* $\tilde{E}$ with respect to trial functions $\tilde{\Psi}$, we can find a solution. Or, at least, a value for the ground-state energy since our derivation does not ensure that the $\tilde{\Psi}$ which minimises $\tilde{E}$ is actually *identical* to Ψ, the true solution. However, since the ground state of any Schrödinger equation is always *non-degenerate*, there can be no other function associated with energy E so the minimising $\tilde{\Psi}$ must be identical to Ψ.

Unfortunately $\tilde{E}$ is a *functional* of $\tilde{\Psi}$ and the standard manipulations involved in the functional minimisation of the expression for $\tilde{E}$ simply lead us

---
[10] With appropriate boundary conditions.

1.9. THE VARIATION PRINCIPLE

back to the Schrödinger equation; we need to make a further step to show progress.

In fact we have already prepared the way in the earlier sections; if we make a specific *type* of systematic approximation to Ψ, that is, we choose a $\tilde{\Psi}$ which is of *known* mathematical form containing some undetermined *parameters*, the expression for $\tilde{E}$ becomes a *function* of these parameters and we know how to find the minimum in a function!

In the spirit of the last section we therefore expand $\tilde{\Psi}$ as a linear combination of a finite number (N) of determinants (Φ_K) composed of known spin-orbitals:

$$\tilde{\Psi}(\vec{x}_1, \vec{x}_2, \ldots, \vec{x}_n,) = \sum_{K=1}^{N} D_K \Phi_K(\vec{x}_1, \vec{x}_2, \ldots, \vec{x}_n,)$$

then evaluate the corresponding $\tilde{E}$ and minimise it with respect to the undetermined parameters, the linear coefficients D_K. The energy expression is easily evaluated:

$$\tilde{E} = \frac{\sum_{K,L=1}^{N} D_K^* D_L \int \Phi_K^* \hat{H} \Phi_L d\tau}{\sum_{K,L=1}^{N} D_K^* D_L \int \Phi_K^* \Phi_L d\tau}$$

where the integrals involving the (known) determinants Φ_K and the (known) Hamiltonian can be evaluated. Using the notation

$$H_{KL} = \int \Phi_K^* \hat{H} \Phi_L d\tau$$

and

$$S_{KL} = \int \Phi_K^* \Phi_L d\tau$$

we have

$$\tilde{E} = \frac{\sum_{K,L=1}^{N} D_K^* D_L H_{KL}}{\sum_{K,L=1}^{N} D_K^* D_L S_{KL}}$$

or

$$\sum_{K,L=1}^{N} D_K^* D_L S_{KL} \tilde{E} = \sum_{K,L=1}^{N} D_K^* D_L H_{KL}$$

Differentiating the expression with respect to each of the D_K and setting the derivatives to zero ensures a turning point in $\tilde{E}$ which we assume to be the absolute minimum.[11] Setting each such derivative to zero gives an equation

$$E \sum_{L=1}^{N} S_{KL} D_L = \sum_{L=1}^{N} H_{KL} D_L$$

---

[11] We also assume that it is, in fact, a minimum and not just any stationary point.

where the minimum value of $\tilde{E}$ has been given the simple symbol E. The whole set of equations may be collected together into a single matrix equation:

$$\begin{pmatrix} H_{11} & H_{12} & \cdots & H_{1N} \\ H_{21} & H_{22} & \cdots & H_{2N} \\ H_{31} & H_{32} & \cdots & H_{3N} \\ \cdots & \cdots & \cdots & \cdots \\ \cdots & \cdots & \cdots & \cdots \\ \cdots & \cdots & \cdots & \cdots \\ H_{N1} & H_{N2} & \cdots & H_{NN} \end{pmatrix} \begin{pmatrix} D_1 \\ D_2 \\ D_3 \\ \vdots \\ D_{N-1} \\ D_N \end{pmatrix} = E \begin{pmatrix} S_{11} & S_{12} & \cdots & S_{1N} \\ S_{21} & S_{22} & \cdots & S_{2N} \\ S_{31} & S_{32} & \cdots & S_{3N} \\ \cdots & \cdots & \cdots & \cdots \\ \cdots & \cdots & \cdots & \cdots \\ \cdots & \cdots & \cdots & \cdots \\ S_{N1} & S_{N2} & \cdots & S_{NN} \end{pmatrix} \begin{pmatrix} D_1 \\ D_2 \\ D_3 \\ \vdots \\ D_{N-1} \\ D_N \end{pmatrix}$$

or, in an obvious notation,

$$\boldsymbol{HD} = E\boldsymbol{SD}$$

which is a matrix problem of a well-researched form for which there are well-developed solution methods.

It must still be pointed out that, although the integrals H_{KL} and S_{KL} are known *in principle* because everything is known about the integrands, they still remain as formidable technical problems since they are integrals of $3N$ spatial variables.[12]

> The variation method is not limited to obtaining optimum linear parameters although this is a common type of parameter. The fact that the variational result was obtained for an *arbitrary* function (guaranteed by the fact that the solutions of the Schrödinger equation are a complete set) means that any approximate function containing parametric dependence of arbitrary non-linear complexity may be optimised by setting the derivatives of the mean value of the Hamiltonian with respect to these parameters to zero, or, indeed, if the evaluation of the derivatives is too time-consuming and complex a task, one may simply minimise the energy as a *function* of the parameters by brute-force numerical methods.

As we shall see, the most common use of the variation method is not to find a set of linear parameters in the determinantal expansion of the wavefunction but to *model* the electronic structure and optimise the parameters contained in the mathematical formulation of that model.

---

[12] Appendix 2.A to chapter 2 shows how these $3N$-dimensional spatial integrals may be reduced to products of three- and six-dimensional integrals. The "spin integrals" are always trivial since $\hat{H}$ contains no spin-dependent terms

1.10 Summary

This has been a rather wide-ranging introductory chapter but its main purpose has been to introduce the so-called *algebraic approximation* or *linear expansion* technique; the use of sets of antisymmetrised products of one-electron functions (orbitals) to form a basis for a systematic attack on the problem of the computation of molecular electronic energetics and structure:

- The molecular Hamiltonian is taken as the standard Born–Oppenheimer, non-relativistic "electrostatic" model.

- Starting with a set of functions of ordinary three-space which are acceptable as one-electron functions, spin-orbitals are formed which are products of these spatial functions with the two "spin functions".

- Determinants of these spin-orbitals are set up, taking the spin-orbitals n at a time.

- The total n-electron wavefunction is approximated as a function which may always be written as a linear combination of these determinants.

- The values of the optimum linear combination coefficients are determined by ordinary, algebraic equations resulting from the application of the variation method to the above approximate function.

In the next chapter we shall make an attack on a specific model, following the method through in detail.

Appendix 1.A

Atomic units

Chemistry and physics often deal with systems of a very different scale from the familiar everyday scale for which the system of SI units was developed; the SI system is basically aimed at the human, "engineering" size of things while chemists may be dealing with tiny lengths and energies at the molecular level and astronomers with vast distances and huge masses.

In these circumstances the occurrence of large powers of ten in any numerical quantities is inconvenient and error-prone and it is usual to use a system of units whose magnitudes are more suitable for the problem in hand. In atomic and molecular physics and chemistry the system of so-called "atomic units" is almost universally used. It consists of the choice of the sizes of enough of the units to be able to fix all the other magnitudes. There are, of course, many ways of making this choice but the most sensible seems to be to use *fundamental constants* as units rather than an essentially arbitrary choice.

The choices are:

- At the atomic and molecular scale we are always dealing with quantum phenomena and it is therefore appropriate to use Planck's constant as one of our units. Planck's constant has the dimensions of "action" (energy×time) or angular momentum and the choice of

$$\hbar = \frac{h}{2\pi}$$

is used for the unit of action or angular momentum.[13]

---

[13] It is commonly said that a theoretician is one who writes Planck's constant as $2\pi\hbar$.

- The mass of the electron (m_e) is a sensible choice of mass unit since, in molecular electronic structure calculations, the electrons are usually the only things which are moving.

- The charges involved in molecular structures are all multiples of the charge on the proton (or electron) but, since the electron is *negatively* charged, the unit of charge is taken to be the charge on the proton (e).

- There is clearly some feeling amongst the theoretical community that any system of units in which the properties of the vacuum are not zero or unity is strange since the atomic unit of permittivity is taken to be

$$\kappa_0 = 4\pi\epsilon_0$$

With this choice of units, for example, the electrostatic interaction between a nucleus of atomic number Z and an electron separated by a distance r is:

$$\frac{-Z}{r}$$

while in SI units this quantity would be

$$\frac{-(Ze)(e)}{4\pi\epsilon_0 r}$$

When the units of action, charge, mass and permittivity are fixed this fixes the sizes of all (non-magnetic) quantities; in particular the units of length and energy are fixed. The unit of length becomes, rather satisfyingly, about half an Ångstrom and the unit of energy is also of the correct "atomic" proportions, being twice the ionisation energy of the ground state of the hydrogen atom.

The derived unit of velocity is very large compared to everyday experience but that is also commensurate with the speeds of electron in atoms and molecules; it is $c/137$ (where c is the speed of light[14]).

A table of the values of some Atomic Units in SI units is given below:

Quantity	Atomic unit	Value in SI	Symbol
mass	electron mass	9.1091×10^{-31} kg	m
charge	proton charge	1.6021×10^{-19} C	e
action	Planck's constant/2π	1.0545×10^{-34} Js	$\hbar$
permittivity	$4\pi\epsilon_0$	1.1127×10^{-10} Fm^{-1}	none
length	radius of 1st Bohr orbit	0.52917×10^{-11} m	a_0
energy	2× Ionisation energy of H	4.3594×10^{-18} J	E_h

[14] Astronomers and some high-energy physicists use the speed of light as a unit; this choice is possible, of course, but the resulting system of units is not compatible with atomic units.

The symbol h for the atomic unit of energy commemorates the pioneering work of Hartree in this field.

Accurate and up-to-date values of these constants can be found in the publications of the National Bureau of Standards and in derivative tabulations.

We shall *not* use the method of "quantity calculus" which had a period of popularity in the experimental chemistry community; writing, for example,

$$\frac{\exp\left[-\alpha(r/a_0 - R_A/a_0)\right]}{a_0^{-3/2}}$$

for a basis function is too cumbersome compared with the perfectly clear

$$\exp\left[-\alpha(r - R_A)\right]$$

standard notation even though the units of r may be a_0 and the units of a basis functions are $a_0^{-3/2}$.

Appendix 1.B

Standard Notation for Quantum Chemistry

There are a very large number of concepts, symbols and quite terse notations used in quantum chemistry. Most of the standard notation used throughout this work is collected here. All of this appendix should be duplicated somewhere in the text, this is merely a collection and, in some sense, a glossary.

Contents

1.B.1	Introduction	27
1.B.2	The Hamiltonian	28
1.B.3	Many-electron wavefunctions	29
1.B.4	Spin-orbitals	29
1.B.5	Linear expansions for the spatial orbitals	30
1.B.6	Primitive Gaussians	31
1.B.7	Single determinant energy expression	32
1.B.8	Notation for repulsion integrals	33
1.B.9	Spatial orbital repulsion integrals	34
1.B.10	Basis function repulsion integrals	35

1.B.1 Introduction

Much of this notation has been encountered already in Chapter 1 while some of it will not make sense until much later, particularly those definitions which

28 APPENDIX 1.B. STANDARD NOTATION FOR QUANTUM CHEMISTRY

deal with particular types of primitive used to expand the basis functions. In every case, the notation is defined where it first occurs in the text, but it is often tiresome to have to search to find the first use of a particular item of notation.

The notation is fairly standard throughout quantum chemistry although there is one particular departure from the *de facto* standard in that the primitive Gaussian functions are called η_i here rather than g_i; this is a return to the early "POLYATOM" usage rather than the "GAUSSIAN" standard. It has the attraction of using Greek symbols for all levels of expansion function used in the linear expansion technique.

1.B.2 The Hamiltonian

The molecular Hamiltonian used in the energy calculations of quantum chemistry is always the standard non-relativistic, Born–Oppenheimer "electrostatic" Hamiltonian, which is, in atomic units:

$$\hat{H}(\vec{r_1}, \vec{r_2}, \ldots, \vec{r_n}) = \sum_{i=1}^{n} \hat{h}(\vec{r_i}) + \sum_{i=1}^{n} \sum_{j<i} \frac{1}{r_{ij}} \qquad (1.\text{B}.6)$$

Here, there are assumed to be n electrons in the molecule and their position vectors are $\vec{r_i}$. Each electron has a *one-electron Hamiltonian* of identical form:

$$\hat{h}(\vec{r_i}) = -\frac{1}{2}\nabla^2(\vec{r_i}) - \sum_{A=1}^{N} \frac{Z_A}{|\vec{r_i} - \vec{r_A}|} \qquad (1.\text{B}.7)$$

where the position vectors of the nuclei are $\vec{r_A}$ and their charges are Z_A, and it is assumed that there are N of them.

The electron repulsion terms are simply the Coulomb repulsions between unit like charges:

$$\frac{1}{r_{ij}} = \frac{1}{|\vec{r_i} - \vec{r_j}|}$$

the summation is over all *distinct* pairs and $i \neq j$ of course.

The associated Schrödinger equation is:

$$\hat{H}(\vec{r_1}, \vec{r_2}, \ldots, \vec{r_n})\Psi(\vec{x_1}, \vec{x_2}, \ldots, \vec{x_n}) = E\Psi(\vec{x_1}, \vec{x_2}, \ldots, \vec{x_n}) \qquad (1.\text{B}.8)$$

in which the many-electron wavefunction $\Psi(\vec{x_1}, \vec{x_2}, \ldots, \vec{x_n})$ depends on the *spatial* and *spin* "coordinates" of the electrons. The collection of three spatial coordinates $(\vec{r_i})$ and one spin variable is written as $\vec{x_i}$. This equation cannot be solved and *ab initio* methods are designed to generate *approximate and model* solutions of eqn (1.B.8) by a variety of variational and perturbation techniques.

1.B.3 Many-electron wavefunctions

The many-electron wavefunction is approximated by a linear combination of **Slater determinants** $\Phi_K(\vec{x}_1, \vec{x}_2, \ldots, \vec{x}_n)$, each of which is the antisymmetrised product of n **spin-orbitals** $\chi(\vec{x}_i)$ depending on the space and spin variables of just one electron; for example a determinant constructed from the first n spin-orbitals $\chi_1 \ldots \chi_n$ is:

$$\Phi(\vec{x}_1, \vec{x}_2, \ldots, \vec{x}_n) = \frac{1}{\sqrt{n!}} \begin{vmatrix} \chi_1(\vec{x}_1) & \chi_1(\vec{x}_2) & \cdots & \chi_1(\vec{x}_n) \\ \chi_2(\vec{x}_1) & \chi_2(\vec{x}_2) & \cdots & \chi_2(\vec{x}_n) \\ \chi_3(\vec{x}_1) & \chi_3(\vec{x}_2) & \cdots & \chi_3(\vec{x}_n) \\ \cdots & \cdots & \cdots & \cdots \\ \cdots & \cdots & \cdots & \cdots \\ \cdots & \cdots & \cdots & \cdots \\ \chi_{n-2}(\vec{x}_1) & \chi_{n-2}(\vec{x}_2) & \cdots & \chi_{n-2}(\vec{x}_n) \\ \chi_{n-1}(\vec{x}_1) & \chi_{n-1}(\vec{x}_2) & \cdots & \chi_{n-1}(\vec{x}_n) \\ \chi_n(\vec{x}_1) & \chi_n(\vec{x}_2) & \cdots & \chi_n(\vec{x}_n) \end{vmatrix}$$

The factor $1/\sqrt{n!}$ normalises Φ if the χ_i are an orthonormal set.

The number and construction of these determinants defines a **model** of molecular electronic structure and the accuracy with which the spin-orbitals may be computed is defined by practical factors in the system. In general

$$\Psi(\vec{x}_1, \vec{x}_2, \ldots, \vec{x}_n) \approx \sum_{K=0}^{M} D_K \Phi_K(\vec{x}_1, \vec{x}_2, \ldots, \vec{x}_n) \qquad (1.B.9)$$

The **Hartree–Fock** model throws all its effort into obtaining the best possible *one term* expansion; $D_0 = 1$, $D_K = 0$ for $K > 0$. The **Configuration Interaction** and **Møller–Plesset** methods improve on this single-term model by extending the expansion using the *virtual* orbitals generated as a byproduct of the Hartree–Fock variational procedure.

1.B.4 Spin-orbitals

There are just two possible spin "functions" conventionally written α and β and usually the generation of spin-orbitals which are *mixtures* of these two functions is not considered. Thus the computational problem is the determination of the **spatial orbitals** ($\psi_{i'}$) which are the spatially dependent factors of the spin-orbitals χ_i:

$$\chi_i(\vec{x}) = \psi_{i'}(\vec{r})\alpha$$
or
$$\chi_i(\vec{x}) = \psi_{i'}(\vec{r})\beta$$

30 APPENDIX 1.B. STANDARD NOTATION FOR QUANTUM CHEMISTRY

In many applications it is useful to have a more compact notation for the relationship between the spin-orbitals and the spatial orbitals since the spin function is just a label; the "bar" and "no bar" notation is used:

$$\psi_i = \psi_i \alpha$$
$$\overline{\psi}_i = \psi_i \beta$$

That is, the notation ψ_i may mean the spatial orbital *or* the α spin-orbital with spatial factor ψ_i according to context. Care must be taken to distinguish the two cases.

1.B.5 Linear expansions for the spatial orbitals

Each spatial molecular orbital ψ_i, a function of ordinary three-dimensional space, is expanded as a linear combination of **basis functions** which are fixed for a particular calculation and are chosen on a variety of theoretical and (mostly) practical grounds. These basis functions ($\phi_k(\vec{r})$) are key elements in the success of any calculation of molecular electronic structure:

$$\psi_i(\vec{r}) = \sum_{k=1}^{m} \phi_k(\vec{r}) C_{ki} \qquad (1.\text{B}.10)$$

where there are m basis functions with which to expand the n optimum molecular orbitals. In view of the relationship between spatial and spin orbitals, not all the n spatial molecular orbitals need be different; sometimes there will be *pairs* which are the same and the associated spin-orbitals only differ in spin factor. The Hartree–Fock variational method optimises the linear coefficients C_{ki} to ensure the best possible (lowest energy) description of the molecular system.

For reasons which are entirely practical, the basis functions are invariably taken to be **Gaussian functions**, functions which have a factor

$$exp(-\alpha|\vec{r}|^2)$$

as part of their functional form. Since the "natural" atomic orbitals have a dependence like $\exp(-\zeta|\vec{r}|)$ there has to be a much longer expansion in terms of Gaussians to ensure that an accurate molecular orbital is computed, typically two or three times the length of a Slater orbital expansion.

The use of Gaussian functions *directly* in eqn (1.B.10) would therefore make excessive requirements of storage for the electron-repulsion integrals and so a compromise is used whereby the length of the explicit expansion in eqn (1.B.10) is restricted by taking the basis functions themselves to be *fixed* linear combination of so-called **Primitive** Gaussians η_j:

$$\phi_k(\vec{r}) = \sum_{j=1}^{n_k} \eta_j(\vec{r}) d_{jk} \qquad (1.\text{B}.11)$$

1.B.6 Primitive Gaussians

The general form of a primitive Gaussian function is usually chosen to be the product of a Cartesian factor and an exponential:

$$\eta(\vec{r}) = N x^\ell y^m z^n exp(-\alpha r^2) \tag{1.B.12}$$

where $r = |\vec{r}|$ and ℓ, m, n are integers which characterise the *type* or *order* of the Gaussian function. N is a numerical factor chosen to *Normalise* the function to unity, clearly depending on α, ℓ, m, and n.

If a Gaussian primitive is expressed in terms of a global coordinate system, the components of the position vector of the centre on which it is based appear in an obvious way:

$$\eta_j(\vec{r}) = N(\ell_j, m_j, n_j; \alpha_j)(x - x_A)^{\ell_j}(y - y_A)^{m_j}(z - z_A)^{n_j} exp(\alpha_j |\vec{r} - \vec{r_A}|^2) \tag{1.B.13}$$

where the explicit dependence of the primitive on the position of its nucleus is given; $\vec{r_A} = (x_A, y_A, z_A)$ is the position vector of centre A. The subscript j serves to identify this particular η_j among the many.

The type of this primitive is given by

$$t = \ell_j + m_j + n_j$$

and a terminology related to the familiar atomic orbitals is used:

t = 0 an *s*-type Gaussian or zeroth-order Gaussian

t = 1 a *p*-type Gaussian or first-order Gaussian

t = 2 a *d*-type Gaussian or second-order Gaussian

t = 3 an *f*-type Gaussian or third-order Gaussian

In the first two cases (*s* and *p*) there is a direct correspondence between the Gaussians and the real atomic orbitals. For d, f and higher Gaussians there are more Cartesian factors of a given type than real atomic orbitals of the corresponding angular momentum (t is equal to the total angular momentum quantum number usually written as ℓ in atomic theory).[15]

---
[15]Which is not, unfortunately, the same as ℓ used above in our definition of the *type* of a Cartesian Gaussian. Such a collision of notation in such a confined area is unfortunate.

This technique of retaining some of the Gaussian primitives in *fixed* linear combinations is called **contraction**. Of course, the calculations using all the primitives still have to be performed, but the advantage gained by using so-called contracted basis functions is that *storage* is saved. The price to be paid for the contraction technique is loss of variational flexibility; only the linear combination coefficients of the *basis functions* are optimised, not the coefficients of each primitive.

1.B.7 Single determinant energy expression

The mean value of the energy of a single determinant of orthonormal (orthogonal and normalised) molecular orbitals

$$\int \chi_i(\vec{x})\chi_j(\vec{x})d\tau = \delta_{ij}$$

is

$$E[\Phi] = \int \Phi^*(\vec{x_1},\vec{x_2},\ldots,\vec{x_n})\hat{H}(\vec{r_1},\vec{r_2},\ldots,\vec{r_n})\Phi(\vec{x_1},\vec{x_2},\ldots,\vec{x_n})d\tau_1 d\tau_2 \ldots d\tau_n \tag{1.B.14}$$

> Integration over spin *and* space variables is denoted by
>
> $$\int \ldots d\tau \quad \text{or} \quad \int \ldots d\tau_1$$
>
> while integration over space *only* is written
>
> $$\int \ldots dV \quad \text{or} \quad \int \ldots dV_1$$
>
> That is:
>
> $$\int \ldots d\tau = \int\int \ldots dV\, ds$$
>
> where s is the spin "variable". That is, dV refers to the three spatial variables in $\vec{r}$ and $d\tau$ to the four variables in $\vec{x}$.

Completion of the integration, by separation and use of the specific form of $\hat{H}$ (eqn (1.B.6)), together with the orthonormality relationships reduces this integral (over $3n$ spatial variables and n spin "variables") to:

$$E = \sum_{i=1}^{n} h_{ii} + \sum_{i=1}^{n}\sum_{j\leq i}^{n}(J_{i,j} - K_{ij}) \tag{1.B.15}$$

1.B.8. NOTATION FOR REPULSION INTEGRALS

Where

$$h_{ii} = \int \chi_i^*(\vec{x})\hat{h}(\vec{r})\chi_i(\vec{x})d\tau$$

$\hat{h}$ is the one-electron Hamiltonian, eqn (1.B.7), and

$$J_{ij} =< ij|ij > = \int d\tau_1 \int d\tau_2 \chi_i^*(\vec{x_1})\chi_j^*(\vec{x_2}) \left(\frac{1}{r_{12}}\right) \chi_i(\vec{x_1})\chi_j(\vec{x_2}) \quad (1.B.16)$$

$$K_{ij} = \int d\tau_1 \int d\tau_2 \chi_i^*(\vec{x_1})\chi_j^*(\vec{x_2}) \left(\frac{1}{r_{12}}\right) \chi_j(\vec{x_1})\chi_i(\vec{x_2}) \quad (1.B.17)$$

The physical interpretation of the terms is straightforward. The integrals h_{ii} are the energies of an electron moving in the attractive field of the nuclei *alone* (i.e. in the absence of the other electrons). The integrals J_{ij} are the mean repulsions between electrons occupying χ_i and χ_j; they are called "Coulomb" terms for this reason. The integrals K_{ij} arise from the antisymmetry of the wavefunction and have no strict classical analogue but their principle function in the energy expression is to cancel out the "self-repulsion" which would be included if they were not present (notice the summation *includes* the term $i = j$). The K_{ij} are called "exchange integrals" because of the way they arise in the mathematics of the expansion of the determinant.

1.B.8 Notation for repulsion integrals

It is usual to write the electron repulsion terms in the single-determinant energy expression as special cases of a more general repulsion integral, e.g.

$$K_{ij} =< ij|ji > = \int d\tau_1 \int d\tau_2 \chi_i^*(\vec{x_1})\chi_j^*(\vec{x_2}) \left(\frac{1}{r_{12}}\right) \chi_j(\vec{x_1})\chi_i(\vec{x_2}) \quad (1.B.18)$$

The notation $< ij|ij >$ and $< ij|ji >$ has been introduced in anticipation of a more general electron repulsion integral $< ij|k\ell >$:

$$< ij|k\ell > = \int d\tau_1 \int d\tau_2 \chi_i^*(\vec{x_1})\chi_j^*(\vec{x_2}) \left(\frac{1}{r_{12}}\right) \chi_k(\vec{x_1})\chi_\ell(\vec{x_2}) \quad (1.B.19)$$

Integrals of this type appear in the energy expression of multi-determinant wave functions in the "cross terms" between different determinants and are included here to define the notation.

The complex-conjugate notation has been given explicitly since, although the primitives and basis functions are obviously all *real*, the expansion coefficients C_{ki} may well be complex in some applications; leading to complex χ_i.

There are a variety of notations for the repulsion integrals either singly or in "standard combinations", each has its own logic and there is no overwhelming reason to choose one or the other. It is sometimes more intuitively acceptable to use the physical interpretation of these integrals as the net repulsion between a charge distribution $\chi_i(\vec{x_1})\chi_k(\vec{x_1})$ and $\chi_j(\vec{x_2})\chi_\ell(\vec{x_2})$ as a justification for the "charge-cloud" notation:

$$(ik, j\ell) = <ij|k\ell> \tag{1.B.20}$$

where *round brackets* and removal of the vertical bar distinguish between the two notations. The charge-cloud notation is particularly useful when the integrals are over *basis functions* not molecular orbitals since these are always real and the various possible permutations of i, j, k, ℓ which do not change the value of $(ik, j\ell)$ are easier to see. In fact, if the functions *are* real then interchanging i and k, or j and ℓ, or the *pairs* (ik) and $(j\ell)$ do not change the value of $(ik, j\ell)$ as can be seen from the definition (eqn (1.B.19)).

Integrals like eqn (1.B.19) usually occur in *pairs* in the evaluation of energy integrals from determinantal wavefunctions and it is often convenient to use a single symbol to mean a Coulomb integral and its corresponding exchange integral although the difference between the two terms is no longer useful if $i \neq k$ and $j \neq \ell$:

$$<ij||k\ell> = <ij|k\ell> - <ij|\ell j> \tag{1.B.21}$$

The double bar serving to indicate that the "exchange" term has been included.

1.B.9 Spatial orbital repulsion integrals

Throughout the previous two sections the electron-repulsion integrals have been expressed in terms of the *spin-orbitals* $\chi_i(\vec{x})$, but, since the "integration" over the spin is trivial, it is always possible to reduce integrals like eqn (1.B.19) to a "genuine" integration over space (six dimensional since there are two particles involved). The electron-repulsion operator $1/r_{12}$ does not involve spin so that the spin integration can always be separated into factors which are zero or one depending on the spin factors involved:

$$\int \alpha^* \alpha \, ds = 1 \tag{1.B.22}$$

$$\int \beta^* \beta \, ds = 1 \tag{1.B.23}$$

$$\int \alpha^* \beta \, ds = 0 \tag{1.B.24}$$

$$\int \beta^* \alpha \, ds = 0 \tag{1.B.25}$$

1.B.10. BASIS FUNCTION REPULSION INTEGRALS

So that any integral $<ij|k\ell>$ which contains spin-orbitals χ_i and χ_k which have *different* spin factors will be *zero* as will any integral with different spin factors in χ_j and χ_ℓ.

Thus, if the spin factors in the pairs χ_i, χ_k and χ_j, χ_ℓ are the *same* (and only then) the integral

$$<ij|k\ell> = \int d\tau_1 \int d\tau_2 \chi_i^*(\vec{x_1}) \chi_j^*(\vec{x_2}) \left(\frac{1}{r_{12}}\right) \chi_k(\vec{x_1}) \chi_\ell(\vec{x_2}) \quad (1.\text{B}.26)$$

reduces to

$$<i'j'|k'\ell'> = \int dV_1 \int dV_2 \psi_{i'}^*(\vec{r_1}) \psi_{j'}^*(\vec{r_2}) \left(\frac{1}{r_{12}}\right) \psi_{k'}(\vec{r_1}) \psi_{\ell'}(\vec{r_2}) \quad (1.\text{B}.27)$$

where $\psi_{i'}$ is the spatial factor in the spin-orbital χ_i etc.

This fact reduces the number of repulsion integrals, particularly exchange integrals.

1.B.10 Basis function repulsion integrals

In the calculation of molecular electronic structure by the basis function expansion method it is necessary to calculate the *molecular orbital* repulsion integrals by calculating the corresponding repulsion integrals involving the *basis functions* and using the linear combination coefficients to generate the *molecular orbital* integrals. That is we must compute the integrals

$$<ij|k\ell> = \int dV_1 \int dV_2 \phi_i^*(\vec{x_1}) \phi_j^*(\vec{x_2}) \left(\frac{1}{r_{12}}\right) \phi_k(\vec{x_1}) \phi_\ell(\vec{x_2}) \quad (1.\text{B}.28)$$

where the same notation $<ij|k\ell>$ has been used to denote an electron repulsion integral over the basis functions as we used in the last section for these integrals over molecular orbitals. This is standard practice and it is usually clear from the context which integral is meant. If there is doubt then the more explicit notation

$$<\phi_i\phi_j|\phi_k\phi_\ell> = \int dV_1 \int dV_2 \phi_i^*(\vec{r_1}) \phi_j^*(\vec{r_2}) \left(\frac{1}{r_{12}}\right) \phi_k(\vec{r_1}) \phi_\ell(\vec{r_2}) \quad (1.\text{B}.29)$$

may be used for the basis-function integrals and

$$<\chi_i\chi_j|\chi_k\chi_\ell> = \int d\tau_1 \int d\tau_2 \chi_i^*(\vec{x_1}) \chi_j^*(\vec{x_2}) \left(\frac{1}{r_{12}}\right) \chi_k(\vec{x_1}) \chi_\ell(\vec{x_2}) \quad (1.\text{B}.30)$$

for the molecular orbital integrals. In practice, it is usually the basis-function integrals which are denoted by the simpler form $<ij|k\ell>$ and, if necessary, the molecular orbital integrals by the more explicit notation.

Notice that, since the basis functions are *spatial* functions the basis-function electron-repulsion integrals involve no spin integration; if one of them is zero it is because of symmetry or simply a numerical accident because the two charge-clouds are very remote.

Chapter 2

The Hartree–Fock Method

The very special case of modelling a many-electron structure by a single determinant is developed here. Since the only freedom within such a single-determinant wavefunction is in the forms of the spin-orbitals of which the determinant is composed, optimisation of this wavefunction naturally involves the optimisation of these orbitals. The equation which determines these optimum spin-orbitals is the Hartree–Fock equation.

Contents

2.1	Introduction	37
2.2	The variational method	38
2.3	The differential Hartree–Fock equation	40
2.4	Canonical form	47
2.5	Orbital energies	49
2.6	Physical interpretation	50
2.7	Direct parametric minimisation	51
2.8	Summary	53

2.1 Introduction

The general expansion method, using a linear combination of several determinants, outlined in the last chapter is far *too* general both in overall structure and in computational demands to form the basis for either a set of concepts for the understanding of molecular electronic structure or for a practical approach to the computation of the energetics and electron distributions of molecules.

The multi-determinant expansion

$$\Psi = \sum_{K=1}^{\infty} D_K \Phi_K$$

even when truncated to a finite number of terms, has neither the intuitive appeal or the computational accessibility of the most common model of molecular electronic structure, the *single-determinant* or Hartree–Fock model.

In the Hartree–Fock model the expansion is replaced by a *single term* and all the effort is thrown into obtaining the *best possible orbitals* within that single determinant. That is, the idea of a "given" set of expansion functions to be used in forming many determinants is replaced by a fixed (one-term) expansion length within which the component orbitals are to be determined by the variational method.

The computational advantages of this approach will emerge as the derivation and computer implementation proceed but the advantages of *physical interpretation* are also enormous. We have seen in Chapter 1 that, from a physical point of view, it is quite useful to stress the *product* nature of the single-determinant wavefunction while keeping in mind the effects of antisymmetry as complicating factors. If, temporarily, we think of the single-determinant approximation as a single-product approximation its physical interpretation is obvious:

> The distributions of the n electrons in a product of n spin-orbitals is simply the *product* of the n separate one-electron distributions $|\chi_i|^2$.

This is precisely the model of the electronic structure of atoms and molecules which we carry around in our heads all the time; one talks about "a $3d$ electron in a transition-metal atom" or "a π electron in ethene" always perhaps aware of the limitations of this type of mode of expression but, equally, in no doubt about the value of these concepts which form a basic picture; other concepts are usually seen as *corrections* to these ideas rather than replacements for them.

Much of this simple interpretation carries over directly from the product to the determinant, with the principle differences appearing at the level of the distributions of *pairs* of electrons.

2.2 The variational method

Using our established notation, the Hartree–Fock wavefunction (Φ) is a single determinant of n one-electron orbitals χ_i, and the best set of χ_i are to be

2.2. THE VARIATIONAL METHOD

determined by the turning points in the Hartree–Fock Energy Functional:

$$E[\Phi] = \frac{\int \Phi^* \hat{H} \Phi d\tau}{\int \Phi^* \Phi d\tau}$$

Since Φ is completely fixed by the component χ_i, it is more meaningful to express E as a functional[1] of the χ_i:

$$E[\chi_i] = \frac{\int \Phi^* \hat{H} \Phi d\tau}{\int \Phi^* \Phi d\tau} \tag{2.1}$$

So long as this is a *functional* variation principle, we can do nothing practical with it, we must *either* use the fact that the *turning points* of $E[\chi_i]$ correspond to sets of optimum orbitals in the Hartree–Fock model and obtain an *equation* for these χ_i *or* minimise the variational expression in some parametrised form directly. The first of these is the most flexible but the second is also widely used.[2]

> Before setting out on the derivation of an equation which, at least in principle, determines the optimum orbitals χ_i we should come to terms with the *existence* of such functions. In fact, it is possible to prove, at least for uncharged systems, that there *are minimising* solutions of (2.1). That is, for neutral species, the existence of a set of orbitals χ_i which generate a minimum in the energy functional is guaranteed.
> In what follows, we shall tend to make the stronger (but unproven) assumption that, for each turning point in E, i.e. every point for which
>
> $$\delta E = 0$$
>
> (for variations in the χ_i) and E is below the ionisation potential of the system there exists a set of orbitals χ_i *which corresponds to an objectively real electronic state of the system.*
> That is, we assume that it is possible to use the Hartree–Fock method to calculate (approximations to) the electronic structure of the ground *and* excited states of atoms and molecules. However, we make no assumption about the possibility of the computation of the electronic structure of unbound states or states degenerate with unbound states ("resonances").

---

[1] Here, as elsewhere in the text, we use the standard (sloppy) technique of using the same symbol for the physical quantity and the dependence of that quantity on its variables; we also use the more reprehensible method of *not changing* the symbols when dependence on a new set of variables is introduced. Thus, E is used to represent the total energy *and* the functional dependence of the total energy on Φ *and* the functional dependence of the energy on the χ_i etc.

[2] Mainly as the basis of the "Floating Spherical Gaussian Orbital model which we will consider later.

2.3 The differential Hartree–Fock equation

We must now use the results of Appendix 2.A for the expression of the numerator and denominator of eqn (2.1) in terms of integrals involving the χ_i explicitly and investigate the behaviour of the resulting expression when the χ_i are subject to arbitrary variations and finally look for the points at which the quotient has a stationary value.

It is a well known theorem that that the value of a determinant is unchanged by unimodular linear transformations among its rows or columns and, of course, the *quotient* eqn (2.1) is unchanged by *any non-singular* linear transformation among the χ_i. In particular, we may choose the χ_i to be an *orthonormal* set and therefore take advantage of the enormous simplification in the expressions for the numerator and denominator of eqn (2.1) in this case:

$$\int \Phi^* \hat{H} \Phi d\tau = \sum_{i=1}^{n} \int \chi_i^* \hat{h} \chi_i d\tau + \frac{1}{2} \sum_{i,j=1}^{n} \{(\chi_i\chi_i, \chi_j\chi_j) - (\chi_i\chi_j, \chi_i\chi_j)\} \quad (2.2)$$

and

$$\int \Phi^* \Phi d\tau = 1$$

But we wish to subject the energy functional $E[\chi_i]$ to *arbitrary* variations in the functions χ_i, to ensure that any turning points are not artifacts of a constrained variational procedure. This means, of course, that when we allow a small change in one of the orbitals (χ_1, say):

$$\chi_1 \to \chi_1 + \delta\chi_1$$

this variation will, in general, *disturb* the orthonormality of the χ_i and invalidate our use of the energy expression in terms of orthonormal orbitals.

There are two ways forward:

- Use the general formulae for the expansion of eqn (2.1) in terms of a non-orthogonal, un-normalised set of orbitals and we can then be sure that our variations may be arbitrary and the turning points are genuine.

- Use some method to *ensure* that the expansion in terms of orthogonal, normalised orbitals is always correct.

The first of these options is really too awful to contemplate, the expressions are so unwieldy that, even though the derivation can be carried through, it is not at all clear if some terms may have been omitted.

The second option looks, at first sight, as if it carries the danger of locating only *some* of the turning points; those for which the allowed variations only form

2.3. THE DIFFERENTIAL HARTREE–FOCK EQUATION

a restricted class which maintain orthogonality of the χ_i. In fact, it *is* possible to restrict variations to those which maintain orthogonality and obtain the correct result because *arbitrary* variations in *all* the χ_i are extremely *redundant* in a variational sense.

The point is, that for any linearly independent set of χ_i whatever, the single-determinant function is identical to an equivalent set of orthonormal orbitals, related to the χ_i by a linear transformation. Thus, if we assume that *at every point in the variational procedure* this has been done; we may at once avoid searching for turning points for which the determinants are identical [3] *and* stay within the orthogonal orbital expression for eqn (2.1). [4]

In summary, the approach we will take will be to restrict ourselves to variations of an *orthonormal* set of orbitals χ_i *which preserve orthonormality*, not because the optimum orbitals which generate turning points in eqn (2.1) must be orthonormal (far from it!) but for our own convenience in the derivation (to ensure the validity of the simple expressions eqn (2.2)). This is easy enough to say but *how* are we to constrain the orbitals to be orthonormal?

The method universally used in constrained variational problems is the method of Lagrangian multipliers. The basic idea is so simple that, on meeting it for the first time, one is immediately suspicious that it works so well.

The idea is

1. Express the required condition as the variation in some functional; in our case this is simply
$$\delta E = 0$$
for all variations $\delta\chi_i$, $\delta\chi_i^*$ in the orbitals and their complex conjugates (note that $\delta\chi_i^*$ means $\delta(\chi_i^*)$ "a variation in the complex conjugate of χ_i" *not* $(\delta\chi_i)^*$ " the complex conjugate of a variation in χ_i", it is these *two* variations which are linearly independent).

2. Express the constraint on the variations itself as a variation in some other functional; again, in our case this is easy since if it is required that
$$N_{ji} = \int \chi_j^* \chi_i d\tau = \delta_{ji}$$

---

[3] For a given choice of χ_i, there are infinitely many identical determinants related by linear transformations amongst these χ_i and clearly searching for turning points in "directions" determined by these transformations is fruitless.

[4] Of course, there are still many identical determinants, related to the determinant of a set of orthogonal χ_i by *unitary* (orthonormality- preserving) transformations; for the moment we will simply live with this. It will be shown later that this freedom of choice *within* orthonormal sets can be used to both simplify the resulting equations *and* define a *unique* set of optimising orbitals.

then this implies
$$\delta N_{ij} = 0$$
for $i, j = 1, n$.

3. Now insist that, since the variations in all the functionals are to be simultaneously zero for variations in the orbitals, any linear combination of these variations must also be zero; in our case

$$a\delta E + \sum_{j=1}^{n} b_{ji}\delta N_{ij} = 0$$

for variations in each χ_i with an analogous expression for each χ_i^*. The multipliers a and b_{ij} are arbitrary (complex) numbers.

In this way the constraints are incorporated into the variation automatically, and the variations on the linear combination may be *arbitrary*.

We may, without loss of generality, divide throughout the above equation by a and, following convention, put $-b_{ij}/a = \epsilon_{ij}$ to give as a "master equation"

$$\delta E - \sum_{j=1}^{n} \epsilon_{ji}\delta N_{ij} = 0 \qquad (2.3)$$

which must hold for *all* the ($2n$) linearly independent variations $\delta\chi_i^*$, $\delta\chi_i$ in the n functions χ_i.

The variation in the orthonormality constraint induced by the variation

$$\chi_i^* \to \chi_i^* + \delta\chi_i^*$$

is obtained from
$$N_{ij} + \delta N_{ij} = \int (\chi_i^* + \delta\chi_i^*)\chi_j d\tau$$

i.e.
$$\delta N_{ij} = \int \delta\chi_i^* \chi_j d\tau = 0 \qquad (2.4)$$

since the original unvaried orbitals are orthonormal
$$\int \chi_j^* \chi_i d\tau = \delta_{ji}$$

Derivation of the changes in E induced by variations $\delta\chi_i^*$ (and $\delta\chi_j$) is a little more involved. The starting point is eqn (2.2); suppose we make a variation $\delta\chi_i^*$

2.3. THE DIFFERENTIAL HARTREE–FOCK EQUATION

in the function χ_i^* in this expression, then if $\chi_i^* \to \chi_i^* + \delta\chi_i^*$ then the variation induced in the one-electron term depending on χ_i^* is

$$\int (\chi_i^* + \delta\chi_i^*)\hat{h}\chi_i d\tau = \int \chi_i^* \hat{h}\chi_i d\tau + \int \delta\chi_i^* \hat{h}\chi_i d\tau$$

that is

$$\delta \int \chi_i^* \hat{h}\chi_i d\tau = \int \delta\chi_i^* \hat{h}\chi_i d\tau$$

The changes induced in the electron-repulsion terms are more numerous since both the terms $(\chi_i\chi_i, \chi_j\chi_j)$ and $(\chi_j\chi_j, \chi_i\chi_i)$ appear, for example in the Coulomb sum. The corresponding terms in the exchange sum are $(\chi_i\chi_j, \chi_j\chi_i)$ and $(\chi_j\chi_i, \chi_i\chi_j)$. In both these cases the two terms are identical and the change in the total electron-repulsion sum is thus

$$\sum_{j=1}^{n}[(\delta\chi_i\chi_i, \chi_j\chi_j) - (\delta\chi_i\chi_j, \chi_j\chi_i)]$$

remembering that complex conjugation is part of the definition of the $(\ldots,\ldots)$ symbol for a repulsion integral. The doubling of the number of terms associated with a particular $\delta\chi_i^*$ exactly cancels the $1/2$ appearing in eqn (2.2).

Thus, the final expression for δE, the first-order change in the orthonormal expression for the energy of a single determinant function is

$$\delta E = \int \delta\chi_i^* \hat{h}\chi_i d\tau + \sum_{j=1}^{n}[(\delta\chi_i\chi_i, \chi_j\chi_j) - (\delta\chi_i\chi_j, \chi_j\chi_i)] \tag{2.5}$$

So that we may combine this expression with the orthonormality constraint eqn (2.4) to give

$$\int \delta\chi_i^* \hat{h}\chi_i d\tau + \sum_{j=1}^{n}[(\delta\chi_i\chi_i, \chi_j\chi_j) - (\delta\chi_i\chi_j, \chi_j\chi_i)] - \sum_{j=1}^{n}\epsilon_{ji}\int \delta\chi_i^* \chi_j d\tau = 0 \tag{2.6}$$

which may be rearranged and cast into expanded notation to display its essential structure as follows:

$$\int \delta\chi_i^*(\vec{x}_1) \ \{ \ \hat{h}\chi_i(\vec{x}_1)$$
$$+ \sum_{j=1}^{n}[\int \chi_i(\vec{x}_1)\frac{1}{|\vec{r}_1 - \vec{r}_2|}\chi_j^*(\vec{x}_2)\chi_j(\vec{x}_2)d\tau_2$$
$$- \int \chi_j(\vec{x}_1)\frac{1}{|\vec{r}_1 - \vec{r}_2|}\chi_i^*(\vec{x}_2)\chi_j(\vec{x}_2)d\tau_2]$$
$$- \sum_{j=1}^{n}\epsilon_{ji}\chi_j(\vec{x}_1) \ \} \ d\tau_1 = 0$$

where the electron-repulsion integrals have been written out in full and the integrations over the coordinates of the two electrons distinguished explicitly.

Now, if our method is correct, we may treat the $\delta\chi_i^*$ as *independent, arbitrary variations*; thus if the integral above is to vanish for a turning point in the energy functional, it must vanish for arbitrary $\delta\chi_i^*$. That is, the factor in the integrand *which multiplies $\delta\chi_i^*$ must vanish*. This generates a differential equation for χ_i: The expression contained in the $\{\ldots\}$ brackets must vanish for a turning point in the Hartree–Fock energy functional.

This expression is far too unwieldy to repeat several times, let us establish some more convenient notation and some physical interpretation of the terms.

The vanishing of this expression generates a *differential equation* because the one-electron operator $\hat{h}$ contains the (Laplacian) kinetic energy operator:

$$\hat{h} = -\frac{1}{2}\nabla^2 - \sum_{A=1}^{N} \frac{Z_A}{|\vec{r} - \vec{R}_A|}$$

The term arising from the variation of the Coulomb repulsion integrals is quite easy to interpret

$$\sum_{j=1}^{n} \int \chi_i(\vec{x_1}) \frac{1}{|\vec{r_1} - \vec{r_2}|} \chi_j^*(\vec{x_2}) \chi_j(\vec{x_2}) d\tau_2 = \chi_i(\vec{x_1}) \left[\sum_{j=1}^{n} \int \frac{1}{|\vec{r_1} - \vec{r_2}|} \chi_j^*(\vec{x_2}) \chi_j(\vec{x_2}) d\tau_2 \right]$$

and the summed expression is just the electrostatic potential arising from the charge distribution of n electrons, each with its own charge distribution

$$\chi_i^*(\vec{x_2}) \chi_i(\vec{x_2}) = |\chi_i(\vec{x_2})|^2$$

to give a total distribution ρ given by

$$\rho(\vec{x_2}) = \sum_{i=1}^{n} |\chi_i(\vec{x_2})|^2$$

We can define a "Coulomb" operator $\hat{\mathcal{J}}$ by its action on any one of the χ_i:

$$\hat{\mathcal{J}} \chi_i(\vec{x_1}) = \left\{ \int \frac{\rho(\vec{x_2})}{|\vec{r_1} - \vec{r_2}|} d\tau_2 \right\} \chi_i(\vec{x_1})$$

The corresponding exchange sum is harder to interpret physically since its action on χ_i is to produce a combination of mixtures of the other spin-orbitals.

However, it may be cast into each of two forms:

1. The exchange sum can be expressed in a way which enables a single operator $\hat{\mathcal{K}}$ to be defined using a permutation operator $\hat{P}(i,j)$ which permutes

2.3. THE DIFFERENTIAL HARTREE–FOCK EQUATION

the *spin-orbitals* χ_i and χ_j which are functions of the *same* coordinate ($\vec{x}_1$ in the case below):

$$\begin{aligned}
\hat{\mathcal{K}}\chi_i(\vec{x}_1) &= \sum_{j=1}^n \int \chi_j(\vec{x}_1) \frac{1}{|\vec{r}_1 - \vec{r}_2|} \chi_i^*(\vec{x}_2)\chi_j(\vec{x}_2) d\tau_2 \\
&= \sum_{j=1}^n \chi_j(\vec{x}_1) \int \frac{1}{|\vec{r}_1 - \vec{r}_2|} \chi_i^*(\vec{x}_2)\chi_j(\vec{x}_2) d\tau_2 \\
&= \sum_{j=1}^n \left\{ \int \frac{1}{|\vec{r}_1 - \vec{r}_2|} \chi_i^*(\vec{x}_2)\chi_j(\vec{x}_2) d\tau_2 \right\} \chi_j(\vec{x}_1) \\
&= \left[\sum_{j=1}^n \left\{ \int \frac{1}{|\vec{r}_1 - \vec{r}_2|} \chi_i^*(\vec{x}_2)\chi_j(\vec{x}_2) d\tau_2 \right\} \hat{P}(i,j) \right] \chi_j(\vec{x}_1)
\end{aligned}$$

That is, removing one layer of brackets as there is no danger of confusion:

$$\hat{\mathcal{K}} = \left\{ \sum_{j=1}^n \int \frac{1}{|\vec{r}_1 - \vec{r}_2|} \chi_i^*(\vec{x}_2)\chi_j(\vec{x}_2) d\tau_2 \hat{P}(i,j) \right\}$$

where the permutation operator $\hat{P}(i,j)$ interchanges *orbitals* χ_i and χ_j in the integration and the resulting operator is formally independent of the χ_i. This gives a "formal" definition of the *exchange operator* but it is difficult to interpret since it is a *non-local* operator; its action on a function produces effects which do not depend just on the function and its infinitesimal neighbourhood. It is easy to see that

$$J_{ii} = \int \chi_i^* \hat{\mathcal{J}} \chi_i d\tau = \int \chi_i^* \hat{\mathcal{K}} \chi_i d\tau = K_{ii}$$

The "diagonal" exchange term exactly cancels the spurious Coulomb "self-interaction" term if *both* Coulomb and exchange summations are complete. This is the real reason why it is possible to generate a *single* equation for *all* the orbitals in the single-determinant wavefunction.[5]

2. In the second interpretation of the exchange term we take the first form and multiply it by unity in the form

$$1 = \frac{\chi_i(\vec{x}_1)}{\chi_i(\vec{x}_1)}$$

(where $\chi_i(\vec{x}_1) \neq 0$) to give a set of n *different* exchange operators:

$$\hat{\mathcal{K}}_i \chi_i(\vec{x}_1) = \sum_{j=1}^n \left\{ \int \frac{1}{|\vec{r}_1 - \vec{r}_2|} \chi_i^*(\vec{x}_2)\chi_j(\vec{x}_2) d\tau_2 \right\} \chi_j(\vec{x}_1) \frac{\chi_i(\vec{x}_1)}{\chi_i(\vec{x}_1)}$$

---

[5]This cancellation would not occur for a simple product wavefunction as there are *no* exchange terms to do the job. In his original work with characteristic robust good sense, Hartree simply ignored the self-repulsion terms as unphysical.

$$= \sum_{j=1}^{n} \left\{ \int \frac{1}{|\vec{r}_1 - \vec{r}_2|} \chi_i^*(\vec{x}_2) \chi_j(\vec{x}_2) d\tau_2 \frac{\chi_j(\vec{x}_1)}{\chi_i(\vec{x}_1)} \right\} \chi_i(\vec{x}_1)$$

Which gives:

$$\hat{\mathcal{K}}_i = \sum_{j=1}^{n} \left\{ \int \frac{1}{|\vec{r}_1 - \vec{r}_2|} \chi_i^*(\vec{x}_2) \chi_j(\vec{x}_2) d\tau_2 \frac{\chi_j(\vec{x}_1)}{\chi_i(\vec{x}_1)} \right\}$$

in which a *local form* for the exchange operator has been recovered at the expense of having to retain a *separate* exchange operator for each spin-orbital. There are obvious formal disadvantages to this local form but, being local and what is more *multiplicative*, it can be (for example) *plotted* for particular spin-orbitals to get a feel for the nature and size of exchange effects.

With the first of these forms, a *single* exchange operator, we may write down the *unique* equation which the optimising orbitals must satisfy in order that there be a turning point in the Hartree–Fock energy functional eqn (2.1):

$$(\hat{h} + \hat{\mathcal{J}} - \hat{\mathcal{K}})\chi_i = \sum_{j=1}^{n} \epsilon_{ji} \chi_j \qquad (2.7)$$

If we investigate the condition that the Hartree–Fock energy functional must be stationary with respect to variations $\delta\chi_i$ in the functions χ_i as well as variations in $\delta\chi_i^*$ we can use the fact that $\hat{h}$ and the repulsion operators $1/|\vec{r}_i - \vec{r}_j|$ are Hermitian:

$$\int \chi_i^*(\hat{h}\chi_j) d\tau = \int \chi_j^*(\hat{h}\chi_i) d\tau$$

This follows because the *variations* $\delta\chi_i^*$ and $\delta\chi_i$ have the same boundary conditions as the functions χ_i, which ensures the Hermiticity of $\hat{h}$ and the repulsion operators are simply multiplicative.

A completely analogous procedure generates an equation which must be obeyed by the χ_i^*:

$$(\hat{h} + \hat{\mathcal{J}} - \hat{\mathcal{K}})\chi_i^* = \sum_{j=1}^{n} \epsilon_{ij} \chi_j^* \qquad (2.8)$$

These two equations determine the orbitals χ_i at a stationary point in the energy functional. However, taking the complex conjugate of the last equation shows it to be identical to eqn (2.7) *if*

$$\epsilon_{ij}^* = \epsilon_{ji}$$

We can therefore collapse the two equations into one simply by insisting that the ϵ_{ij} are the elements of a $n \times n$ *Hermitian* matrix.

2.4. CANONICAL FORM

Writing
$$\hat{h}^F = \hat{h} + \hat{\mathcal{J}} - \hat{\mathcal{K}}$$
the differential Hartree–Fock Equation becomes

$$\hat{h}^F \chi_i = \sum_{j=i}^{n} \epsilon_{ji} \chi_j \qquad (2.9)$$

where
$$\int \chi_i^* \chi_j d\tau = 0$$

for $i, j = 1, n$ (because we required this in the derivation) and so, by multiplying from the left by χ_j^* and integrating,

$$\epsilon_{ji} = \int \chi_i^* \hat{h}^F \chi_j d\tau$$

This clearly fixes the values of the Lagrange multipliers which were introduced as arbitrary parameters at the start of the derivation.

2.4 Canonical form

The mathematical interpretation of the action of the operator $\hat{h}^F$ on one of the orbitals χ_i is straightforward; the n-dimensional function space spanned by the χ_i is *closed* under the action of $\hat{h}^F$; $\hat{h}^F$ sends any one of the functions into a linear combination of the whole set, the combination coefficients being the ϵ_{ij}. The point here is that the ϵ_{ij} are *constants*; once the Hartree–Fock equations have been solved, the ϵ_{ij} are *independent* of the orbitals χ_i. The interpretation of $\hat{h}^F$ as a transformation within the set of χ_i is clearly related to the determinantal *form* of the Hartree–Fock wavefunction; the component orbitals of any single-determinant wavefunction define a function space and any set of n *linearly independent* elements of this function space (formed from non-singular linear transformations of the original set) form an *identical* determinant (apart, possibly, from multiplication by a constant).

In these circumstances, our derivation has chosen one possible specially convenient kind of linearly independent set; an orthonormal set. As we noted at the outset of the derivation, this orthonormal set is not unique, it merely defines a unique *determinant*. The question naturally arises:

> Are there, amongst all these sets of orthonormal χ_i, any particularly interesting sets and, in particular is there any *unique* basis which has some special mathematical and/or physical interpretation and properties?

The easiest way to see that there is a one basis of particular interest is to use matrix notation to make the action of $\hat{h}^F$ somewhat clearer. The possibility of other physically interesting bases will be examined later.

If the n functions solving eqn (2.9) are written as a row matrix

$$\boldsymbol{\chi} = (\chi_1, \chi_2, \chi_3, \ldots, \chi_n)$$

then all n eqns (2.9) may be written concisely as

$$\hat{h}^f \boldsymbol{\chi} = \boldsymbol{\chi} \boldsymbol{\epsilon}$$

Now suppose that a matrix $\boldsymbol{U}$ is unitary (orthonormality preserving)

$$\boldsymbol{U}^\dagger \boldsymbol{U} = \boldsymbol{U} \boldsymbol{U}^\dagger = 1$$

then the above equation may be multiplied from the right by $\boldsymbol{U}$ and have the unit matrix $\boldsymbol{U}\boldsymbol{U}^\dagger$ inserted between the two factors on the right-hand-side without changing the equation:

$$\hat{h}^f \boldsymbol{\chi} \boldsymbol{U} = \boldsymbol{\chi} \boldsymbol{U} \boldsymbol{U}^\dagger \boldsymbol{\epsilon} \boldsymbol{U}$$

which becomes, if

$$\boldsymbol{\chi}' = \boldsymbol{\chi} \boldsymbol{U}$$

and

$$\boldsymbol{\epsilon}' = \boldsymbol{U}^\dagger \boldsymbol{\epsilon} \boldsymbol{U}$$

an equation of identical form in the primed quantities:

$$\hat{h}^f \boldsymbol{\chi}' = \boldsymbol{\chi}' \boldsymbol{\epsilon}'$$

Now ϵ is constrained to be a Hermitian matrix and $\boldsymbol{U}$ is unitary, so we may *choose* $\boldsymbol{U}$ to be that matrix which reduces ϵ to its (unique) diagonal form;

$$\epsilon'_{ij} = \delta_{ij} \epsilon'_i$$

(say). Thus, the n eqns (2.9) become individual *eigenvalue* equations:

$$\hat{h}^F \chi'_i = \epsilon'_i \chi'_i$$

and we can omit the primes and define the standard canonical form of the differential Hartree–Fock equation as

$$\hat{h}^F \chi_i = \epsilon_i \chi_i \tag{2.10}$$

This equation is made deceptively attractive and simple-looking by the extreme contraction of notation for the Hartree–Fock operator $\hat{h}^F$. The fact is that $\hat{h}^F$ *is dependent on the whole set of solutions* χ_i through the electron-repulsion operators $\hat{\mathcal{J}}$ and $\hat{\mathcal{K}}$, making even the eqns (2.10) highly non-linear and incapable of direct solution.

2.5. ORBITAL ENERGIES

> Equations (2.10) are a set of strongly coupled, highly non-linear, pseudo-eigenvalue, three-dimensional partial differential equations for the orbitals determining the stationary points of the Hartree–Fock energy functional. As such, there is not the slightest hope of solving them by analytical *or* numerical methods for any but the simplest atoms and molecules. We shall turn to practical (approximate) methods of solving them in Chapter 3.

2.5 Orbital energies

The Hartree–Fock equation has the look of a Schrödinger equation when written in compact form and the quantities ϵ_i, originally introduced as Lagrangian multipliers, must have the *dimensions* of energy, but what is their physical interpretation? One would like ϵ_i to be the energy of an electron whose orbital is χ_i, in the same way that the energies of the *states* of the hydrogen atom are analogous quantities in a one-electron equation where there is no distinction between state energy and orbital energy.

The easiest way to verify this suspicion is to evaluate the energy of the single-determinant with χ_i removed, using the now-familiar single-determinant energy formula. For simplicity let us assume that we remove the "last" χ_i i.e. χ_n. The total electronic energy of the determinant of $(n-1)$ orbitals (Φ_n^+, say) is just

$$E_n^+ = \int \Phi_n^{+*} \hat{H} \Phi_n^+ d\tau = \sum_{i=1}^{n-1} \int \chi_i^* \hat{h} \chi_i d\tau + \frac{1}{2} \sum_{i,j=1}^{n-1} \{(\chi_i\chi_i, \chi_j\chi_j) - (\chi_i\chi_j, \chi_i\chi_j)\}$$

But

$$\epsilon_n = \int \chi_n^* \hat{h} \chi_n d\tau + \sum_{j=1}^{n} \{(\chi_n\chi_n, \chi_j\chi_j) - (\chi_n\chi_j, \chi_n\chi_j)\}$$

so that

$$E_n^+ = E - \epsilon_n$$

where E is the total energy of the full determinant of n orbitals which was our starting point in the derivation.

Apart from the convenience of expression, there is nothing special about ϵ_n in this result, so we may quote the general result:

$$\epsilon_i = E - E_i^+$$

The ϵ_i are, indeed, the energies associated with the orbitals χ_i in the sense that ϵ_i is the energy required to remove the electron with orbital χ_i from the total n-electron system *maintaining the other orbitals χ_j ($j \neq i$) unchanged.*

Clearly these numbers are approximations to the ionisation energies of the system and, equally clearly, the deeper in energy the electrons in χ_i lie, the more unrealistic will be the assumption that the other orbitals do not relax on removing an electron from χ_i. One might reasonably expect that the highest ϵ_i might be a reasonable approximation to the first ionisation energy.

At first sight, one might expect that the constraint that the orbitals are not allowed to relax might be compensated by some mixing of the "ionised wavefunctions"; the Φ_i^+ might not be pure approximations to the states of the ion, they may mix so that a more reasonable approximation to any ion state might be

$$\Phi^+ = \sum_{i=1}^{n} A_i \Phi_i^+$$

for some numerical coefficients A_i. It is easy to show, using the results of Appendix 2.A, that *all* the integrals

$$\int \Phi_i^{+*} \hat{H} \Phi_j^+ d\tau$$

for $i \neq j$, are *zero* so no mixing actually occurs; a result due to Koopmans. It is this latter result, rather than the rather obvious energy summation, which enables the identification of the ϵ_i with separate approximate ionisation energies.

2.6 Physical interpretation

In using a single determinant form for the variational approximate solution of the Schrödinger equation we have used the only freedom available to optimise the determinant, the forms of the individual orbital of which the determinant is composed. Unlike the full variational principle, our procedure does not allow small changes in the determinant by adding infinitesimal amounts of *determinants* since a sum of determinants is not necessarily a determinant and we would be "outside" our variational choice.

The optimum orbitals have an interesting physical interpretation. Obviously, by construction, they are the components of the best possible single determinant and, as such, they are the solution of a *single-particle* Schrödinger-like equation, so that an examination of the terms in the Hartree–Fock "Hamiltonian" will tell us a lot about their interpretation. What we might call the "parent Hamiltonian" of the HF Hamiltonian — the one used in the single-determinant variational method — induces the appearance of the one-particle kinetic energy and nuclear attraction terms in the HF Hamiltonian; the difference between the parent and the HF lies in the way in which electron repulsion is represented.

2.7. DIRECT PARAMETRIC MINIMISATION

In the full Hamiltonian the electron repulsion is "exact", it is simply the sum of all the inter-particle repulsions. In the HF Hamiltonian this is replaced by Coulomb and exchange terms. The Coulomb term is simply the net *average* repulsion field due to all the electrons in the molecule and the exchange term removes the "self-interaction" term included in this average sum plus some further small corrections.

Thus, in the Hartree–Fock (single-determinant) model of electronic structure each electron only "sees" the other electrons "on the average". The details of the distribution of the electrons which depend on the point-by-point repulsions between the electrons cannot be obtained from the energetics and electron distributions obtained within the HF model.

If, therefore, we have the spatial distributions of two electrons, the distribution of the electron *pair* is just the (antisymmetrised) *product* of these two single-particle distributions; by analogy with the product rule in probability theory we say that the distributions of the particles are *uncorrelated*.[6]

The difference in energy between the Hartree–Fock model and the exact[7] energy is called the correlation energy and the driving force for this change in energy is the improved treatment of electron repulsion in a many-determinant model.

2.7 Direct parametric minimisation

In the previous sections we have generated a differential equation which generates (at least in principle) a set of orbitals which ensure a stationary point in the single-determinant energy functional eqn (2.1). We have proceeded from an algebraic expression to a differential equation; a considerable increase in complexity. There is an alternative approach which does not involve differential equations and attacks the problem of finding the turning points of eqn (2.1) directly. In fact, this method is only practicable for *local minima* in the functional because of the technical difficulties associated with the location of maxima, saddle points etc.

Suppose that we can make a guess at the *form* of the orbitals which minimise eqn (2.1) and further suppose that this functional form of each of the approximate orbitals depends on a number of *parameters*. An obvious example is the

---

[6]The fact that the distribution is the *antisymmetrised* product rather than just the simple product means that the electron distributions are, strictly, correlated but this definition of electron correlation is now standard.

[7]"Exact" here means, of course, an exact solution of the full Schrödinger equation using the parent Hamiltonian, not "experimental".

case of the electronic structure of atoms which are all the same "shape" and there is a familiar "model" for the electronic structure: the hydrogen atom. The parameters on which this model of atomic electronic structure might depend are the effective nuclear charges of each orbital; i.e. replacing the hydrogenic Z/n by ζ_i (say) in the exponential radial factor in each orbital.

Then with these explicitly known functional forms for the orbitals, we can actually *evaluate* the Hartree–Fock energy functional as a *function* of the parameters. If these parameters are collectively called α_i we have

$$E[\Phi] = \frac{\int \Phi^* \hat{H} \Phi d\tau}{\int \Phi^* \Phi d\tau} = E(\alpha_i) \tag{2.11}$$

where the symbol E has been used for the *function* as well as the *functional* and different types of brackets distinguish the two.[8]

There are now two technical methods we can use to find the stationary points of this *parametric* variational expression;

1. If possible, evaluate the derivatives and find the points at which

$$\frac{\partial E(\alpha_i)}{\partial \alpha_k} = 0$$

and, if necessary examine the second derivatives.

2. If the explicit evaluation of the derivatives is too difficult, simply minimise eqn (2.11) by standard numerical techniques.

There are two general types of application of this direct method:

- Applications to *systems* of similar structure: the atomic example is the most familiar one.

- Applications using simple *functions* in different types of system: the Floating Gaussian Orbital method is the best-known of these.

The single most important difficulty with this, as with any other parametric method, is that, once the energy *functional* has been replaced by the parametric energy *function*, a solution always *exists*. Unless one is extremely inept or unlucky, the energy function will always have at least one turning point and one cannot know if this corresponds to some objective electronic state or is simply an artifact of the parametrisation scheme. To give a familiar example, one can quite easily use this method to calculate the energies and electronic structures of atomic anions, even when these energies turn out to be higher than those of the uncharged systems; the model is simply not flexible enough to let the system lose electrons, the parametrisation has not allowed for this fact.

---

[8]Again, without changing the symbol E.

> In practice *almost all* methods of solution of the Hartree–Fock equations are parametric and we must be alert to the trap of the guaranteed existence of "solutions" in all work on the electronic structure of systems with weakly bound electrons. In particular, the "matrix Hartree–Fock" equations developed in the next chapter result from the application of a parametric variation method.

2.8 Summary

The Hartree–Fock (single-determinant) model of the structure of an n-electron system leads to a *single* one-electron equation. All the spin-orbitals which comprise the determinant are solutions of this equation which is a highly non-linear but Hermitian integro-differential equation. There is clearly no hope of obtaining solutions to this equation for systems of chemical interest by classical analytical methods. However, the very existence of this equation is the basis of numerical and algebraic approximation techniques for the generation of spin-orbitals which will provide a quantitative foundation for the majority of our chemical concepts.

The apparent absence of any *boundary conditions* to be imposed on the Hartree–Fock equation is the result of the rather careless derivation. The variation method should determine the solutions completely; it should give an equation satisfied by them in the interior of the relevant space *and* some boundary conditions. The single-determinant energy functional is actually a functional of the spin-orbitals, their complex conjugates *and* the gradients of these functions (or the Laplacian of them). If *these* variations had been included we would have obtained a definite integral which must vanish, generating the required boundary conditions. Since we are not proposing to *solve* the equation as a differential equation, we may impose (implicitly or explicitly) the boundary conditions which we might expect; mainly that the trial functions should vanish at infinity.

Appendix 2.A

Single-determinant energy expression

This appendix uses the most elementary and pedestrian method for the reduction of the many-electron integrals involved in the single-determinant energy expression to integrals involving the coordinates of one and two electrons.

Contents

2.A.1	Introduction	**54**
	2.A.1.1 Vanishing determinants	56
2.A.2	The normalisation integral	**56**
	2.A.2.1 Summary	60
2.A.3	One-electron terms	**61**
2.A.4	Two-electron terms	**65**
2.A.5	Summary	**70**
	2.A.5.1 General case; Løwdin's rules	71
	2.A.5.2 The orthonormal case: Slater's rules	72

2.A.1 Introduction

The expression for the mean value of the energy of the (possibly unnormalised) single-determinant function Φ is

$$E = \frac{\int \Phi^* \hat{H} \Phi d\tau}{\int \Phi^* \Phi d\tau} \tag{2.A.12}$$

2.A.1. INTRODUCTION

In order to be able to use the variation method *in practice* we need to be able to evaluate this expression; i.e. to be able to *reduce* the expression to integrals over the n individual one-electron functions χ_i of which Φ is the antisymmetrised product. In this appendix the most "elementary" possible derivation of this energy expression is given in the sense that no *specifically developed* techniques are used. Only the standard method for the expansion of a determinant is assumed. There are many more advanced methods for the reduction of these many-dimensional integrals to sums and products of integrals of smaller dimension but the elementary method is chosen because:

- It is in line with our general philosophy of making the simplest and most direct approach to any task.

- Most important, special techniques have to be *developed* and have to be shown to be efficient and relevant for the task in hand. The length of the manipulations presented here should go a long way to convincing that more powerful techniques are needed here.

- Long experience shows that advanced mathematical techniques are only taken to heart if they *systematise* and *summarise* existing knowledge; the motivation for their use must be a reorganisation of material which is *already familiar* in a more elementary context.[9]

In the above equation and in what follows, integration over spin *and* space variables is denoted by

$$\int \ldots d\tau \quad or \quad \int \ldots d\tau_1$$

while integration over space *only* is written

$$\int \ldots dV \quad or \quad \int \ldots dV_1$$

In general, $d\tau$ or dV (without a subscript) means integration over the variables of all the particles in the integrand while $d\tau_i$ or dV_i means integration over the coordinates of the ith particle. That is:

$$\int \ldots d\tau = \int \int \ldots dV ds$$

where s is the spin "variable(s)". That is, dV refers to the spatial variables typically denoted by the $\vec{r}_i$ and $d\tau$ to these variables plus the spin variable(s) the whole set denoted, typically, by the $\vec{x}_i$.

---

[9]Mathematics texts tend to ignore these elementary facts; they are mostly written "upside down" with the motivation and application presented *after* the general theorems. This method of presentation is often called the "lapidary" approach; beautifully polished objects (theorems) are presented with no clue about their origin.

56   APPENDIX 2.A.   SINGLE-DETERMINANT ENERGY EXPRESSION

2.A.1.1 Vanishing determinants

A certain amount of laxness with the derivations is often evident because of the combination of two of the elementary properties of determinants:

- The value of a determinant is unchanged by linear combinations of its rows (or columns); in the case of determinants of orbitals this means that any (invertible) linear transformation among the orbitals χ_i does not change the value of the determinant. This property will be used many times in the theory of the Hartree–Fock method and its implementation.

- The value of a determinant with an entire row (or column) of zeroes is zero; in our case an identically vanishing orbital will ensure that the entire single-determinant wavefunction is zero.

If, therefore, there is *linear dependence* amongst the orbitals comprising a single-determinant function it will be possible to transform this linear dependence into the form

$$\chi_1 = \sum_{k=2}^{m} d_k \chi_k$$

by renumbering the χ_i and appropriate choice of some numerical coefficients. One more step of subtraction will ensure that one column of the determinant is identically zero. Thus *if the orbitals of a single-determinant function are linearly dependent* then this function is identically zero.

In this case both the numerator and denominator of the energy expression are zero leading to a meaningless expression. In everything which follows *unless explicitly mentioned* it will be assumed that the orbitals comprising a single-determinant wavefunction are linearly independent.

Clearly, one can take advantage of the invariance of the determinant against linear combinations of the orbitals to choose particularly simple expressions for the energy and normalisation integrals in special cases but the general case will be considered below.

2.A.2 The normalisation integral

It is convenient to deal with the denominator of eqn (2.A.12) first and use this result in approaching the reduction of the numerator which is complicated by the presence of the operator $\hat{H}$.

We shall proceed with the evaluation of this integral by elementary methods which may seem a little pedestrian but this approach needs no apology since:

2.A.2. THE NORMALISATION INTEGRAL

- There is no harm at all in making sure that the reduction of these very basic integrals is clearly understood.
- The "normalisation integral" which forms the denominator of eqn (2.A.12) is the simplest case of a whole class of "many-electron overlap integrals" which occur in the reduction of the energy integral which forms the numerator of eqn (2.A.12)

The integral is

$$\int \Phi^* \Phi d\tau \qquad (2.A.13)$$

where

$$\Phi(\vec{x_1}, \vec{x_2}, \ldots, \vec{x_n}) = \frac{1}{\sqrt{n!}} \begin{vmatrix} \chi_1(\vec{x_1}) & \chi_1(\vec{x_2}) & \cdots & \chi_1(\vec{x_n}) \\ \chi_2(\vec{x_1}) & \chi_2(\vec{x_2}) & \cdots & \chi_2(\vec{x_n}) \\ \chi_3(\vec{x_1}) & \chi_3(\vec{x_2}) & \cdots & \chi_3(\vec{x_n}) \\ \cdots & \cdots & \cdots & \cdots \\ \cdots & \cdots & \cdots & \cdots \\ \cdots & \cdots & \cdots & \cdots \\ \chi_{n-2}(\vec{x_1}) & \chi_{n-2}(\vec{x_2}) & \cdots & \chi_{n-2}(\vec{x_n}) \\ \chi_{n-1}(\vec{x_1}) & \chi_{n-1}(\vec{x_2}) & \cdots & \chi_{n-1}(\vec{x_n}) \\ \chi_n(\vec{x_1}) & \chi_n(\vec{x_2}) & \cdots & \chi_n(\vec{x_n}) \end{vmatrix} \qquad (2.A.14)$$

In this expression, each orbital χ_i is a function of the coordinate of just *one* electron; typically three spatial coordinates and one spin "coordinate". In the following reduction the aim is simply to reduce the many-dimensional integration indicated symbolically by "$d\tau$" to one-electron integrations.

Clearly, since Φ is a determinant of n orbitals, the expansion of Φ will generate $n!$ products of n orbitals which only differ from each other by *which* electron is associated with ("occupies") *which* orbital. The simplest non-trivial case is for two particles

$$\Phi(\vec{x_1}, \vec{x_2}) = \frac{1}{\sqrt{2!}} \begin{vmatrix} \chi_1(\vec{x_1}) & \chi_1(\vec{x_2}) \\ \chi_2(\vec{x_1}) & \chi_2(\vec{x_2}) \end{vmatrix} = \frac{1}{\sqrt{2}} [\chi_1(\vec{x_1})\chi_2(\vec{x_2}) - \chi_1(\vec{x_2})\chi_2(\vec{x_1})] \qquad (2.A.15)$$

Each of these two products represents an identical physical picture: two electrons occupying two orbitals. The terms only differ in the permutation of (physically identical) electrons among the orbitals.

In the "normalisation integral" which forms the denominator of eqn (2.A.12) there are obviously, then, $(n!)^2$ terms to consider, in the 2×2 case

$$\int \Phi^* \Phi d\tau = \frac{1}{2} \int d\tau_1 \int d\tau_2 \, [\chi_1^*(\vec{x_1})\chi_2^*(\vec{x_2}) - \chi_1^*(\vec{x_2})\chi_2^*(\vec{x_1})]$$
$$\times [\chi_1(\vec{x_1})\chi_2(\vec{x_2}) - \chi_1(\vec{x_2})\chi_2(\vec{x_1})]$$

APPENDIX 2.A. SINGLE-DETERMINANT ENERGY EXPRESSION

$$= \frac{1}{2}\int d\tau_1 \int d\tau_2 \, [\chi_1^*(\vec{x_1})\chi_2^*(\vec{x_2})\chi_1(\vec{x_1})\chi_2(\vec{x_2})$$
$$-\chi_1^*(\vec{x_1})\chi_2^*(\vec{x_2})\chi_2(\vec{x_1})\chi_1(\vec{x_2})$$
$$-\chi_2^*(\vec{x_1})\chi_1^*(\vec{x_2})\chi_1(\vec{x_1})\chi_2(\vec{x_2})$$
$$+\chi_2^*(\vec{x_1})\chi_1^*(\vec{x_2})\chi_2(\vec{x_1})\chi_1(\vec{x_2}) \,]$$

This expression can be simplified considerably by noting that, in any definite integral, the integration variable is a "dummy"; *i.e.* for any function f:

$$\int f(t)dt = \int f(x)dx = \int f(y)dy \ \ldots$$

Thus, in our particular case,

$$\int d\tau_1 \int d\tau_2 \, [\chi_1^*(\vec{x_1})\chi_2^*(\vec{x_2})\chi_1(\vec{x_1})\chi_2(\vec{x_2})]$$
$$= \int d\tau_1 \int d\tau_2 \, [\chi_2^*(\vec{x_1})\chi_1^*(\vec{x_2})\chi_2(\vec{x_1})\chi_1(\vec{x_2})]$$

That is, in the expansion of the normalisation integral for the 2×2 case above, there are only *two* distinct terms:

$$\int d\tau_1 \int d\tau_2 \, [\chi_1^*(\vec{x_1})\chi_2^*(\vec{x_2})\chi_1(\vec{x_1})\chi_2(\vec{x_2})]$$

and

$$\int d\tau_1 \int d\tau_2 \, [\chi_1^*(\vec{x_1})\chi_2^*(\vec{x_2})\chi_2(\vec{x_1})\chi_1(\vec{x_2})]$$

(say)

A little experimentation with the 3×3 case and thinking about the results is enough to convince that this result is general.

Of the $(n!)^2$ terms in the expansion of the normalisation integral associated with an $n \times n$ determinant, only $n!$ are actually *distinct*.

In practice, we can choose *just one* of the permutations of the electronic coordinates in one of the appearances of Φ in the integrand (say in Φ^*) and then use *all* the possible permutations in the other (Φ) and be sure that we have included all the distinct terms. Of course, the result of this procedure must be multiplied by $n!$ to generate the correct *value* of the integral.

In the 3×3 case ($n! = 6$) we have

$$\int \Phi^*\Phi d\tau = 6 \times \left(\frac{1}{\sqrt{6}}\right)^2 \int \ [\ \ \chi_1(\vec{x_1})\chi_2(\vec{x_2})\chi_3(\vec{x_3}) \,] \times$$

2.A.2. THE NORMALISATION INTEGRAL

$$\begin{aligned}
[&\chi_1(\vec{x_1}) & \chi_2(\vec{x_2}) & \quad \chi_3(\vec{x_3}) \\
-&\chi_1(\vec{x_2}) & \chi_2(\vec{x_1}) & \quad \chi_3(\vec{x_3}) \\
-&\chi_1(\vec{x_1}) & \chi_2(\vec{x_3}) & \quad \chi_3(\vec{x_2}) \\
+&\chi_1(\vec{x_2}) & \chi_2(\vec{x_3}) & \quad \chi_3(\vec{x_1}) \\
-&\chi_1(\vec{x_3}) & \chi_2(\vec{x_2}) & \quad \chi_3(\vec{x_1}) \\
+&\chi_1(\vec{x_3}) & \chi_2(\vec{x_1}) & \quad \chi_3(\vec{x_2}) \,]\, d\tau
\end{aligned}$$

The factor of $n!$ exactly cancelling the $(1/\sqrt{6})^2$ occurring in the definition of the single-determinant function.

It now only remains to evaluate each of the six different terms to generate the value of the normalisation integral.

Taking a typical one:

$$\int [\chi_1^*(\vec{x_1})\chi_2^*(\vec{x_2})\chi_3^*(\vec{x_3})] [\chi_1(\vec{x_2})\chi_2(\vec{x_3})\chi_3(\vec{x_1})] \, d\tau$$

This is clearly a *product* of integrals of functions of $\vec{x_1}$, $\vec{x_2}$ and $\vec{x_3}$ *separately*, so writing

$$d\tau = d\tau_1 d\tau_2 d\tau_3$$

this integral becomes

$$\left(\int \chi_1^*(\vec{x_1}) \chi_3(\vec{x_1}) d\tau_1 \right) \left(\int \chi_2^*(\vec{x_2}) \chi_1(\vec{x_2}) d\tau_2 \right) \left(\int \chi_3^*(\vec{x_3}) \chi_2(\vec{x_3}) d\tau_3 \right)$$

which may, without ambiguity, be written

$$\left(\int \chi_1^* \chi_3 d\tau \right) \left(\int \chi_2^* \chi_1 d\tau \right) \left(\int \chi_3^* \chi_2 d\tau \right)$$

because, once separated, there is no danger of confusion of coordinates and $d\tau$ is used for all the coordinates involved in the integrand; this time just those of one particle.

These "one-electron" integrals are the orbital overlap integrals which we can denote by S_{ij}

$$S_{ij} = \int \chi_i^* \chi_j d\tau = \left(\int \chi_j^* \chi_i d\tau \right)^* = S_{ji}^*$$

and our typical term in the expansion of the normalisation integral is

$$S_{13} S_{21} S_{32}$$

Using this technique to evaluate the remaining five terms gives

$$\begin{aligned}
\int \Phi^* \Phi d\tau &= (S_{11}S_{22}S_{33} - S_{12}S_{21}S_{33} - S_{11}S_{32}S_{23} \\
&+ S_{12}S_{32}S_{13} - S_{13}S_{22}S_{31} + S_{12}S_{23}S_{13}) \\
&= \begin{vmatrix} S_{11} & S_{12} & S_{13} \\ S_{21} & S_{22} & S_{23} \\ S_{31} & S_{32} & S_{33} \end{vmatrix} = D
\end{aligned}$$

(say). This is the general result

The normalisation integral of a single determinant of one-electron orbitals is just the determinant of the orbital overlap matrix.

In particular, if the orbitals are *orthonormal*, that is normalised to unity and orthogonal

$$S_{ij} = \delta_{ij}; \quad \boldsymbol{S} = \boldsymbol{1}$$

then the single-determinant function defined by eqn (2.A.14) is already normalised to unity since the determinant of a unit matrix is just 1. In general, if the orbitals of which a single-determinant function is composed are not orthonormal, it is usual to normalise the function by multiplication by $D^{-1/2}$ where D is the determinant of the orbital overlap matrix.

2.A.2.1 Summary

The normalisation integral

$$\int \Phi^* \Phi d\tau$$

of a single-determinant function Φ is

$$D = det|S_{ij}|$$

where

$$S_{ij} = \int \chi_i^* \chi_j d\tau = \left(\int \chi_j^* \chi_i d\tau \right)^* = S_{ji}^*$$

are the orbital overlap integrals. Thus the single determinant function is normalised by

$$\frac{1}{\sqrt{det|S_{ij}|}}$$

Note that these integrations are all over *both* space and spin coordinates.

This result is easily extended to the integral

$$\int \Phi_A^* \Phi_B d\tau \tag{2.A.16}$$

which is the (many-electron) overlap integral analogous to the (one-electron) orbital overlap integrals. Of course, both functions must be functions of the same number of coordinates. The result is identical to the normalisation integral except that the orbital overlap integrals are between the n orbitals of which Φ_A is composed and the n (possibly different) orbitals which make up Φ_B.

So, if

$$S_{ij}^{AB} = \int (\chi_i^A)^* \chi_j^B d\tau$$

2.A.3. ONE-ELECTRON TERMS

where χ_i^A is a function from Φ_A and χ_j^B is one from Φ_B, the general result is

$$\int \Phi_A^* \Phi_B d\tau = \frac{det|S_{ij}^{AB}|}{\sqrt{D^A D^B}}$$

which will prove useful in the expansion of the energy integral in terms of orbital integrals. D^A and D^B are the normalisation determinants of the two functions Φ_A and Φ_B.

2.A.3 One-electron terms

The rather lengthy considerations of the last section are in preparation for the more involved expressions of the numerator of the variational energy expression. The energy integral is

$$\int \Phi^* \hat{H} \Phi dV$$

and the complications arise because of the number and nature of the terms in $\hat{H}$.

Let us take the simplest possible case when $\hat{H}$ is simply a sum of one-electron operators; that is the electrons really *are* independent particles:

$$\hat{H}_1 = \sum_{i=1}^{n} \hat{h}(i)$$

where $\hat{h}(i)$ means that the operator $\hat{h}$ depends only on the (space and spin) coordinates of electron i. Since electrons are identical, it goes without saying that the *form* of each $\hat{h}(i)$ is *the same*, the particles' Hamiltonians only differ by their *labels*.

This special case of using *only* one-electron operators has been distinguished by calling the total n-electron Hamiltonian $\hat{H}_1$ in place of the more usual $\hat{H}$.

We can now proceed to evaluate

$$\int \Phi^* \hat{H}_1 \Phi dV$$

using the same general approach as in the last section: explicit consideration of the simplest non-trivial case.

In the three-electron case

$$\hat{H}_1 = \hat{h}(1) + \hat{h}(2) + \hat{h}(3)$$

so there are $3 \times (3!)^2$ terms in the energy integral when it is expanded out in terms of orbital integrals. However, as before, because many of the terms only differ by permutations of the coordinates of identical particles, there are simplifications to be made.

Because $\hat{H}_1$ is *symmetrical* with respect to permutations amongst the coordinates of the electrons we can, in fact, use the same simplification as before; of the $(n!)^2$ terms due to expansion of the determinant (twice), only $n!$ are distinct and they can be written in the same way. But this time the operator $\hat{H}_1$ must also be included, giving a total of $n \times n!$ terms which may be written

$$\int \Phi^* \hat{H}_1 \Phi d\tau = 6 \times \left(\frac{1}{\sqrt{6}}\right)^2 \int \quad [\quad \chi_1^*(\vec{x_1})\chi_2^*(\vec{x_2})\chi_3^*(\vec{x_3}) \,]$$
$$\times \quad (\hat{h}(1) + \hat{h}(2) + \hat{h}(3))$$
$$\begin{array}{rccc}
[& \chi_1(\vec{x_1}) & \chi_2(\vec{x_2}) & \chi_3(\vec{x_3}) \\
- & \chi_1(\vec{x_2}) & \chi_2(\vec{x_1}) & \chi_3(\vec{x_3}) \\
- & \chi_1(\vec{x_1}) & \chi_2(\vec{x_3}) & \chi_3(\vec{x_2}) \\
+ & \chi_1(\vec{x_2}) & \chi_2(\vec{x_3}) & \chi_3(\vec{x_1}) \\
- & \chi_1(\vec{x_3}) & \chi_2(\vec{x_2}) & \chi_3(\vec{x_1}) \\
+ & \chi_1(\vec{x_3}) & \chi_2(\vec{x_1}) & \chi_3(\vec{x_2}) \,] \, d\tau
\end{array}$$

by analogy with the expansion of the normalisation integral for the case with $n = 3$ in the last section.

Again taking a typical one of the terms

$$\int [\chi_1^*(\vec{x_1})\chi_2^*(\vec{x_2})\chi_3^*(\vec{x_3})] \left[\hat{h}(1) + \hat{h}(2) + \hat{h}(3)\right] [\chi_1(\vec{x_2})\chi_2(\vec{x_3})\chi_3(\vec{x_1})] \, d\tau$$

separation of the coordinates of the three electrons in the integrand gives

$$\left(\int \chi_1^*(\vec{x_1})\hat{h}(1)\chi_3(\vec{x_1})d\tau_1\right) \left(\int \chi_2^*(\vec{x_2})\chi_1(\vec{x_2})d\tau_2\right) \left(\int \chi_3^*(\vec{x_3})\chi_2(\vec{x_3})d\tau_3\right)$$
$$+ \left(\int \chi_1^*(\vec{x_1})\chi_3(\vec{x_1})d\tau_1\right) \left(\int \chi_2^*(\vec{x_2})\hat{h}(2)\chi_1(\vec{x_2})d\tau_2\right) \left(\int \chi_3^*(\vec{x_3})\chi_2(\vec{x_3})d\tau_3\right)$$
$$+ \left(\int \chi_1^*(\vec{x_1})\chi_3(\vec{x_1})d\tau_1\right) \left(\int \chi_2^*(\vec{x_2})\chi_1(\vec{x_2})d\tau_2\right) \left(\int \chi_3^*(\vec{x_3})\hat{h}(3)\chi_2(\vec{x_3})d\tau_3\right)$$

where we have used the fact that both the orbitals *and* the operators $\hat{h}$ only involve the coordinates of a single particle, thus maintaining the "product" form of the integral.

Defining

$$h_{ij} = \int \chi_i^* \hat{h} \chi_j d\tau = \left(\int \chi_j^* \hat{h} \chi_i d\tau\right)^* = h_{ji}^*$$

because $\hat{h}$ is symmetrical in the electron's coordinates and therefore need not have its electron label when there is no ambiguity *and* is Hermitian, and using

2.A.3. ONE-ELECTRON TERMS

the existing notation for the orbital overlap integrals, we have for our typical term
$$h_{13}S_{21}S_{32} + S_{13}h_{21}S_{32} + S_{13}S_{21}h_{32}$$
and all the other five product terms contribute similar sets of three terms, giving $3 \times 3! = 18$ terms altogether.

Collecting these terms together with their signs and regrouping them shows them to be the expansion of *three* determinants. Each determinant has one row (or column) of the determinant of orbital overlap integrals replaced by a row (or column) of elements of the matrix of h_{ij} The final result is

$$\begin{vmatrix} h_{11} & h_{12} & h_{13} \\ S_{21} & S_{22} & S_{23} \\ S_{31} & S_{32} & S_{33} \end{vmatrix} + \begin{vmatrix} S_{11} & S_{12} & S_{13} \\ h_{21} & h_{22} & h_{23} \\ S_{31} & S_{32} & S_{33} \end{vmatrix} + \begin{vmatrix} S_{11} & S_{12} & S_{13} \\ S_{21} & S_{22} & S_{23} \\ h_{31} & h_{32} & h_{33} \end{vmatrix}$$

This result is clearly capable of being expressed in terms of the *minors* of the orbital overlap determinant: the sub-determinants obtained by striking out rows and columns of this determinant.

If each of the n determinants is expanded about the row of h_{ij}, we get a sum of terms each of which is one of the h_{ij} multiplied by an $(n-1) \times (n-1)$ determinant D_{ij} (say) which is the minor of the $n \times n$ original determinant with the ith row and jth column removed, all multiplied by a sign determined in the usual way from the expansion: $(-1)^{i+j}$. Obviously *all* the possible h_{ij} occur so that the general result is

$$\int \Phi^* \hat{H}_1 \Phi d\tau = \sum_{i,j=1}^{n} (-1)^{i+j} h_{ij} D_{ij}$$

It is worthwhile to give the result in the special case of an orthonormal set of orbitals; when

$$\boldsymbol{S = 1}$$

This case is particularly simple because, of the $n \times n!$ terms (18 in our case), only three do not involve at least one S_{ij} with $i \neq j$ which are zero. For example, the typical term which we have just evaluated becomes

$$h_{13} \times 0 \times 0 + 0 \times h_{21} \times 0 + 0 \times 0 \times h_{32} = 0$$

Evaluating all the remaining terms shows that the final result is

$$h_{11}S_{22}S_{33} + S_{11}h_{22}S_{33} + S_{11}S_{22}h_{33}$$

which, because the orbitals are *normalised* as well as orthogonal, is just

$$h_{11} + h_{22} + h_{33}$$

So that, in the case of orthonormal orbitals we have

$$\int \Phi^* \hat{H}_1 \Phi d\tau = \sum_{i=1}^{n} h_{ii}$$

This result is also easily seen from the expansion of the integral in terms of the minors of the orbital overlap determinant. If $S = 1$ then all the D_{ij} are zero if $i \neq j$, the $D_{ii} = 1$ and $(-1)^{2i} = 1$ giving exactly the same result.

In view of the technique to be used to evaluate the electron repulsion terms, it is worthwhile to use this general expansion technique for the result we have already obtained.

We may expand the determinant of order n in terms of the first *column* to show its dependence on the coordinates of particle 1 explicitly (omitting the conventional "normalisation factor" $1/\sqrt{n!}$ for simplicity):

$$\Phi(\vec{x_1}, \vec{x_2}, \ldots, \vec{x_n}) = \begin{vmatrix} \chi_1(\vec{x_1}) & \chi_1(\vec{x_2}) & \cdots & \chi_1(\vec{x_n}) \\ \chi_2(\vec{x_1}) & \chi_2(\vec{x_2}) & \cdots & \chi_2(\vec{x_n}) \\ \chi_3(\vec{x_1}) & \chi_3(\vec{x_2}) & \cdots & \chi_3(\vec{x_n}) \\ \cdots & \cdots & \cdots & \cdots \\ \cdots & \cdots & \cdots & \cdots \\ \cdots & \cdots & \cdots & \cdots \\ \chi_{n-2}(\vec{x_1}) & \chi_{n-2}(\vec{x_2}) & \cdots & \chi_{n-2}(\vec{x_n}) \\ \chi_{n-1}(\vec{x_1}) & \chi_{n-1}(\vec{x_2}) & \cdots & \chi_{n-1}(\vec{x_n}) \\ \chi_n(\vec{x_1}) & \chi_n(\vec{x_2}) & \cdots & \chi_n(\vec{x_n}) \end{vmatrix}$$

$$= \sum_{i=1}^{n} (-1)^{i+1} \chi_i(\vec{x_1}) \times \begin{vmatrix} \chi_1(\vec{x_2}) & \chi_1(\vec{x_3}) & \cdots & \chi_1(\vec{x_n}) \\ \chi_2(\vec{x_2}) & \chi_2(\vec{x_3}) & \cdots & \chi_2(\vec{x_n}) \\ \chi_3(\vec{x_2}) & \chi_3(\vec{x_3}) & \cdots & \chi_3(\vec{x_n}) \\ \cdots & \cdots & \cdots & \cdots \\ \cdots & \cdots & \cdots & \cdots \\ \cdots & \cdots & \cdots & \cdots \\ \chi_{n-2}(\vec{x_2}) & \chi_{n-2}(\vec{x_3}) & \cdots & \chi_{n-2}(\vec{x_n}) \\ \chi_{n-1}(\vec{x_2}) & \chi_{n-1}(\vec{x_3}) & \cdots & \chi_{n-1}(\vec{x_n}) \\ \chi_n(\vec{x_2}) & \chi_n(\vec{x_3}) & \cdots & \chi_n(\vec{x_n}) \end{vmatrix}$$

$$= \sum_{i=1}^{n} (-1)^{i+1} \chi_i(\vec{x_1}) \times \Phi_{1i}$$

where Φ_{1i} is the determinant of order $n - 1$ obtained by striking out the i'th row and first column of Φ *but without the factor of* $1/\sqrt{(n-1)!}$.

If we now evaluate the integral

$$\int \Phi^* \hat{h}(1) \Phi d\tau = \frac{1}{n!} \sum_{i=1}^{n} \sum_{j=1}^{n} \int (-1)^{i+j} \chi_i(\vec{x_1}) \Phi_{1i} \hat{h}(1) \chi_j(\vec{x_1}) \Phi_{1j} d\tau$$

2.A.4. TWO-ELECTRON TERMS

we see that it separates neatly into an integral involving $\vec{x}_1$ which is h_{ij} and an integral involving the remaining coordinates $(\vec{x}_2 \ldots \vec{x}_n)$ which is an inter-determinant overlap integral. Performing the same integration with the other $(n-1)$ operators $\hat{h}_i$ and summing the results generates exactly our result above obtained by generalising the 3×3 case.[10]

2.A.4 Two-electron terms

The only remaining terms to evaluate in the energy integral

$$\int \Phi^* \hat{H} \Phi dV$$

are the ones arising from the so-called two-electron terms in $\hat{H}$:

$$\frac{1}{r_{ij}}$$

where r_{ij} is the distance between electrons i and j. Obviously the two terms

$$\frac{1}{r_{ij}} \quad \text{and} \quad \frac{1}{r_{ji}}$$

are the same and there are no "self-repulsion" terms $1/r_{ii}$. The whole set of terms can be called $\hat{H}_2$, where

$$\hat{H}_2 = \frac{1}{2} \sum_{i,j=1}^{n} \frac{1}{r_{ij}}$$

and where there are actually only $n(n-1)/2$ distinct repulsion operators and the summation must be understood to exclude the term with $i = j$.

In this case, the straightforward approach which we have used for the normalisation integral and the one-electron terms would work, but it turns out to be very unwieldy to extract the general result from a simple case. If we try to evaluate the result for the simplest manageable case explicitly, $n = 3$ generating six terms, the outcome is not typical because only individual orbital overlaps remain after integration over the coordinates of the two electrons involved in the repulsion integral. The next case $n = 4$ with twenty-four terms is too unwieldy to handle explicitly.

We must, therefore, use a general method; a method of expansion of the determinant which is adapted to the problem in hand. That is, expand the

---
[10]There is some careful counting of the numbers of terms involved here!

APPENDIX 2.A. SINGLE-DETERMINANT ENERGY EXPRESSION

original determinant of orbitals in terms of its minors instead of relying on recognising this expansion in the final result.

Since the repulsion operators involve the coordinates of *two* electrons, it is obvious that, in any one of the basic product integrals which result from an expansion of the $n \times n$ determinant, there will be an integral over the coordinates of two electrons (involving the product of *four* orbitals) and $(n-2)$ orbital overlap integrals (each involving *two* orbitals).

It is therefore useful to be able to expand the determinant in a way which gives us a separation of *two* of the variables from the set of n. This technique is elementary but tedious to demonstrate explicitly; it involves expanding the $n \times n$ determinant in terms of the column of functions of one of the electrons (electron 1, coordinates $\vec{x_1}$, say) and then expanding each of the resulting n determinants of order $(n-1)$ in terms of the column of the functions of another electron (electron 2, coordinates $\vec{x_2}$, say). It requires a vivid imagination and a willingness to take things on trust!

The full determinant of order n is:

$$\Phi(\vec{x_1}, \vec{x_2}, \ldots \vec{x_n}) = \begin{vmatrix} \chi_1(\vec{x_1}) & \chi_1(\vec{x_2}) & \cdots & \chi_1(\vec{x_n}) \\ \chi_2(\vec{x_1}) & \chi_2(\vec{x_2}) & \cdots & \chi_2(\vec{x_n}) \\ \chi_3(\vec{x_1}) & \chi_3(\vec{x_2}) & \cdots & \chi_3(\vec{x_n}) \\ \cdots & \cdots & \cdots & \cdots \\ \cdots & \cdots & \cdots & \cdots \\ \cdots & \cdots & \cdots & \cdots \\ \chi_{n-2}(\vec{x_1}) & \chi_{n-2}(\vec{x_2}) & \cdots & \chi_{n-2}(\vec{x_n}) \\ \chi_{n-1}(\vec{x_1}) & \chi_{n-1}(\vec{x_2}) & \cdots & \chi_{n-1}(\vec{x_n}) \\ \chi_n(\vec{x_1}) & \chi_n(\vec{x_2}) & \cdots & \chi_n(\vec{x_n}) \end{vmatrix}$$

where the normalisation constant $1/\sqrt{n!D}$ has been omitted for clarity; it will be restored at the end of the derivation.

The first expansion, involving the functions of $\vec{x_1}$, has terms like:

$$\Phi = \chi_1(\vec{x_1}) \left\{ \begin{vmatrix} \chi_2(\vec{x_2}) & \chi_2(\vec{x_3}) & \cdots & \chi_2(\vec{x_n}) \\ \chi_3(\vec{x_2}) & \chi_3(\vec{x_3}) & \cdots & \chi_3(\vec{x_n}) \\ \cdots & \cdots & \cdots & \cdots \\ \cdots & \cdots & \cdots & \cdots \\ \cdots & \cdots & \cdots & \cdots \\ \chi_{n-2}(\vec{x_2}) & \chi_{n-2}(\vec{x_3}) & \cdots & \chi_{n-2}(\vec{x_n}) \\ \chi_{n-1}(\vec{x_2}) & \chi_{n-1}(\vec{x_3}) & \cdots & \chi_{n-1}(\vec{x_n}) \\ \chi_n(\vec{x_2}) & \chi_n(\vec{x_3}) & \cdots & \chi_n(\vec{x_n}) \end{vmatrix} \right\}$$

2.A.4. TWO-ELECTRON TERMS

$$-\chi_2(\vec{x_1}) \left\{ \begin{vmatrix} \chi_1(\vec{x_2}) & \chi_1(\vec{x_3}) & \cdots & \chi_2(\vec{x_n}) \\ \chi_3(\vec{x_2}) & \chi_3(\vec{x_3}) & \cdots & \chi_3(\vec{x_n}) \\ \cdots & \cdots & \cdots & \cdots \\ \cdots & \cdots & \cdots & \cdots \\ \cdots & \cdots & \cdots & \cdots \\ \chi_{n-2}(\vec{x_2}) & \chi_{n-2}(\vec{x_3}) & \cdots & \chi_{n-2}(\vec{x_n}) \\ \chi_{n-1}(\vec{x_2}) & \chi_{n-1}(\vec{x_3}) & \cdots & \chi_{n-1}(\vec{x_n}) \\ \chi_n(\vec{x_2}) & \chi_n(\vec{x_3}) & \cdots & \chi_n(\vec{x_n}) \end{vmatrix} \right\} + \cdots$$

Then these two "typical" terms of this expansion may each be expanded in terms of the remaining $(n-1)$ functions of the coordinates $\vec{x_2}$. Each large bracket is such an expansion and only the first term of each is given.

$$\Phi = \chi_1(\vec{x_1}) \left\{ \chi_2(\vec{x_2}) \begin{vmatrix} \chi_3(\vec{x_3}) & \cdots & \chi_3(\vec{x_n}) \\ \chi_4(\vec{x_3}) & \cdots & \chi_4(\vec{x_n}) \\ \cdots & \cdots & \cdots \\ \cdots & \cdots & \cdots \\ \chi_{n-1}(\vec{x_3}) & \cdots & \chi_{n-1}(\vec{x_n}) \\ \chi_n(\vec{x_3}) & \cdots & \chi_n(\vec{x_n}) \end{vmatrix} + \cdots \right\}$$

$$-\chi_2(\vec{x_1}) \left\{ \chi_1(\vec{x_2}) \begin{vmatrix} \chi_3(\vec{x_3}) & \cdots & \chi_3(\vec{x_n}) \\ \chi_4(\vec{x_3}) & \cdots & \chi_4(\vec{x_n}) \\ \cdots & \cdots & \cdots \\ \cdots & \cdots & \cdots \\ \chi_{n-1}(\vec{x_3}) & \cdots & \chi_{n-1}(\vec{x_n}) \\ \chi_n(\vec{x_3}) & \cdots & \chi_n(\vec{x_n}) \end{vmatrix} + \cdots \right\}$$

$$+ \left\{ \cdots \begin{vmatrix} \cdots & \cdots & \cdots \\ \cdots & \cdots & \cdots \\ \cdots & \cdots & \cdots \end{vmatrix} \right.$$

$$+ \cdots \cdots \cdots$$

These two terms have the determinant of order $(n-2)$ in common and may be collected together to give a single term:

$$\{\chi_1(\vec{x_1})\chi_2(\vec{x_2}) - \chi_2(\vec{x_1})\chi_1(\vec{x_2})\} \times \begin{vmatrix} \chi_3(\vec{x_3}) & \cdots & \chi_3(\vec{x_n}) \\ \chi_3(\vec{x_3}) & \cdots & \chi_3(\vec{x_n}) \\ \cdots & \cdots & \cdots \\ \cdots & \cdots & \cdots \\ \chi_{n-1}(\vec{x_3}) & \cdots & \chi_{n-1}(\vec{x_n}) \\ \chi_n(\vec{x_3}) & \cdots & \chi_n(\vec{x_n}) \end{vmatrix}$$

Which itself is the product of a 2×2 determinant and a determinant of order $(n-2)$.

$$\begin{vmatrix} \chi_1(\vec{x_1}) & \chi_1(\vec{x_2}) \\ \chi_2(\vec{x_1}) & \chi_2(\vec{x_2}) \end{vmatrix} \times \begin{vmatrix} \chi_3(\vec{x_3}) & \cdots & \chi_3(\vec{x_n}) \\ \chi_3(\vec{x_3}) & \cdots & \chi_3(\vec{x_n}) \\ \cdots & \cdots & \cdots \\ \cdots & \cdots & \cdots \\ \chi_{n-1}(\vec{x_3}) & \cdots & \chi_{n-1}(\vec{x_n}) \\ \chi_n(\vec{x_3}) & \cdots & \chi_n(\vec{x_n}) \end{vmatrix}$$

Now, what is perhaps reasonable but not obvious is that the whole determinant may be expanded in this way to give a collection of products of determinants of order 2 and $(n-2)$ with no terms excluded and no extra terms. This is a special case of a theorem which treats the general case of the expansion of a determinant as products of determinants of lower order. The general result is:

$$\Phi(\vec{x}_1, \vec{x}_2, \ldots \vec{x}_n) = \frac{1}{\sqrt{n!D}} \sum_{i<j=1}^{n} (-1)^{(i+j+1)} \begin{vmatrix} \chi_i(\vec{x}_1) & \chi_i(\vec{x}_2) \\ \chi_j(\vec{x}_1) & \chi_j(\vec{x}_2) \end{vmatrix} \times \Phi_{1i2j}$$

(2.A.17)

where Φ_{1i2j} is the determinant of the n orbitals χ_k *with the two columns labelled by orbitals* χ_i and χ_j and the two rows labelled by coordinates $\vec{x}_1$, $\vec{x}_2$ removed (as before, *without the factor of* $1/\sqrt{(n-2)!}$).

If we now proceed with the original objective of evaluating the electron-repulsion energy, we can see that this expansion is exactly what is required. Since all electronic coordinates are equivalent, we can take the term involving $1/r_{12}$ and the above expansion which separates the coordinates of electrons 1 and 2, which are involved in the energy operator, from all the others which will just generate products (determinants, actually) of overlap integrals.

The only remaining problem is one of book-keeping; having an adequate notation to identify the large number of products of repulsion integrals and overlap determinants of order $(n-2)$.

Using the above expansion in terms of products of determinants of order 2 and $(n-2)$ a typical one of the electron-repulsion terms becomes:

2.A.4. TWO-ELECTRON TERMS

$$\int \Phi^* \frac{1}{r_{12}} \Phi d\tau = $$

$$\frac{1}{n!D} \sum_{i,j<i=1}^{n} \sum_{k,\ell<k=1}^{n} (-1)^{i+j+k+\ell} \int d\tau_3 \int \ldots d\tau_n \Phi_{ij} \Phi_{k\ell}$$

$$\times \int d\tau_1 \int d\tau_2 \begin{vmatrix} \chi_i(\vec{x_1}) & \chi_i(\vec{x_2}) \\ \chi_j(\vec{x_1}) & \chi_j(\vec{x_2}) \end{vmatrix} \left(\frac{1}{r_{12}} \right) \begin{vmatrix} \chi_k(\vec{x_1}) & \chi_k(\vec{x_2}) \\ \chi_\ell(\vec{x_1}) & \chi_\ell(\vec{x_2}) \end{vmatrix}$$

Where the integration over the coordinates of particles 1 and 2 has been separated from the remaining coordinates $3\ldots n$ since these are the ones which involve the electron repulsion operator. The integration over $d\tau_3 \ldots d\tau_n$ is simply a special case of our earlier integral for the overlap of two determinants. The integration over $d\tau_1$ and $d\tau_2$ can be reduced to our standard notation for repulsion integrals: simply by expanding the determinants of order 2 and collecting terms.

There are, of course, the remaining repulsion operators $1/r_{ij}$ to be used in exactly the same way and if we add together all $n(n-1)/2$ sets of terms we obtain the general result for the electron-repulsion contribution to the energy of a single-determinant wavefunction:

$$\int \Phi^* \left(\sum_{i,j<i=1}^{n} \frac{1}{r_{ij}} \right) \Phi d\tau = \sum_{i,j<i=1}^{n} \sum_{k,\ell<k=1}^{n} D_{ijkl}$$

$$\times \left[\int \chi_i^*(\vec{x}_1)\chi_j^*(\vec{x}_2) \frac{1}{r_{12}} \chi_k(\vec{x}_1)\chi_\ell(\vec{x}_2) d\tau_1 d\tau_2 \right.$$

$$\left. - \int \chi_i^*(\vec{x}_1)\chi_j^*(\vec{x}_2) \frac{1}{r_{12}} \chi_\ell(\vec{x}_1)\chi_k(\vec{x}_2) d\tau_1 d\tau_2 \right]$$

There are considerable simplifications in this enormously long expression if the orbitals χ_k form an orthonormal set. In this case the vast majority of the D_{ijkl} are zero and the remaining ones are unity, giving a final expression of:

$$\int \Phi^* \left(\sum_{i,j<i=1}^{n} \frac{1}{r_{ij}} \right) \Phi d\tau = \sum_{i,j<i=1}^{n} \left[\int \chi_i^*(\vec{x}_1)\chi_j^*(\vec{x}_2) \frac{1}{r_{12}} \chi_i(\vec{x}_1)\chi_j(\vec{x}_2) d\tau_1 d\tau_2 \right.$$

$$\left. - \int \chi_i^*(\vec{x}_1)\chi_j^*(\vec{x}_2) \frac{1}{r_{12}} \chi_j(\vec{x}_1)\chi_i(\vec{x}_2) d\tau_1 d\tau_2 \right]$$

involving only a twofold summation in place of the fourfold sum in the nonorthogonal case.

It is useful to have a more convenient notation for the electron-repulsion integrals. As we see above, each one of them involves *four* spin-orbitals and the integration is over the coordinates of *two* electrons. If we use the *order* of the orbitals in the integrand to indicate which orbital is a function of which electron's coordinates we can dispense with the explicit integration notation altogether and write:

$$< \chi_i \chi_j | \chi_k \chi_\ell > = \int d\tau_1 \int d\tau_2 \chi_i^*(\vec{x_1}) \chi_j^*(\vec{x_2}) \left(\frac{1}{r_{12}} \right) \chi_k(\vec{x_1}) \chi_\ell(\vec{x_2})$$

where the pointed brackets are reminiscent of P. A. M. Dirac's notation ("bra"s and "ket"s forming a "bracket" on integration) and the repulsion operator is omitted altogether, being understood by the inclusion of the central "bar" (|). In this notation the order of the electrons' coordinates is $1,2|1,2$ emphasising their pedigree since they arise from the reduction of the product of determinants. However, for the purposes of physical interpretation and computational simplicity it is more convenient to use:

$$(\chi_i \chi_k, \chi_j \chi_\ell) = < \chi_i \chi_j | \chi_k \chi_\ell >$$

where ordinary (round) brackets replace the Dirac-type ones and a comma replaces the bar. In this form the order of the electrons' coordinates is $11,22$ which emphasises the physical interpretation of the integral as the repulsion between an electronic "charge cloud" $\chi_i^*(\vec{x_1}) \chi_k(\vec{x_1})$ and $\chi_j^*(\vec{x_2}) \chi_\ell(\vec{x_2})$. Where there is no danger of confusion, the generic symbol representing the spin-orbital may be dropped giving an extremely compact notation for these integrals:

$$(ik, j\ell) = < ij | k\ell >$$

In fact, although the use of orthonormal spin-orbitals reduces the expression to a twofold sum, it is not this reduction which is the greatest difference between the orthonormal and non-orthonormal cases. The largest difference lies in the trivial nature of the $D_{ijk\ell}$ in the orthogonal case. In the non-orthogonal case each of these "coefficients" is made up of a product of about $(n-2)!$ orbital overlap integrals. Now, although as we shall see these orbital overlaps are rather simple to compute, the huge numbers of products of them quickly become overwhelming in any computation. As usual in computer implementations one has to decide either to store them or to compute them as needed, leading to being either overwhelmed by filestore use or by computing requirements.

There have been very significant advances over the past several years in the technology of handling these problems but they are still a limiting factor in the use of a set of *non-orthogonal* orbitals in the quantum theory of many-electron systems.

2.A.5 Summary

There are the two distinct cases to summarise; the full expressions where no assumption of orthogonality is made about the spin-orbitals of the determinants and the orthogonal case.

2.A.5. SUMMARY

2.A.5.1 General case; Løwdin's rules

The evaluation of the energy integrals involving determinants of *non-orthonormal* orbitals were first systematically derived by P-O. Løwdin and are known as *Løwdin's Rules*.

With the techniques which we have outlined above there are only trivial differences between the evaluation of the integrals

$$\int \Phi^* \hat{H} \Phi d\tau$$

and the more general case involving *two* determinants of n orbitals:

$$\int \Phi_A^* \hat{H} \Phi_B d\tau$$

where $\hat{H}$ is a typical zero-, one- or two-electron operator.

All that is necessary is to use a suitable notation to *identify* which quantities belong to which determinant. The superscripts A and B will be used for this purpose when the existing quantities have subscripts and subscripts A and B used where there are existing superscripts. This leads to a rather over-full notation but it is unambiguous.

Overlap.

We already have the general expression:

$$\int \Phi_A^* \Phi_B d\tau = \frac{det|D_{ij}^{AB}|}{\sqrt{D^A D^B}}$$

One-electron Hamiltonian.

For

$$\hat{H} = \sum_{i=1}^{n} \hat{h}(i)$$

we have

$$\int \Phi_A^* \hat{H} \Phi_B d\tau = \sum_{i,j=1}^{n} (-1)^{(i+j)} h_{ij} D_{ji}^{AB}$$

where

$$h_{ij} = \int \chi_i^{A*} \hat{h} \chi_j^B d\tau$$

and D_{ji}^{AB} is the j,i minor of the overlap determinant D^{AB}.

APPENDIX 2.A. SINGLE-DETERMINANT ENERGY EXPRESSION

Electron Repulsion.

For
$$\hat{H} = \sum_{j<i=1}^{n} \frac{1}{|\vec{r}_i - \vec{r}_j|}$$
the energy integral becomes
$$\int \Phi_A^* \hat{H} \Phi_B d\tau = \sum_{i>j=1}^{n} \sum_{k>\ell}^{n} (-1)^{(i+j+k+\ell)} D_{ijk\ell}^{AB}$$
$$\times \ \{<\chi_i^A \chi_j^A | \chi_k^B \chi_\ell^B> - <\chi_i^A \chi_j^A | \chi_\ell^B \chi_k^B>\}$$

where the electron-repulsion integrals are written in standard form and the $D_{ijk\ell}^{AB}$ are the minors of the overlap determinant D^{AB} with the rows i, j and columns k, ℓ removed.

2.A.5.2 The orthonormal case: Slater's rules

Historically, the mean values of the various operators were evaluated first for determinants of *orthonormal* orbitals (the derivation being performed by J C Slater) and the expressions are now universally known as *Slater's Rules*.[11]

Overlap.

$$\int \Phi_A^* \Phi_B d\tau = \delta_{AB}$$

where δ_{AB} is the Kronecker delta; zero if $A \neq B$, and unity if $A = B$.

One-electron Hamiltonian

There are only three possibilities which are better enumerated than put into a general formula:

1.
$$\int \Phi_A^* \hat{H} \Phi_A d\tau = \sum_{i=1}^{n} h_{ii}$$

---

[11] There are another set of rules (for the evaluation of approximate exponential atomic orbitals) which are also known as Slater's rules but the contexts are sufficiently different to avoid serious confusion.

2.A.5. SUMMARY

where
$$h_{ii} = \int \chi_i^* \hat{h} \chi_i d\tau$$

2. If Φ_A and Φ_B differ by only one spin-orbital; say χ_i in Φ_A is replaced by $\chi_{i'}$ in Φ_B then:
$$\int \Phi_A^* \hat{H} \Phi_B d\tau = h_{ii'}$$
where
$$h_{ii'} = h_{i'i}^* = \int \chi_i^* \hat{h} \chi_{i'} d\tau$$

3. If Φ_A and Φ_B differ by two or more spin-orbitals then:
$$\int \Phi_A^* \hat{H} \Phi_B d\tau = 0$$

Electron Repulsion.

This time
$$\hat{H} = \sum_{i,j<i=1}^{n} \frac{1}{r_{ij}}$$

Again there are a small number of cases:

1.
$$\int \Phi_A^* \hat{H} \Phi_A d\tau = \sum_{i,j<i=1}^{n} <\chi_i \chi_j | \chi_i \chi_j> - <\chi_i \chi_j | \chi_j \chi_i>$$

2. If Φ_A and Φ_B differ by only one spin-orbital; say χ_i in Φ_A is replaced by $\chi_{i'}$ in Φ_B then:
$$\int \Phi_A^* \hat{H} \Phi_B d\tau = \sum_{k \neq i=1}^{n} (<\chi_{i'} \chi_k | \chi_i \chi_k> - <\chi_{i'} \chi_k | \chi_k \chi_i>)$$

3. If Φ_A and Φ_B differ by two spin-orbitals; say χ_i in Φ_A is replaced by $\chi_{i'}$ in Φ_B, and χ_j in Φ_A is replaced by $\chi_{j'}$ in Φ_B then:
$$\int \Phi_A^* \hat{H} \Phi_B d\tau = <\chi_{i'} \chi_{j'} | \chi_i \chi_j> - <\chi_{i'} \chi_{j'} | \chi_j \chi_i>$$

4. If Φ_A and Φ_B differ by three or more spin-orbitals;
$$\int \Phi_A^* \hat{H} \Phi_B d\tau = 0$$

APPENDIX 2.A. SINGLE-DETERMINANT ENERGY EXPRESSION

The Orthogonal-Orbital Energy Expression.

Easily the most widespread use to which Slater's Rules are put is in the evaluation of the mean value of the energy for a system with the standard electrostatic Hamiltonian. This simply means adding the one- and two-electron terms in the above expressions; the result is

$$E = \sum_{i=1}^{n} h_{ii} + \sum_{i,j<i=1}^{n} <\chi_i\chi_j|\chi_i\chi_j> - <\chi_i\chi_j|\chi_j\chi_i>$$

which, in the more usual "charge-cloud" notation is

$$E = \sum_{i=1}^{n} h_{ii} + \sum_{i,j<i=1}^{n} (\chi_i\chi_i, \chi_j\chi_j) - (\chi_i\chi_j, \chi_i\chi_j) \qquad (2.A.18)$$

This is the result we need for the derivations in the present chapter.

Chapter 3

The matrix SCF equations

The Hartree–Fock equation of the previous chapter is a non-separable partial differential equation in three dimensions. In this chapter we derive an equation which is satisfied by an approximation to the differential Hartree–Fock equation. This matrix equation provides the techniques and concepts for the vast majority of quantum chemistry.

Contents

3.1	Introduction	75
3.2	Notation	77
3.3	The expansion	78
3.4	The energy expression	80
3.5	The numerator: Hamiltonian mean value	81
3.6	The denominator: normalisation condition	84
3.7	The Hartree–Fock equation	86
3.8	"Normalisation": the Lagrangian	87
3.9	Preliminary summary	88
3.10	Some technical manipulations	89
3.11	Canonical orbitals	92
3.11.1	The full MO matrix	94
3.12	The total energy	95
3.13	Summary	96

3.1 Introduction

It has been shown earlier (Chapter 2) that it is possible to use the variation principle to derive a single *differential* equation whose solutions are the optimum orbitals of a single-determinant wavefunction. We now wish to carry this derivation forward in order to be able to have a practical method of obtaining (approximations to) these optimum orbitals. In particular, since the solution of a highly non-linear differential equation by numerical methods for molecules of arbitrary geometry is out of the question, it is desirable to transform the *differential* equation into an *algebraic* equation which will be more amenable to a systematic method of solution independently of the size and shape of the molecule, radical, ion or group of molecules.

The differential Hartree–Fock equation

$$\hat{h}^F \chi_i = \epsilon_i \chi_i$$

has the n self-consistent molecular orbitals χ_i as its solutions, together with their orbital energies ϵ_i. The differential equation is, of course, for the *spin-orbitals* which include the spatial component(s) of each as a factor. However, the essentially *numerical* problem is to determine the *spatial* parts of these orbitals. Nevertheless, we shall retain the full spin-orbital formalism since it presents no complications and turn to special cases later.

The most useful method of approximation of functions is by expressing the unknown function as a *linear* combination of a set of known functions of the same number of variables and having similar behaviour (existence of gradients, boundary conditions etc.).[1] The mathematical and physical advantages of the use of atomic orbitals[2] to expand molecular orbitals are obvious enough:

- Mathematically, the set of "Atomic Orbitals" (atom-centred functions) may be extended indefinitely so that the approximation may be made arbitrarily accurate in principle.

- Physically, the electron distribution in molecules is not so very different from the electron distribution in the component atoms and so the orbitals of the separate atoms is an obvious (to chemists) place to start the expansion of the molecular orbitals.

- Not all many-electron systems are molecules and, in some contexts, it is useful to expand the orbitals of parts of an electronic structure in terms of functions better adapted to those structures; plane waves are often used in this application. Although the derivations given below do not depend on the nature of the functions used, certain assumptions about the finite nature of the integrals are implicit and, in the case of plane waves, these assumptions may not be true without some additional transformations.

---

[1] Here we see why it was not necessary to worry about the boundary conditions on the solutions of the Hartree–Fock equations; if we use a linear expansion of this type, they are automatically satisfied.

[2] Notice that it has not yet been indicated how we would, in fact, calculate these atomic orbitals and we seem to be locked into a regress; this point is addressed in Appendix 3.A to this chapter; "Atomic Orbitals"

3.2. NOTATION

It is not possible to simply *substitute* an approximation to the MOs χ_i into the above equation and attempt a solution since we would then be violating the variational conditions under which the equations have been derived. In the derivation of the last chapter we obtained the solution by optimisation with respect to *arbitrary* variations in the functional form of the χ_i, not simply one form of parametric optimisation. We must start the variational derivation anew under the circumstances of a special kind of parametric variation procedure: the linear expansion of the orbitals in terms of a set of known functions.

To see this more clearly it is only necessary to look at the effect of a *differential* operator on a finite (and therefore incomplete) linear expansion of a molecular orbital in terms of some known functions; if

$$\chi = \sum_{i=1}^{m} c_i \varphi_i$$

then

$$\hat{h}\chi = \sum_{i=1}^{m} c_i \hat{h}\varphi_i$$

but, unless we have been very wise or very lucky in our choice of expansion functions φ_i, the right-hand-side of this expression will *not be capable of being expanded in terms of those φ_i*:

$$\hat{h}\chi \neq \sum_{i=1}^{m} d_i \varphi_i$$

for *any* choice of d_i. Thus the whole simple scheme will not work. We shall have to return to this point later when we look at "semi-numerical" methods which try to solve the differential equation directly by use of the linear expansion technique without resort to the linear variation method. This is taken up in Chapter 34.

3.2 Notation

The molecular orbitals are denoted by χ_i and are functions of space and spin for a single electron

$$\chi_i = \chi_i(\vec{x})$$

In the overwhelming majority of cases the spin dependence of all of these MOs is a simple spin "factor" α or β, and the *essential* functional dependence of the MO on the variables of the electron is through the spatial factor. That is, in these cases the MO χ_i is given by

$$\chi_i(\vec{x}) = \psi_{i'}(\vec{r})(\alpha(s) \quad \text{or} \quad \beta(s))$$

where the "spin variable" s is written explicitly for emphasis.

The symbol ψ (lower-case psi) will always be used for the *spatial* part of a molecular orbital (almost exclusively one which is the result of some variational calculation).

However, it is by no means necessary that the spin part of an individual MO be entirely α or β; the more general form

$$\chi_i(\vec{x}) = \psi_{i'}^{\alpha}(\vec{r})\alpha(s) + \psi_{i'}^{\beta}\beta(s)$$

is possible where $\psi_{i'}^{\alpha}$ and $\psi_{i'}^{\beta}$ are different.

In every case, however, the expansion of the MOs in terms of some known basis functions is essentially the use of a known set of *spatial* functions to expand the unknown *spatial* parts of the MOs. If, therefore, we use a particular set of spatial *basis functions* we must "double" this spatial basis by multiplication of the two possible spin factors to be sure of using a basis which is fully balanced for expansion of the general MOs.

In the first instance we shall assume that this has been done and when we speak of "a basis of m functions" we shall mean $m/2$ spatial functions multiplied by the spin factors. No assumption is made about the *order* in which these basis functions are numbered but, of course, when the equations come to be implemented there are certain obvious conventions and manipulational devices which will be used.

Although it is useful to distinguish between the *spin-orbital* MOs $\chi_i(\vec{x})$ and the *spatial* MOs $\psi_i(\vec{r})$ by usin a completely different symbol for each, the different types of *basis function* are more closely related and merit only a change of style of the same symbol. Where it is necessary, the symbol $\varphi_k(\vec{x})$ will be used for a typical spin-orbital basis function, which are spatial basis functions $\phi_k(\vec{r})$ multiplied by one or other of the spin factors. This has a slight danger of confusion, but there are only so many symbols available.

3.3 The expansion

The linear expansion ("Algebraic") technique consists of expanding each molecular orbital $\chi_i(\vec{x})$ as a linear combination of the basis functions $\varphi_r(\vec{x})$:

$$\chi_i = \sum_{r=1}^{m} \varphi_r C_{ri} \tag{3.1}$$

In this expansion we begin to try to make a distinction amongst the increasing profusion of sub- and super-scripts by using the familiar "integer" indexing

3.3. THE EXPANSION

labels $i, j, k, \ell, \ldots$ to label MOs (or MO-related quantities) and the less familiar indexing quantities $r, s, t, u, \ldots$ to label the basis functions (and quantities related to basis functions).[3]

The quantities C_{ri} are the expansion coefficients which are to be determined variationally. There are n expansions like eqn (3.1): one for each of the n electrons in the system. Clearly, all these n expansions can be conveniently collected into a single matrix equation by the definition of a *row* of χ_is, a *row* of φ_rs and an $m \times n$ *matrix* $\boldsymbol{C}$:

$$\boldsymbol{\chi} = (\chi_1, \chi_2, \chi_3, \ldots, \chi_n) \quad (3.2)$$

is the $1 \times n$ matrix (row) of MOs,

$$\boldsymbol{\varphi} = (\varphi_1, \varphi_2, \varphi_3, \ldots, \varphi_m) \quad (3.3)$$

is the $1 \times m$ matrix (row) of basis functions and

$$\boldsymbol{C} = \begin{pmatrix} C_{11} & C_{12} & \ldots & C_{1n} \\ C_{21} & C_{22} & \ldots & C_{2n} \\ C_{31} & C_{32} & \ldots & C_{3n} \\ \ldots & \ldots & \ldots & \ldots \\ \ldots & \ldots & \ldots & \ldots \\ \ldots & \ldots & \ldots & \ldots \\ C_{m1} & C_{m2} & \ldots & C_{mn} \end{pmatrix} \quad (3.4)$$

is the $m \times n$ matrix of expansion coefficients. So that eqn (3.1) becomes

$$\boldsymbol{\chi} = \boldsymbol{\varphi} \boldsymbol{C} \quad (3.5)$$

It is sometimes useful to indicate the *functional dependence* of the columns of functions in an obvious way

$$\boldsymbol{\varphi}(\vec{x}) \quad \text{and} \quad \boldsymbol{\chi}(\vec{x})$$

in order to facilitate the definition of *matrices* of certain integrals appearing in the theory. Thus the matrix of orbital overlap integrals

$$S_{rs} = \int \varphi_r^* \varphi_s d\tau$$

can be written

$$\boldsymbol{S} = \int \boldsymbol{\varphi}^\dagger \boldsymbol{\varphi} d\tau$$

where $\boldsymbol{\varphi}^\dagger$ is the Hermitian conjugate of $\boldsymbol{\varphi}$ (transposition and complex conjugation).

---
[3] Of course, the actual *values* of the quantities i, j, k, l and r, s, t, u can only be integers so that this distinction serves only as an heuristic aid during derivations.

Similarly, the matrix of one-electron integrals may be written

$$h = \int \varphi^\dagger \hat{h} \varphi d\tau$$

The relationship between the overlap matrix involving the basis functions φ_r and the MOs χ_i is therefore, using eqn (3.5),

$$\begin{aligned} \int \chi^\dagger \hat{h} \chi d\tau &= \int (\varphi C)^\dagger \hat{h} \varphi C d\tau \\ &= C^\dagger \left(\int \varphi^\dagger \hat{h} \varphi d\tau \right) C \\ &= C^\dagger h C \end{aligned}$$

Of course we cannot use the same symbol h, for the matrix of one-electron integrals over both the functions χ_i and φ_r; there are different numbers of them ($n \times n$ in the MO case, $m \times m$ in the case of the basis functions) and they have different values. If there is a danger of confusion, we can distinguish between them by a label:

$$h^\chi \quad \text{or} \quad h^\varphi$$

for example.

3.4 The energy expression

The mean value of the energy associated with a single determinant of orthonormal orbitals χ_i is a special case of Slater's Rules obtained in the last section of Appendix 2.A:

$$E = \frac{\int \Phi^* \hat{H} \Phi d\tau}{\int \Phi^* \Phi d\tau} \tag{3.6}$$

where

$$\int \Phi^* \Phi d\tau = 1$$

and

$$\begin{aligned} \int \Phi^* \hat{H} \Phi d\tau &= \sum_{i=1}^n \int \chi_i^* \hat{h} \chi_i d\tau \\ &+ \sum_{i=1}^n \sum_{j=1}^n [(\chi_i \chi_i, \chi_j \chi_j) - (\chi_i \chi_j, \chi_i \chi_j)] \end{aligned} \tag{3.7}$$

in the notation of Appendix 1.B.

The Hamiltonian operator is a sum of one- and two-electron operators which is symmetrical in the coordinates of all the particles:

$$\hat{H}(\vec{r}_1, \vec{r}_2, \ldots, \vec{r}_n) = \sum_{i=1}^{n} \hat{h}(\vec{r}_i) + \sum_{i=1}^{n}\sum_{i\neq j=1}^{n} \frac{1}{|\vec{r}_i - \vec{r}_j|}$$

Notice that it has been assumed that the Hamiltonian operator depends only on the *spatial* coordinates of the n electrons; there is no explicit dependence of the energy *operator* on electron spin. There is no difficulty of *principle* in including Hamiltonians which have spin-dependence; indeed the derivation is, line for line, the same. However, the *size* of the spin-dependent energies is so small that is is not sensible to include them in a *variational* treatment where their effects would be swamped by the electrostatic terms.[4]

3.5 The numerator: Hamiltonian mean value

The one-electron terms in eqn (3.6) are simply the diagonal elements of the matrix of one-electron integrals derived in the last section:

$$\int \chi_i^* \hat{h} \chi_i d\tau = \left(\boldsymbol{C}^\dagger \boldsymbol{h} \boldsymbol{C}\right)_{ii} \tag{3.8}$$

Thus the sum of them is just

$$\sum_{i=1}^{n} \int \chi_i^* \hat{h} \chi_i d\tau = \sum_{i=1}^{n} \left(\boldsymbol{C}^\dagger \boldsymbol{h} \boldsymbol{C}\right)_{ii} = \mathrm{tr}\left(\boldsymbol{C}^\dagger \boldsymbol{h} \boldsymbol{C}\right) \tag{3.9}$$

where the "trace" of a matrix is the sum of the elements along its principle diagonal and is denoted by "tr" in mathematical notation:

$$\mathrm{tr}\boldsymbol{A} = \sum_{i=1}^{n} A_{ii}$$

The electron-repulsion contribution to the energy is more complicated because the transformation from the basis-function repulsion integrals to the MO repulsion integrals involves *four* appearances of the elements C_{ri} of the coefficient matrix.

A typical "Coulomb"-type integral

$$(\chi_i \chi_i, \chi_j \chi_j) = \int d\tau_1 \int d\tau_2 \chi_i^*(\vec{x}_1) \chi_i(\vec{x}_1) \frac{1}{|\vec{r}_1 - \vec{r}_2|} \chi_j^*(\vec{x}_2) \chi_j(\vec{x}_2)$$

---
[4]There will be certain advantages of organisation which result from the omission of spin-dependent terms from the Hamiltonian when we come to consider the implementation of the solution method.

can be expanded in terms of the basis-function repulsion integrals

$$(\varphi_r\varphi_s, \varphi_t\varphi_u) = \int d\tau_1 \int d\tau_2 \varphi_r^*(\vec{x_1})\varphi_s(\vec{x_1})\frac{1}{|\vec{r_1}-\vec{r_2}|}\varphi_t^*(\vec{x_2})\varphi_u(\vec{x_2})$$

as

$$(\chi_i\chi_i, \chi_j\chi_j) = \sum_{r,s,t,u=1}^{m} C_{ri}^*C_{si}C_{tj}^*C_{uj}(\varphi_r\varphi_s, \varphi_t\varphi_u) \qquad (3.10)$$

so that the sum of such integrals over all i and j is:

$$\begin{aligned}\sum_{i,j=1}^{n}(\chi_i\chi_i, \chi_j\chi_j) &= \sum_{i,j=1}^{n}\sum_{r,s,t,u=1}^{m} C_{ri}^*C_{si}C_{tj}^*C_{uj}(\varphi_r\varphi_s, \varphi_t\varphi_u) \qquad (3.11)\\ &= \sum_{r,s,t,u}^{m}\left(\sum_{i=1}^{n}C_{ri}^*C_{si}\right)\left(\sum_{j=1}^{n}C_{tj}^*C_{uj}\right)(\varphi_r\varphi_s, \varphi_t\varphi_u)\\ &= \sum_{r,s,t,u}^{m}\left(CC^\dagger\right)_{rs}\left(CC^\dagger\right)_{tu}(\varphi_r\varphi_s, \varphi_t\varphi_u)\\ &= \sum_{r,s=1}^{m}\left(CC^\dagger\right)_{rs}J_{sr}\end{aligned}$$

where the so-called Coulomb matrix J has elements defined by

$$J_{rs} = \sum_{t,u=1}^{m}\left(CC^\dagger\right)_{tu}(\varphi_r\varphi_s, \varphi_t\varphi_u) \qquad (3.12)$$

The equation can be further simplified by noting that

$$\sum_{r,s=1}^{m}\left(CC^\dagger\right)_{rs}J_{sr} = \text{tr}\left(CC^\dagger J\right) = \text{tr}\left(C^\dagger JC\right)$$

The trace of the product of two $(m \times m)$ matrices is simply

$$\text{tr}\,AB = \sum_{i,j=1}^{m} A_{ij}B_{ji} = \text{tr}\,BA$$

therefore the trace of a product of matrices is unchanged by *cyclic* permutation of the factors in the product.

A similarly straightforward but lengthy analysis shows that the sum of the so-called exchange integrals leads to a similar expression

$$(\chi_i\chi_j, \chi_i\chi_j) = \int d\tau_1 \int d\tau_2 \chi_i^*(\vec{x_1})\chi_j(\vec{x_1})\frac{1}{|\vec{r_1}-\vec{r_2}|}\chi_i^*(\vec{x_2})\chi_j(\vec{x_2})$$

3.5. THE NUMERATOR: HAMILTONIAN MEAN VALUE 83

and
$$\sum_{i,j=1}^{n}(\chi_i\chi_j,\chi_i\chi_j) = \mathrm{tr}\left(\boldsymbol{C}^\dagger \boldsymbol{K}\boldsymbol{C}\right) \qquad (3.13)$$

where
$$K_{rs} = \sum_{t,u=1}^{m}\left(\boldsymbol{C}\boldsymbol{C}^\dagger\right)_{tu}(\varphi_r\varphi_u,\varphi_t\varphi_s) \qquad (3.14)$$

We are now, at last, in a position to write down the numerator of the energy expression eqn (3.6) for the MO linear expansion method in terms of the (assumed known) integrals involving the basis functions φ_r and the coefficients $\boldsymbol{C}$ which are to be determined.

$$E = \mathrm{tr}\left[\boldsymbol{C}^\dagger \boldsymbol{h}\boldsymbol{C}\right] + \frac{1}{2}\mathrm{tr}\left[\boldsymbol{C}^\dagger\left(\boldsymbol{J} - \boldsymbol{K}\right)\boldsymbol{C}\right] \qquad (3.15)$$

In order to optimise eqn (3.6) with respect to the coefficients in $\boldsymbol{C}$ we need to know how the expression varies when changes are made in $\boldsymbol{C}$. Clearly, we can find expressions for

$$\frac{\partial E}{\partial C_{ri}} \quad \text{and} \quad \frac{\partial E}{\partial C^*_{ri}}$$

for $1 \leq r \leq m$ and $1 \leq i \leq n$ and use the familiar conditions for a minimum. This proves to be rather unwieldy; it is clearer and involves a good deal less involved algebra if we use the idea of the "small change" in a *whole matrix* rather than derivatives with respect to the matrix *elements*.

We therefore look for the changes induced in E by small changes[5] ($\delta\boldsymbol{C}$ and $\delta\boldsymbol{C}^\dagger$) in the matrices $\boldsymbol{C}$ and $\boldsymbol{C}^\dagger$, separate the changes according to their *order* in $\delta\boldsymbol{C}$, $\delta\boldsymbol{C}^\dagger$ and define a turning point (minimum, usually) when the *first order* change in E induced by

$$\boldsymbol{C} \to \boldsymbol{C} + \delta\boldsymbol{C} \qquad (3.16)$$
$$\boldsymbol{C}^\dagger \to \boldsymbol{C}^\dagger + \delta\boldsymbol{C}^\dagger \qquad (3.17)$$

is zero. The *sign* of the *second order* change in E will confirm that we are at a minimum in the energy expression (or not!).

The changes ($\delta\boldsymbol{C}$ and $\delta\boldsymbol{C}^\dagger$) in the matrices are assumed to be *linearly independent* since the matrices ($\boldsymbol{C}$ and $\boldsymbol{C}^\dagger$) are, in general, complex.

The notation
$$E \to E + \delta E + \delta^{(2)}E + \ldots \qquad (3.18)$$

will be used for the change in E induced by the change shown in eqn (3.17). Thus,

$$(E + \delta E + \delta^{(2)}E + \ldots) = \mathrm{tr}\left(\boldsymbol{C} + \delta\boldsymbol{C}\right)^\dagger\left[\boldsymbol{h} + \frac{1}{2}\left(\boldsymbol{J} - \boldsymbol{K}\right)\right]\left(\boldsymbol{C} + \delta\boldsymbol{C}\right) \qquad (3.19)$$

---
[5]Note, as in the previous chapter that the notation $\delta\boldsymbol{C}^\dagger$ means $\delta(\boldsymbol{C}^\dagger)$ not $(\delta\boldsymbol{C})^\dagger$, there are two *independent* variations.

It is straightforward to extract the first-order terms:

$$\begin{aligned} \delta E &= \operatorname{tr}\left[\delta C^\dagger \left(h + \frac{1}{2}(J - K)\right) C\right] \\ &+ \operatorname{tr}\left[C^\dagger \delta \left(h + \frac{1}{2}(J - K)\right) C\right] \\ &+ \operatorname{tr}\left[C^\dagger \left(h + \frac{1}{2}(J - K)\right) \delta C\right] \end{aligned}$$

Now the matrix h is a constant so that

$$\delta\left(h + \frac{1}{2}(J - K)\right) = \frac{1}{2}\delta(J - K)$$

but J and K *do* depend on C and $C^\dagger$. In fact, it is a lengthy but simple exercise (done in Appendix 3.C) to show that

$$\operatorname{tr}\left(C^\dagger \delta J C\right) = \operatorname{tr}\left(\delta C^\dagger J C\right) + \operatorname{tr}\left(C^\dagger J \delta C\right)$$

and

$$\operatorname{tr}\left(C^\dagger \delta K C\right) = \operatorname{tr}\left(\delta C^\dagger K C\right) + \operatorname{tr}\left(C^\dagger K \delta C\right)$$

These expressions simplify the first-order change in E induced by the changes δC and $\delta C^\dagger$ to:

$$\begin{aligned} \delta E &= \operatorname{tr}\left[\delta C^\dagger (h + (J - K)) C\right] \\ &+ \operatorname{tr}\left[C^\dagger (h + (J - K)) \delta C\right] \end{aligned} \quad (3.20)$$

This completes the expression for the numerator of the single-determinant energy expression of eqn (3.6).

3.6 The denominator: normalisation condition

We have seen earlier that a single determinant of orbitals is normalised if the individual orbitals of which it is composed are themselves normalised:

$$\int \Phi^* \Phi d\tau = 1 \quad \text{if} \quad \int \chi_i^* \chi_j d\tau = \delta_{ij}$$

That is

$$\sum_{r,s=1}^{m} C_{ri}^* C_{sj} \int \varphi_r^* \varphi_j d\tau = \delta_{ij}$$

3.6. THE DENOMINATOR: NORMALISATION CONDITION

or, more compactly,

$$C^\dagger S C = 1 \qquad (3.21)$$

If now (small) changes are made in the MO coefficients C, $C^\dagger$ and the normalisation condition is to be retained, then the possible changes (δC and $\delta C^\dagger$) are constrained by

$$\delta\left(C^\dagger S C\right) = 0$$

or, since the matrix S is a constant:

$$\delta C^\dagger S C + C^\dagger S \delta C = 0 \qquad (3.22)$$

The variational problem is to combine the two expressions of eqn (3.20) and eqn (3.22) and to look for the condition where the original energy expression, eqn (3.6), is a turning point.

The "direct" method of obtaining the change in the *quotient* eqn (3.6) has not even been considered as it is far too involved to be manageable. The traditional method is to use Lagrange's method of undetermined multipliers; to form a *linear combination* of the two expressions which are required to vanish, and require this linear combination to vanish *for each degree of variational freedom*. In our case this is to combine eqns (3.20) and (3.22) using a linear combination coefficient *for each of the linear degrees of freedom* C_{ri}. Obviously this can be done most transparently by the continuation of our use of matrix methods. The initial problem is that one of the equations to be satisfied, eqn (3.20), is a scalar equation and the other, eqn (3.22), is a matrix equation.

However, the matrix equation eqn (3.22) has the form

$$A = 0$$

and, if ϵ is an *arbitrary* matrix (of the correct $n \times n$ size) then this is equivalent to

$$\text{tr}\left(\epsilon A\right) = \text{tr}\left(A\epsilon\right) = 0$$

because *each* element of A must be zero if the sum is zero for arbitrary ϵ. This means that eqn (3.22) is equivalent to

$$\text{tr}\left(\delta C^\dagger S C \epsilon\right) + \text{tr}\left(C^\dagger S \delta C \epsilon\right) = 0$$

or, since $\text{tr} AB = \text{tr} BA$,

$$\text{tr}\left(\delta C^\dagger S C \epsilon\right) + \text{tr}\left(\epsilon C^\dagger S \delta C\right) = 0 \qquad (3.23)$$

where ϵ is a single general arbitrary matrix with as many degrees of freedom as necessary.

3.7 The Hartree–Fock equation

Now that we have two *scalar* equations, both of which are required to be zero, we may combine them; subtracting eqn (3.23) from eqn (3.20) together gives

$$\text{tr}\left[\delta C^\dagger \left(h + (J - K)\right) C\right]$$
$$+\text{tr}\left[C^\dagger \left(h + (J - K)\right) \delta C\right]$$
$$-\text{tr}\left(\delta C^\dagger S C \epsilon\right) - \text{tr}\left(\epsilon C^\dagger S \delta C\right) = 0$$

which may be re-arranged into a more compact and transparent form:

$$\text{tr}\delta C^\dagger \left[(h + J - K) C - S C \epsilon\right]$$
$$+\text{tr}\left[C^\dagger (h + J - K) - \epsilon C^\dagger S\right] \delta C = 0$$

Now we may take the step equivalent to requiring the *partial* derivatives of a function with respect to several variables to be *separately* zero. Since the variations δC and $\delta C^\dagger$ are *independent* and *arbitrary*, requiring the above expression to vanish implies that the *multipliers* of these arbitrary independent variations must vanish. That is,

$$(h + J - K) C - S C \epsilon = 0$$

from variations $\delta C^\dagger$, and

$$C^\dagger (h + J - K) - \epsilon C^\dagger S = 0$$

from corresponding variations δC.

The Hermitian conjugate of the latter equation is (remember that h, J and K are Hermitian)

$$(h + J - K) C - S C \epsilon^\dagger = 0$$

which is clearly identical to the first equation *if ϵ is Hermitian*, i.e. if

$$\epsilon^\dagger = \epsilon$$

Thus, the requirement that the *two* independent equations be solved can be replaced by the solution of the single (matrix) equation

$$(h + J - K) C = S C \epsilon \tag{3.24}$$

plus the requirement that ϵ be a Hermitian matrix. This, then, is the equation satisfied by the variationally optimum linear combination coefficients C in the best single-determinant wavefunction.

3.8 "Normalisation": the Lagrangian

It is worth mentioning one point which may well turn up later. In using the variation induced in the normalisation condition on the MOs

$$\delta\left(C^\dagger SC\right) = 0$$

ostensibly derived from the normalisation of those MOs,

$$C^\dagger SC = 1$$

it should have been mentioned that this result can, in fact, be generated from the condition

$$C^\dagger SC = A$$

where A is *any constant matrix*; in particular A may be any constant diagonal matrix. Thus the optimum MOs χ_i may well not be normalised, which is also obvious from eqn (3.24), where the matrix C may be multiplied by any scalar without disturbing the solution.

The question of the *orthogonality* of the MOs is not so easily disposed of and hinges on the properties of determinants which we will address later; for the moment we simply note that the *actual* effect of the normalisation condition is not so much to ensure that the MOs are orthonormal but to take care of the fact that the elements of the matrix C are not, in fact, independently variable in the variational procedure. The inclusion of the constraint ensures that the correct number of *independent* variables are used; in a sense it is pure good luck that the apparently *physical* condition which we imposed on the MOs does in fact have the correct mathematical properties to ensure a unique variationally optimised single determinant of MOs.

The true role of the Normalisation Condition is to generate an expression in which the matrix C may be varied unconditionally. The number of linear coefficients in the matrix C is, of course, $m \times n$ but the number of *independent* linear degrees of freedom in the Hartree–Fock calculation by the linear expansion method is $n(m-n)$, the size of the matrix C *minus* the n^2 constraints implied by the normalisation condition.

The Lagrangian method is precisely to set up this expression, which we may call the "Lagrangian"[6]

$$L\left(C\right) = \mathrm{tr}\left[C^\dagger\left(h + \frac{1}{2}(J-K)\right)C\right] - \mathrm{tr}\left[C^\dagger SC\epsilon\right] \quad (3.25)$$

---

[6]It is a little unfortunate that this name is also used in mechanics for a completely different function related to the classical Hamiltonian, but Lagrange was a multi-faceted theoretician!

This function will prove useful in any applications where it is important to treat the elements of C as independent; derivatives of E with respect to the C_{ri}, for example.

For the moment we simply note that the Hartree–Fock equation (3.24) could, more elegantly, have been derived from the vanishing of the first-order variation (δL) in L induced by *arbitrary* variations in the matrix C.

3.9 Preliminary summary

The linear expansion technique eqn (3.5)

$$\chi = \varphi C$$

involves the generation (computation, estimation etc.) of the one-electron energy integrals

$$h = \int \varphi^\dagger \hat{h} \varphi d\tau$$

and the electron-repulsion integrals

$$(\varphi_r \varphi_s, \varphi_t \varphi_u) = \int d\tau_1 \int d\tau_2 \varphi_r^*(\vec{x_1}) \varphi_s(\vec{x_1}) \frac{1}{|\vec{r_1} - \vec{r_2}|} \varphi_t^*(\vec{x_2}) \varphi_u(\vec{x_2})$$

as well as the more straightforward overlap integrals

$$S = \int \varphi^\dagger \varphi d\tau$$

When these numbers have been obtained, the optimum molecular orbital coefficient matrix is the solution of eqn (3.24):

$$(h + J - K)\,C = SC\epsilon$$

which may be written more compactly as

$$h^F C = SC\epsilon$$

with the definition

$$h^F = (h + J - K)$$

where the matrices J and K are defined by

$$J_{rs} = \sum_{t,u=1}^{m} \left(CC^\dagger\right)_{tu} (\varphi_r \varphi_s, \varphi_t \varphi_u)$$

and

$$K_{rs} = \sum_{t,u=1}^{m} \left(CC^\dagger\right)_{tu} (\varphi_r \varphi_u, \varphi_t \varphi_s)$$

The basis functions φ_r are products of spatial functions and an α or β spin factor, and the elements of the coefficient matrix are (in general) complex numbers.

3.10 Some technical manipulations

The Hartree–Fock equation, eqn (3.24), regarded as a set of linear equations, is a set of m equations for m unknowns and so completely soluble. However, the m unknowns present themselves in an unusual way; essentially as various *ratios* between the interdependent elements of the C matrix and the $n(n+1)/2$ elements of a Hermitian matrix ϵ.

It is both conceptually and computationally useful to cast the equation into a form which can be interpreted more easily. This possibility hinges on a simple property of determinants; the fact that the value of a determinant is invariant against a unimodular transformation amongst its rows (or columns). The molecular orbitals of which the single-determinant wavefunction is composed may be transformed by any linear transformation of unit determinant *without changing the wavefunction* although the orbitals themselves (the coefficients in the matrix C and the elements of the matrix ϵ) are changed.

Thus, we can form a new set of n MOs χ' (say) by the matrix D (say):

$$\chi' = \chi D$$

Repeating the derivation would yield the corresponding equation for the optimum coefficients C' (say) as

$$(h + J - K)C' = SC'\epsilon'$$

where
$$C' = CD$$

and
$$\epsilon' = D^{\dagger}\epsilon D$$

But ϵ is *Hermitian* and D is arbitrary (if non-singular) so that we may *choose* D so that ϵ takes its *simplest possible form*. That is, we may choose D so that as few as possible of the mn independent variables are contained in the matrix ϵ.

We choose D to be the unitary matrix that *diagonalises* ϵ. This choice, which uses up $n(n-1)/2$ of the degrees of freedom in the system of equations, reduces the number of unknowns associated with the matrix ϵ from $n(n+1)/2$ complex numbers to n *real* numbers which, of course, exactly compensates (the eigenvalues of a Hermitian matrix are always *real*).

The effect of reducing the number of degrees of freedom in ϵ is to *concentrate* the number of independent variables of the system into C; what we have done is to make *all but one* of the elements of each of the n columns of C independent, leaving the remaining one of each column (n in total) to be determined by the normalisation requirement on the columns.

Let us therefore assume in what follows that this choice has always been made;[7] the matrix ϵ is *always (real and) diagonal*. We simply take over eqn (3.24), together with this simplification and call this *the* Hartree–Fock Equation.

With this choice, the form of the Hartree–Fock equation begins to look more like the form of a Schrödinger equation, particularly if we write out the full equation as a set of n equations for the column of coefficients for a particular MO; that is if

$$C = (c^1, c^2, \ldots, c^n)$$

where each c^i is a column of the m coefficients defining χ_i in terms of the basis functions φ_r:

$$\chi_i = \sum_{r=1}^{m} \varphi_r c_r^i = \varphi c^i$$

(for $i = 1, n$) Then, with diagonalised ϵ, the equations become

$$h^F c^i = \epsilon_i S c^i$$

where we have written ϵ_i in place of ϵ_{ii} and used

$$h^F = h + J - K$$

for the "Fock matrix" to emphasise the similarity to a Schrödinger equation. This similarity is even more striking if the basis functions form an *orthonormal set*:

$$S_{rs} = \delta_{rs}$$

since, in this case the equations become:

$$h^F c^i = \epsilon_i c^i$$

for $i = 1, n$. In fact, it proves possible to reduce the Hartree–Fock equation to this orthonormal form, by a simple matrix multiplication technique, *in all cases* where the basis set is non-redundant.

The equation can be written in the new compact form as

$$h^F C = SC\epsilon$$

We can, of course, insert a unit matrix anywhere in the equation

$$h^F 1 C = S 1 C \epsilon$$

---

[7]It may seem paradoxical to assume that we can insist on the *form* of what are, after all, unknown matrices. What, in fact, we do when we come to *implement* a method of solution is to use a *numerical method* which solves the equations by *generating* a diagonal ϵ.

3.10. SOME TECHNICAL MANIPULATIONS

and we choose a particular representation of **1**:

$$1 = VV^{-1}$$

(for a non-singular matrix V) to give

$$h^F VV^{-1} C = SVV^{-1} C\epsilon$$

Multiplying this equation from the right by $V^\dagger$ and grouping the terms gives

$$\left(V^\dagger h^F V\right)\left(V^{-1} C\right) = \left(V^\dagger S V\right)\left(V^{-1} C\right)\epsilon$$

If we now define

$$V^\dagger h^F V = \bar{h}^F$$

and

$$V^{-1} C = \bar{C}$$

and noting that

$$V^\dagger S V = \bar{S}$$

the Hartree–Fock equation takes an identical form to the original

$$\bar{h}^F \bar{C} = \bar{S}\bar{C}\epsilon$$

Notice that, as expected, the elements of the (diagonal) matrix ϵ are unchanged by this transformation but, of course, the elements of the MO coefficient matrix are changed *because they refer to a new basis*.

Now, we may choose V in any way that is convenient for simplification of the equations (provided it is non-singular). Obviously if we choose V so that $\bar{S} = 1$, i.e.

$$V^\dagger S V = 1$$

the Hartree–Fock equation takes the simple orthonormal-basis form

$$\bar{h}^F \bar{C} = \bar{C}\epsilon$$

This transformation is of much more than cosmetic importance; it is this step which makes the technical implementation of the matrix Hartree–Fock equations possible. We shall return to this point in detail later.[8]

---

[8]The Hartree–Fock equation can be reduced to a simple form simply by multiplying from the left by the inverse of the overlap matrix to give

$$(S^{-1} h^F) C = C\epsilon$$

but this form is not, in practice, useful because $(S^{-1} h^F)$ is not a *symmetrical* matrix which makes implementation difficult.

The requirement that $V^\dagger SV = 1$ does not fix V uniquely since, if V is replaced by the product VU for any *unitary* matrix U, we have

$$U^\dagger V^\dagger SVU = U^\dagger 1 U = 1$$

since

$$U^\dagger U = 1$$

for any unitary matrix. This freedom corresponds to the freedom to make unitary transformations amongst the orthonormal orbitals of a determinant.

It is sometimes technically useful to have the capability to use *different* sets of orthonormal orbitals in different situations, as we shall see, but for the moment we can take the *simplest* solution of the problem

$$V^\dagger SV = 1$$

i.e.

$$VV^\dagger = S^{-1}$$

namely

$$V = V^\dagger = S^{-1/2}$$

which always exists if the basis set has a non-singular overlap matrix S.

3.11 Canonical orbitals

The particular set of MO coefficients C (the ones which are associated with a diagonal ϵ) are, of course, *unique* and, as such, have an important physical interpretation which we can begin to investigate now.

Clearly the elements of the diagonal matrix ϵ have the dimensions of energy; each ϵ_i is an energy quantity associated with the orbital χ_i (coefficients c^i). Simply by analogy with the Schrödinger equation, one might well expect that these ϵ_i are energies of electrons in the associated MOs χ_i.

In order to confirm this conjecture we need to know the actual relationship between the ϵ_i and the MO coefficients and the basis-function energy integrals. The most direct way to obtain an expression for (and hence a physical interpretation) of the ϵ_i is from the Hartree–Fock equation for each column of C in the orthonormal basis:

$$\bar{h}^F \bar{c}^i = \epsilon_i \bar{c}^i$$

If the column c^i is normalised, ($\bar{c}^{i\dagger}\bar{c} = 1$), then

$$\epsilon_i = \bar{c}^{i\dagger} \bar{h}^F \bar{c}$$

3.11. CANONICAL ORBITALS

or
$$\epsilon_i = \text{tr} \bar{\boldsymbol{h}}^F (\bar{\boldsymbol{c}}^i \bar{\boldsymbol{c}}^{i\dagger})$$

But, since the orthonormal basis coefficients are related to the original basis coefficients by
$$\bar{\boldsymbol{c}}^i = \boldsymbol{S}^{1/2} \boldsymbol{c}^i$$
that is,
$$\boldsymbol{c}^i = \boldsymbol{S}^{-1/2} \bar{\boldsymbol{c}}^i$$

the expression for ϵ_i in terms of the original (non-orthogonal) basis is
$$\epsilon_i = \text{tr} \bar{\boldsymbol{h}}^f \boldsymbol{S}^{1/2} \boldsymbol{c}^i \boldsymbol{c}^{i\dagger} (\boldsymbol{S}^{1/2})^\dagger$$

which, because $\boldsymbol{S}$ (and therefore $\boldsymbol{S}^{1/2}$) is Hermitian, and
$$\boldsymbol{S}^{1/2} \bar{\boldsymbol{h}}^f \boldsymbol{S}^{1/2} = \boldsymbol{h}^F$$
becomes
$$\epsilon_i = \text{tr} \boldsymbol{h}^F \boldsymbol{c}^i \boldsymbol{c}^{i\dagger}$$

an expression of identical form to the one involving the orthogonalised basis.

Written out in full, this expression becomes
$$\epsilon_i = \text{tr} \left(\boldsymbol{h} + \boldsymbol{J} - \boldsymbol{K} \right) \boldsymbol{c}^i \boldsymbol{c}^{i\dagger}$$

Now it is shown in Appendix 3.C that
$$\text{tr} \boldsymbol{J} \boldsymbol{c}^i \boldsymbol{c}^{i\dagger} = \sum_{j=1}^{m} (\chi_j \chi_j, \chi_i \chi_i)$$
and
$$\text{tr} \boldsymbol{K} \boldsymbol{c}^i \boldsymbol{c}^{i\dagger} = \sum_{j=1}^{m} (\chi_i \chi_j, \chi_i \chi_j)$$

While, of course,
$$\text{tr} \boldsymbol{h} \boldsymbol{c}^i \boldsymbol{c}^{i\dagger} = \boldsymbol{c}^{i\dagger} \boldsymbol{h} \boldsymbol{c}^i = \int \chi_i^* \hat{h} \chi_i dV$$

so that the full expression for ϵ_i in terms of integrals over the molecular orbitals χ_i is
$$\epsilon_i = \int \chi_i^* \hat{h} \chi_i dV + \sum_{j=1}^{m} [(\chi_j \chi_j, \chi_i \chi_i) - (\chi_i \chi_j, \chi_i \chi_j)] \quad (3.26)$$

We can now investigate the physical interpretation of the ϵ_i simply by evaluating the energy of the single determinant of $(n-1)$ MOs with one electron removed. If, for convenience of notation, we remove orbital χ_n from the determinant we obtain an energy expression for the remaining system precisely analogous to the original n-electron expression:

$$\sum_{i=1}^{n-1} \int \chi_i^* \hat{h} \chi_i d\tau + \sum_{i=1}^{n-1} \sum_{j=1}^{n-1} [(\chi_i\chi_i, \chi_j\chi_j) - (\chi_i\chi_j, \chi_i\chi_j)]$$

And the difference between this $(n-1)$-electron energy and the n-electron original energy is just ϵ_n, the energy of the "removed" electron:

$$E_n = E_{n-1} + \epsilon_n$$

Clearly there is nothing particular about the value of n; it was simply used for notational convenience.

> The result is that the values of the (diagonal) ϵ_i are just the energy of an electron in orbital χ_i in the restricted sense that the electron distribution is assumed unchanged on removing the electron. It is, in this sense that the ϵ_i may be called "orbital energies."

We shall see later that a much stronger result may be derived which reinforces this interpretation of the ϵ_i.

It is important to remember that this result only applies to the *unique* set of orbitals for which the matrix ϵ is diagonal. These energy-optimised molecular orbitals are called the *canonical* molecular orbitals.

3.11.1 The full MO matrix

As we shall shortly see when we come to actually implement the solution of the matrix SCF equations, the technique is to solve

$$\bar{h}^F \bar{U} = \bar{U} \epsilon'$$

where U is the *square* matrix of which the first n columns are the matrix C. This solution is equivalent to the solution of the target equation

$$\bar{h}^F \bar{C} = \bar{C} \epsilon$$

in the sense of generating the same C and ϵ, if the form of ϵ' is "block diagonal":

$$\epsilon' = \begin{pmatrix} \epsilon & 0 \\ 0 & \epsilon'' \end{pmatrix}$$

That is, depending on whether or not we are interested in canonical orbitals, ϵ may or may not be diagonal; similarly ϵ'' may or may not be diagonal. But, *there must be no non-zero elements in the "off-diagonal blocks" connecting the two*. This is, in fact, another way of defining the solution of the SCF equations:

3.12. THE TOTAL ENERGY

The matrix U' must bring the square Hartree–Fock matrix into block-diagonal form in which the matrix ϵ' has no non-zero elements ϵ'_{ij} or ϵ'_{ji} for $(i = 1, n), (j = n + 1, m)$.

This condition is completely equivalent to the matrix eigenvalue statement but is of more formal value as we shall see later when considering the restricted open shell SCF equations. However it cannot be made the basis of a strategy for obtaining the actual solutions of the SCF equations.

3.12 The total energy

Having generated an equation for the orthonormal orbitals, we can use the equation eqn (3.15) as the energy expression and it is useful to put it into a form which involves the matrix h^F. The energy is given by

$$E = \operatorname{tr}\left[C^\dagger h C\right] + \frac{1}{2}\operatorname{tr}\left[C^\dagger \left(J - K\right) C\right]$$

which may be rearranged by using

$$h^F = h + J - K$$

to

$$E = \operatorname{tr}\left[C^\dagger h^F C\right] - \frac{1}{2}\operatorname{tr}\left[C^\dagger \left(J - K\right) C\right]$$

and the first term is simply the sum of the orbital energies:

$$\operatorname{tr} C^\dagger h^F C = \sum_{i=1}^{n} \epsilon_i$$

so that the total electronic energy is *not* simply the sum of the individual electron energies but must be corrected by a amount exactly equal to the total inter-electron repulsion energy.[9]

$$E = \sum_{i=1}^{n} \epsilon_i - \frac{1}{2}\operatorname{tr}\left[C^\dagger \left(J - K\right) C\right]$$

The reason for this is clear; the energy of each electron is calculated *in the field of all the others* so that the sum of these energies includes the electron repulsion energy exactly *twice*. For computational purposes it is useful to recast the total energy into a form which only involves *two* matrices:

$$E = \frac{1}{2}\left[\operatorname{tr} C^\dagger h C + \operatorname{tr} C^\dagger h^F C\right] \tag{3.27}$$

---

[9] This means that it is not completely obvious that the ground state of an n-electron system is obtained by occupying the n lowest-lying MOs; it is possible that the additional electron-repulsion terms could upset this simple picture.

3.13 Summary

Section 3.9 gives what might be called a mathematical summary of the Hartree–Fock equations; we are now in a position to give a "physical" summary. For convenience some of the preliminary summary is repeated.

The linear expansion technique (3.5)

$$\chi = \varphi C$$

involves the acquisition of the overlap and one-electron integrals

$$S = \int \varphi^\dagger \varphi d\tau \quad \text{and} \quad h = \int \varphi^\dagger \hat{h} \varphi d\tau$$

and the electron-repulsion integrals

$$(\varphi_r \varphi_s, \varphi_t \varphi_u) = \int d\tau_1 \int d\tau_2 \varphi_r^*(\vec{x_1}) \varphi_s(\vec{x_1}) \frac{1}{|\vec{r_1} - \vec{r_2}|} \varphi_t^*(\vec{x_2}) \varphi_u(\vec{x_2})$$

When these numbers have been obtained, the optimum molecular orbital coefficient matrix is the solution of:

$$h^F C = SC\epsilon$$

with the definition

$$h^F = (h + J - K)$$

where the matrices J and K are defined by

$$J_{rs} = \sum_{t,u=1}^{m} \left(CC^\dagger\right)_{tu} (\varphi_r \varphi_s, \varphi_t \varphi_u)$$

and

$$K_{rs} = \sum_{t,u=1}^{m} \left(CC^\dagger\right)_{tu} (\varphi_r \varphi_u, \varphi_t \varphi_s)$$

The basis functions φ_r are products of spatial functions and either an α or β spin factor and the elements of the coefficient matrix are (in general) complex numbers.

When the matrix ϵ is diagonal, the resulting molecular orbitals are the canonical set for which the diagonal elements of ϵ, $\epsilon_{ii} = \epsilon_i$ are the orbital energies. The total electronic energy is then given by

$$E = \frac{1}{2} \left[\text{tr} C^\dagger h C + \text{tr} C^\dagger h^F C \right]$$

Appendix 3.A

Atomic orbitals

The term "orbital" is used extremely loosely in quantum chemistry sometimes unintentionally through long-established abbreviations and acronyms. Thus LCAO originally meant Linear Combination of Atomic Orbitals and STO, Slater-Type Orbital; but, in each of these acronyms, "orbital" means "atom-centred function with a family resemblance to an orbital". This usage is clarified here.

The word "orbital" is usually taken to mean "the solution of the Schrödinger equation for a one-particle system". This definition in practice is, then, limited to functions which are solutions of the Schrödinger equation for:

1. The free particle or the "particle in a box" (no potential).
2. The one-electron atom (central potential; Coulomb's law).
3. The harmonic oscillator (central potential; Hooke's law).
4. The one-electron diatomic (cylindrical potential; Coulomb's law).

since these are the only cases where the one-particle Schrödinger equation is actually soluble.

It is physically reasonable to expect that these orbitals may be well-adapted to the approximate solution of other problems with similar potentials.

Thus, one might expect the one-electron atom orbitals (AOs, say, the familiar s, p, d etc. orbitals) to be the ones to use in expanding the orbitals in a single-determinant model of the electronic structure of (at least) the rare-gas atoms which (presumably) have a central potential which is a combination of many interactions all described by Coulomb's law. The forms of these orbitals have been determined by the action of Coulomb's law and there are a huge number of them of all possible symmetry types, with which to make an expansion of apparently arbitrary accuracy.

The solutions of any one-electron atom (or ion) form a complete set and so the expectation that the orbitals of, for example, the neon atom could be expanded as linear combinations of them is both physically reasonable and mathematically under-pinned. But there are some practical considerations:

- The radial (r) dependence of the spatial form of the one-electron AOs contains the factor
$$\exp(-Zr/n)$$
where Z is the atomic number and n is the principle quantum number. Taking a particular symmetry-type, the s-functions become very diffuse very quickly as n increases. If we take the AOs for a particular Z, they are quite unsuitable for the expansion of the AOs of an atom of atomic number $5Z$ (say), and yet the AOs form a complete set!

- The deficiencies in the *bound* s-type AOs compensated by the millions of s-type *unbound* (scattering) orbitals which, although they form part of the complete set are not at all suitable for use in the theory of bound molecular electrons.

- As a complete set, the AOs of (*e.g.*) the hydrogen atom could be used to expand the AOs of an atom *anywhere* in space; or for any molecular system whatsoever. This would clearly require expansions of enormous length, of little or no practical use.

What is clear is that the formal property of completeness, even when compounded with an ostensibly suitable potential, is no guarantee of utility. The choice of expansion functions to use in any calculation has to be made on the basis of practical considerations. For a molecule (or atom) this choice would normally include:

1. Functions centred on each nucleus of the molecule.

2. Spatial forms which may be scaled to "fill" the space expected to be occupied by electrons around each nucleus.

3. As short an expansion length as practical.

4. If possible, functions which have a "family relationship" to the solutions of the Schrödinger equation with a Coulomb potential.

5. Functions which have the boundary conditions expected for the orbitals of the molecule.

6. Functions which give rise to tractable integrals occurring in the theory.

The practical points raised by the choice of expansion functions will be taken up later. For the moment we will simply call a suitable set of atom-centred expansion functions, with as many of the above properties as possible, a set of "atomic orbitals".

> Thus, in line with current practice and in defiance of rigour, we use the term "orbital" sometimes to mean simply "any function of the coordinates of one electron which is normalisable, single-valued and has derivatives almost everywhere" and sometimes in its original sense; the context should enable the distinction to be made.

Appendix 3.B

Charge density

*Perhaps the simplest electronic quantity of chemical interest is the total n-electron probability density—the "charge density". The **R** matrix provides an algebraic representation of the total electronic charge density.*

The probability distribution for electron i in the n-electron wave function Φ is obtained by integrating over the coordinates of all the other $(n-1)$ electrons in the total probability distribution

$$\Phi^*(\vec{x}_1, \vec{x}_2, \ldots, \vec{x}_n)\Phi(\vec{x}_1, \vec{x}_2, \ldots, \vec{x}_n)$$

that is:

$$\rho_i(\vec{x}_i) = \int \Phi^*\Phi dV_1 dV_2 \ldots dV_{i-1}dV_{i+1}\ldots dV_n$$

and the *total* probability distribution for the presence of *any* of the n electrons described by Φ is just the sum of all these expressions:

$$\rho(\vec{x}) = \sum_{i=1}^{n} \rho_i(\vec{x})$$

dropping the subscript now that the function is a function of ordinary three-space and spin. But the function Φ is antisymmetric with respect to all the electrons, so that any one of the $\rho_i(\vec{x}_i)$ is the same as all the others. Thus, if we pick the first variable and drop its subscript:

$$\rho(\vec{x}) = n \times \int \Phi^*(\vec{x}, \vec{x}_2, \ldots, \vec{x}_n)\Phi(\vec{x}, \vec{x}_2, \ldots, \vec{x}_n dV_2 \ldots dV_n$$

If Φ is a single determinant of orthonormal orbitals, we can use the earlier result for the normalisation integral of a single determinant. The expression above is the same as the normalisation integral except that integration over the coordinates of the first (now unlabelled) electron is missing, the result is

$$\rho(\vec{x}) = \sum_{i=1}^{n} \chi_i^*(\vec{x})\chi_i(\vec{x}) \tag{3.B.28}$$

Although it is strictly a *probability distribution*, it has become common to call $\rho(\vec{x})$ the *electron density*. This terminology is obviously more realistic for the Uranium atom than for the helium atom but the meaning is clear.

If the MOs χ_i are expanded in the usual way in terms of a finite number of basis functions

$$\boldsymbol{\chi} = \boldsymbol{\varphi C}$$

then

$$\rho = \boldsymbol{\chi}\boldsymbol{\chi}^\dagger = \sum_{i=1}^n \chi_i^* \chi_i$$

and

$$\rho = \boldsymbol{\chi}\boldsymbol{\chi}^\dagger = \boldsymbol{\varphi C C^\dagger \varphi^\dagger} = \boldsymbol{\varphi R \varphi^\dagger} \quad \text{say}$$

where

$$\boldsymbol{R} = \boldsymbol{C C^\dagger}$$

The physical interpretation of the elements of $\boldsymbol{R}$ may be inferred either from the above expression or from the result of integrating it

$$\int \rho(\vec{x}) d\tau = \sum_{r,s=1}^{m} R_{ij} \int \varphi_j^* \varphi_i d\tau = \text{tr}\boldsymbol{RS}$$

In particular, if the expansion functions are orthonormal, $\boldsymbol{S} = \boldsymbol{1}$ and

$$\int \rho(\vec{x}) d\tau = \text{tr}\boldsymbol{R} = \sum_{i=1}^{m} R_{ii}$$

Now, if the orbitals are orthonormal,

$$\text{tr}\boldsymbol{C C^\dagger} = \text{tr}\boldsymbol{C^\dagger C} = \text{tr}\boldsymbol{1} = n$$

the number of orbitals (electrons) in the system. Thus, in this case,

$$\text{tr}\boldsymbol{R} = n$$

and in the general case

$$\text{tr}\boldsymbol{RS} = n$$

In some sense, therefore, the diagonal elements of $\boldsymbol{R}$ are the *occupation numbers* of the corresponding basis functions in the total electron density and the off-diagonal elements are the occupation numbers of the overlap distributions between the basis functions. At the very least, any analysis of the electron density of a single-determinant wave function is bound to involve the matrix $\boldsymbol{R}$. We shall take up this analysis in due course.

It has been noted earlier that the single-determinant wavefunction is invariant against any non-singular linear transformation of the occupied molecular orbitals (provided that the orbitals remain normalised which is always possible). Now that we are dealing with an *orthonormal* set of MOs χ_i, the transformation to a new orthonormal set will be the special case of a *unitary* matrix. So, if a new set of MOs is formed by such a linear transformation:

$$\chi' = \chi D$$

where $\boldsymbol{D}$ is unitary, the new MOs can be expressed in terms of the basis functions as

$$\chi' = \chi D = \varphi C D$$

and the new electron density is

$$\boldsymbol{R}' = \boldsymbol{CD}\left(\boldsymbol{CD}\right)^\dagger = \boldsymbol{CDD^\dagger C^\dagger} = \boldsymbol{C1C^\dagger} = \boldsymbol{CC^\dagger} = \boldsymbol{R}$$

since $\boldsymbol{D}$ is unitary:

$$\boldsymbol{DD^\dagger} = \boldsymbol{D^\dagger D} = 1$$

which is precisely what one would expect. If the wavefunction is unchanged by a unitary transformation among the component orbitals, the associated electron probability distribution must be unchanged. In the language of the expansion method, the matrix $\boldsymbol{R}$ is a fundamental invariant of the single-determinant wavefunction.

It is useful to invoke this matrix in the definition of the electron-repulsion matrices involved in the Hartree–Fock matrix. The matrices $\boldsymbol{J}$ and $\boldsymbol{K}$ which depend on $\boldsymbol{C}$ and $\boldsymbol{C}^\dagger$ do so only through the single matrix $\boldsymbol{R} = \boldsymbol{CC}^\dagger$:

$$J_{rs} = \sum_{t,u=1}^{m} \left(\boldsymbol{CC^\dagger}\right)_{tu} (\varphi_r\varphi_s, \varphi_t\varphi_u) = \sum_{t,u=1}^{m} R_{tu}(\varphi_r\varphi_s, \varphi_t\varphi_u)$$

and

$$K_{rs} = \sum_{t,u=1}^{m} \left(\boldsymbol{CC^\dagger}\right)_{tu} (\varphi_r\varphi_u, \varphi_t\varphi_s) = \sum_{t,u=1}^{m} R_{tu}(\varphi_r\varphi_u, \varphi_t\varphi_s)$$

emphasising the physical interpretation of these operators as the *total* electron interaction operators in the single-determinant case.

In fact, the $\boldsymbol{R}$-matrix has very useful *formal* properties as well as its interpretation as a numerical measure of the MO charge distribution. These properties will be explored when we meet the theory of "shells" of electrons.

Appendix 3.C

Properties of the J and K matrices

The two matrices summarising the effects of electron repulsion in SCF theories have important physical interpretations and will soon be seen to dominate the implementation of the SCF method. Some of their most important properties are summarised here.

Contents

3.C.1	Mathematical properties	103
3.C.2	Physical interpretation	106
3.C.3	Supermatrices	107

3.C.1 Mathematical properties

The electron-repulsion matrices $\boldsymbol{J}$ and $\boldsymbol{K}$ are so important in the theory of the single-determinant model of electronic structure that it is worthwhile summarising some of their mathematical properties and their links with physically interpretable quantities.

The definitions are

$$J_{rs} = \sum_{t,u=1}^{m} \left(\boldsymbol{CC}^\dagger\right)_{tu} (\varphi_r \varphi_s, \varphi_t \varphi_u)$$

and

$$K_{rs} = \sum_{t,u=1}^{m} \left(\boldsymbol{CC}^\dagger\right)_{tu} (\varphi_r \varphi_u, \varphi_t \varphi_s)$$

where the functions φ_r are the basis functions and the matrix C is a matrix defining a set of molecular orbitals in terms of these basis functions

$$\chi = \varphi C$$

Using the notation

$$R = CC^\dagger$$

the definitions become a little more compact:

$$J_{rs} = \sum_{t,u=1}^{m} R_{tu}(\varphi_r\varphi_s, \varphi_t\varphi_u)$$

and

$$K_{rs} = \sum_{t,u=1}^{m} R_{tu}(\varphi_r\varphi_u, \varphi_t\varphi_s)$$

and we may emphasise this "functional" dependence by writing $J(R)$ and $K(R)$.

It is easy to show that both matrices have the property

$$\mathrm{tr}\,J(A)B = \mathrm{tr}\,J(B)A$$

and

$$\mathrm{tr}\,K(A)B = \mathrm{tr}\,K(B)A$$

since, for example,

$$\begin{aligned}
\mathrm{tr}\,J(A)B &= \sum_{t,u=1}^{m} B_{tu}\,[A_{ut}(\varphi_r\varphi_s, \varphi_t\varphi_u)] \\
&= \sum_{t,u=1}^{m} A_{tu}\,[B_{ut}(\varphi_r\varphi_s, \varphi_t\varphi_u)] \\
&= \mathrm{tr}\,J(B)A
\end{aligned}$$

with a similar expression for the K formula.

This property is useful in reducing the number of terms in the variational derivation of the Hartree–Fock equations. But, when the physical interpretation of the J and K is known, these properties quickly become intuitively obvious.

It is often useful to separate the MO coefficient matrix into columns of coefficients defining each orbital:

$$C = (c^1, c^2, c^3, \ldots, c^n)$$

and to define a separate "electron distribution" matrix for each MO:

$$R^i = c^i c^{i\dagger}$$

3.C.1. MATHEMATICAL PROPERTIES

where, of course, each $\boldsymbol{R}^i$ is $m \times m$ so, in contrast to the formation of $\boldsymbol{C}$ from the $\boldsymbol{c}^i$, the total $\boldsymbol{R}$ matrix is obtained from the individual $\boldsymbol{R}^i$ by simple addition:

$$\boldsymbol{R} = \sum_{i=1}^{n} \boldsymbol{R}^i$$

Using this notation we can look at the physical interpretation of, for example,

$$\left(\boldsymbol{J}(\boldsymbol{R}^i)\right)_{rs} = \sum_{t,u=1}^{m} R_{tu}(\varphi_t \varphi_u, \varphi_r \varphi_s)$$

Expanding $\boldsymbol{R}^i$ in terms of the MO coefficients we have

$$\left(\boldsymbol{J}(\boldsymbol{R}^i)\right)_{rs} = \sum_{t,u=1}^{m} c_t^i c_u^i (\varphi_t \varphi_u, \varphi_r \varphi_s)$$

but this sum on the left-hand-side is nothing more than the expansion of the repulsion integral $(\chi_i \chi_i, \varphi_r \varphi_s)$ so that

$$\left(\boldsymbol{J}(\boldsymbol{R}^i)\right)_{rs} = (\chi_i \chi_i, \varphi_r \varphi_s) \tag{3.C.29}$$

Because the full matrix $\boldsymbol{R}$ is just the sum of the individual orbital $\boldsymbol{R}$ matrices, the corresponding expression involving $\boldsymbol{R}$ alone is:

$$\left(\boldsymbol{J}(\boldsymbol{R})\right)_{rs} = \sum_{i=1}^{n} (\chi_i \chi_i, \varphi_r \varphi_s)$$

In a similar manner it is easy to show that, for example,

$$\text{tr} \boldsymbol{J}(\boldsymbol{R}^i) \boldsymbol{R}^j = \text{tr} \boldsymbol{J}(\boldsymbol{R}^j) \boldsymbol{R}^i = (\chi_i \chi_i, \chi_j \chi_j) = (\chi_j \chi_j, \chi_i \chi_i) \tag{3.C.30}$$

Using the full $\boldsymbol{R}$ matrix in eqn (3.C.29) we obtain

$$\left(\boldsymbol{J}(\boldsymbol{R})\right)_{rs} = \sum_{i=1}^{n} (\chi_i \chi_i, \varphi_r \varphi_s) \tag{3.C.31}$$

Entirely analogous expressions can be easily obtained for the repulsion integrals associated with the $\boldsymbol{K}$ matrix:

$$\left(\boldsymbol{K}(\boldsymbol{R}^i)\right)_{rs} = (\chi_i \varphi_s, \varphi_r \chi_i) \tag{3.C.32}$$

$$\text{tr} \boldsymbol{K}(\boldsymbol{R}^i) \boldsymbol{R}^j = (\chi_i \chi_j, \chi_i \chi_j) \tag{3.C.33}$$

$$\left(\boldsymbol{K}(\boldsymbol{R})\right)_{rs} = \sum_{i=1}^{n} (\chi_i \varphi_s, \varphi_r \chi_i) \tag{3.C.34}$$

3.C.2 Physical interpretation

We will concentrate, for simplicity, on the physical interpretation of the quantities associated with the $\boldsymbol{J}$ matrices eqn (3.C.29), eqn (3.C.30) and eqn (3.C.31).

The elements, $\boldsymbol{J}(\boldsymbol{R})_{rs}$, of the full matrix are simply the repulsion energies experienced by an electron having a probability distribution given by the product $\varphi_r^*\varphi_s$ of two of the basis functions *due to all n electrons* occupying the *molecular* orbitals χ_i. This total repulsion energy is, of course, the sum of the repulsions between such a basis-function-product distribution and the occupied molecular orbitals χ_i: the quantities $\boldsymbol{J}(\boldsymbol{R}^i)_{rs}$

Taking the trace of the matrix $\boldsymbol{J}(\boldsymbol{R}^i)$ with a matrix $\boldsymbol{R}^j$ generates the repulsion integral between two of the molecular orbitals. The sum of all these MO repulsions is contained in

$$\mathrm{tr}\boldsymbol{J}(\boldsymbol{R})\boldsymbol{R}$$

and it is easy to see that this sum must contain the repulsion of each electron *with itself*, since the sum implicitly includes the terms with $i = j$. This is obviously unphysical corresponding, as it does, to repulsions between different parts of the distribution of the *same* electron.[10] Fortunately, as we noted in the derivation of the Hartree–Fock equations, this spurious self-repulsion is exactly cancelled by the corresponding "diagonal" terms in the expression

$$\mathrm{tr}\boldsymbol{K}(\boldsymbol{R})\boldsymbol{R}$$

because

$$\mathrm{tr}\boldsymbol{J}(\boldsymbol{R}^i)\boldsymbol{R}^i = \mathrm{tr}\boldsymbol{K}(\boldsymbol{R}^i)\boldsymbol{R}^i$$

In fact, this is the principle role of the "exchange" term; to cancel the unphysical self-repulsion in the Coulomb sum. It is the difference between the Hartree and the Hartree–Fock methods, and the reason why all the MOs are the eigenfunctions of the *same* Hartree–Fock operator, while a separate Hartree operator is needed for each MO which excludes the self-repulsion *for that MO*.

This explicit removal of the *self-interaction* amongst the electrons is a great strength of the (algebraic approximation to the) Hartree–Fock equations. We shall see later that *separate* approximations to parts of the total interaction energy of a system of electrons do *not* have this convenient property and may often include spurious energies of interaction between "different parts of a given electron". The so-called self-interaction correction (SIC) must be invoked in such cases.

---

[10]Clearly, this is one of the pitfalls of the "electron density" picture of the probability distribution since an actual density *does* have self-repulsion but a probability distribution does not.

3.C.3 Supermatrices

For products of two matrices the "trace" operation is reminiscent of a scalar product; it is a sum of m^2 indexed quantities just as the scalar product is a sum of m indexed quantities. The two-index quantities J_{rs} and R_{sr} may easily be numbered by the single index:

$$(rs) = m(s-1) + r$$

and

$$(sr) = m(r-1) + s$$

so that, if $J'_{(rs)} = J_{rs}$ and $R'_{(rs)} = R_{sr}$,

$$\text{tr}\boldsymbol{JR} = \sum_{i=1}^{m^2} J'_i R'_i$$

a fact which is often useful in implementations.

In the same spirit it is possible to collapse four subscripts into two so that four-index quantities become analogues of *matrices* just as two-index quantities have been treated as *vectors*. "Vectors" and "matrices" of this type are often called supervectors and supermatrices, respectively. We shall use this notation later when molecular symmetry transformations are encountered.

Appendix 3.D

An artifact of expansion

There is an artifact which may occur when using the linear expansion technique which does not arise in the general functional *variation method. If we use a basis set which contains no functions of the same symmetry as the exact ground state, we cannot reach that ground state. This leads, paradoxically, to an extension of the linear variation method.*

Contents

3.D.1   Lowest state of a given symmetry 108

3.D.1 Lowest state of a given symmetry

The solution of the Schrödinger equation for the hydrogen atom is known to be the familiar 1s function:

$$\psi_{1s}(r,\theta,\varphi) = \exp(-r)$$

apart from normalisation.

If, however, we did not know this and used a linear expansion of the ground state function in terms of a set of p- and d-type functions in the variation method we would not, of course, obtain the correct ground-state solution (or even a reasonable approximation to it). This is simply because all the p functions and the d functions are *orthogonal* to the ground-state function *by symmetry*.

3.D.1. LOWEST STATE OF A GIVEN SYMMETRY 109

If we used a moderately inappropriate set of functions which included some s functions which were not orthogonal to the ground-state $1s$ function but which had exponential factors very different to $\exp(-r)$, then we would get a very bad approximation to the ground-state function.

These two cases are *qualitatively* different. In the latter case we are suffering from poor judgement which is rewarded by a very bad approximate wavefunction. In the former case we are suffering from bad luck which is rewarded by an approximate solution, which is not a poor approximation to the *ground state*, but which is an approximation to the lowest state which can be reached by functions of the given symmetry type. In the case quoted we would obtain an approximation to the $2p$ state of the H atom.

A little thought shows that this result must be general; if there are *constraints* applied to the action of the variation principle, it will respond by seeking out the lowest solution which can be found *consistent* with those constraints. These constraints may be of several types explicit or implicit:

- Spatial symmetry constraints of which the above example is the simplest. In a many-electron case, the constraint may be expressed as a *product* of orbital symmetries.

- Spin symmetry constraints; a two-electron single determinant may have the electrons with parallel or anti-parallel spins. In the first case the variation principle will find the lowest triplet, in the second the lowest singlet.

- More subtle constraints due to the *model* being used; the most common is the UHF/ROHF example to be discussed later.

In summary, when a parametric variation method is used, the action of the variation principle may be constrained by the scope of the parameters used. In particular this result enables us to make the generalisation that the variation method can be used to find the lowest state of a given symmetry for a many-electron system.

Appendix 3.E

Single determinant: choice of orbitals

Contents

3.E.1	Orthogonal invariance	110
3.E.2	Koopmans' theorem	112
3.E.3	Localised orbitals	113
3.E.4	"Zeroth-order" perturbed orbitals	113

3.E.1 Orthogonal invariance

It has been stressed already that the individual orbitals in the single-determinant wavefunction are not uniquely defined. The properties of determinants ensure that any (non-singular) linear transformation amongst these orbitals does not change the determinant. Our derivation of the *optimum* single-determinant wavefunction concentrated, for practical reasons, on *orthonormal* orbitals and, if we wish to retain this simplification, this restricts allowed transformations amongst the orbitals to be unitary or orthogonal if we deal only with *real* orbitals. Thus, any set of orbitals λ_k given by

$$\boldsymbol{\lambda} = \boldsymbol{\chi}\boldsymbol{O}$$

(where $\boldsymbol{O}$ is an $n \times n$ orthogonal matrix) generates the same determinant.

This transformation induces the transformation

$$\boldsymbol{\epsilon}' = \boldsymbol{O}^T \boldsymbol{\epsilon} \boldsymbol{O}$$

3.E.1. ORTHOGONAL INVARIANCE

on the elements of the original matrix ϵ associated with the set χ_i.

Now the electron density due to the set of orbitals forming the determinant is just

$$\rho(\vec{x}) = \sum_{i=1}^{n} \chi_i^2(\vec{x}) = \sum_{i=1}^{n} \lambda_i^{\,2}(\vec{x}) \qquad (3.\text{E}.35)$$

so that the use of the freedom in the definition of the orbitals in the determinant offers some obvious possibilities:

- Choose the orbitals to have particularly convenient energetics. We have seen that the *diagonalisation* of ϵ generates the most useful form for the orbital energies; approximations to the vertical ionisation energies of the molecule.

- Choose the orbitals to reflect some particular partitioning of the spatial distribution of the electrons.

Of course, if we make the first choice then the resulting orbitals *do* define a particular partitioning of the spatial distribution of the electrons, but this is a *by-product* of the energy-determined choice rather than the main reason for that choice. In fact, it is well known that (for valence electrons, at least) these canonical orbitals are *delocalised* over the whole molecule in general.

In chemistry we are often interested in the behaviour of *parts* of the electron distribution and it would be extremely useful to *concentrate* the interest on a few occupied orbitals or even just one or two. Thus the idea of obtaining a set of orbitals that are adapted to the description of physical and chemical properties other than the ionisation energy is well worth exploring.

For example, if the single determinant were an approximation to the electronic structure of a closed-shell system, then one might want to choose a set of orbitals λ_i for which:

- Each orbital represented a *localised* pair of electrons; an electron-pair bond or a lone pair. Clearly in this case the summation over the occupied orbitals represented by eqn (3.E.35) would be mainly a summation over *separate* regions of the molecule rather than the *superposition* of a set of (ionisation-energy-optimal) *delocalised* orbitals.

- One (or more) of the orbitals was particularly well adapted to the donation of a pair of electrons in a nucleophilic substitution reaction. In this case both energy and spatial form might well be important.

- The (assumed localised) orbitals are assumed to be constructed by pairwise addition of pairs of "hybrid" orbitals; the ϵ matrix can be transformed to a "tri-diagonal form" to reflect this model of the electronic structure.

3.E.2 Koopmans' theorem

The unique properties of the orbitals associated with *diagonal* ϵ have been mentioned before; they are routinely called *the* molecular orbitals and the diagonal elements of this ϵ *the* orbital energies. Let us now establish this result.

If we take an optimum[11] single-determinant wavefunction and remove one electron from each one of a chosen set of occupied orbitals we can generate n different $(n-1)$-electron single-determinant wavefunctions, and these wavefunctions should bear some relationship to the states of the corresponding molecular ion. In fact, they can be taken as a basis for the linear expansion of the states of the ion, and the diagonalisation of the associated linear variation problem would give the best approximation to these states available with this restricted class of function.

If we denote by Φ_i the single determinant obtained by removing χ_i from the n-electron determinant Φ, then the matrix elements of $\hat{H}$ between different Φ_i are:

$$\int \Phi_i \hat{H} \Phi_j d\tau = h_{ij} + \sum_k [(\chi_i\chi_i, \chi_k\chi_k) - (\chi_i\chi_k, \chi_i\chi_k)]$$
$$= \epsilon_{ij}$$

where, as usual, ϵ_{ij} is the off-diagonal element of the HF matrix expressed in the basis of MOs (χ_i). Clearly, if the χ_i are chosen to make ϵ *diagonal* ($\epsilon_{ij} = 0$) then there is no variational problem to solve and the corresponding orbital energies are the best available approximations to the ionisation energies available by this route. If, however, ϵ is not diagonal, then each diagonal element still represents the energy required to remove an electron from the orbital in question, but the resulting wavefunction is not the optimum one for the molecular ion; there is a non-trivial linear variation problem to solve which is, naturally, entirely equivalent to the diagonalisation of ϵ.

The conclusion is obvious:

> The energy associated with the removal of an electron from an orbital of a single determinant wavefunction is the diagonal element of the associated ϵ matrix of the Hartree–Fock equation. This energy is only an optimum description of the ionisation energy of the system if the matrix ϵ is *diagonal*.

This result is Koopmans' Theorem. As we see there is a good case for referring to these orbital energies as *the* orbital energies but the associated orbitals have no preferred status *other* than their association with ionisation processes.

---
[11] UHF for convenience of notation.

3.E.3 Localised orbitals

The best-known sets of orbitals that are alternatives to the diagonal-ϵ set are *localised* molecular orbitals which have been produced by some localisation criterion defining an orthogonal transformation of the energy-canonical set.

There are various kinds of criteria which may be used to generate a localised set:

- Maximise the "self-repulsion" of two electrons in the same orbital.
- Minimise the inter-orbital repulsions.
- Minimise the inter-orbital exchange interactions.
- Minimise the overlap *density* between orbitals.

all subject to the orthogonality of the resulting orbitals, of course.

3.E.4 "Zeroth-order" perturbed orbitals

Formally, since (for real orbital coefficients) the matrix ϵ is symmetric, any orthogonal transformation of the χ_i can be represented as the diagonalisation of some matrix $\boldsymbol{A}$ (say) with elements

$$A_{ij} = \int \chi_i \hat{A} \chi_j d\tau$$

for some (Hermitian) operator $\hat{A}$. The idea would be to choose some physical or chemical criterion which could be represented by $\hat{A}$ and hence obtain a set of orbitals that are well adapted to the description of the phenomenon under investigation.

It must be stressed, of course, that the determinant is *unchanged* by these transformations. However, it might be useful to draw an analogy between these transformations and another, familiar, transformation in quantum theory. If an energy level is *degenerate* (has several wavefunctions associated with it) then in the application of perturbation theory to such a system it is necessary to establish the so-called "zeroth-order" function; the particular linear combination of wavefunctions of the same energy to which the system reverts if the perturbation is smoothly "turned off". This is done by simply setting up the linear variation problem using the matrix elements of the perturbation operator in the basis of the degenerate functions and diagonalising the resulting matrix.

APPENDIX 3.E. SINGLE DETERMINANT: CHOICE OF ORBITALS

As we have seen, the freedom in the definition of the orbitals in a single determinant can be expressed as the orthogonal transformation associated with the diagonalisation of a symmetric matrix so that there is a tempting analogy here. If we choose an operator which is of modelling value, diagonalising the representation of that operator in the basis of orbitals defining a single determinant may well produce a set of orbitals which are adapted to the description of the phenomenon being modelled by the operator. For example, if we diagonalise the matrix of the operator representing a point charge approaching the molecule the resulting orbitals might well be adapted to a description of the $S_N 2$ process.

It must be stressed that this is only an analogy; the determinants formed from each set of orbitals related by these transformations are not degenerate, they are *identical*; only their compositions in terms, for example, of individual orbital contributions to the total electron density differ.

Chapter 4

A special case: closed shells

The vast majority of stable molecules are of "closed-shell" electronic structure; having an even number of electrons and no net electron-spin properties. For this class of molecule, the theory of the last chapter is a little redundant. In the present chapter the theory of the special case of a single determinant of "doubly occupied spatial orbitals" is obtained from the general single-determinant theory.

Contents

4.1	Introduction	115
4.2	Notation for the closed-shell case	116
4.3	Closed-shell expansion	116
4.4	The closed-shell "HF" equation	117
4.5	Closed-shell summary	121

4.1 Introduction

The linear expansion Hartree–Fock eqn (3.24) of the last chapter was derived under very general conditions; no constraints were applied except those implicit in the single-determinant model of electronic structure and the quality of the basis functions used in the expansion. Of course, in any *parametric* variation method the important question of the very *existence* of the solutions of the corresponding "real " (functional) Hartree–Fock equation is *assumed*.

For many purposes this equation is *too* general; for example, when there is some physical or chemical knowledge or suspicion of the *qualitative* form of the electronic structure. In these cases the solutions of the Hartree–Fock equations are often required simply to confirm this knowledge and to provide *quantitative* information about the energetics and distributions of the electrons.

The most common of these special cases is the so-called closed-shell model of the electronic structure of systems with an even number of electrons which have no evidence of magnetic behaviour. The assumption in these cases is that the electrons are "paired"[1], i.e. there are two electrons of opposite spin in each of $n/2$ spatial orbitals which are themselves entirely *real* combinations of the spatial factors in the basis functions.

In this case there are considerable simplifications to be made, not to the *form* of eqn (3.24) but to the details of its implementation and to the amount of numerical work involved in its solution.

4.2 Notation for the closed-shell case

The n electrons in the closed shell case occupy $n/2$ spatial orbitals in pairs; the occupied MOs are

$$\chi_1(\vec{x}_1) = \psi_1(\vec{r}_1)\alpha(s_1)$$
$$\chi_2(\vec{x}_2) = \psi_1(\vec{r}_2)\beta(s_2)$$
$$\chi_3(\vec{x}_3) = \psi_2(\vec{r}_3)\alpha(s_3)$$
$$\chi_4(\vec{x}_4) = \psi_2(\vec{r}_4)\beta(s_4)$$
$$\vdots$$
$$\chi_{n-1}(\vec{x}_{n-1}) = \psi_{n/2}(\vec{r}_{n-1})\alpha(s_{n-1})$$
$$\chi_n(\vec{x}_n) = \psi_{n/2}(\vec{r}_n)\beta(s_n)$$

where the $\psi_i(\vec{r}_i)$ are the *spatial* components of the molecular orbitals.

Thus each spatial MO ψ_i can be expanded directly in terms of the spatial components of the basis functions; recall that the basis functions were

$$\varphi_i(\vec{x}_i) = \phi_k(\vec{r}_i) \times (\quad \alpha(s_i) \quad \text{or} \quad \beta(s_i) \quad)$$

We can therefore divide the number of basis functions *and* the number of MOs by two and work entirely in terms of spatial functions. This is possible, as we shall see, because we can carry out the spin "integrations" explicitly.

4.3 Closed-shell expansion

With this in mind, we can change the definition of the *numbers* of basis functions and MOs to avoid having to use division by two throughout.

---

[1] But not in the sense of valence bond theory, we mean here that every spatial orbital is occupied by two electrons of opposite spin.

4.4. THE CLOSED-SHELL "HF" EQUATION

> In this chapter we shall use m to mean the number of *spatial* basis functions $\phi_s(\vec{r})$ (in contrast to its earlier use as the number of space-spin basis functions $\varphi_r(\vec{x})$). We shall also use n to mean the number of (doubly occupied) *spatial* MOs $\psi_i(\vec{r})$ (again in contrast to its use as the number of singly occupied spin-space MOs $\chi_i(\vec{x})$). Thus the number of *electrons* becomes $2n$ in place of the earlier n.

Thus, we may make a similar definition of the *spatial* MOs, the *spatial* basis functions and the matrix of coefficients that relates the two and is to be determined by a restriction of the form of eqn (3.24) derived earlier.

$$\boldsymbol{\psi} = (\psi_1, \psi_2, \psi_3, \ldots, \psi_n) \tag{4.1}$$

is the $1 \times n$ matrix (row) of spatial MOs,

$$\boldsymbol{\phi} = (\phi_1, \phi_2, \phi_3, \ldots, \phi_m) \tag{4.2}$$

is the $1 \times m$ matrix (row) of basis spatial functions and

$$\boldsymbol{C} = \begin{pmatrix} C_{11} & C_{12} & \ldots & C_{1n} \\ C_{21} & C_{22} & \ldots & C_{2n} \\ C_{31} & C_{32} & \ldots & C_{3n} \\ \ldots & \ldots & \ldots & \ldots \\ \ldots & \ldots & \ldots & \ldots \\ \ldots & \ldots & \ldots & \ldots \\ C_{m1} & C_{m2} & \ldots & C_{mn} \end{pmatrix} \tag{4.3}$$

is the $m \times n$ matrix of expansion coefficients relating the spatial (doubly-occupied) MOs to the spatial basis functions.

Thus, the spatial expansion is

$$\boldsymbol{\psi} = \boldsymbol{\phi} \boldsymbol{C} \tag{4.4}$$

and the spatial analogue of the $\boldsymbol{R}$ matrix of the last chapter is

$$\boldsymbol{R} = \boldsymbol{C}\boldsymbol{C}^T$$

(for real $\boldsymbol{C}$) but, since the spatial orbitals are *doubly occupied*, the spatial charge density is defined by $2\boldsymbol{R}$

$$\rho(\vec{r}) = \boldsymbol{\phi}(2\boldsymbol{R})\boldsymbol{\phi}^T$$

4.4 The closed-shell "HF" equation

It is obviously possible to repeat the derivation of the linear expansion Hartree-Fock equations under these new conditions but it is much more transparent to note the nature of the constraints on our original, more general, eqn (3.24) and simply write down the new equation.

The simple key piece of information which we have to use is the fact that the two spin factors form an orthonormal set in spin "space":

$$\int \alpha^*(s)\alpha(s)ds = 1$$
$$\int \alpha^*(s)\beta(s)ds = \int \beta^*(s)\alpha(s)ds = 0$$
$$\int \beta^*(s)\beta(s)ds = 1$$

and, since each spatial MO occurs multiplied by *both* of these two factors, we can carry through the spin integration explicitly in a way which was not possible in general when the space-spin MOs χ_i were allowed to contain a more general dependence on the spin factors.

For example, of the $2n$ one-electron integrals involving the MOs, only n are distinct since they differ only by integration over α or β spin factors. The one-electron Hamiltonian does not depend on spin so that

$$\int \psi_i^*(\vec{r})\alpha^*(s)\hat{h}\psi_i(\vec{r})\alpha(s)dVds = \int \psi_i^*(\vec{r})\beta^*(s)\hat{h}\psi_i(\vec{r})\beta(s)dVds \qquad (4.5)$$

and separation of the spin integration means that both these integrals are both equal to

$$\int \psi_i^*(\vec{r})\hat{h}\psi_i(\vec{r})dV$$

which means that the one-electron contribution to the total energy of a single determinant of n doubly occupied spatial orbitals is just

$$2\sum_{i=1}^{n}\int \psi_i^*(\vec{r})\hat{h}\psi_i(\vec{r})dV$$

The situation with the electron repulsion integrals is a little more involved since there are *two* sets of spin integration involved in each term. However in the case of the Coulomb integral the two pairs of functions are necessarily the same since they both arise from the same spin-orbital involving either an α or β spin-factor:

$$(\psi_i\psi_i,\psi_j\psi_j) = \int dV_1 \int ds_1 \int dV_2 \int ds_2 \psi_i^*(\vec{r_1})\sigma^*(s_1)\psi_i(\vec{r_1})\sigma(s_1)$$
$$\times \frac{1}{|\vec{r_1}-\vec{r_2}|}\psi_j^*(\vec{r_2})\mu^*(s_2)\psi_j(\vec{r_2})\mu(s_2)$$

where *each of* σ and μ are α or β. This enables the spin integration to be carried out as in the one-electron case (the electron-repulsion operator does not

4.4. THE CLOSED-SHELL "HF" EQUATION

depend on spin) to give

$$(\psi_i\psi_i, \psi_j\psi_j) = \int dV_1 \int dV_2 \psi_i^*(\vec{r_1})\psi_i(\vec{r_1}) \frac{1}{|\vec{r_1} - \vec{r_2}|} \psi_j^*(\vec{r_2})\psi_j(\vec{r_2}) \quad (4.6)$$

Of course, the energy expression will contain each of these Coulomb integrals *twice* because of the double occupancy of the spatial MOs.

In the case of the exchange integral we have a slightly more complicated situation since the spin factors involving each electron come from *different* space-spin MOs. A typical integral is

$$(\psi_i\psi_j, \psi_i\psi_j) = \int dV_1 \int ds_1 \int dV_2 \int ds_2 \psi_i^*(\vec{r_1})\sigma^*(s_1)\psi_j(\vec{r_1})\mu(s_1)$$
$$\times \frac{1}{|\vec{r_1} - \vec{r_2}|} \psi_i^*(\vec{r_2})\sigma^*(s_2)\psi_j(\vec{r_2})\mu(s_2)$$

with the same understanding about σ and μ as before.

This time, however, there are *two* possibilities in place of the single possibility in the case of the Coulomb integral; σ and μ may be the same *or* different. Both may be equal to α or β or one may be α and the other β.

If the ψ_i are all doubly occupied, *all* possibilities will necessarily occur and so, exactly *half* of the integrals will be zero, leading to the occurrence of each of the *spatial* exchange integrals exactly once (note that when $i = j$ the Coulomb and exchange integrals are the same). These considerations lead to the two-electron contribution to the total energy of a single determinant of doubly occupied MOs of

$$\sum_{i=1}^{n}\sum_{j=1}^{n}[2(\psi_i\psi_i, \psi_j\psi_j) - (\psi_i\psi_j, \psi_i\psi_j)]$$

and a total energy mean value of

$$E = 2\sum_{i=1}^{n}\left[\int \psi_i^* \hat{h}\psi_i dV\right] + \sum_{i=1}^{n}\sum_{j=1}^{n}[2(\psi_i\psi_i, \psi_j\psi_j) - (\psi_i\psi_j, \psi_i\psi_j)] \quad (4.7)$$

Obviously the normalisation condition on the spatial molecular orbitals has a form precisely analogous to the general case, and the variational derivation of the Hartree–Fock equations determining the optimum *spatial* MOs ψ_i for the real closed-shell special case can be carried through in exactly the same way as the general case. It is only necessary to give a summary of the results of this derivation since it is so similar to the general case of Section 3.13.

It is important to keep in mind the difference in the definition of the general (space-spin) quantities for the general case and the special (space-only) quantities for the closed shell case. In particular the expression for the total electronic energy in the closed-shell case is

$$E = \text{tr}\boldsymbol{h}^F\boldsymbol{R} + \text{tr}\boldsymbol{h}\boldsymbol{R} \quad (4.8)$$

the factor of one half exactly cancelling the factor of two arising from the double occupation of the spatial orbitals in R.

4.5 Closed-shell summary

> For the real closed-shell case, the linear expansion technique
> $$\boldsymbol{\psi} = \boldsymbol{\phi C}$$
> involves the generation (computation, estimation etc.) of the one-electron energy integrals
> $$\boldsymbol{h} = \int \boldsymbol{\phi}^T \hat{h} \boldsymbol{\phi} dV$$
> and the electron-repulsion integrals
> $$(\phi_r\phi_s, \phi_t\phi_u) = \int dV_1 \int dV_2 \phi_r(\vec{r_1})\phi_s(\vec{r_1}) \frac{1}{|\vec{r_1} - \vec{r_2}|} \phi_t(\vec{r_2})\phi_u(\vec{r_2})$$
> as well as the overlap integrals
> $$\boldsymbol{S} = \int \boldsymbol{\phi}^T \boldsymbol{\phi} dV$$
> over the *spatial* basis functions.
> When these numbers have been obtained, the optimum real Molecular Orbital coefficient matrix is the solution of the equation:
> $$\boldsymbol{h}^F \boldsymbol{C} = \boldsymbol{SC}\boldsymbol{\epsilon}$$
> where the *real closed-shell* Hartree–Fock matrix $\boldsymbol{h}^F$ is given by
> $$\boldsymbol{h}^F = (\boldsymbol{h} + 2\boldsymbol{J} - \boldsymbol{K}) = \boldsymbol{h} + \boldsymbol{G}$$
> and the matrix $\boldsymbol{G}$ is defined by
> $$G_{rs} = \sum_{t,u=1}^{m} R_{tu} \left[2(\phi_r\phi_s, \phi_t\phi_u) - (\phi_r\phi_u, \phi_t\phi_s)\right]$$
> where $\boldsymbol{R} = \boldsymbol{CC}^T$. The basis functions ϕ_r are assumed to be *real* functions and the elements of the closed-shell coefficient matrix $\boldsymbol{C}$ are also real (complex conjugation has been dropped from the definition of the repulsion integrals and † changed to T to denote simple transposition of the matrix).

For the moment, we will take the *theory* of the single-determinant model no further; it is time to think about the problems of implementation which we will begin with the closed-shell case.

Chapter 5

Implementation of the closed-shell case

This is a long chapter in which a preliminary full implementation of the solution Hartree–Fock equations for the closed-shell case is developed. Many ideas are introduced and not every point of detail is explained at this initial stage; it is hoped that the general structure of the approach will be apparent and individual points of detail (particularly of computer language) will be taken largely on trust. It has been said that the generation of software is not about languages but about design decisions and concepts; let us see.

Contents

5.1	Preview		124
5.2	Vectors, matrices and arrays		125
	5.2.1	Matrix and array storage	126
5.3	The implementation: getting started		131
	5.3.1	A skeleton code	132
	5.3.2	One iteration	142
	5.3.3	The iterative procedure	144
5.4	The implementation: repulsion integral access		148
	5.4.1	General considerations	148
	5.4.2	Obtaining the integrals	152
	5.4.3	Processing the repulsion integrals	154
5.5	Building a testbench: conventional SCF		158
	5.5.1	A test getint	159
	5.5.2	The testbench	160

	5.5.3 Running the testbench 162	
5.6	Another testbench: direct SCF	**165**
	5.6.1 The π-electron approximation 165	
	5.6.2 A direct SCF `getint` 167	
	5.6.3 A direct SCF testbench 169	
5.7	Summary	**173**
5.8	What next?	**174**

5.1 Preview

There are several *types* of decision to be made when embarking on the implementation of the solution of the finite expansion Hartree–Fock equations for molecules because we know in advance that the finished system will be

- very demanding of computer resources and therefore should be efficient in the use of those resources and
- constantly undergoing revisions and developments as new scientific projects arise.

The requirements of "efficiency" and "easy maintenance" are basically contradictory as we noted at the very start of this work and, where this conflict becomes evident, the choice of functional clarity and structured programming techniques will always be preferred over numerical efficiency and computational speed. The basic philosophy will be to rely on hardware development for improved efficiency and present a reasonably clean implementation which itself can always be modified if increased performance is required; if there is a choice it will always be to let the machine not the human researcher do the work.

However, there are also much more detailed decisions to make:

- Which numerical method to use for a particular task.
- How to organise the storage of the intermediate numbers generated during the calculation.
- If a set of numbers is used several times during the calculation, should they be computed once and stored or computed as they are needed?
- How to structure the program; the identification of the basic modules of the problem.

5.2. VECTORS, MATRICES AND ARRAYS

- How to keep the maximum number of options for extension and change open.

- How to minimise the possibility of obsolescence by developments in computer technology (hardware and software).

- Perhaps the most important of all and intimately related to the last point; how to choose a computer language and related software tools for the implementation and *ensure that one has full control over its future*.

Some of these decisions cannot be made *in general* and must be decided for particular cases. Some of them are dictated to us by the cost and performance of hardware. All of them are difficulties faced by all Software Engineers and we must not be paralysed by them.

We shall adopt the general approach of what has come to be called "top-down design"; writing the software as if the lower levels existed and deducing from that process what the low-level code *must* do. In this way hard decisions are pushed out of sight until they absolutely must be faced and the algorithms coded. The language used and the software tools available for the generation of the system have been chosen with this in mind. These and other design considerations will be taken up in detail in Chapter 11.

There is much material to be presented here which can only be presented basically *all at once*; a protracted discussion of each of the points above in the absence of a particular implementation soon becomes diffuse and boring. It is hoped that the presentation is not *too* dense with new ideas and material.

5.2 Vectors, matrices and arrays

It is obvious from the last chapter that much of the work of any implementation of the Hartree–Fock method consists of operations with matrices which are stored in a computer as (usually contiguous) strings of numbers. However, it is organisationally convenient to distinguish between a matrix (or a vector) and a string or array of numbers, since this distinction affects the way in which matrices are stored and manipulated.

The term "matrix" will be reserved for a rectangular array of numbers which forms a particular *representation* of some mathematical object with respect to some particular choice of frame of reference (coordinates, basis functions etc.). That is, a matrix is not just an array of numbers. In the same spirit, a "vector" is a linear array of numbers which represent some mathematical object in a particular frame. Thus,

- The set of three numbers (x, y, z) giving the position of a point in space is a vector (change the coordinate system and the numbers transform in a well-known way).

- The set of three points $(p_x p_y, p_z)$ specifying the momentum of an electron with respect to a Cartesian coordinate system is a vector (change the reference frame and the numbers change in a well-defined way).

- The set of numbers (p, V, T) giving the pressure, volume and temperature of a point in space is not a vector, merely an array of scalars (none of the values depends on a reference frame).

- The set of n^2 numbers
$$H_{ij} = \int \phi_i^* \hat{h} \phi_j dV$$
is a matrix, the representation of the operator $\hat{h}$ in the basis ϕ_i (change the basis and the numbers change)

- The set of nine numbers representing a rotation about a given axis by a given angle is a matrix since the numbers transform in a well-defined way when the coordinate system is changed.

- The set of $4N$ numbers (x, y, y, Z) for a set of N nuclei giving the coordinates and nuclear charges for those nuclei is not a matrix but a rectangular array (the coordinates depend on the reference frame but the nuclear charges do not).

The distinction between matrices, vectors and arrays is important in the computer implementation of our equations because *matrices* are normally treated, accessed and manipulated *as a whole* since they are an internally connected collection of numbers; while arrays are simply convenient places to keep sets of related but heterogeneous data: orbital specifications, descriptions of atoms *etc*.

5.2.1 Matrix and array storage

Matrices and arrays are stored in computers as contiguous strings of numbers and are accessed by knowing the address of the start of the array and the offset within that array which is generated from the subscripts giving the particular element. Most computer languages provide subscripting for at least single and double subscripts so that one can, for example use A(I) or B(K,7) to access particular elements of an array.

When an array is used *as* an array, then the use of these subscripts is very convenient for random access of the elements of the array. However, when an

5.2. VECTORS, MATRICES AND ARRAYS

array is used to store a *matrix* the elements will normally be accessed in a systematic way and for this and other reasons which will become clear later we will choose to store the elements of a matrix in a singly subscripted array and use a simple algorithm to locate particular elements.

In practice, we shall assume the existence of a set of matrix manipulation routines (which will be coded later) and only rarely have to access the individual elements of *matrices*.

> Thus the mn numbers comprising a $m \times n$ matrix $\boldsymbol{A}$ (m rows, n columns) will be stored "by columns": the first column of the matrix is stored starting at the first location of the array, followed by the second column and so on until all n columns of length m are stored contiguously. Thus, the number A_{ij} will be located at A(m*(j-1) + i). That is, the so-called method of *single subscripting* will be employed. Note that, if a matrix is $m \times n$ the length of a *column* is m and the length of a *row* is n.

For example the code to evaluate the trace of the product of two symmetric, real $n \times n$ matrices:

$$s = \sum_{i,j=1}^{n} A_{ij} B_{ji}$$

is particularly simple because, if the matrices are *symmetric* the elements A_{ij} and A_{ji} are the same so that;

$$s = \sum_{i,j=1}^{n} A_{ij} B_{ji} = \sum_{i,j=1}^{n} A_{ij} B_{ij}$$

and, whatever method is used to locate the elements of the two matrices, their elements are in corresponding locations:

```
s = ZERO;
do i = 1,n*n
    s = s + A(i)*B(i);
```

or

```
s = ZERO;
for (i = 1; i <= n*n; i = i + 1)
    s = s + A(i)*B(i);
```

or, more compactly,

```
for (s = ZERO, i = 1; i <= n*n; i = i + 1)
    s = s + A(i)*B(i);
```

or, possibly,

```
s = ZERO;
i = 1;
while ( i <= n*n )
    {
    s = s + A(i)*B(i);
    i = i + 1;
    }
```

or even

```
s = ZERO;
i = 1;
repeat
    {
    s = s + A(i)*B(i);
    i = i + 1;
    }
until (i == n*n +1)
```

Each of these fragments illustrates, in a self-evident way, two points:

1. A particular looping construct of a programming language which has the general form:

    ```
    initialise
    looping construct
        statement
    ```

2. The way in which a **statement** may be replaced by a group of statements in *braces;* ({.....}) is treated as a *single statement*. This simple device, together with care about code presentation and indentation goes a long way to making the local structure of codes clear.

The **for** construct is particularly useful since it combines the "initialise", "test" and "increment" facilities in one place. Its general form is

`for(initialise thing(s); test thing(s); change thing(s))`

where **thing(s)** is a comma-separated list.

For the most part, these elements of the programming language are not in need of any further explanation but a full description of where the specifications of the language can be found is given in Chapter 35 later in this work.

5.2. VECTORS, MATRICES AND ARRAYS

In particular:

- Access of the elements of a matrix is usually hidden from the user by using procedures to handle matrices as a whole.
- However, the numbers in an *array* will be accessed by explicit double (or triple or whatever) subscripting.

Thus the code to extract the Cartesian coordinates and nuclear charge of the ith atom from an *array* Zlist might look like

```
for ( i = 1; i <= number_of_centres; i = i + 1)
    {
    x(i) = Zlist(i,1);   y(i) = Zlist(i,2);
    z(i) = Zlist(i,3);
    bigZ(i) = Zlist(i,4);
    }
```

No explanation of the elements of the programming language is necessary; they are obvious. This will be the general approach, the elements of the computer language will be introduced as needed and only rarely will any explanation be necessary; the full specification will be given later.

Turning to the management and manipulation of matrices, a sensible way to implement a foolproof matrix manipulation system might look like

```
double precision A(MATRIX_SIZE), B(MATRIX_SIZE)
double precision R(MATRIX_SIZE)
integer reply, matrix_multiply
.
.
reply = matrix_multiply(A, B, R,...)
if ( reply != OK) call error_out(reply)
.
.
```

Where **A, B** and **R** are matrices, **reply** is an error code and **OK** is a "satisfactory completion" code (zero, maybe). **matrix_multiply** does the required manipulations and returns **OK** or an error code which is analysed by **error_out**. However, in most of our applications, this is a little over-cautious and constant error analysis will prove unnecessary.

One point about these program fragments *is* worth mentioning, upper-case strings have been used for some of the "constants" in the code: **ZERO, OK** and **MATRIX_SIZE**. This is deliberate. One of the principles to be insisted on in the generation of clean, comprehensible code is:

> There shall be no "magic numbers" in the code

Any constants are given symbolic names which are associated with actual numerical values by a simple replacement facility *at compilation time* once and for all. A given symbolic name shall have the same significance *throughout the code* So we require statements like

```
define ( ZERO, 0.0d00)
define (OK,0)
define (MATRIX_SIZE,81)
```

or, even better,

```
define (BASIS_SIZE,9)
define (MATRIX_SIZE, BASIS_SIZE*BASIS_SIZE)
```

to make the required associations, the last example would ensure that the basis size was consistent with the size of any matrices used.

Notice that `define (OK,0)` is not the same as `OK = 0`, the latter is an *arithmetic statement* while the former is a *literal replacement*; wherever `OK` occurs it is replaced by the *character* "0" not by the *number* zero.

This elementary string replacement facility is very useful to make programs more readable and, equally important, easy to maintain and change; if the notation is consistent throughout it is only necessary to make *one change* to a `define` statement.

This is the simplest example of a *macro*, one of the commonest tools in programming to avoid repetitive work. Macros are familiar to all FORTRAN programmers as "inline functions" like `ABS` which is expanded *at compile time* (in this case) to the code necessary to give the modulus of the argument.

Although it is not given a thorough presentation here, a macro processor is capable of extending a computer language almost indefinitely, from the use of simple string replacement facilities (the simple definitions

```
define (depuis, while)
define (hasta, until)
define (cierra, close)
define (pisz, write)
```

have obvious attractions for some programmers), to general language extensions. In fact, the disciplined use of macros enables the generation of a particular "application-orientated language" with no run-time overheads.

5.3 The implementation: getting started

It is a very important empirical psychological fact that people find it much easier and more rewarding to improve a working system than to spend large amounts of time planning to generate an efficient program first time. In carrying out an implementation of the solution of the matrix equations of quantum chemistry, the general plan will be

> Get the overall design right, build something which works as quickly as possible then improve, improve, improve.

If the design is right improvements can be made in a controlled and modular way with a minimum of interactions.

The first thing to be said about the solution of the closed-shell linear expansion Hartree–Fock equation

$$\boldsymbol{h}^F \boldsymbol{C} = \boldsymbol{S} \boldsymbol{C} \boldsymbol{\epsilon} \qquad (5.1)$$

is that the technique used must be *iterative* since, although it is disguised in the above equation by the compact notation, the (assumed known) matrix $\boldsymbol{h}^F$ actually depends on $\boldsymbol{C}$ because of the dependence of $\boldsymbol{h}^F$ on $\boldsymbol{R}$ through $\boldsymbol{G}$:

$$\boldsymbol{h}^F = \boldsymbol{h} + \boldsymbol{G}(\boldsymbol{R})$$

so that the problem falls naturally into three areas:

- the setting up of the equations to be solved for a given matrix $\boldsymbol{R}$,
- the use of numerical techniques to solve these equations,
- the use of an iterative method to generate the final solution.

Clearly, the matrix equations have a solution (i.e. the original energy functional has a turning point) when the matrix $\boldsymbol{C}$ which solves eqn (5.1) is identical (to some numerical tolerance) to the matrix $\boldsymbol{C}$ used to form $\boldsymbol{R}$ and hence $\boldsymbol{h}^F$. That is, equation eqn (5.1 has a solution when $\boldsymbol{C}$ is *self consistent*.

This property of the solution matrix $\boldsymbol{C}$ gives the associated molecular orbitals their familiar name of the *Self-Consistent-Field* (SCF) Molecular Orbitals and the whole technique has come to be known as the SCF method. Thus,

> The "SCF method" is shorthand for "The solution of the matrix equations associated with the approximation of the orbitals of the Hartree–Fock model within a finite expansion approach".

Notation and Matrix Storage Requirements

The number of basis functions which we have called m will appear in the code as m and the number of electrons ($2n$ in the closed-shell case) will be nelec.

In the course of the calculation of the matrix C we shall need storage for at least the following matrices:

- The one-electron Hamiltonian matrix h: H
- Either the overlap matrix S or a related matrix which will orthogonalise the basis functions ($S^{-1/2}$, for example): V
- The Hartree–Fock matrix h^F: HF
- The matrix C which is to be found: C
- The matrix R used to form h^F: R
- One matrix to store either C or R for comparisons between iterations: Rold
- A smaller array for the orbital energies ϵ_i: epsilon

as well as working storage in which to perform matrix multiplications etc.

In spite of the fact that some of these matrices are *symmetrical* and therefore only have $m(m+1)/2$ distinct elements, the whole of each matrix of size m^2 will be stored and used since additional logic is required to access the elements in the former case and, more important, the product of two symmetrical matrices is not necessarily symmetrical.

5.3.1 A skeleton code

Let us assume that the storage locations are as defined in the previous section and that, initially, we have arranged for the location H to contain the one-electron Hamiltonian matrix over the basis functions (the "core" matrix), the location V to contain a matrix which orthogonalises the basis functions and that there exists a method for obtaining the electron-repulsion integrals from wherever they are stored. We will, of course, have to provide all these facilities in due course.[1]

The steps in one iterative step of the calculation are as follows:

---
[1] Here, "top-down design" looks more like "procrastination".

5.3. THE IMPLEMENTATION: GETTING STARTED

1. Form the charge and bond-order matrix $\mathbf{R}$ from the matrix of MO coefficients, the number of basis functions (m) and the number of electrons ($2n$).

2. Using this $\mathbf{R}$ matrix, form the electron-repulsion matrix $\mathbf{G}$ from the stored repulsion integrals.

3. Add this repulsion matrix to the one-electron Hamiltonian matrix to give a "Hartree–Fock" matrix defined over the (non-orthogonal) basis orbitals.[2]

4. Use the orthogonalising matrix in $\mathbf{V}$ to generate a "Hartree–Fock" matrix over an orthogonal basis.

5. Solve the eigenvalue problem

$$\bar{h}^F \bar{C} = \bar{C}\epsilon$$

in the orthogonal basis.

6. Using the orthogonal basis $\bar{C}$ and the matrix in $\mathbf{V}$, generate the corresponding matrix of MO coefficients over the non-orthogonal basis.

$$C = V\bar{C}$$

This process has, of course, to be *started* in some way by an initial guess of the MO coefficient matrix C.

There is one obvious point to be made about this series of steps; one might ask

> Why perform all these transformations between the two bases? Why not work entirely in the orthogonalised basis?

This is a valid point and the answer to it hinges on the technical question which always dominates electronic structure calculations. The one-electron Hamiltonian and the repulsion integrals are *computed* over the non-orthogonal ("ordinary") basis functions and, while it is not a problem to perform the matrix multiplications to transform the one-electron Hamiltonian to the orthogonal basis:

$$\bar{h} = V^T h V$$

the transformation of the repulsion integrals to a new basis is a major task, typically requiring more time than the whole SCF calculation.

---
[2] Hartree–Fock is in quotes here since the matrix generated has the *form* of the Hartree–Fock matrix but, of course, since the MO coefficients are not yet self-consistent, it is not *the* Hartree–Fock matrix in this basis.

As we shall see as the implementation develops, it is precisely the possibility of avoiding this transformation of repulsion integrals which gives the SCF method its appeal and value.

We can now consider each of the above steps in turn and either develop an implementation or defer it as appropriate.

1. **The R matrix: subroutine scfR.**

The charge and bond-order matrix $\boldsymbol{R}$ is defined in terms of the (real) MO coefficient matrix by

$$\boldsymbol{R} = \boldsymbol{C}\boldsymbol{C}^T$$

that is, in component form

$$R_{ij} = \sum_{k=1}^{m} C_{ik} C_{jk}$$

so that only a single matrix multiplication is involved.

Here is a code to do the job and a manual page:

```
subroutine scfR (C, R, m, nocc);
double precision C(MATRIX_SIZE), R(MATRIX_SIZE);
integer m, nocc;
{
double precision sum, zero;
integer i, j, k, ij, ji, kk, ik, jk  ;
data zero/0.0d00/;
do i = 1, m;
   {
   do j = 1, i;           #   R matrix is symmetric
      {
      sum = zero;         #   initialise accumulator
      do k = 1, nocc;     #   Sum over columns of C
         {
         kk = m*(k-1); ik = kk + i; jk = kk + j;
         sum = sum + C(ik)*C(jk);
         }
      ij = m*(j-1) + i; ji = m*(i-1) + j;
      R(ij) = sum;
      R(ji) = sum;
      }
   }
return;
}
```

5.3. THE IMPLEMENTATION: GETTING STARTED

NAME scfR
  Form the density matrix R from the occupied columns of U.

SYNOPSIS

   subroutine scfR(C,R,m,nocc)
   double precision R(ARB), C(ARB)
   integer m, noccc

DESCRIPTION
  C contains (as columns) the MO coefficients and the R-matrix is formed from the first nocc columns.

ARGUMENTS

  C The eigenvectors are stored as columns of length m.

  R The charge and bond order matrix generated is stored in R by columns ($m \times m$)

  m The dimension of the matrices (basis set size)

  nocc The number of occupied orbitals (half of the number of electrons in the closed-shell case)

DIAGNOSTICS
  None

This code requires very little comment; sum is used for accumulation rather that accumulating directly into R(ij) to avoid unnecessary subscripting overheads (the central statement sum =.... is executed m*(m+1)*nocc/2 times).

The subscripts of the matrices are calculated explicitly each time which is not necessary since one can get a new value from the last value simply by a knowledge of the "stride" through the elements (either down a column, where the elements are adjacent, or along a row, where they are m apart). These improvements will be incorporated into the matrix multiplication routines discussed below.

Three further elements of the programming language have been introduced here:

1. The concept of a subroutine with an explicit *interface*; the *arguments* of the subroutine are a comma-separated list in parentheses following the subroutine's name. They are "dummies" in the sense that the subroutine

136   CHAPTER 5. IMPLEMENTATION OF THE CLOSED-SHELL CASE

takes whatever objects it finds in these locations and performs the operations on them as if they were C, R, m and nocc. Obviously the objects passed to scfR should be of the correct type (array, integer or whatever).

2. *comments* can be placed anywhere in the code, anything following a hash (sharp) sign (#) on a line is ignored by the compiler.

3. statements are ended by *semicolons* not by the end of a line.

2. and 3. The Hartree–Fock Matrix: subroutine scfGR.

The organisational problems associated with the acquisition and processing of electron-repulsion integrals represent the bulk of the work in the generation of an SCF program. We shall, for the moment, simply assume the existence of a subroutine scfGR which does the work and return to this more complex problem later since it is not just a technical problem; it involves decisions on how to store repulsion integrals (and, indeed, the decision *whether* to store these integrals). Furthermore, steps 2 and 3 in the above scheme can easily be done at the same time. The usual way is to arrange for the repulsion integral matrix to be generated and *added* to a storage location which already contains the one-electron Hamiltonian to form the Hartree–Fock matrix.

Here is a suitable interface:

```
subroutine scfGR(R, G, m, nfile)
```

the code to use it in our case would look like

```
        .
        .
do i = 1, m*m;
   HF(i) = H(i);
call scfGR(R, HF, m, nfile);
        .
        .
```

The subroutine scfGR takes the matrix in R, which is m by m, and, using the repulsion integrals assumed to be available on a file nfile, forms and adds the repulsion matrix G to the matrix in G. In the fragment of code above the one-electron Hamiltonian is put into HF before calling scfGR.

Section 5.4 of this chapter will address the problem of the organisation of repulsion integrals and a suitable scfGR will be outlined there.

5.3. THE IMPLEMENTATION: GETTING STARTED

4. Orthogonalisation: subroutine gmprd , subroutine gtprd

The transformation of the Hartree–Fock matrix from the original basis to an orthogonalised basis involves two matrix multiplications:

$$\bar{h} = (V^T h)V = BV \quad (say)$$

rather than take up space to store the transpose of a matrix, we provide two matrix multiplication routines: one to form the simple product (**gmprd**) and another which *uses* the matrix to form the transposed product (**gtprd**).[3] Here is gmprd:

```
subroutine gmprd(A, B, R, n, m, l);
double precision A(MATRIX_SIZE), B(MATRIX_SIZE);
double precision R(MATRIX_SIZE);
integer n, m, l;
{
double precision zero;
integer k, ik, j, ir, ji, ib;
data zero/0.0d00/;
ir = 0; ik = -m; # initialise the "stride" counters
do k = 1, l;
   {
   ik = ik + m;    # use the row stride
   do j = 1, n;
      {
      ir = ir + 1; ji = j - n; ib = ik;
      R(ir) = zero;
      do i = 1, m;
         {
         ji = ji + n;  ib = ib + 1;
         R(ir) = R(ir) + A(ji)*B(ib);
         }
      }
   }
return;
}
```

Although apparently more complex than scfR, this program does a very similar job; matrix multiplication. There are more statements but less work is done: all the indexing of the one-dimensional arrays is done without multiplications. The positions of the elements of the matrix within the storage array are worked out with reference to the position of the previous elements and the row and column stride which are fixed integers for a given matrix.

---
[3] These simple routines are converted versions of the IBM Scientific Subroutine Package versions.

This general-purpose matrix multiplication routine has *six* arguments: storage space for the matrices *and* information about their sizes. In fact, so far, we only require to perform matrix multiplications with *square* matrices but the routine is general enough to allow for future changes and the code is only trivially different from the square case.

However, we shall have to remember things like the *order* in which the matrix arguments go and the order of the matrix dimensions. In short, it is time to think about *documentation* for this and the other routines before things get out of hand.

The Unix "Manual Page" method is the most widely-used method of presenting the specification of the interface, function and limitations of a program; although our matrix multiplication routine is rather simple it still deserves the full treatment.

> Documentation is at least as important as the actual code. Given the documentation one can always regenerate the program; the opposite is much more difficult.

Here, then, is a manual page for **gmprd**.

5.3. THE IMPLEMENTATION: GETTING STARTED

NAME gmprd
  Matrix multiplication primitive

SYNOPSIS

```
subroutine gmprd(A,B,R,n,m,l)
double precision A(ARB), B(ARB), R(ARB)
integer n, m, l
```

DESCRIPTION
  The matrix R is formed as $AB = R$

ARGUMENTS

  A The first factor in the product $(n \times m)$, unchanged on output

  B Second factor in product $(m \times \ell)$, unchanged on output

  R Product matrix $(n \times \ell\ )$

  n number of rows in A (and in R)

  m number of columns in A and rows in B

  l number of columns in B (and in R)

SEE ALSO
  gtprd

DIAGNOSTICS
  None

Notice that the dimensions of the arrays used to store the matrices are all ARB. This is another global constant and, depending on the particular installation/compiler, one might have

```
define(ARB,*)
define(ARB,1)
define(ARB,10000000)
```

or something to indicate that the routine can be used with matrices of arbitrary size.

The code for a routine to pre-multiply a given matrix by the transpose of a matrix (**gtprd**) is very similar to that of **gmprd**, only the subscript of the first matrix is changed; here it is:

```
subroutine gtprd(A, B, R, n, m, l);
double precision A(MATRIX_SIZE), B(MATRIX_SIZE);
double precision R(MATRIX_SIZE);
integer n, m, l;
{
double precision zero;
integer k, ik, j, ir, ij, ib;
data zero/0.0d00/;
ir = 0; ik = -n; # initialise the "stride" counters
do k = 1, l;
   { ij = 0;
   ik = ik + m;    # use the row stride
   do j = 1, m;
      {
      ir = ir + 1; ib = ik;
      R(ir) = zero;
      do i = 1, n;
         {
         ij = ij + 1; ib = ib + 1;
         R(ir) = R(ir) + A(ij)*B(ib);
         }
      }
   }
return;
}
```

The manual page is similar and, to avoid too much detail, we omit it.

We can now turn to the application of these two simple routines to transform the Hartree–Fock matrix ($m \times m$) to an orthogonal basis. Since the storage R, which was used to form and use the R matrix is now freed, we may use it as work space for an intermediate matrix product and write

```
call gtprd(V,HF,R,m,m,m);
call gmprd(R,V,HF,m,m,m);
```

which uses the matrix in V to form the orthogonalised Hartree–Fock matrix and over-write the old non-orthogonal basis h^F matrix in HF, using the space where R was stored (R) as working space for the intermediate product $V^T h^F$.

We are now in a position to solve the matrix MO coefficient problem.

5.3. THE IMPLEMENTATION: GETTING STARTED

5. The eigenvalue problem: `eigen`.

The solution of the so-called matrix eigenvalue problem

$$AB = Ba$$

where A is a symmetric matrix and a is a real diagonal matrix is a very well-researched numerical problem and a number of stable methods are available.

The solution of the problem is available for the case when B is *square* as well as for rectangular B. That is, we can generate a total of m linearly independent columns of coefficients which individually satisfy the equation

$$Ab^i = a_i b^i$$

where $a_i = a_{ii}$ are the diagonal elements of a and the b^i are the corresponding columns of B.

In our case, we only need the n columns, $\bar{c}^i$, of $\bar{C}$ that solve the matrix equation

$$\bar{h}^F \bar{C} = \bar{C} \epsilon$$

in order to form the R matrix for the next iteration of the SCF process. It is only the n molecular orbitals associated with these columns which are specified by the variational method.

However, it is common to use one of the numerical methods which generates all m columns of coefficients and to *use* only the required n. If we are interested in the *ground state* of the $2n$ electron system, then, presumably,[4] we require the n columns which are associated with the n lowest values of the eigenvalue $\epsilon_i = \epsilon_{ii}$.

The actual numerical technique of obtaining the eigenvalues and eigenvectors (as the numbers ϵ_i and associated columns $\bar{c}^i$ are called) is quite simple but, again to avoid too much interference with the "flow" of the presentation, a discussion is deferred to Appendix 5.A at the end of this chapter.

In the same way as we treated the formation of the electron-repulsion matrix G, we simply assume the existence of a subroutine which does the work:

```
subroutine eigen(H, U, m)
double precision H(ARB), U(ARB);
integer m;
```

---

[4]We will examine this presumption in due course since it is basic to the whole approach to ground-state calculations.

i.e. it takes the matrix stored in H (which is assumed to be symmetric) of dimension m and generates the eigenvalues and eigenvectors. The eigenvalues are sorted into ascending order and placed on the diagonal positions of H (i.e. at locations H(m*(i-1) + i)) and the eigenvectors are placed in the corresponding columns of U. Thus, for example, H(1) contains the lowest eigenvalue and the column of coefficients which is the associated eigenvector is stored in locations U(1) ...U(m), H(m+2) contains the next-lowest eigenvalue and the associated eigenvector is stored as U(m+1) ...U(2*m) etc.

6. The C matrix: gmprd again

Having obtained the MO coefficients with respect to the orthogonalised basis as the first n columns of the matrix $\bar{C}$ in storage Cbar, we can simply multiply by the orthogonalising matrix stored in V to generate the new set of MO coefficients in C over the original (non-orthogonal) basis; the code is simply:

```
call gmprd(V, Cbar, C, m,m,m);
```

and we are back to where we started with a new current approximation to the optimum (variational) MO coefficients. In fact, the code has destroyed *both* the original C matrix and its associated R matrix, so that we cannot tell if our new C (or R) is different from the one we started with. We can take care of that now by arranging to save a copy of the R matrix as it is formed in the storage location Rold.

5.3.2 One iteration

We can now put together the codes we have either developed or specified to form the centre of an LCAOMOSCF code, the steps involved in each iteration of the self-consistent procedure. Remember that H contains the one-electron Hamiltonian, V contains an orthogonalising matrix (which we have not yet seen how to calculate) and the repulsion integrals are assumed to be accessed from a file nfile.

Here is the fragment:

5.3. THE IMPLEMENTATION: GETTING STARTED

```
      double precision H(ARB), HF(ARB), C(ARB), V(ARB), R(ARB);
      double precision Cbar(ARB), epsilon(ARB);
      integer m, n, i, nfile;
.
#  Assume that C has been generated in a previous iteration
.
.
      do i = 1, m*m;
        HF(i) = H(i);
      call scfR(C, R, m, n);              # Form the R matrix
      call scfGR(HF, R, m, nfile);        # Use it to form HF
      do i = 1, mm;                       # Save a copy of R
        Rold(i) = R(i);                   # for comparison
      call gtprd(V, HF, R, m, m, m);      # Transform HF to
      call gmprd(R, V, HF, m, m, m);      # orthogonal basis
      call eigen(HF, Cbar, m);            # Diagonalise HF bar
      do i = 1, n;
        epsilon(i) = HF(m*(i-1)+i);       # Get eigenvalues
      call gmprd(V, Cbar, C, m, m, m);    # Transform coefficients
                                          # back to basis
.
.
```

Even this small piece of code could be made more compact by using the storage space R as work-space once more in place of the storage space Cbar which is only used once.

One thing which it is useful to do during the course of each SCF iteration is to calculate the current value of the total electronic energy and report it. This total energy takes no part in the iterative process and is far too coarse a criterion to be used to test self-consistency, but is useful as a reassurance that all is well if the total energy is lowered as the iterations proceed.

The total electronic energy of a closed-shell single determinant of (doubly occupied spatial) orbitals is given in terms of the MO coefficients by

$$E = 2\mathrm{tr}\left[\boldsymbol{C}^T(\boldsymbol{h} + \frac{1}{2}\boldsymbol{G})\boldsymbol{C}\right]$$

which, because $\boldsymbol{R} = \boldsymbol{C}\boldsymbol{C}^T$ and $\mathrm{tr}\boldsymbol{A}\boldsymbol{B} = \mathrm{tr}\boldsymbol{B}\boldsymbol{A}$, may be written

$$E = \mathrm{tr}\boldsymbol{h}\boldsymbol{R} + \mathrm{tr}\boldsymbol{h}^F\boldsymbol{R}$$

which is nothing more than the sum of the trace of the product of the matrices stored in R and HF before and after the call scfGR statement. The code becomes:

```
      double precision H(ARB), HF(ARB), C(ARB), V(ARB);
      double precision Cbar(ARB), epsilon(ARB);
      double precision E, zero;
      integer m, n, i, nfile, mm;
      data zero/0.0d00/;
.
#  Assume that C has been generated in a previous iteration
.
.
      mm = m*m;
      E = zero;                       # Initialize E
      do i = 1, mm;
         HF(i) = H(i);
      call scfR(C, R, m, n);          # Form the R matrix
      do i = 1, mm;                   # Accumulate the total
         E = E + R(i)*HF(i);          # energy
      call scfGR(HF, R, n, nfile);    # Use R to form HF
      do i = 1, mm;                   # Accumulate rest
         E = E + R(i)*HF(i);          # of energy
      write(OUTPUT_UNIT,200) E;
 200  format(" Current Electronic Energy = ", f12.6)
      do i = 1, mm;                   # Save a copy of R
         Rold(i) = R(i);              # for comparison
      call gtprd(V, HF, R, m, m, m);  # Transform HF to
      call gmprd(R, V, HF, m, m, m);  # orthogonal basis
      call eigen(HF, Cbar, m);        # Diagonalise HF bar
      do i = 1, n;
         epsilon(i) = HF(m*(i-1)+i);  # Get eigenvalues
      call gmprd(V, Cbar, C, m, m, m); # Transform coefficients
                                      # back to basis
.
.
```

OUTPUT_UNIT is some output channel number.

5.3.3 The iterative procedure

All that is necessary now is to take the code of the last section, form some condition based on the difference between successive C or R matrices and do

```
repeat
   {
   code
   }
until (condition)
```

5.3. THE IMPLEMENTATION: GETTING STARTED

and we have a working LCAOMOSCF skeleton.

In setting up this implementation of the solution of the finite-expansion Hartree–Fock equation we are, of course, solving an approximation to the original parametric variational problem; the calculation of the minimum energy of the variational expression.[5]

At first sight, therefore, a suitable criterion for termination of the iterative process might be the minimisation of the total electronic energy E. In fact, at the solution E *will* be a minimum but experience shows that the total electronic energy is a rather insensitive criterion of self-consistency. What this means is that, frequently, the total electronic energy has "settled down" to its final value (to machine accuracy) long before the coefficients have themselves stopped changing. The `condition` will, therefore, be based on a criterion involving the differences between successive values of the elements of the R matrix.

A suitable code might look like:

```
      double precision Rsum, term, crit;
      integer icon;
      data crit/1.0d-06/;
    .
    .
      icon = 0;              #  initialize counter and accumulator
      Rsum = zero;
      do i =1, mm;
        {
        term = abs(R(i) - Rold(i));   #  Difference in each element
        Rold(i) = R(i);               #  Set up Rold for next time
#       Count the offenders
        if ( term > crit ) icon = icon + 1;
        Rsum = Rsum + term;
        }
      write(OUTPUT_UNIT,201) Rsum, icon;
  201 format (" Sum of differences in R = ", f12.5,i6,
              "  Changing");
```

This code generates `icon`, which is zero if there is no change (to the tolerance `crit`) in the elements of R between successive iterations or non-zero if the R matrix has not converged. `Rsum` is merely information, it is not used in the

---

[5]In fact, of course, we are only guaranteed a turning point in the energy functional which may not be a minimum. This point is related to the choice of *which* columns of coefficients to use in forming the R matrix (which orbitals are occupied in the single determinant). This point will be addressed later.

criterion but, presumably, will become smaller and smaller as the calculation proceeds.

The opportunity has been taken, during this do loop, to put the latest R matrix into Rold in preparation for the next cycle. Since the criterion is based on the R matrix we may as well make an initial assumption of a suitable R rather than an initial C and take the opportunity to reorganise the code slightly to give our initial LCAOMOSCF code, presented on one page for convenience:

5.3. THE IMPLEMENTATION: GETTING STARTED

```
      double precision H(ARB), HF(ARB), C(ARB), V(ARB), R(ARB);
      double precision Cbar(ARB), epsilon(ARB), Rold(ARB);
      double precision E, zero, Rsum, term, crit;
      integer m, n, i, nfile, icon,mm;
      data zero/0.0d00/,crit/1.0d-06/;
      mm = m*m;
# Set initial R to zero, i.e. start from H not HF
      do i = 1, mm;
        { R(i) = zero; Rold(i) = zero}
#  Assume that the first iteration will change R to be non-zero!
      repeat
        {
        E = zero ; icon = 0;      # Initialize E and counter
        do i = 1, m*m;
          HF(i) = H(i);
        do i = 1, mm;                   # Accumulate the total
          E = E + R(i)*HF(i);           # energy
        call scfGR(HF, R, m, nfile);    # Use R to form HF
        do i = 1, mm;                   # Accumulate rest
          E = E + R(i)*HF(i);           # of energy
        write(OUTPUT_UNIT,200) E;
   200 format(" Current Electronic Energy = ", f12.6)
        call gtprd(V, HF, R, m, m, m);  # Transform HF to
        call gmprd(R, V, HF, m, m, m);  # orthogonal basis
        call eigen(HF, Cbar, m);        # Diagonalise HF bar
        do i = 1, n;
          epsilon(i) = HF(m*(i-1)+i);      # Get eigenvalues
        call gmprd(V, Cbar, C, m, m, m); # Transform coefficients
        call scfR(C, R, m, n);          # Form the R matrix
        Rsum = zero;
        do i =1, mm;
          {
          term = abs(R(i) - Rold(i));  # Difference in each element
          Rold(i) = R(i);              # Set up Rold for next time
          if ( term > crit ) icon = icon + 1;  # Count the offenders
          Rsum = Rsum + term;
          }
        write(OUTPUT_UNIT,201) Rsum, icon;
   201 format (" Sum of differences in R = ", f12.5,i6,
               " Changing")
        }
      until (icon == 0)
```

The assumptions behind this code are:

1. There are **m** basis functions and **n** (doubly) occupied orbitals.
2. The **m*m** numbers forming the one-electron Hamiltonian are present in **H**.
3. The **m*m** numbers forming a matrix which will generate an orthogonal basis from the given basis are present in **V**.
4. The electron repulsion integrals over the basis functions are accessible by `scfGR` (on a file `nfile`).

The tools we still have to build to complete the implementation are

1. A repulsion-integral storage and access method and the code for `scfGR`.
2. A matrix diagonalisation method and the code for `eigen`.
3. A method of computing the orthogonalisation matrix V.

Not forgetting, of course, attacking the all-important but logically distinct task of actually *calculating* the overlap, one-electron and electron-repulsion integrals themselves.

However, although in the approach to the problem of molecular electronic structure so far it has been assumed that these integrals will be *computed from an explicitly defined basis of functions*, it is clear from our implementation that the *source* of the integrals is logically distinct from their *use* in the SCF method. There is nothing to prevent us from (*e.g.*) taking the values of some integrals from experimental measurements, from neglecting others assumed on physical grounds to be small and approximating others. In particular, there is nothing to prevent us from *assuming* that we may work in an (unspecified) orthogonal basis and therefore being able to omit the calls to `gmprd` and `gtprd` in the above code. In short, with the removal of the calls to these two routines, our code is the basis of a semi-empirical SCF technique.

For the time being, however, we must turn to the building of the tools mentioned above.

5.4 The implementation: repulsion integral access

5.4.1 General considerations

We are considering the special case of a single determinant of *doubly occupied* real spatial orbitals; that is, the MOs are to be expanded in terms of a set of

5.4. THE IMPLEMENTATION: REPULSION INTEGRAL ACCESS

real basis functions ϕ_r with *real* expansion coefficients C. In the more general case of complex singly occupied spin-orbitals, the *basis functions* will always be real; only the expansion coefficients may be allowed to be complex. This is not a restriction of principle but one of practice, since there is only slight advantage to be gained by the use of complex orbitals[6] in those cases where there is an axis of cylindrical symmetry (linear molecules and atoms) and this symmetry suggests the use of complex functions. We shall assume, therefore, that the spatial basis functions are *always* real and therefore (with the usual electrostatic Hamiltonian) so are the energy integrals and overlaps arising from the basis.

In particular, the spatial electron-repulsion integrals

$$(\phi_r\phi_s, \phi_t\phi_u) = \int dV_1 \int dV_2 \phi_r(\vec{r_1})\phi_s(\vec{r_1}) \frac{1}{|\vec{r_1}-\vec{r_2}|} \phi_t(\vec{r_2})\phi_u(\vec{r_2})$$

are real numbers and the use of complex conjugate notation has been dropped. Clearly, since the four basis functions defining any electron-repulsion integral may be any of the m basis functions, there are potentially m^4 repulsion integrals to calculate, store and manipulate during an SCF calculation. Equally clearly, since the electron-repulsion "operator" $1/r_{12}$ is merely a multiplying factor, certain *permutations* amongst the four basis functions do not change the value of the integral.

In the general case where the four indices r, s, t, u are all different, there are *eight* possible permutations of the ϕ_r that lead to integrals with the same value; we may:

- interchange r and s since these two indices simply serve to describe the "charge density" of electron 1;
- interchange t and u since these two indices simply serve to describe the "charge density" of electron 2;
- interchange the *pairs* (rs) and (tu), since the value of the integral does not depend on the numbering of the electrons.

Thus, in the general case, we have

$$\begin{aligned}(\phi_r\phi_s, \phi_t\phi_u) &= (\phi_s\phi_r, \phi_t\phi_u) = (\phi_r\phi_s, \phi_u\phi_t) = (\phi_s\phi_r, \phi_u\phi_t) = \\ (\phi_t\phi_u, \phi_r\phi_s) &= (\phi_t\phi_u, \phi_s\phi_r) = (\phi_u\phi_t, \phi_r\phi_s) = (\phi_u\phi_t, \phi_s\phi_r)\end{aligned}$$

The notation involving the generic symbol ϕ is a little redundant when discussing electron-repulsion integrals only over the *basis functions* and there is no danger of confusion with integrals over other functions, so we drop the symbol ϕ and

---
[6]In the absence of the complication of magnetic terms in the Hamiltonian.

refer to the integrals over basis functions by the simpler notation (rs,tu). In this notation the above equalities become simply

$$(rs,tu) = (sr,tu) = (rs,ut) = (sr,ut) =$$
$$(tu,rs) = (tu,sr) = (ut,rs) = (ut,sr)$$

There are only $M = m(m+1)/2$ distinct pairs of products of the functions $\phi_r \phi_s$ and so there are only $m\$ = M(M+1)/2$ distinct repulsion integrals arising from a basis of m real basis functions. It is worth noting and emphasising here how quickly m^4 increases with m and how much of a saving this simple "permutational" symmetry effects. The following table gives some typical numbers arising from the simplest possible calculation on the named molecules[7]

Molecule	m	$M = m(m+1)/2$	$m\$ = M(M+1)/2$	m^4
H_2O	7	28	406	2401
C_2H_6	16	136	9316	65,536
C_6H_6	36	666	222,111	1,679,616
$Fe(C_5H_5)_2$	88	3916	7,669,486	59,969,536

Of course in the case of molecules having some elements of spatial symmetry, many thousands of these integrals may be zero or equal to spatially equivalent integrals. This is a separate problem which will be addressed later in Chapter 17.

If we assume that the "typical" molecule will be associated with a number of repulsion integrals which cannot be kept in the computer's main high-speed memory we are left with just two choices for an attack on the storage problems generated by such huge quantities of data:

1. Compute and store the integrals on an external medium (disk file) *once and for all* and read and process them as they are required during the SCF (and other) process.

2. Compute the integrals *as they are needed* and do not store them at all.

Let us assume here and now that, in either case, we shall only deal with the $m\$$ permutationally distinct integrals and not the entire set of m^4 numbers. This simple decision has some consequences however, since, in addition to the *value* of a particular repulsion integral, we need to know *which one it is*, i.e. for a particular electron-repulsion integral we need to know

- the value of the integral (the value, say),
  - the four indices r, s, t and u (the label, say),

---

[7] A "minimal basis" calculation; using as a basis only those atomic orbital shells which are occupied in the ground state of the atoms of which molecule is composed.

5.4. THE IMPLEMENTATION: REPULSION INTEGRAL ACCESS

- which one of the eight potential orderings of the labels is to be used as *the* label of that integral.

The second point can be met in basically two possible ways:

1. The values can be stored *in a particular order* in an array or file so that the label may be inferred from the position of the value in the file.
2. The label may be stored *explicitly* along with the value in the file or array.

The first of these choices seems to be more efficient, since only half the storage space is needed. But, in fact, it is much more effective to use the second choice since it is often desirable to

- Omit integrals which are zero or fall below some numerical threshold from the file to save space and processing time.
- Compute and store the integrals in an *order* dictated by numerical convenience rather than in an order dictated by some labelling scheme.

The upshot is that, if the repulsion integrals are to be *stored* then the $m\$$ (at most) permutationally distinct integrals will be stored together with their labels and any code which accesses the file so generated must assume that the integrals occur in random order.

The third point above, that is *which* label to use for a permutationally equivalent set is simply a matter of convention: if

$$(ij) = \frac{i(i-1)}{2} + j$$

then the label (rs, tu) for which

$$r \geq s; \quad t \geq u; \quad (rs) \geq (tu)$$

will be the one chosen to do duty for the whole set of labels. There may be only the one (if $r = s = t = u$), two, four or eight depending on equalities among r, s, t and u.

The programs which access these file must make sure that all the integrals are generated and used from the single representative present in the file.

5.4.2 Obtaining the integrals

In the latter part of the last section we have silently made the assumption that the repulsion integrals will be computed and stored rather than computed as they are needed. This assumption will be continued in what follows but, in designing any codes, it will be kept in mind that the other option may be more valid in different circumstances and the codes should be flexible enough to accommodate both. The only way to ensure this is to push the details of the acquisition of the repulsion integrals out of sight into a "primitive" segment which will have different implementations for the two basis possibilities: "compute and store" or "compute on demand".

This primitive will be **getint** with the following specification:

```
integer function getint(file,i,j,k,l,mu,val,pointer)
integer file, i, j, k, l, mu, pointer
double precision val
```

DESCRIPTION
   This is the repulsion integral primitive which generates repulsion integrals one at a time on successive "calls" for processing by any of the procedures in the system. Returns OK if it finds an integral or END_OF_FILE when the integrals are exhausted.

ARGUMENTS

   file The logical number of a sequential file containing the electron repulsion integrals (if the files are on a file, otherwise undefined). `file` must be rewound before the first entry to **getint** for a particular pass over the integrals.

   i,j,k,l The labels of the repulsion integral currently being generated or requested.

   mu This is a spare argument for use perhaps if the integrals are marked in some way or divided into classes for special purposes.

   val The numerical value of the current repulsion integral.

   pointer The book-keeping pointer for the integrals. It must be set to zero before the first entry to **getint** and is maintained by **getint** thereafter.

The duty of getint is to hand out electron-repulsion integrals on

5.4. THE IMPLEMENTATION: REPULSION INTEGRAL ACCESS 153

demand, one at a time, whatever the circumstances.

It may be that these integrals are already computed and reside on an external file, in which case use of `getint` will tend to look like

```
rewind file; pointer = 0;
while ( getint(file,i,j,k,l,mu,val,pointer) != END_OF_FILE)
   {
   process (i,j,k,l,val)
   }
```

Or, it may be that `getint` is required to *compute* the electron-repulsion integrals as required. In this case usage is different and will look like

```
do i = 1, m;
   {
   do j = 1, i;
      {
      do k = 1, i;
         { ltop = k         # get upper limit for "l" right
         if ( i == k ) ltop = j;
         do l = 1, ltop;
            {
            status = getint (file,i,j,k,l,mu,val,pointer);
               process (i,j,k,l,val)
            }
         }
      }
   }
```

Where `file` and `pointer` are not used. Of course, in the latter case, `getint` must have access to the details of the basis functions etc. from another source in order to compute the repulsion integrals.

In the first case `getint` *generates* the integers i, j, k, l and in the second they are *supplied*, but in each case `getint` is required to ensure that it can provide all the electron repulsion integrals. It is sometimes clearer to have both uses of `getint` called by similar coding structures. The generation of the four integral label loops can be pushed down out of sight by coding them as a small function[8] (`next_label`, say) so that the use of `getint` in a direct SCF situation can be written

---

[8]Coded in Appendix 5.D

```
while (next_label(i,j,k,l,m) == YES)
  {
  status = getint (file,i,j,k,l,mu,val,pointer)
  process (i,j,k,l,val)
  }
```

Implementations of **getint** are deferred until the file structure for storage of repulsion integrals has been finalised and the details of **getint**'s partner **putint** have been decided.

5.4.3 Processing the repulsion integrals

The most direct and transparent way to illustrate the way in which the eight possible integrals arising from the permutations of the (possibly) different indices forming the label of a repulsion integral is by a program fragment (where "USE" is simply shorthand for any code which might be used to process the indices, label and value of the repulsion integral).

The indices are **i**, **j**, **k** and **l** and the value of the repulsion integral is **val**.

```
double precision val;
integer i, j, k, l, ij, kl;
USE (i,j,k,l,val);                           # 1
if ( i != j ) USE (j,i,k,l,val);             # 2
if ( k != l )
   {
   USE (i,j,l,k,val);                        # 3
   if ( i != j ) USE (j,i,l,k,val);          # 4
   }
ij = (i*(i-1))/2 + j; kl = (k*(k-1))/2 + l;
if ( ij != kl )
   {
   USE (k,l,i,j,val);                        # 5
   if ( i != j) USE (k,l,j,i,val);           # 6
   if ( k != l )
      {
      USE (l,k,i,j,val);                     # 7
      if ( i != j ) USE (l,k,j,i,val);       # 8
      }
   }
```

The code has been written in this way on purpose since we can make it into a useable fragment simply by a suitable macro definition of USE. If we **define**

5.4. THE IMPLEMENTATION: REPULSION INTEGRAL ACCESS   155

USE in the following way:

```
define(USE, { irs = m*($2-1) + $1; itu = m*($4-1) + $3;
              iru = m*($4-1) + $1; its = m*($2-1) + $3;
              G(irs) = G(irs) + 2.0d00*R(itu)*$5;
              G(iru) = G(iru) - R(its)*$5; })
```

then, provided the locations G and R are allocated in the program, this combination will compute the contribution of a particular repulsion integral to the electron-repulsion matrix.

It is, perhaps, worth noting at this point the relative advantages and disadvantages of macros and subroutines (or functions) since we could obviously have written a subroutine USE(i,j,k,l,val) to accomplish the same task.

- There is a unavoidable runtime "overhead" associated with the transmission of arguments of functions or subroutines which is *relatively* large the smaller the "body" of the subroutine. Of course, once compiled, the subroutine is available generally without recompilation.

- A macro is *expanded inline* by the compiler and so the result is exactly equivalent to (laboriously) writing out the code for each separate case. There is no run-time penalty of the transmission of arguments.

- The attractiveness of the use of macros is rapidly attenuated the more complex the task becomes; the macro may become illegible if it is more than about six to ten lines and uses a large number of arguments.

Efficiency is not yet an issue here, the use of the macro USE is simply for purposes of illustration of the possibilities available to us and to make the program look more and more like its *description*.

We now have the essentials of an implementation of scfGR. It only remains to assemble the pieces. Here is the *first model* of the code for scfGR:

CHAPTER 5. IMPLEMENTATION OF THE CLOSED-SHELL CASE

```
subroutine scfGR(R, G, m, nfile)
double precision G(*), R(*);
double precision val;
integer irs, itu, iru, its;
integer i, j, k, l, mu, ij, kl, getint;
integer m, nfile, pointer;

define(USE, { irs = m*($2-1) + $1; itu = m*($4-1) + $3;
              iru = m*($4-1) + $1; its = m*($2-1) + $3;
              G(irs) = G(irs) + 2.0d00*R(itu)*$5;
              G(iru) = G(iru) - R(its)*$5; })

rewind nfile;
pointer = 0;              # initialise "file"
while( getint(nfile, i, j, k, l, mu, val, pointer) != END_OF_FILE)
   {
   USE (i,j,k,l,val);                     # 1
   if ( i != j )
      USE (j,i,k,l,val);                  # 2
   if ( k != l )
      {
      USE (i,j,l,k,val);                  # 3
      if ( i != j )
         USE (j,i,l,k,val);               # 4
      }
   ij = (i*(i-1))/2 + j; kl = (k*(k-1))/2 + l;
   if ( ij != kl )
      {
      USE (k,l,i,j,val);                  # 5
      if ( i != j)
         USE (k,l,j,i,val);               # 6
      if ( k != l )
         {
         USE (l,k,i,j,val);               # 7
         if ( i != j )
            USE (l,k,j,i,val);            # 8
         }
      }
   }        # end of while
return
end
```

This implementation will work, however slowly, and we are deferring all considerations of efficiency until we have a working program.

The manual entry is preliminary but should not change too much since it reflects, not the detailed coding *within* scfGR, but the interface.

5.4. THE IMPLEMENTATION: REPULSION INTEGRAL ACCESS

NAME scfGR

Formation of two-electron HF contributions

SYNOPSIS

```
subroutine scfGR(R,G,m,nfile)
double precision R(ARB), G(ARB)
integer n, nfile
```

DESCRIPTION

scfGR processes the electron repulsion integrals to be found on file nfile.

ARGUMENTS

R Input: contains the "charge and bond order" matrix on input, unchanged on output

G Input/Output: this routine adds (repeat adds) the electron repulsion matrix $G(R)$ to the contents on input of G. Thus, the matrix G will normally contain a one-electron Hamiltonian on entry and the Hartree–Fock matrix on output.

m Input: the number of basis functions.

nfile Input: if the repulsion integrals are on a file, this is its number.

SEE ALSO

getint

DIAGNOSTICS

None

One point is worth mentioning; the definition of the electron-repulsion matrix is:

$$G_{rs} = 2J_{rs} - K_{rs} = \sum_{t,u=1}^{m} 2R_{tu}(rs, tu) - \sum_{t,u=1}^{m} R_{tu}(ru, ts)$$

but we are accessing the repulsion integrals *serially*. That is, in contrast with the definition above which says:

> I am forming a particular matrix element G_{rs}, I need *all* the repulsion integrals to do so.

requiring the repulsion integral file to be processed m^2 times (once for each

G_{rs}), we are saying

> I have a particular repulsion integral (rs, tu) to hand, to which matrix elements of G does it make a contribution?

requiring the file to be processed just once.

Thus a more useful piece of information is obtained by inverting the above sum and expressing the contributions to J and K in terms of a *given* repulsion integral:

$$J_{rs} = \sum_{t,u=1}^{m} 2R_{tu}(rs, tu)$$

$$K_{ru} = \sum_{t,s=1}^{m} R_{ts}(rs, tu)$$

these are the formulae implemented in the macro USE.

5.5 Building a testbench: conventional SCF

It is time to try out the codes developed in previous sections. To do this we need just two things:

- Access to a matrix eigenvalue routine (`eigen` above).
- Values of the elements of the one-electron Hamiltonian matrix (h), the orthogonalising matrix (V) and the electron repulsion integrals (ij, kl) for some particular case.

Methods of finding the eigenvalues and eigenvectors are discussed in Appendix 5.A, but methods of computing the various basis-function integrals will be discussed *much* later. So, we can build a testbench for the SCF codes by taking the values of these integrals as data and building them into the program. This gives us the opportunity of

- Writing a simple version of `getint` to supply the repulsion integrals.
- Testing the logic of the ideas outlined in this chapter.
- Calculating the MOs for closed-shell single-determinant functions for the molecule and its ions.

5.5. BUILDING A TESTBENCH: CONVENTIONAL SCF

Perhaps the simplest non-trivial example is a "minimal basis" calculation on the ground electronic state of the water molecule at a nuclear geometry close to its equilibrium value. The minimal basis is just those atomic orbitals which are occupied in the ground states of the component atoms of the molecule: 7 AOs, the $1s$, $2s$, $2p_x$, $2p_y$ and $2p_z$ orbitals of oxygen and a $1s$ AO for each hydrogen.[9] The integrals arising from this basis have been computed by the routines shown later in this work and are here simply put into **data** statements.

5.5.1 A test getint

The function **getint** is simply required to hand out repulsion integrals one at a time when called and to signal OK or END_OF_FILE to the calling program to indicate whether or not there are any more integrals to be had. The mechanism by which **getint** obtains these integrals is invisible to the calling program, so that they may be computed or stored somewhere; there need not be any physical file involved.

In the case of our testbench, **getint** has all the integrals and their labels stored as part of itself and passes them out on demand. The full listing of this very specialised **getint** is given in Appendix 5.C to this chapter, the essence of it is that the labels and values of the repulsion integrals are stored in

```
ir(pointer),jr(pointer),kr(pointer),lr(pointer),
   v(pointer),pointer=1, 228
```

and are handed out on demand until **pointer** $>$ 228 at which point END_OF_FILE is signalled. Here is the code excluding the tedious **data** statements:

```
# Number of integrals for H2O
   define(NUMBER,228)
# Dummy getint for the special case of H2O minimal basis
   integer function getint(file, i, j, k, l, mu, val, pointer)
   save;     # make sure that the data is not lost
   integer file, i, j, k, l, mu, pointer
   double precision val;
# here is storage space for the integrals and labels

   integer ir(NUMBER),is(NUMBER),it(NUMBER),iu(NUMBER);
   double precision v(NUMBER);

#  Not all the repulsion integrals are present, the zeroes are omitted.
```

---
[9]The detailed specification of the basis functions will be given later when various basis sets have been discussed; for the moment the simple intuitive picture of the minimal basis is enough.

```
#   The full number, including all zeroes, is 406
#   The integrals are in "standard order" but they could be
#      in any order

#   ''data'' statements for ir,jr,kr,lr and v go here

#
    pointer = pointer + 1;
    if (pointer > NUMBER ) return (END_OF_FILE);
    i = ir(pointer);
    j = is(pointer);
    k = it(pointer);
    l = iu(pointer);
    val = v(pointer);
#
    return(OK);
    end
```

This function is rather boring, but it is just a test version. Notice that, in common with "genuine" versions of **getint**, this function must have **pointer** initialised to zero before the first call and **pointer** must not be changed thereafter by the calling program; **pointer** is managed entirely by **getint**.

Used with **scfGR** above, this **getint** will calculate the repulsion matrix G for the seven basis functions.

5.5.2 The testbench

We simply have to insert **data** statements with the values of the elements of the matrices h and V into our closed-shell SCF program to have a working testbench. These data are also relegated to Appendix 5.C.

The full program is repeated below for convenience with a slight modification to allow for the fact that the iterative process may not converge. The string **MAX_ITERATIONS** is **define**d to be a fairly large number, the iterations counted in **kount** and the test for completion modified to

$$\text{until (icon == 0 | kount == MAX\_ITERATIONS)}$$

In common with standard logical usage | is "or" (the f77 .OR.).

```
    define(MAX_ITERATIONS,50)
#   Matrices of suitable size for 7 basis functions
        double precision H(49), HF(49), C(49), V(49), R(49);
```

5.5. BUILDING A TESTBENCH: CONVENTIONAL SCF 161

```
      double precision Cbar(49), epsilon(7), Rold(49);
      double precision E, zero, Rsum, term, crit;
      integer m, n, i, nfile, icon, mm, kount;
      data zero/0.0d00/,crit/1.0d-06/;

# Start of special data for H2O
#     THE DATA STATEMENTS FOR V AND H ARE INSERTED HERE
      m = 7;         # Basis size for H2O
      n = 5;         # 10 electrons for H2O

      irite= ERROR_OUTPUT_UNIT
# End of data for H2O

      mm = m*m;
# Set initial R to zero, i.e. start from H not HF
      do i = 1, mm;
         { R(i) = zero; Rold(i) = zero; }
# Assume that the first iteration will change R to be non-zero!
      kount = 0;
      repeat
         {
         kount = kount + 1;       # count the iterations
         E = zero ; icon = 0;     # Initialize E and counter
         do i = 1, m*m;
            HF(i) = H(i);
         do i = 1, mm;                 # Accumulate the total
            E = E + R(i)*HF(i);        # energy
         call scfGR( R, HF, m, nfile); # Use R to form HF
         do i = 1, mm;                 # Accumulate rest
            E = E + R(i)*HF(i);        # of energy
         write(ERROR_OUTPUT_UNIT,200) E;
  200    format(" Current Electronic Energy = ", f12.6)
         call gtprd(V, HF, R, m, m, m);   # Transform HF to
         call gmprd(R, V, HF, m, m, m);   # orthogonal basis
         call eigen(HF, Cbar, m);         # Diagonalise HF bar
         do i = 1, n;
            epsilon(i) = HF(m*(i-1)+i);   # Get eigenvalues
         call gmprd(V, Cbar, C, m, m, m); # Transform coefficients
         call scfR(C, R, m, n);           # Form the R matrix
         Rsum = zero;
         do i =1, mm;
            {
            term = abs(R(i) - Rold(i)); # Difference in each element
            Rold(i) = R(i);             # Set up Rold for next time
            if ( term > crit ) icon = icon + 1; # Count the offenders
            Rsum = Rsum + term;
            }
         write(ERROR_OUTPUT_UNIT,201) Rsum, icon;
  201 format (" Sum of differences in R = ", f12.5,i6,
             " Changing")
```

```
        }
    until (icon == 0 | kount == MAX_ITERATIONS)
    write(ERROR_OUTPUT_UNIT,202) kount;
202 format(" SCF converged in",i4," iterations")
    write(ERROR_OUTPUT_UNIT,203) (epsilon(i), i=1,n);
203 format(" Orbital Energies ", (7f10.5))
    STOP;
end
```

5.5.3 Running the testbench

Compiling and running this program produces the following output:

```
Current Electronic Energy =     0.000000
Sum of differences in R =       9.12580    31 Changing
Current Electronic Energy =    -82.438005
Sum of differences in R =       7.87478    30 Changing
Current Electronic Energy =    -84.152331
Sum of differences in R =       0.69720    30 Changing
Current Electronic Energy =    -84.168228
Sum of differences in R =       0.14597    30 Changing
Current Electronic Energy =    -84.168951
Sum of differences in R =       0.05467    30 Changing
Current Electronic Energy =    -84.169045
Sum of differences in R =       0.02207    30 Changing
Current Electronic Energy =    -84.169061
Sum of differences in R =       0.00934    30 Changing
Current Electronic Energy =    -84.169064
Sum of differences in R =       0.00398    28 Changing
Current Electronic Energy =    -84.169065
Sum of differences in R =       0.00171    28 Changing
Current Electronic Energy =    -84.169065
Sum of differences in R =       0.00074    28 Changing
Current Electronic Energy =    -84.169065
Sum of differences in R =       0.00032    24 Changing
Current Electronic Energy =    -84.169065
Sum of differences in R =       0.00014    21 Changing
Current Electronic Energy =    -84.169065
Sum of differences in R =       0.00006    17 Changing
Current Electronic Energy =    -84.169065
Sum of differences in R =       0.00003    13 Changing
Current Electronic Energy =    -84.169065
Sum of differences in R =       0.00001     2 Changing
Current Electronic Energy =    -84.169065
Sum of differences in R =       0.00000     0 Changing
SCF converged in  16 iterations
Orbital Energies  -20.24147  -1.26907  -0.61856  -0.45321  -0.39137
```

5.5. BUILDING A TESTBENCH: CONVENTIONAL SCF

These numbers are correct; the total electronic energy of -84.16907 has to have the (constant) nuclear repulsion energy added to give the total energy of the water molecule. The orbital energies are the energies of the five doubly occupied MOs.

The scope for further calculations with this program is rather limited! If we ignore the physics of the molecule, we can perform calculations on ions with an even number of electrons up to 14 (all seven MOs doubly occupied). It is, perhaps, worth trying calculation on $(H_2O)^{2+}$ and $(H_2O)^{2-}$ just to see what happens.

For $(H_2O)^{2+}$, changing n = 5 to n = 4 generates the output

```
Current Electronic Energy =     0.000000
Sum of differences in R =       8.12580      30  Changing
Current Electronic Energy =   -82.014408
Sum of differences in R =       5.17110      30  Changing
Current Electronic Energy =   -82.806664
Sum of differences in R =       0.57795      30  Changing
Current Electronic Energy =   -82.818777
Sum of differences in R =       0.02663      30  Changing
Current Electronic Energy =   -82.818808
Sum of differences in R =       0.00559      30  Changing
Current Electronic Energy =   -82.818809
Sum of differences in R =       0.00187      30  Changing
Current Electronic Energy =   -82.818810
Sum of differences in R =       0.00064      30  Changing
Current Electronic Energy =   -82.818810
Sum of differences in R =       0.00022      25  Changing
Current Electronic Energy =   -82.818810
Sum of differences in R =       0.00008      21  Changing
Current Electronic Energy =   -82.818810
Sum of differences in R =       0.00003      13  Changing
Current Electronic Energy =   -82.818810
Sum of differences in R =       0.00001       0  Changing
SCF converged in  11 iterations
Orbital Energies  -21.83551  -2.44325  -1.78156  -1.65429
```

The things to notice are both the consequence of the electrons being held much more tightly in the dipositive ion:

1. The calculation converges in a smaller number of iterations.

2. The orbital energies are all more negative; the electrons are more tightly bound.

In the case of $(H_2O)^{2-}$, changing n = 5 to n = 6 generates

164 CHAPTER 5. IMPLEMENTATION OF THE CLOSED-SHELL CASE

```
Current Electronic Energy =     0.000000
Sum of differences in R =      14.21867    31  Changing
Current Electronic Energy =   -81.098730
Sum of differences in R =       5.20071    25  Changing
Current Electronic Energy =   -82.127178
Sum of differences in R =       0.56812    25  Changing
Current Electronic Energy =   -82.142615
Sum of differences in R =       0.15446    25  Changing
Current Electronic Energy =   -82.143384
Sum of differences in R =       0.05532    25  Changing
Current Electronic Energy =   -82.143465
Sum of differences in R =       0.01999    25  Changing
Current Electronic Energy =   -82.143476
Sum of differences in R =       0.00724    25  Changing
Current Electronic Energy =   -82.143477
Sum of differences in R =       0.00263    25  Changing
Current Electronic Energy =   -82.143477
Sum of differences in R =       0.00095    25  Changing
Current Electronic Energy =   -82.143477
Sum of differences in R =       0.00035    25  Changing
Current Electronic Energy =   -82.143477
Sum of differences in R =       0.00013    24  Changing
Current Electronic Energy =   -82.143477
Sum of differences in R =       0.00005    14  Changing
Current Electronic Energy =   -82.143477
Sum of differences in R =       0.00002     6  Changing
Current Electronic Energy =   -82.143477
Sum of differences in R =       0.00001     0  Changing
SCF converged in  14 iterations
Orbital Energies  -18.79697  -0.18382   0.37157   0.58737   0.69482   1.2810
```

Again, the salient points are, perhaps, obvious

1. The electronic energy is much higher than that of the molecule showing that the dinegative ion would (at least in this approximation) spontaneously emit electrons.

2. Except for the very low "core" orbital energy and one rather loosely bound orbital, the orbital energies are *all positive*, again emphasising the unstable nature of this electronic structure.

3. The convergence is also quicker than that of the neutral molecule which is not simply explicable on physical grounds (one would expect loosely bound electrons to be more difficult to bring to self-consistency). However, in this case there are not many degrees of freedom since six of the seven orbitals are occupied so that the relatively rapid convergence (and, as we shall see later, the very existence of the solution) is an artifact of the minimal basis.

5.6. ANOTHER TESTBENCH: DIRECT SCF

This last point can be made even more strongly by proceeding to the unphysical limit of the basis. A calculation on $(H_2O)^{4-}$, obtained by replacing n = 5 by n = 7 in the testbench code generates the following output:

```
Current Electronic Energy =     0.000000
Sum of differences in R =      18.38466     31   Changing
Current Electronic Energy =   -78.215430
Sum of differences in R =       0.00000      0   Changing
SCF converged in    2 iterations
Orbital Energies  -16.95016    0.99590    1.46425    1.66517
                    1.97417    2.33073    2.44065
```

There is no freedom here for the variation principle to operate, *all* the MOs are doubly occupied and, since a single determinant is invariant against linear transformations among the orbitals, the transformation induced by diagonalising the h^F matrix has no effect on the total wavefunction and therefore no effect on the R matrix. The calculation "converges" after one genuine iteration.

In the last two examples the program has clearly been used in an entirely inappropriate way, we have used the *formalism* way beyond its range of physical applicability. This type of point will be discussed in a later chapter.

5.6 Another testbench: direct SCF

The testbench in the last section was an implementation of "conventional SCF"; the one-electron and repulsion integrals were computed *once* and *stored* for use during the iterations of the SCF procedure. It is just as easy to set up a testbench to demonstrate the other extreme; the *calculation* of the energy integrals *as they are required* during the SCF iterations; the so-called "direct" SCF method.

However, in order to give such a testbench we would have to be able to actually *code* the formulae for the repulsion integrals (and the one-electron and overlap integrals). This problem will be attacked in due course but, for the moment, this would be a large detour from the essence of the implementation of the direct method, which is simply to stress the *organisation* of the SCF method when integrals are computed on demand rather than being recovered on demand. Strictly speaking, the intention is to stress how little difference there is between the two cases if the design of the implementation is good.

5.6.1 The π-electron approximation

Since we have not yet learned how to evaluate the energy integrals, and we are still concentrating on the logic of the SCF process, the testbench is a special

166   CHAPTER 5. IMPLEMENTATION OF THE CLOSED-SHELL CASE

case for which very simple *approximations* are available for the integrals which may be coded in a few lines. SCF calculations of the electronic structure of the π-electron system of conjugated polyenes have a venerable history and their peculiarly simple properties (one AO per atom, low overlap between AOs) has made them suitable candidates for semi-empirical approaches to the calculation of electronic structure. The fact that this example is a semi-empirical one and involves, not only the approximation of the energy integrals, but the *neglect* of most of the repulsion integrals and the assumption of an *orthogonal* basis simply shows the flexibility of the SCF design.

The repulsion integrals for this system are particularly easy to approximate. There is only one basis function ($2p^\pi$ atomic orbital) per atom and the single one-centre repulsion integral

$$(2p_i^\pi 2p_i^\pi, 2p_i^\pi 2p_i^\pi) = \gamma_{ii} = 11.4002 eV$$

is taken from a study of spectroscopic experiments. Its value is given here in electron-Volts for historical reasons. The other repulsion integrals of the "diagonal" type

$$(2p_i^\pi 2p_i^\pi, 2p_j^\pi 2p_j^\pi) = \gamma_{ij} = \gamma_{ji}$$

are divided into two classes depending on the inter-atomic (inter-orbital) distance r:

1. Those involving AOs up to two C—C bonds apart, which are given by the empirical formula

$$\gamma_{ij} = 10.528 - 2.625r + 0.2157r^2 \quad eV$$

2. Those between AOs further apart than this are computed by a charged-sphere formula, which for carbon-atom AOs is

$$\gamma_{ij} = \frac{7.2}{r}\left(1 + \left(1 + \frac{2}{r}\right)^{\frac{1}{2}}\right)$$

These are very simple formulae to implement and it is easy to see that it is scarcely worthwhile to store the resulting integrals.

The one-electron Hamiltonian integrals are also parametrised. In fact, for hydrocarbons the formulae are as follows:

1. The diagonal elements are given by

$$h_{ii} = \omega - \sum_{j=1}^{m} \gamma_{ij}$$

where the atomic one-centre parameter ω is taken from experiment ("valence-state ionisation energy", -11.16 eV) and the γ_{ij} is doing duty for nuclear attraction integral as well as electron repulsion integral!

5.6. ANOTHER TESTBENCH: DIRECT SCF

2. The off-diagonal values are zero if the two AOs are not nearest neighbours, and take the empirical value of 2.395 eV if they are nearest neighbours.

5.6.2 A direct SCF getint

Here is a **getint** which implements the approximation schemes for the electron-repulsion integrals outlined in the last section.

```
# Dummy getint for the special case of PPP integrals
    integer function getint(file, i, j, k, l, mu, val, pointer)
    integer file, i, j, k, l, mu, pointer
    double precision val
    integer status, ppptwo
    save;     # for id, jd, kd, ld

    if (pointer == 0)
       { id = 0; kd = 0;}  # Initialise labels
    pointer = pointer + 1;
    kd = kd + 1;            # increment labels
    if ( kd > id )
       { kd = 1; id = id + 1}
    jd = id; ld = kd;       # ZDO approximation for integrals

    status = ppptwo(id,kd,val);  # PPP repulsion integral generator
    { i = id; j = jd; k = kd; l = ld;}  # regenerate labels

    return(status);
    end

#  PPP ZDO integrals, values from Tom Peacock's book p97
    integer function ppptwo(i,j,val)
    integer i, j;
    double precision val;
    double precision r, two;

    C_DATA             #  Common block for geometric data
                       #  and number of atoms/orbitals
    data two/2.0d00/;

    if ( i > m ) return(END_OF_FILE);
    if ( i == j )
       {
       val = 11.4d00;     # One-centre value in eV
       return(OK);
       }
    r = sqrt((xy(i,1)-xy(j,1))**2 + (xy(i,2) - xy(j,2))**2);
    if ( abs(r - two* RCC) < 0.1d00 );          # Next nearest-neighbours
```

168   CHAPTER 5. IMPLEMENTATION OF THE CLOSED-SHELL CASE

```
      val = 10.528d00 -2.625d00*r + 0.2157*r*r;
   else
      {
      val = (1.0d00 + 2.0d00/r)**(-0.5d00);   # Charged-sphere formula
      val = 7.2d00*(1.0d00 + val)/r;
      }
              # values in eV
   return(OK);
   end
```

This code requires little in the way of explanation. Some points worth remarking on are

1. The actual calculation of the repulsion integrals is pushed down into ppptwo, leaving getint with the sole responsibility of generating the integral *labels*.

2. The function ppptwo needs to know the geometry of the molecule; this is provided by the macro C_DATA:

   ```
   define(C_DATA,    double precision xy(20,2);
                     common /xydata/xy,m;)
   ```

   which must, of course, be present in the main program to enable this geometric data to be read in.

3. The fact that pointer is initialised to zero before the first call to getint enables the loops which generate the labels to be set up.

4. Although only the integrals with $i = j$ and $k = l$ are non-zero, the algorithm in scfGR still works properly provided that all four labels are supplied.

Since the one-electron Hamiltonian is so closely related to the repulsion integral formulae, a code to calculate these numbers is given here:

```
   double precision function pppone(xy,m,i,j)
   double precision xy(20,2);
   integer i, j, m, junk;
   double precision zero, r, val, gamma, half;
   integer ppptwo;       # Repulsion integral function
   data zero/0.0d00/, half/0.5d00/;

   val = zero;
```

```
if ( i == j )
    {
    #    diagonal values
    val = -11.16d00;
    do k = 1,m;
        {
        if ( i == k ) next;
        junk = ppptwo(i,k,gamma);
        val = val - gamma;              # Nuclear attraction modelled
                                        # by electron repulsion
        }
    }
else
    {
    #   off-diagonal integrals
    r = sqrt((xy(i,1)-xy(j,1))**2 + (xy(i,2) - xy(j,2))**2);
    if ( abs(r - 1.414d00) < 0.1d00 );    # C-C distance
        val = -2.395d00;    # Beta in eV
    }
return(val);
end
```

5.6.3 A direct SCF testbench

The principle idea behind these testbenches is that the various special cases used to display the capabilities of the code developed earlier should be, as far as is reasonable, *invisible* to the design of the overall system; no major restructuring should be necessary to the basic code. In the case of the electron-repulsion integrals, the function **getint** is changed from case to case and no other changes are needed.

However, in the case in question here, it would be a little perverse to insist on the full machinery when there are some obvious simplifications which can be implemented *without* affecting the main design. It must be emphasised, though, that these simplifications are contingent on the particular case being used (the π-electron model) and are *not* part of the general idea of direct SCF.

The $2p^\pi$ AOs are assumed to form an *orthogonal* basis so that no orthogonalising matrix V is needed. We can therefore either

- set $V = 1$ in our SCF code which would involve no changes whatsoever

170   CHAPTER 5. IMPLEMENTATION OF THE CLOSED-SHELL CASE

in the SCF code at the expense of a few matrix multiplications by unit matrices or

- omit the calls to **gmprd** and **gtprd** which are necessary if the basis is not orthogonal and make corresponding slight changes to the code to make it consistent

$$\text{call scfR(C, R, m, n);}$$

must be replaced by

$$\text{call scfR(Cbar, R, m, n);}$$

and some matrices are no longer needed (H, C, V.)

Here, then, is the direct SCF testbench:

```
    define(MAX_ITERATIONS,50)
    define(C_DATA,    double precision xy(20,2);
                      common /xydata/xy,m;)

        double precision  HF(400), R(400);
        double precision Cbar(400), epsilon(7), Rold(400);
        double precision E, zero, Rsum, term, crit;
        double precision pppone;
        integer m, n, i, nfile, icon,mm;
        C_DATA
        data zero/0.0d00/,crit/1.0d-06/;

        irite= ERROR_OUTPUT_UNIT;
        iread = STDINUNIT;
        read(iread,*) m,n;
        read (iread,*) (xy(i,1), xy(i,2), i = 1,m);
        mm = m*m;
# Set initial R to zero, i.e. start from H not HF
        do i = 1, mm;
            { R(i) = zero; Rold(i) = zero; }
#   Assume that the first iteration will change R to be non-zero!
        kount = 0;
            repeat
                {
        kount = kount + 1;
            E = zero ; icon = 0;  # Initialize E and counter
#   Compute the one-electron Hamiltonian each time
            do i = 1, m;
                {
                do j = 1,i;
                    { ij = m*(j-1) + i; ji = m*(i-1) + j;
```

5.6. ANOTHER TESTBENCH: DIRECT SCF

```
            val = pppone(xy,m,i,j);
            HF(ij) = val; HF(ji) = val;
            }
         }
      do i = 1, mm;                       # Accumulate the total
         E = E + R(i)*HF(i);              # energy
      call scfGR( R, HF, m, nfile);       # Use R to form HF
      do i = 1, mm;                       # Accumulate rest
         E = E + R(i)*HF(i);              # of energy
      write(ERROR_OUTPUT_UNIT,200) E;
  200 format(" Current Electronic Energy = ", f12.6)
      call eigen(HF, Cbar, m);            # Diagonalise HF bar
      do i = 1, m;
         epsilon(i) = HF(m*(i-1)+i);      # Get eigenvalues
      call scfR(Cbar, R, m, n);           # Form the R matrix
      Rsum = zero;
      do i =1, mm;
         {
         term = abs(R(i) - Rold(i));  #  Difference in each element
         Rold(i) = R(i);              # Set up Rold for next time
         if ( term > crit ) icon = icon + 1; # Count the offenders
         Rsum = Rsum + term;
         }
      write(ERROR_OUTPUT_UNIT,201) Rsum, icon;
  201 format (" Sum of differences in R = ", f12.5,i6,
                " Changing")
      }
   until (icon == 0 | kount == MAX_ITERATIONS);
   write(ERROR_OUTPUT_UNIT,202) kount;
  202 format(" SCF converged in",i4," iterations")
   write(ERROR_OUTPUT_UNIT,203) (epsilon(i), i=1,n);
  203 format(" Orbital Energies ", (6f10.5))
   STOP;
   end
```

The code is very similar to the conventional SCF code; statements to read the data for a particular (planar) molecule appear at the start of the executable code and the simplifications due to an orthogonal basis have been included.

Compiling and running this code with the appropriate **getint** and support with the data for (e.g.) butadiene:

```
  4 2
0.0    0.0
1.224  0.707
2.448  0
3.672  0.707
```

generates the output:

CHAPTER 5. IMPLEMENTATION OF THE CLOSED-SHELL CASE

```
Current Electronic Energy =     0.000000
Sum of differences in R =       4.14689   16  Changing
Current Electronic Energy =   -86.095766
Sum of differences in R =       1.72713   16  Changing
Current Electronic Energy =   -90.338850
Sum of differences in R =       0.46040   16  Changing
Current Electronic Energy =   -90.770393
Sum of differences in R =       0.12311   16  Changing
Current Electronic Energy =   -90.806990
Sum of differences in R =       0.03501   16  Changing
Current Electronic Energy =   -90.810049
Sum of differences in R =       0.01040   16  Changing
Current Electronic Energy =   -90.810306
Sum of differences in R =       0.00313   16  Changing
Current Electronic Energy =   -90.810327
Sum of differences in R =       0.00096   16  Changing
Current Electronic Energy =   -90.810329
Sum of differences in R =       0.00030   16  Changing
Current Electronic Energy =   -90.810329
Sum of differences in R =       0.00010   16  Changing
Current Electronic Energy =   -90.810329
Sum of differences in R =       0.00003   12  Changing
Current Electronic Energy =   -90.810329
Sum of differences in R =       0.00001    8  Changing
Current Electronic Energy =   -90.810329
Sum of differences in R =       0.00000    0  Changing
SCF converged in  13 iterations
Orbital Energies  -13.76105 -10.58504
```

Note that the energy quantities are in eV (not atomic units) this time. A similar calculation on the π-system of benzene with geometrical data:

```
6 3
0 1.414
1.22456 0.707
1.22456 -0.707
0 -1.414
-1.22456 -0.707
-1.22456 0.707
```

generates, perhaps surprisingly, the following output:

5.7. SUMMARY

```
Current Electronic Energy =      0.000000
Sum of differences in R  =      8.00000    24  Changing
Current Electronic Energy =   -176.411887
Sum of differences in R  =      0.00000     0  Changing
SCF converged in    2 iterations
Orbital Energies   -15.02436 -11.44957 -11.44957
```

a result similar to the calculation on $(H_2O)^{4-}$, only one iteration generating a self-consistent set of MOs.

This time, however, it is not the fact that all possible MOs are filled (obviously not, there are six AOs and therefore the possibility of six MOs). Benzene is so symmetrical that (in a minimal basis of six AOs) the self-consistent MOs are determined *completely by symmetry*; there is no variational problem to solve. A simpler case is provided by the H_2 molecule with a minimal basis of two $1s$ AOs; the lowest orbital *must* be equal contributions of the two AOs thus no variational solution is necessary.

The use of symmetry in molecular calculations will be discussed later.

5.7 Summary

This has been a lengthy chapter and a lot of material has been covered. The considerations would have been much more irksome if it had been necessary to describe, in detail, the elements of the programming language used. It is hoped that the structures used are more-or-less self-evident to anyone with the slightest acquaintance with any programming language; the structures look a little like a mixture of FORTRAN and the language C. This is intentional and the detailed definitions necessary for a confident use of the programming language will be forthcoming. For the moment it is certainly enough that the fragments of code be *understood* in much the same way as it is possible to understand a foreign language much more readily than it is to speak it. However, unlike foreign languages, it is possible, in computing, to *develop* any structures which you would like to be present by the use of macros.

What has been done in this chapter is to show that the most straightforward possible design can lead to:

- Clean implementations of both conventional and direct SCF.
- The logic of the SCF process being cleanly separated from the technical problems of integral generation.
- Any or all of the procedures may be replaced without changing the structure of the program provided only that the *interface* is retained.

5.8 What next?

For the moment, we can simply congratulate ourselves that the design for solving the LCAMOSCF equations works; in view of the remarks at the end of the last section "works" means "works as a *program*" even if not (in some cases) as a scientifically meaningful *project*.

There are now several obvious and pressing problems and several possible directions to go from here; do we

- Improve and clean up the various segments of the above closed-shell program; there are very obvious places where improvements are needed (mainly in `scfGR`)
- Extend the above program to the general single-determinant case and generate an unrestricted Hartree–Fock (actually unrestricted LCAOMOSCF) UHF program
- Go on and generate some integral calculation utilities so that we can do closed-shell calculations on any system

All these problems *will* be solved, it is simply a question of which is the most logical next step.

The problem of the calculation of the basis-function integrals is both self contained and quite involved; it is deferred until later.

The two remaining problems can be tackled to some extent together and to some extent separately. The next chapter is devoted to some improvements to the existing codes in preparation for the development of a UHF program which is very similar to the closed-shell case. In fact, very crudely, the UHF program is just the closed-shell program with the basis doubled and a new `scfGR`.

Appendix 5.A

Jacobi diagonalisation

The simplest and most easily implemented method of computing the eigenvalues and eigenvectors of a symmetric matrix is developed and implemented here.

Contents

5.A.1	Introduction	175
5.A.2	The problem	176
5.A.3	The solution	177
5.A.4	Implementation	179
5.A.5	Other diagonalisation methods	182

5.A.1 Introduction

Once the matrix SCF equations have been set up and transformed to an orthogonal basis, the only numerical problem in their solution is the calculation of the eigenvalues and eigenvectors of a symmetric matrix. This is a problem which occurs in many branches of science, particularly those involving optimisation of some kind and has, consequently, received much attention.

All the methods used to "diagonalise" a symmetric matrix depend on a series of transformations to obtain the matrix in an equivalent but simpler form; one with more zeroes. The most commonly used methods apply a series of transformations to the elements which:

- Transform each pair of off-diagonal elements to zero in turn, subsequent transformations may regenerate elements, the method is iterative: the Jacobi method.

- Transform each pair of off-diagonal elements to zero in such a way that they are not regenerated by later transformations: the Givens method.

- Transform a whole row (and column) to zero in one step which is not regenerated by subsequent transformations: the Householder method.

In fact, it is not possible to diagonalise a matrix completely by the Givens and Householder methods as we shall see; these transformation can only reduce the matrix to *tri-diagonal* form (a matrix with non-zero principal diagonal elements plus the adjacent diagonals non-zero). The latter two methods must be used in conjunction with a method for the diagonalisation of such a tri-diagonal matrix.

The processes involved in all these methods can be appreciated by a detailed look at the Jacobi method which is extremely simple in concept and implementation.

5.A.2 The problem

The equation to be solved is of the form

$$hC = C\epsilon$$

where h is $m \times m$, C is $m \times n$ and ϵ is $n \times n$ and *diagonal*; $\epsilon_{ij} = \epsilon_{ii}\delta_{ij}$.

In fact, the methods for matrix diagonalisation are all capable of solving the larger problem

$$hU = U\epsilon$$

where *all* the matrices are $m \times m$ so this is the one which will be addressed.

If the above equation is multiplied from the left by U^{-1} we obtain

$$U^{-1}hU = \epsilon$$

and the transpose of this equation is

$$U^T h \left(U^{-1}\right)^T = \epsilon$$

since both h and ϵ are symmetric:

$$h^T = h \quad \epsilon^T = \epsilon$$

5.A.3. THE SOLUTION

If
$$U^T = U^{-1}$$
these are versions the same equation so that we may restrict attention to a search for *orthogonal* matrices U that solve the equation
$$U^T h U = \epsilon$$
for diagonal ϵ.

Considered as a problem in algebra, the equation
$$h u^i = u^i \epsilon_{ii}$$
for one column of U (or C) looks, at first sight, like a set of m equations for $(m+1)$ unknowns since there are m elements of the column u^i plus the eigenvalue ϵ_{ii}. It is the constraint of *orthogonality* on the full matrix U which provides the additional piece of information to complete the solution. In the case of just one column, the coefficients are required to be *normalised* to unity by the orthogonality of the full matrix.

5.A.3 The solution

The most obvious way to proceed is to consider the simplest possible case; 2×2 matrices. An orthogonal 2×2 matrix has only *one* degree of freedom: the four elements are constrained by two requirements of normalisation of the columns plus the additional constraint of orthogonality *between* the columns. The most convenient way to incorporate this information into the form of the matrix is to *parametrise* the matrix elements by a single parameter.

Writing
$$c = \cos\theta$$
$$s = \sin\theta$$
we may write the matrix U as
$$U = \begin{pmatrix} c & -s \\ s & c \end{pmatrix}$$
and look for a value of the parameter θ such that the 2×2 problem
$$\begin{pmatrix} c & s \\ -s & c \end{pmatrix} \begin{pmatrix} h_{11} & h_{12} \\ h_{21} & h_{22} \end{pmatrix} \begin{pmatrix} c & -s \\ s & c \end{pmatrix} = \begin{pmatrix} \epsilon_{11} & 0 \\ 0 & \epsilon_{22} \end{pmatrix}$$
has a solution. Recall that $h_{12} = h_{21}$ since h is symmetric.

Clearly, the main point is that the two (equal!) off-diagonal elements should be zero; this is enough to fix θ. Performing the multiplication for the element $\epsilon_{12} = 0$ gives:
$$h_{12}(c^2 - s^2) + sc(h_{22} - h_{11}) = 0$$
If we now define
$$C = c^2 - s^2 = \cos(2\theta)$$
$$S = 2sc = \sin(2\theta)$$

which means
$$c^2 = \frac{1+C}{2}$$
$$s^2 = \frac{1-C}{2}$$

The equation becomes
$$Ch_{12} + \frac{1}{2}S(h_{22} - h_{11}) = 0$$
that is
$$\tan(2\theta) = \frac{S}{C} = \frac{2h_{12}}{h_{22} - h_{11}}$$
Thus, we can obtain values of θ which diagonalise a 2×2 matrix in one step.

In the case of matrices of larger dimension, it is quickly obvious by experiment that the elimination of one off-diagonal element h_{ij} is not "final" since, in eliminating (e.g.) h_{ik}, a new non-zero h_{ij} is generated. However, if the process is continued — eliminating each off-diagonal element in turn — it is found that the matrix can be brought into diagonal form.

Traditionally, the *largest* off-diagonal element was eliminated at each step until the matrix was diagonalised to some numerical tolerance. However, in practice the time taken to locate the largest of the off-diagonal elements of a large matrix is much more than that required to eliminate several elements so that one simply sweeps through the off-diagonal elements eliminating them in order until they are all below some tolerance.

The matrix U which solves the eigenvalue problem is then the product of all the individual "2×2" matrices used in the process. Of course, all these individual matrices involved in the diagonalisation of an $m \times m$ matrix are $m \times m$ but they have only the four non-zero elements of a 2×2 matrix. The matrix h is transformed into the matrix ϵ as the process proceeds.

The 2 × 2 matrices
$$\begin{pmatrix} c & -s \\ s & c \end{pmatrix}$$
have the form of a *rotation* by an angle θ in the plane and so the Jacobi procedure is known as the method of successive rotations.

5.A.4 Implementation

The implementation of the Jacobi method is straightforward; all that is necessary are the steps:

1. Initialise the U matrix.
2. Set up loops on the two indices of the off-diagonal elements.
3. Calculate the value of θ to eliminate the "current" off-diagonal element.
4. Transform the current element away and change the diagonal elements with the current rotation.
5. Update the U matrix by multiplying by the current rotation matrix.
6. Continue until the matrix is diagonal.

Here is the code to perform the task.

APPENDIX 5.A. JACOBI DIAGONALISATION

```
      subroutine eigen(H,U,n)
      implicit double precision (a-h,o-z)
      double precision H(ARB), U(ARB)
      integer n
#
      data zero,eps,one,two,four,big/0.0d0,1.0d-20,1.0d0,2.0d0,4.0d0,1.0d20/;

      define(loch,($2-1)*n + $1)    # Macro for matrix subscripting

#   Initialise the U matrix to unity
      do i = 1,n;
         { ii = loch(i,i);
         do j = 1,n;
            { ij = loch(i,j);
            U(ij) = zero;
            }
         U(ii) = one;
         }
      }
# start sweep through off-diagonal elements
      repeat
      {
      hmax = zero;
      do i = 2,n;
         { jtop = i-1;
         do j = 1,jtop;
            { ii = loch(i,i); jj = loch(j,j); ij = loch(i,j);
            ji = loch(j,i);           # positions of matrix elements
            hii = H(ii); hjj = H(jj); hij = H(ij); hsq = hij*hij;

            if ( hsq > hmax ) hmax = hsq;
            if ( hsq < eps )   next;           # omit zero H(ij)
            del = hii - hjj; sign = one;
            if ( del < zero )
               { sign = -one; del = -del; }
            denom = del + dsqrt(del*del + four*hsq);
#   Here is the expression for tan (2 theta)
            tan = two*sign*hij/denom;
#   Now get cos and sin of theta by trigonometry
            c = one/dsqrt(one + tan*tan); s = c*tan;
#       Update the U matrix with the current 2 by 2 rotation
            do k = 1,n;
               { kj = loch(k,j); ki = loch(k,i);
                 jk = loch(j,k); ik = loch(i,k);
               temp = c*U(kj) - s*U(ki);
               U(ki) = s*U(kj) + c*U(ki); U(kj) = temp;
               if ( (i == k) | (j == k) ) next;
#   Update the parts of the H matrix affected by the rotation
               temp = c*H(kj) - s*H(ki);
               H(ki) = s*H(kj) + c*H(ki);
```

5.A.4. IMPLEMENTATION

```
                H(kj) = temp; H(ik) = H(ki); H(jk) = H(kj);
                }
#       Now transform the four elements explicitly targeted by theta
                H(ii) = c*c*hii + s*s*hjj + two*c*s*hij;
                H(jj) = c*c*hjj + s*s*hii - two*c*s*hij;
                H(ij) = zero; H(ji) = zero;
                }
            }
        }
        until ( hmax < eps );   # Finish when largest off-diagonal
                                #   is small enough
    return
    end
```

In practice, it is useful to have the eigenvalues sorted into ascending order with the associated columns of U in the same order, of course. The following code can be inserted before the **return** statement to effect this sorting and pass the sorted eigenvalues back as the diagonals of H and the associated eigenvectors as the corresponding columns of U.

```
# now sort eigenvectors into eigenvalue order
    iq = -n;
    do i = 1,n;
        { iq = iq + n; ii = loch(i,i);   jq = n*(i-2);
        do j = i,n;
            { jq = jq + n; jj = loch(j,j);
            if ( H(ii) < H(jj) ) next; # this means H(1) is lowest!
            temp = H(ii);   H(ii) = H(jj);   H(jj) = temp;
            do k = 1,n;
                { ilr = iq + k;   imr = jq + k;
                temp = U(ilr);   U(ilr) = U(imr);   U(imr) = temp;
                }
            }
        }
```

A manual entry for the Jacobi diagonalisation routine **eigen** would look like:

> **NAME** eigen
> Jacobi diagonalisation of a symmetric matrix
>
> **SYNOPSIS**
>
> ```
> subroutine eigen(H,U,n)
> double precision H(ARB), U(ARB)
> integer n
> ```
>
> **DESCRIPTION**
> This is a straightforward implementation of the old faithful Jacobi method. The eigenvalues and eigenvectors are generated in order of lowest eigenvalue first.
>
> **ARGUMENTS**
>
> **H** Input; symmetric square matrix to be diagonalised stored as columns. On output the eigenvalues are on the diagonal of H in increasing order (lowest first). Note input matrix H is destroyed.
>
> **U** Output; orthogonal matrix of ordered eigenvectors - each vector is a column of U.
>
> **n** Input; dimension of square matrices H and U (number of basis functions in SCF applications).
>
> **DIAGNOSTICS**
> None

5.A.5 Other diagonalisation methods

The advantages of the Jacobi method, in addition to its simplicity are obvious:

- It requires no working space; all the transformations are done in the two matrices H and U.
- It is very stable and reliable; no precautions have to be taken with degenerate eigenvalues.

Its main disadvantage is equally obvious:

5.A.5. OTHER DIAGONALISATION METHODS

- It is a potentially infinite process, there is no explicit guarantee that it will always "converge". In fact the time consumption of the process is proportional to N^4.

The Givens and Householder methods are both proportional to N^3 in their time consumption.

Most scientific program libraries will contain implementations of these methods as well as another class of methods based on the transformation of symmetric matrices to *triangular* form; the so-called QR and LU algorithms.

Appendix 5.B

Orthogonalisation

In order to be able to carry through the implementation of the solution of the SCF equations the ability to transform the equations to a basis of orthogonal functions is required. The simplest way to carry this out is the so-called "symmetrical" or Løwdin orthogonalisation method.

Contents

5.B.1	Introduction	184
5.B.2	Functions of a matrix	186
5.B.3	Implementation	187

5.B.1 Introduction

The matrix diagonalisation methods described in Appendix 5.A assume that the basis is an orthonormal one:

$$\int \phi_i \phi_j dV = \delta_{ij}$$

and, of course, the "naturally occurring" bases of atomic functions are not orthogonal. In order to be able to use the diagonalisation methods in our implementation of the LCAO method we must have a way of (reversibly) generating an orthogonal basis from an arbitrary basis.

The matrix representation of any operator $\hat{A}$ in the basis ϕ_i is given by

$$\boldsymbol{A} = \int dV \boldsymbol{\phi}^T \hat{A} \boldsymbol{\phi}$$

5.B.1. INTRODUCTION

where ϕ is the row matrix of basis functions ϕ_i.

In particular, the overlap matrix ($\hat{A} = 1$) is given by

$$S = \int dV \phi^T \phi$$

(where T means "transpose" and would be replaced by $\dagger$ for Hermitian conjugate in the case of complex ϕ_i).

If new basis functions ($\bar{\phi}$ are formed from the ϕ_i by the action of a linear transformation matrix V, say,

$$\bar{\phi} = \phi V$$

then the associated overlap matrix is, of course,

$$\bar{S} = \int dV \bar{\phi}^T \bar{\phi} = \int dV V^T \phi^T \phi V = V^T S V$$

If we require this transformation matrix to generate an orthonormal basis then

$$\bar{S} = V^T S V = 1$$

or

$$VV^T = S^{-1}$$

Any matrix which satisfies this equation will generate an orthonormal basis $\bar{\phi}$ from the original set ϕ.

Obviously, there are many matrices which satisfy the equation since any orthogonal transformation among the members of an orthogonal set does not disturb the orthogonality; this fact may be expressed concisely by inserting

$$OO^T = 1$$

into the equation where O is any orthogonal matrix

$$O^{-1} = O^T$$
$$V\left(OO^T\right)V^T = (VO)\left(O^T V^T\right) = S^{-1}$$

and taking the matrix

$$V' = VO$$

as the alternative orthogonalising matrix.

For the purposes of the solution of the matrix SCF equations it is most convenient to ask for the *simplest possible* solution. This is accomplished by requiring V to be symmetrical

$$V^T = V$$

so that the equation to solve is

$$V^2 = S^{-1}$$

or

$$V = S^{-1/2}$$

Clearly, there are some constraints on the matrix S before this operation can be performed, principally the matrix must not be *singular*, i.e. it must be capable of being inverted. Physically, this means that the basis functions should not be *redundant*; there should not be sufficiently large overlap amongst them to enable one of the functions to be removed from the basis *without affecting the quality of the calculation*.

5.B.2 Functions of a matrix

The generation of the inverse square root of a matrix is a special case of the calculation of a general function of a matrix

$$f(S)$$

This is particularly easily achieved if the function is sufficiently well behaved to have a convergent series expansion. That is, if

$$f(x) = \sum_{i=0}^{\infty} a_i x^i$$

we can evaluate the function because, in this case, all we have to do is to be able to form *powers* of the matrix which is always possible. Thus, in these cases, the function of the matrix is well defined.

However, it is not exactly convenient to have to carry through the power series expansion to *evaluate* the function explicitly.

If we have the matrix which *diagonalises* S

$$d = E^T S E$$

obtained for example with the routine `eigen` of Appendix 5.A, then

$$S = E d E^T$$

where d is diagonal $d_{ij} = \delta_{ij} d_{ii}$, then any power of S may be expressed as

$$S^2 = \left(E d E^T\right) = E d E^T E d E^T = E d^2 E^T$$

because
$$EE^T = E^T E = 1$$
In general, then
$$S^n = E d^n E^T$$
and, using the series definition of $f(S)$

$$f(S) = E \left(\sum_{i=0}^{\infty} a_i d^i \right) E^T$$

The advantage of this transformation is, of course, that powers of a *diagonal* matrix are just the diagonal matrices of the powers of those *diagonal elements*. So, we can write the inner sum of the above equation as the diagonal matrix whose elements are the functions of the diagonal matrix elements $f(d_{ii})$:

$$f(S) = E(f(d)) E^T$$

where the notation $f(d)$ has been used as shorthand for

$$\begin{pmatrix} f(d_{11}) & 0 & 0 & \ldots \\ 0 & f(d_{22}) & 0 & \ldots \\ 0 & 0 & f(d_{33}) & \ldots \\ \ldots & \ldots & \ldots & \ldots \\ \ldots & \ldots & 0 & f(d_{mm}) \end{pmatrix}$$

which is easily evaluated.

In the particular case we are dealing with

$$S^{-1/2} = E \left(d^{-1/2} \right) E^T$$

The matrix $d^{-1/2}$ is just the diagonal matrix with elements $d_{ii}^{-1/2}$. The diagnostic test for the existence of the required solution is then clear:

The overlap matrix S should have no zero eigenvalue in order that the basis may be orthogonalised.

5.B.3 Implementation

The implementation of the calculation of $S^{-1/2}$ is almost trivial. It uses the matrix diagonalisation routine `eigen` and the two matrix multiplication routines introduced earlier (`gmprd` and `gtprd`). Here is the code and a manual entry.

```
      subroutine shalf(S,U,W,m)
#        Routine to replace S by S**(-1/2)
      implicit double precision (a-h,o-z);
      double precision S(ARB), U(ARB), W(ARB);
      integer m;
#
      data crit, one/1.0d-10,1.0d0/;
      call eigen(S,U,m);
#   Transpose the eigenvectors of S for convenience
#   of using gmprd and gtprd
      do  i=1,m;
         {
         do  j=1,i;
            {    ij=m*(j-1)+i  ;    ji=m*(i-1)+j  ;    d=U(ij);
            U(ij)=U(ji) ;    U(ji)=d;
            }
         }
#   Get the inverse square root of the eigenvalues
      do  i=1,m;
         {
         ii = (i-1)*m+i;
         if ( S(ii) < crit )        # Is the basis redundant?
            { write( ERROR_OUTPUT_UNIT ,200) ; STOP ;}
         S(ii)=one/sqrt(S(ii));
         }
#  Do the transformation with the matrix multipliers
      call gtprd(U,S,W,m,m,m);
      call gmprd(W,U,S,m,m,m);
#
      return;
      200 format ( " Basis is linearly dependent; S is singular ")
      end
```

NAME shalf

Compute the (orthonormalising) matrix $S^{-\frac{1}{2}}$ from the overlap matrix in S.

SYNOPSIS

```
subroutine shalf(S,U,W,m)
double precision S(ARB), U(ARB), W(ARB)
integer m
```

DESCRIPTION

shalf computes $S^{-\frac{1}{2}}$ by diagonalising S and transforming the diagonal form with the eigenvectors of S. The matrix S is overwritten by the output.

ARGUMENTS

S Input/output: on input this array contains the overlap matrix as columns, the matrix is overwritten by the output $S^{-\frac{1}{2}}$

U Workspace (as large as S)

W Workspace (as large as S)

m The dimension of the square matrices S, W, U

SEE ALSO
eigen

DIAGNOSTICS

If the overlap matrix is singular (the basis is redundant) a message is printed on ERROR_OUTPUT_UNIT and control is returned to the operating system.

Appendix 5.C

getint and data for H_2O

Here is the special **getint** used in the first testbench program. It *contains* all the repulsion integrals and labels for the minimal-basis calculation on H_2O. There are 228 of them and their values are kept as (v(i), i=1,228) and their labels as (ir(i),jr(i),kr(i),lr(i), i = 1,228). Thus, there is no file to be read but, of course, all this is invisible to the calling program; all that the calling program knows is that, if it calls **getint** an integral will be handed out or an end-of-file condition signalled.

```
# Number of integrals for H2O
    define(NUMBER,228)
# Dummy getint for the special case of H2O minimal basis
    integer function getint(file, i, j, k, l, mu, val, pointer)
    save      # make sure that the data is not lost
    integer file, i, j, k, l, mu, pointer
    double precision val
    integer ir(NUMBER),is(NUMBER),it(NUMBER),iu(NUMBER)
    double precision v(NUMBER)
#   Not all the repulsion integrals are present, the zeroes are omitted
#   the full number, including all zeroes, is 406
#   The integrals are in "standard order" but they could be
#      in any order
    data ir /_
1, 2, 2, 2, 2, 2, 3, 3, 3, 3, 3, 3, 3, 4, 4, 4, 4, 4, 4, 4,
4, 4, 5, 5, 5, 5, 5, 5, 5, 5, 5, 5, 6, 6, 6, 6, 6, 6, 6, 6,
6, 6, 6, 6, 6, 6, 6, 6, 6, 6, 6, 6, 6, 6, 6, 6, 6, 6, 6, 6,
6, 6, 6, 6, 6, 6, 6, 6, 6, 6, 6, 6, 6, 6, 6, 6, 6, 6, 6, 6,
6, 6, 6, 6, 6, 6, 6, 6, 6, 6, 6, 6, 6, 6, 6, 6, 6, 6, 6, 6,
6, 6, 6, 6, 6, 6, 6, 6, 7, 7, 7, 7, 7, 7, 7, 7, 7, 7, 7, 7,
7, 7, 7, 7, 7, 7, 7, 7, 7, 7, 7, 7, 7, 7, 7, 7, 7, 7, 7, 7,
7, 7, 7, 7, 7, 7, 7, 7, 7, 7, 7, 7, 7, 7, 7, 7, 7, 7, 7, 7,
```

7, 7, 7, 7, 7, 7, 7, 7, 7, 7, 7, 7, 7, 7, 7, 7, 7, 7,
7, 7, 7, 7, 7, 7, 7, 7, 7, 7, 7, 7, 7, 7, 7, 7, 7, 7,
7, 7, 7, 7, 7, 7, 7, 7, 7, 7, 7, 7, 7, 7, 7, 7, 7, 7,
7, 7, 7, 7, 7, 7, 7, 7/
    data is /_
1, 1, 1, 2, 2, 2, 1, 2, 2, 3, 3, 3, 3, 1, 2, 2, 3, 4, 4, 4,
4, 4, 1, 2, 2, 3, 4, 5, 5, 5, 5, 5, 5, 1, 1, 1, 1, 1, 1, 1,
1, 1, 1, 1, 1, 2, 2, 2, 2, 2, 2, 2, 2, 2, 2, 2, 2, 2, 3, 3,
3, 3, 3, 3, 3, 3, 3, 3, 3, 3, 3, 3, 4, 4, 4, 4, 4, 4, 4, 4,
4, 4, 4, 4, 4, 4, 4, 5, 5, 5, 5, 5, 6, 6, 6, 6, 6, 6, 6, 6,
6, 6, 6, 6, 6, 6, 6, 6, 1, 1, 1, 1, 1, 1, 1, 1, 1, 1, 1, 1,
1, 1, 1, 1, 1, 2, 2, 2, 2, 2, 2, 2, 2, 2, 2, 2, 2, 2, 2, 2,
2, 2, 2, 3, 3, 3, 3, 3, 3, 3, 3, 3, 3, 3, 3, 3, 3, 3, 3, 3,
3, 3, 4, 4, 4, 4, 4, 4, 4, 4, 4, 4, 4, 4, 4, 4, 4, 4, 4, 4,
4, 4, 5, 5, 5, 5, 5, 5, 6, 6, 6, 6, 6, 6, 6, 6, 6, 6, 6, 6,
6, 6, 6, 6, 6, 6, 7, 7, 7, 7, 7, 7, 7, 7, 7, 7, 7, 7, 7, 7,
7, 7, 7, 7, 7, 7, 7, 7/
    data it /_
1, 1, 2, 1, 2, 2, 3, 3, 3, 1, 2, 2, 3, 4, 4, 4, 4, 1, 2, 2,
3, 4, 5, 5, 5, 5, 5, 1, 2, 2, 3, 4, 5, 1, 2, 2, 3, 3, 3, 4,
4, 4, 4, 5, 6, 1, 2, 2, 3, 3, 3, 4, 4, 4, 4, 5, 6, 6, 1, 2,
2, 3, 3, 3, 4, 4, 4, 4, 5, 6, 6, 6, 1, 2, 2, 3, 3, 3, 4, 4,
4, 4, 5, 6, 6, 6, 6, 5, 5, 5, 5, 6, 1, 2, 2, 3, 3, 3, 4, 4,
4, 4, 5, 6, 6, 6, 6, 6, 1, 2, 2, 3, 3, 3, 4, 4, 4, 4, 5, 6,
6, 6, 6, 6, 7, 1, 2, 2, 3, 3, 3, 4, 4, 4, 4, 5, 6, 6, 6, 6, 6, 7,
7, 7, 1, 2, 2, 3, 3, 3, 4, 4, 4, 4, 5, 6, 6, 6, 6, 6, 7, 7,
7, 7, 5, 5, 5, 5, 6, 7, 1, 2, 2, 3, 3, 3, 4, 5, 6, 6, 6, 6,
6, 7, 7, 7, 7, 7, 1, 2, 2, 3, 3, 3, 4, 4, 4, 4, 5, 6, 6, 6,
6, 6, 7, 7, 7, 7, 7, 7/
    data iu /_
1, 1, 1, 1, 1, 2, 1, 1, 2, 1, 1, 2, 3, 1, 1, 2, 3, 1, 1, 2,
3, 4, 1, 1, 2, 3, 4, 1, 1, 2, 3, 4, 5, 1, 1, 2, 1, 2, 3, 1,
2, 3, 4, 5, 1, 1, 1, 2, 1, 2, 3, 1, 2, 3, 4, 5, 1, 2, 1, 1,
2, 1, 2, 3, 1, 2, 3, 4, 5, 1, 2, 3, 1, 1, 2, 1, 2, 3, 1, 2,
3, 4, 5, 1, 1, 2, 1, 2, 3, 1, 2,
3, 4, 5, 1, 2, 3, 4, 6, 1, 1, 2, 1, 2, 3, 1, 2, 3, 4, 5, 1,
2, 3, 4, 6, 1, 1, 1, 2, 1, 2, 1, 2, 3, 1, 2, 3, 4, 5, 1, 2, 3, 4,
6, 1, 2, 1, 1, 2, 1, 2, 3, 1, 2, 3, 4, 5, 1, 2, 3, 4, 6, 1,
2, 3, 1, 1, 2, 1, 2, 3, 1, 2, 3, 4, 5, 1, 2, 3, 4, 6, 1, 2,
3, 4, 1, 2, 3, 4, 5, 5, 1, 1, 2, 1, 2, 3, 4, 5, 1, 2, 3, 4,
6, 1, 2, 3, 4, 6, 1, 1, 2, 1, 2, 3, 1, 2, 3, 4, 5, 1, 2, 3,
4, 6, 1, 2, 3, 4, 6, 7/
    data (v(i), i=1,108) /_

4.78507,	0.74138,	0.13687,	1.11895,	0.25663,	0.81721,
0.02448,	0.03781,	0.18052,	1.11581,	0.25668,	0.81702,
0.88016,	0.02448,	0.03781,	0.18052,	0.04744,	1.11581,
0.25668,	0.81702,	0.78527,	0.88016,	0.02448,	0.03781,
0.18052,	0.04744,	0.04744,	1.11581,	0.25668,	0.81702,
0.78527,	0.78527,	0.88016,	0.17229,	0.03152,	0.05868,

APPENDIX 5.C. GETINT AND DATA FOR H_2O

```
          0.00089,    0.00166,    0.05870,    0.00115,    0.00214,    0.00016,
          0.05878,    0.05857,    0.00736,    0.40448,    0.09387,    0.32923,
          0.00430,    0.03234,    0.32979,    0.00555,    0.04176,    0.00779,
          0.33382,    0.32376,    0.02201,    0.16198,    0.17444,    0.04102,
          0.15367,    0.01165,    0.07192,    0.16296,    0.00213,    0.01855,
          0.01232,    0.15124,    0.14664,    0.01006,    0.09184,    0.07502,
          0.22530,    0.05298,    0.19847,    0.00213,    0.01855,    0.19295,
          0.01275,    0.08152,    0.01138,    0.21284,    0.18939,    0.01299,
          0.11861,    0.06650,    0.10942,    0.01000,    0.05756,    0.00678,
          0.00876,    0.02354,    0.53256,    0.12560,    0.50371,    0.00768,
          0.07878,    0.50603,    0.00992,    0.10175,    0.02950,    0.52129,
          0.48319,    0.02974,    0.30701,    0.19885,    0.25682,    0.77461/
          data (v(i),i=109,216)/ _
          0.17229,    0.03152,    0.05868,    0.00089,    0.00166,    0.05870,
         -0.00115,   -0.00214,   -0.00016,    0.05878,    0.05857,    0.00724,
          0.02133,    0.00975,    0.01146,    0.02846,    0.00736,    0.40448,
          0.09387,    0.32923,    0.00430,    0.03234,    0.32979,   -0.00555,
         -0.04176,   -0.00779,    0.33382,    0.32376,    0.02133,    0.13255,
          0.07310,    0.06662,    0.20792,    0.02201,    0.16198,    0.17444,
          0.04102,    0.15367,    0.01165,    0.07192,    0.16296,   -0.00213,
         -0.01855,   -0.01232,    0.15124,    0.14664,    0.00975,    0.07310,
          0.05836,    0.03588,    0.12589,    0.01006,    0.09184,    0.07502,
         -0.22530,   -0.05298,   -0.19847,   -0.00213,   -0.01855,   -0.19295,
          0.01275,    0.08152,    0.01138,   -0.21284,   -0.18939,   -0.01146,
         -0.06662,   -0.03588,   -0.01526,   -0.09163,   -0.01299,   -0.11861,
         -0.06650,    0.10942,    0.01000,    0.05756,    0.00678,   -0.00876,
          0.01866,    0.02354,    0.16138,    0.03778,    0.14317,    0.00275,
          0.02432,    0.14674,    0.14271,    0.13990,    0.00877,    0.06953,
          0.04432,    0.04383,    0.13492,    0.00877,    0.06953,    0.04432,
         -0.04383,    0.03609,    0.53256,    0.12560,    0.50371,    0.00768,
          0.07878,    0.50603,   -0.00992,   -0.10175,   -0.02950,    0.52129/
          data (v(i),i=217,228)/ _
          0.48319,    0.02846,    0.20792,    0.12589,    0.09163,    0.34338,
          0.02974,    0.30701,    0.19885,   -0.25682,    0.13492,    0.77461/
#
          pointer = pointer + 1
          if (pointer > NUMBER ) return (END_OF_FILE)
          i = ir(pointer)
          j = is(pointer)
          k = it(pointer)
          l = iu(pointer)
          val = v(pointer)
#
          return(OK)
          end
```

The remaining data are just the matrices of one-electron integrals and a matrix which will generate an orthogonal basis from the minimal basis of "AOs". Here are the **data** statements which supply these data to the testbench program:

```
# Start of special data for H2O. Here is the orthogonalising matrix
    data V /_
    1.02421,  -0.14351,  -0.01030,   0.00000,   0.00000,   0.02205,   0.02205,
   -0.14351,   1.24886,   0.11273,   0.00000,   0.00000,  -0.29740,  -0.29740,
   -0.01030,   0.11273,   1.05600,   0.00000,   0.00000,  -0.14822,  -0.14822,
    0.00000,   0.00000,   0.00000,   1.11451,   0.00000,  -0.23271,   0.23271,
    0.00000,   0.00000,   0.00000,   0.00000,   1.00000,   0.00000,   0.00000,
    0.02205,  -0.29740,  -0.14822,  -0.23271,   0.00000,   1.21199,  -0.09117,
    0.02205,  -0.29740,  -0.14822,   0.23271,   0.00000,  -0.09117,   1.21199/
# Now the matrix of one-electron integrals
    data H /_
  -32.72233,  -7.61313,  -0.01906,   0.00000,   0.00000,  -1.75465,  -1.75465,
   -7.61313,  -9.33609,  -0.22404,   0.00000,   0.00000,  -3.74917,  -3.74917,
   -0.01906,  -0.22404,  -7.55265,   0.00000,   0.00000,  -1.64732,  -1.64732,
    0.00000,   0.00000,   0.00000,  -7.61542,   0.00000,  -2.03290,   2.03290,
    0.00000,   0.00000,   0.00000,   0.00000,  -7.45869,   0.00000,   0.00000,
   -1.75465,  -3.74917,  -1.64732,  -2.03290,   0.00000,  -5.08179,  -1.61434,
   -1.75465,  -3.74917,  -1.64732,   2.03290,   0.00000,  -1.61434,  -5.08179/
```

Appendix 5.D

Coding the standard index loops

There are many occasions during the computation of molecular electronic structure when electron-repulsion integrals must be processed: computed, used, transformed etc. In many cases this will involve the setting up of, typically, four nested loops over the labels i, j, k, ℓ of these integrals. If standard practice of indentation and closing of program loops by braces (}) is used this has the disadvantage of throwing all the body of the code too far over to the right of the listing and, in a nest of loops, it is not always clear where particular loops end.

It would be nice to hide the coding of the loops completely in a macro or a function so that one could say something like:

```
while(next_label(i,j,k,l,n) == YES)
    {
    DO SOMETHING WITH i, j, k, l
    }
```

which requires `next_label` to generate the appropriate sets of values of `i`, `j`, `k`, `l` and make them available for the computation in hand.

Of course, this code is a lot more transparent than the alternative which we have used before:

```
do i = 1,n;
   {
   do j = 1,i;
      {
      do k = 1,i;
         {
         ltop = k;
         if ( i == k ) ltop = j ;
         do l = 1,ltop;
            {
            DO SOMETHING WITH i, j, k, l
            }
         }
      }
   }
```

This type of usage matches the construction we have used to *access* the repulsion integrals *e.g.* from a disk file:

```
while(getint(nfile,i,j,k,l,mu,val,pointer) != END_OF_FILE)
   {
   DO SOMETHING WITH i, j, k, l, val
   }
```

and brings together the constructs used for the generation and use of repulsion integrals.

What remains is to code **next_label**; here is a **function** to do the job since a macro might appear too long and convoluted:

APPENDIX 5.D. CODING THE STANDARD INDEX LOOPS

```
integer function next_label(i,j,k,l,n);
integer i, j, k, l, n;
{
integer ltop;

next_label = YES;
 ltop = k;
 if ( i == k ) ltop = j ;
if(l < ltop) l = l + 1;
else
   {
     l = 1;
     if (k < i) k = k + 1;
     else
        {
        k = 1;
        if ( j < i ) j = j + 1;
        else
           {
           j = 1;
           if (i < n ) i = i + 1;
           else
              {
              next_label = NO;
              }
           }
        }
   }
return;
 }
end
```

The whole process has to be initialised by setting

i = 1; j = 1; k = 1; l = 0;

since **next_label** (of course!) generates the *next* label when given the current one.

This process of initialisation does not have to induce **next_label** to produce the whole list; the usage encourages the idea of *restarting* a calculation from a particular point simply by replacing the starting initialisation by the values of i, j, k, l which were the last ones reached in the process. This restart facility, if introduced into the original full **do**-loop structure, makes the code very messy.

Obviously, one could code a general version which replaced the generation of the standard canonical order of the integral labels given above by a more general looping construct which generated the next label for some particular algorithm; see, for example, next_mp2_label in Appendix 29.A.

Chapter 6

Improvements: tools and methods

> *Having developed a working SCF implementation, we should now begin to prepare for the tasks involved in coding a program system which will never be "stable". The codes will constantly need modifying, updating and improving. There are software tools available for making these tasks bearable, even enjoyable.*

Contents

6.1	Introduction	199
6.2	Versions: conditional compilation	200
6.3	Improved diagonalisation	207
6.4	Simple interpolation	210
6.5	Improving the formation of G(R)	212
6.6	Summary	215

6.1 Introduction

The design and codes developed in the last chapter work when tested with the two representative cases. But these cases are both rather trivial compared with the tasks which might be expected of a program for the calculation of molecular electronic structure; six or seven basis functions is hardly typical.

If the program were tested with a more realistic set of data, it would be found to be unreasonably slow (compared, for example, with other "mature" programs to perform the same task). For example, it it relatively straightforward to write a `getint` which will read the files generated by the GAUSSIAN series of programs and compare the SCF runs of our pilot program and the program which is part of the GAUSSIAN suite. This chapter is devoted to improving those parts of the code which are responsible for the most obvious inefficiencies, consistent with maintaining a modular design.

There are only two parts of the code which make any significant demands on the computing system:

1. The formation of the electron-repulsion matrix G in `scfGR`
2. The diagonalisation of the h^F matrix in `eigen`

In both cases there are two *kinds* of question to ask about our implementation:

- Is the *design* right? Can the algorithm be replaced by a better one?
- Is the *implementation* right? Can the code be improved to effect a better implementation of the chosen method?

Both of the two problems will be analysed in the light of these criteria.

But before the code is re-examined, it is, perhaps, not too early to introduce one important *organisational* point; using these two testbench programs as an illustration.

6.2 Versions: conditional compilation

The two testbench programs of the last chapter have a great deal in common; they are *functionally* identical and only differ in a few details which depend on the use to which the code is to be put. There is an important general rule here which gives the opportunity of demonstrating the value of *conditional compilation*.

> Never, never keep multiple copies of programs which do the same (or very closely related) tasks

If more than one copy of the source code is kept, they will quickly diverge as changes are not kept "in step" and, very quickly, two (or more) documented, working programs will become two (or more) inadequately documented programs which do different things and do not work.

There are two methods of getting to grips with this problem of "version control" :

6.2. VERSIONS: CONDITIONAL COMPILATION

1. Use a software tool which has been specifically developed for this purpose; a system which keeps all related versions of a program suite on a single file and can generate specific versions on demand.

2. Use conditional compilation; arrange for the compiler to be capable of compiling only certain parts of a given source file depending on the definition of some global symbol(s).

In this context "version" might mean:

- One of several related programs which perform closely related tasks like the two related testbenches of the last chapter.
- One of several functionally identical programs which are system or compiler dependent (UNIX version, DOS version, Windows version etc.).

A particular tool for version control will be outlined later (the RCS, Revision Control System), for the moment we take the opportunity to introduce the conditional compilation facility of a macro processor which is typical of Unix-based systems.

The main difference between the two testbenches is the fact that the second of the two uses an orthogonal basis and so some of the steps in the SCF method are omitted. The relevant program fragment in the full calculation is

```
   kount = 0
     repeat
       {
       kount = kount + 1;
       E = zero ; icon = 0;    # Initialize E and counter
       do i = 1, m*m;
          HF(i) = H(i);
       do i = 1, mm;                        # Accumulate the total
          E = E + R(i)*HF(i);               # energy
       call scfGR( R, HF, m, nfile);        # Use R to form HF
       do i = 1, mm;                        # Accumulate rest
          E = E + R(i)*HF(i);               # of energy
       write(ERROR_OUTPUT_UNIT,200) E;
       200 format(" Current Electronic Energy = ", f12.6)
       call gtprd(V, HF, R, m, m, m);       # Transform HF to
       call gmprd(R, V, HF, m, m, m);       # orthogonal basis
       call eigen(HF, Cbar, m);             # Diagonalise HF bar
       do i = 1, m;
          epsilon(i) = HF(m*(i-1)+i);          # Get eigenvalues
       call gmprd(V, Cbar, C, m, m, m); # Transform coefficients
       call scfR(C, R, m, n);               # Form the R matrix
       Rsum = zero;
       do i =1, mm;
          {
          term = abs(R(i) - Rold(i));  #  Difference in each element
          Rold(i) = R(i);              # Set up Rold for next time
          if ( term > crit ) icon = icon + 1; # Count the offenders
          Rsum = Rsum + term;
          }
       write(ERROR_OUTPUT_UNIT,201) Rsum, icon;
201 format (" Sum of differences in R = ", f12.5,i6,
                   " Changing")
       }
     until (icon == 0 | kount == MAX_ITERATIONS);
```

The orthogonal basis version does not need the calls to **gmprd** and **gtprd** and the call to **eigen** generates C directly not via the intermediate Cbar:

6.2. VERSIONS: CONDITIONAL COMPILATION

```
    kount = 0;
      repeat
        {
        kount = kount + 1;
        E = zero ; icon = 0;   # Initialize E and counter
        do i = 1, m*m;
          HF(i) = H(i);
        do i = 1, mm;                   # Accumulate the total
          E = E + R(i)*HF(i);           # energy
        call scfGR( R, HF, m, nfile);   # Use R to form HF
        do i = 1, mm;                   # Accumulate rest
          E = E + R(i)*HF(i);           # of energy
        write(ERROR_OUTPUT_UNIT,200) E;
    200 format(" Current Electronic Energy = ", f12.6)
        call eigen(HF, C, m);           # Diagonalise HF bar
        do i = 1, m;
          epsilon(i) = HF(m*(i-1)+i);   # Get eigenvalues
        call scfR(C, R, m, n);          # Form the R matrix
        Rsum = zero;
        do i =1, mm;
          {
          term = abs(R(i) - Rold(i));   # Difference in each element
          Rold(i) = R(i);               # Set up Rold for next time
          if ( term > crit ) icon = icon + 1; # Count the offenders
          Rsum = Rsum + term;
          }
        write(ERROR_OUTPUT_UNIT,201) Rsum, icon;
    201 format (" Sum of differences in R = ", f12.5,i6,
               "  Changing")
        }
      until (icon == 0 | kount == MAX_ITERATIONS);
```

Most macro processors have the facility to process source code (or not) depending on the existence of a global symbol in their table of **defined** strings, thus

```
ifdef(ORTHOGONAL_BASIS)
   {
     Orthogonal basis code
   }
elsedef
   {
     Other code
   }
enddef
```

will only process **Orthogonal basis code** if there exists earlier in the code the statement

$$\text{define(ORTHOGONAL\_BASIS,...)}$$

If this statement is not present, **Other code** will be processed. As usual, **Orthogonal basis code** or **Other code** may be one statement (in which case the braces {...} are redundant) or many statements (in which cases the braces are required).

The **elsedef** and therefore the associated **enddef** are optional so that if it is required simply to add code conditionally on the presence of a **define**d symbol one may simply say:

```
ifdef(ORTHOGONAL_BASIS)
   {
     Orthogonal basis code
   }
```

Note that it is not strictly necessary to actually **define** the symbol to be a "real" (non-empty) string, thus

$$\text{define(ORTHOGONAL\_BASIS,)}$$

places ORTHOGONAL_BASIS in the symbol table manipulated by the macro processor with the value NULL. It is useful to distinguish between symbols **define**d to be used in this way for conditional compilation and other, "genuine" definitions

6.2. VERSIONS: CONDITIONAL COMPILATION

by using the NULL form; the definitions used to control conditional compilation may be then seen at a glance.

It is sometimes more convenient to test for the *non-existence* of a definition; the obvious construct is:

```
ifnotdef(ORTHOGONAL_BASIS)
    {
     Non-orthogonal basis code
    }
elsedef
    {
    Orthogonal basis code
    }
enddef
```

Again, the **elsedef** and **enddef** are optional.

With this tool we can amalgamate the two versions of the SCF program:

```
    kount = 0;
      repeat
         {
         kount = kount + 1;
         E = zero ; icon = 0;    # Initialize E and counter
         do i = 1, m*m;
            HF(i) = H(i);
         do i = 1, mm;                        # Accumulate the total
            E = E + R(i)*HF(i);               # energy
         call scfGR( R, HF, m, nfile);        # Use R to form HF
         do i = 1, mm;                        # Accumulate rest
            E = E + R(i)*HF(i);               # of energy
         write(ERROR_OUTPUT_UNIT,200) E;
         200 format(" Current Electronic Energy = ", f12.6)
   ifnotdef(ORTHOGONAL_BASIS)
      {
         call gtprd(V, HF, R, m, m, m);       # Transform HF to
         call gmprd(R, V, HF, m, m, m);       # orthogonal basis
         call eigen(HF, Cbar, m);             # Diagonalise HF bar
      }
     elsedef
         call eigen(HF, C, m);                # HF is orthogonal basis
     enddef
         do i = 1, m;
            epsilon(i) = HF(m*(i-1)+i);       # Get eigenvalues
   ifnotdef(ORTHOGONAL_BASIS)
      {
         call gmprd(V, Cbar, C, m, m, m);  # Transform coefficients
      }
         call scfR(C, R, m, n);               # Form the R matrix
         Rsum = zero;
         do i =1, mm;
            {
            term = abs(R(i) - Rold(i));  #  Difference in each element
            Rold(i) = R(i);                   # Set up Rold for next time
            if ( term > crit ) icon = icon + 1; # Count the offenders
            Rsum = Rsum + term;
            }
         write(ERROR_OUTPUT_UNIT,201) Rsum, icon;
  201 format (" Sum of differences in R = ", f12.5,i6,
                " Changing")
         }
       until (icon == 0 | kount == MAX_ITERATIONS);
```

Here, both types of construct are used, the full **ifnotdef**, **elsedef**, **enddef** and the short form, simply **ifnotdef**.

The second **ifnotdef** controls only one statement and the braces are superfluous but they make the conditional statement stand out from the rest of the code.

6.3. IMPROVED DIAGONALISATION

Of course, our two testbenches are merely that, *examples* to be discarded as experience is gained with the implementation, and the use of conditional compilation here is rather out-of-place; but it does provide a convenient introduction to the technique and its use.

6.3 Improved diagonalisation

The method which has been implemented to generate the eigenvalues and eigenvectors of a real symmetric matrix is not, in fact, the fastest method; there are methods which have asymptotic dependence on m^3 floating point operations, while the Jacobi method depends asymptotically on m^4. f77 implementations of these methods (the Givens and Householder methods) are available for most computers and calls to `eigen` may be simply replaced by corresponding calls to the other routine.

What is more important from the point of view of the general design of codes for the SCF method is the *way* in which `eigen` (or its replacement) is used. If we can make an improvement in the *design* of the SCF process, this will make the use of *any* diagonalisation method more effective.

There are two important considerations which may be brought together to use the matrix diagonalisation technique more effectively.

- The matrix diagonalisation routine `eigen` (or some equivalent), if it is a general-purpose routine, cannot *assume* anything about the solution eigenvector matrix; it will be used in contexts quite different from the L-CAMOSCF method. It must, therefore, always start from the unit matrix and generate the eigenvector matrix by suitable transformations of the data matrix (h^F in our case).

  Now as the SCF calculation proceeds and if the iterative procedure is converging, consecutive cycles of the process will generate eigenvector matrices which are more and more similar to each other; the MO coefficients will "settle down". It would be sensible, therefore, to try to make use of this fact to reduce the time taken to diagonalise the matrix during each SCF iteration. If the information about the successful eigenvectors from the previous cycle can be passed to `eigen`, then `eigen` can *start* from that point to diagonalise a matrix which is only slightly different from that of the previous cycle and therefore, presumably, has eigenvectors rather close to the previous iteration.

- From a computational point of view, the only necessary property of the orthogonalising matrix V is that it should represent a linear transformation which generates an orthogonal set of functions ($\bar{\phi}$) from the basis functions (ϕ). But the matrix C, formed by the product of $\bar{C}$ and V, does this just as well as the matrix V does. That is nothing more than a statement of the obvious fact that we have, for convenience, chosen to assume a single determinant of *orthogonal* MOs. Thus, at each iteration of the SCF process we generate a matrix which defines an orthogonal set of functions; the current approximation to the MOs. This means, among other things, that the matrix C can be used in place of V to transform h^F to an orthogonal basis to be diagonalised by `eigen`.

The conclusion is obvious; instead of using a *fixed* orthogonalising matrix V we *update* the orthogonalising matrix to be the current eigenvector matrix. Thus the transformation to the orthogonal basis using the previous C matrix generates an $\bar{h}^F$ over a basis of orbitals (presumably) quite close to those which actually diagonalise it. Therefore we can make the "unit matrix starting point" for the next eigenvector matrix more and more realistic. The calculation will converge when the eigenvector matrix generated by this procedure is, in fact, the unit matrix; subsequent updates to the "orthogonalising" matrix do not change it.

It is wise to add a cautionary note here; any method which depends on the *cumulative* updating of a set of numbers may well accumulate errors due to machine rounding etc. It is sometimes wise to check every now and again that the cumulative process is not being adversely affected in this way. The natural check is to verify that the orbitals defined by the cumulated matrix are, in fact, orthogonal.

The changes required are surprisingly small, being confined to changes in the arguments of some of the routines called:

6.3. IMPROVED DIAGONALISATION

```
      kount = 0;
        repeat
          {
          kount = kount + 1;
          E = zero ; icon = 0;     # Initialize E and counter
          do i = 1, m*m;
            HF(i) = H(i);
          do i = 1, mm ;                   # Accumulate the total
            E = E + R(i)*HF(i);            # energy
          call scfGR( R, HF, m, nfile);    # Use R to form HF
          do i = 1, mm;                    # Accumulate rest
            E = E + R(i)*HF(i);            # of energy
          write(ERROR_OUTPUT_UNIT,200) E;
  200 format(" Current Electronic Energy = ", f12.6)
***       call gtprd(C, HF, R, m, m, m);   # Transform HF to
***       call gmprd(R, C, HF, m, m, m);   # orthogonal 'MO' basis
          call eigen(HF, Cbar, m);         # Diagonalise HF bar
          do i = 1, m;
            epsilon(i) = HF(m*(i-1)+i);         # Get eigenvalues
***       call gmprd(C, Cbar, V, m, m, m); # Update orthogonaliser
***       call scfR(V, R, m, n);           # Form the R matrix
          Rsum = zero;
          do i =1, mm;
            {
            term = abs(R(i) - Rold(i));  # Difference in each element
            Rold(i) = R(i);              # Set up Rold for next time
***         C(i) = V(i);                 # Restore C as orthogonaliser
            if ( term > crit ) icon = icon + 1; # Count the offenders
            Rsum = Rsum + term;
            }
          write(ERROR_OUTPUT_UNIT,201) Rsum, icon;
  201 format (" Sum of differences in R = ", f12.5,i6,
                  " Changing")
          }
        until (icon == 0 | kount == MAX_ITERATIONS);
```

The changed statements have been marked with three asterisks (***). Also, this new program assumes that an initial orthogonalising matrix is present in C; it will normally be the matrix originally supplied in V whose role is simply to ensure an initial orthogonal set.

Running this changed program with either of the two getints produces results which can only be called visually unspectacular; the output is exactly the same as the original version. But, of course, this is exactly as it should be. The calculation follows the same path as before but the *time taken* in the diagonalisation steps decreases as the MOs approach self-consistency. If we were to insert calls to some system-specific timing routines we could get an estimate of the improved performance. In fact, the improvement in speed is deceptively good in the testbench because the diagonalisation process is the rate-determining step in this simple case as there is no input/output-bound file-reading step for the repulsion integrals. In general, if a more realistic case were to be studied, the improvement in timing would be much less marked since, in both the case of integrals read from a file and that of integrals computed as needed, it is `getint` which is the rate-determining step not the diagonalisation process.

6.4 Simple interpolation

An examination of the information generated by either of the testbench programs shows that there are very large changes in the elements of the R matrix between the first few iterations and subsequent differences are much less marked.

The reason for this is very obvious. If we think about the calculation on the water molecule for concreteness, we have arranged for the calculation to start from an initial "approximation" to the R matrix of 0. This means that the first set of "MO" coefficients generated will be those for a set of ten electrons in the field of the oxygen and hydrogen atoms *with no electron repulsion at all*. Thus this first set of MO coefficients will represent a set of five doubly occupied MOs which are all concentrated around the relatively highly charged oxygen nucleus in some kind of approximation to the neon configuration since there is no repulsion between the electrons to force them away from the oxygen nucleus. Equally obviously, once electron repulsion is "turned on" by the R matrix associated with this very unrealistic electron configuration, this Neon configuration will be very high in energy due to the crowded electron configuration; the next set of MO coefficients will reflect this by representing an equally violent swing away from a large oxygen atom population.

As the calculation proceeds these oscillations will settle down but it is clear that some steps should be taken to avoid them in the first place. There are two *sorts* of solution to this problem:

- A physical method: try to start the calculation from a more realistic "guess" at the equilibrium electron distribution.

- A numerical method: try to "damp" the oscillations caused by the poor initial guess.

In a more complete program both techniques would be used; in fact as experience is gained it quickly becomes obvious that oscillations in the R matrix are a

6.4. SIMPLE INTERPOLATION

general problem in the SCF method and this problem will have to be addressed by more careful techniques than the crude method detailed below.

The problem of a good starting guess for the electronic structure of a molecule of general geometry has received a lot of attention and has become something of a "black art". Most modern methods rely on the idea that the diagonal elements of the self-consistent h^F matrix tend to have values which do not differ enormously from one molecular environment to another, and so a "library" of such atomic values can be used as the guess for the diagonal elements of the molecular h^F matrix. A numerical recipe is used to generate the largest off-diagonal elements from the atomic diagonal ones and (typically) the AO overlap integrals. Since we have no experience at all of the values of SCF h^F matrix elements, this approach is deferred!

A simple interpolation technique, based on the values of just the consecutive R matrices, is easily implemented. If we assume that for the first few cycles in the SCF process each R matrix will overshoot the target which lies somewhere between these successive cycles we can write

$$R^i \rightarrow R^i - \lambda \left(R^i - R^{i-1} \right)$$

where λ is some (positive, presumably) numerical parameter lying between zero and unity. Obviously, if $\lambda = 0$, no interpolation is performed and if $\lambda = 1$, the cycle is cancelled and the matrix is returned to its previous value. It is a matter of numerical experiment to find a suitable value for λ, if possible valid for a range of cases.

Once the initial oscillations have been damped, the interpolation procedure can be turned off since it may become a hindrance to convergence; again at a point determined by experience. Putting this simple algorithm into the test-bench

```
double precision lambda, turm;
integer interp;
interp = 4;
lambda = 0.2d00;
    .
    .
    do i =1, mm;
      {
      turm = (R(i) - Rold(i));    # Save the change in R
      term = abs(turm);
      Rold(i) = R(i);             # Set up Rold for next time
      C(i) = V(i);                # Restore C
      if ( term > crit ) icon = icon + 1; # Count the offenders
      Rsum = Rsum + term;              if ( kount < interp )
         R(i) = R(i) - lambda*turm;  # Damping for the first few
      }                              #         iterations
```

and trying out various values of `lambda` and `interp` enables the total number of cycles in the SCF process to be reduced. In the case of the testbench for the ground state of the H_2O molecule, the values of `lambda` and `interp` quoted give a saving of two (out of 16) iterations in the calculation.

Later, it will prove possible to base a whole class of interpolation and extrapolation methods on the formal properties of the solutions to the matrix SCF equations.

As was noted above, the matrix diagonalisation step is not usually the rate-determining step in an SCF calculation. However, methods of improved diagonalisation and interpolation are still important since for many systems the calculation is not nearly so well-behaved as our testbench and the iterative procedure is often either very poorly convergent or even oscillatory after many cycles. In these cases it is wise to retain the interpolation procedure for the whole calculation.

6.5 Improving the formation of G(R)

The routine presented in the last chapter for the formation of the matrix $\boldsymbol{G}(\boldsymbol{R})$, the electron repulsion contribution to the Hartree–Fock matrix, was coded to be as direct as possible and to illustrate the possibilities of the use of macros to simplify repetitive coding. No consideration at all was given to questions of speed and efficiency. In particular some attention should be given to:

- The matrix $\boldsymbol{G}$ is *symmetrical*, $G_{ij} = G_{ji}$, so that it is only necessary to generate *half* (approximately) of the matrix and use this symmetry to generate the other half.

- The equalities among the integrals due to permutations of the labels can be used more efficiently than permuting the labels and processing them in the same way as the originals; it is often a question of multiplying by a factor of two or four rather than re processing the whole set of labels and value.

- The repulsion integrals are processed many times during an SCF calculation (once per iteration) and the processing logic is *always identical* (all that differs are the *values* of the $\boldsymbol{R}$ matrix elements). It should therefore be possible to somehow code this logic *once and for all* so that the tortuous path through the `if` statements for each integral is *stored* and used in each iteration rather than having to be discovered at each pass.

6.5. IMPROVING THE FORMATION OF G(R)

In this section we can tackle the first two of these points, the third one, although quite straightforward to implement, does involve some additional decisions to be taken which are deferred for the moment. The possibility of labelling the "path" through the logic of the formation of $G(R)$ is the reason for carrying the redundant integer parameter mu throughout the routines. In due course mu will be used to label the *type* of the integral, in the sense of the particular set of equalities and inequalities which obtain between the labels i, j, k, ℓ for each integral.

To take care of the first two points we can drop the use of the macro USE from subroutine scfGR and simply write down the decisions necessary to evaluate the contributions to the distinct elements of the G matrix, using the equalities amongst the labels to decide whether or not to double the relevant contribution. This leads to some very unappealing looking code but it is a good deal more efficient than the original.

```
        subroutine scfGR(R, G, m, nfile)
        double precision G(*), R(*);
        integer m, nfile;
#
        double precision val, two, coul1, coul2, coul3, exch;
        integer i, j, k, l, mu, ij, kl, il, ik, jk, jl;
        integer getint;
        integer pointer;
        data two/2.0d00/;
        define(locGR,($2-1)*m+$1)   # macro for subscripting
#       rewind nfile;               # There is still no real file to be read
        pointer = 0;                # initialise "file"
        while( getint(nfile, i, j, k, l, mu, val, pointer) != END_OF_FILE)
           {
           ij = locGR(i,j); kl = locGR(k,l);   # Set up all possible sets
           il = locGR(i,l); ik = locGR(i,k);   # of subscripts for given
           jk = locGR(j,k); jl = locGR(j,l);   # i j k l.
           if ( j < k ) jk = locGR(k,j);       # Notice convention for
           if ( j < l ) jl = locGR(l,j);       # label storage.

           coul1 = two * R(ij)*val; coul2 = two*R(kl)*val; exch = val;
#       coul1 and coul2 are the possible "Coulomb" contributions,
#          exch is the "Exchange" type
           if ( k != l )
              {
              coul2 = coul2 + coul2;           # Double the term if k != l
              G(ik) = G(ik) - R(jl)*exch
              if ( ( i != j ) & ( j >= k ) ) G(jk) = G(jk) - R(il)*exch;
              }
           G(il) = G(il) - R(jk)*exch; G(ij) = G(ij) + coul2;
           if ( ( i != j) & ( j >= l ) ) G(jl) = G(jl) - R (ik)*exch;
           if ( ij != kl )
              {
              coul3 = coul1;
```

```
              if ( i != j ) coul3 = coul3 + coul1;    # Doubled if i != j
              if( j <= k )
                 {
                 G(jk) = G(jk) - R(il)*exch;
                 if (( i!= j ) & ( i <= k )) G(ik) = G(ik) - R(jl)*exch;
                 if (( k!= l ) & ( j <= l )) G(jl) = G(jl) - R(ik)*exch;
                 }
                 G(kl) = G(kl) + coul3;
              }
       }    # end of while
 # now symmetrise the G matrix
    do i = 1, m;
       { do j = 1, i;
           { ij = locGR(i,j);    ji = locGR(j,i);
             if ( i == j ) next;
             G(ji) = G(ij);
           }
       }
    return;
    end
```

Again, to improve readability, a macro (locGR) has been used to perform the subscripting tasks.

The manual entry requires no changes; the interface is the same, a small triumph for "top-down design".

These changes to the codes for the subroutines and the testbench program provide us with a working LCAOSCFMO system which is by no means as efficient as it might be, but it has had the worst of the original defects removed. It is now time to turn to the task of making the system work *in general*, that is to replace the testbench with a program which will work for *any* molecular electronic structure. What this means in practical terms is

- The design of a general structure for integer function getint; something to replace the two special cases of the testbench. This involves a decision about the file organisation for the repulsion integrals and the implementation of a partner for getint to generate the file which getint reads; putint, say.

- Facing the large task of coding the numerical methods for the non-empirical evaluation of the various energy integrals involved in the electronic structure calculation. As usual, this task will be approached gently, doing the simplest and most obvious things first.

6.6 Summary

Three improvements have been made to the first implementation of the SCF code:

1. Information about the "eigenvectors" has been passed forward to the next iteration so that the diagonalisation procedure of each SCF iteration may start from the eigenvectors of the previous SCF iteration.

2. Arrangements have been made to avoid violent oscillations in the R matrix during the SCF process, especially during the initial steps.

3. Some improvements to the handling of the electron-repulsion integrals during the formation of $G(R)$ have been made.

There are much more significant changes to be made as more experience is gained.

Chapter 7

Molecular integrals: an introduction

Here we begin to look at the most technically demanding (mathematically and computationally) part of the finite-basis expansion method; the computation of the various energy integrals over the basis functions. The derivations are lengthy and are only outlined here. The computer codes are voluminous and dull, and present qualitatively new problems of documentation which will be addressed by using the WEB programming system.

Contents

7.1	Introduction		**218**
7.2	Basis functions		**219**
7.3	AOs and atom-centred-functions		**221**
7.4	Multi-dimensional integral evaluation		**222**
7.5	Molecular integrals over STOs		**223**
	7.5.1	Overlap integrals	228
	7.5.2	Kinetic energy integrals	229
	7.5.3	Nuclear attraction integrals	229
	7.5.4	Electron-repulsion integrals	231
	7.5.5	Summary for STOs	232
7.6	Basis functions of convenience		**232**
7.7	Gaussian basis functions		**234**
	7.7.1	Overlap integrals	235
	7.7.2	Kinetic energy integrals	238
	7.7.3	Nuclear-attraction integrals	239

	7.7.4	Electron-repulsion integrals 245
	7.7.5	Summary for GTFs 250
	7.7.6	More efficient methods 250
7.8		**The contraction technique 251**

7.1 Introduction

The most pressing problem at this point in our development of the theory and implementation of the LCAOMO method is a way of actually *computing* the various energy integrals which are required in the calculation. The two test-benches of previous chapters avoided this problem in order to concentrate on the actual solution of the matrix Hartree–Fock equations.

If we assume for the present that we will be computing the various molecular integrals just *once* and storing them for use by the LCAOMO program (analogous to the H_2O testbench), then we must also design a method for the *storage* of the integrals. The integrals required are:

- The overlap integrals over the basis functions ("atomic orbitals") in order to generate a matrix which will orthogonalise the basis.

- The one-electron integrals over the basis functions

$$\int \phi_i \hat{h} \phi_j dV$$

- The electron repulsion integrals over the basis functions

$$\int dV_1 \int dV_2 \phi_i(\vec{r_1}) \phi_j(\vec{r_1}) \frac{1}{r_{12}} \phi_k(\vec{r_2}) \phi_\ell(\vec{r_2})$$

As we have seen, matrix storage for the first two of these sets of integrals must be allocated in the LCAOMO program so that they are simply read into the appropriate place in the program and kept as static storage. There are only about m^2 of them and this storage is not a penalty.

There are, however, about $m^4/8$ repulsion integrals and, for any molecule of reasonable size, this number is far too large to store all the integrals permanently. These integrals must be kept in a file on some external medium.

7.2 Basis functions

In deciding that the only practical way to compute the molecular orbitals of an arbitrary polyelectronic system was to *expand* these MOs as a *linear* combination of some basis functions we did not give any thought to the form which these all-important functions might take. The testbench programs used to test the codes do contain some implicit assumptions on this point, but the point has not so far received any attention. We must now give the matter some consideration if we are to develop methods of calculation suitable for any molecule, radical or ion.

The expansion method itself places some mathematical and physical constraints on the basis functions; mathematically, the basis functions ϕ_i should:

- be functions of the same number and type of variables as the target MOs; functions of three-dimensional space, possibly multiplied by a spin function;

- have the same kind of asymptotic behaviour as the MOs for extreme values of the coordinates; in particular they should go to zero sufficiently quickly (and smoothly) as the coordinates go to infinite distance from any nucleus;

- be sufficiently smooth to be at least twice differentiable everywhere except possibly at a finite number of points;

- be of such a functional form as to "fill out" the form of the MOs;[1] strictly speaking, the basis functions should be sufficient for the expansion of any function of the same number of variables and boundary conditions as the MOs.

These mathematical conditions can be interpreted and made a good deal less strict by their physical counterparts:

- The first two of the above conditions can be met by using what can only be called "physically sensible" basis functions; functions which can be expected on general physical grounds to describe the electron distribution in molecules sensibly. Use of some form of approximate atomic orbitals as basis functions satisfy these requirements.

- Use of atomic orbitals as basis functions also goes a good deal of the way to satisfying the third condition. If the basis functions are well adapted to the expansion of the MOs, then the *length* of the expansions can be limited. In short, the "mathematical completeness" of the basis set is an impractical ideal and we shall have to be satisfied with a well-chosen (on physical grounds, that is) incomplete basis.

---

[1]This is rather loose terminology for the concept of "near completeness" which will be discussed later.

These considerations would seem to point unequivocally to the use of AOs as the natural expansion functions for MOs and, indeed, this is the assumption implicit in the two testbench programs.

However, there are two points which must be made in mitigation of this non-rigorous approach:

- In practice, the AOs *themselves* are calculated as linear combinations of some more primitive basis functions and all the considerations which we have given to the expansion of MOs apply with equal force to the expansion of AOs.

- The linear expansion method using AOs as a basis allows the AOs to "mix" to form MOs and therefore allows for some *inter*-atomic redistribution of electrons to occur on molecule formation. But there are other processes occuring on molecule formation, one of these is *intra*-atomic electron redistribution and this cannot be represented very well if AOs are used as basis functions.

These two factors are decisive in fixing the usual basis set not as atomic orbitals but as a set of *atom-centred functions* which are adapted to the *expansion* of the AOs of each of the component atoms of the molecule under study. It will also be useful from time to time to augment these basis functions with additional atom-centred functions that allow the description of aspects of the molecular electron distribution which are specific to the molecule. For example, in any satisfactory description of the H_2 molecule one would use those sets of spherically symmetric atom-centred functions which are used to expand the $1s$ AOs of the hydrogen atoms. But one might also add to the basis one or more p_σ functions on each atom to allow for the *polarisation* of the electron distribution *on each atom* on molecule formation; functions which do not take part in the expansion of the AOs of the ground state of the component atoms of the molecule.

These considerations lead naturally to the decision to use, as basis functions for the expansion of MOs, sets of atom-centred functions which are:

- Well adapted to the expansion of the AOs of the atoms of which the molecule is composed.

- Numerous enough to enable the changes suffered by the electron distribution on molecule formation to be reflected in the expansion of the MOs.

The detailed form of these basis function must now be determined by a study of the properties of known atomic orbitals.

7.3 AOs and atom-centred-functions

As usual in any theory of atomic and molecular electronic structure, we must turn to the hydrogen atom for guidance as the only exactly soluble model.

The AOs of the hydrogen atom are of the form:

$$\chi(r, \theta, \phi) = \text{(polynomial in } r) \times \text{(function of } \theta, \phi) \times exp(-\zeta r)$$

in polar coordinates (r, θ, ϕ) based on the nucleus as origin. The parameter ζ is *fixed* by the quotient of the nuclear charge and the principal quantum number.

In Cartesian coordinates with origin at the nucleus the AOs can be written (without changing the symbol for the AO):

$$\chi(x, y, z) = \text{(polynomial in } x, y, z) \times exp(-\zeta r)$$

Thus, if we make the reasonable assumption (based on atomic spectra and atomic properties) that the AOs of polyelectronic atoms are at least *qualitatively* similar to those of the hydrogen atom, then we might hope to expand the AOs as

$$\chi = \sum_{i=1}^{m} c_i x^{\ell_i} y^{m_i} z^{n_i} \exp(-\zeta_i r)$$

where the place of the (fixed) polynomial of the hydrogen atom has been taken by an expansion involving the optimisable parameters c_i and ζ_i. The powers ℓ_i, m_i, and n_i are assumed to be fixed by the "symmetry" of the AO (s, p, d, ...)

In fact, it is usual to make the "spherical approximation" for atoms (assume that the same radial function will do duty for a whole shell of electrons) and write:

$$\chi = \text{(spherical harmonic in } \theta, \phi) \times \sum_{i=1}^{m} c_i r^{n_i} exp(-\zeta_i r)$$

Again, the particular type of spherical harmonic is fixed by the s, p, d etc. nature of the AO, and the parameters to be optimised (by the variational process) are the c_i and the ζ_i.

This method has proved extremely effective for the computation of AOs for atoms, it is certainly competitive in accuracy with the much more atom-specific method of solution of the radial equation by numerical quadrature. In particular the energy and overlap integrals associated with these basis functions *for atoms* are quite trivial to evaluate.

These functions, which one might call the "monomial equivalent" of the hydrogenic polynomial functions, were introduced as approximation functions for AOs very early by J. C. Slater and have become universally known as *Slater-Type*

Orbitals (STOs). From what we have said, this notation is slightly unfortunate since they are not orbitals but *basis functions* and, indeed, they are sometimes referred to as Slater-Type Functions (STFs). This may be a convenient place to remind ourselves of the usage of the term "orbital" in quantum chemistry. Strictly speaking:

> An orbital is any solution of a one-particle Schrödinger equation.

But usage is much more diffuse and "orbital" has come to mean:

> An Orbital is any normalisable function of three-dimensional space

in the context of the expansion of AOs or MOs.

Thus, there is a strong *prima facie* case for the use of the functions

$$\phi_i = x^{\ell_i} y^{m_i} z^{n_i} exp(-\zeta_i r)$$

as basis functions for molecular calculations. They are capable of giving a good description of the electron distribution and energies of atoms.

7.4 Multi-dimensional integral evaluation

In the case of the spherical approximation applied to atoms, the three-dimensional overlap and energy integrals which must be evaluated *separate* very neatly into an integral involving a product of spherical harmonics and a radial integral involving the exponential function. Both types of integral are very well researched and the results are well known.

However, the methods involved are general; if a multi-dimensional integral is to be evaluated it must be

- separable into a product (or sum of products) of integrals of smaller dimension and only becomes really tractable if it separates into a product (or sum of products) of three one-dimensional integrals; and
- the separate products of functions of one (or, rarely, two) variables must be themselves tractable; if the integrals are not analytically tractable then they may be attacked by direct numerical quadrature, but this is useful only if they are one-dimensional.

7.5. MOLECULAR INTEGRALS OVER STOS

Numerical quadrature of multi-dimensional integrals is difficult and very time consuming.

It is easy to see that, for atoms, it is the centro-symmetric nature of the functions that facilitates the separation of the three- and six-dimensional energy integrals into manageable one-dimensional forms. In general, when there is no symmetry at all, it is not easy to see how the integrals will separate. Indeed, the existence of some form of symmetry would seem to be a prerequisite for the separation of the integrals. For diatomic molecules the existence of an axis of symmetry and the possibility of using prolate spheroidal coordinates as a basis for a separation does mean that the basis functions can be used successfully for diatomics.

Of course, any set of nuclei arbitrarily distributed in space can be considered as a set of "diatomics" by taking the nuclei a pair at a time. Although this technique *does* allow some of the more elementary integrals to be calculated for the hydrogen-like basis functions (overlap, kinetic energy and a minority of the nuclear-attraction and electron-repulsion integrals), the basic difficulties due to lack of symmetry remain in the general case.

That is the limit of separation for these functions; even linear triatomics require some numerical quadrature. If we consider a general case it is easy to see what the essential difficulties are.

7.5 Molecular integrals over STOs

The general features of the calculation of the various integrals using a basis of STOs can best be reviewed by taking a particular case; the H$_2$O molecule with a so-called "minimal" basis is a typical example. The basis functions of the minimal basis comprise just *one* function of each of the types that are occupied in the ground states of the atoms of which the molecule is composed; in our case

$$1s_O, 2s_O, 2p_{xO}, 2p_{yO}, 2p_{zO}$$
$$1s_{H_1} \text{ and } 1s_{H_2}$$

where

$$1s_X = N_{1s} \exp(-\zeta_{1s} r_X)$$
$$2s_X = N_{2s} r_X \exp(-\zeta_{2s} r_X)$$
$$2px_X = N_{2p} x_X \exp(-\zeta_{2p} r_X)$$
$$2py_X = N_{2p} y_X \exp(-\zeta_{2p} r_X)$$
$$2pz_X = N_{2p} z_X \exp(-\zeta_{2p} r_X)$$

here, (x_X, y_X, z_X) is a Cartesian coordinate system *based on nucleus X* and

$$r_X = \sqrt{x_X^2 + y_X^2 + z_X^2}$$

is the radial distance from nucleus X. It is usual to write the STOs in this "mixed" form involving both local Cartesian and local polar coordinates:

$$\begin{aligned} x_X &= r_X \sin\theta_X \cos\phi_X \\ y_X &= r_X \sin\theta_X \sin\phi_X \\ z_X &= r_X \cos\theta_X \end{aligned}$$

In one sense it is immediately obvious that it will be difficult to separate integrals involving these functions; the explicit appearance of r_X with its square root definition does not bode well for separation into a product (or sum) of separate functions of x, y and z. However, for *pairs* of such STOS it is possible to find a coordinate system into which the transforms of the STOs do separate; the prolate spheroidal system, usually denoted by (ξ, η, ϕ) [2]

The relationships are:

$$\begin{aligned} \xi &= \frac{1}{R}(r_A + r_B) \\ \eta &= \frac{1}{R}(r_A - r_B) \\ \phi &= \phi_A = \phi_B \end{aligned}$$

where

$$R = r_{AB}$$

the (fixed) distance between the two centres on which the STOs are based. That is:

$$\begin{aligned} r_A &= \frac{R}{2}(\xi + \eta) \\ r_B &= \frac{R}{2}(\xi - \eta) \end{aligned}$$

These definitions enable the various quantities related to the *Cartesian and polar* systems on each nucleus to be written:

$$\cos\theta_A = \frac{1 + \xi\eta}{\xi + \eta}$$

---

[2] Notice, during these integral discussions, the unfortunate collision of notation; ϕ is often used as an angular coordinate and we have used it as a symbol denoting a basis function!

7.5. MOLECULAR INTEGRALS OVER STOS

$$\cos\theta_B = \frac{-1+\xi\eta}{\xi-\eta}$$

$$\sin\theta_A = \frac{(\xi^2-1)(1-\eta^2)}{\xi+\eta}$$

$$\sin\theta_B = \frac{(\xi^2-1)(1-\eta^2)}{\xi-\eta}$$

$$x_A = x_B = \frac{R}{2}\sqrt{(\xi^2-1)(1-\eta^2)}\cos\phi$$

$$y_A = y_B = \frac{R}{2}\sqrt{(\xi^2-1)(1-\eta^2)}\sin\phi$$

$$z_A = z_B + R = \frac{R}{2}(1+\xi\eta)$$

$$1 \leq \xi \leq \infty \quad -1 \leq \eta \leq 1$$

and the volume element is

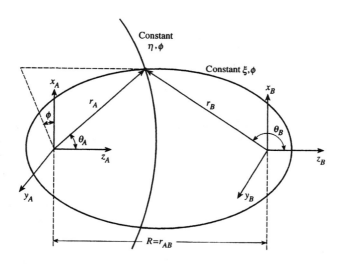

Figure 7.1: The "diatomic" coordinate system

$$dV = \frac{R^3}{2}(\xi^2-\eta^2)d\xi d\eta d\phi$$

The surfaces of constant ξ are prolate ellipsoids, those of constant η are hyperbolic sheets. The system is illustrated in Fig. 7.1.

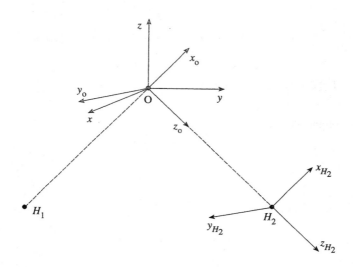

Figure 7.2: The H$_2$O molecule; axis systems

If we now take a particular pair of the STO functions comprising the minimal basis for H$_2$O, say $1s_{H_1}$ and one of the $2p_O$ functions we can transform them to the new coordinate system and see what is involved in the integral calculation.

The first thing to notice is that the diatomic coordinate system *defines* the local coordinate axes with respect to the "diatomic" axis which will not, in general, coincide with the "global" Cartesian axis system used to set up the molecular geometry and, therefore, the individual basis functions. In our case this means that we need a diagram (like Fig. 7.2) of the H$_2$O molecule involving some likely choice of overall Cartesian coordinate system. There are three local diatomic coordinate choices here (H$_1$—H$_2$, H$_1$—O and H$_2$—O) each differently orientated with respect to a global system and so the non-spherically-symmetric basis functions in each of the three cases will be *different*. Fortunately, of course, the sets of basis functions are related to each other by rotations through some Euler angles and, computing the integrals over the local basis is sufficient to determine the corresponding integrals over the global basis.

7.5. MOLECULAR INTEGRALS OVER STOS

The s-type functions are unaffected by these considerations, the set of three $2p$ functions behave like a standard vector in three-space. Sets of d and higher functions are more complicated to deal with but the calculation is always possible.

What remains, then, is to actually compute some of the molecular integrals over the local diatomic basis. Let us choose $1s_{H_1}$ and the $2p$ functions of the oxygen atom so that

$$\begin{aligned} 1s_H &= N_{1s} \exp(-\zeta_{1s} r_H) \\ &= N_{1s} \exp\left[-\zeta_{1s}\left(\frac{R(\xi+\eta)}{2}\right)\right] \\ 2p_{\sigma O} &= N_{2p}\sigma \exp(-\zeta_{2p} r_O) \\ &= \left(\frac{R}{2}\right) N_{2p}(1+\xi\eta) \exp\left[-\zeta_{2p}\left(\frac{R(\xi-\eta)}{2}\right)\right] \\ 2p_{\pi O} &= N_{2p}\pi \exp(-\zeta_{2p} r_O) \\ &= \left(\frac{R}{2}\right) N_{2p} \left[(\xi^2-1)(1-\eta^2)\right]^{\frac{1}{2}} \cos(\phi) \exp\left[-\zeta_{2p}\left(\frac{R(\xi-\eta)}{2}\right)\right] \\ 2p_{\bar{\pi}O} &= N_{2p}\bar{\pi} \exp(-\zeta_{2p} r_O) \\ &= \left(\frac{R}{2}\right) N_{2p} \left[(\xi^2-1)(1-\eta^2)\right]^{\frac{1}{2}} \sin(\phi) \exp\left[-\zeta_{2p}\left(\frac{R(\xi-\eta)}{2}\right)\right] \end{aligned}$$

where the "local diatomic" set of Cartesian axes is denoted by $(\pi, \bar{\pi}, \sigma)$ in order to distinguish it from the global Cartesian coordinate system in which the geometry of the molecule is specified. It is the local system $(\pi, \bar{\pi}, \sigma)$ which is related to the global system (x, y, z) by the Euler angle rotations.

Some of the integrals over this local diatomic set of basis functions can be "calculated" by inspection even before the basis functions are translated into the (ξ, η, ϕ) coordinates. It is obvious, for example, that the overlap integrals

$$\int 1s_H 2p_\pi dV \qquad \int 1s_H 2p_{\bar{\pi}} dV$$

are both zero since the two functions have different symmetry properties with respect to the local diatomic axis; the positive and negative regions in the overlap density exactly cancel when summed. Similar considerations apply to the corresponding kinetic energy integrals over the same pairs of functions and to the nuclear attraction integrals *involving the two nuclei of the local system* (and any collinear nuclei). *Some* of the electron repulsion integrals may be seen to be zero by similar "local symmetry" considerations; the integral

$$\int dV_1 \int dV_2 1s_H(\vec{r_1}) 1s_H(\vec{r_1}) \frac{1}{r_{12}} 1s_H(\vec{r_2}) 2p_{\pi O}(\vec{x_2})$$

is zero, for example.

7.5.1 Overlap integrals

The only non-zero two-centre overlap integral between the basis functions under consideration is

$$\int 1s_H 2p_{\sigma O} dV$$

which may be written out explicitly in prolate spheroidal coordinates as

$$N_{1s}N_{2p}\frac{R^4}{2}\int_1^\infty d\xi \int_{-1}^{+1} d\eta \int_0^{2\pi} d\phi(\xi^2 - \eta^2)$$
$$\times \exp\left[-\zeta_{1s}\left(\frac{R(\xi+\eta)}{2}\right)\right](1+\xi\eta)\exp\left[-\zeta_{2p}\left(\frac{R(\xi-\eta)}{2}\right)\right]$$

which can be simplified by defining

$$\rho = \frac{R}{2}(\zeta_{1s} + \zeta_{2p})$$
$$\tau = \frac{(\zeta_{1s} - \zeta_{2p})}{(\zeta_{1s} + \zeta_{2p})}$$

and dropping the numerical factors from the front of the integral:

$$\int_1^\infty d\xi \int_{-1}^{+1} d\eta \int_0^{2\pi} d\phi(\xi^2 - \eta^2)(1+\xi\eta)\exp\left[-\rho(\xi+\tau\eta)\right]$$

It is easily seen that the factors involving powers of ξ and η may be expanded out to form a polynomial and the exponential factor can be separated as a product of exponentials of ξ and η, and so the whole expression may be written as a sum of products of integrals of functions of ξ, η (and, of course, ϕ where the integration is trivial).

The relevant forms are

$$A_n(\rho) = \int_1^\infty d\xi \xi^n \exp(-\xi\rho)$$
$$= \frac{1}{\rho}[\exp(-\rho) + nA_{n-1}(\rho)]$$
$$A_0(\rho) = \frac{\exp(-\rho)}{\rho}$$
$$B_n(\rho\tau) = \int_{-1}^1 d\eta \eta^n \exp(-\rho\tau\eta)$$
$$= -A_n(\rho\tau) - (-1)^n A_n(-\rho\tau)$$

7.5.2 Kinetic energy integrals

The kinetic energy integrals have similar symmetry properties to the overlap integrals; the only non-zero one between the functions of our choice is:

$$\int 1s_H \left(-\frac{1}{2}\nabla^2\right) 2p_{\sigma O} dV = \int 2p_{\sigma O} \left(-\frac{1}{2}\nabla^2\right) 1s_H dV$$

a typical term of which is

$$\int 2p_{\sigma O} \left(\frac{\partial^2}{\partial x^2}\right) 1s_H dV$$

Obviously, the kinetic energy operator can be transformed into the prolate spheroidal coordinate system and the whole analysis carried out consistently in the (ξ, η, ϕ) system, but it is easier to see the way the derivation goes if the $1s$ function is first differentiated in Cartesians.[3]

Since

$$\begin{aligned}\frac{\partial}{\partial x}\exp(-\zeta_{1s}r_H) &= -\frac{\partial}{\partial x}\exp\left(-\zeta_{1s}(x_H^2+y_H^2+z_H^2)^{\frac{1}{2}}\right) \\ &= \frac{-x_H}{r_H}\zeta_{1s}\exp(-\zeta_{1s}r_H)\end{aligned}$$

and a further differentiation is similar, it is clear that the operation of the kinetic energy operator on an STO is to produce a linear combination of STO-like functions. Thus the kinetic energy integrals reduce to the evaluation of overlap integrals.

7.5.3 Nuclear attraction integrals

The nuclear attraction integrals over a local diatomic basis in a molecule of general geometry fall into three types:

1. One-centre integrals; both basis functions *and* the attracting centre all share a common origin. These integrals are *atomic* and are trivial to evaluate in the local polar coordinates.

2. Two-centre integrals: the two basis functions and the attracting centre involve at most two nuclei. All the nuclear-attraction integrals in a diatomic molecule are either one- or two-centre, for example.

---
[3]Since the various local and global Cartesian coordinate systems are related by addition of constants, it is of no importance *which* Cartesian system is used in the differentiations.

3. Three centre integrals: the two basis functions are on different nuclei and the attracting centre is different from both. This case occurs in the water molecule, for example.

It is fairly clear from the earlier discussion that the second of these two categories might well be evaluated exactly in the spheroidal coordinate system, since the distances involved are all easily expressed in the (ξ, η, ϕ) system. But the third category of integral introduces the three coordinates of the attracting nucleus which destroys the "symmetry" of the system rendering separation impossible in general.

The central problem in both the STO nuclear-attraction integrals and the STO electron-repulsion integrals is the fact that the expression for the inverse-distance operator is intractable. If $\vec{r}_a$ and $\vec{r}_b$ are two position vectors (either for one electron and a nucleus or for two electrons) then the inter-particle distance may be expressed:

- by the familiar trigonometric formula:

$$|\vec{r}_{ab}|^2 = |\vec{r}_a|^2 + |\vec{r}_b|^2 - 2|\vec{r}_a||\vec{r}_b|$$

It is straightforward to put these magnitudes into the prolate spheroidal system of coordinates but the resulting expressions will not separate even in these coordinates which are well adapted to the separation of the STO products:

$$r_{ab}^2 = 2\left(\frac{R}{2}\right)^2 (Q - S\cos(\phi_a - \phi_b))$$
$$Q = \xi_a^2 + \xi_b^2 + \eta_a^2 + \eta_b^2 - 2\xi_a\xi_b\eta_a\eta_b - 2$$
$$S = 2\left[(\xi_a^2 - 1)(\xi_b^2 - 1)(1 - \eta_a^2)(1 - \eta_b^2)\right]^{\frac{1}{2}}$$

- by the Neumann expansion:

$$\frac{1}{r_{ab}} = \left(\frac{2}{R}\right) \sum_{\ell=0}^{\infty} \sum_{m=0}^{\ell} (-1)^m (2 - \delta_{m0})(2\ell + 1)$$
$$\times \left(\frac{(\ell-m)!}{(\ell+m)!}\right)^2 P_\ell^m(\xi_<) Q_\ell^m(\xi_>) P_\ell^m(\eta_a) P_\ell^m(\eta_b)$$
$$\times \cos m(\phi_a - \phi_b)$$

($\xi_<$ and $\xi_>$ are the lesser and greater respectively of ξ_a and ξ_b). However, except for very untypical circumstances (equality amongst the STO exponents), this expansion does not lead to a *terminating* expansion for the molecular integrals.[4] The P_ℓ^m and the Q_ℓ^m are the Legendre polynomials of the first and second kinds.

---

[4]Contrasting with the *one*-centre expansion for the inter-particle distance which always terminates when used with AOs because of the orthogonality properties of the spherical harmonics of the first kind.

7.5. MOLECULAR INTEGRALS OVER STOS

The evaluation of these integrals has to depend on either some numerical integration[5] or the use of different techniques for different types of integral and different types of STO product (charge distribution).

7.5.4 Electron-repulsion integrals

Again these integrals fall into several categories depending on the number of nuclei involved in the "charge distributions"; products of basis functions:

1. One-centre integrals: both charge distributions are on one centre and this centre is common. These are atomic integrals and their evaluation is straightforward.

2. Two-centre "Coulomb" integrals: each charge distribution is one-centre and the two centres are different. These integrals are diatomic in nature and can, in fact, be evaluated by use of the prolate spheroidal coordinate transformation.

3. Two-centre "hybrid and exchange" integrals: integrals involving at least one charge distribution which is a product of two basis functions on different centres but all basis functions are confined to just two centres. These are the most general integrals involved in the diatomic case and are tractable with a good deal more effort (some methods involve numerical quadrature in one dimension; all involve truncation of an expansion).

4. Three- and four-centre integrals: the most general type where the charge distributions may both involve products of basis functions on different centres and up to four centres may be involved. All the methods of computing electron-repulsion integrals over an STO basis are extremely complex and involve some numerical quadrature and/or truncation of expansions. These integrals are *hard* to compute for the reasons we outlined in discussing the nuclear-attraction integrals.

There is no doubt, however, that work will continue on the evaluation of integrals involving STOs as basis functions; they have superior formal and physical properties to the Gaussians and after all, they are the functions which solve the Schrödinger equation for the Coulomb potential. Perhaps it is not too optimistic to think that the speed of hardware and the power of software will enable a newer approach to problems like this. The traditional method of numerical computing has been to "do the analysis" and then "code the formulae" (as we shall do for the Gaussians). But the tools are now available for a more "algorithmic" approach in which the machine *derivation* and computing of effective numerical techniques will be possible. This kind of approach using recursion relationships has already begun to appear.[6]

---

[5]This is not always a disadvantage, see Chapters 33 and 34.
[6]See J. Fernandez Rico, R. Lopez and G. Ramirez *J. Chem. Phys.*, **97**, 7613 (1992), for example.

7.5.5 Summary for STOs

The use of Slater-type orbitals as basis functions is currently feasible only for atoms, diatomics and, with effort, linear polyatomic molecules. However, research into more powerful methods continues and the physical attractiveness of these functions is such that the development of effective integral evaluation methods would be a major breakthrough in quantum chemistry.

There is one point from our outline which will prove useful in the efficient evaluation of integrals over other basis functions:

> The use of a "local diatomic" coordinate system maximises the number of integrals of all types *which are zero by local symmetry*. Thus of the 256 possible electron-repulsion integrals which arise from a set of four ($2s$ and $2p$) basis functions on two centres when the axes are arbitrarily orientated, only 88 are non-zero in the local diatomic system.

This simple fact has been made the basis of a general method for the rapid evaluation of integrals.

7.6 Basis functions of convenience

The natural attractiveness of the STO basis is countered by the numerical difficulties associated with the computation of the various molecular integrals for polyatomic molecules of general geometry. An enormous amount of work has been done over the past thirty years to try to solve these computational problems but despite this effort it is still true to say that the use of the STO basis is only practical for atoms and linear molecules. Even here we are being slightly optimistic; it is only atoms, diatomics and linear triatomics which are well within reach for the STO basis.

Faced with this impasse there is only one way forward. We must approach the problem from the other direction; look for functions over which the molecular integrals *are* accessible and choose the most physically realistic of these for a basis.

If the chosen basis is less well adapted to the description of the structure of a system of particles in mutual Coulomb interaction we must expect that the expansion lengths of MOs in terms of the basis function will be *longer*.

We need the following criteria:

7.6. BASIS FUNCTIONS OF CONVENIENCE

- The molecular integrals shall be tractable for polyatomic molecules of general geometry.

- The basis functions should be atom-centred if possible to enable a physical interpretation to be made of the results.

- The basis functions should have appropriate boundary conditions; vanishing at large distance from the atoms.

- In view of their presumed lack of natural physical appeal, the basis functions should have some known "completeness" properties. If they were the solutions of a single-particle Schrödinger equation, this would be ideal.

The set of Gaussian functions (Gaussian-Type Functions, GTFs) fulfils all of these requirements admirably.

- As *required*, all the molecular integrals can be evaluated analytically in terms of standard functions after separation into products of integrals of single variables.

- They can be chosen to be centred on each atom in an analogous way to the STO basis, only the radial (exponential) factor is different.

- An atom-centred GTF vanishes asymptotically at large r. In fact it vanishes rather *too* quickly.

- The GTFs are directly related to the solutions of the one-particle Schrödinger equation for the simple harmonic oscillator and so have well-defined completeness properties.

It is easy to see from this qualitative description the penalty to be paid for the benefits of easy integral evaluation.

Electrons and nuclei interact with each other via Coulomb's law not via Hooke's law and so the expansion of a function dominated by the *electrostatic* interactions in terms of functions well adapted to the description of particles connected together by *springs* might well be a lengthy one.

There are, of course, various attitudes which might be taken to this choice of basis; on the one hand we cannot get very far *without* the use of a basis of convenience, on the other we must be careful of the interpretation of our results with an "unphysical" basis.[7]

---

[7] Perhaps the most extreme view has been expressed by M. Randić: "... but tomorrow they [GTFs] may be thought of as bastard surrogates, which served their purpose in the transition period... " in *"ETO Multicenter Molecular Integrals"* Eds. C. A. Weatherford and H. W. Jones p. 142 (D Reidel, 1982)

7.7 Gaussian basis functions

The Gaussian basis is identical in *general form* to the STO basis, but the exponential factor is different:

$$\eta_i = x_A^{\ell_i} y_A^{m_i} z_A^{n_i} \exp(-\alpha_i r_A^2)$$

is the *Cartesian* form (which is the most widely used form in molecules) of a GTF centred on nucleus A or

$$\eta_i = r_A^{n_i} P_\ell^m(cos\theta_A) \exp(-\alpha_i r_A^2)$$

in a less familiar complex spherical polar form.

The notation η_i has been used in place of our standard of ϕ_i deliberately, not in order to distinguish GTFs from STOs, but in order to be able to make the distinction between *basis functions* and *Gaussian primitives* which will prove one of the consequences of the use of the GTF basis.[8]

The only new feature here is the crucial one: the appearance of the radial distance *squared* in the exponential factor which, of course, means that, unlike the STO exponential factor, the GTF exponential factor splits into a *product* of separate functions of the (local) Cartesian coordinates. Further, since the relationship between any two (parallel) Cartesian coordinate systems is simply addition of a constant to each coordinate, this separation process is possible in the global coordinate system.

$$\begin{aligned}\eta_i &= \left(x_A^{\ell_i} \exp(-\alpha_i x_A^2)\right) \\ &\times \left(y_A^{m_i} \exp(-\alpha_i y_A^2)\right) \\ &\times \left(z_A^{n_i} \exp(-\alpha_i z_A^2)\right)\end{aligned}$$

with clear mathematical advantages.[9]

If it is necessary to have an explicit knowledge of the parameters $(\ell_i, m_i, n_i, \vec{A}, \alpha_i)$ that characterise a particular GTF, the notation

$$\eta_i(\vec{A}, \alpha_i, \ell_i, m_i, n_i) = x_A^{\ell_i} y_A^{m_i} z_A^{n_i} \exp(-\alpha_i r_A^2)$$

will be used.

---

[8]The technique of "contraction", used to avoid being overwhelmed by the sheer numbers of GTFs needed to expand MOs.

[9]Indeed, it is obviously not necessary that we use the same α for each of the Cartesian factors, leading to further possibilities for basis choice.

7.7. GAUSSIAN BASIS FUNCTIONS

The relationship between the local Cartesian systems on each nucleus (x_A, y_A, z_A) (say) and the global Cartesian system (x, y, z) over which the calculation is ultimately performed is:

$$x_A = x - A_x$$
$$y_A = y - A_y$$
$$z_A = z - A_z$$

where the position vector of nucleus A with respect to the global origin is (A_x, A_y, A_z).

7.7.1 Overlap integrals

The central result in the calculation of the Cartesian GTF overlap integrals is the overlap between the x, y, or z *factors* in each GTF. Typically:

$$\int_{-\infty}^{\infty} \exp\left(-\alpha_1(x-A)^2\right) \exp\left(-\alpha_2(x-B)^2\right) dx$$
$$= \int_{-\infty}^{\infty} \exp\left(-\alpha_1(x-A)^2 - \alpha_2(x-B)^2\right) dx$$

The exponent can be expanded to

$$x^2(\alpha_1 + \alpha_2) - 2x(A\alpha_1 + B\alpha_2) + (\alpha_1 A^2 + \alpha_2 B^2)$$

which can be rearranged to

$$(\alpha_1 + \alpha_2)\left(x^2 - 2x\frac{A\alpha_1 + B\alpha_2}{\alpha_1 + \alpha_2} + \frac{\alpha_1 A^2 + \alpha_2 B^2}{\alpha_1 + \alpha_2}\right)$$

which, on dividing by $(\alpha_1 + \alpha_2)$ becomes

$$x^2 - 2Px + P^2 + C = (x-P)^2 + C$$

where

$$P = \frac{A\alpha_1 + B\alpha_2}{\alpha_1 + \alpha_2}$$

and

$$C = \frac{\alpha_1\alpha_2(A-B)^2}{\alpha_1 + \alpha_2}$$

which is *a constant* independent of x.

> Thus the original integrand becomes
>
> $$\exp\left(-\alpha_1(x-A)^2\right)\exp\left(-\alpha_2(x-B)^2\right)$$
> $$= \exp\left(\frac{\alpha_1\alpha_2(A-B)^2}{\alpha_1+\alpha_2}\right)\times\exp\left((\alpha_1+\alpha_2)(x-P)^2\right)$$
>
> which (apart from the constant term) is just the Cartesian factor of another GTF centered at another point on the x-axis.

This "product rule" has very important ramifications[10] for the nuclear-attraction and electron-repulsion integrals as we shall emphasise later.

The only integral required for the evaluation of the GTF overlap integrals is then

$$\int_{-\infty}^{\infty}\exp(-\gamma t^2)dt = \sqrt{\frac{\pi}{\gamma}}$$

In the case of the more general (non s-type) GTF overlap integral the integrand will involve a product of factors from the powers of the local Cartesian coordinates:

$$(x-A_x)^{\ell_1}(x-B_x)^{\ell_2}$$

Clearly, these terms can be expanded out as polynomials in powers of x and collected together to give a factor in the overall GTF overlap which is composed of a sum of terms like

$$\int_{-\infty}^{\infty}t^{2n}\exp(-\gamma t^2)dt = \frac{(2n-1)!!}{2^{n+1}}\sqrt{\frac{\pi}{\gamma^{2n+1}}}$$

where $2n$ has been used intentionally to indicate that the power of the coordinate must be even for a non-zero integral.

The notation $(2n-1)!!$ (the so-called *double factorial*) means

$$(2n-1)!! = 1\times 3\times 5\times\ldots\times(2n-1)$$

which product is guaranteed to terminate since $(2n-1)$ is always odd.

The terms arising from the Cartesian factors of each GTF may be written in terms of the position vectors of the nuclei A and B with respect to the point P (the centre of the product GTF). Since

$$\vec{r}_A = \vec{r}-\vec{A} = (\vec{r}-\vec{P})+(\vec{P}-\vec{A}) = \vec{r}_P+(\vec{P}-\vec{A}) = \vec{r}_P-\vec{PA}$$

---
[10]This is just a restatement of the fact, familiar from elementary statistics, that the product of two normal (Gaussian) distributions is normal.

7.7. GAUSSIAN BASIS FUNCTIONS

with similar expressions for the y and z terms we have:

$$\begin{aligned} x_A^{\ell_1} x_B^{\ell_2} &= (x + \vec{PA}_x)^{\ell_1}(x + \vec{PB}_x)^{\ell_2} \\ &= \sum_{j=0}^{\ell_1+\ell_2} f_j(\ell_1, \ell_2, \vec{PA}_x, \vec{PB}_x) x_P^j \end{aligned}$$

which is the *definition* of the expression $f_j(\ell, m, a, b)$ as the coefficient of x^j in the expansion of $(x+a)^\ell (x+b)^m$. It is easy to show that

$$f_j(\ell, m, a, b) = \sum_{k=max(0, j-m)}^{min(j, \ell)} \binom{\ell}{k} \binom{m}{j-k} a^{\ell-k} b^{m+k-j}$$

Therefore the product of two general GTFs:

1.
$$\eta_1(\vec{A}, \alpha, \ell_1, m_1, n_1)$$

centred at $\vec{A}$, exponent α, Cartesian factor $x_A^{\ell_1} y_A^{m_1} z_A^{n_1}$ and

2.
$$\eta_2(\vec{B}, \beta, \ell_2, m_2, n_2)$$

centred at $\vec{B}$, exponent β, Cartesian factor $x_B^{\ell_2} y_B^{m_2} z_B^{n_2}$

is, expressed in terms of the Cartesian system at $\vec{P}$:

$$\begin{aligned} \eta_1 \eta_2 &= \exp\left(\frac{-\alpha\beta|\vec{AB}|^2}{\gamma}\right) \\ &\times \sum_{i=0}^{\ell_1+\ell_2} f_i(\ell_1, \ell_2, \vec{PA}_x, \vec{PB}_x) x_P^i \exp(-\gamma x_P^2) \\ &\times \sum_{j=0}^{m_1+m_2} f_j(m_1, m_2, \vec{PA}_y, \vec{PB}_y) y_P^j \exp(-\gamma y_P^2) \\ &\times \sum_{k=0}^{n_1+n_2} f_k(n_1, n_2, \vec{PA}_z, \vec{PB}_z) z_P^k \exp(-\gamma z_P^2) \end{aligned}$$

where
$$\gamma = \alpha + \beta$$

The integral of this form is the product of three integrals over the individual Cartesian coordinates multiplied by the factor arising from expression of a product of GTFs as a single GTF:

$$\int \eta_1 \eta_2 dx dy dz = \exp\left(\frac{-\alpha\beta|\vec{AB}|^2}{\gamma}\right) S_x S_y S_z$$

where, for example,

$$S_x = \int_{-\infty}^{\infty} \sum_{i=0}^{\ell_1+\ell_2} f_i(\ell_1, \ell_2, \vec{PA}_x, \vec{PB}_x) x_P^i \exp(-\gamma x_P^2) dx$$

The symmetry of the integral means that odd powers of x_P give zero contributions and the terms with even powers of x_P may be evaluated in terms of the Γ function used for s-type GTFs above.

$$\int_{-\infty}^{\infty} t^j \exp(-\alpha t^2) dt = \frac{\Gamma\left(\frac{j+1}{2}\right)}{\alpha^{(j+1)/2}}$$

so that

$$S_x = \sum_{j=0}^{[(\ell_1+\ell_2)/2]} f_{2j}(\ell_1, \ell_2, \vec{PA}_x, \vec{PB}_x) \frac{\Gamma\left(\frac{2j+1}{2}\right)}{\gamma^{(2j+1)/2}}$$

It was seen above, the Γ function with argument of the form $(2j+1)/2$ is of a particularly simple form:

$$\Gamma\left(\frac{2j+1}{2}\right) = \sqrt{\pi} \frac{(2j-1)!!}{2^j}$$

and therefore the integral I_x becomes

$$S_x = \sqrt{\frac{\pi}{\gamma}} \sum_{j=0}^{[(\ell_1+\ell_2)/2]} f_{2j}(\ell_1, \ell_2, \vec{PA}_x, \vec{PB}_x) \frac{(2j-1)!!}{(2\gamma)^j}$$

with exactly analogous expressions for S_y and S_z.

7.7.2 Kinetic energy integrals

The calculation of the GTF kinetic energy integrals involves exactly the same simplification as the STO case; the derivative of a GTF is another GTF:

$$\frac{\partial}{\partial x} \exp(-\alpha x_A^2) = 2x_A \alpha \exp(-\alpha x_A^2)$$

in fact, for the general GTF $\eta(\vec{A}, \alpha, \ell, m, n)$

$$\frac{\partial \eta(\vec{A}, \alpha, \ell, m, n)}{\partial x} = \ell \eta(\vec{A}, \alpha, \ell-1, m, n) - 2\alpha \eta(\vec{A}, \alpha, \ell+1, m, n)$$

7.7. GAUSSIAN BASIS FUNCTIONS

so that the kinetic energy operator $\nabla^2/2$ generates a sum of three possible GTFs (in general):

$$\begin{aligned}
\frac{-1}{2}\nabla^2 \eta(\vec{A}, \alpha, \ell, m, n) = & \ \alpha[2(\ell + m = n) + 3]\eta(\vec{A}, \alpha, \ell, m, n) \\
& - \ 2\alpha^2[\eta(\vec{A}, \alpha, \ell + 2, m, n) \\
& + \ \eta(\vec{A}, \alpha, \ell, m + 2, n) \\
& + \ \eta(\vec{A}, \alpha, \ell, m, n + 2)] \\
& - \ \frac{1}{2}[\ell(\ell - 1)\eta(\vec{A}, \alpha, \ell - 2, m, n) \\
& + \ m(m - 1)\eta(\vec{A}, \alpha, \ell, m - 2, n) \\
& + \ n(n - 1)\eta(\vec{A}, \alpha, \ell, m, n - 2)]
\end{aligned}$$

The terms involving $\ell - 2$, $m - 2$, or $n - 2$ are, of course, omitted unless $\ell \geq 2$ etc.

Thus the kinetic energy integrals over the GTFs reduce to a sum of overlap integrals.

7.7.3 Nuclear-attraction integrals

Because of the important product property of the GTFs which we discovered in the evaluation of overlap integrals, the GTF nuclear-attraction integrals fall into just *two* essential classes:

1. One-centre integrals: the attracting nucleus coincides with the point in space on which the product of the two GTFs generates a GTF function.

2. Two-centre integrals: the attracting nucleus is at a different point in space from the point on which the product of the two GTFs generate a GTF function.

Note that we are now no longer talking about the *two* GTFs which describe the distribution of the electron but about the *one* product function generated by the rule in the section above. Obviously, case 1 is just a special case of 2.

The evaluation of the nuclear-attraction integrals like

$$\int\int\int \eta_i \left(-\frac{1}{r_C}\right) \eta_j \, dx dy dz$$

still has to resolve the problem of the inverse-distance operator $1/r_C$. We have seen that the problems of *actual quadrature* are made very much simpler in the GTF case by the "separation" in Cartesian coordinates of the GTF form. We can, therefore, afford to introduce a transform which generates *another* separable quadrature in order to resolve the problem.

Before introducing this transformation, it is as well to sketch the general line of attack on the evaluation of the nuclear-attraction integrals since both they and the more complicated electron-repulsion integrals involve a morass of detailed analysis and algebra and it is important to be able to appreciate the general direction of the derivation.

The basic steps are as follows:

- Form the product Gaussian from the GTF functions and express it as a product of Cartesian factors (exactly as in the overlap integrals).

- Use an integral transform to express the nuclear-attraction operator $1/r_C$ in a form which will also separate into Cartesian factors.

- Carry through the quadrature of the integrable factors.

- Use an additional transform to enable the remaining one-dimensional integral to be reduced to a standard form for which stable numerical methods are available.

There are, in fact, two candidates for the second step and both are used:

1. The identity:
$$\frac{1}{r} = \frac{1}{\sqrt{\pi}} \int_0^\infty s^{-\frac{1}{2}} \exp(-sr^2) ds$$

    transforms the nuclear-attraction operator into a GTF-like form (with all the advantages this has) at the expense of another integration.

2. The more familiar Fourier transform of $1/r$:
$$\frac{1}{r} = \frac{1}{2\pi} \int_{-\infty}^\infty \frac{d\vec{k}}{|\vec{k}|^2} \exp(i\vec{k} \cdot \vec{r})$$

    is also composed of a product of Cartesian factors; and when used to transform $1/r_C$ generates integrals of the type
$$\int_{-\infty}^\infty x^n \exp(ax^2 + bx) dx$$

    which are also standard forms since they may be reduced, by a change of variable, to Gaussian form provided that $n \geq 0$.

7.7. GAUSSIAN BASIS FUNCTIONS

With this overall plan in mind we can begin the rather lengthy operations involved in computing the nuclear attraction integrals.

The general case is:

$$\begin{aligned} V &= \int \eta_1(\vec{A}, \alpha, \ell_1, m_1, n_1) \frac{1}{r_C} \eta_2(\vec{B}, \beta, \ell_2, m_2, n_2) dV \\ &= \exp\left(\frac{-\alpha\beta|\vec{AB}|^2}{\gamma}\right) \\ &\times \sum_{\ell=0}^{\ell_1+\ell_2} f_\ell(\ell_1, \ell_2, \vec{PA}_x, \vec{PB}_x) \\ &\times \sum_{m=0}^{m_1+m_2} f_m(m_1, m_2, \vec{PA}_y, \vec{PB}_y) \\ &\times \sum_{n=0}^{n_1+n_2} f_n(n_1, n_2, \vec{PA}_z, \vec{PB}_z) \\ &\times \int x_P^\ell y_P^m z_P^n \frac{1}{r_C} \exp(-\gamma r_P^2) \end{aligned}$$

In the above equation the (dummy) summation indices have been changed from i, j, k (used during the overlap integral evaluation) to ℓ, m, n since the danger of confusion between (e.g.) ℓ and ℓ_1 is less than the confusion caused by using i, j, k when the Fourier transform uses i for $\sqrt{-1}$ and k for the transformation variable![11]

Now, using the Fourier transform of $1/r_C$ and expressing r_C in terms of r_P:

$$\vec{r}_C = \vec{r} - \vec{C} = (\vec{r} - \vec{P}) + (\vec{P} - \vec{C}) = \vec{r}_P + \vec{PC}$$

(say, $\vec{PC} = \vec{P} - \vec{C}$)

$$\frac{1}{r_C} = \frac{1}{2\pi^2} \int \frac{d\vec{k}}{k^2} \exp(i\vec{k} \cdot \vec{PC}) \exp(i\vec{k} \cdot \vec{r}_P)$$

and making the substitution in the expression for V the integrand can be expressed in a form containing the separate Cartesian factors:

$$\begin{aligned} V &= \exp\left(\frac{-\alpha\beta|\vec{AB}|^2}{\gamma}\right) \frac{1}{2\pi^2} \\ &\times \sum_{\ell=0}^{\ell_1+\ell_2} f_\ell(\ell_1, \ell_2, \vec{PA}_x, \vec{PB}_x) \end{aligned}$$

---

[11] The original paper by H. Taketa, S. Huzinaga and K. O-Ohata (*J. Phys. Soc. Japan*, **21**, 2313 (1966)) is marred by typographical errors compounded the double use of i, so that i^i occurs in some of the formulae with the two i's meaning different things!

$$\times \sum_{m=0}^{m_1+m_2} f_m(m_1, m_2, \vec{PA}_y, \vec{PB}_y)$$

$$\times \sum_{n=0}^{n_1+n_2} f_n(n_1, n_2, \vec{PA}_z, \vec{PB}_z)$$

$$\times \int \left(k^{-2} \exp(i\vec{k} \cdot \vec{PC}) V_\ell^x V_m^y V_n^z \right) d\vec{k}$$

where, as usual, $d\vec{k}$ is written for $dk_x dk_y dk_z$ ands $|\vec{k}|$ has been abbreviated to simply k.

The Cartesian factors of V which are represented by V_ℓ^x are given by

$$V_\ell^x = \int_{-\infty}^{\infty} x_P^\ell \exp(-\gamma x_P^2 + ik_x x_P) dx_P$$

Now this integral may be evaluated by the standard form:

$$\int_{-\infty}^{\infty} t^n \exp(-at^2 + ibt) dt = i^n n! \left(\frac{\pi}{a}\right)^{1/2} \left(\frac{1}{2\sqrt{a}}\right)^n \exp(-c^2)$$

$$\times \sum_{j=0}^{[n/2]} \frac{(-1)^j}{j!} \frac{(2c)^{n-2j}}{(n-2j)!}$$

where $c = b/2\sqrt{a}$ and the notation $[n/2]$ means the largest integer less than or equal to $n/2$.

Using this form to evaluate V_ℓ^x we have

$$V_\ell^x = i^\ell \left(\frac{\pi}{\gamma}\right)^{1/2} \ell! \epsilon^{\ell/2} \exp(-\epsilon k_x^2)$$

$$\times \sum_{r=0}^{[\ell/2]} \frac{(-1)^r (2\sqrt{\epsilon} k_x)^{\ell-2r}}{r!(\ell-2r)!}$$

where

$$\epsilon = \frac{1}{4\gamma}$$

This completes only a *partial* solution as the integrals V_ℓ^x, V_m^y and V_n^z appear *inside* the integration (over $d\vec{k}$) introduced by the Fourier transform and the completed *spatial* integrals still contain exponentials and powers of the components k_x, k_y, k_z of $\vec{k}$.

Examining the above expression for V_ℓ^x and using the analogy with the other Cartesian factors it is straightforward to see that the product of Cartesian factors is

$$V_\ell^x V_m^y V_n^z = \text{(factors independent of } \vec{k} \text{)} \times \exp(-\epsilon k^2) k_x^{\ell-2r} k_y^{m-2s} k_z^{n-2t}$$

7.7. GAUSSIAN BASIS FUNCTIONS

where s and t are the summation indices analogous to r in the Cartesian factors V_m^y and V_n^z. Combining this residual from the Cartesian integration with the factor involving $\vec{k}$ from the Fourier transform gives a $\vec{k}$-dependent integral of

$$\int k^{-2} k_x^{\ell-2r} k_x^{m-2s} k_x^{n-2t} \exp(i\vec{k}\cdot\vec{PC}) d\vec{k}$$

to be evaluated.

The compact notation $d\vec{k}$ disguises the fact that this is still a *triple* integral over the components of $\vec{k}$. The problem is the awkward appearance of the factor

$$\frac{1}{|\vec{k}|^2} = \frac{1}{k_x^2 + k_y^2 + k_z^2}$$

which precludes a complete separation.

This factor can be removed (at the expense of yet another integration!) by the identity:

$$\exp(\epsilon k^2) = 2\epsilon k^2 \int_0^1 u^{-3} \exp\left(-\epsilon \frac{k^2}{u^2}\right) du$$

This identity replaces the exponential factor

$$\exp(-\epsilon k^2 + i\vec{k}\cdot\vec{PC})$$

in the above integral by

$$\exp(-\epsilon k^2 u^{-2} + i\vec{k}\cdot\vec{PC})$$

and removes the factor k^2 thus enabling the integrals over the components k_x, k_y, k_z of $\vec{k}$ to be evaluated separately. The resulting (now quadruple) integral is:

$$\begin{aligned}\int k^{-2} k_x^{\ell-2r} k_x^{m-2s} k_x^{n-2t} \exp(i\vec{k}\cdot\vec{PC}) d\vec{k} &= 2\epsilon \int_0^1 \frac{du}{u^3} \\ &\times \int_{-\infty}^{\infty} dk_x k_x^{\ell-2r} \exp\left(-\epsilon u^{-2} k_x^2 + i\vec{PC}_x k_x\right) \\ &\times \int_{-\infty}^{\infty} dk_y k_y^{m-2s} \exp\left(-\epsilon u^{-2} k_y^2 + i\vec{PC}_y k_y\right) \\ &\times \int_{-\infty}^{\infty} dk_z k_z^{n-2t} \exp\left(-\epsilon u^{-2} k_z^2 + i\vec{PC}_z k_z\right)\end{aligned}$$

When the integrations over the components of $\vec{k}$ are carried through, a dependence on the new variable u is left in (what is now, at last) the final integration.

Each of the integrations over the components of $\vec{k}$ is of the same standard form that was used to evaluate the original *spatial* Cartesian factors. The only remaining technical problem is to find enough letters to express the summations.

After using the standard integration formula the remaining integrals over u are of the form:

$$\int_0^1 u^{2\nu} \exp(-\gamma|\vec{PC}|^2 u^2) du = F_\nu(\gamma|\vec{PC}|^2)$$

(say) where

$$\nu = (\ell - 2r - i + m - 2s - j + n - 2t - k) = \ell + m + n - 2(r+s+t) - (i+j+k)$$

The indices i, j, and k are generated by the integration over k_x, k_y and k_z, respectively, and their ranges are

$$\begin{aligned} 0 \leq\ & i\ \leq [(\ell - 2r)/2] \\ 0 \leq\ & j\ \leq [(m - 2r)/2] \\ 0 \leq\ & k\ \leq [(n - 2r)/2] \end{aligned}$$

The function $F_n(x)$ is a well-known one from applied mathematics and techniques for its evaluation are well researched.

Thus, in order to evaluate the nuclear-attraction integrals it is necessary to be able to calculate $F_n(x)$ *and* to keep track of all the algebra arising from the other *six* (spatial and Fourier-transform) integrations. The latter is a large "book-keeping" problem.

The terms arising in the various summations may be regrouped to give a final expression for the integral:

$$\begin{aligned} V =\ & \int \eta_1(\vec{A}, \alpha, \ell_1, m_1, n_1) \frac{1}{r_C} \eta_2(\vec{B}, \beta, \ell_2, m_2, n_2) dV \\ =\ & \frac{2\pi}{\gamma} \exp\left(\frac{-\alpha\beta|\vec{AB}|^2}{\gamma}\right) \\ & \times \sum_{\ell=0}^{\ell_1+\ell_2} \sum_{r=0}^{[\ell/2]} \sum_{i=0}^{[(\ell-2r)/2]} A_{\ell,r,i}(\ell_1, \ell_2, \vec{A}_x, \vec{B}_x, \vec{C}_x, \gamma) \\ & \times \sum_{m=0}^{m_1+m_2} \sum_{s=0}^{[m/2]} \sum_{j=0}^{[(m-2s)/2]} A_{m,s,j}(m_1, m_2, \vec{A}_y, \vec{B}_y, \vec{C}_y, \gamma) \\ & \times \sum_{n=0}^{n_1+n_2} \sum_{t=0}^{[n/2]} \sum_{k=0}^{[(n-2t)/2]} A_{n,t,k}(n_1, n_2, \vec{A}_z, \vec{B}_z, \vec{C}_z, \gamma) \\ & \times F_{\ell+m+n-2(r+s+t)-(i+j+k)}(\gamma|\vec{PC}|^2) \end{aligned}$$

7.7. GAUSSIAN BASIS FUNCTIONS

Where the separation into Cartesian factors is clearly shown by the symmetrical appearance of the A factors which are defined by

$$A_{\ell,r,i}(\ell_1, \ell_2, \vec{A}_x, \vec{B}_x, \vec{C}_x, \gamma) = (-1)^\ell f_\ell(\ell_1, \ell_2, \vec{PA}_x, \vec{PB}_x) \frac{(-1)^i \ell! \vec{PC}_x^{\ell-2r-2i} \epsilon^{r+i}}{r! i! (\ell - 2r - 2i)!}$$

with expressions of identical form for the y and z factors.

Recall that

$$\gamma = \alpha + \beta$$
$$\epsilon = \frac{1}{4\gamma}$$

$\vec{A}$, $\vec{B}$ and $\vec{C}$ are the position vectors of the centres of η_1, η_2 and the attracting nucleus, respectively, and $\vec{P}$ is the position vector of the point at which the "product Gaussian" formed from the exponential factors of η_1 and η_2 is centred.

This formula is straightforward to implement, provided a convenient way to evaluate the function $F_\nu(x)$ is available. Methods of evaluating this integral are outlined later.

7.7.4 Electron-repulsion integrals

The electron-repulsion integrals which arise in a molecular calculation are integrals over the coordinates of *two* electrons; that is *six-dimensional* integrals involving, in general, four basis functions and the inverse inter-electronic distance operator $1/r_{12}$. Because the computation of these integrals is complex and relatively time-consuming, and because the integrals are so numerous, an enormous amount of effort has been spent on developing efficient algorithms for their evaluation. In this section only the "direct" method of evaluation is discussed; no attempt is made to optimise the resulting formulae for speed. However, most of the methods which are available for the efficient and speedy evaluation of electron-repulsion integrals take the formulae derived below as a starting point and achieve their savings by regrouping the summations or by using special axis systems to maximise the number of integrals which are zero by symmetry or by avoiding duplicate calculations etc.[12]

Fortunately, the method of evaluating these numerous integrals follows the derivation of the nuclear-attraction integrals very closely; the amount of manipulation is, however, *doubled* because of the four orbitals involved. In the

---

[12] The method of Rys polynomials is an exception to this generalisation.

evaluation of the integral

$$G = \int \eta_1(\vec{A}, \alpha_1, \ell_1, m_1, n_1)(\vec{r}_{A1})\eta_2(\vec{B}, \alpha_2, \ell_2, m_2, n_2)(\vec{r}_{B1})$$
$$\times \frac{1}{r_{12}}\eta_3(\vec{C}, \alpha_3, \ell_3, m_3, n_3)(\vec{r}_{C2})\eta_4(\vec{D}, \alpha_4, \ell_4, m_4, n_4)(\vec{r}_{D2})dV_1 dV_2$$

the new notation used is obvious:

$$\begin{aligned}
\vec{r}_{A1} &= \vec{r}_1 - \vec{A} \\
\vec{r}_{B1} &= \vec{r}_1 - \vec{B} \\
\vec{r}_{C2} &= \vec{r}_2 - \vec{C} \\
\vec{r}_{D2} &= \vec{r}_2 - \vec{D} \\
r_{12} &= |\vec{r}_1 - \vec{r}_2|
\end{aligned}$$

The steps in the evaluation of this integral should now be familiar.

Use the GTF product.

Use the Gaussian product theorem to reduce the two basis-function products to linear combinations of GTFs centred on just two points

$$\begin{aligned}
\exp\left(-\alpha_1 r_{A1}^2\right)\exp\left(-\alpha_2 x_{B1}^2\right) &= \exp\left(\frac{\alpha_1 \alpha_2 |\vec{AB}|^2}{\alpha_1 + \alpha_2}\right)\exp\left(\gamma_1 r_{P1}^2\right) \\
&= K_1 \exp\left(\gamma_1 r_{P1}^2\right) \quad (say) \\
\exp\left(-\alpha_3 r_{C2}^2\right)\exp\left(-\alpha_4 x_{D2}^2\right) &= \exp\left(\frac{\alpha_3 \alpha_4 |\vec{CD}|^2}{\alpha_3 + \alpha_4}\right)\exp\left(\gamma_2 r_{Q2}^2\right) \\
&= K_2 \exp\left(\gamma_2 r_{Q2}^2\right) \quad (say)
\end{aligned}$$

Again the notation is almost self-explanatory; the points with position vectors $\vec{P}$ and $\vec{Q}$ are the centres of the "product" GTFs formed from $\eta_1\eta_2$ and $\eta_3\eta_4$, respectively

$$\begin{aligned}
\gamma_1 &= \alpha_1 + \alpha_2 \\
\gamma_2 &= \alpha_3 + \alpha_4 \\
\vec{P} &= \frac{(\alpha_1 \vec{A} + \alpha_2 \vec{B})}{\gamma_1} \\
\vec{Q} &= \frac{(\alpha_3 \vec{C} + \alpha_4 \vec{D})}{\gamma_1}
\end{aligned}$$

7.7. GAUSSIAN BASIS FUNCTIONS

and

$$\begin{aligned}
\vec{AB} &= \vec{A} - \vec{B} \\
\vec{CD} &= \vec{C} - \vec{D} \\
\vec{r}_{P1} &= \vec{r}_1 - \vec{P}; \quad x_{P1} = x_1 - P_x \quad \text{etc.} \\
\vec{r}_{Q2} &= \vec{r}_2 - \vec{Q}; \quad x_{Q2} = x_2 - Q_x \quad \text{etc.}
\end{aligned}$$

It is also useful to have the compact notation

$$\vec{P} - \vec{Q} = \vec{p}$$

Expand the Cartesian Products.

The expansion of the Cartesian product factors in terms of the $f_j(i,k,a,b)$ factors introduced in the derivation of the nuclear-attraction integral enables the integral G to be expanded as a sum of six-dimensional integrals involving the Cartesian factors referred to the two points $\vec{P}$ and $\vec{Q}$:

$$\begin{aligned}
G &= K_1 K_2 \sum_{\ell=0}^{\ell_1+\ell_2} \sum_{m=0}^{m_1+m_2} \sum_{n=0}^{n_1+n_2} f_\ell(\ell_1, \ell_2, \vec{PA}_x, \vec{PB}_x) \\
&\times f_m(m_1, m_2, \vec{PA}_y, \vec{PB}_y) f_n(n_1, n_2, \vec{PA}_z, \vec{PB}_z) \\
&\times \sum_{\ell'=0}^{\ell_3+\ell_4} \sum_{m'=0}^{m_3+m_4} \sum_{n'=0}^{n_3+n_4} f_{\ell'}(\ell_3, \ell_4, \vec{QC}_x, \vec{QD}_x) \\
&\times f_{m'}(m_3, m_4, \vec{QC}_y, \vec{QD}_y) f_{n'}(n_3, n_4, \vec{QC}_z, \vec{QD}_z) \\
&\times \iint x_{P1}^\ell y_{P1}^m z_{P1}^n x_{Q2}^{\ell'} y_{Q2}^{m'} z_{Q2}^{n'} \frac{1}{r_{12}} \exp(-\gamma_1 r_{P1}^2 - \gamma_2 r_{Q2}^2) dV_1 dV_2
\end{aligned}$$

where, of course,

$$\begin{aligned}
dV_1 &= dx_{P1} dy_{P1} dz_{P1} \\
dV_2 &= dx_{Q2} dy_{Q2} dz_{Q2}
\end{aligned}$$

Use the Fourier Transform.

The introduction of the Fourier transform for the operator $1/r_{12}$ in terms of r_{P1} and r_{Q2} is straightforward:

$$\vec{r}_{12} = \vec{r}_1 - \vec{r}_2 = (\vec{r}_1 - \vec{P}) - (\vec{r}_2 - \vec{Q}) = \vec{r}_{P1} - \vec{r}_{Q2} + \vec{p}$$

so that
$$\frac{1}{r_{12}} = \frac{1}{2\pi^2} \int \frac{d\vec{k}}{k^2} \exp(i\vec{r}_{P1} \cdot \vec{k}) \exp(i\vec{r}_{Q2} \cdot \vec{k}) \exp(i\vec{p} \cdot \vec{k})$$

Substitution of this transform into the expression for G gives the integral as an integral over $\vec{k}$ involving *six* spatial integrals *each of which* has some $\vec{k}$-dependence in an way exactly analogous to the nuclear-attraction integral.

Perform the Quadratures.

The reductions now parallel those of the earlier integral:

1. Carry through each of the six Cartesian-factor integrals using the standard form used in the treatment of the nuclear-attraction integral.

2. Collect together the resulting terms involving powers and exponentials of the components, (k_x, k_y, k_z), of $\vec{k}$ separately.

3. Use the integral identity to remove the appearance of the $\vec{k}^2$ factor and introduce the new integral.

4. Perform the integrations over the components of $\vec{k}$ to give the final integral as a sum of multiples of the function $F_\nu(x)$ as before.

Collect the Terms.

These steps involve considerable pedestrian manipulations which contain no steps not discussed in the treatment of the nuclear-attraction integral and so are not given in detail here. Instead we simply quote the result for the general electron-repulsion integral over four GTFs:

$$\begin{aligned}
G &= \int \eta_1(\vec{A}, \alpha_1, \ell_1, m_1, n_1)(\vec{r}_{A1}) \eta_2(\vec{B}, \alpha_2, \ell_2, m_2, n_2)(\vec{r}_{B1}) \\
&\times \frac{1}{r_{12}} \eta_3(\vec{C}, \alpha_3, \ell_3, m_3, n_3)(\vec{r}_{C2}) \eta_4(\vec{D}, \alpha_4, \ell_4, m_4, n_4)(\vec{r}_{D2}) dV_1 dV_2 \\
&= \Omega \sum_{\ell=0}^{\ell_1+\ell_2} \sum_{r=0}^{[\ell/2]} \sum_{i=0}^{[(\ell-2r)/2]} \sum_{\ell'=0}^{\ell_3+\ell_4} \sum_{r'=0}^{[\ell'/2]} \\
&\quad B_{\ell,\ell',r_1,r_2,i}(\ell_1, \ell_2, \vec{A}_x, \vec{B}_x, \vec{P}_x, \gamma_1; \ell_3, \ell_4, \vec{C}_x, \vec{D}_x, \vec{Q}_x, \gamma_2) \\
&\times \sum_{m=0}^{m_1+m_2} \sum_{s=0}^{[m/2]} \sum_{j=0}^{[(m-2s)/2]} \sum_{m'=0}^{m_3+m_4} \sum_{s'=0}^{[m'/2]} \\
&\quad B_{m,m',s_1,s_2,j}(m_1, m_2, \vec{A}_y, \vec{B}_y, \vec{P}_y, \gamma_1; m_3, m_4, \vec{C}_y, \vec{D}_y, \vec{Q}_y, \gamma_2)
\end{aligned}$$

7.7. GAUSSIAN BASIS FUNCTIONS

$$\times \sum_{n=0}^{n_1+n_2} \sum_{t=0}^{[n/2]} \sum_{k=0}^{[(n-2t)/2]} \sum_{n'=0}^{n_3+n_4} \sum_{t'=0}^{[n'/2]}$$
$$B_{n,n',t_1,t_2,k}(n_1, n_2, \vec{A}_z, \vec{B}_z, \vec{P}_z, \gamma_1; n_3, n_4, \vec{C}_z, \vec{D}_z, \vec{Q}_z, \gamma_2)$$
$$\times F_\nu(\vec{p}^2/4\delta) \qquad (7.1)$$

where

$$\nu = \ell + \ell' + m + m' + n + n' - 2(r + r' + s + s' + t + t') - (i + j + k)$$
$$\delta = \frac{1}{4\gamma_1} + \frac{1}{4\gamma_2}$$

and the "Gaussian product factor" Ω, involving the original factors K_1 and K_2, is given by

$$\Omega = \frac{2\pi^2}{\gamma_1 \gamma_2} \left(\frac{\pi}{\gamma_1 + \gamma_2}\right)^{1/2} \exp\left(-\frac{\alpha_1 \alpha_2 \vec{AB}^2}{\gamma_1} - \frac{\alpha_3 \alpha_4 \vec{CD}^2}{\gamma_2}\right)$$

One thing is clear from this rather awesome formula; the electron-repulsion integral is simply a weighted sum of the integrals $F_\nu(x)$. The coefficients which multiply the $F_\nu(x)$ involve

1. The powers of x, y, and z in the Cartesian factors of each of the GTFs.
2. The exponents of each of the GTFs.
3. The components of the position vectors of each GTF.

The B terms can be simplified by the definition of

$$\theta(\ell, \ell_1, \ell_2, a, b, r, \gamma) = f_\ell(\ell_1, \ell_2, a, b) \frac{\ell! \gamma^{r-\ell}}{r!(\ell - 2r)!}$$

Then the B involving the "x-components" is

$$B_{\ell,\ell',r_1,r_2,i}(\ell_1, \ell_2, \vec{A}_x, \vec{B}_x, \vec{P}_x, \gamma_1; \ell_3, \ell_4, \vec{C}_x, \vec{D}_x, \vec{Q}_x, \gamma_2)$$
$$= (-1)^{\ell'} \theta(\ell, \ell_1, \ell_2, \vec{PA}_x, \vec{PB}_x, r, \gamma_1) \theta(\ell', \ell_3, \ell_4, \vec{QC}_x, \vec{QD}_x, r', \gamma_2)$$
$$\times \frac{(-1)^i (2\delta)^{2(r+r')}(\ell + \ell' - 2r - 2r')! \delta^i \vec{p}_x^{\ell+\ell'-2(r+r'+i)}}{(4\delta)^{\ell+\ell'} i! [\ell + \ell' - 2(r + r' + i)]!}$$

With completely analogous expressions for the other two B factors.

Thus, with some effort, the electron-repulsion integrals involving GTFs can be obtained in closed form; again if the functions $F_\nu(x)$ can be computed.

7.7.5 Summary for GTFs

This completes our sketches of the derivations of the molecular integrals involving the GTF basis. Unlike the case of an STO basis, *all* the overlap and energy integrals can be obtained in a closed form giving the GTF basis an overwhelming *practical* superiority over the STO basis.

In the next chapter we must address the daunting problem of the computational implementation of these expressions. At present, the expressions have been derived in their full generality; in particular, the GTFs may have any Cartesian factor

$$x_A^\ell y_A^m z_A^n$$

for arbitrary (positive integer or zero) values of ℓ, m, and n. In practice, of course, we shall be concerned with values of ℓ, m, and n typically 0, 1 and 2. This means that many of the summations in the general formulae range from 0 to 0 or 1; that is, the summations may well be just one or two terms. Also, many of the factorials and double factorials will be 0!, 1!, 0!! or 1!! so that, for many practical applications the "overheads" involved in setting up the summations may be prohibitive. All these kind of considerations must be taken into account when implementing these lengthy formulae.

7.7.6 More efficient methods

The complexity and numbers of repulsion integrals in any electronic structure calculation places them at the centre of any consideration of *efficiency* in an implementation. There are now many techniques available for the rapid evaluation of these integrals; no-one would use the eqn (7.1) as it stands for the *routine* calculation of repulsion integrals. Broadly speaking, there are two types of technique which are in routine use:

- GTFs may be grouped together in "shells"[13] In particular, a shell of GTFs may be chosen to have common primitive exponents and may be centred on a common centre so that the geometrical and other factors in eqn (7.1) are the same or closely related and are calculated only once. This method, together with the use of a local coordinate system to maximise the number of integrals which are zero by local symmetry has proved very powerful. The small penalty to be paid for efficiency here is the *constraint* that the expansion of a whole shell of GTFs is done by the same primitives, involving a small loss of freedom in the expansion. This method was pioneered by J. A. Pople and W. J. Hehre (*Journal of Computational Physics*, **27**, 161, 1978) and is a prominent feature of the GAUSSIAN system of programs.

---

[13] A concept closely related to but quite distinct from the idea of a "shell" of electrons in a many-electron structure.

7.8. THE CONTRACTION TECHNIQUE

- Notwithstanding the apparent complexity of eqn (7.1), it is nothing more than a *finite* linear combination of values of the function $F_\nu(t)$:

$$G = \sum_{\nu=0}^{L} C_\nu(x) F_\nu(x)$$

where $x = \vec{p}^{\,2}/4\delta$ and $C_\nu(x)$ is simply a compact symbol for the linear combination of B-factors in eqn (7.1) which themselves depend on x; L, the upper limit of the summation, is fixed by the actual GTFs in eqn (7.1). Of course

$$F_\nu(x) = \int_0^1 t^{2\nu} \exp(-xt^2) dt$$

Which means that G is a one-dimensional integral of a product of a polynomial in even powers of x with coefficients C_ν:

$$G = \int_0^1 P_L(t) \exp(-xt^2) dt$$
$$P_L(t) = \sum_\nu C_\nu t^{2\nu}$$

Now, if the integral may be written as a finite sum of values of a function in this way, it is capable of being evaluated *exactly* by a weighted sum of $L/2 + 1$ values of the polynomial P_L in the integrand:

$$G = \sum_{i=0}^{L/2+1} P_L(t_i) W_i$$

where the points (t_i) and the associated weight factors (W_i) are obtained by the solutions of a polynomial equation. This is the basic theorem at the heart of the various Gaussian[14] numerical quadrature methods. There are methods for the evaluation of the zeroes of these Rys polynomials; the method is described in "Evaluation of molecular integrals over Gaussian basis functions" by M. Dupuis, J. Rys and H. F. King (*Journal of Chemical Physics* **65**, 111, 1976).

- There are various ways of using interpolation and extrapolation as well as purely programming techniques for speeding up integral calculation. The best way to keep abreast of these methods is to consult the documentation of the GAUSSIAN and other large suites.

7.8 The contraction technique

The central problem with the use of a GTF basis is the "wrong" potential for which they are well adapted. Functions closely-related to the GTF basis

---
[14] Another mathematical method due to the titan Gauss, nothing to do with the GTFs.

252   CHAPTER 7. MOLECULAR INTEGRALS: AN INTRODUCTION

arise "naturally" in the solution of the Schrödinger equation for the Hooke's law potential. This means that, if a satisfactory result is to be obtained for calculations on systems where the potential is Coulomb's law, then many such basis functions will have to be used in the expansion of the orbitals involved. In practice, a factor of at least three over the STO basis has been found.

In one sense there is no way round this problem, the molecular integrals simply *have* to be computed. But recall that the number of electron repulsion integrals is about $m^4/8$ for m basis functions. This means that, if the basis has to be at least $3m$ for GTFs, then the number of repulsion integrals goes up by a factor of $3^4 = 81$. Numbers of this kind place a strain on computing power but, more particularly, on *file storage*. It quickly becomes out of the question to *store* all the integrals arising from the use of a large GTF basis.

It has been found, in practice, that the most convenient compromise between use of many GTF basis functions and finite file storage is the so-called *contraction* technique.

The GTFs themselves are regarded as *primitive* functions which are used to form basis functions which are *linear combinations* of the primitive GTFs:

$$\phi_i = \sum_{k=1}^{K} \eta_k d_{ki}$$

where the d_{ki} are the contraction coefficients involved in forming the basis function ϕ_i from the primitives η_k.

Thus the actual GTFs are "pushed down" by one more level of linear combinations.

In this way the final variationally determined MOs are linear combinations of the GTFs which occur in *fixed* linear combinations as the basis functions.

The name "contraction" for a method of *expanding* the basis functions needs a little explanation, since one would naturally call the contraction coefficients expansion coefficients! The point is simply that the contraction scheme is viewed from the point of view of the *primitives* rather than from that of the basis functions. The *number* of functions which have to be manipulated explicitly in integral storage and retrieval is *reduced* by this technique and so the term "contraction" seems more appropriate.

The next question to consider is: how do we determine the particular sets of linear combinations of GTFs to use to generate a useful set of basis functions? There are two general methods:

1. Use a least-squares fitting procedure to some other, less computationally tractable, basis functions (STOs, perhaps).

7.8. THE CONTRACTION TECHNIQUE

2. Perform atomic SCF calculations with contracted GTFs and *optimise* the contraction coefficients for lowest atomic energy.

Both of these methods are widely used and will be described later; for the moment we will simply take the existence of contracted GTF basis functions as given.

A start will be made on the problem of actually *coding* these formulae in the next chapter. It is, surely, already clear from the complexity of the formulae that we need more powerful tools to keep track of the coding and its documentation than we have been using for the relatively simple SCF codes.

Chapter 8

Molecular integrals: implementation

The computer implementation of the general formulae for the molecular integrals over Gaussian functions presents problems of an entirely new kind. The codes are long and tedious to code and even more demanding to maintain. Codes of this potential complexity must be planned with the appropriate software tools in mind; our previous method of "clear code, comments and a manual page" is simply not adequate here. The WEB *method of structured, documented code is introduced as one solution to these problems.*

Contents

8.1	Introduction		**256**
8.2	Data structures		**256**
8.3	Normalisation		**260**
8.4	Overview; the general structure		**263**
8.5	Complex code management: the WEB system		**269**
	8.5.1	Sample	274
	8.5.2	INDEX	275
8.6	A working WEB		**276**
	8.6.1	The integral $F_\nu(t)$	278
	8.6.2	fmch	280
	8.6.3	INDEX	286
8.7	Some comments on the WEB		**287**
	8.7.1	Cosmetic features of weave	287
8.8	The full integral codes		**288**

CHAPTER 8. MOLECULAR INTEGRALS: IMPLEMENTATION

8.1 Introduction

In this chapter the "simple-minded" implementation of the formulae of the last chapter is planned and the new programming problems are discussed. The approach is simple-minded in the sense that, apart from some very obvious devices, the formulae are to be programmed *as they stand* with no particular regard for speed of execution or savings due to symmetry or attempts to avoid duplication of calculations. The reasons for this are:

- There is always a need for a benchmark programs which can be relied upon to generate, however slowly, the correct values of the molecular integrals for use when writing newer and faster codes.
- Only by coding the full formulae can experience be gained for future use.

The implementation of these formulae present problems which have not yet been encountered in the essentially matrix manipulations of the SCF method:

1. Data structures must be devised to contain a full description of the basis functions and primitives; positions in space, orbital exponents, specification of the Cartesian factors (powers of x, y and z) contraction coefficients etc.

2. Some decisions must be taken about the *flexibility* of the contraction scheme; are we to allow GTFs of *arbitrary* type be combined or restrict the linear combinations to be all of one type?

3. Most important of all, the description of the steps involved in the calculation is more difficult when general mathematical formulae are being implemented (integrals, Greek letters, mathematical symbols etc.).

The first two of these are related and the third one may involve a complete rethink about the tools used to present a documented program.

8.2 Data structures

Since we have taken the decision to use the contraction technique to define our basis functions in terms of primitive GTFs let us incorporate this decision into the design of the implementation.

The basis functions are given by

$$\phi_i = \sum_{k=1}^{K} \eta_k d_{ki}$$

8.2. DATA STRUCTURES

where each of the η_k depends on a number of parameters:

$$\eta_k(\vec{A}_k, \alpha_k, \ell_k, m_k, n_k)$$

The parameters in a typical η_k are:

- $\vec{A}_k$ is the position vector of the centre of η_k (usually a nucleus on which basis functions are centred),
- α_k is the orbital exponent of the primitive η_k,
- ℓ_k, m_k, and n_k are the powers of x, y and z in the local axis system of the Cartesian factor in η_k

and the d_{ki} are the contraction coefficients forming the basis function from the primitive GTFs.

Although there is absolutely no reason why the basis functions should not be allowed to be combinations of primitives of *arbitrary* type, it is more usual to make the Cartesian factor common to all the primitives of a given basis function, so that one can talk about "an s-type basis function" or "a p_x-type basis function" etc. Similarly, convenience usually dictates that all the primitives in a given basis function should share a common origin so that (again usually) one may refer to "a d_{xy}-type basis function *on a given nucleus*". With these restrictions (which prove to be rather mild ones in practice) the relationship between basis functions and primitives may be written more explicitly as:

$$\begin{aligned}\phi_i(\vec{A}_i, \ell_i, m_i, n_i) &= \sum_{k=1}^{K} \eta_k(\vec{A}_k, \alpha_k, \ell_k, m_k, n_k) d_{ki} \\ &= x_{A_i}^{\ell_i} y_{A_i}^{m_i} z_{A_i}^{n_i} \sum_{k=1}^{K} \exp(-\alpha_k r_{A_i}^2)\end{aligned}$$

That is, the contraction technique is, strictly, applied to the exponential factors only and any implementation of the formulae can take advantage of common geometrical terms in the Cartesian factors.

The sum of the powers of x, y and z in the Cartesian factor classifies the basis functions in a way which is *almost* the same as the classification of the real hydrogenic AOs:

1.
$$\ell_i + m_i + n_i = 0$$

defines an s basis function,

2.
$$\ell_i + m_i + n_i = 1$$
defines an p basis function,

3.
$$\ell_i + m_i + n_i = 2$$
defines an d basis function,

4.
$$\ell_i + m_i + n_i = 3$$
defines an f basis function,

etc. There are, however, *more* types of d and f (and higher) basis functions of this "Cartesian monomial" form than the hydrogenic AOs; six d types and ten f types compared to five and seven in the AO case. In other words the geometrical factors in the familiar AOs are linear combinations of the Cartesian monomials. This complication will be dealt with in due course, for the moment attention is focussed on the integral computation.

The data necessary to specify the basis functions are, then:

1. The type of function (the Cartesian factor); this information can be coded by placing the possible types in some standard order and simply numbering the types in that standard order. One such method is

1	x	y	z	x^2	y^2	z^2	xy	xz	yz
1	2	3	4	5	6	7	8	9	10
x^3	y^3	z^3	x^2y	x^2z	xy^2	y^2z	xz^2	yz^2	xyz
11	12	13	14	15	16	17	18	19	20

for the first twenty types which are the most commonly used basis functions (s, p, d and f functions).

2. The position of the centre of the function (the nucleus); easily specified by three Cartesian coordinates in some global system.

3. The number of primitives it contains; say the number of the first and of the last primitive in a list of primitives.

4. The exponents of the primitives; a list of values.

5. The contraction coefficients; a list of values.

8.2. DATA STRUCTURES

6. Slightly less obviously, a normalisation constant for each primitive which only need be computed *once*.

In the implementation discussed here, a slight compromise has been adopted between full generality and the usual contraction method; the coordinates of each *primitive* are stored rather than the coordinates of each basis function.

Looking at the above list it is clear that some form of data structure (provided by the language C, for example) would be the most elegant way of storing this mixture of real numbers and integers. If we are to remain in contact with FORTRAN we must simple define the relevant real and integer arrays, which we do as follows using a suitably modified manual-page format.

STRUCTURES.

Data structures used during the evaluation of molecular integrals. These structures are constant throughout the calculation and will, typically, be formed from a more compact form of input at an early stage in the calculation.

vlist(MAX_CENTRES,4) (vlist(n,kk), kk = 1,3) are the Cartesian coordinates of the nth nucleus of the molecule in the global system.
   vlist(n,4) is the atomic number of the nth nucleus.

eta(MAX_PRIMITIVES,5) (eta(k,kk), kk=1,3) are the Cartesian coordinates of the centre of the kth primitive.
   eta(k,4) is the orbital exponent of this primitive.
   eta(k,5) is the contraction coefficient by which this primitive is multiplied in its appearance in a basis function. This presumes, of course, that the same primitive cannot appear in different basis functions with different coefficients. During the computation this coefficient will be multiplied by the normalisation constant of the kth primitive to give the *total* multiple of that primitive in the basis function in which it appears.

nfirst(MAX_BASIS_FUNCTIONS) nfirst(i) is the number of the first primitive (in the numbering scheme used within the eta array) in the contraction of the ith basis function.

nlast(MAX_BASIS_FUNCTIONS) nlast(i) is the number of the last primitive in the contraction of the ith basis function.

ntype(MAX_BASIS_FUNCTIONS) ntype(i) is the type of the ith basis function using notation of the above table. It is used in conjunction with nr, into which it points, to generate the Cartesian factor of a basis function.

ncntr(MAX_BASIS_FUNCTIONS) ncntr(i) is the nuclear centre on which the ith basis function is based (i.e. ncntr(i) points into the array vlist). This information is clearly also contained in the eta array, but it is so much quicker to be able to simply compare two integers rather than compute a distance from Cartesian coordinates and see if it is zero (to some tolerance), that the redundancy of information is tolerable.

nr(NO_OF_TYPES,3) nr(ntype(i),1) nr(ntype(i),2) and nr(ntype(i),3) are, respectively the powers of x, y and z in the ith basis function (and therefore of all the primitives which it contains); these integers are ℓ, m and n in the usual notation. This is a *fixed* array, independent of the particular basis used.

The information in these structures is sufficient to complete the description of a molecule and a set of contracted basis functions and so the calculation of the molecular integrals is just a question of manipulating these data and the formulae of the previous chapter.

There is the much smaller problem of constructing these structures from a more compact form of input data; this matter can be left to the user's preferences.

8.3 Normalisation

There is much less danger of error in the preparation of data if, in the presentation of sets of contraction coefficients, both the primitives and the basis functions are *normalised*. That is each primitive is of the form

$$\eta = N(\alpha, \ell, m, n) x^\ell y^m z^n \exp(\alpha r^2)$$

where $N(\alpha, \ell, m, n)$ is calculated so that

$$\int \eta^2 dV = 1$$

However, there are savings to be made if the computations are done over *unnormalised* primitives of the type discussed so far. Thus, it is necessary to be clear about which set of contraction coefficients is to be used. We will assume that any provided contraction coefficients are defined for normalised primitives and so need some additional factor to refer to the "raw" primitives.

The normalisation factor $N(\alpha, \ell, m, n)$ depends only on the "local" form of the primitive, not on its position in space so may be computed just once. The

8.3. NORMALISATION

formula for $N(\alpha, \ell, m, n)$ is obtained from the overlap formulae of the previous chapter as a simple special case:

$$N(\alpha, \ell, m, n) = \sqrt{\frac{(2\ell - 1)!!(2m - 1)!!(2n - 1)!!}{\alpha^{\ell+m+n}}} \left(\frac{\pi}{2\alpha}\right)^{3/2}$$

This number may be computed and stored in `eta(i,5)`.

The normalisation of the basis functions is nearly as simple and involves the (one-centre) overlap integral between two primitives of the same type but different exponents:

$$\int \eta(\vec{A}, \alpha, \ell, m, n)\eta(\vec{A}, \beta, \ell, m, n)dV = \alpha\beta \frac{(2\ell - 1)!!(2m - 1)!!(2n - 1)!!}{(2\alpha + 2\beta)^{\ell+m+n}}$$
$$\times \left(\frac{\pi}{\alpha + \beta}\right)^{3/2}$$

With these two formulae we can take a given set of contraction coefficients and generate a new set which define the basis function as a *normalised* linear combination of individually *un-normalised* primitive GTFs. The products of the normalisation factors for the primitives and the normalised contraction coefficients are stored in `eta(i,5)` which are then the most basic contraction coefficients; those with which the primitives

$$x^\ell y^m z^n \exp(\alpha r^2)$$

must be multiplied to form the normalised basis function.

The following fragment of code implements this procedure assuming the above data structures and that `dfact(i+1)` supplies $(2i - 1)!!$ and the original contraction coefficients are in the array `c`.

```
    pitern = 5.568327997d00 # pi**1.5;
#   Normalise the primitives
    for ( j = 1; j <= nbasis; j = j+1) # Loop over Basis functions
       {
# Get information about primitives in current basis function
       jtyp = ntype(j); js = nfirst(j); jf = nlast(j);
       l = nr(jtyp,1); m = nr(jtyp,2); n = nr(jtyp,3);
# Loop over primitives in the current basis function
       for ( i = js; i <=  jf; i = i+1)
           {
           alpha = eta(i,4); S00 = pitern*(half/alpha)**1.5;
           t1 = dfact(l+1)/alpha**l;
           t2 = dfact(m+1)/alpha**m;
           t3 = dfact(n+1)/alpha**n;
           eta(i,5) = one/dsqrt(S00*t1*t2*t3)
           }
       }

#   Now normalise the basis functions
    for ( j = 1; j <= nbasis; j = j + 1)
       {
       jtyp = ntype(j); js = nfirst(j); jf = nlast(j);
       l = nr(jtyp,1); m = nr(jtyp,2); n = nr(jtyp,3);

       sum = zero;
       for ( ii = js; ii <= jf; ii = ii+1)
          {
          for ( jj = js; jj <= jf; jj = jj+1)
              {
              t = one/(eta(ii,4)+eta(jj,4));
              S00 = pitern*(t**onep5)*eta(ii,5)*eta(jj,5);
              t = half*t;
              t1 = dfact(l+1)*t**l;
              t2 = dfact(m+1)*t**m;
              t3 = dfact(n+1)*t**n;
              sum = sum + c(ii)*c(jj)*S00*t1*t2*t3;
              }
          }
       sum = one/sqrt(sum);
       for ( ii = js; ii <= jf; ii = ii+1)
           c(ii) = c(ii)*sum;
       }
    for ( ii = 1; ii <= ngauss; ii = ii+1)
       eta(ii,5) = eta(ii,5)*c(ii);
```

This code is not a complete program segment but the meaning is clear enough. This procedure prepares the structure **eta** for the actual integral generation

8.4 Overview; the general structure

step.

8.4 Overview; the general structure

The general approach to the computation of molecular integrals is quite simple; we:

- set up loops over the basis functions,
- access the data in the data structures above and
- implement the formulae of the last chapter.

However complex the details of the coding, the outline is simple. Let us start from the top and see what is required. A suitable "top-level" code is given below, togther with a manual page.

```
# genint routine to evaluate S, T,V i.e. H
#          and a two-electron integral file
#
    subroutine genint (ngmx,nbfns,eta,ntype,ncntr,nfirst,
                       nlast,vlist,ncmx,noc,S,H,nfile)
    integer ngmx, nbfns, noc, ncmx;
    double precision eta(ngmx,5),vlist(ncmx,4);
    double precision S(ARB),H(ARB);
    integer ntype(ARB),nfirst(ARB),nlast(ARB),ncntr(ARB),nfile;
    {
    integer i, j, k, l, ltop, ij, ji, mu;
    double precision generi,genoei;
    integer pointer, last, ovltot,kintot;
    double precision val,crit;
    integer nr(NO_OF_TYPES,3);
    data nr /
         0,1,0,0,2,0,0,1,1,0,3,0,0,2,2,1,0,1,0,1,
         0,0,1,0,0,2,0,1,0,1,0,3,0,1,0,2,2,0,1,1,
         0,0,0,1,0,0,2,0,1,1,0,0,3,0,1,0,1,2,2,1/;
    data crit/1.0d-08/;

    mu = 0;   #  tag  for use later perhaps

#
#   One-electron integrals
#
    for ( i = 1; i <= nbfns; i = i+1)
      {
```

```
            for ( j = 1; j <= i; j = j+1)
              {
              ij = (j-1)*nbfns+i;   ji = (i-1)*nbfns+j;
              H(ij) = genoei (i,j,eta,ngmx,nfirst,nlast,ntype,
                              nr,NO_OF_TYPES,vlist,noc,ncmx,ovltot,kintot);
              H(ji) = H(ij);
              S(ij) = ovltot; S(ji) = ovltot;
              }
        }
#
# H now contains T + V

    rewind nfile;  pointer = 0;         # initialisation for putint
    last = NO;                          #   ditto
    i = 1; j = 1; k = 1; l = 0;         # initialise next_label
    while(next_label(i,j,k,l,nbfns) == YES)
        {
        if ( l == nbfns ) last = YES;        # last integral
        val = generi (i,j,k,l,0,eta,ngmx,nfirst,nlast,
                      ntype,nr,NO_OF_TYPES);
        if(abs(val) < crit ) next;   # this assumes that the
                                     # last integral is never zero
        call putint(nfile,i,j,k,l,mu,val,pointer,last);
        }
#   all done
    return;
    }
    end
```

8.4. OVERVIEW; THE GENERAL STRUCTURE

NAME genint
Routine to drive the calculation of molecular integrals.

SYNOPSIS

```
subroutine genint (ngmx,nbfns,eta,ntype,ncntr,nfirst,
    nlast,vlist,ncmx,noc,S,H,nfile)
integer ngmx, nbfns, noc, ncmx
double precision eta(ngmx,5),vlist(ncmx,4)
double precision S(ARB),H(ARB)
integer ntype(ARB),nfirst(ARB),nlast(ARB)
integer ncntr(ARB),nfile
```

DESCRIPTION
This routine takes the data describing the basis functions and the molecular geometry and sets up the logic to calculate all the molecular integrals needed for a molecular wavefunction calculation.

ARGUMENTS
Consult the **STRUCTURES** information for a detailed description of the data structures used by the integral generation routines; only arguments specific to this routine are described here.

ngmx Input: The explicit first dimension of the array `eta` in the program which calls `genint`.

nbfns Input: The number of basis functions.

ncmx Input: The explicit first dimension of `vlist` in the calling program.

noc Input: The number of nuclei in the system.

S Output: The matrix of overlap integrals.

H Output: The matrix of one-electron integrals.

nfile Input: The logical channel number of a (sequential) file on which the electron-repulsion integrals are to be placed.

SEE ALSO
STRUCTURES, `genoei`, `generi`

DIAGNOSTICS
None; underflow messages may appear, they are harmless.

There are a few points to note in this segment:

- All the "real work" has been pushed down out of sight into two **functions**:
    1. **genoei** which is given the task of calculating the overlap, kinetic energy and nuclear-attraction integrals, given the information about two basis functions and
    2. **generi** which is required to calculate an electron-repulsion integral, given the data for four basis functions.
- The task of deciding what to do with the electron-repulsion integrals once they are computed has been passed on to **putint** (which is, of course, a companion to **getint**, introduced in the implementation of the SCF method). **putint** has to be given the labels and value of an integral and the information whether this integral is the last one or not. Given this information **putint** stores the integrals on **nfile** and remembers where they are. Full details of **putint** are given in the next chapter on the storage of molecular integrals.
- A small point; integrals are not even offered to **putint** if they are less than **crit**; they do not appear on the final file.

The specification of the two working **functions**; **genoei** and **generi** is given in the relevant manual pages below. Of course, meeting these specifications involves coding the formulae derived in Chapter 7.

8.4. OVERVIEW; THE GENERAL STRUCTURE

NAME genoei
   One electron integral function

SYNOPSIS

```
double precision function genoei (i,j,eta,ngmx,nfirst,
   nlast,ntype,nr,ntmx,vlist,noc,ncmx,ovltot,kintot)
double precision eta(ngmx,5), vlist(ncmx,4)
double precision ovltot, kintot
integer i, j, ngmx, ntmx, ncmx, noc
integer nfirst(ARB), nlast(ARB), ntype(ARB), nr(ntmx, 3)
```

DESCRIPTION
   genoei generates the overlap, kinetic energy, and total one- electron Hamiltonian integrals for two basis functions.

ARGUMENTS
   Consult the **STRUCTURES** information for a detailed description of the data structures used by the integral generation routines; only arguments specific to this routine are described here.

   i,j Input: the labels of the one-electron integrals to be computed. i and j label *basis functions* (not primitives)

   ngmx Input: the dimension of eta, the maximum number of primitive GTFs

   ntmx Input: the dimension of nr; the maximum number of types. Currently this is 20 allowing up to f-type GTFs.

   noc Input: the number of centres (charged and dummies)

   ncmx Input: the dimension of vlist; the maximum number of centres.

   ovltot Output: the output overlap integral between basis functions i and j

   kintot Output: the kinetic energy integral between basis functions i and j. (The one-electron Hamiltonian integral; kinetic energy plus nuclear attraction is returned in genoei.)

SEE ALSO
   genints, STRUCTURES

DIAGNOSTICS
   None; but floating point underflows may occur which are harmless.

NAME generi
general two electron integral function

SYNOPSIS

```
double precision function generi(i,j,k,l,xyorz,
    eta,ngmx,nfirst, nlast,ntype,nr,ntmx)
integer i, j, k, l, xyorz, ngmx, ntmx
integer nfirst(ARB), nlast(ARB), ntype(ARB),
    nr(ntmx, 3)
double precision eta(ngmx,5)
```

ARGUMENTS
Consult the **STRUCTURES** entry in the manual for a detailed description of the data structures used by the integral generation routines; only arguments specific to this routine are described here.

i,j,k,l Input: the labels of the two-electron integral to be computed.
i,j,k,l label basis functions (not primitives)

xyorz Input: for ordinary repulsion integrals this is zero, for the three possible spin-orbit coupling integrals this is 1, 2 or 3 for the x, y or z component.

ngmx Input: the dimension of **eta**, the maximum number of primitive GTFs

ntmx Input: the dimension of **nr**; the maximum number of types. Currently this is 20 allowing up to f-type GTFs.

SEE ALSO
genints, STRUCTURES

DIAGNOSTICS
None; but floating point underflows may occur which are harmless.

It is only necessary to glance back at the complexity of the formulae for the nuclear-attraction and electron-repulsion integrals to be reminded of the likely complexity of the codes which implement these formulae. If we proceed to simply code these formulae as they stand, with no changes in our methods, the programs will quickly become illegible and impossible to read, understand or maintain. Both from the point of view of ease of maintenance and to maintain interest in this basically boring task we need to make a qualitative change in our approach to coding and documentation.

Just as in the case of the Revision Control System for version control, there is a public domain system available which is the best that there is; the **WEB** system of program development.

8.5 Complex code management: the WEB system

The implementations which have been presented so far have been codes which, with a few interspersed comments, can be made to be almost self-evident; the codes do not require much explanation and this explanation has been just a few remarks. In the case of the implementation of the formulae for the molecular integrals we are faced with problems of quite a different order. There are two *kinds* of problem:

1. The procedures are necessarily rather long and contain much coding that is tedious and, however necessary to the implementation, does tend to get in the way of the general flow of the program.

2. In order to have a comprehensible code, it is highly desirable to have the capability of annotating the code with *mathematical* comments. Unlike the earlier implementations, simple indications of the form that the implementation takes is not adequate.

The first of these problems has been met before but in a quite different context; it has proved possible in the past to break up a long implementation into smaller procedures which each perform a quite specific task and have a definite interface. However, although this is possible to some extent in the implementation of the molecular integral formulae, it is often quite artificial to write procedures which do not perform a well-defined task and the names of these procedures are simply *mnemonics* for parts of what is essentially a monolithic program.

For example, to extract the data to be used for the computation of a particular electron-repulsion integral from the data structures defined earlier would require a very special procedure which might have as many as a dozen arguments and would require its own manual page which would never be read because it would never be used outside that particular context; such a procedure would not have the natural cohesiveness of, for example, a matrix multiplication routine.

In short, we require the ability to write the *logical outline structure* of the program on (at most) two pages (so that it can all be seen at once) and be able to supply the *details* in the form of blocks of coding[1] as required, in much the same way as one would with an implementation which breaks up naturally into independent tasks which can be implemented as procedures (like **integer function scf**).

---
[1] Which may be separate procedures, of course.

The second problem is easily illustrated by reference to the material in this book.

The theory of the various codes generated in this work has been developed and the codes themselves are embedded in the text of the book which exists both as a physical (paper) object and as a series of disk files. If it is required to use the codes, one must use a text editor (say) to remove them from these files, compile, link and run them. In a similar way, if the theory of a particular implementation or a manual page or some item of descriptive text is required, it must be extracted from the files with a text editor and processed by LaTeX. It is, of course, very easy to write a program to perform these tasks: to arrange for the automatic searching of the text files for a particular segment of code and extract it for use.

> Thus, the whole set of files of which this text is composed may be viewed from another angle; perhaps they are nothing more than a very comprehensively annotated set of programs and the techniques one uses to extract either the "programs" or the "documentation" are simply various preprocessors used to facilitate the use of this annotated program.

In this view the whole book and the programs contained in it are logically *identical*; the text is just the complete annotated program for human use and the extracted codes are just those parts of the programs which are required for machine compilation.

Basically, this is the point of view (a little exaggerated, perhaps) of the WEB system of programming invented by Knuth[2] for the generation of structured and documented programs in the Pascal language. Thanks to the efforts of Silvio Levy and John Krommes, the WEB system is available to FORTRAN, C and Ratfor programmers since 1990.

The WEB system uses one or more programming languages (FORTRAN, Ratfor or C) together with the LaTeX typesetting system to enable the generation of structured programs with documentation of arbitrary complexity and comprehensiveness. The basic ideas are those outlined above which are necessary for the integral implementations:

- The facility to break down a large implementation into **modules** which can be arranged to display the functionality of the implementation in an optimum way. Modules are simply **named pieces of coding;** they may be self-contained procedures (`subroutines` and `functions`) or, more usually, groups of related statements in the chosen language which perform a specific task but are not of sufficient generality to be made into a separate, named, procedure.

---

[2]see, for example, "Literate Programming" by Donald E. Knuth, *The Computer Journal*, **27:2**, (1984) 97-111.

8.5. COMPLEX CODE MANAGEMENT: THE WEB SYSTEM

- Macro processing capability.

- The ability to give as full a documentation as desired using the full power of LaTeX.

- The generation of an index of modules and of all symbols used in the coding.

In practice, one prepares a file (**application.web**, say) which contains the sources for the coding, structure and documentation and processes this file with two tools (**weave** and **tangle**).

The first of which (**weave**) produces a file (**application.tex**, say) which can be processed by the LaTeX typesetting system [3] to generate a high-quality "listing" of the application which displays the structure and documentation together with the index and cross-references. In view of what has been said, this "listing" is much more than a simple "pretty printing" of the code. The second tool (**tangle**) generates a file (**application.f**, say) which contains the material to be compiled.

By far the best way to give the flavour of the WEB system is to give an example of the output of **weave** and LaTeX. But in order to get the most out of such an example a few notes on terminology are in order.

- **Modules** are given names and these names are displayed in the listing between pointed brackets like

$$<\text{Declarations}>$$
$$<\text{Special Case}>$$

or even

$$<\text{What to do when x is zero}>$$

weave supplies these module identifiers with an additional number which is used in the cross-referencing.

The "contents" of the modules are placed in the listing, typically after they have been referred to by name and are introduced by a self-explanatory sequence like:

$$<\text{Declarations}> =$$

or

---
[3] TeX is itself written in WEB, of course.

<What to do when x is zero> =

followed by the relevant code in the chosen programming language. The names of the modules are, therefore, mnemonics for blocks of code.

Thus, the general appearance of a procedure (or other major structure in the program) in the file will, typically, be a logical "skeleton" which has calls to other procedures and the names of modules in the order in which they are used. tangle makes sure that, when the actual compilable program is produced, these module names are replaced by the pieces of code comprising those modules.

- weave is not a compiler, it does not know the *rules* of the programming languages, but it can recognise the keywords of all the languages which it can process (things like double precision, for and exp etc. in FORTRAN or Ratfor) and prints them in bold type.

- weave has its own ideas about how code should be formatted but these are easily over-ruled by the user.

Leaving aside the macro-processing capability and other features the following artificial example gives the flavour of the system; it is hoped that it is self-explanatory!

Here is the actual WEB code:

```
\def\title{--- EXAMPLE ---}

@r

@* SAMPLE. This is a silly little program to illustrate WEB.
Notice the overall structure; the main named module
(|integer function sample|) is a skeleton which simply
shows the overall plan of the code. The short individual
modules of actual code are referred to (by name) and
can be referred to where they occur. The coding is broken
down in a modular way similar to the use of many small
procedures.

@a
/* Comments are ''C-style'' */

    integer function sample(theta);

    {

    @<Declarations@>
```

8.5. COMPLEX CODE MANAGEMENT: THE WEB SYSTEM

```
      if ( theta < zero )

         @<Do This@>    /* Use one module of code if x is negative  */

      else

         @<Do That@>    /* Use another module if x is not negative  */

      return(i);

      }
```

@ Code modules can be defined anywhere, even after they are used.
It is usual, in fact, to define them after they are used in the
same manner as |subroutines| and |functions| in f77.

```
@<Declarations@>=
   integer i;
   real theta, zero;
   data zero/0.0/;
```

@ If |theta| is negative, compute $2 cos(\theta)$ and truncate
it (note that \LaTeX\ enables one to use θ rather
than |theta| and we have the full power of \LaTeX\ for comments.
Normally, the module names would be more sensible than these
and would be a shorthand description of the task which they
perform.

```
   @<Do This@>=
   i = int (2.0*cos(theta));
```

@ If |theta| is not negative, return zero.

```
   @<Do That@>=
   i = 0;
```

@* INDEX.

Here is the output from **weave** and LaTeX:

8.5.1 Sample

This is a silly little program to illustrate WEB. Notice the overall structure; the main named module (**integer function** *sample*) is a skeleton which simply shows the overall plan of the code. The short individual modules of actual code are referred to (by name) and can be referred to where they occur. The coding is broken down in a modular way similar to the use of many small procedures.

"example.f" 8.5.1 ≡
    /∗ Comments are "C-style" ∗/
  integer function *sample*(*theta*);
  {
  ⟨ Declarations 8.5.1.1 ⟩
  if (*theta* < *zero*)
  ⟨ Do This 8.5.1.2 ⟩   /∗ Use one module of code if x is negative ∗/
  else
  ⟨ Do That 8.5.1.3 ⟩   /∗ Use another module if x is not negative ∗/
  return (*i*);
  }

Code modules can be defined anywhere, even after they are used. It is usual, in fact, to define them after they are used in the same manner as *subroutines* and *functions* in f77.

⟨ Declarations 8.5.1.1 ⟩ ≡
  integer *i*;
  real *theta*, *zero*;
  data *zero*/0.0/;

This code is used in section 8.5.1.

If *theta* is negative, compute $2cos(\theta)$ and truncate it (note that LaTeX enables one to use θ rather than *theta* and we have the full power of LaTeX for comments. Normally, the module names would be more sensible than these and would be a shorthand description of the task which they perform.

⟨ Do This 8.5.1.2 ⟩ ≡
  $i = \textbf{\textit{int}}(2.0 * \textbf{\textit{cos}}(theta))$;

This code is used in section 8.5.1.

- providing the full power of the most advanced document-preparation system available for "comments".

The demands of discipline are still present but those demands are concentrated on the maintenance of *one* file and there is always the aesthetic pleasure in developing something which looks better and better all the time instead of getting messier and messier.

Perhaps the best introduction to the system is to give a simple example which illustrates some of the most obvious facilities. We can choose to start with a "real" example and illustrate the code for the integral central to the evaluation of GTF nuclear-attraction and electron-repulsion integrals; $F_\nu(t)$.

The WEB source file is relegated to Appendix 8.A to this chapter, the output from running **weave** on the WEB file followed by LaTeX on **weave**'s output is listed below in the hope that it requires no further justification; if successful, the text below will both explain the theory of the calculation of these integrals and document the code to do the job, while simultaneously convincing the reader of the value of the WEB system.

For our purposes, the use of WEB is simply to use the output from **weave** to illustrate the methods used to code the molecular integrals. A full description of the use of WEB would take us too far out of our way since the facilities available are quite sophisticated and, of course, presume that the user is familiar with the LaTeX document-preparation system. This is one of the main advantages of WEB; it provides a very "programming-language-insensitive" form of code which is legible to humans who may have little or no programming experience; it is rather like the typical experience of a (latin-based!) foreign language, it is much easier to understand than it is to generate.[5]

8.6.1 The integral $F_\nu(t)$

The following pages are an example of WEB applied to the coding of the **double precision function fmch**, the source for the WEB is given as Appendix 8.A. It is, perhaps, a little over-explicit in documentation. The central idea behind the actual code is to present the **function** itself as just bare bones:

- Interface
- Declarations
- Decision between two expansion methods depending on the size of one argument.

---

[5]The manuals, primers, information systems, "Frequently-Asked-Question" lists (FAQs) for WEBs are all freely available, sources are listed elsewhere.

8.6. A WORKING WEB

- The discipline required to keep the actual code of a program "in step" with what the documentation says the program will do is very demanding particularly if, as is usually the case, the person(s) generating and maintaining the program are not full-time professional programmers but are scientific research workers. There is no doubt that a working environment in which scientific workers generate, share and maintain their own programs is much more rewarding to work in than one in which the "theory" and "application" are separated and, in these environments, the assumption has been that one should be able to understand and modify the programs that one works with.

- The tools available in "ordinary" programming languages for scientific numerical work are simply not adequate to enable satisfactory documentation ("comments") to be included in the code. For example, the standard ASCII character set is simply not capable of describing the jobs which mathematical software do. Compare

```
C Compute F sub nu (x) =
C          integral from 0 to 1 t**(2*nu)exp(-x*t**2) dt
```

to
```
/* Compute
```
$$F_\nu(x) = \int_0^1 t^{2\nu} \exp(-xt^2) dt$$
```
*/
```
We want to be able to say θ_j and Λ^n not `theta(j)` and `LAMBDA**n` in the comments and to be able to use all the standard notations for Laplace transforms ($\mathcal{L}[f(t)]$) and spherical harmonics ($Y_\ell^m(\theta, \varphi)$, for example, and not to be confined to the pidgin-notation provided by ASCII.

The WEB system of programming makes a radical attack on these problems[4] by (among many other things):

- abolishing the distinction between the program and its documentation and

---

[4]These ideas were clearly anticipated by Frederick Brooks as early as 1975 in his account of a large software project (The mythical man-month, Addison-Wesley, 1975): "A basic principle of data processing teaches the folly of trying to maintain independent files in synchronism. It is far better to combine them into one file with each record containing all the information both files held concerning a given key.

Yet our practice in programming documentation violates our own teaching. We typically attempt to maintain a machine-readable form of a program and an independent set of human-readable documentation, consisting of prose and flow charts...

The solution, I think, is to merge the files, to incorporate the documentation in the source program... The time has come to devise radically new approaches and methods for program documentation."

The modules in this example are trivially small; the executable ones are one statement each which might just as well have been placed in the body of the program. But they *can* be arbitrarily large and complex codes. The reason for grouping a particular set of statements together as a module is always to structure the program and to separate the essential from the inessential in the display of the main program components.

The *essence* of the simple `function sample` is to do one thing if $\theta < 0$ and something else if not. What is done in both cases may be safely relegated to a lower level of the overall structure and, depending on what is to be done, the code may be itself another procedure containing other modules or simply, as here, a module of code not grand enough to be given a separate existence and interface.

Notice that, in viewing the WEB source and the **weave** output as *the* program, we take a particular point of view towards the output; of **tangle** these files should, ideally, *never be looked at* by the programmer. They are an intermediate between the programmer and the final working program; only as a last resort should debugging occur at this primitive level.

The point which was emphasised when the version control system RCS was described —one must **never** keep duplicate copies of source files— applies with even more force when using the WEB system.

Only the WEB source file must have a permanent existence in filestore; the files generated by weave and tangle (LATEX and FORTRAN, Ratfor or C) must be discarded once they have been processed.

The confusions possible here are enormous if discipline is not used; one could easily come to have **weave** and **tangle** output which were not "in step".

In fact, the ideal approach is to program in the WEB system and use RCS for version control *of the* WEB *source files* in the same way that RCS was used earlier for version control of the ratfor files.

Let us now turn to the use of WEB to think about meeting our specifications for the integral-generation codes.

8.6 A working WEB

As we have seen, there are several well-known problems associated with "scientific programming" and all of them have in common the difficulties of documentation and maintenance of programs which can become very large:

8.5. COMPLEX CODE MANAGEMENT: THE WEB SYSTEM

If *theta* is not negative, return zero.

⟨ Do That 8.5.1.3 ⟩ ≡
   $i = 0$;

This code is used in section 8.5.1.

8.5.2 INDEX

cos: 8.5.1.2.

functions: 8.5.1.1.

i: <u>8.5.1.1</u>.

int: 8.5.1.2.

sample: <u>8.5.1</u>.

subroutines: 8.5.1.1.

theta: 8.5.1, <u>8.5.1.1</u>, 8.5.1.2, 8.5.1.3.

zero: 8.5.1, <u>8.5.1.1</u>.

8.6. A WORKING WEB

Then each of these "modules" is presented as a self-contained piece of code with appropriate comments.

The modules may be anywhere in the file, it is one of the tasks of **tangle** to assemble them into the correct place in the machine-readable code in much the same way as the linker sorts out inter-procedure references when generating an executable from a set of compiled procedures. Conventionally, one refers to the module before it is listed since "top-down" design requires that the *structure* be visible before its detailed implementation.

The implementation of **fmch** is just one major module, the function itself, with the other modules being accessed explicitly simply by module name. As we said above, these module names are always of the form

```
<Text describing the function of a module>
```

while the body of the module itself is identified, elsewhere in the listing, by

```
<Text describing the function of a module> =
    code statements
    .
    .
```

in a manner reminiscent of, but more general than, the coding of **functions** and **subroutines** in Fortran. Of course a module may be a **subroutine**, a part of a **subroutine** or, indeed, a part of several **subroutines**. The problems of finding one's way around a WEB is deferred until a particular one has been examined.

In the original descriptions of WEB, modules were said to be optimum if they were around a dozen statements but, in numerical work, there is often much self-evident, repetitive coding and this rule is a little too strict.

Here is the output from **weave** after being processed by LaTeX.

> It must be emphasised that this output is simply the following subsections showing how simply the idea of a piece of software as a "book" can be implemented with the WEB system.[6] The output from **tangle** on the same source file is not listed and, in any case, is a scarcely legible set of FORTRAN statements containing comments to link it to the human-legible form given below. The only function of this file is to be a compiler-legible *equivalent* of the output of weave/LaTeX.

---

[6] Strictly speaking, this example shows the power of John Krommes' FWEB system. The very fact that the system has the capability to become completely continuous with the text by having sequential section numbers which appear in the contents *without* any author intervention is just one convenience.

8.6.2 fmch

This code is for the oldest and most general and reliable of the methods of computing

$$F_\nu(x) = \int_0^1 t^{2\nu} \exp(-xt^2) dt$$

One of two possible series expansions is used depending on the value of x.

For $x \leq 10$ (Small x Case) the (potentially) infinite series

$$F_\nu(x) = \frac{1}{2} \exp(-x) \sum_{i=0}^{\infty} \frac{\Gamma(\nu + \frac{1}{2})}{\Gamma(\nu + i + \frac{3}{2})} x^i$$

is used.

The series is truncated when the value of terms falls below 10^{-8}. However, if the series seems to be becoming unreasonably long before this condition is reached (more than 50 terms), the evaluation is stopped and the function aborted with an error message on *ERROR_OUTPUT_UNIT*.

If $x > 10$ (Large x Case) a different series expansion is used:

$$F_\nu(x) = \frac{\Gamma(\nu + \frac{1}{2})}{2x^{(\nu + \frac{1}{2})}} - \frac{1}{2} \exp(-x) \sum_{i=0}^{\infty} \frac{\Gamma(\nu + \frac{1}{2})}{\Gamma(\nu - i + \frac{3}{2})} x^{-i}$$

This series, in fact, diverges but it diverges so slowly that the error obtained in truncating it is always less than the last term in the truncated series. Thus, to obtain a value of the function to the same accuracy as the other series, the expansion is terminated when the last term is less than the same criterion (10^{-8}).

It can be shown that the minimum term is always for i close to $\nu + x$, thus if the terms for this value of i are not below the criterion, the series expansion is abandoned, a message output on *ERROR_OUTPUT_UNIT* and the function aborted.

The third argument, y, is $\exp(-x)$, since it is assumed that this function will only be used *once* to evaluate the function $F_\nu(x)$ for the maximum value of ν required and other values of $F_\nu(x)$ will be obtained by downward recursion:

$$F_{\nu-1}(x) = \frac{\exp(-x) + 2xF_\nu(x)}{2\nu - 1}$$

which also requires the value of $\exp(-x)$ to be available.

8.6. A WORKING WEB

NAME
    fmch

SYNOPSIS

    double precision function fmch(nu,x,y);

    implicit double precision (a-h,o-z);
    double precision x, y;
    integer nu;

DESCRIPTION
    Computes
$$F_\nu(x) = \int_0^1 t^{2\nu} \exp(-xt^2) dt$$
    given ν and x. It is used in the evaluation of GTF nuclear-attraction and electron-repulsion integrals.

ARGUMENTS

    nu Input: The value of ν in the explicit formula above (**integer**).

    x Input: x in the formula (**double precision**).

    y Input: $\exp(-x)$, assumed to be available.

DIAGNOSTICS
    If the relevant series expansions used do not converge to a tolerance of 10^{-8}, an error message is printed on standard output and the computation aborted.

"fmch.f" 8.6.1 ≡

    @m *ERROR_OUTPUT_UNIT* 6

    double precision function *fmch*(*nu*, *x*, *y*);

    implicit double precision ($a - h$, $o - z$);
    double precision x, y;
    integer *nu*;

```
{
    /* First, make the variable declarations; notice that comments, even in
       Fortran, are "C-style" */
    ⟨ Internal Declarations 8.6.1.1 ⟩
    /* Temporary integer and real values for nu = ν */
m = nu;
a = dble(float(m));
    /* Now do the computation of $F_\nu(x)$ */
    if (x ≤ ten)
    {
        /* Use the expansion for the smaller value of x */
        ⟨ Small x Case 8.6.1.2 ⟩
    }
    else
    {
        /* Use the other expansion for the larger value of x */
        ⟨ Large x Case 8.6.1.3 ⟩
    }
```

Here are the declarations and **data** statements which are entirely internal to *fmch*. Nomenclature is fairly obvious except possible *rootpi4* which is $\sqrt{(\pi/4)}$.

⟨ Internal Declarations 8.6.1.1 ⟩ ≡
    double precision *ten*, *half*, *one*, *zero*, *rootpi4*, *xd*, *crit*;
    double precision *term*, *partialsum*;
    integer *m*, *i*, *numberofterms*, *maxone*, *maxtwo*;
    data *zero*, *half*, *one*, *rootpi4*, *ten*/0.0, 0.5, 1.0, 0.88622692, 10.0/;

    /* *crit* is the required accuracy of the series expansion */
    data $crit/1.0 \cdot 10^{-08}/$;

    /* *maxone* is the upper limit of the summation in the small-x
       summation; *maxtwo* is the corresponding limit in the large-x
       summation */
    data *maxone*/50/, *maxtwo*/200/;

This code is used in section 8.6.1.

8.6. A WORKING WEB 283

This is the case of x less than 10 where the series is:

$$F_\nu(x) = \frac{1}{2}\exp(-x)\sum_{i=0}^{\infty}\frac{\Gamma(\nu+\frac{1}{2})}{\Gamma(\nu+i+\frac{3}{2})}x^i$$

The series is summed by using a recursion relationship between succesive terms in the series. The sum is accumulated in *partialsum* and the terms are obtained from previous ones by the ratio x/a with a incremented by 1 for each new *term*.

⟨ Small x Case 8.6.1.2 ⟩ ≡
```
    a = a + half;
    term = one/a;
        /* Remember that a is ν */
    partialsum = term;
    for (i = 2; i ≤ maxone; i = i + 1)
    {
    a = a + one;
    term = term * x/a;
    partialsum = partialsum + term;
        /* Here is the possibility of convergence */
    if (term/partialsum < crit)
        break;
    }
        /* If the upper summation limit is reached, it means that the series has
           failed to converge, so say so and exit. */
    if (i ≡ maxone)
    {
    write (ERROR_OUTPUT_UNIT, 200) ;
200:
    format ('␣i␣>␣50␣in␣fmch') ;
    STOP;
    }
        /* Otherwise return silently */
    return (half * partialsum * y);
```
This code is used in section 8.6.1.

This is the other case for x greater than ten:

$$F_\nu(x) = \frac{\Gamma(\nu + \frac{1}{2})}{2x^{(\nu+\frac{1}{2})}} - \frac{1}{2}\exp(-x)\sum_{i=0}^{\infty}\frac{\Gamma(\nu + \frac{1}{2})}{\Gamma(\nu - i + \frac{3}{2})}x^{-i}$$

again, the terms in the sum are obtained by recursion from the previous ones.

⟨ Large x Case 8.6.1.3 ⟩ ≡

```
b = a + half;
a = a - half;
    /* Once again, remember that a is ν */
xd = one/x;
approx = rootpi4 * dsqrt(xd) * xd^m;
if (m > 0)
   {
   for (i = 1; i ≤ m; i = i + 1)
      {
      b = b - one;
      approx = approx * b;
      }
   }
fimult = half * y * xd;
partialsum = zero;
    /* Take care of special case */
if (fimult ≡ zero)
   return (approx);
    /* Otherwise continue */
fiprop = fimult/approx;
term = one;
partialsum = term;
numberofterms = maxtwo;
for (i = 2; i ≤ numberofterms; i = i + 1)
   {
   term = term * a * xd;
   partialsum = partialsum + term;
       /* See if the criterion is satisfied, return silently if so */
   if (dabs(term * fiprop/partialsum) ≤ crit)
      return (approx - fimult * partialsum);
   a = a - one;       /* or carry on */
   }
    /* If i gets as far as numberofterms then the expansion has failed, print
       a message and quit. */
```

8.6. A WORKING WEB

    write(*ERROR_OUTPUT_UNIT*, 201);
201:
    format('␣␣numberofterms␣reached␣in␣fmch');
    STOP; /∗ no convergence ∗/
    }

This code is used in section 8.6.1.

8.6.3 INDEX

approx: 8.6.1.3.

crit: <u>8.6.1.1</u>, 8.6.1.2, 8.6.1.3.

dabs: 8.6.1.3.

dble: 8.6.1.

dsqrt: 8.6.1.3.

ERROR_OUTPUT_UNIT: 8.6.1, 8.6.1.2, 8.6.1.3.

fimult: 8.6.1.3.

fiprop: 8.6.1.3.

float: 8.6.1.

fmch: <u>8.6.1</u>, 8.6.1.1.

half: <u>8.6.1.1</u>, 8.6.1.2, 8.6.1.3.

i: <u>8.6.1.1</u>.

m: <u>8.6.1.1</u>.

maxone: <u>8.6.1.1</u>, 8.6.1.2.

maxtwo: <u>8.6.1.1</u>, 8.6.1.3.

nu: <u>8.6.1</u>.

numberofterms: <u>8.6.1.1</u>, 8.6.1.3.

one: <u>8.6.1.1</u>, 8.6.1.2, 8.6.1.3.

partialsum: <u>8.6.1.1</u>, 8.6.1.2, 8.6.1.3.

rootpi4: <u>8.6.1.1</u>, 8.6.1.3.

STOP: 8.6.1.2, 8.6.1.3.

ten: 8.6.1, <u>8.6.1.1</u>.

term: <u>8.6.1.1</u>, 8.6.1.2, 8.6.1.3.

x: <u>8.6.1</u>.

xd: <u>8.6.1.1</u>, 8.6.1.3.

y: <u>8.6.1</u>.

zero: <u>8.6.1.1</u>, 8.6.1.3.

8.7 Some comments on the WEB

We said earlier that perhaps this WEB is a little over-documented; it is shown here primarily to show the sort of thing which can be done. The code and its general commentary should be self-explanatory. However, there are some points which were not mentioned in the introduction:

- weave produces a contents page listing the major modules (only one, fmch, in this case but, as we shall see, a more complete WEB would have many major modules; one per separate procedure (function or subroutine).
- weave automatically produces an index[7] of symbols, variable names and procedure names, with the index references pointing to lines within modules and, if relevant, the place where the symbol is defined is underlined.

Altogether, the document provides a very legible and complete listing of the program. It is easy to read both in the coding sense of typeface, symbols and cross-referencing and in the sense of comprehensive documentation. The "manual page" at the start of the document cannot be produced automatically, of course, since this depends on the *meaning* of the implementation; nevertheless LaTeX provides the capability to include this material within the WEB. Normally (a copy of) this manual page would reside in another place along with all other such items.

It is obvious that the index produced by weave will often be far too detailed, listing the occurrences of symbols which are simply used as working variables in addition to those symbols which are important to track; remember, weave does not know what these symbols *mean* to us. It is possible to manipulate this index by asking weave to omit chosen symbols from the index or to forcibly include some symbols which would not normally be included (language keywords, for example). We would not normally want to see where all occurrences of i are in a WEB if it is regularly used as a subscripting variable; nor would a listing of all occurrences of the keyword do be particularly illuminating in scanning a WEB.

8.7.1 Cosmetic features of weave

Although weave does not know the meaning of all the language structures, it does know the meaning of several of the common features so that, for example:

- Constants like 27.3D+01 are recognised and converted to $27.3 \cdot 10^1$.

---
[7] The references in the index are local to the particular WEB, it would be too much to expect weave to remember the whole of the pagination of this book.

288   CHAPTER 8. MOLECULAR INTEGRALS: IMPLEMENTATION

- The logical operators (||, .OR., .LE. etc.) are converted to their standard mathematical form.

- Powers (like x**m) are recognised and converted to x^m.

In general, the ASCII form of programming-language elements is converted to standard mathematical notation. Of course, the output from `tangle` contains the ASCII equivalents for machine consumption.

It is not claimed here that this example is perfect or even particularly admirable but what it *does* do is illustrate the quality of "coding" which can be obtained by the WEB system; its legibility and its relative independence of a detailed knowledge of the particular computer language being used.

8.8 The full integral codes

If we are convinced of the value of the WEB system then one byproduct of the above example is the sheer *size* of the codes generated by the rather simple case addressed above.

It is clearly not practical to give the output from `weave` for the nuclear-attraction and electron-repulsion integrals here since they require many pages of listings which, basically repeat the formulae given in Chapter 7. The WEBs are clear but boring and some compromise is necessary. The whole of the material for the WEB system (source, makefiles, manuals etc. together with the WEBs for the integral generation routines and many of the other materials in the text are available by anonymous `ftp` from a site listed at the end of the text.

Appendix 8.A

Source for the WEB of fmch

Here is the raw source text for the **function fmch** printed in Chapter 8. Without giving the whole of the FWEB manual[8] just a few points to make the text comprehensible:

- The character "@" is used to prefix the main "commands". In particular
    1. **@r** means the language is ratfor.
    2. **@m** introduces a macro (the equivalent of the more explicit **define** in our earlier codes).
    3. **@** means that a new module is being started (terminated by the next **@** or **@***).
    4. **@*** means a major module is being started (one which appears in the Table of Contents generated by **weave**).
    5. **@a** signals the start of the program code in a module.
    6. **@<** and **@>** delineate the names of other (minor) modules.
- Text bracketed by "|" characters in descriptive comments is set in the same font as it will be in the program code.
- All the non-code text is LATEX.

One source of possible confusion is the designation of "comments" in the code. For Unix users, there is always the dilemma, does one use:

- a hash (#) symbol to make all the text until the end of the current line a comment as one does for comments in scripts or

---
[8] Available from the **ftp** site listed.

- bracket the comment between the pair /* ... */ as one does in the C language?

As we have already seen, The convention for `ratfor` is the first but for the WEB system it is the second.

@r

@* fmch. This code is for the oldest and most general
and reliable of the methods of computing
\[
  F_\nu (x) = \int_0^1 t^{2 \nu} \exp (-x t^2) dt
\]
One of two possible series expansions is used depending on the
value of |x|.

For |x <= 10| (Small |x| Case) the (potentially) infinite series
\[
F_\nu (x) = \frac{1}{2} \exp(-x) \sum_{i=0}^{\infty}
   \frac{\Gamma (\nu + \frac{1}{2})}
   {\Gamma (\nu + i + \frac{3}{2}) }
     x^i
\]
is used.

The series is truncated when the value of terms falls below
10^{-8}.
However, if the series seems to be becoming unreasonably
long before this condition is reached (more than 50 terms), the
evaluation is stopped and the function aborted with an
error message
on |ERROR_OUTPUT_UNIT|.

If |x > 10 | (Large |x| Case) a different series
expansion is used:
\[
F_\nu (x) = \frac{\Gamma (\nu + \frac{1}{2})}
         {2x^{(\nu + \frac{1}{2}) }}
- \frac{1}{2} \exp(-x) \sum_{i=0}^{\infty}
   \frac{\Gamma (\nu + \frac{1}{2}) }
   {\Gamma (\nu - i + \frac{3}{2}) }
     x^{-i}
\]
This series, in fact, diverges but it diverges
so slowly that the error
obtained in truncating it is always less than
the last term in the
truncated series. Thus, to obtain a value of the function to the
same accuracy as the other series,
the expansion is terminated when the
last term is less than the same criterion (10^{-8}).

It can be shown that the minimum term is always for
|i| close to $\nu + x$, thus if the terms for
this value of |i| are not below the criterion,
the series expansion is abandoned, a message output

on |ERROR_OUTPUT_UNIT|
and the function aborted.

The third argument, |y|, is $\exp(-x)$, since it
is assumed that this
function will only be used {\it once} to evaluate the function
$F_\nu (x) $ for the maximum value
of ν required and other values of
$F_\nu (x)$ will be obtained by downward recursion:
\[
F_{\nu-1}(x) = \frac{\exp(-x) + 2 x F_\nu (x) }{2 \nu -1 }
\]
which also requires the value of $\exp(-x)$ to be available.
%
\ \\ \ \\
\begin{boxedminipage}{4.5in}
  \ \\
\begin{description}
\item[NAME] \   \\
  fmch

\item[SYNOPSIS] \ \\
  {\tt double precision function fmch(nu,x,y); \\
    \ \\
    implicit double precision (a-h,o-z); \\
    double precision x, y; \\
    integer nu; \\
    }
\item[DESCRIPTION] \ \\
  Computes
\[
F_\nu (x) = \int_0^1 t^{2 \nu} \exp (-x t^2) dt
\]
given ν and x. It is used in the evaluation of GTF
nuclear-attraction and electron-repulsion integrals.
\item[ARGUMENTS] \ \\
\begin{description}
\item[nu] Input: The value of ν in the explicit formula above
  ({\tt integer}).

\item[x] Input: x in the formula ({\tt double precision}).

\item[y] Input: $\exp(-x)$, assumed to be available.

\end{description}

\item[DIAGNOSTICS] \ \\
  If the relevant series expansions used do not converge to a

    tolerance of 10^{-8}, an error message is printed on standard
    output and the computation aborted.
\end{description}
\ \\ \ \\
\end{boxedminipage}
\ \\ \ \\

@a @m ERROR_OUTPUT_UNIT 6

double precision function fmch(nu,x,y);

implicit double precision (a-h,o-z); double precision x, y; integer nu;

{

  /* First, make the variable declarations; notice that comments, even
  in Fortran, are ``C-style'' */

  @<Internal Declarations@>

  /* Temporary integer and real values for |nu =| ν */

  m = nu; a = dble(float(m));

  /* Now do the computation of $F_\nu (x)$ */

  if (x <= ten)
  {

    /* Use the expansion for the smaller value of |x| */

    @<Small x Case@>

  }
  else
  {

    /* Use the other expansion for the larger value of |x| */

    @<Large x Case@>

  }

@ Here are the declarations and |data| statements which are entirely
internal to |fmch|. Nomenclature is fairly obvious except possible
|rootpi4| which is $\sqrt{(\pi/4)}$.

APPENDIX 8.A. SOURCE FOR THE WEB OF FMCH

@<Internal Declarations@> = double precision ten, half, one, zero,
rootpi4, xd, crit; double precision term, partialsum; integer m, i,
numberofterms, maxone, maxtwo; data
zero,half,one,rootpi4,ten/0.0,0.5,1.0, 0.88622692,10.0/;

/* |crit| is the required accuracy of the series expansion */

data crit/1.0e-08/;

/* |maxone| is the upper limit of the summation in the small-x
summation; |maxtwo| is the corresponding limit in the large-x
summation */

data maxone/50/, maxtwo/200/;

@ This is the case of |x| less than 10 where the series is:
\[
F_\nu (x) = \frac{1}{2} \exp(-x) \sum_{i=0}^{\infty} \frac{\Gamma (\nu
  + \frac{1}{2})}{\Gamma (\nu + i + \frac{3}{2}) } x^i
\]

The series is summed by using a recursion relationship between
successive terms in the series. The sum is accumulated in |partialsum|
and the terms are obtained from previous ones by the ratio |x/a| with
|a| incremented by 1 for each new |term|.

@<Small x Case@> = a = a + half; term = one/a;

/* Remember that |a| is ν */

partialsum = term; for (i = 2; i <= maxone; i=i+1) { a = a + one;
  term = term*x/a; partialsum = partialsum + term; /* Here is the
  possibility of convergence */ if (term/partialsum < crit) break; }

/* If the upper summation limit is reached, it means that the series
has failed to converge, so say so and exit. */ if (i == maxone) {
  write(ERROR_OUTPUT_UNIT,200); 200: format (' i > 50 in fmch'); STOP;
  }

/* Otherwise return silently */

return (half*partialsum*y);

@ This is the other case for |x| greater than ten:
\[
F_\nu (x) = \frac{\Gamma (\nu + \frac{1}{2})} {2x^{\nu + \frac{1}{2}}
) }} - \frac{1}{2} \exp(-x) \sum_{i=0}^{\infty} \frac{\Gamma (\nu
  + \frac{1}{2} } {\Gamma (\nu - i + \frac{3}{2}) } x^{-i}
\]

again, the terms in the sum are obtained by recursion from the
previous ones.

@<Large x Case@> =

b = a + half; a = a - half;

/* Once again, remember that |a| is ν */ xd = one/x; approx =
rootpi4*dsqrt(xd)*xd^m ; if (m > 0) { for (i = 1; i <= m; i=i+1) {
    b = b - one; approx = approx*b; } } fimult = half*y*xd; partialsum
= zero;

/* Take care of special case */

if (fimult == zero) return (approx);

/* Otherwise continue */

fiprop = fimult/approx; term = one; partialsum = term; numberofterms =
maxtwo; for (i = 2; i <= numberofterms; i=i+1) { term = term*a*xd;
  partialsum = partialsum + term;

  /* See if the criterion is satisfied, return silently if so */

  if (dabs(term*fiprop/partialsum) <= crit) return (approx -
  fimult*partialsum); a = a - one; /* or carry on */ }

/* If |i| gets as far as |numberofterms| then the expansion has
failed, print a message and quit. */

write(ERROR_OUTPUT_UNIT,201); 201: format (' numberofterms reached in
fmch'); STOP; /* no convergence */ }

@* INDEX.

Chapter 9

Repulsion integral storage

Having seen how to compute the one- and two-electron integrals over a basis of contracted GTFs, we must now address the problems associated with the very large number of repulsion integrals. If these numbers are computed as they are required then there is no storage problem. But if they are computed just once and stored, we must design and implement a suitably flexible algorithm for the storage of real numbers labelled by several (at least four) integer labels.

Contents

9.1	Introduction	297
9.2	A storage algorithm	298
9.3	Implementation: putint	300
	9.3.1 Details of putint	303
9.4	A partner for putint; getint	307
9.5	Conclusion	309

9.1 Introduction

The most common way of dealing with the problem of manipulating the large numbers of electron-repulsion integrals which arise in any calculation of molecular electronic structure is to compute them and *store* them for future use in a file on the computing systems backing store (a disk file). We have earlier introduced the function `getint` whose duty it is to obtain these integrals and make them available for processing. Clearly, in the case of stored repulsion integrals, `getint` must be taught the algorithm involved in the storage and retrieval of repulsion integrals. The testbench `getint`s were strictly for demonstration purposes.

If we have access to a function which *computes* arbitrary electron-repulsion integrals then, with a suitable design, we can develop `putint` (say) which is the partner of `getint` and stores the integrals in a way which is meaningful to `getint`. There is another possibility; we may have access to *pre-existing* files of repulsion integrals computed by someone else's program. In this case it is perfectly possible to write a particular `getint` for each format of file with the same interface to our codes.

9.2 A storage algorithm

If we assume that we are intending to generate a file of repulsion integrals for *general* use (not just for SCF purposes, for example) then we shall need to store all the (non-zero) integrals separately, not particular linear combinations of them. In this case there are two general kinds of possibility:

- Store all integral values *and* their four integer labels

$$i, j, k, \ell \text{ and value}$$

characterising

$$\int dV_1 \int dV_2 \phi_i(\vec{r}_1)\phi_j(\vec{r}_1)\frac{1}{r_{12}}\phi_k(\vec{r}_2)\phi_\ell(\vec{r}_2)$$

- Store just the values in such a way that the values of the four labels may be inferred from the *position* of the value in the file; i.e. store the integral values in a standard *order*.

Each of these two methods has its own advantages and disadvantages:

1. The first method requires more storage since *five* quantities must be saved for each integral.
2. The second method constrains the computation to be carried out in a particular order which may not be the most efficient.
3. If an integral is zero (either exactly or to some tolerance) it may be omitted from the first storage method but not from the second since this would disturb the order of the integrals.

9.2. A STORAGE ALGORITHM

The third of these points tends to mitigate the higher storage requirement of the first method, particularly in view of the fact that the four labels are small integers which may be stored in a small space.

Overall, the universal choice for flexibility is the first method; the storage of the value of an electron-repulsion integral *and* its four labels.

Of course, it is only necessary to store *one* of the (up to) eight integrals with the same value which differ only in allowed permutations of the labels. By convention, the one with "reverse dictionary order" in the labels is stored:

The labels and value of the integral

$$(ij, k\ell)$$

for which

$$i \geq j \quad k \geq \ell \quad (ij) \geq (k\ell)$$

is stored and must be used for all integrals

$$(ij, k\ell) \quad (ji, k\ell) \quad (ij, \ell k) \quad (ji, \ell k)$$
$$(k\ell, ij) \quad (k\ell, ji) \quad (\ell k, ij) \quad (\ell k, ji)$$

With this convention, it is only necessary to decide on a general method of storage for the unique integrals. Obviously, the integral file must consist of a number of *records* each of which is small enough to fit into a main storage buffer in the machine in order that they can be processed (by, e.g. scfGR). The simplest approach is to have the file consist of a number of records of equal length plus a "last" record which will be, in general, of different length from the others. A program processing these records simply needs each record to contain the information:

- How many integrals in this record.
- Is this the last record or are there more to follow?
- The values of the integrals in this record and their labels.

Conversely, a program *generating* the integrals needs to:

- Ensure the labels obey the above convention.

- Compute integrals and store them and their labels in an internal buffer until it is full.

- Output to file the buffer of integral values and labels together with the two other pieces of information; number of integrals and last-record indicator.

The most direct way of illustrating this algorithm is by coding it.

9.3 Implementation: putint

Let us use our newly-coded
subroutine genint(.....)
which calculates electron-repulsion integrals; this routine organises the implementation of the formulae of the last chapter.

It will be recalled that this **subroutine** called **subroutine putint** which was not explained at that time. However, it was clear what an implementation of **putint** would have to do:

- Given the four labels (in the correct reverse-dictionary order) and the value of an electron-repulsion integral plus the information whether this particular integral is the last one in the file or not, **putint** had to arrange to put the integral into a file.

- The details of the structure and updating of the file are all invisible to the calling program and are the exclusive business of **putint** and its partner **getint**.

Here is one possible way of coding the specification of **putint** contained in the following manual page.

9.3. IMPLEMENTATION: PUTINT

```
subroutine putint(nfile,i,j,k,l,mu,val,pointer,last)
implicit double precision (a-h,o-z);
save;  #  Make sure nothing is lost; labels and value are
       #  local to this routine
integer nfile,i,j,k,l,mu,pointer,last;
double precision labels(INT_BLOCK_SIZE),value(INT_BLOCK_SIZE);
double precision val;
data max_pointer/INT_BLOCK_SIZE/,id/0/;
if ( last == ERR) return;
iend = NOT_LAST_BLOCK
if ( pointer == max_pointer )
   {            #  The buffer is full - write it out
   write (nfile) pointer,iend,labels,value;
   pointer = 0;  #   Re-initialise the buffer pointer
   }
pointer = pointer + 1;
call pack(labels(pointer),i,j,k,l,mu,id);
value(pointer) = val;
if ( last == YES )
   {  #  write out the last block
   iend = LAST_BLOCK;
   last = ERR;
   write (nfile) pointer,iend,labels,value;
   }
return;
end
```

NAME putint

Primitive for putting out repulsion integrals to file; converse of `getint`.

SYNOPSIS

```
subroutine putint(nfile,i,j,k,l,mu,val,pointer,last)
integer nfile, i, j, k, l, mu, pointer, last
double precision val
```

DESCRIPTION

During the generation of the repulsion integrals, this routine is called when each integral is completed. It arranges to store the integral and its labels in the file `nfile` by buffered output.

ARGUMENTS

nfile Input: the logical number of a (sequential, unformatted) file which is to receive the repulsion integrals.

It must be REWOUND and `pointer` set to zero before this routine is called for the first time for a particular integral file.

i,j,k,l Input: the labels of the current integral being added to the file

mu Input: a spare argument for possibly labelling the integrals according to some scheme in the future.

val Input: the value of the current integral.

pointer Input/Output: this integer is maintained by `putint` - it points to the next integral position in a file record. It must be set to zero before the routine is first called for a particular file

last Input: this integer is set to NO until the last integral is reached when it must be set to YES

SEE ALSO

getint

DIAGNOSTICS

None, it is the user's responsibility to open `nfile`, set `pointer` to zero and to set `last` correctly to ensure file termination

---

It is easy to see at a glance how `putint`'s job is done; it uses `pointer` to keep

9.3. IMPLEMENTATION: PUTINT

track of how many integrals have been put into the current `labels` and `value` buffers and writes the buffers out when they are full (contain `INT_BLOCK_SIZE` integrals), resetting `pointer`.

From the point of view of the calling program, the trickiest part is to know when the last integral is reached (`last = YES`). In the case of `genint`, this is easy since the last integral is always (mm, mm) (for m basis functions) which, being a one-centre Coulomb integral, is never zero and so will always need to be stored. If the integrals are computed in some non-standard order, as they often are when symmetry is used or the integrals are computed by some fast algorithm, then the occurrence of the last integral is harder to establish. This point will be met later when we introduce the use of molecular symmetry to avoid redundant computations.

9.3.1 Details of putint

The only point in this implementation of `putint` which has not been mentioned is the use of the routine `pack` and some mnemonic symbols not explicitly `defined` but whose meanings are obvious.

It was mentioned above that the labels of an electron-repulsion integral are likely to be small (positive) integers (the maximum label is, of course, m the number of basis functions. It is therefore convenient to use a method of storage which takes advantage of that fact. The traditional method is to store these integers in the space allocated to one `double precision` variable (eight 8-bit bytes on most machines). A working hypothesis might be that the maximum number of basis functions in a calculation will generally[1] be less than 256 and so there is room in such a variable for eight such integers; it is just a matter of arranging a convenient way of "packing" the integers into the `double precision` storage. There are two portable ways of doing this in ratfor (i.e. in FORTRAN):

1. Simply use a power law:

    `label = `$256^3 \times$` i + `$256^2 \times$` j + `$256 \times$` k + l`

    and use the fact that one can recover the integers by taking advantage of integer truncation:

$$i = \mathtt{label}/(256^3)$$
$$j = (\mathtt{label} - 256^3 \times \ i \ )/(256^2)$$
$$\text{etc.}$$

---

[1] This is likely to be superceded!

2. Use the fact that a **double precision** variable may be **equivalenced** to an array of **character*1** variables and manipulate the individual characters.

Neither of these two methods is very elegant and one may make use of system-dependent bit-manipulation routines as an alternative. Here is an example of the second approach and a manual entry.

```
      subroutine pack(a,i,j,k,l,m,n)
      double precision a;
      integer i, j, k, l, m, n;
#
      double precision word;   #   Working storage
      integer id(6);
      CHARACTER*1 chr1(8), chr2(24);
      equivalence (word,chr1(1)), (id(1),chr2(1));
      id(1) = i; id(2) = j; id(3) = k;
      id(4) = l; id(5) = m; id(6) = n;
      do ii = 1,6;
         chr1(ii) = chr2((ii-1)*BYTES_PER_INTEGER + LEAST_BYTE);
      a = word;
      return;
      end
```

9.3. IMPLEMENTATION: PUTINT

NAME pack
Store six integers in a **double precision** location, usage is for repulsion integral labels.

SYNOPSIS

```
subroutine pack(label,i,j,k,l,mu,id)
integer  i, j, k, l, mu, id
double precision label
```

DESCRIPTION
Stores the least significant 8 bits (up to 255) of the six integers i, j, k, l, mu, id in the double precision variable label

ARGUMENTS

label Output: the integers are stored here.

i,j,k,l,mu,id Input: the labels to be packed.

SEE ALSO
unpack

DIAGNOSTICS
None, it is the user's responsibility to ensure that the integers lie in the range 0 to 255.

Of course, one needs a complementary facility in order to regenerate the labels from the **double precision** value; here it is

```
    subroutine unpack(a,i,j,k,l,m,n)
    double precision a;
    integer i, j, k, l, m, n;
#
#
    double precision word;
    integer id(6);
    CHARACTER*1 chr1(8), chr2(24);
    equivalence (word,chr1(1)), (id(1),chr2(1));
    do ii = 1,6;
       id(ii) = 0;
    word = a;
    do ii = 1, 6;
       chr2((ii-1)*BYTES_PER_INTEGER + LEAST_BYTE) = chr1(ii);
    i = id(1); j = id(2); k = id(3);
    l = id(4); m = id(5); n = id(6);
    return;
    end
```

NAME unpack

Recovers 6 integers in a **double precision** location, usage is for repulsion integral labels.

SYNOPSIS

```
        subroutine unpack(label,i,j,k,l,mu,id)
        integer  i, j, k, l, mu, id
        double precision label
```

DESCRIPTION

Recovers the least significant 8 bits (up to 255) of the six integers i, j, k, l, mu, id from the **double precision** variable **label**

ARGUMENTS

label Input: the integers are stored here.

i,j,k,l,mu,id Output: the labels to be unpacked.

SEE ALSO
pack

DIAGNOSTICS

None, it is the user's responsibility to have ensured that the integers lie in the range 0 to 255 in the original call to **pack**

9.4. A PARTNER FOR PUTINT; GETINT

The two symbols BYTES_PER_INTEGER and LEAST_BYTE are likely to be system dependent.

- BYTES_PER_INTEGER might, typically be formed by one of

    ```
    define(BYTES_PER_INTEGER,2)
    define(BYTES_PER_INTEGER,4)
    ```

    or whatever.

- While LEAST_BYTE tells pack which *end* of an integer variable is the least significant; is the integer stored left-to-right or right-to-left? Typically one of

    ```
    define(LEAST_BYTE,1)
    define(LEAST_BYTE,4)
    ```

    would be appropriate if BYTES_PER_INTEGER were 4.

This pair of routines stores and recovers six integers "just in case" we need more than four for a particular integral.

9.4 A partner for putint; getint

The implementations of the SCF method in earlier chapters were "testbenches" since they use precomputed molecular integrals or simple semi-empirical values supplied by very special "one-off" getints. Having now the full capability to *compute* all the molecular integrals for any system and use putint to store them we can now write the general partner to putint which will read a file of repulsion integrals generated by putint and make sense of them. Here it is:

CHAPTER 9. REPULSION INTEGRAL STORAGE

```
#       getint: hand out repulsion integrals one at a time
        integer function getint(file,i,j,k,l,mu,val,pointer)
        integer file, i, j, k, l, mu, pointer;
        double precision val;
        {
        save;           # Make sure that getint remembers where it is!
        integer max_pointer, id, iend;
        double precision  zero;
        double precision labels(INT_BLOCK_SIZE), value(INT_BLOCK_SIZE);
        data max_pointer/0/, iend/NOT_LAST_BLOCK/, zero/0.0d00/;
#
#   file must be rewound before first use of this function
#   and pointer must be set to 0
#
        if (pointer == max_pointer)    # pointer must be initialised to 0
          {
          if ( iend == LAST_BLOCK)
             {
             val = zero; i = 0; j = 0;
             k = 0; l = 0; max_pointer = 0; iend = NOT_LAST_BLOCK
             return(END_OF_FILE);
             }
          read (file) max_pointer, iend, labels, value;
          pointer = 0;
          }
        pointer = pointer + 1;
        call unpack(labels(pointer),i,j,k,l,mu,id); # id is unused
        val = value(pointer);
        return(OK);
        }
        end
```

This function simply reads a block of integrals and passes them out to the calling program one at a time on subsequent calls until that buffer is exhausted and then reads another block etc. If it finds an integral it returns OK, if it hits the end of the integrals it returns END_OF_FILE. This implementation satisfies the specification given in the manual page earlier which simply referred (at that time) to the testbench program for the closed-shell case.

Of course, this *particular* getint is only useable with the putint implemented above. But any program written to use getint will be useable *unchanged* if the getint interface is retained and a new getint coded to meet different circumstances; using a file of repulsion integrals computed by another program, for example.

9.5 Conclusion

We now have all the pieces of code needed to implement all the steps in the calculation of molecular electronic structure by the expansion method:

- Computation of molecular overlap and energy integrals over a basis of contracted GTF primitives.
- Storage and retrieval of the repulsion integrals.
- Calculation of the molecular wavefunction at the single-determinant level (closed-shell or UHF).

There are, of course, many improvements which we might make to the particular implementations and other models of molecular electronic structure to be investigated. But the tools developed so far provide a basis for further study.

Chapter 10

"Virtual orbitals"

For reasons of convenience, the implementation of the LCAO SCF equations has been arranged to compute a set of m molecular orbitals from a basis of m expansion functions even though only $n < m$ are required to form the single-determinant wavefunction. In this chapter we take a look at the remaining $(m - n)$ "orbitals" and try to interpret them.

Contents

10.1	Introduction	**311**
10.2	Virtual orbitals in practice	**312**
	10.2.1 Virtuals for H_2O	312
	10.2.2 Objections	315
10.3	The virtual space in LCAO	**317**
	10.3.1 Occupying the virtuals	317
	10.3.2 Convergence aids	318
	10.3.3 Level shifters and damp factors	319
10.4	Conclusions	**322**

10.1 Introduction

The Hartree–Fock equations are differential equations for the orbitals which span a certain space defined by the optimum single-determinant wavefunction. When we implemented the algebraic approximation for these orbitals (LCAO) we found that, by diagonalising the Hartree–Fock *matrix*, which is the representation of the Hartree–Fock operator in the chosen basis, we were able to generate m molecular orbitals, the lowest n of which were (approximations to) the self-consistent molecular orbitals which span the single-determinant space. No "use" was made of the remaining $m - n$ orbitals during the calculation insofar as the density matrix was constructed from just the n orbitals sought for the optimum single determinant.

It is, however, natural to ask if these orbitals have any meaning; do they have any physical interpretation; for example, are they approximations to any orbitals associated with the system under investigation or are they just an artifact of the use of the LCAO technique? In the case of the orbitals of the single determinant — the "occupied" orbitals — the LCAO expansion is a more or less good approximation to the solutions of the differential Hartree–Fock equation depending on the quality and length of the "AO" expansion. As the quality and size of the basis is improved, the occupied MOs presumably become better and better approximations to these orbitals. What about the "unoccupied" molecular orbitals?

There are two general approaches which we might take to the interpretation of these unoccupied or "virtual" orbitals:

1. An empirical investigation; do the orbitals and their energy eigenvalues converge smoothly to some limiting functions as the occupied orbitals do. If they do (bearing in mind that as the size of the basis increases, so does the *number* of the virtuals), then there might be some hope that they are physically interpretable and a correlation with some properties of the system under investigation can be sought.

2. A more theoretical investigation; what formal properties do these virtual orbitals have and how do they relate to any corresponding properties of the (differential) Hartree–Fock equation.

Needless to say, since we now have the technology to hand, an empirical investigation is the easiest approach; we will try this first, extract some comments from the work and then begin a more theoretical approach.

10.2 Virtual orbitals in practice

10.2.1 Virtuals for H_2O

Using the closed-shell program developed earlier and the integral generation programs it is possible to make an empirical computational investigation of the

10.2. VIRTUAL ORBITALS IN PRACTICE

effect of increasing basis quality on the virtual orbitals of a trial molecule, say H_2O. Some standard basis sets from the literature have been used, starting with the minimal basis used in the testbench calculations (seven basis functions, each comprising a linear combination of three GTFs) and proceeding to a moderately comprehensive basis consisting of a good $1s$ basis function for the oxygen atom to be sure of a good total energy (six GTFs), two sets of $2s$ and $2p$ functions on the oxygen (three GTFs and one GTF) plus a set of $3d$ functions and two $1s$ functions and a set of $2p$ functions on each hydrogen atom.

The details of the basis functions are not so important as the fact that they are in order of increasing quality.

In the table below the highest two occupied MO energies are listed together with the lowest two virtual MOs.

	Molecular orbital energies: H_2O		
	Minimal basis ("STO-3G")	Intermediate basis ("3-21G")	Good basis ("6-31G")
Total Energy	-74.962757	-75.585330	-75.984001
ϵ_7	0.74377	0.36257	0.30019
ϵ_6	0.60691	0.26414	0.20419
ϵ_5	-0.39134	-0.47968	-0.50150
ϵ_4	-0.45320	-0.53781	-0.56800

The first and most obvious thing to note is that the virtual orbital eigenvalues are *all positive*. This immediately makes one suspicious of the physical interpretation of the virtual orbitals; how can "molecular orbitals" which lie in the energy continuum beyond the ionisation energy of the molecule have any physical meaning?

However, both the virtual orbital energies and the occupied orbital energies do seem to be converging to some kind of limit[1] and one could be forgiven for thinking that this indicated some kind of physical interpretation for the virtuals.

The central difference between the occupied orbitals and the virtuals in an LCAO calculation is that the variation theorem has acted on the occupied MOs but not on the virtuals, and this provides a clue to a further investigation of the virtuals. The basis functions in the literature are, of course, optimised for the description of the actual electronic structure of atoms and molecules; they are the result of the application of the variation method either directly or indirectly. Thus, if we include any arbitrary basis function in a calculation it will be used

---
[1] In the particular case of the water molecule used here as an example the convergence is, like that of the total energy, from above as the quality of the basis improves, but this is just coincidence as there is no variation theorem for the *orbital* energies corresponding to that for the *total* energy.

by the calculation if it improves the total energy but not otherwise. That is, arbitrary basis functions will be included in the form of the occupied MOs if this improves the energetic properties of the total wavefunction.

Since the variation principle does not act on the virtual orbitals, it is of interest to see how arbitrary "unsuitable" basis functions are incorporated into the virtual orbitals.

If we take the minimal basis calculation on the water molecule and add a single $1s$ function to the basis centred on a point which is not one of the nuclei, say along the bisector of the H-O-H angle (the reflex angle) and move it further and further away it should be irrelevant to the calculation of the occupied MOs.

The Table below summarise the results of these calculations.

Virtual orbital energies: H_2O with single remote $1s$ function			
R_{O-1s}	ϵ_6	ϵ_7	ϵ_8
2.0	0.01460	0.61573	0.74313
5.0	0.01185	0.60940	0.74360
10.0	0.00942	0.60715	0.74375
20.0	0.00819	0.60691	0.74377
30.0	0.00012	0.60691	0.74377

In this case we do see a surprising result: the more remote the "unphysical" basis function, the lower the energy of the lowest virtual orbital. In fact, at large distances, the lowest virtual *is* this remote basis function with an eigenvalue which becomes lower and lower the more remote it is from the molecule and the more diffuse it is.

The interpretation of these facts is straightforward and is only surprising if we were *expecting* a physical interpretation for the virtuals.

The key to the understanding of what the virtual orbitals represent lies in the fact that they are *positive*, not just in the case studied here but in almost all cases. If basis functions are added which are either remote from the molecule or very diffuse or both, then there appear in the calculation low-lying virtual orbitals which are composed of these unrealistic basis functions. What is clearly happening is the combinations of unrealistic basis functions are "trying" to mimic the behaviour of the continuum functions of positive energy. The more remote and diffuse the functions, the less interaction they have with potential generated by the molecule and the lower their (positive) kinetic energy.[2]

The results of this, admittedly rather limited, empirical investigation suggest the following conjectures:

---

[2] The kinetic energy integral of a basis function is proportional to a power of the exponent which itself determines the size of the function.

10.2. VIRTUAL ORBITALS IN PRACTICE

- The virtual orbitals generated by the solution of the LCAO approximation to the Hartree–Fock equations are indeed an artifact of the LCAO technique and do not have any physical interpretation except as a residue of those features of the basis functions which are not suitable for the description of the single-determinant model of the electronic structure of the molecule.

- The solution of the (differential) Hartree–Fock equations (if this were possible) would yield:
    1. A set of functions which span the space of the optimum single-determinant wavefunction: the occupied MOs.
    2. Continuum functions.

The LCAO approximation with the usual kind of basis functions (i.e. those optimised for the description of the electronic structure of the separate atoms) gives good approximations to the occupied MOs and very bad approximations to the continuum functions.

Physically, one might expect that the solutions of the Hartree–Fock equations of a very electronegative atom or, *a fortiori*, a highly charged cation which, experimentally, would bind an additional electron would have a virtual orbital with a negative eigenvalue. This is certainly true for cations and may well be true for very electronegative molecules or molecules with a high dipole moment[3] but the above conjectures will certainly be the rule rather than the exception for neutral species. The question of the solutions of the Hartree–Fock equations for anions carries its own difficulties, as we shall see in due course.

10.2.2 Objections

The most immediate reaction to the results of the last section might be:

1. But molecules *do have* bound excited states and, presumably, at least some of these excited states may be approximated by single-determinant wavefunctions composed of orbitals, and the highest of these orbitals must contain electrons which are *bound* to the system.
2. That is, is it not the case *experimentally* that molecules have genuine, bound "excited orbitals" and the failure of the Hartree–Fock method to find them is a failure in the model?

---
[3]There is a critical value of a point-dipole which will bind an electron, the value is about 0.64 atomic units.

The answer to this apparent paradox is that, because of electron repulsion, the solutions of the Hartree–Fock equations are not at all like those of the Schrödinger equation; the Hartree–Fock equations are non-linear.

When the Schrödinger equation is solved for any system, the result is a set of pairs of eigenvalues and eigenfunctions which are the quantum mechanical description of *all* the allowed states of the system. So, for example, the fourth eigenvalue and eigenfunction is a description of the third excited state of the system etc. The essential fact which ensures that this is the case (in the context here) is that the Hamiltonian operator $\hat{H}$ of the time-independent Schrödinger equation

$$\hat{H}\Psi_k = E_k \Psi_k$$

is *independent* of the functions Ψ_k.

The *apparent* similarity of the Hartree–Fock equation:

$$\hat{h}^F \psi_i = \epsilon_i \psi_i$$

to the Schrödinger equation perhaps encourages the view that the *orbitals* ψ_i might share this particular property of the solutions Ψ_k of the Schrödinger equation.

But the similarity is only apparent: the operator $\hat{h}^F$ depends on the ψ_i via electron repulsion and so, although there may well be a Hartree–Fock equation for the orbitals comprising *any state* of a particular molecule, this Hartree–Fock equation will be different for different states so that the orbitals comprising a single-determinant function for each state will *depend on that state*.

In short, in the presence of electron repulsion, there is no *single* set of molecular orbitals which solve a single Hartree–Fock equation from which one can construct an optimum single-determinant wavefunction for an arbitrary state of a molecule.

The situation is different in the Hückel model, for example, since no electron repulsion is included and therefore the equations are linear, the orbitals and their energies are independent of the number of electrons in the molecule or its state and therefore, *within the confines of this model* one may construct approximations to any state of the molecule. This is not to say that such functions are realistic, just that the model is consistent with their construction!

We assumed, without proof, very early in this work that, for every turning point in the Hartree–Fock energy functional there is a Hartree–Fock equation which corresponds in some way to some solution of the associated Schrödinger equation; the empirical investigation above just re-emphasises this assumption

10.3. THE VIRTUAL SPACE IN LCAO

and, perhaps, stresses that these Hartree–Fock equations only produce interpretable *occupied* orbitals.

However, these considerations, no matter how reasonable and sound they may seem, do not seem to satisfy the nagging doubt that there must be *something* physical in the forms of the virtual MOs; after all, there has been a whole theory of chemical reactivities and related matters based on the forms of the highest occupied MO (HOMO) and the lowest *unoccupied* MO (LUMO). Theories like this have had a good deal of success and are the basis of much qualitative chemical thinking. The use of the LUMO (or, indeed, any virtual MO) in this physically meaningful way can only have one possible interpretation:

> Using a basis of "AOs" which is adapted to the description of the low-lying states of the constituent atoms of a molecule, the general qualitative and semi-quantitative form of the LUMO is similar to the HOMO of the system in an excited state or to the HOMO of the anion of the system.

That is, if a calculation is performed on the anion or on a low-lying excited state of the molecule, then the new orbital occupied is similar to the LUMO of the ground state of the neutral system but this HOMO will now be *bound*. There are all kinds of provisos which one can make on this conclusion, the most obvious being that it seems to depend on the good luck of original basis choice; the exclusion of diffuse functions on the (economic) grounds that they do not contribute significantly to the low-lying orbitals automatically excludes the possibility of the lowest virtuals being approximations to continuum functions.

What is certainly the case is that qualitative theories based on the form of the LUMO and other virtuals using a minimal basis will not be supported by quantitative calculations using more and more extended bases.

With these empirical conjectures in mind we can turn to a more theoretical examination of the virtual orbitals and their relationship to the implementations of the LCAO method.

10.3 The virtual space in LCAO

10.3.1 Occupying the virtuals

The fact remains that, in the solution of the matrix Hartree–Fock equations, virtual orbitals *are* generated and, whatever we might think or even prove about the physical interpretation of these orbitals, they can be used to form determinants and a sum of determinants is always a better approximation to the true wavefunction than a single determinant if the linear parameters it contains are optimised. Further, these virtual orbitals are *orthogonal* to the occupied orbitals; the whole set, occupied plus virtual, is orthonormal and so they have many convenient properties. In particular, we can write an approximation to the wavefunction of any system for which we have a solution of the LCAO Hartree–Fock equations as:

$$\Psi = \sum_{k=0}^{N} C_k \Phi_k$$

where the Φ_k are determinants of the orbitals which solve the LCAO equations. Typically, one might take Φ_0 as the determinant of occupied orbitals and the other Φ_k as determinants in which one or more of the occupied orbitals are replaced by the virtuals.

Intuitively, because Φ_0 is the *best* single-determinant approximation to Ψ, and the physical interpretation of the virtuals, and therefore determinants containing them, is suspect one might expect in these circumstances that

- $C_0 \approx 1.0$
- C_k very small for $k \geq 1$

and therefore a large number of terms might have to be included in the sum to obtain a significant improvement over the best single-determinant approximation.

This method, known as Configuration Interaction (CI) or superposition of configurations is a widely-used model for the calculation of molecular electronic structures beyond the Hartree–Fock model. We shall look at this model in Chapter 20.

10.3.2 Convergence aids

A formal view of the process of solving the LCAO Hartree–Fock equations is that one takes a set of functions which span a certain function space and, by means of a linear transformation amongst the members of this set, divides the space into two mutually orthogonal and exclusive spaces: that of the occupied orbitals and that of the virtuals. The single-determinant function is, of course, invariant against linear transformations amongst the occupied orbitals and so it is the occupied *space* which is defined by the solution of the LCAO equations, not any particular set of functions which span that space. The virtual space is just the complement of the occupied space; the space of functions which completes the original space.

10.3. THE VIRTUAL SPACE IN LCAO

During the iterative solution of the LCAO equations, at each iteration, the diagonalisation of the "current" approximation to the self-consistent Hartree–Fock matrix generates such a partition of the total function space, i.e. a current (non-self-consistent) set of occupied orbitals and a set of current virtuals: a current occupied space and a current virtual space. These current spaces share some of the properties of the final self-consistent spaces; in particular the current single-determinant is invariant against linear transformations within the current occupied space.

Thus, when the next iteration is performed, if the single-determinant wavefunction is to be improved, mixing must occur between the current occupied and the current virtual spaces; the current occupied orbitals can only be changed in the correct way by addition of linear combinations of the current virtual orbitals. That is, the corrections to the new occupied vectors at the kth iteration must be a linear combination of the virtual eigenvectors of the previous $((k-1)$th non-self-consistent) HF matrix:

$$\delta \boldsymbol{c}^i_{(k+1)} = \sum_{a=n+1}^{m} A_a \boldsymbol{c}^a_{(k)}$$

for some choice of the numbers A_a.

We can make use of this simple fact to attempt to improve the convergence properties of the LCAO method. Recall that our earlier attempts in this direction were based on simple interpolation or extrapolation by "brute force"; there was no underlying theory.

10.3.3 Level shifters and damp factors

In Appendix 10.A to this chapter the matrix analogue of perturbation theory is developed, and that theory can be applied to analyse the process of iteration to self-consistency by treating the difference between successive Hartree–Fock matrices as a perturbation on the current matrix.

During any iteration of the solution of the matrix Hartree–Fock equations, it is computationally convenient to transform the Hartree–Fock matrix into the basis of orbitals defined by the diagonalisation of the previous Hartree–Fock matrix. When this has been done, and if the calculation is converging, the new set of orbitals is close to the previous set.

Thus, at a particular kth cycle, the new matrix equation for any one of the orbitals is

$$\boldsymbol{h}^F \boldsymbol{c}^i_{(k)} = \epsilon^{(k)}_i \boldsymbol{c}^i_{(k)} \tag{10.1}$$

where the results of the previous iteration is just the set of functions being used to expand the current set.

This situation fulfils all the conditions of Appendix 10.A:

- The functions are all the solutions of the same matrix equation.
- The expansion of the old functions is just unity in *one* place in the expansion matrix c^i.
- The functions are orthogonal.

As a result of changes in the density matrix and therefore in the matrices J and K, the next cycle of the process will have a different h^F matrix. Let us take this difference (in the basis of previous eigenvectors)

$$h^F_{(k+1)} - h^F_{(k)} = h_1$$

as a perturbation on the original non-self-consistent equation of the previous cycle.

Under these conditions, the first-order change to the new expansion c^i is given by the final expression in Appendix 10.A:

$$\delta c^i_{(k+1)} = \sum_{j=1}^{m} \left(\frac{c^j_{(k)} h_1 c^i_{(k)}}{\epsilon^{(k)}_i - \epsilon^{(k)}_j} \right) c^j_{(k)}$$

except that, in addition to excluding the term with $i = j$ from the perturbation theory, we may also exclude from the summation which corrects any *occupied* orbital all the contributions from the *other occupied orbitals*.

This is not because these terms do not contribute to the change in any particular occupied orbital but because, as we noted above, the single-determinant nature of the total wavefunction means that all these contributions will cancel; giving no *net* change to the total wavefunction. So the correction to each *occupied* orbital coefficient vector may be written:

$$\delta c^i_{(k+1)} = \sum_{a=n+1}^{m} \left(\frac{c^a_{(k)} h_1 c^i_{(k)}}{\epsilon^{(k)}_i - \epsilon^{(k)}_a} \right) c^a_{(k)}$$

(if the lowest n of the m orbitals are occupied). The summation index j has been changed to a to emphasise this change.

Since, during the iterative process, we transform to a basis of the eigenvectors of the previous cycle this expression may be written as

$$\delta c^i_{(k+1)} = \sum_{a=n+1}^{m} \left(\frac{h^F_{ai}}{h^F_{ii} - h^F_{aa}} \right) c^a_{(k)}$$

entirely in terms of the current HF matrix in the basis of the eigenvectors of the previous HF matrix.

10.3. THE VIRTUAL SPACE IN LCAO

Thus we now have an explicit expression for the A_a in the above expansion

$$c^i_{(k+1)} = \sum_{a=n+1}^{m} A_a c^a_{(k)}$$

This equation gives an explicit expression for the largest changes in the coefficients as the iterative procedure progresses. If the iterative procedure oscillates or diverges we may use our knowledge of this expansion to improve the convergence properties.

Roughly speaking, there are two *kinds* of reason why an SCF calculation does not converge:

- Successive cycles are alternating between two different *"states"* of the system; there are two possible sets of orbital occupations which lie close in energy. Often the two states are of different symmetry thus precluding any "averaging" process during the iterations.

- The SCF process generates changes which are extremely small or too large to "settle down".

There is always the possibility that, using some "artificial" convergence aids, we may "lock" the iterative calculation into a state that is not the lowest state and we must always be alert to this possibility. However, we can always improve the convergence of a calculation by manipulating sizes of the correction coefficients A_a by artificially changing the elements h^F_{ai} and h^F_{aa}.

As we saw in our empirical attempts to improve the convergence of the iterative procedure when running the testbench, the "natural" changes to the matrices c^i are often too large, leading to successive overshooting of the "best" correction per cycle. If, during the SCF process, we monitor the progress of convergence by, for example, checking the value of the sum of differences or the monotone decrease of the total energy we can intervene. In the case of non-convergence, the simplest thing to do is to add a (positive) constant to all the elements h^F_{aa} (the diagonals of h^F associated with the virtuals of the previous SCF cycle) which will increase the energy denominators in the perturbation expression and have the effect of reducing the size of the "steps" taken into the virtual space during the iterations. That is, the convergence of the calculation will be improved by the addition of a *level shifter* to the current virtual orbitals. In fact, we could add different level shifters to different virtual eigenvalues in really difficult cases but usually it is sufficient to simply exaggerate the energy gap between the occupied and virtual MOs to obtain convergence.

A very similar effect can be obtained by reducing the size of the numerators in the perturbation expression; multiplying the elements h^F_{ai} by a positive number less than unity will have this effect. The use of such so-called *damp factors* as this process is called is less widely used. Again it is possible to scale different h_{ai}^F by different damp factors.

It is clear that (ignoring the possibility of lower states) the use of level shifters and damp factors can *ensure* the convergence of a calculation (in the sense that the convergence of a perturbation series is ensured), but usually at the expense of an increased *number* of iterations since many small steps may have to be taken.

The actual calculation is unaffected by the arbitrary changes to the virtual orbital diagonals which become, on convergence, the virtual orbital energies since they take no part in the final single-determinant wavefunction. If the virtual orbitals and their eigenvalues are to be *used* after the SCF calculation (e.g. in a CI calculation) one must, of course, remember to *remove* the level shifter from the virtual eigenvalues!

10.4 Conclusions

In making the algebraic (LCAO) approximation for molecular orbitals and using standard matrix techniques for the solution of the resulting equations we generate a set of virtual orbitals, usually with positive energies. There are no corresponding solutions of the differential HF equations since any orbital with positive energy is degenerate with a continuum solution.

Because of the invariance property of the single-determinant wavefunction, it is possible to use the virtual orbitals during the SCF process to manipulate the convergence properties of the computer implementation.

Perhaps the best way to regard the virtual space is as providing additional degrees of freedom for the action of the variation principle beyond the MO model. Certainly any attempt to provide a physical interpretation of these orbitals is fraught with paradox.

Appendix 10.A

Perturbation theory

This is an appendix about simple quantum-mechanical perturbation theory; it is in a rather strange place because the first time perturbation theory is needed is in the introduction of level shifters in this chapter.

Contents

10.A.1 Introduction **323**
10.A.2 Perturbation theory **324**
    10.A.2.1 Solution of the first-order equation 325
    10.A.2.2 The second-order energy 328
10.A.3 Perturbation theory for matrix equations **329**

10.A.1 Introduction

This small appendix is devoted to providing the barest minimum of perturbation theory necessary to support the material in the previous chapter on the use of convergence aids in SCF theory. The "genuine" (self-consistent) perturbation theory used to evaluate molecular properties from an SCF wavefunction is developed and implemented later. We begin with an elementary exposition of the standard (Rayleigh–Schrödinger) perturbation theory for the Schrödinger equation and continue with its matrix analogue.

10.A.2 Perturbation theory

The central idea in the perturbation technique is to use knowledge of the solutions of a given Schrödinger equation to obtain approximate solutions of another Schrödinger equation which is, in some way, "nearby" the original equation but which is not soluble.

Thus, if we know the solutions of an *unperturbed* or so-called "zeroth-order" Schrödinger equation:

$$\hat{H}_0 \Psi^{(0)} = E^{(0)} \Psi^{(0)} \tag{10.A.2}$$

we might want to find the solutions of a related Schrödinger Equation:

$$\hat{H}\Psi = E\Psi \tag{10.A.3}$$

For example, if we know the solutions of the equation for a hydrogen atom we might like to find approximate solutions for the hydrogen atom in an electric or magnetic field or in the presence of a point charge.

We write the Hamiltonian of the insoluble system in terms of that of the soluble system (10.A.2):

$$\hat{H} = \hat{H}_0 + \lambda(\hat{H} - \hat{H}_0) = \hat{H}_0 + \lambda \hat{H}_1 \tag{10.A.4}$$

where $\lambda = 1$ and has been included as a parameter for mathematical reasons which will become obvious shortly.

We now make the basic assumption of perturbation theory:

> If the solutions of the eqn (10.A.3) are "close enough" to those of eqn (10.A.2), then they can be expanded as power series in the parameter λ:
>
> $$(\hat{H}_0 + \lambda \hat{H}_1)\Psi = E\Psi \tag{10.A.5}$$
> $$\Psi = \Psi^{(0)} + \lambda \Psi^{(1)} + \lambda^2 \Psi^{(2)} + \ldots \tag{10.A.6}$$
> $$E = E^{(0)} + \lambda E^{(1)} + \lambda^2 E^{(2)} + \ldots \tag{10.A.7}$$

That is, we assume that the (unknown) perturbed energies and wavefunctions are capable of being expanded *as power series* in the so-called "perturbation parameter" λ.

We must now insert eqn (10.A.7) into eqn (10.A.3) to obtain equations satisfied by the energies $E^{(i)}$ and functions $\Psi^{(i)}$.

$$(\hat{H}_0 + \lambda \hat{H}_1)(\Psi^{(0)} + \lambda \Psi^{(1)} + \lambda^2 \Psi^{(2)} + \ldots)$$
$$= (E^{(0)} + \lambda E^{(1)} + \lambda^2 E^{(2)} + \ldots)(\Psi^{(0)} + \lambda \Psi^{(1)} + \lambda^2 \Psi^{(2)} + \ldots)$$

10.A.2. PERTURBATION THEORY

The key to any further progress with this intractable-looking equation is to realise that *from the point of view of* λ it is *an identity*; it must be true *for all values of* λ. If this is so then we can equate like powers of λ on both sides of the equation:

$$(\lambda^0) \quad \hat{H}_0 \Psi^{(0)} = E^{(0)} \Psi^{(0)} \quad (10.\text{A}.8)$$
$$(\lambda^1) \quad \hat{H}_0 \Psi^{(1)} + \hat{H}_1 \Psi^{(0)} = E^{(0)} \Psi^{(1)} + E^{(1)} \Psi^{(0)} \quad (10.\text{A}.9)$$
$$(\lambda^2) \quad \hat{H}_0 \Psi^{(2)} + \hat{H}_1 \Psi^{(1)} = E^{(0)} \Psi^{(2)} + E^{(1)} \Psi^{(1)} + E^{(2)} \Psi^{(0)} \quad (10.\text{A}.10)$$
$$(\lambda^n) \quad \hat{H}_0 \Psi^{(n)} + \hat{H}_1 \Psi^{(n-1)} = E^{(0)} \Psi^{(n)} + E^{(1)} \Psi^{(n-1)} + E^{(2)} \Psi^{(n-2)}$$
$$+ \ldots + E^{(n)} \Psi^{(0)} \quad (10.\text{A}.11)$$

The first of these is not new; it is simply the unperturbed equation whose solutions we assume to be known. The second equation (eqn (10.A.9)), determines the first-order energy $E^{(1)}$ and the first-order correction to the wavefunction $\Psi^{(1)}$. The third equation determines the corresponding second-order quantities, and the last one is the general case. Note that there are only *two* terms on the left-hand-side in every case but that there are more and more on the right as n increases. This is simply because the perturbed *Hamiltonian* expansion only goes as far as the *first* power of λ, although one could envisage a longer power series for $\hat{H}$ with correspondingly more complicated equations.

10.A.2.1 Solution of the first-order equation

The first thing to say about eqn (10.A.9) is that it is a *differential* equation; $\hat{H}$ contains differential operators. However the energy quantity $E^{(1)}$ is a *number* and our aim is to get at this number by *algebraic* methods. The way to get numbers from differential equations is to *integrate* over all the variables involved in the equation;

$$\int \cdots dV$$

is used to indicate integration over all the variables of all the electrons involved (space and spin, if necessary). If eqn (10.A.9) is multiplied from the left by $\Psi^{(0)}$ and integrated over all space we obtain

$$\int \Psi^{(0)} \hat{H}_0 \Psi^{(1)} dV + \int \Psi^{(0)} \hat{H}_1 \Psi^{(0)} dV = E^{(0)} \int \Psi^{(0)} \Psi^{(1)} dV + E^{(1)} \int \Psi^{(0)} \Psi^{(0)} dV$$
$$(10.\text{A}.12)$$

since $E^{(0)}$ and $E^{(1)}$ are just numbers. Now, the unperturbed wavefunction $\Psi^{(0)}$ can always be *normalised* and we will assume that this has been done so that:

$$\int \Psi^{(0)} \Psi^{(0)} dV = 1$$

and we obtain

$$E^{(1)} = \int \Psi^{(0)} \hat{H}_0 \Psi^{(1)} dV + \int \Psi^{(0)} \hat{H}_1 \Psi^{(0)} dV - E^{(0)} \int \Psi^{(0)} \Psi^{(1)} dV \quad (10.\text{A}.13)$$

We must now use the fact that the Hamiltonian $\hat{H}_0$ is a *Hermitian* operator:

$$\int \Psi^{(0)} \hat{H}_0 \Psi^{(1)} dV = \int \Psi^{(1)} \hat{H}_0 \Psi^{(0)} dV$$

But the integrand on the right of the above equation can be simplified by the unperturbed equation:

$$\hat{H}_0 \Psi^{(0)} = E^{(0)} \Psi^{(0)}$$

so that

$$\int \Psi^{(0)} \hat{H}_0 \Psi^{(1)} dV = \int \Psi^{(1)} \hat{H}_0 \Psi^{(0)} dV = E^{(0)} \int \Psi^{(1)} \Psi^{(0)} dV$$

Thus two of the terms on the right of eqn (10.A.13) cancel and we obtain:

$$\boxed{E^{(1)} = \int \Psi^{(0)} \hat{H}_1 \Psi^{(0)} dV \quad (10.\text{A}.14)}$$

The *first-order* correction to the energy is determined completely by the unperturbed wavefunction and the perturbing Hamiltonian.

In order to calculate the first-order correction to the wavefunction ($\Psi^{(1)}$), recall the first-order equation eqn (10.A.9):

$$\hat{H}_0 \Psi^{(1)} + \hat{H}_1 \Psi^{(0)} = E^{(0)} \Psi^{(1)} + E^{(1)} \Psi^{(0)}$$

which rearranges to

$$(\hat{H}_0 - E^{(0)}) \Psi^{(1)} = E^{(1)} \Psi^{(0)} - \hat{H}_1 \Psi^{(0)}$$

$E^{(0)}$, $E^{(1)}$ and $\Psi^{(0)}$ are all now known and so this is a differential equation for $\Psi^{(1)}$.

As we have seen above, it is always easier to solve ordinary, algebraic equations than to solve differential equations so, again, we use a device to obtain an algebraic equation from this differential equation. A standard technique is to *expand* the unknown function in terms of a set of *known* functions:

$$\Psi^{(1)} = \sum_i C_i f_i$$

where the f_i are a set of known functions and the C_i are a set of numbers to be determined from the differential equation for $\Psi^{(1)}$.

Obviously, the f_i must be *suitable* in some way to expand $\Psi^{(1)}$, that is

10.A.2. PERTURBATION THEORY

- They must have the same number of variables and similar boundary conditions.
- There must be enough of them to be sure of getting a good description of $\Psi^{(1)}$.

A suitable set of functions are the *excited* solutions of the unperturbed Schrödinger equation. Why not correct the ground state function in the presence of a perturbation by addition of pieces of the excited state functions? The unperturbed equation has, in addition to the solution $\Psi^{(0)}$ of eqn (10.A.2), lots of higher-energy solutions

$$\hat{H}_0 \Psi_i^{(0)} = E_i^{(0)} \Psi_i^{(0)} \qquad (10.A.15)$$

where i is used to label the different solutions and we might take $i = 0$ to be the ground state, i.e. use a slightly more explicit notation for the ground-state function when we are dealing with *many* solutions of the unperturbed equation:

$$\Psi_0^{(0)} = \Psi^{(0)}$$

Using these functions as the f_i we have:

$$\Psi^{(1)} = \sum_{i=1}^{\infty} C_i \Psi_i^{(0)} \qquad (10.A.16)$$

Now it is only necessary to insert this expansion into the equation which determines $\Psi^{(1)}$ to obtain the values of the coefficients in terms of integrals involving *known* functions (the $\Psi_i^{(0)}$) and operators ($\hat{H}_1$). Equation (10.A.9) is (using the new notation $\Psi_0^{(0)}$ in place of $\Psi^{(0)}$):

$$\hat{H}_0 \Psi^{(1)} + \hat{H}_1 \Psi_0^{(0)} = E^{(0)} \Psi^{(1)} + E^{(1)} \Psi_0^{(0)}$$

which can be rearranged to

$$(\hat{H}_0 - E^{(0)})\Psi^{(1)} = E^{(1)} \Psi_0^{(0)} - \hat{H}_1 \Psi_0^{(0)}$$

Now, using the expansion eqn (10.A.16), the left-hand-side of this equation becomes:

$$(\hat{H}_0 - E^{(0)})\Psi^{(1)} = \sum_{i=1}^{\infty} (\hat{H}_0 - E^{(0)}) C_i \Psi_i^{(0)}$$
$$= \sum_{i=1}^{\infty} C_i (\hat{H}_0 - E^{(0)}) \Psi_i^{(0)}$$
$$= \sum_{i=1}^{\infty} C_i (E_i^{(0)} - E^{(0)}) \Psi_i^{(0)}$$

because the functions $\Psi_i^{(0)}$ solve the unperturbed Schrödinger equation

$$\hat{H}_0\Psi_i^{(0)} = E^{(0)}\Psi_i^{(0)}$$

If we now multiply from the left by any *one* of the unperturbed solutions $\Psi_k^{(0)}$ and integrate over all space we obtain the value of the coefficient C_k, and clearly we can do this for *all* C_k:

$$(E_k^{(0)} - E_0^{(0)})C_k = \int \Psi_k^{(0)} \hat{H}_1 \Psi_0^{(0)} dV$$

In obtaining this expression we use the fact that the *overlap* integrals between *different* solutions of the *same* Schrödinger equation are zero:

$$\int \Psi_i^{(0)} \Psi_k^{(0)} dV = \delta_{ik}$$

Writing

$$H_{k0} = \int \Psi_k^{(0)} \hat{H}_1 \Psi_0^{(0)} dV$$

we have

$$C_k = \frac{H_{k0}}{(E_k^{(0)} - E_0^{(0)})}$$

and so, finally, the expression for $\Psi^{(1)}$:

$$\boxed{\Psi^{(1)} = \sum_{k=1}^{\infty} \frac{H_{k0}}{(E_k^{(0)} - E_0^{(0)})} \Psi_k^{(0)}} \qquad (10.\text{A}.17)$$

10.A.2.2 The second-order energy

The *second-order* eqn (10.A.10) is:

$$\hat{H}_0\Psi^{(2)} + \hat{H}_1\Psi^{(1)} = E^{(0)}\Psi^{(2)} + E^{(1)}\Psi^{(1)} + E^{(2)}\Psi_0^{(0)}$$

and, without going through the steps in as minute detail as before, we can use the same technique to obtain the second-order *energy* $E^{(2)}$ since we know the zeroth-order function, the first-order energy and the first-order wavefunction. The technique is exactly analogous to the calculation of the first-order energy:

1. Multiply the equation from the right by $\Psi_0^{(0)}$ and integrate over all space.
2. Use the *normalisation* of the unperturbed wavefunction.

10.A.3. PERTURBATION THEORY FOR MATRIX EQUATIONS

3. Use the *Hermiticity* of the Hamiltonian to evaluate

$$\int \Psi_0^{(0)} \hat{H}_0 \Psi^{(2)} dV = \int \Psi^{(2)} \hat{H}_0 \Psi_0^{(0)} dV = E_0^{(0)} \int \Psi^{(2)} \Psi_0^{(0)} dV$$

and cancel the resulting simplified term.

4. The result is the expression

$$E^{(2)} = \int \Psi^{(1)} \hat{H}_1 \Psi_0^{(0)} dV = \int \Psi_0^{(0)} \hat{H}_1 \Psi^{(1)} dV$$

writing in the expression for $\Psi^{(1)}$ in terms of the $\Psi_k^{(0)}$ eqn (10.A.17) the final expression is obtained:

$$E^{(2)} = \sum_{k=1}^{\infty} \frac{H_{k0}^2}{(E_k^{(0)} - E_0^{(0)})} \qquad (10.A.18)$$

remember the definition of the H_{k0}:

$$H_{k0} = \int \Psi_k^{(0)} \hat{H}_1 \Psi_0^{(0)} dV$$

10.A.3 Perturbation theory for matrix equations

If we work in a fixed finite basis (assumed orthogonal for simplicity) the matrix eigenvalue problem is:

$$\boldsymbol{h}_0 \boldsymbol{c}_i^0 = \epsilon_i^0 \boldsymbol{c}_i^0$$

(for $i = 1, m$).

If the matrix $\boldsymbol{h}_0$ is changed to

$$\boldsymbol{h} = \boldsymbol{h}_0 + \lambda \boldsymbol{h}_1$$

and we assume a power series in the perturbed eigenvalue and eigenvector:

$$\epsilon_i = \epsilon_i^0 + \lambda \epsilon_i^1 + \ldots$$
$$\boldsymbol{c}_i = \boldsymbol{c}_i^0 + \lambda \boldsymbol{c}_i^1 + \ldots$$

Then we may use the fact that the equation resulting from the substitution of these expansions is an *identity* in λ to equate terms in like powers of λ to generate the equation of first order in λ as

$$(\boldsymbol{h}_0 \boldsymbol{c}_i^1 - \epsilon_i^0 \boldsymbol{c}_i^1) = (\epsilon_i^1 \boldsymbol{c}_i^0 - \boldsymbol{h}_1 \boldsymbol{c}_i^0)$$

In the fixed basis we may use the eigenvectors of the original eigenvalue problem to expand any vector so, in particular, we may use these eigenvectors to expand the first-order correction to the ith eigenvector c_i^1.

$$c_i^1 = \sum_{j=1}^{m} d_j c_j^0$$

where the d_j are the required expansion coefficients (the subscript i has been suppressed to avoid to much notational confusion; in fact there is a set of d_j for each i of course) and the summation should be understood to exclude the term with $j = i$.

We can now insert this expansion into the first-order equation and multiply the resulting equation by each c_k^0 in turn. The orthonormality of the c_k^0 and the fact that they are eigenvectors of the original unperturbed equation gives an expression for each d_k:

$$d_k = \frac{c_k^0 h_1 c_i^0}{\epsilon_i^0 - \epsilon_k^0}$$

so that the first-order approximate eigenvector for the perturbed matrix eigenvalue problem is:

$$c_i^1 = \sum_{j=1}^{m} \left(\frac{c_j^0 h_1 c_i^0}{\epsilon_i^0 - \epsilon_j^0} \right) c_j^0$$

where the summation excludes the term with $i = j$.

Chapter 11

Choice of tools

This chapter is a pause in the development of the theories to think again about the large problems which loom over the generation and maintenance of large software suites and the need to interact with existing sets of programs.

Contents

11.1	Existing software	**331**
11.2	Why ratfor?	**335**
11.3	The Revision Control System: RCS	**337**
	11.3.1 The system	337
	11.3.2 Use of the system	338

11.1 Existing software

Here are a few facts in random order about existing quantum chemistry software:

- Quantum chemistry programs are large in size and *very* demanding of resources; time and storage.

- The programs have "grown up" over the years by generations of graduate research workers, by mergers and accretions and have been developed with a view to operational efficiency (speed and storage).

- There are a number of basically equivalent program packages which do the same *kind* of computation with different emphases. Hardly any of these programs have anything approaching adequate documentation. Indeed, it is rather rare to have moderately complete *operating instructions* let alone documentation.[1]

- These programs are now so large, unstructured and complex as to be almost unmaintainable.

- It has been estimated recently that any of these large packages would require 25 man-years of effort to recreate.

- All are written in a mixture of FORTRAN66, f77 various system-specific "FORTRANs" and various assemblers. *None* will pass the ANSI f77 standard.[2]

Without exception, the large suites of quantum chemistry programs have been written by research workers and enthusiasts in quantum chemistry and not by software developers. They have not been through the design, planning and specification routes which are now routine in software development. The interests of these researchers (who were and are their own systems analysts, specification writers and programmers) has always been to get their systems up and running in the shortest possible time to enable them to do scientific research.

Furthermore, until quite recently it was standard practice for computational research workers to make their programs freely available to their colleagues to use wherever those colleagues happened to be. This led to a very fertile sequence of program enhancements as pieces of existing program were inserted into programs being generated.

The unspoken assumption behind all this admirable cooperation was that the programs would always be used by people who:

- are familiar with the basic theories underlying the software
- can look at the source code of the programs and understand how to correct or change it.

That is to say, the documentation associated with the programs was and is always minimal and usually amounts to little more than operating instructions; the programs were intended for expert user maintenance. Typically, such documentation as exists has been written some time after the programs themselves

---

[1] There does not seem to be, in fact, in the quantum chemistry community a clear understanding of the difference between program documentation and operating instructions.
[2] And certainly not the the f90 standard.

11.1. EXISTING SOFTWARE

have been written. As programs were interfaced together in *ad hoc* manners, the idea that a research worker in the field could manage these large suites of programs became more and more unrealistic and so the opposite course of action was taken: to explicitly *hide* the inner workings of the software by the provision of a user-friendly shell which can be used with only a knowledge of the operating instructions.

Essentially this has meant that the idea of an explicit specification and documentation for the procedures comprising the programs has been abandoned.

Thus, the more and more widespread *use* of quantum chemistry software has not been matched by an increase in the understanding of the theoretical models and methods of implementation of these models which are implicit in that software.

It is the central conviction of the work presented here that a constantly refreshed familiarity with theoretical models and mathematical methods *and* the way in which they are implemented are absolutely essential to the vitality of scientific research. It is simply not good science to be able to "press the buttons" and generate the results.

However, there is nothing more debilitating for any activity than to call a halt to all production while tools are redesigned. It is simply absurd to ask for systematic specifications of existing programs which have been designed for speed of execution rather than quality of structure and documentation. It is even unrealistic to ask for a modular and structured approach to the development of *new* software since these programs are so hungry for computer resources (time and storage) that considerations of efficiency of execution will always take priority over design and portability.

What is being proposed here is a compromise.

The methods and techniques will be developed and implemented in a modular and structured manner which will serve as a didactic skeleton for the understanding of existing software and the generation of new programs; a framework whose sole reason for existence is to be improved upon.

In the system derived, developed and presented here:

- All the theories will be developed from first principles and fully implemented to generate a working program which is also supplied in machine-readable form.

- Each program segment will have a specification (a "manual entry", in the Unix sense).

- There will be explicit communication between program segments via fully specified argument lists; there will be no "COMMON blocks".

- Programming techniques will be fully structured; there will be no labelled statements and no "goto" statements.

- There will be no "magic numbers" (explicit, mysterious constants) in the code; all symbols will be explicitly defined with mnemonics which are valid throughout the code.

- Above all, the code is designed to reflect the mathematical formulae in both structure and nomenclature rather than being organised for efficiency of execution.

The specification and generation of structured programs is not feasible without access to a certain minimum of program development tools. By far the best working environment available to software developers at this time is Unix together with the excellent tools from the Free Software Foundation and the associated *aficionados*.[3] The very minimum requirements are:

- A suitable programming language which has the structure to encourage good programming practice,

- A macro processor,

- A system for version control,

- Methods for the generation and display of both on-line and hard copy documentation,

- Ideally, an automatic method of generating programs, structured and informative listings and program documentation from the same master file (a system similar to Knuth's WEB, see Chapter 8).

Some of these tools are provided with the system and are available by anonymous ftp.

It must be stressed that without a macro processor and a version control system, the generation of portable software is scarcely feasible, and, if there is no on-line manual the use of the system is severely restricted. However, the most important decision to be taken is the choice of programming language.

---

[3] emacs is *the* editor and gcc, g77 are rapidly becoming *the* compilers.

11.2 Why ratfor?

All the fragments of code in Chapter 5 were in a computer language which looks, perhaps, familiar to anyone with any experience of the use of high-level languages (FORTRAN, Pascal, C etc.) but is not recognisable as any of these. The code is rat77, a version of ratfor ("rational fortran") which is processed into f77 by a precompiler, one of the many preprocessors for FORTRAN. This choice is not just a personal quirk on the part of the author; it has been made on the basis of many years of experience in the development and maintenance of large suites of programs.

There are several factors to be taken into consideration when choosing a programming language to use in a major software generation project:

1. Availability: will any operating system offer the language?
2. Compatibility: will one be able to access existing software from the language?
3. Flexibility: does the language offer the sorts of constructs and structures which are best for the project?
4. Longevity: will the language continue to be supported?

The traditional scientific programmer's language, FORTRAN, certainly offers all of these except flexibility and will certainly continue to be used for the foreseeable future. The candidates for a modern replacement for FORTRAN are, perhaps, PASCAL, ADA and C with C being very much the preferred choice by enthusiasts and *aficionados*. Some operating systems offer a "common object format" as output from their compilers; meaning that programs may be written in a mixture of languages and linked together. We are really looking for a new way of writing scientific software but tied to the very valuable heritage of existing FORTRAN. The main conclusion is that we are in the position of the motorist who asked the yokel the way to the capital and received the classic reply "if I were you, I wouldn't start from here".

While it is very desirable to write software in a more flexible and structured language than FORTRAN, it is unrealistic to lose contact with the vast quantity of FORTRAN software which has been honed for efficiency over the years.

It is this fact which points to rat77 which is simply a version of the Ratfor introduced by Kernighan and Plauger[4] in their now-classic work *Software Tools*. Ratfor is particularly appropriate for scientific programming although it was introduced by Kernighan and Plauger as a general-purpose language more

---
[4] *Software Tools* by B. W. Kernighan and P. J. Plauger (Addison-Wesley, 1976).

orientated to system programming than scientific applications. The idea is very simple; one writes in Ratfor and a "preprocessor" takes this code and emits a FORTRAN equivalent which is then compiled in the usual way. Of course, rat77 is just Ratfor which emits f77-compatible code and there will be a rat90[5] in due course.

The pre-processor is, of course, itself written in rat77 thus one is completely free from compiler restrictions and can be sure that the language will be supported indefinitely; or, at least, as long as FORTRAN continues to be supported. rat77 itself is very "C-like", the main exception being the lack of structures and the associated facilities: pointers etc.

In using rat77, we are treating FORTRAN in much the same way as other compilers (FORTRAN, C etc.) treat the specific machine's assembly language; rat77 emits FORTRAN code while other compilers emit assembler code. This similarity goes deeper than a mere analogy. In the same way that a FORTRAN programmer would never dream of looking at the assembler output by the compiler, so the rat77 programmer would not normally look at the FORTRAN. De-bugging would be done entirely at the rat77 level and it is anticipated that there would be less of it since the structure of rat77 is better than that of FORTRAN. More important, the assembler code emitted by a FORTRAN compiler will only work on a specific family of operating systems. But the FORTRAN code emitted by rat77 will work *anywhere where FORTRAN is supported*. In deciding to work with rat77 we can regard ourselves as working on a "FORTRAN virtual machine".

With rat77 we are writing programs for a machine whose assembler is FORTRAN. That is, we are writing programs which will run anywhere; rat77 is a "cross-compiler" for all computers. We can, for example, write rat77 on a personal computer, use the rat77 preprocessor to generate the FORTRAN and send it directly to a supercomputer and be sure that it will go. What is more, since we have control of the source of the preprocessor, we can enhance it with statements and structures of our own choice in the certain knowledge that these enhancements will not become obsolete because they will always run on the FORTRAN virtual machine.

Access to existing FORTRAN software (for molecular integrals and other utilities) is also guaranteed; this software, in general, does not run on the FORTRAN virtual machine but on particular FORTRANs all of which are accessible.

Details of the rat77 language are given in material available by ftp.

---

[5] The FWEB system of John Krommes described in Chapter 8 is such a system.

11.3 The Revision Control System: RCS

When a document (source code, program write-up, lecture notes) has reached a "mature stage" it will, typically, require *modification* rather than major changes and will require less continuous attention than during its development stage. In this context "mature" might mean "working" in the case of a program; "moderately comprehensive" in the case of a program write-up and "issued at least once" in the case of lecture notes.

Modifications might be:

- The removal of bugs or minor enhancements to source code,
- Associated changes to a program write-up and
- Corrections or clarifications in the case of lecture notes.

In all these cases it will sooner or later happen that bugs "corrected" will prove to be bugs added or "clarifications" prove to be obfuscations and it will prove necessary to try to go back to an earlier version to start again. In this situation it is obviously necessary to get to grips with the problem of controlling the number of "versions" of a document.

The Revision Control System (RCS) is a set of tools which enables the recovery of any previous version of a changing document from a single file which contains the initial version and *all* the changes to which it has been subject in its development history since its introduction to the RCS system.

11.3.1 The system

The procedure is simple in concept and consists of one initial step and three stages per change:

(A) The text of a mature document (`program.f`, say) is admitted to the filestore of the RCS system; say it is stored by RCS as `program..f,v` (v for "version"). Normally, RCS deletes the original to avoid confusion.

(Bi) Subsequently, a "working copy" of the document (at a particular version level) is obtained from the RCS filestore; say the working copy which is ultimately destined to become the new version is generated from `program.f,v` as `program.f`.

(Bii) This working copy is edited and changed (compiled and tested in the case of a program) until a new version is generated — this may take minutes or months; in due course the modifications to `program.f` are finished and working.

(Biii) The changes which were necessary to generate the new version (`program.f`) are *incorporated into the same* RCS file (`program.f,v`). The changes are incorporated by a file comparison algorithm. RCS, when incorporating changes to make a new version, will ask for a set of comments to describe the essence of the changes made since the last version.

Steps (Bi)–(Biii) are repeated as often as necessary. In both the original step (A) and the final step (Biii) the user is prompted for comments, reasons for the changes etc.

What actually happens at step (Biii) is that *lines* which differ between the new version and the version generated in (Bi) are inserted with an appropriate code and version number to enable the new version to be regenerated. The original lines are not deleted but are simply marked as belonging to the earlier version.

11.3.2 Use of the system

With the combined use of RCS and, optionally, the conditional compilation facility of `rat77` (or C) it is *never* necessary to keep multiple copies of any document.[6]

Any version of the source document can be regenerated from a single master RCS file and so one can work on one's programs in the certain knowledge that an earlier (working!) version can always be regenerated.

For an individual user the system is excellent as it stands, for a *group* of people working on the same programs there clearly has to be some voluntary or compulsory discipline, typically:

- Once a given version of the document has been generated from `program.f,v`, then all other users are "locked out" from that version until the revisions have been completed.

A more detailed account of RCS is given in Appendix 11.A which is actually extracts from the author of the system's own description.

---

[6]Except backup or archive copies, of course, which should themselves be RCS files.

Appendix 11.A

RCS: version control

This appendix consists of selections (with the kind permission of Walter F. Tichy) from:

> RCS A System for Version Control
> by
> Walter F. Tichy
> Department of Computer Sciences
> Purdue University
> West Lafayette, Indiana 47907

Contents

11.A.1	Motivation	**339**
11.A.2	Introduction	**340**
11.A.3	Getting started with RCS	**341**

11.A.1 Motivation

An important problem in program development and maintenance is version control, i.e. the task of keeping a software system consisting of many versions and configurations well organized. The Revision Control System (RCS) is a software tool that assists with that task. RCS manages revisions of text documents, in particular source programs, documentation, and test data. It automates the storing, retrieval, logging and identification of revisions, and it provides selection mechanisms for composing configurations. This paper introduces basic version control concepts and discusses the practice of version control using RCS. For conserving space, RCS stores deltas, i.e. differences between successive revisions. Several delta storage methods are discussed. Usage statistics show that RCS's delta storage method is space and time efficient.

11.A.2 Introduction

Version control is the task of keeping software systems consisting of many versions and configurations well organized. The Revision Control System (RCS) is a set of UNIX commands that assist with that task.

RCS' primary function is to manage **revision groups**. A revision group is a set of text documents, called **revisions**, that evolved from each other. A new revision is created by manually editing an existing one. RCS organizes the revisions into an ancestral tree. The initial revision is the root of the tree, and the tree edges indicate from which revision a given one evolved. Besides managing individual revision groups, RCS provides flexible selection functions for composing configurations. RCS may be combined with **make**, resulting in a powerful package for version control.

RCS also offers facilities for merging updates with customer modifications, for distributed software development, and for automatic identification. Identification is the "stamping" of revisions and configurations with unique markers. These markers are akin to serial numbers, telling software maintainers unambiguously which configuration is before them.

RCS is designed for both production and experimental environments. In production environments, access controls detect update conflicts and prevent overlapping changes. In experimental environments, where strong controls are counterproductive, it is possible to loosen the controls.

Although RCS was originally intended for programs, it is useful for any text that is revised frequently and whose previous revisions must be preserved. RCS has been applied successfully to store the source text for drawings, VLSI layouts, documentation, specifications, test data, form letters and articles.

This paper discusses the practice of version control using RCS. It also introduces basic version control concepts, useful for clarifying current practice and designing similar systems. Revision groups of individual components are treated in the next three sections, and the extensions to configurations follow. Because of its size, a survey of version control tools appears at the end of the paper.

11.A.3 Getting started with RCS

Suppose a text file f.c is to be placed under control of RCS. Invoking the check-in command

```
ci    f.c
```

creates a new revision group with the contents of f.c as the initial revision (numbered 1.1) and stores the group into the file f.c,v. Unless told otherwise, the command deletes f.c. It also asks for a description of the group. The description should state the common purpose of all revisions in the group, and becomes part of the group's documentation. All later check-in commands will ask for a log entry, which should summarize the changes made. (The first revision is assigned a default log message, which just records the fact that it is the initial revision.)

Files ending in ,v are called RCS files (v stands for versions); the others are called working files. To get back the working file f.c in the previous example, execute the check-out command:

```
co    f.c
```

This command extracts the latest revision from the revision group f.c,v and writes it into f.c. The file f.c can now be edited and, when finished, checked back in with ci:

```
ci    f.c
```

ci assigns number 1.2 to the new revision. If ci complains with the message

```
ci error: no lock set by <login>
```

then the system administrator has decided to configure RCS for a production environment by enabling the 'strict locking feature'. If this feature is enabled, all RCS files are initialized such that check-in operations require a lock on the previous revision (the one from which the current one evolved). Locking prevents overlapping modifications if several people work on the same file. If locking is required, the revision should have been locked during the check-out by using the option -l:

```
co   -l   f.c
```

Of course it is too late now for the check-out with locking, because f.c has already been changed; checking out the file again would overwrite the modifications. (To prevent accidental overwrites, co senses the presence of a working file and asks whether the user really intended to overwrite it. The overwriting check-out is sometimes useful for backing up to the previous revision.) To be able to proceed with the check-in in the present case, first execute

```
rcs   -l   f.c
```

This command retroactively locks the latest revision, unless someone else locked it in the meantime. In this case, the two programmers involved have to negotiate whose modifications should take precedence.

If an RCS file is private, i.e. if only the owner of the file is expected to deposit revisions into it, the strict locking feature is unnecessary and may be disabled. If strict locking is disabled, the owner of the RCS file need not have a lock for check-in. For safety reasons, all others still do. Turning strict locking off and on is done with the commands:

```
rcs   -U   f.c     and      rcs   -L   f.c
```

These commands enable or disable the strict locking feature for each RCS file individually. The system administrator only decides whether strict locking is enabled initially.

To reduce the clutter in a working directory, all RCS files can be moved to a subdirectory with the name RCS. RCS commands look first into that directory for RCS files. All the commands presented above work with the RCS subdirectory without change.

Pairs of RCS and working files can actually be specified in three ways: a) both are given, b) only the working file is given, c) only the RCS file is given. If a pair is given, both files may have arbitrary path prefixes; RCS commands pair them up intelligently.

It may be undesirable that ci deletes the working file. For instance, sometimes one would like to save the current revision, but continue editing. Invoking

```
ci   -l   f.c
```

checks in f.c as usual, but performs an additional check-out with locking afterwards. Thus, the working file does not disappear after the check-in. Similarly, the option -u does a check-in followed by a check-out without locking. This option is useful if the file is needed for compilation after the check-in.

Chapter 12

Open shells: implementing UHF

We now return to the problem of implementing our earlier result; the solution of the equations which determine the orbitals of the optimum single-determinant wavefunction for an n-electron system. This necessarily involves electronic "open shells".

Contents

12.1	Introduction	344
12.2	Choice of constraints	344
12.3	Organising the basis	347
12.4	Integrals over the spin-basis	348
12.4.1	One-electron integrals	349
12.4.2	Electron-repulsion integrals	349
12.5	Implementation	350
12.6	J and K for GUHF	351
12.7	The GUHF testbench	357
12.8	Interpreting the MO coefficients	360
12.9	DODS or GUHF?	363
12.10	Version 1 of the SCF code	363
12.10.1	Manual page for function scf	365
12.10.2	Simple coding of function scf	367
12.11	WEB output for function scf	368
12.11.1	scf	369
12.11.2	INDEX	374
12.12	Comments	376

CHAPTER 12. OPEN SHELLS: IMPLEMENTING UHF

12.1 Introduction

In carrying through the implementation of the closed-shell LCAOMO method we have, in fact, had to solve most of the design problems associated with all LCAOMO-type models:

- General flow of control
- Formation of $\boldsymbol{R}$ and $\boldsymbol{G}(\boldsymbol{R})$
- Matrix diagonalisation
- Repulsion integral storage
- Convergence control

therefore the implementation of a new *model* of molecular electronic structure will not require nearly so much new material as we had to consider in the closed-shell case. Attention can be concentrated on just those items in which a new model *differs* from the closed-shell model.

The calculation of the electronic structure of systems with unpaired electrons within the MO model can be done in two ways, by:

1. Retaining the single-determinant model: the Unrestricted Hartree–Fock (UHF) approach.
2. Using a model of sets of doubly occupied and singly occupied orbitals: the Restricted Hartree–Fock (RHF) model.

The second of these methods is best treated as a special case of the "shell model" of molecular electronic structure which will be dealt with later since it introduces some new formal techniques which are to be introduced shortly. For the moment the single-determinant UHF model will be considered.

12.2 Choice of constraints

In the derivation of the single-determinant (Hartree–Fock) model of electronic structure given earlier, no *constraints* were placed on the form of the orbitals or on the values of any of the properties which might be computed from the resulting single-determinant function. In deciding to use the algebraic approximation (the expansion of each orbital as a linear combination of basis functions), an

12.2. CHOICE OF CONSTRAINTS

implicit constraint is placed on the orbitals; they can only be as good as the size and quality of the basis allows. For the moment, this constraint will be ignored and we will assume that any basis used is capable of indefinite accuracy.

We might consider the following hierarchy of models as placing the single-determinant under increasingly severe *constraints*; and, since a constraint on the freedom of the variational principle to act will necessarily mean that a resulting wavefunction may be less than optimum, the hierarchy will be that of *increasing* total energy.

1. No constraints at all; each MO has the full freedom of the basis and the coefficient field (the *complex numbers*). In this most general case each MO is a linear combination of spatial basis functions with both an α and a β spin factor and the coefficients are allowed to to be complex. Compared to our implementation of the closed-shell model there are *four times* as many (real) degrees of freedom. This model will be called the Complex General Unrestricted Hartree–Fock model (CGUHF). In this context "general" is used to mean "having (or being allowed to have) both α and β spin factors in one MO". "Unrestricted" should not be necessary but it is put in for emphasis and for contrast with the (historically) more usual restricted models.

2. The field of expansion coefficients is constrained to be the *real numbers* while retaining the possibility that each MO has an α and β spin component. This is the Real General Unrestricted Hartree–Fock model; (R)GUHF. The use of complex coefficients is sufficiently rare that this model is often called simply GUHF.

3. The field of real numbers is retained for the coefficients, but now each MO is formed by linear combinations of basis functions having *either* an α spin factor *or* a β spin factor. The most graphic name of this model is "Different Orbitals for Different Spins" (DODS). However, historically, this was the first Hartree–Fock method to be used which had any of the common constraints removed and so has also come to be known as simply the Unrestricted Hartree–Fock model (UHF). Obviously this name should really be used for CGUHF.

4. Finally, the (real) coefficients may be constrained so that the MOs occur in *pairs* in which a given spatial orbital appears twice in the determinant multiplied by an α spin factor *and* a β spin factor. The closed-shell model falls into this category since this single determinant has *only* such paired MOs.

It should be noted here that this hierarchy is not strict; for example a single-determinant model which has DODS but the expansion coefficients are allowed to be complex is perfectly possible. But this series of models does show the possibilities which arise from the application of various constraints *on the form of the expansion* of each MO.

The Hartree–Fock equations which arise from the application of the variation principle to any of these constrained models guarantees a minimum in the energy of the associated single-determinant wavefunction. What they do *not* guarantee is that this wavefunction is physically meaningful. This point will be taken up in detail later; for the moment simply recall the single-determinant solution of the closed-shell model for $(H_2O)^{2-}$ obtained with the testbench which had *positive* orbital energies!

The variation principle guarantees the best *energy* possible within a given model but the quality of any other computed molecular properties is not guaranteed (except indirectly *via* the best-energy solution). Suppose we would like to *constrain* the model so that it does produce some property other than energy to be optimum or even to have its experimental value. It is clear that, in principle at least, such a constraint would limit the freedom of the variation principle to act and generate a solution with an energy higher than the corresponding unconstrained function.

It is not clear whether the errors introduced by this process of what one might call *physical* constraint would be merely *quantitative* (a numerical increase in the computed energy) or *qualitative* (an incorrect description of the electronic structure). This can only be discovered by (numerical) experiment.

It is traditional to *impose* three kinds of constraint on computed MOs:

1. The spin-eigenfunction constraint: the form of the MOs is constrained so that the single-determinant function is an eigenfunction of the *total* spin operators $\hat{S}^2$ and $\hat{S}_z$.

2. Symmetry constraints: the form of the MOs is constrained so that they carry an irreducible representation of the molecular point group.

3. In atoms the spherical constraint or the angular momentum eigenfunction constraint: all the orbitals of a given angular momentum shell are required to have the same radial function, this may have the effect of forcing the total single-determinant function to be an eigenfunction of the *total* angular momentum operators $\hat{L}^2$ and $\hat{L}_z$.

One could consider that *all* these constraints are types of symmetry constraint and, clearly, case 3 is a special case of 2. Also, these constraints are not completely distinct from the earlier hierarchy of constraints, but it is useful to think of them in this convenient group.

It is not always possible to fulfil symmetry constraints of this kind and stay within the single-determinant model. In fact the general case involves a wavefunction of several determinants which form a linear combination with coefficients *fixed* by the symmetry constraint. This fact means, unfortunately, that many of the familiar properties of the single-determinant model no longer apply; in particular the idea that the orbital energies are approximations to the ionisation energies of the molecule is lost. In general, therefore, it will be necessary to use a multi-determinant approach to these constrained models of electronic structure, and explicit treatment is deferred until the shell model of molecular electronic structure is discussed.[1]

In choosing a particular *one* of these single-determinant models, technical considerations are never far away, in particular the question of complex MOs involves:

- Do we want to use *complex basis functions*?
- Can we write a procedure for the diagonalisation of complex matrices?

In fact, there is no new freedom to be had by using *both* complex basis functions *and* complex expansion coefficients so we shall always use real basis functions.[2] For the moment we will aim at somewhere in the "middle" of the hierarchy of models and begin to implement the RGUHF method (real basis, real coefficients, mixtures of different spin components) which will be called simply GUHF.

12.3 Organising the basis

A GUHF calculation will involve *twice* as many basis functions as (e.g.) a closed-shell calculation because the spin-basis functions are used, not simply the spatial basis functions. However, unless one wants to be *very* eccentric and use different *spatial basis* functions in the definition of the spin-basis functions involving different spin factors, the basis is simply *doubled*. Each spatial basis function is simply multiplied by the two spin factors to form a convenient spin-basis set.

There are just two rational possibilities for the enumeration of the basis functions:

1. Have each α/β pair adjacent:

$$\varphi_1 = \phi_1 \alpha$$

---

[1] Sometimes, by good luck, a constrained model will turn out to be a single-determinant case.
[2] In the usual restricted model of *atomic* electronic structure the opposite choice is made, since the symmetry-adapted functions are complex.

$$\varphi_2 = \phi_1\beta$$
$$\ldots$$
$$\varphi_{2m-1} = \phi_m\alpha$$
$$\varphi_{2m} = \phi_m\beta$$

2. Make all the α functions first and all the β functions follow:

$$\varphi_1 = \phi_1\alpha$$
$$\varphi_2 = \phi_2\alpha$$
$$\ldots$$
$$\varphi_m = \phi_m\alpha$$
$$\varphi_{m+1} = \phi_1\beta$$
$$\varphi_{m+2} = \phi_2\beta$$
$$\ldots$$
$$\varphi_{2m} = \phi_m\beta$$

The second method will be used as it is much easier to implement and, as we shall see, enables the program to be used for the DODS model with only trivial changes.

12.4 Integrals over the spin-basis

The familiar Born–Oppenheimer non-relativistic molecular Hamiltonian contains no terms that depend on electron spin. The overlap and energy integrals therefore only involve spin through the orthonormality properties of the spin factors:

$$\int \alpha^*\alpha\, ds = \int \beta^*\beta\, ds = 1$$
$$\int \alpha^*\beta\, ds = \int \beta^*\alpha\, ds = 0$$

and so it is not sensible to store any more integrals than the ones we have used for the implementation of the closed-shell case.

Of course, this single set of *spatial*-basis-function integrals must do duty for the whole set of *spin*-basis-function integrals and so our codes will be more complicated.

12.4.1 One-electron integrals

Because of the orthonormality of the spin factors the overlap and one-electron Hamiltonian integral matrices over the spin-basis will be "diagonal blocked". Two (identical) $m \times m$ blocks on the diagonal of a $2m \times 2m$ matrix with two $m \times m$ blocks making up the full matrix:

$$S = \begin{pmatrix} S_{\alpha\alpha} & 0 \\ 0 & S_{\beta\beta} \end{pmatrix}$$

and

$$h = \begin{pmatrix} h_{\alpha\alpha} & 0 \\ 0 & h_{\beta\beta} \end{pmatrix}$$

where, of course,

$$S_{\alpha\alpha} = S_{\beta\beta} \qquad h_{\alpha\alpha} = h_{\beta\beta}$$

and the subscripts are added for mnemonic purposes only.

There is no "mixing" between the spin-basis functions induced by the one-electron Hamiltonian.

12.4.2 Electron-repulsion integrals

Any molecular integral that involves the integration over one or more pairs of *different* spin factors is zero. Conversely all those integrals which involve the integration of only pairs of identical spin factors are non-zero (at least by spin integration!). The one-electron integrals only involve one pair of spin factors and so the situation is very simple. In the case of the electron-repulsion integrals, and the associated electron-repulsion matrices J and K, it is both physically and mathematically obvious that there must be interaction between electrons of different spin

- Electrons of different spin repel each other with the same law of force as electrons of the same spin.

- There must be non-zero off-diagonal blocks in the electron-repulsion matrix

$$J = \begin{pmatrix} J_{\alpha\alpha} & J_{\alpha\beta} \\ J_{\beta\alpha} & J_{\beta\beta} \end{pmatrix}$$

at least, otherwise the GUHF method would simply be two copies of the closed-shell method.

In fact, of course, all the integrals of the type

$$\int dV_1 \int dV_2 \varphi_i(\vec{r_1})\varphi_j(\vec{r_1})\frac{1}{r_{12}}\varphi_k(\vec{r_2})\varphi_\ell(\vec{r_2})$$

are non-zero if the spin factors of φ_i and φ_j are the same *and* the spin factors of φ_k and φ_ℓ are the same.

This means that the electron-repulsion integrals are doubled, as we would expect, *plus* all the integrals of the above type *between* the two different spin-factor sets, provided φ_i and φ_j belong to the same set *and* φ_k and φ_ℓ belong to the same set.

So that *one* spatial-basis integral may have to do duty for as many as *four* spin-basis integrals (each of which does duty for several by the usual rule of saving only one of the set of permutation-related integrals). This must all be taken care of in our code to generate the electron-repulsion matrices J and K.

$$\int dV_1 \int dV_2 \varphi_2(\vec{r_1})\varphi_1(\vec{r_1})\frac{1}{r_{12}}\varphi_3(\vec{r_2})\varphi_4(\vec{r_2})$$
$$= \int dV_1 \int dV_2 \varphi_{m+2}(\vec{r_1})\varphi_{m+1}(\vec{r_1})\frac{1}{r_{12}}\varphi_{m+3}(\vec{r_2})\varphi_{m+4}(\vec{r_2})$$
$$= \int dV_1 \int dV_2 \varphi_{m+2}(\vec{r_1})\varphi_{m+1}(\vec{r_1})\frac{1}{r_{12}}\varphi_3(\vec{r_2})\varphi_4(\vec{r_2})$$
$$= \int dV_1 \int dV_2 \varphi_2(\vec{r_1})\varphi_1(\vec{r_1})\frac{1}{r_{12}}\varphi_{m+3}(\vec{r_2})\varphi_{m+4}(\vec{r_2})$$

making use of the numbering convention established earlier.

12.5 Implementation

The implementation of the GUHF method is surprisingly similar to that of the closed-shell MO method. Both of them are concerned with the calculation of a set of orbitals which optimise the energy of a single-determinant wavefunction. The only real difference, apart from the doubling of the number of basis functions, is in the processing of the *spatial* basis functions to form the electron-repulsion matrices.

The codes for

- matrix multiplication etc.,
- formation of the R-matrix,

12.6. J AND K FOR GUHF

- matrix diagonalisation,
- convergence testing etc.

can be used unchanged. Indeed,

> If the basis size is doubled, the whole of the testbench program can be used **as it stands** for a GUHF program *if we write a new* **scfGR** *procedure* and arrange for the orthogonalisation and one-electron Hamiltonian matrices to be doubled and stored correctly.

This fact holds out the possibility of using *the same program* to do both types of calculation depending on some coded "option" supplied as data to the program.

12.6 J and K for GUHF

In the GUHF case the procedure **scfGR** must do three types of action:

1. Obtain the spatial repulsion integrals; either from an external file (conventional SCF) or by computing them (direct SCF).

2. Ensure that a given integral is used in the different "spin blocks" in the output matrix.

3. Use the permutation symmetries of the labels of an integral to make sure that all integrals are used.

Since the last two of these activities interact (the renumbering associated with adding m to some labels changes the ordering convention) it is best to separate the two by pushing the more primitive operation, the third, down into a separate procedure.

As a preliminary, the closed-shell **scfGR** is repeated below with this change implemented The **while** loop no longer needs the braces { } since it is only one statement; but the braces are retained as the loop is to be extended very soon.

```
    subroutine scfGR(R, G, m, nfile)
    double precision G(*), R(*);
    integer m, nfile;
#
    double precision val, two,one;
    integer i, j, k, l, mu, ij, ji;
    integer getint;
    integer pointer;
    data one,two/1.0d00,2.0d00/;
#   rewind nfile;
    pointer = 0;                    # initialise "file"
    while( getint(nfile, i, j, k, l, mu, val, pointer) != END_OF_FILE)
      {
      call GofR(R,G,m,two,one,i,j,k,l,val);
      }
# now symmetrise the G matrix
    do i = 1, m;
      { do j = 1, i;
         { ij = locGR(i,j);  ji = locGR(j,i);
           if ( i == j ) next;
           G(ji) = G(ij);
         }
      }
    return;
    end
```

12.6. J AND K FOR GUHF

```
      subroutine GofR(R,G,m,a,b,i,j,k,l,val)
      double precision R(*), G(*), val, a, b;
      integer i, j, k, l, m;
      integer ij, kl, il, ik, jk, jl;
      double precision coul1, coul2, coul3, exch;
      define(locGR,($2-1)*m+$1)   # macro for subscripting
         ij = locGR(i,j); kl = locGR(k,l);  # Set up all possible sets
         il = locGR(i,l); ik = locGR(i,k);  # of subscripts for given
         jk = locGR(j,k); jl = locGR(j,l);  # i j k l.
         if ( j < k ) jk = locGR(k,j);      # Notice convention for
         if ( j < l ) jl = locGR(l,j);      # label storage.

         coul1 = a * R(ij)*val; coul2 = a*R(kl)*val; exch = b*val;
#     coul1 and coul2 are the possible "Coulomb" contributions,
#     exch is the "Exchange" type
         if ( k != l )
            {
            coul2 = coul2 + coul2       # Double the term if k != l
            G(ik) = G(ik) - R(jl)*exch;
            if ( ( i != j ) & ( j >= k ) ) G(jk) = G(jk) - R(il)*exch;
            }
         G(il) = G(il) - R(jk)*exch; G(ij) = G(ij) + coul2;
         if ( ( i != j) & ( j >= l ) ) G(jl) = G(jl) - R (ik)*exch;
         if ( ij != kl )
            {
            coul3 = coul1;
            if ( i != j ) coul3 = coul3 + coul1;    # Doubled if i != j
            if( j <= k)
               {
               G(jk) = G(jk) - R(il)*exch;
               if (( i!= j ) & ( i <= k )) G(ik) = G(ik) - R(jl)*exch;
               if (( k!= l ) & ( j <= l )) G(jl) = G(jl) - R(ik)*exch;
               }
               G(kl) = G(kl) + coul3;
            }
      return;
      end
```

The operation of the new subroutine GofR has been very slightly generalised from the code of the closed-shell case; it is now possible to calculate a more general electron-repulsion matrix:

$$aJ(R) - bK(R)$$

for arbitrary a and b in anticipation of the GUHF case. These changes do not affect the operation of the call to scfGR of course.

What is necessary now is simply to insert the code to take care of the four possible spin cases into the while loop

354 CHAPTER 12. OPEN SHELLS: IMPLEMENTING UHF

```
while( getint(nfile, i, j, k, l, mu, val, pointer) != END_OF_FILE)
   {
   call GofR(R,G,m,two,one,i,j,k,l,val)
   }
```

The natural way to do this is to use an integer index (**spin**, say) in a loop and use the **switch/case** statements:

```
while( getint(nfile, i, j, k, l, mu, val, pointer) != END_OF_FILE)
   {
   for ( spin = 1; spin <= 4; spin = spin+1)
      {
      switch (spin)
         {
         case 1:    # Code for alpha - alpha case

         case 2:    # Code for beta - beta case

         case 3:    # Code for beta - alpha case

         case 4:    # Code for alpha - beta case

         default: STOP;  # Can't happen!!
         }          # End of switch
#   Arrange for i,j,k,l to be renumbered
#   and correct values of a and b
         call GofR(R,G,m,a,b,i,j,k,l,val)
      }             # End of for
   }                # End of while
```

12.6. J AND K FOR GUHF

There is only one small complication which prevents the coding being completely obvious. We must take steps to avoid counting the $\alpha - \beta$ and $\beta - \alpha$ contributions *twice* in the case when the labels are such that $(ij) = (k\ell)$.

Here is the code to do the job; explanation follows:

```
      subroutine scfGR(R, G, m, nfile)
      double precision G(*), R(*);
      integer m, nfile;
#
      integer mby2;
      double precision val;
      integer i, j, k, l, is, js, ks, ls, ijs, kls, mu;
      integer pointer, spin, skip;
      integer getint;    # integral reading function
      double precision zero, one, a, b;
      data zero,one/0.0d0,1.0d0/;
      define(locGR,($2-1)*m + $1)

      mby2 = m/2;
      rewind nfile;
      pointer = 0;
      while ( getint(nfile,is,js,ks,ls,mu,val,pointer) != END_OF_FILE )
         {
         ijs = is*(is-1)/2+js;   kls=ks*(ks-1)/2+ls;
         for ( spin = 1; spin <= 4; spin = spin+1)
            {   # runs over four possible spin combinations
            skip = NO;  # skip is a code used to avoid counting some
                        #        contributions twice: see case 4:
            switch ( spin )
              {
              case 1: { i=is; j=js; k=ks; l=ls; break;}
                         # alpha - alpha, ordering not changed
              case 2: {i=is+mby2; j=js+mby2; k=ks+mby2; l=ls+mby2; break;}
                         # beta - beta, ordering not changed
              case 3: {i=is+mby2; j=js+mby2; k=ks; l=ls; break;}
                         # beta - alpha, ordering not changed
              case 4: { if (ijs == kls )   skip= YES;
                         i=is; j=js; k=ks+mby2; l=ls+mby2;
                         call order(i,j,k,l); break;}
                         # alpha - beta, ordering may be changed
              } #end of "switch"
#
            if ( skip == YES ) next;
            a = one; b = one;
            if ( spin >= 3 )  b = zero;  # drop alpha-beta exchange
            call GofR(R,G,m,a,b,i,j,k,l,val);
         }  # end of "spin" loop
```

```
    } # end of while loop
# now symmetrise the G matrix
  do i = 1,m;
    { do j = 1,i;
      { ij = locGR(i,j);   ji = locGR(j,i);
        if ( i == j ) next;
        G(ji) = G(ij);
      }
    }
  return;
  end
```

The following notes explain the new pieces of code

- `mby2` is the spatial basis size, so that `m = 2*mby2` the spin-basis size.

- `is`, `js` etc. are used for the original spatial integral labels, `i`, `j` etc. are the "current" spin-basis labels.

- Adding `mby2` to each of the labels only destroys the canonical order of the labels in `case 4` where `k`, `l` may then be greater than `i`, `j`. In this case the labels must be reordered; the existence of `subroutine order` which does this has been assumed; here it is

  ```
      subroutine order(i,j,k,l)
      integer i, j, k, l, int
  #
      i = abs(i); j = abs(j); k = abs(k); l = abs(l);   # just in case
      if ( i < j )
        { int = i; i = j; j = int; }
      if ( k < l )
        { int = k; k = l; l = int; }
      if ( (i < k) | ((i == k) & (j < l)) )
        { int = i; i = k; k = int; int = j; j = l; l = int; }
      return;
      end
  ```

  The use of this routine is slight overkill since only the order in the *pairs* is disturbed by adding `mby2` to `k` and `l`.

- `skip` is used to avoid counting things twice in `case 3:` and `case 4:`

- In `case 3:` and `case 4:` there is no "exchange" contribution to the repulsion matrix so `b` is set to zero in these cases. This is slightly wasteful; the calculation should be omitted not multiplied by zero.

12.7 The GUHF testbench

Before putting together a GUHF testbench we need one utility to facilitate the generation of initial matrices for the calculation; a procedure to generate, from an $m \times m$ matrix (stored in a one-dimensional array as usual), a $2m \times 2m$ matrix of block form with the two $m \times m$ diagonal blocks equal to the given matrix and the two off-diagonal $m \times m$ blocks zero. This output matrix to be stored by rows in a one-dimensional array. Here is an example **spinor**: the code and a manual entry.

```
      subroutine spinor(H,m)
      double precision H(*);
#  Matrix comes in as contiguous columns of length m

      double precision zero;
      integer m;
#
      integer i, j, ij, ip, jp, ijp, nl,  n;
      data zero/0.0d0/;
      n=2*m;    nl=m+1;
#  Copy original into lower diagonal block
#  in columns of length 2*m
      do    i=1,m;
        {
          do    j=1,m;
            {
              ij=m*(j-1)+i ;   ip=i+m;   jp=j+m;
              ijp=n*(jp-1)+ip ;   H(ijp)=H(ij);
            }
        }
#  Now re-store the top diagonal block in columns of length 2*m
      do    i=1,m;
        {
          do    j=1,m;
            {
            ip=i+m ;    jp=j+m ;    ijp = n*(jp-1)+ip;
            ijd=n*(j-1)+i ;   H(ijd)=H(ijp);
            }
        }
#  Now put zeros into the off-diagonal blocks
      do    i=1,m;
        {
          do    j=nl,n;
            {
            ij=n*(j-1)+i ;    ji=n*(i-1)+j ;   H(ij)=zero ;   H(ji)=zero;
            }
        }
      return;
      end
```

NAME spinor
> Expand out a matrix over the spatial basis to the full spin-orbital basis

SYNOPSIS

```
subroutine spinor(H,m)
double precision H(ARB)
integer m
```

DESCRIPTION
> spinor puts two copies of the input data matrix (m x m) into H in 'blocked form' with off-diagonal blocks of zeroes. Usage is to set up the one-electron matrices for GUHF calculation. That is, H goes into spinor with m*m elements and emerges with (2m)*(2m) elements.

ARGUMENTS

> **H** Input/output: the matrix to be expanded. It is stored by columns of m elements on input and expanded out to be columns of (2*m) on output.
>
> **m** the dimension of the (square) input matrix H

DIAGNOSTICS
> None
> Take care that there is enough room in H for expansion.

With this utility and the new codes for scfGR we have all the materials for a GUHF testbench to hand.

In fact, the changes to the closed-shell program are minimal; all the changes of substance have been done in the new scfGR and its support GofR. All that is necessary is to

- Increase the size of all the arrays to allow for the doubling of the basis (and set these arrays to zero).
- Use spinor to double the h and V matrices stored in H and C and double the basis size.
- Make a trivial change by halving the total electronic energy since the orbitals are now *singly* occupied.

Nothing else is changed. Here is the new part of the code which is inserted at the start of the closed-shell program:

12.7. THE GUHF TESTBENCH

```
    mby2 = 7;    # Half the number of spin-basis functions
    n = 9;       # Number of electrons in H2O+
    interp = 4;
    irite= ERROR_OUTPUT_UNIT;
#   End of data for H2O

#   Double the basis and initial matrices
    call spinor(H,mby2);  call spinor(C,mby2);
    m = mby2 + mby2;
```

The simplest test of the program is to run it with the water data to recalculate the electronic structure of the ground state of the (closed-shell) H_2O system. If we do that we get the same total energy and the same MOs and MO energies (appearing twice in each case as both a α and a β spin-orbital).

A more interesting case is the genuine open-shell electronic structure of the H_2O^+ molecule for which the data is displayed in the above fragment of code (14 spin-basis functions, 9 electrons).

The output from the program is given below (with some of the intermediate iterations removed for conciseness.

```
Current Electronic Energy =      0.000000
Sum of differences in R =       17.25160     61  Changing
Current Electronic Energy =    -72.093987
Sum of differences in R =        8.19765     60  Changing
Current Electronic Energy =    -83.353555
....
....
Sum of differences in R =        0.00004     16  Changing
Current Electronic Energy =    -83.861532
Sum of differences in R =        0.00002      8  Changing
Current Electronic Energy =    -83.861532
Sum of differences in R =        0.00002      0  Changing
SCF converged in  23 iterations
Orbital Energies  -21.03405 -21.00316  -1.92935  -1.75931
   -1.20191  -1.15965  -1.11564  -1.06896  -1.00938  -0.22745
    0.08696   0.11939   0.20099   0.22133
```

This output needs a little explanation since it contains the numerical illustration of some new concepts.

- First of all, of course, there are now *nine* occupied MOs since the orbitals are all singly occupied. In fact, there are ten bound MOs (MOs with negative orbital energies) since the positively charged system holds all the electrons more tightly than the neutral molecule.

- Secondly, the MO energies show that the orbital energies of those orbitals which would be described as "doubly occupied" in the naive description of the system are not, in fact, the same; there is a group of MO energies from about -1.2 to about -1.0 which cannot be easily correlated with the MOs of the neutral molecule *on the basis of MO energies alone*.

- The total energy of the ion can be subtracted from the total energy of the molecule (computed with the closed-shell testbench earlier) to give an estimate of the ionisation energy of the water molecule: $84.16907 - 83.86153 = 0.30754$ (Hartrees). Notice that the MO energy of the highest occupied MO in the neutral water molecule calculation is -0.39137.

This estimate of the ionisation energy obtained by subtracting the total energies of the two species *at the same geometry, with the same orbital basis* is, at best, an approximation to the "vertical" ionisation energy; no allowance is made for molecular relaxation although the *electrons* have been allowed to attain self-consistency in each separate case.

- The tenth MO energy and the remaining four (positive) "MO energies" are just the result of diagonalising the $\boldsymbol{h}^F$ matrix in the iterative procedure; they are not involved in the wavefunction and the variation principle has not acted on them. These orbitals will not have any *physical* interpretation.

12.8 Interpreting the MO coefficients

We have obtained the (real) GUHF molecular orbitals of a simple open-shell system and seen that the result is a set of n (number of electrons) different MO energies. It is useful at this stage to take a look at the MO coefficients and make a preliminary interpretation of and explanation of the form that they take. The tables below list the MO coefficients of the minimal basis (real) GUHF calculation on the ground state of H_2O^+; the notation is standard:

$$\psi^i = \sum_{r=1}^{14} c_r^i \phi_r$$

with the associated MO energy ϵ_i.

Molecular orbital coefficients: H_2O^+ real GUHF

12.8. INTERPRETING THE MO COEFFICIENTS

ϵ_i	-21.03405	-21.00316	-1.92935	-1.75931	-1.20191
r of ϕ_r	c^1	c^2	c^3	c^4	c^5
1	0.000000	0.994666	0.000000	-0.232408	0.000000
2	0.000000	0.024204	0.000000	0.847403	0.000000
3	0.000000	0.003944	0.000000	0.180987	0.000000
4	0.000000	0.000000	0.000000	0.000000	0.000000
5	0.000000	0.000000	0.000000	0.000000	0.000000
6	0.000000	-0.005440	0.000000	0.138437	0.000000
7	0.000000	-0.005440	0.000000	0.138437	0.000000
8	0.994070	0.000000	-0.245296	0.000000	0.000000
9	0.026796	0.000000	0.915123	0.000000	0.000000
10	0.004273	0.000000	0.173334	0.000000	0.000000
11	0.000000	0.000000	0.000000	0.000000	0.741188
12	0.000000	0.000000	0.000000	0.000000	0.000000
13	-0.006002	0.000000	0.086242	0.000000	0.321110
14	-0.006002	0.000000	0.086242	0.000000	-0.321110

Molecular orbital coefficients: H_2O^+ real GUHF

ϵ_i	-1.15965	-1.11564	-1.06896	-1.00938
r of ϕ_r	c^6	c^7	c^8	c^9
1	0.000000	0.000000	0.000000	0.097348
2	0.000000	0.000000	0.000000	-0.476989
3	0.000000	0.000000	0.000000	0.840537
4	0.697694	0.000000	0.000000	0.000000
5	0.000000	0.000000	0.000000	0.000000
6	0.363371	0.000000	0.000000	0.192749
7	-0.363371	0.000000	0.000000	0.192749
8	0.000000	0.000000	0.084520	0.000000
9	0.000000	0.000000	-0.421485	0.000000
10	0.000000	0.000000	0.858440	0.000000
11	0.000000	0.000000	0.000000	0.000000
12	0.000000	1.000000	0.000000	0.000000
13	0.000000	0.000000	0.186231	0.000000
14	0.000000	0.000000	0.186231	0.000000

The first and most obvious comment to make about the MO coefficients is that they are, in fact, DODS not GUHF. That is, any one MO is composed entirely of α-spin basis functions (numbered 1 to 7) *or* entirely of β-spin basis functions (numbered 8 to 14). This is not an error, although the question of whether we found the *absolute* minimum in the electronic energy is still open; we have not *constrained* the MOs to be DODS, that is the way they came out. Further comment on this point is required and will be given shortly.

The physical interpretation of the MO coefficients is made easier if we have the information from the closed-shell calculation on the ground state of the neutral molecule available; here it is in the same format. Notice that, in this case, there are only seven spatial basis functions in the expansion.

Molecular Orbital Coefficients: H_2O

ϵ_i	-20.24147	-1.26907	-0.61856	-0.45321	-0.39137
r of ϕ_r	c^1	c^2	c^3	c^4	c^5
1	0.994124	-0.232734	0.000000	0.103260	0.000000
2	0.026582	0.833031	0.000000	-0.537344	0.000000
3	0.004353	0.130094	0.000000	0.776946	0.000000
4	0.000000	0.000000	0.606515	0.000000	0.000000
5	0.000000	0.000000	0.000000	0.000000	1.000000
6	-0.005991	0.158803	0.444933	0.277359	0.000000
7	-0.005991	0.158803	-0.444933	0.277359	0.000000

There are some striking similarities and some equally striking surprises:

1. The lowest two GUHF MOs are very similar to the doubly occupied MO of the closed shell.

2. MOs 3 and 4 are, again, rather similar in qualitative form to MO 2 of the closed shell (dominated by oxygen atom s orbital and hydrogen atom $1s$ contributions).

3. MOs 5 and 6 are qualitatively similar to MO 3 of the closed-shell case.

4. MO 7 is the same as MO 5 of the closed-shell case. It is the same because, in this limited basis, there is only *one* spatial basis function of π-type symmetry; if there were more basis functions of this symmetry then, presumably, orbital 7 would differ quantitatively from MO 5 of the neutral molecule. That is:

   The only MO of H_2O^+ which could be said to be "singly occupied", in the usual qualitative sense of the term, is the π-type MO which is the *highest* occupied MO in the ground state of the neutral molecule but is below *two* occupied MOs in the ion.

5. MOs 8 and 9 are qualitatively similar to MO 4 of the closed-shell case.

If we interpret the *qualitative* similarities noted in 1, 2, 3 and 5 as indicating some correlation with the simple idea of a doubly occupied spatial MO, we are left with the qualitative picture of the H_2O^+ ion which we might have expected: an unpaired electron in the highest occupied MO of the neutral species but *it is not the highest MO in energy*, which is in sharp contrast to the simple picture.

In fact this picture of the electronic structure of open-shell systems is the rule rather than the exception; the "unpaired electron" is usually in a MO which is *not* the highest in energy.

12.9 DODS or GUHF?

The physical interpretation of the reason why the spatial part of the molecular orbitals should differ for orbitals which differ in their spin factors is easy.

Although there are no specifically *spin-dependent* terms in the usual molecular Hamiltonian (Born–Oppenheimer, non-relativistic), it is easy to see that the actual energy expression is different for electrons with different spins if there are different *numbers* of α and β spins. Taking the simplest possible case of three electrons, there must be two of one kind of spin (α, say) and one of the other (β, say) if all the electrons are not to have the same spin. In this case, it is clear that the energy expression for the α electrons contains an additional non-zero term when written out in terms of spatial integrals; there is an additional "exchange"-type term in the energy expression involving the electrons of like spin which does not occur for the electrons of opposite spin. The effect is, of course, that when the variation principle acts during the solution of the SCF equations this difference is reflected in a difference in the detailed form of the α and β spatial orbitals; a phenomenon known as "spin polarisation".

The absence of any "mixing" between the two types of basis function — ones with α spin factor and ones with β — is also explicable in terms of the particular form of the molecular Hamiltonian. Since the Hamiltonian does not contain any α/β interaction terms, it *cannot* initiate any mixing between the two types of basis function in the generation of the MOs. Thus, starting, as the program does, from a zero initial $\boldsymbol{R}$ matrix, the separation of the α and β basis functions persists throughout the calculation leading to a DODS result.

What is more, if the calculation is started from an $\boldsymbol{R}$ matrix which does contain some $\alpha - \beta$ mixing, the standard Hamiltonian cannot *unmix* the two types of basis function for precisely the same reason as it cannot mix them: no spin interaction terms.

Thus the calculation of a "genuine" GUHF wavefunction is more difficult than our simple theory would suggest. For the time being we can preserve our GUHF program and use it for DODS calculations.

12.10 Version 1 of the SCF code

We now have code for the calculation of the optimum single-determinant wavefunction for the closed-shell and for the GUHF (or DODS) cases. These programs are so similar that it is not worthwhile to preserve separate versions for the two cases. We can amalgamate the two very easily and use a "switch" in the program to distinguish between the two cases. That is, we can "freeze" the development at this stage and call the result Version 1 and enter it into some version-control system as such. Of course, we shall go on to develop the methods and the implementation; but at the moment we have an SCF program which is capable of useful work.

The changes to the code to make it work for *both* the closed-shell *and* the GUHF case are very small; the program has to be taught to recognise which case to work on and take appropriate action when forming the J and K matrices in scfGR. This is done by the slightly clumsy method of "overloading" one of the arguments to the SCF function; the number of electrons is passed as positive if a GUHF calculation is required, and negative to request a closed-shell calculation.

At the risk of boredom by repetition, the complete codes for **integer function** scf and scfGR, together with manual pages are given below. After the manual page there are no less then *three* listings given since this is the time try and compare the merits and demerits of:

1. A straightforward listing of the code with comments.

2. The output from **weave** and LaTeX applied to the **WEB** source.

3. As in Chapter 8, the **WEB** source listing is given as an appendix (Appendix 12.A).

These are the codes for Version 1 of the SCF program, one or other of them should now be entered into the Revision Control System (RCS) described in Chapter 11 and Appendix 11.A.

12.10.1 Manual page for function scf

NAME scf
    LCAOMOSCF calculations

SYNOPSIS

```
integer function scf(H,C,nbasis,nelec,nfile,
            irite,damp,E,HF,V,R,Rold,Ubar,eps)
integer nbasis, nelec,  nfile , irite
double precision damp, E
double precision H(ARB), C(ARB), HF(ARB), V(ARB), R(ARB)
double precision Rold(ARB), Ubar(ARB), eps(ARB)
```

DESCRIPTION
    scf performs LC'AO' MO SCF calculations of either closed-shell RHF type or general (real) UHF type. The method is the traditional Roothaan repeated diagonalisations of the Hartree–Fock matrix until self-consistency is achieved.

ARGUMENTS

H Input: the one-electron Hamiltonian, `nbasis*nbasis` i.e. integrals over spatial orbitals

C Input/Output: an initial MO matrix - it must at least orthogonalise the basis. Normally, it is simply the orthogonalisation matrix `S**(-1/2)` on output the SCF eigenvector matrix U is in here

nbasis Input: the number of spatial orbitals in the basis - i.e. half the final spin-orbital basis size if `nelec` $\geq$ 0.

nelec Input: the number of electrons in the system - these electrons will be placed in the lowest `nelec` spin-orbitals. If `nelec` is less than zero it is assumed to be even and a standard closed shell calculation is done using only the `nbasis` spatial orbitals and occupying the lowest `abs(nelec)/2` spatial orbitals with two electrons each.

nfile the electron repulsion integrals are assumed to be stored in a file numbered `nfile` in the format of blocks of INT_BLOCK_SIZE integrals.

irite Channel number for convergence information or zero if this information is not required.

damp The value of the "Hartree damping" parameter: zero gives straight iteration, $0 \leq \text{damp} \leq 1$ gives damping $-1 \leq \text{damp} \leq 0$ gives extrapolation (not recommended!) try `damp` around 0.2 as a first guess.

This completes the input required - the rest of the argument list is workspace or output.

HF Workspace: for use as the Fock matrix

V Workspace: for the eigenvector matrix will have the eigenvectors on exit (as well as `U`)

R Workspace: for the density (charge and bond order) matrix and contains this matrix on exit

Rold Workspace: for retaining the penultimate density matrix for convergence testing.

Ubar Workspace: contains junk on exit

eps Output: contains the orbital energies on exit (first `nelec` are the occupied ones)

E Output: total electronic energy on exit.

Note that all the output matrices are (`2*nbasis`)-dimensional in the UHF case; even those put in as `nbasis`-dimensional.

RETURNS

YES if the calculation converged in `MAX_ITERATIONS`, NO if not. Typical useage is:

```
if ( scf(.....) == YES )
    output successful calculation
```

SEE ALSO

scfR , scfGR , eigen and support

DIAGNOSTICS

A message is printed on ERROR_OUTPUT_UNIT if the process does not converge after `MAX_ITERATIONS` iterations.

12.10. VERSION 1 OF THE SCF CODE

12.10.2 Simple coding of function scf

```
      integer function scf(H,C,nbasis,nelec,nfile,
                        irite,damp,E,HF,V,R,Rold,Cbar,epsilon)
      implicit double precision (a-h,o-z)
      integer nbasis, nelec, nfile
      integer irite
      double precision damp, E
      double precision H(ARB), C(ARB), HF(ARB), V(ARB), R(ARB)
      double precision Rold(ARB), Cbar(ARB), epsilon(ARB)
      {
      integer scftype, kount, maxit, nocc, m, mm, i
      double precision term, turm, Rsum
      double precision zero, half, crit
      data zero,half,crit/0.0d0,0.5d0,1.0d-06/
      data maxit/MAX_ITERATIONS/
      if ( nelec < 0 )              # test for closed shell or UHF
            { scftype = CLOSED_SHELL_CALCULATION;
              nocc = abs( nelec/2 );  # closed shells - even no. electrons
              m = nbasis;             # basis size is as given
            }
      else
            {
              scftype = UHF_CALCULATION;
              nocc = nelec;           # number of spin-orbitals
              m = 2 * nbasis;         # double the spatial basis; spin-orbitals
              call spinor (H, nbasis);   # double the one-electron Hamiltonian
              call spinor (C, nbasis);   # and the orthogonaliser
            }
      mm = m*m;
# Set initial R to zero, i.e. start from H not HF
      do i = 1, mm
            { R(i) = zero; Rold(i) = zero; }
#  Assume that the first iteration will change R to be non-zero!
      scf = YES;                    # successful return code
      kount = 0;
         repeat
            {
              kount = kount + 1;
              E = zero ; icon = 0;   # Initialize E and counter
              do i = 1, m*m
                HF(i) = H(i);
              do i = 1, mm                      # Accumulate the total
                E = E + R(i)*HF(i);             # energy
              call scfGR( R, HF, m, nfile,scftype);     # Use R to form HF
              do i = 1, mm                      # Accumulate rest
                E = E + R(i)*HF(i);             # of energy
              if (scftype == UHF_CALCULATION) E = half*E;
```

```
        write(ERROR_OUTPUT_UNIT,200) E;
        200 format(" Current Electronic Energy = ", f12.6);
        call gtprd(C, HF, R, m, m, m);    # Transform HF to
        call gmprd(R, C, HF, m, m, m);    # orthogonal basis
        call eigen(HF, Cbar, m);          # Diagonalise HF bar
        do i = 1, m
            epsilon(i) = HF(m*(i-1)+i);       # Get eigenvalues
        call gmprd(C, Cbar, V, m, m, m); # Transform coefficients
        call scfR(V, R, m, nocc);         # Form the R matrix
        Rsum = zero;
        do i =1, mm
            {
            turm = (R(i) - Rold(i));    #  Difference in each element
            term = abs(turm);
            Rold(i) = R(i);             # Set up Rold for next time
            C(i) = V(i);                # Restore C
            if ( term > crit ) icon = icon + 1; # Count the offenders
            Rsum = Rsum + term;
            R(i) = R(i) - damp*turm;    # damping
            }
        write(ERROR_OUTPUT_UNIT,201) Rsum, icon;
    201 format (" Sum of differences in R = ", f12.5,i6,
                " Changing");
    }
    until (icon == 0 | kount == maxit)
if ( kount == maxit )             # were iterations exhausted?
        {
        scf = NO;       # failure return code
        write( ERROR_OUTPUT_UNIT , 202 );
        202 format (" CONVERGENCE FAILURE, CYCLE LIMIT REACHED");
        }
return
}
end
```

Now here is the WEB output after processing by LaTeX.

12.11 WEB output for function scf

This is the output from WEB/LaTeXon the WEB source text for our function scf.

12.11.1 scf

This is Version 1 of the closed-shell/DODS SCF code; it returns *YES* if it converges, *NO* if not. However, the output is still supplied in the output arguments in the list.

Note that if *nelec* is positive the output matrices will be of the spin-orbital dimension i.e. $2 * nbasis$.

§. @I "defns.hweb" *Section(s) skipped...*

@m *MAX_ITERATIONS* 50

"scf.f" 12.11.1 ≡

 integer function *scf*(*H*, *C*, *nbasis*, *nelec*, *nfile*, *damp*, *E*, *HF*, *V*, *R*, *Rold*, *Cbar*, *epsilon*);

 ⟨Interface declarations 12.11.1.1⟩

 {

 ⟨scf local declarations 12.11.1.2⟩

 ⟨Establish type of SCF and set up basis 12.11.1.4⟩

 ⟨Initialise R matrix and counters 12.11.1.3⟩

 repeat

 {

 ⟨Single SCF Iteration 12.11.1.5⟩

 }

 until (*icon* ≡ 0 ∨ *kount* ≡ *maxit*)

 ⟨Evaluate convergence and finish off 12.11.1.9⟩

 return;

 }

These declarations are for the function arguments, the interface parameters passed to **function** *scf*.

⟨Interface declarations 12.11.1.1⟩ ≡
    implicit double precision $(a - h, \; o - z)$;
    integer *nbasis*, *nelec*, *nfile*;
    double precision *damp*, *E*;
    double precision $H(*), C(*), HF(*), V(*), R(*)$;
    double precision $Rold(*), Cbar(*), epsilon(*)$;

This code is used in section 12.11.1.

These are declarations for variables local to *scf*.

⟨scf local declarations 12.11.1.2⟩ ≡
    integer *scftype*, *kount*, *maxit*, *nocc*, *m*, *mm*, *i*;
    double precision *term*, *turm*, *Rsum*;
    double precision *zero*, *half*, *crit*;
    data *zero*, *half*, *crit*/$0.0 \cdot 10^0$D, $0.5 \cdot 10^0$D, $1.0 \cdot 10^{-06}$D/;
    data *maxit*/MAX_ITERATIONS/;

This code is used in section 12.11.1.

Various initialisations, set up the initial density matrix and the "previous" density matrix and initial information about convergence.

⟨Initialise R matrix and counters 12.11.1.3⟩ ≡
    $mm = m * m$;   /∗ Set initial R to zero, i.e. start from H not HF ∗/
    do $i = 1, \; mm$;
      {
      $R(i) = zero$;
      $Rold(i) = zero$;
      }   /∗ Assume that the first iteration will change R to be non-zero! ∗/
    $scf = YES$;   /∗ successful return code ∗/
    $kount = 0$;

This code is used in section 12.11.1.

12.11. WEB OUTPUT FOR FUNCTION SCF

Use information in *nelec* to decide between closed-shell and DODS and set up labels, basis sizes and matrices accordingly.

⟨Establish type of SCF and set up basis 12.11.1.4⟩ ≡
   if ($nelec < 0$)   /∗ test for closed shell or UHF ∗/
   {
     $scftype = CLOSED\_SHELL\_CALCULATION$;
     $nocc = \boldsymbol{abs}(nelec/2)$;   /∗ closed shells - even no. electrons ∗/
     $m = nbasis$;   /∗ basis size is as given ∗/
   }
   else
   {
     $scftype = UHF\_CALCULATION$;
     $nocc = nelec$;   /∗ number of spin-orbitals ∗/
     $m = 2 * nbasis$;   /∗ double the spatial basis; spin-orbitals ∗/
     call $spinor(H, nbasis)$;   /∗ double the one-electron Hamiltonian ∗/
     call $spinor(C, nbasis)$;   /∗ and the orthogonaliser ∗/
   }

This code is used in section 12.11.1.

This is the code for a single SCF iteration. It is broken down into smaller modules.

⟨ Single SCF Iteration 12.11.1.5 ⟩ ≡

   $kount = kount + 1$;

   ⟨ Initialise and one-electron energy 12.11.1.6 ⟩

      /* Use current R to form Hartree-Fock matrix in HF */
      call $scfGR(R, HF, nbasis, nfile, scftype)$;

   ⟨ Accumulate two-electron energy contribution 12.11.1.7 ⟩

      /* Transform HF to an orthogonal basis */
      call $gtprd(C, HF, R, m, m, m)$;
      call $gmprd(R, C, HF, m, m, m)$;

      /* Diagonalise the orthogonal-basis HF */
      call $eigen(HF, Cbar, m)$;

      /* Recover the eigenvalues from the diagonal of HF */
      do $i = 1, m$;
         $epsilon(i) = HF(m * (i - 1) + i)$;

      /* Transform current MO coefficients back to original basis */
      call $gmprd(C, Cbar, V, m, m, m)$;

      /* Use these coefficients to form next R-matrix */
      call $scfR(V, R, m, nocc)$;

   ⟨ Test for convergence and transfer matrices 12.11.1.8 ⟩

This code is used in section 12.11.1.

Put the one-electron Hamiltonian into HF and start the energy accumulation.

⟨ Initialise and one-electron energy 12.11.1.6 ⟩ ≡
   $E = zero$;
   $icon = 0$;   /* Initialise E and counter */
   do $i = 1, m * m$;
      $HF(i) = H(i)$;
   do $i = 1, mm$;   /* Accumulate the total energy */
      $E = E + R(i) * HF(i)$;

This code is used in section 12.11.1.5.

12.11. WEB OUTPUT FOR FUNCTION SCF

Add the two-electron contributions to the total energy and sort out the closed-shell/DODS difference.

⟨ Accumulate two-electron energy contribution 12.11.1.7 ⟩ ≡
   do $i = 1$, mm;   /∗ Accumulate rest of Energy ∗/
   $E = E + R(i) * HF(i)$;
   /∗ for DODS, the energy is double its true value ∗/
   if ($scftype \equiv UHF\_CALCULATION$)
   $E = half * E$;
   write ($ERROR\_OUTPUT\_UNIT$,
      200) E; 200 **format** ("␣Current␣Electronic␣Energy␣=␣", $f12.6$) ;

This code is used in section 12.11.1.5.

Compare the current and previous R-matrices and prepare for a new iteration by swapping old and new R matrices and replacing the orthogonalisation matrix.

⟨ Test for convergence and transfer matrices 12.11.1.8 ⟩ ≡
   $Rsum = zero$;
   do $i = 1$, mm;
   {
   $turm = (R(i) - Rold(i))$;
   $term = \mathbf{abs}(turm)$;
      /∗ Transfer R to $Rold$ for next iteration ∗/
   $Rold(i) = R(i)$;
   $C(i) = V(i)$;   /∗ Restore C ∗/
      /∗ Count the offenders ∗/
   if ($term > crit$)
      $icon = icon + 1$;
   $Rsum = Rsum + term$;
   $R(i) = R(i) - damp * turm$;   /∗ damping ∗/
   }
   write ($ERROR\_OUTPUT\_UNIT$, 201) $Rsum$, $icon$; 201
      format ("␣Sum␣of␣differences␣in␣R␣=␣", $f12.5$, $i6$,
      "␣␣Changing") ;

This code is used in section 12.11.1.5.

Check to see if things are OK. If run has converged return *YES* if not, return *NO* and cooment on this non-convergence.

⟨ Evaluate convergence and finish off 12.11.1.9 ⟩ ≡
   if ($kount \equiv maxit$)   /* were iterations exhausted? */
   { $scf = NO$;   /* failure return code */
   write ($ERROR\_OUTPUT\_UNIT$,
       202) ; 202 **format** ("␣␣CONVERGENCE␣FAILURE,\
   ␣CYCLE␣LIMIT␣REACHED") ; }

This code is used in section 12.11.1.

12.11.2 INDEX

abs: 12.11.1.4, 12.11.1.8.

C: 12.11.1.1.

Cbar: 12.11.1, 12.11.1.1, 12.11.1.5.

CLOSED_SHELL_CALCULATION: 12.11.1.4.

crit: 12.11.1.2, 12.11.1.8.

damp: 12.11.1, 12.11.1.1, 12.11.1.8.

E: 12.11.1.1.

eigen: 12.11.1.5.

epsilon: 12.11.1, 12.11.1.1, 12.11.1.5.

ERROR_OUTPUT_UNIT: 12.11.1.7, 12.11.1.8, 12.11.1.9.

gmprd: 12.11.1.5.

gtprd: 12.11.1.5.

H: 12.11.1.1.

half: 12.11.1.2, 12.11.1.7.

HF: 12.11.1, 12.11.1.1, 12.11.1.5, 12.11.1.6, 12.11.1.7.

i: 12.11.1.2.

icon: 12.11.1, 12.11.1.6, 12.11.1.8.

kount: 12.11.1, 12.11.1.2, 12.11.1.3, 12.11.1.5, 12.11.1.9.

m: 12.11.1.2.

MAX_ITERATIONS: 12.11.1, 12.11.1.2.

maxit: 12.11.1, 12.11.1.2, 12.11.1.9.

mm: 12.11.1.2, 12.11.1.3, 12.11.1.6, 12.11.1.7, 12.11.1.8.

nbasis: 12.11.1, 12.11.1.1, 12.11.1.4, 12.11.1.5.

nelec: 12.11.1, 12.11.1.1, 12.11.1.4.

nfile: 12.11.1, 12.11.1.1, 12.11.1.5.

NO: 12.11.1, 12.11.1.9.

nocc: 12.11.1.2, 12.11.1.4, 12.11.1.5.

12.11. WEB OUTPUT FOR FUNCTION SCF

R: <u>12.11.1.1</u>.

Rold: 12.11.1, <u>12.11.1.1</u>, 12.11.1.3, 12.11.1.8.

Rsum: <u>12.11.1.2</u>, 12.11.1.8.

scf: <u>12.11.1</u>, <u>12.11.1.1</u>, 12.11.1.2, 12.11.1.3, 12.11.1.9.

scfGR: 12.11.1.5.

scfR: 12.11.1.5.

scftype: <u>12.11.1.2</u>, 12.11.1.4, 12.11.1.5, 12.11.1.7.

spinor: 12.11.1.4.

term: <u>12.11.1.2</u>, 12.11.1.8.

turm: <u>12.11.1.2</u>, 12.11.1.8.

UHF_CALCULATION: 12.11.1.4, 12.11.1.7.

V: <u>12.11.1.1</u>.

YES: 12.11.1, 12.11.1.3, 12.11.1.9.

zero: <u>12.11.1.2</u>, 12.11.1.3, 12.11.1.6, 12.11.1.8.

12.12 Comments

The only way in which these codes differ from our earlier versions are in the definitions of CLOSED_SHELL_CALCULATION and UHF_CALCULATION which are defined to be two arbitrary (but different) integers. A small modification to the calculation of the total energy is required since, in the closed-shell case, the orbitals are doubly-occupied while in the UHF case they are just singly-occupied hence the multiplication of the total energy E by 0.5.

Appendix 12.A

WEB Source for the scf code

Here is the source for the WEB presented in Chapter 12. The simple outline of some of the WEB commands is available in Appendix 8.A which contains WEB source for another program.

The manual page is normally part of this WEB source but it is omitted from here since it is listed in Chapter 12.

```
@r

@* scf.
This is Version 1 of the closed-shell/DODS SCF code; it
returns |YES| if it converges, |NO| if not. However, the
output is still supplied in the output arguments in the list.

Note that if |nelec| is positive the output matrices will
be of the spin-orbital dimension  i.e. |2*nbasis|.

@I defns.hweb
@m MAX_ITERATIONS 50

@a
    integer function scf(H,C,nbasis,nelec,nfile,
                    damp,E,HF,V,R,Rold,Cbar,epsilon);

@<Interface declarations@>

    {

@<scf local declarations@>
```

@<Establish type of SCF and set up basis@>

@<Initialise R matrix and counters@>

    repeat

        {

@<Single SCF Iteration@>

        }

    until (icon == 0 || kount == maxit)

@<Evaluate convergence and finish off@>

    return;

}

@ These declarations are for the function arguments, the interface parameters passed to |function scf|.

```
@<Interface declarations@> =
   implicit double precision (a-h,o-z);
   integer nbasis, nelec, nfile;
   double precision damp, E;
   double precision H(*), C(*), HF(*), V(*), R(*);
   double precision Rold(*), Cbar(*), epsilon(*);
```

@ These are declarations for variables local to |scf|.

```
@<scf local declarations@> =
   integer scftype, kount, maxit, nocc, m, mm, i;
   double precision term, turm, Rsum;
   double precision zero, half, crit;
   data zero,half,crit/0.0d0,0.5d0,1.0d-06/;
   data maxit/MAX_ITERATIONS/;
```

@ Various initialisations, set up the initial density matrix and the ''previous'' density matrix and initial information about convergence.

```
@<Initialise R matrix and counters@> =
      mm = m*m;
/* Set initial R to zero, i.e. start from H not HF */
      do i = 1, mm;
         { R(i) = zero; Rold(i) = zero; }
```

```
/*  Assume that the first iteration will change |R|
    to be non-zero! */
    scf = YES;          /* successful return code */
    kount = 0;
```

@ Use information in |nelec| to decide between closed-shell
and DODS and set up labels, basis sizes and matrices accordingly.

```
@<Establish type of SCF and set up basis@> =
    if ( nelec < 0 ) /* test for closed shell or UHF */
        { scftype = CLOSED_SHELL_CALCULATION;
          nocc = abs( nelec/2 ); /* closed shells - even no. electrons */
          m = nbasis; /* basis size is as given */
        }
    else
        {
          scftype = UHF_CALCULATION;
          nocc = nelec;    /* number of spin-orbitals */
          m = 2 * nbasis;  /* double the spatial basis; spin-orbitals */
          call spinor (H, nbasis);  /* double the one-electron Hamiltonian */
          call spinor (C, nbasis);  /* and the orthogonaliser */
        }
```

@ This is the code for a single SCF iteration. It is broken down
into smaller modules.

```
@<Single SCF Iteration@> =

        kount = kount + 1;

@<Initialise and one-electron energy@>

/* Use current |R| to form Hartree--Fock matrix in |HF| */

        call scfGR( R, HF, nbasis, nfile, scftype);

@<Accumulate two-electron energy contribution@>

/* Transform |HF| to an orthogonal basis */

        call gtprd(C, HF, R, m, m, m);
        call gmprd(R, C, HF, m, m, m);

/* Diagonalise the orthogonal-basis |HF| */

        call eigen(HF, Cbar, m);

/* Recover the eigenvalues from the diagonal of |HF| */

        do i = 1, m;
```

```
            epsilon(i) = HF(m*(i-1)+i);

/* Transform current MO coefficients back to original basis */

        call gmprd(C, Cbar, V, m, m, m);

/* Use these coefficients to form next R-matrix */

        call scfR(V, R, m, nocc);


@<Test for convergence and transfer matrices@>


@ Put the one-electron Hamiltonian into |HF| and start
the energy accumulation.

@<Initialise and one-electron energy@> =
        E = zero; icon = 0;     /* Initialise E and counter */
        do i = 1, m*m;
            HF(i) = H(i);
        do i = 1, mm;           /* Accumulate the total energy */
            E = E + R(i)*HF(i);

@ Add the two-electron contributions to the total energy
and sort out the closed-shell/DODS difference.

@<Accumulate two-electron energy contribution@> =
        do i = 1, mm;   /* Accumulate rest of Energy */
            E = E + R(i)*HF(i);

/* for DODS, the energy is double its true value */

        if (scftype == UHF_CALCULATION) E = half*E;
        write(ERROR_OUTPUT_UNIT,200) E;
  200   format(" Current Electronic Energy = ", f12.6);

@ Compare the current and previous R-matrices and
prepare for a new iteration by swapping old and new R matrices
and replacing the orthogonalisation matrix.

@<Test for convergence and transfer matrices@> =
        Rsum = zero;
        do i =1, mm;
          {
          turm = (R(i) - Rold(i));
          term = abs(turm);

/* Transfer |R| to |Rold| for next iteration */
```

```
          Rold(i) = R(i);
          C(i) = V(i);  /*    Restore C */
/* Count the offenders */
          if ( term > crit ) icon = icon + 1;
          Rsum = Rsum + term;
          R(i) = R(i) - damp*turm;  /*   damping */
          }
       write(ERROR_OUTPUT_UNIT,201) Rsum, icon;
    201 format (" Sum of differences in R = ", f12.5,i6,
                " Changing");
```

@ Check to see if things are OK. If run has converged return
|YES| if not, return |NO| and comment on this non-convergence.

```
@<Evaluate convergence and finish off@> =
   if ( kount == maxit ) /*  were iterations exhausted? */
          {
          scf = NO;   /*  failure return code */
          write( ERROR_OUTPUT_UNIT , 202 );
       202 format (" CONVERGENCE FAILURE, CYCLE LIMIT REACHED");
          }
```

@* INDEX.

Appendix 12.B

Blocking the Hartree–Fock matrix

Almost all implementations of the GUHF equations are, in fact, Different Orbitals for Different Spins (DODS). If this model is used, the Hartree–Fock matrix will always be capable of being written in blocked form. The procedure used to diagonalise such a blocked matrix can be made to take advantage of this fact.

Contents

12.B.1 The block form of the HF matrix	**382**
12.B.2 Implementation	**384**
12.B.2.1 beigen	385
12.B.2.2 epsort	392
12.B.2.3 INDEX	392

12.B.1 The block form of the HF matrix

In this chapter, Version 1 of the SCF code has been presented which will perform closed-shell or DODS single-determinant wavefunction calculations. The only reason why the code, as presented, will not perform GUHF calculations is because of certain properties of the *equations*, not the codes. The fact that the conventional one-electron Hamiltonian has no spin-dependent terms, together with the fact that one normally uses the eigenvectors of this one-electron Hamiltonian to form an initial R-matrix, means that there is no mechanism within the iterative procedure to induce α–β "mixing" in forming optimum spin-orbitals. The "GUHF" orbitals will always turn out to be DODS in the absence of some

12.B.1. THE BLOCK FORM OF THE HF MATRIX

explicit procedure to form orbitals which are linear combinations of basis functions of both spin types.

What this means in practice is that, when we diagonalise the Hartree–Fock matrix, we spend a lot of time "transforming to zero" matrix elements which are already zero; the $\alpha\beta$ blocks of the matrix. Now this is not a lot of *numerical* work (it is trivial to test each element to see if it is zero before transforming it away) but there is a lot of wasted book-keeping involved here and it would be better not done at all. There are two kinds of approach we could take to this problem and make our GUHF program explicitly specialised to DODS:

1. Treat the α and β blocks of $\boldsymbol{h}^F$ separately and perform two linked SCF calculations, each of the dimension of the spatial basis. The calculations are linked because of the α–β repulsion in the $\boldsymbol{G}$ matrix.

2. Change the diagonalisation routine so that it knows that the matrix consists of two blocks which can be diagonalised separately, ignoring the zero off-diagonal blocks:

$$\boldsymbol{h}^F = \begin{pmatrix} \boldsymbol{h}^F_{\alpha\alpha} & \boldsymbol{0} \\ \boldsymbol{0} & \boldsymbol{h}^F_{\beta\beta} \end{pmatrix}$$

The first method has the advantage that no storage is wasted on the zero off-diagonal blocks at the expense of some increase in the complexity of the code.

The second method involves storing the zero blocks but trivial changes to the coding; all that is necessary is to teach the diagonalisation program (`eigen`) to recognise blocked matrices.

In fact, the second method will be used, because it is simpler *and* because we shall meet the need to deal with blocked Hartree–Fock matrices later when we try to simplify the calculations by use of molecular symmetry.

There is just one small complication; normally, in an SCF calculation we shall be requiring the *lowest state* of the molecule and we shall assume, as usual, that this lowest state is obtained by filling up the n lowest spin-orbitals. Thus, we shall require that the eigenvectors resulting from the blocked diagonalisation be in *"global"*, overall, energy order, so that `scfR` will generate the correct $\boldsymbol{R}$ matrix from the occupied orbitals. In fact, it is sometimes useful to be able to generate molecular states other than the ground state by, for example, calculating the SCF orbitals for a system of six α and four β electrons rather than the closed-shell of five electrons of each spin. This is possible when there is no state lower than the target state; clearly the single-determinant here would be a triplet spin state and there are no lower-lying single-determinants of this spin multiplicity *even though there are lower* single-determinant wavefunctions of different overall spin multiplicity. As we shall see, the same phenomenon occurs with molecular states classified by molecular symmetry.

The upshot is that it will be useful for now and for the future to have a blocked matrix diagonaliser and, possibly, a new scfR which, together, will enable DODS calculations to be performed more efficiently and to enable the calculation of a wider class of DODS wavefunctions. For the moment, we concentrate on the blocked diagonalisation code and associated eigenvector sorting.

12.B.2 Implementation

Implementation is simplicity itself; if we have a program for diagonalising a matrix all that is necessary is to arrange for that program to be used to deal with several matrices along the diagonal of a larger matrix.

Additional information required by the replacement for eigen is just:

1. The number of diagonal blocks comprising the overall matrix (nblock, say).

2. The dimensions of each of these blocks
   (iblock(i), i = 1, nblock, say)

Of course
$$\sum_{i=1}^{nblock} iblock(i) = n$$
The opportunity is also taken here to given the block diagonaliser the possibility of having an approximation to the matrix which is being sought passed to it *via* the argument:

- If init is zero, the diagonalisation is started from scratch (a unit U matrix).

- If init is non-zero, the diagonalisation process starts from the matrix supplied and merely updates this matrix until the process is complete.

Here is a WEB to perform this task, the manual page at the start is very similar to the eigen page.

12.B.2. IMPLEMENTATION

12.B.2.1 beigen

Old faithful Jacobi diagonisation routine. This version is modified to diagonalise a "blocked" matrix. The matrix is assumed to be *nblock* sub-matrices along the principal diagonal of dimension $iblock(1)$, $iblock(2)$, ... $iblock(nblock)$ with zeroes off these blocks. Of course, if $nblock = 1$ then $iblock(1) = n$.

NAME
    beigen

SYNOPSIS

```
subroutine beigen(H,U,n,init,nblock,iblock);
implicit double precision (a-h,o-z);
double precision H(*), U(*);
integer n, init, nblock, iblock(*);
```

DESCRIPTION
    Diagonalises a symmetric matrix which consists of *nblock* symmetric matrices along the diagonal of H. The eigenvalues and eigenvectors are returned in order of ascending eigenvalue.

ARGUMENTS

  H Input/Output Matrix to be diagonalised. On output, the eigenvalues are on the main diagonal of H.

  U Input/Output: If *init* is non-zero U contains a guess at the eigenvector matrix, if not it is initialised to the unit matrix. Ouput is the eigenvectors.

  n Input: The size of the matrices overall,

  init Input: zero if U is a sensible starting-point for the process, non-zero if not.

  nblock Input: The number of sub-matrices in H.

  iblock Input: $iblock(1)$ ... $iblock(nblock)$ are the dimensions of the sub-matrices.

DIAGNOSTICS
    None, possibility of infinite loop but unlikely.

"beigen.f" 12.B.2.1 ≡
   subroutine *beigen*(H, U, n, *init*, *nblock*, *iblock*);
   ⟨ Interface declarations 12.B.2.1.1 ⟩
   {
   data *zero*, *eps*, *one*, *two*, *four* /$0.0 \cdot 10^0$D, $1.0 \cdot 10^{-20}$D, $1.0 \cdot 10^0$D, $2.0 \cdot 10^0$D, $4.0 \cdot 10^0$D/;
   ⟨ Initialize U 12.B.2.1.2 ⟩
   $nmax = 0$;
   do $nsym = 1$, $nblock$;   /∗ Loop over the *nsym* sub-matrices ∗/
   {
      /∗ *nmin* and *nmax* are the limits of the current sub-matrix ∗/
   $nmin = nmax + 1$;
   $nmax = nmax + iblock(nsym)$;
      /∗ Start sweep through off-diagonal elements; the sweep is repeated
         until the largest off-diagonal element of H is less than *eps* ∗/
   repeat
   {
   ⟨ Reduce current off-diagonal to zero 12.B.2.1.3 ⟩ }
   until ($hmax < eps$)
   }   /∗ End of *nsym* loop on blocks ∗/
   ⟨ Sort the eigenvalues and vectors 12.B.2.1.7 ⟩ **return**;
   }

⟨ Interface declarations 12.B.2.1.1 ⟩ ≡
   implicit double precision ($a - h$, $o - z$);
   double precision $H(*)$, $U(*)$;
   integer n, *init*, *nblock*, *iblock*(∗);

This code is used in section 12.B.2.1.

12.B.2. IMPLEMENTATION

This simply sets U to the unit matrix. It is used if *init* is zero. If *init* is not zero, the incoming U is assumed to be a sensible starting-point for the calculation.

⟨ Initialize U 12.B.2.1.2 ⟩ ≡
   if $(init \equiv 0)$
   {
     do $i = 1,\ n;$
      {
      $ii = n * (i - 1) + i;$
      do $j = 1,\ n;$
       {
       $ij = n * (j - 1) + i;$
       $U(ij) = zero;$
       }
      $U(ii) = one;$
      }
   }

This code is used in section 12.B.2.1.

This is the central algorithm which calclautes and uses the "angle" which reduces the current off-diagonal element of H to zero. The effect on the other off-diagonals is computed and the largest off-diagonal element saved for convergence testing

⟨ Reduce current off-diagonal to zero 12.B.2.1.3 ⟩ ≡
    hmax = *zero*;
        /∗ *hmax* keeps track of the largest off-diagonal element of H ∗/

do $i = nmin + 1,\ nmax$;

{

jtop = $i - 1$;

do $j = nmin,\ jtop$;

{

⟨ Calculate Rotation Angle 12.B.2.1.4 ⟩

⟨ Apply Rotation to U 12.B.2.1.5 ⟩

⟨ Apply Rotation to H 12.B.2.1.6 ⟩

}

}

This code is used in section 12.B.2.1.

12.B.2. IMPLEMENTATION

tan is $tan(2\theta)$ where θ is the rotation angle which makes $H(ij)$ vanish. c and s are $cos\theta$ and $sin\theta$ obtained by the usual "half-angle formula" from $tan(2\theta)$

⟨ Calculate Rotation Angle 12.B.2.1.4 ⟩ ≡
    $ii = n * (i - 1) + i;$
    $jj = n * (j - 1) + j;$
    $ij = n * (j - 1) + i;$
    $ji = n * (i - 1) + j;$ /* positions of matrix elements */
    $hii = H(ii);$
    $hjj = H(jj);$
    $hij = H(ij);$
    $hsq = hij * hij;$
    if $(hsq > hmax)$
        $hmax = hsq;$
    if $(hsq < eps)$
        next;   /* omit zero H(ij) */
    $del = hii - hjj;$
    sign $= one;$
    if $(del < zero)$
      {
      sign $= -one;$
      $del = -del;$
      }
    $denom = del + \mathbf{dsqrt}(del * del + four * hsq);$

    tan $= two * \mathbf{sign} * hij / denom;$
    $c = one / \mathbf{dsqrt}(one + \mathbf{tan} * \mathbf{tan});$
    $s = c * \mathbf{tan};$

This code is used in section 12.B.2.1.3.

⟨ Apply Rotation to U 12.B.2.1.5 ⟩ ≡
  do $k = 1, \ n;$
  {
   $kj = n * (j - 1) + k;$
   $ki = n * (i - 1) + k;$
   $jk = n * (k - 1) + j;$
   $ik = n * (k - 1) + i;$
   $temp = c * U(kj) - s * U(ki);$
   $U(ki) = s * U(kj) + c * U(ki);$
   $U(kj) = temp;$
     /* If k is niether i or j then apply the current rotation */
   if $((i \equiv k) \vee (j \equiv k))$
     next;
   $temp = c * H(kj) - s * H(ki);$
   $H(ki) = s * H(kj) + c * H(ki);$
   $H(kj) = temp;$
   $H(ik) = H(ki);$
   $H(jk) = H(kj);$
  } /* This does not make any off-diagonal element zero; in fact it will, in general, re-generate ones which have been zeroized in other cycles */

This code is used in section 12.B.2.1.3.

⟨ Apply Rotation to H 12.B.2.1.6 ⟩ ≡
  $H(ii) = c * c * hii + s * s * hjj + two * c * s * hij;$
  $H(jj) = c * c * hjj + s * s * hii - two * c * s * hij;$
  $H(ij) = zero;$
  $H(ji) = zero;$
    /* This is the key step; it generates one zero off-diagonal element */

This code is used in section 12.B.2.1.3.

12.B.2. IMPLEMENTATION

Now Sort the eigenvalues and eigenvectors into ascending order. OVERALL; i.e. not within each block.

If it is required to sort the eigenvalues and vectors into ascending order *within* each block then this coding must be changed. For example, one may wish to occupy the lowest orbitals of each of several symmetry types to generate specific states of the molecule.

⟨ Sort the eigenvalues and vectors 12.B.2.1.7 ⟩ ≡
    $nmax = 0$;
    do $nsym = 1$, $nblock$;
        {
        $nmin = nmax + 1$;
        $nmax = nmax + iblock(nsym)$;
        call $epsort(H, U, n, nmin, nmax)$;
        }

This code is used in section 12.B.2.1.

12.B.2.2 epsort

Sort eigenvectors from n1 to n2 into eigenvalue order

"beigen.f" 12.B.2.1.8 ≡
```
    subroutine epsort(H, U, n, n1, n2);
    implicit double precision (a − h, o − z);
    double precision H(∗), U(∗);
    integer n, n1, n2;

    {
    double precision temp;
    integer iq, jq, ii, jj, k, ilr, imr;

    iq = (n1 − 2) ∗ n;
    do i = n1, n2;
      {
      iq = iq + n;
      ii = (i − 1) ∗ n + i;
      jq = n ∗ (i − 2);
      do j = i, n2;
        {
        jq = jq + n;
        jj = (j − 1) ∗ n + j;
        if (H(ii) < H(jj))
          next;    /∗ this means H(1) is lowest! ∗/
        temp = H(ii);
        H(ii) = H(jj);
        H(jj) = temp;
        do k = 1, n;
          {
          ilr = iq + k;
          imr = jq + k;
          temp = U(ilr);
          U(ilr) = U(imr);
          U(imr) = temp;
          }
        }
      }
    return;
    }
```

12.B.2.3 INDEX

12.B.2. IMPLEMENTATION 393

beigen: <u>12.B.2.1</u>.

del: 12.B.2.1.4.

denom: 12.B.2.1.4.

dsqrt: 12.B.2.1.4.

eps: <u>12.B.2.1</u>, 12.B.2.1.4.

epsort: 12.B.2.1.7, <u>12.B.2.1.8</u>.

four: <u>12.B.2.1</u>, 12.B.2.1.4.

H: <u>12.B.2.1.1</u>, <u>12.B.2.1.8</u>.

hii: 12.B.2.1.4, 12.B.2.1.6.

hij: 12.B.2.1.4, 12.B.2.1.6.

hjj: 12.B.2.1.4, 12.B.2.1.6.

hmax: 12.B.2.1, 12.B.2.1.3, 12.B.2.1.4.

hsq: 12.B.2.1.4.

iblock: 12.B.2.1, <u>12.B.2.1.1</u>, 12.B.2.1.7.

ii: 12.B.2.1.2, 12.B.2.1.4, 12.B.2.1.6, <u>12.B.2.1.8</u>.

ij: 12.B.2.1.2, 12.B.2.1.4, 12.B.2.1.6.

ik: 12.B.2.1.5.

ilr: <u>12.B.2.1.8</u>.

imr: <u>12.B.2.1.8</u>.

init: 12.B.2.1, <u>12.B.2.1.1</u>, 12.B.2.1.2.

iq: <u>12.B.2.1.8</u>.

ji: 12.B.2.1.4, 12.B.2.1.6.

jj: 12.B.2.1.4, 12.B.2.1.6, <u>12.B.2.1.8</u>.

jk: 12.B.2.1.5.

jq: <u>12.B.2.1.8</u>.

jtop: 12.B.2.1.3.

k: <u>12.B.2.1.8</u>.

ki: 12.B.2.1.5.

kj: 12.B.2.1.5.

n: <u>12.B.2.1.1</u>, <u>12.B.2.1.8</u>.

nblock: 12.B.2.1, <u>12.B.2.1.1</u>, 12.B.2.1.7.

nmax: 12.B.2.1, 12.B.2.1.3, 12.B.2.1.7.

nmin: 12.B.2.1, 12.B.2.1.3, 12.B.2.1.7.

nsym: 12.B.2.1, 12.B.2.1.7.

n1: <u>12.B.2.1.8</u>.

n2: <u>12.B.2.1.8</u>.

one: <u>12.B.2.1</u>, 12.B.2.1.2, 12.B.2.1.4.

sign: 12.B.2.1.4.

tan: 12.B.2.1.4.

temp: 12.B.2.1.5, <u>12.B.2.1.8</u>.

two: <u>12.B.2.1</u>, 12.B.2.1.4, 12.B.2.1.6.

U: <u>12.B.2.1.1</u>, <u>12.B.2.1.8</u>.

zero: <u>12.B.2.1</u>, 12.B.2.1.2, 12.B.2.1.3, 12.B.2.1.4, 12.B.2.1.6.

Appendix 12.C

The Aufbau principle

From the very start, we have silently assumed that the aufbau principle is valid; we have made the reasonable assumption that the lowest-energy single-determinant wavefunction is the one obtained by occupying the n lowest-energy solutions of the Hartree–Fock equations. As the complexity of the systems which we treat becomes greater and particularly the highest occupied MOs crowd more closely together, this assumption looks less and less obvious. Here we examine the matter a little more closely, both theoretically and practically. The details of the investigation have some affinities with the self-consistent perturbation theory to be developed later.

Contents

12.C.3	Introduction	394
12.C.4	The second variation	395
12.C.5	Special case: a single excitation	396

12.C.3 Introduction

The energy of a single-determinant wavefunction is given by

$$E = \sum_{i=1}^{n} <\chi_i|\hat{h}|\chi_i> + \frac{1}{2}\sum_{i=1}^{n}\sum_{j=1}^{n}\left(<\chi_i\chi_j|\frac{1}{r_{12}}|\chi_i\chi_j> - <\chi_i\chi_j|\frac{1}{r_{12}}|\chi_j\chi_i>\right)$$

or, in terms of matrices involved in the SCF theory:

$$E = \mathrm{tr}\,\boldsymbol{h}^F\boldsymbol{R} - \frac{1}{2}\mathrm{tr}G\boldsymbol{R}$$

12.C.4. THE SECOND VARIATION

which may be expressed as

$$E = \sum_{i=1}^{n} \epsilon_i - \frac{1}{2} \sum_{i,j=1}^{n} \left(<\chi_i\chi_j|\frac{1}{r_{12}}|\chi_i\chi_j> - <\chi_i\chi_j|\frac{1}{r_{12}}|\chi_j\chi_i> \right)$$

This expression shows very clearly that, in addition to the involvement of electron repulsion in the values of the orbital energies, additional *electron repulsion* occurs in the expression for the total energy. That is in the absence of a *proof* of the aufbau principle, we cannot simple assume that the first (orbital energy) term is always dominant.

12.C.4 The second variation

In elementary analysis, the *nature* of a turning point in a function (a point at which the derivative is zero) is determined by the *sign* of the second derivative. We must, therefore, look at the analogue of the second derivative in our derivation; the *second-order* variation induced in the total energy by a variation in the MO coefficients.

Working in terms of the invariant density matrix $\boldsymbol{R}$ that characterises the SCF solution we obtain[3]

$$E \longrightarrow E + \delta E + \delta^{(2)} E + \ldots$$

where the so-called second variation $\delta^{(2)}$ is given by

$$\delta^{(2)} E = \text{tr}\left(\delta^{(2)} \boldsymbol{R} \boldsymbol{h}^F\right) + \frac{1}{2}\text{tr}\left[\delta \boldsymbol{R} G(\delta \boldsymbol{R})\right] - \sum_{i=1}^{n} \left(\text{tr}\boldsymbol{R}^i \boldsymbol{h}^F\right)\left(\text{tr}\delta^{(2)} \boldsymbol{R}^i\right)$$

where, as usual,

$$\boldsymbol{R} = \sum_{i=1}^{n} \boldsymbol{R}^i$$

(n of the m MOs are occupied) and, in terms of the MO coefficients of these occupied MOs:

$$\delta \boldsymbol{R} = \delta \boldsymbol{c}^i \boldsymbol{c}^{i\dagger} + \boldsymbol{c}^i \delta \boldsymbol{c}^{i\dagger}$$

and

$$\delta^{(2)} \boldsymbol{R} = \delta \boldsymbol{c}^i \delta \boldsymbol{c}^{i\dagger}$$

If we work in a *canonical basis* (one in which the matrix $\boldsymbol{\epsilon}$ is diagonal):

$$\boldsymbol{h}^F \boldsymbol{c}^i = \epsilon_i \boldsymbol{c}^i$$

---
[3]This result contains the idempotency constraint on the density matrix.

for $i = 1, n$ for the occupied MOs, and

$$\boldsymbol{h}^F \boldsymbol{c}^k = \epsilon_k \boldsymbol{c}^k$$

for the virtual orbitals $k = (n + 1), n$, we may expand the *change* in each $\boldsymbol{c}^i$ entirely in terms of the $\boldsymbol{c}^k$ because of the invariance of the total $\boldsymbol{R}$:

$$\delta \boldsymbol{c}^i = \sum_{k=(n+1)}^{n} \lambda_{ik} \boldsymbol{c}^k$$

for some parameters λ_{ik} which are to be determined. We will be interested in a special case of the latter expression when one occupied MO is *replaced* by a virtual MO in the single determinant, but, for the moment, we derive the general expression.

Substitution of these expansions into the expression for $\delta^{(2)} E$ gives:

$$\begin{aligned}
\delta^{(2)} E &= \sum_{i=1}^{n} \sum_{k=(m+1)}^{m} \left((\epsilon_k - \epsilon_i) |\lambda_{ik}|^2 \right) \\
&- \frac{1}{2} \sum_{i,j=1}^{n} \sum_{k,\ell=(m+1)}^{m} \lambda_{ik} \lambda_{j\ell} \left(2 < \chi_i \chi_j | \frac{1}{r_{12}} | \chi_k \chi_\ell > \right. \\
&- \left. < \chi_i \chi_j | \frac{1}{r_{12}} | \chi_\ell \chi_k > - < \chi_i \chi_\ell | \frac{1}{r_{12}} | \chi_j \chi_k > \right)
\end{aligned}$$

12.C.5 Special case: a single excitation

Now, consider the case when one occupied MO is replaced by occupancy of one of the virtual orbitals in the single determinant; what one might call an *excitation* from χ_i (determined by $\boldsymbol{c}^i$) to χ_k (coefficients $\boldsymbol{c}^k$), that is

$$\lambda_{ik} = 1$$

and all other λs are zero

$$\lambda_{j\ell} = 0$$

In this case, the second variation in the total energy becomes:

$$\delta^{(2)} E = (\epsilon_k - \epsilon_i) - \left(< \chi_i \chi_j | \frac{1}{r_{12}} | \chi_i \chi_j > - < \chi_i \chi_j | \frac{1}{r_{12}} | \chi_j \chi_i > \right)$$

Now, it is physically reasonable that the Coulomb-type integral is always greater than the associated exchange-type integral and, in fact, it has been proved[4] that this is so, thus the bracketed repulsion term is *always positive*.

---

[4] A. Okninski, *Chem. Phys. Letters*, **27**, 603 (1974).

12.C.5. SPECIAL CASE: A SINGLE EXCITATION

So, if the second variation in the energy is to be positive, guaranteeing a *minimum* in the total electronic energy, the orbital energy difference ($\epsilon_k - \epsilon_i$) must be *positive* (and of sufficient value to more than cancel the repulsion term). That is, a minimum can only be guaranteed if

$$\epsilon_k > \epsilon_i$$

for all occupied MOs χ_i and all virtuals χ_k. This is just the aufbau principle; the lowest-energy single determinant of an n-electron system is obtained by occupying the *lowest* n solutions of the Hartree–Fock equation.

There is one silent assumption behind this result: the n lowest solutions must be the ones obtained as *optimum* (unconstrained) solutions. That is, in general, if any constraints are placed on the solution of the Hartree–Fock equation, the aufbau principle cannot be guaranteed. We shall return to this point in due course.

In view of this result it is unlikely that any occupied MO will be degenerate with any unoccupied MO, in particular the familiar "spherically constrained" model of atomic structure in which (*e.g.*) all $3d$ AOs, whether occupied or not, have the same energy and radial function is likely to violate the aufbau principle.

Chapter 13

Population analysis

Now that we have the capability to calculate the electron distributions in both open- and closed-shell systems, we can try to analyse these distributions in terms of the atomic contributions of which they are composed.

Contents

13.1	Introduction	399
13.2	Densities and spin-densities	400
13.3	Basis representations: charges	401
13.4	Basis-function analysis	404
13.5	A cautionary note	406
13.6	Multi-determinant forms	407
13.7	Implementation	408

13.1 Introduction

In an earlier appendix we introduced the idea that the R matrix carries an algebraic representation of the total electron probability distribution, what is loosely called the charge distribution:

$$\rho(\vec{x}) = \boldsymbol{\varphi}(\vec{x})\boldsymbol{R}\boldsymbol{\varphi}^\dagger(\vec{x})$$

where the basis functions φ are spin-basis functions in general.

In the two cases we have met so far the spin-dependence may be either:

1. Eliminated, because a closed-shell system has equal and opposite α and β spin distribution.

   or

2. Separated by writing the overall $\boldsymbol{R}$ matrix in block form in terms of the α- and β-basis functions.

With this in mind let us write the total electron density as

$$\rho^T(\vec{x}) = \rho^\alpha(\vec{r})|\alpha(s)|^2 + \rho^\beta(\vec{r})|\beta(s)|^2$$

writing the space/spin overall density (dependent on $\vec{x}$) as a sum of a pair of *spatial* functions multiplied by the appropriate densities in spin space. In both cases there are no spin "cross-terms" like $\alpha\beta$. The two functions of space only given the *spatial* distribution of the two types of electron: spin α and spin β.

The purely spatial *total* electron distribution may be obtained simply by integrating over the spin variable, generating unity in both cases:

$$\rho(\vec{r}) = \int ds \rho^T(\vec{x}) = \rho^\alpha(\vec{r}) + \rho^\beta(\vec{r})$$

Similarly we may obtain the *spatial* distribution of spin difference by integrating the difference between the two separate spin densities:

$$\rho^S(\vec{r}) = \int ds \left(\rho^\alpha(\vec{r}) - \rho^\beta(\vec{r})\right)$$

Now both of these expressions involve only *spatial* functions and so may be expanded in terms of the basis of spatial functions (ϕ not φ) used in the original SCF calculation.

13.2 Densities and spin-densities

The DODS $\boldsymbol{R}$ matrix[1] has the block form

$$\boldsymbol{R} = \begin{pmatrix} \boldsymbol{R}^\alpha & 0 \\ 0 & \boldsymbol{R}^\beta \end{pmatrix}$$

so that the two densities defined in the last section may be written as

$$\rho(\vec{r}) = \boldsymbol{\phi}(\vec{r})\boldsymbol{R}^\alpha\boldsymbol{\phi}^\dagger(\vec{r}) + \boldsymbol{\phi}(\vec{r})\boldsymbol{R}^\beta\boldsymbol{\phi}^\dagger(\vec{r})$$

and

$$\rho^S(\vec{r}) = \boldsymbol{\phi}(\vec{r})\boldsymbol{R}^\alpha\boldsymbol{\phi}^\dagger(\vec{r}) - \boldsymbol{\phi}(\vec{r})\boldsymbol{R}^\beta\boldsymbol{\phi}^\dagger(\vec{r})$$

---
[1] This matrix is of size $2m \times 2m$ for m spatial basis functions.

13.3. BASIS REPRESENTATIONS: CHARGES

while, for the closed-shell case "$R^\alpha = R^\beta = R$" so that

$$\rho(\vec{r}) = 2\phi(\vec{r})R\phi^\dagger(\vec{r})$$

and

$$\rho^S(\vec{r}) \equiv 0$$

So, if we use the notation P for that matrix which represents the total electron density and Q for the spatial distribution of electron spin, we have in both cases,

$$\begin{aligned} P &= R^\alpha + R^\beta \\ Q &= R^\alpha - R^\beta \end{aligned}$$

in which all the matrices are the size of the *spatial* basis, what we have usually called m. Thus, the two spatial functions become

$$\begin{aligned} \rho(\vec{r}) &= \phi(\vec{r})P\phi^\dagger(\vec{r}) \\ \rho^S(\vec{r}) &= \phi(\vec{r})Q\phi^\dagger(\vec{r}) \end{aligned}$$

The normalisation of the MO coefficients is expressed by

$$\int \rho(\vec{r}) dV = \text{tr}PS = n$$

$$\int \rho^S(\vec{r}) dV = \text{tr}QS = n_\alpha - n_\beta$$

where n is the total number of electrons and n_α, n_β are the numbers of electrons of each spin type and

$$n = n_\alpha + n_\beta$$

and

$$S = M_S = \frac{1}{2}(n_\alpha - n_\beta)$$

13.3 Basis representations: charges

One of the results of the last section may be written in full as

$$\rho(\vec{r}) = \sum_{i,j=1}^{m} P_{ij} \phi_i \phi_j$$

which may have the terms grouped together by expressing the summations over basis functions centred on particular atomic *centres*:

$$\rho(\vec{r}) = \sum_{A,B=1}^{N} \left(\sum_{i \in A, j \in B} P_{ij} \phi_i \phi_j \right)$$

Now the products $\phi_i(\vec{r})\phi_j(\vec{r})$ are themselves[2] components of the electron densities of the separate atoms. Indeed if the basis functions were just the *atomic orbitals* of the separate atoms, these products would simply sum to the electron densities of those atoms:

$$\rho_A(\vec{r}) = \sum_{i \in A} \nu_i \phi_i^2(\vec{r})$$

if the AOs have occupation numbers ν_i in the separate atoms.

Keeping this AO interpretation in mind, we can see that the expression for the total molecular electron density has the form:

$$\begin{aligned}\rho(\vec{r}) &= \sum_{AOs}(Number) \times (AO\ density) \\ &+ \sum_{Pairs\ of\ AOs}(Number) \times (AO\ overlap\ density)\end{aligned}$$

with an expression of identical form for the spin density, where "*Number*" is an element of $\boldsymbol{P}$ or $\boldsymbol{Q}$, as appropriate. If the basis functions are merely atom-centred functions rather than actual atomic orbitals the physical interpretation is similar, but the relationship to the distributions in the separate is atoms not quite so direct.

The detailed expression for ρ is

$$\rho(\vec{r}) = \sum_{i,j=1}^{m} P_{ij}\phi_i(\vec{r})\phi_j(\vec{r})$$

(for real MO coefficients and real basis functions). This expression integrates to

$$\begin{aligned}\int \rho(\vec{r})dV &= \sum_{i,j=1}^{m} P_{ij}S_{ji} \\ &= \sum_{i=1}^{m} P_{ii} + \sum_{i,i \neq j=1}^{m} P_{ij}S_{ji} \\ &= \text{tr}\boldsymbol{PS} = n\end{aligned}$$

If the basis were orthonormal, $\boldsymbol{S} = \boldsymbol{1}$, the second (double) sum would vanish, not because the P_{ij} were zero but because the off-diagonal overlaps are all zero. This would generate the convenient result

$$\sum_{i=1}^{m} P_{ii} = n$$

and the interpretation of the numbers P_{ii} would be straightforward:

---

[2] Assuming *real* basis functions, as usual.

13.3. BASIS REPRESENTATIONS: CHARGES

In an orthonormal basis, the diagonal elements, P_{ii}, of the matrix $\boldsymbol{P}$ can be interpreted as the "occupation numbers" of the basis functions in the molecule.

The definition of $\boldsymbol{P}$ as $2\boldsymbol{R}$ ensures that these numbers all lie between zero and two and therefore satisfy the obvious simplest requirement of occupation numbers. Since the charge on the electron is the magnitude of the unit of charge we use, these numbers are frequently called the "charges" associated with each basis function. Chemical information can be obtained now by comparing the P_{ii} (which are the occupation numbers of the AOs *within* the molecule) with the occupation numbers (ν_i) of the same AOs in the separate atoms of which the molecule is composed.

Clearly information is lost if this terminology were to be taken too literally since it would suggest the electron distribution to be

$$\rho(\vec{r}) = \sum_{i=1}^{m} P_{ii} \phi_i^2(\vec{r})$$

which is incorrect. The difference between this expression, involving only the charges, and the full expression lies in electron distribution contributions which *integrate to zero*; they are present in the distribution but do not contribute to the coarse numerical measure of the distribution given by the P_{ii}.

If the basis functions were AOs, then the amount by which the orbital products $\phi_i \phi_j$ were occupied in the charge density expression is obviously important; the numbers multiplying the distributions $\phi_i(\vec{r})\phi_j(\vec{r})$ *for nearest-neighbours* are contributions to the charge-density involved in bond formation (or not, as the case may be). There is a good case for assuming that, the larger these numbers (in an algebraic sense), the stronger the bond between the atoms on which the two functions ϕ_i and ϕ_j are centred; these quantities P_{ij}, $i \neq j$, are often called "bond orders". As usual, this physical interpretation is at its strongest and most unequivocal when the basis of functions used to expand the MOs are actually the AOs of the separate atoms and not just a set of convenient atom-centered functions.

Perhaps the feature which causes the most unease among these definitions and interpretations is the fact that the bond-orders make no contribution to the overall electron count in an orthonormal basis. However, if we look at the matrices for the simplest possible case, a minimal basis (AO) calculation on a two-orbital model, we see that

$$\boldsymbol{c} = \begin{pmatrix} c \\ s \end{pmatrix}$$

($c = cos\theta$, $s = sin\theta$ for some θ)

$$\boldsymbol{P} = 2\boldsymbol{R} = 2\boldsymbol{cc}^\dagger = \begin{pmatrix} 2c^2 & 2cs \\ 2cs & 2s^2 \end{pmatrix}$$

and it is clear that the bond-orders are actually fully determined by the charges;

$$P_{12} = P_{21} = \sqrt{P_{11} \times P_{22}}$$

This simple result depends on the orthogonality of the basis functions and the simplicity of the two-AO model, but it may go some way to showing that there are fewer degrees of freedom in the P-matrix than might, at first sight, appear.[3] With these initial considerations in mind we can try to interpret the full, non-orthogonal, non-AO-basis P and Q matrices.

13.4 Basis-function analysis

We take the closed-shell $P = 2R$ as the archetype of this analysis for the moment.

Let us suppose that the MOs of a molecule are linear combinations of some basis functions which are divided into two sets (A and B, say) which might be associated with two atoms A and B or grouped together in some more general way, perhaps as molecular fragments:

$$\psi = \phi C = \left(\phi^A, \phi^B\right) \begin{pmatrix} C^A \\ C^B \end{pmatrix}$$

and that

$$P = 2R = 2CC^\dagger = 2 \begin{pmatrix} C^A \\ C^B \end{pmatrix} \left(C^{A\dagger}, C^{B\dagger}\right)$$

That is,

$$P = \begin{pmatrix} P^{AA} & P^{AB} \\ P^{BA} & P^{BB} \end{pmatrix}$$

where

$$P^{AB} = 2C^A C^{B\dagger}$$

etc. The overlap matrix may be partitioned similarly:

$$S = \begin{pmatrix} S^{AA} & S^{AB} \\ S^{BA} & S^{BB} \end{pmatrix}$$

The normalisation condition on the MOs C is

$$\text{tr} PS = n$$

---

[3]It must be stressed that this result is acutely dependent on the single-determinant model; the number of independent elements in a symmetric matrix is $(m(m+1))/2$ (2, in the simple model above), while the number of independent parameters in the MO matrix is $n(m-n)$ which has a maximum of $m^2/4$ (1, as expected, in the simple model).

13.4. BASIS-FUNCTION ANALYSIS

that is, in terms of the partitioned form of these matrices,

$$\mathrm{tr}\begin{pmatrix} \boldsymbol{P}^{AA}\boldsymbol{S}^{AA}+\boldsymbol{P}^{AB}\boldsymbol{S}^{BA} & \boldsymbol{P}^{AA}\boldsymbol{S}^{AB}+\boldsymbol{P}^{AB}\boldsymbol{S}^{BA} \\ \boldsymbol{P}^{AA}\boldsymbol{S}^{AA}+\boldsymbol{P}^{BB}\boldsymbol{S}^{BA} & \boldsymbol{P}^{BA}\boldsymbol{S}^{AB}+\boldsymbol{P}^{BB}\boldsymbol{S}^{BB} \end{pmatrix}=n$$

but, since the trace operation only involves the diagonals,

$$\begin{aligned}\mathrm{tr}\boldsymbol{PS} &= \mathrm{tr}\left(\boldsymbol{P}^{AA}\boldsymbol{S}^{AA}+\boldsymbol{P}^{AB}\boldsymbol{S}^{BA}\right) \\ &+ \mathrm{tr}\left(\boldsymbol{P}^{BB}\boldsymbol{S}^{BB}+\boldsymbol{P}^{BA}\boldsymbol{S}^{AB}\right)\end{aligned}$$

However, the matrices $\boldsymbol{S}$ and $\boldsymbol{P}$ are *symmetrical*:

$$\boldsymbol{S}^{AB} = \boldsymbol{S}^{BA\dagger} \quad \boldsymbol{P}^{AB} = \boldsymbol{P}^{BA\dagger}$$

so that

$$\mathrm{tr}\boldsymbol{P}^{AB}\boldsymbol{S}^{BA} = \mathrm{tr}\boldsymbol{P}^{BA}\boldsymbol{S}^{AB}$$

Taking our cue from the orthonormal case, we write

$$\mathrm{tr}\boldsymbol{P}^{AA}\boldsymbol{S}^{AA} = \sum_{i,j} P_{ij}^{AA} S_{ji}^{AA} = n_A$$
$$\mathrm{tr}\boldsymbol{P}^{BB}\boldsymbol{S}^{BB} = \sum_{i,j} P_{ij}^{BB} S_{ji}^{BB} = n_B$$
$$\mathrm{tr}\boldsymbol{P}^{BA}\boldsymbol{S}^{AB} = \mathrm{tr}\boldsymbol{P}^{AB}\boldsymbol{S}^{BA} = \sum_{i,j} P_{ij}^{BA} S_{ji}^{AB} = n_{AB}$$

where

$$n = n_A + n_B + 2n_{AB}$$

and n_A, n_B are the numbers of electrons unequivocally associated with atom (or group) A and B, respectively, and n_{AB} is a general overlap population. Notwithstanding possible bond polarities, it is common to associate one copy each of n_{AB} with n_A and n_B to generate a partitioning of the electrons between atoms so that they sum to the total:

$$n = (n_A + n_{AB}) + (n_B + n_{AB}) = q_A + q_B$$

(say), where the q_A, q_B are now *atomic* charges, the overlap populations having been absorbed into the atomic partitioning of charge.

Naturally, the *elements* of the product matrices are interpretable as the individual *basis function* charges and "bond orders" and the way in which the *relative* populations of these products change from atom to molecule or from molecule to molecule gives a great deal of detailed information which is absent in the coarse atomic charge analysis.

Everything which has been said about $\boldsymbol{P}$ may also be said about $\boldsymbol{Q}$, the spin-density matrix. The analysis may be carried through in a completely parallel way to generate spin populations and spin overlap populations. There is one significant difference in the actual *numbers*. One conventionally describes an open-shell system as having more α-electrons than β-electrons so that the spin densities will describe an excess of α-spin spatial distribution over β-spin spatial distribution. It often happens that the effects of spin-polarisation are such that, although there is an *overall* excess of α-spin distribution this may be composed of spatial areas where there is an excess of α-spin *and* spatial regions where β-spin is in excess. These regions are known as *negative spin-density* regions. For very obvious reasons there is no analogy with the total electron density, which must be everywhere positive (although negatively charged!).

Similarly, open-shell populations may be computed separately for each type of electron or for each shell, as appropriate.

13.5 A cautionary note

One must not get carried away by the simplicity and convenience of the algebraic approach; after all, the *numbers* P_{ij} must be multiplied by the products of the basis functions in order to get the actual *spatial* distribution of the electrons. The analysis counts the electrons in terms of the occupations of the functions *based* on a particular *centre*, it cannot tell us about the spatial distribution of the functions on that centre.

The ideas of population analysis were all originally developed to help with the physical interpretation of the results of calculations of the electronic structure of molecules of conventional structure formed from atoms of the first row (up to neon) of the periodic table. As we have noted several times, the physical interpretation is at its best when the basis functions are actually AOs.

When the basis is extended to obtain a better description of the electronic structure or when calculations are performed on molecules of less conventional electronic structure, one is bound to come across new problems in the population analysis. For example, in the calculation of the electronic structure of a transition metal compound of the type ML_n, it is often the case that the highest s or p AOs (or basis functions) are quite diffuse and, in a sense, are *larger than the whole molecule*. Thus if one of these functions is occupied to an extent P_{ii}, it is not at all clear that this electron distribution should be assigned to the atom on which it just happens to be *centred* rather than the atom(s) where it has its maximum *values*. Similar considerations apply to the use of any diffuse functions centred on atoms of any atomic number.

This type of difficulty can be overcome to a large extent by transforming the P matrix to the basis of AOs of the separate atoms which have been computed using the same basis functions as are used for the molecule. That is, to *generate*

13.6. MULTI-DETERMINANT FORMS

the AOs of the separate atoms with the same basis and use these AOs to perform the population analysis, demoting the basis functions to a computational tool. The problem here is that there are *fewer* AOs than there are basis functions so that, in order to recover all the information in the basis-function expansion, one must use the full set of "AOs", *including* the unoccupied "virtual atomic orbitals". This necessarily means that the final population analysis will have numbers referring to occupations of the separate-atom virtual AOs as well as to AOs occupied in the ground state of the separate atoms. If one can place any reliance on the physical interpretation of the virtuals, this will provide an interpretation of bond formation in terms of inter- and intra-atomic rearrangement of charge *and* local atomic excitations.

The transformations are simple, if the collection of all the AOs of the separate atoms are λ_i^A and λ_j^B, say, and the coefficients which generate the AOs from the basis functions ϕ_k are collected in D, which is *square* and therefore invertible if it defines *all* the AOs of the (two) separate atoms, occupied and virtual:

$$\boldsymbol{\lambda} = \boldsymbol{\phi} \boldsymbol{D} \quad i.e. \quad \boldsymbol{\phi} = \boldsymbol{\lambda} \boldsymbol{D}^{-1}$$

and, as before, the MO coefficients in terms of the basis functions are C, where

$$\boldsymbol{\psi} = \boldsymbol{\phi} \boldsymbol{C}$$

then

$$\boldsymbol{\psi} = \boldsymbol{\lambda} \left(\boldsymbol{D}^{-1} \boldsymbol{C} \right) = \boldsymbol{\lambda} \boldsymbol{C}^D$$

giving a final expression of the MOs in terms of the AOs of the separate atoms:

$$\boldsymbol{\psi} = \boldsymbol{\lambda} \boldsymbol{C}^D$$

where $\boldsymbol{C}^D = \boldsymbol{D}^{-1} \boldsymbol{C}$, and the whole analysis may be carried through in terms of $\boldsymbol{C}^D$ in place of $\boldsymbol{C}$, generating partitioned *AO-based* $\boldsymbol{P}^D$ matrices and AO-based charges and bond-orders etc.

13.6 Multi-determinant forms

Everything so far in this chapter has assumed a single-determinant wavefunction. It is, however, obvious that the electron density function ρ, when computed using a basis-function expansion method, must always be capable of being written as a sum of quadratic and bilinear terms in the basis functions. That is, for any wavefunction which is expanded as a linear combination of determinants of orbitals expressed as linear combinations of basis functions ϕ_k we must have:

$$\int \Psi^*(\vec{x}, \vec{x}_2, \ldots, \vec{x}_n) \Psi(\vec{x}, \vec{x}_2, \ldots, \vec{x}_n) dV_2 \ldots dV_n ds_1 \ldots ds_n$$

$$= \sum_{i,j=1}^{m} P_{ij} \phi_j(\vec{r}) \phi_i(\vec{r})$$

The difference between the MO expression and the more general expression is *entirely contained in the values* of the elements of $\boldsymbol{P}$.

Many of the simple properties of $\boldsymbol{P}$ (and the associated $\boldsymbol{R} = \boldsymbol{P}/2$) are different for multi-determinant functions but, from the point of view of population analysis, some properties are shared.

The most characteristic difference between an "MO" $\boldsymbol{P}$-matrix and a more general one is revealed by *diagonalisation*. Since the MO $\boldsymbol{P}$ is generated from the MO coefficients, when the $\boldsymbol{P}$ matrix is transformed to the MOs, it is diagonalised by that same matrix, which is most easily seen in the orthogonalised basis:

$$\boldsymbol{C}^\dagger \boldsymbol{P} \boldsymbol{C} = 2\boldsymbol{C}^\dagger \boldsymbol{R} \boldsymbol{C} = 2\boldsymbol{C}^\dagger \boldsymbol{C} \boldsymbol{C}^\dagger \boldsymbol{C} = 2(\boldsymbol{C}^\dagger \boldsymbol{C})(\boldsymbol{C}^\dagger \boldsymbol{C}) = 2\mathbf{1}$$

which expresses the well-known fact that the MOs are each doubly occupied.

It is easy to see that, when a more general $\boldsymbol{P}$ is expressed in an orthogonal basis and diagonalised, there is no set of "doubly occupied" MOs that generate the full electron density. The occupation numbers of the resulting orbitals range from close to 2 to much smaller values depending on the number of determinants and the care with which they have been chosen and optimised.

The orbitals generated by diagonalising $\boldsymbol{P}$ are known as the natural (spin) orbitals and have useful properties in both physical interpretation and the optimum choice of determinants in a more general model wavefunction. We shall return to a more detailed consideration of these natural orbitals when considering many-determinant wavefunctions.

13.7 Implementation

The implementation of the ideas and methods discussed in this chapter is trivial; this simply involves matrix multiplication and some book-keeping to partition the basis functions. Since it is normal to order the basis functions in terms of atomic centres, even the latter is quite simple. We have all the necessary routines elsewhere.

Chapter 14

The general MO functional

The single-determinant energy expression is just one example of a general form of energy expression which involves only one-electron integrals and MO repulsion integrals of the strict "Coulomb" and "exchange" type. In this chapter it is shown that it is possible to develop an effective "Hartree–Fock-like" Hamiltonian matrix equation for the orbitals involved in any such energy expression.

Contents

14.1	A generalisation	409
14.2	Shells of orbitals	410
14.3	The variational method	412
14.4	A single "Hartree–Fock" operator	416
14.5	Non-orthogonal basis	419
14.6	Choice of the arbitrary matrices	420
14.7	Implementation: stacks of matrices	422
14.7.1	rmcz	424
14.7.2	heffz	427
14.7.3	INDEX	432

14.1 A generalisation

The expression for the energy of a single determinant of orthonormal orbitals:

$$E = \int \Phi^* \hat{H} \Phi d\tau = \sum_{i=1}^{n} \int \chi_i^* \hat{h} \chi_i d\tau$$

$$+ \sum_{i=1}^{n}\sum_{j=1}^{n}[(\chi_i\chi_i,\chi_j\chi_j) - (\chi_i\chi_j,\chi_i\chi_j)]$$

when transformed to matrix form became

$$E = \mathrm{tr}\boldsymbol{C}^\dagger(\boldsymbol{h} + \frac{1}{2}\boldsymbol{J} - \frac{1}{2}\boldsymbol{K})\boldsymbol{C}$$

and variations in the orbital coefficients subject to normalisation, generated the Hartree–Fock (SCF) equation

$$(\boldsymbol{h} + \boldsymbol{J} - \boldsymbol{K})\boldsymbol{C} = \boldsymbol{SC}\boldsymbol{\epsilon}$$

The Hartree–Fock matrix resulting from making the energy functional stationary has the contributions from electron repulsion matrices $\boldsymbol{J}$ and $\boldsymbol{K}$ multiplied by a factor of two, cancelling the factor 1/2 in the repulsion energy.

It is not difficult to see that this procedure is capable of generalisation since the *essential* step in the development is that *only MO repulsion integrals of the types*

$$(\chi_i\chi_i,\chi_j\chi_j) \quad \text{and} \quad (\chi_i\chi_j,\chi_i\chi_j)$$

occur in the energy expression; i.e. the electron repulsion terms *must* be capable of being expressed in the form

$$\mathrm{tr}\boldsymbol{J}(\boldsymbol{A})\boldsymbol{B} \quad \text{or} \quad \mathrm{tr}\boldsymbol{K}(\boldsymbol{A})\boldsymbol{B}$$

for some matrices $\boldsymbol{A}$, $\boldsymbol{B}$.

So, suppose that we abandon the *original source* of the energy functional (the single-determinant wavefunction) and find the stationary points in the general energy functional

$$\begin{aligned} E &= \sum_{i=1}^{n} \nu_i \int \chi_i^* \hat{h} \chi_i d\tau \\ &+ \sum_{i=1}^{n}\sum_{j=1}^{n}[\mu_{ij}(\chi_i\chi_i,\chi_j\chi_j) - \lambda_{ij}(\chi_i\chi_j,\chi_i\chi_j)] \end{aligned}$$

(where the ν_i, μ_{ij}, and λ_{ij} are numerical multipliers), and *then* look around for applications for the resulting equations; i.e. wavefunctions which do, in fact, generate an energy functional of this kind. If a wavefunction can be found which generates an energy functional of this form, it will determine the values of the multipliers ν_i, μ_{ij}, and λ_{ij}.

14.2 Shells of orbitals

The most general form involving a different multiplier for *each energy integral* is a little *too* general for actual applications, although the formalism may easily

14.2. SHELLS OF ORBITALS

be carried through. A more useful form involves the conceptual idea that the n electrons may be divided into N groups or *shells* with the same *form* of interaction between the shells rather than the individual electrons. So, if S and R denote these electron shells, we need only include numerical multipliers for intra- and inter-shell energy terms; separate multipliers for the one-electron, Coulomb and exchange integrals. The form of the energy functional involves sums over shells containing sums over orbitals within those shells:

$$\begin{aligned} E & = \sum_{S=1}^{N} \nu_S \{ \sum_{i \in S} \int \chi_i^* \hat{h} \chi_i d\tau \\ & + \frac{1}{2} \sum_{i \in S} \sum_{j \in S} \nu_S \left[\mu_{SS}(\chi_i\chi_i, \chi_j\chi_j) - \lambda_{SS}(\chi_i\chi_j, \chi_i\chi_j) \right] \} \\ & + \sum_{S=1}^{N} \sum_{R \neq S=1}^{N} \nu_S \nu_R \{ \sum_{i \in S} \sum_{k \in R} \left[\mu_{RS}(\chi_i\chi_i, \chi_k\chi_k) - \lambda_{RS}(\chi_i\chi_k, \chi_i\chi_k) \right] \} \end{aligned} \qquad (14.1)$$

where some interpretation and some constraints have been placed on the numerical multipliers in the energy expression:

- It has been assumed that one of the similarities amongst the orbitals comprising a shell is that they should have the same *occupation number* which has been denoted by ν_S. Thus one may speak of the "shell occupation number" of shell S meaning that all the orbitals comprising shell S have the (same) occupation number ν_S. In these cases, the orbitals are *spatial* orbitals and these numbers should be less than 2. Normally, these numbers would be either 1 or 2 but we retain the generality.

- The numerical multipliers of the repulsion integrals $\mu_{RS} = \mu_{SR}$ and $\lambda_{RS} = \lambda_{SR}$ are again characteristic of *inter-shell* rather than inter-orbital interactions.

This energy functional can most easily be transformed into an energy function of the coefficients in an LCAO expansion of the orbitals

$$\chi = \varphi C$$

by using the separate (column) matrices for each orbital introduced in Appendix 3.C:

$$C = (c^1, c^2, c^3, \ldots, c^n)$$

and defining a separate "orbital" R matrix for each MO:

$$R^i = c^i c^{i\dagger}$$

The contributions from the orbitals within a given shell are then added to form a set of "shell" R matrices:

$$R^S = \sum_{i \in S} R^i$$

Finally, the column coefficient matrices associated with a given shell are concatenated to form a set of "shell" C matrices:

$$C^S = (c^{k_1}, c^{k_2}, c^{k_3}, \ldots)$$

($k_i \in S$) so that

$$R^S = C^S C^{S\dagger}$$

With these definitions, and the functional notation for the electron repulsion matrices introduced in Appendix 3.C, the energy function becomes

$$\begin{aligned} E &= \sum_{S=1}^{N} \nu_S \{ \text{tr} R^S h \\ &+ \frac{1}{2}\nu_S [\, \mu_{SS} \text{tr} J(R^S) R^S - \lambda_{SS} \text{tr} K(R^S) R^S \,] \} \\ &+ \sum_{S=1}^{N} \sum_{S \neq R=1}^{N} \nu_S \nu_R [\, \mu_{RS} \text{tr} R^S J(R^R) - \lambda_{RS} \text{tr} R^S K(R^R) \,] \end{aligned} \quad (14.2)$$

This expression can be cast into a more attractive form by taking *half* of each inter-shell repulsion term and adding it to the corresponding intra-shell term. Also, using the results of Appendix 3.C

$$\text{tr} J(A) B = \text{tr} J(B) A$$

and

$$\text{tr} K(A) B = \text{tr} K(B) A$$

the expression assumes a symmetrical form:

$$E = \sum_{S=1}^{N} \nu_S \text{tr} \{ \, h + \frac{1}{2} \sum_{R=1}^{N} \nu_R [\, \mu_{RS} J(R^R) - \lambda_{RS} K(R^R) \,] \, \} R^S \quad (14.3)$$

or, defining

$$h_S = \nu_S h + \frac{1}{2} \nu_S \sum_{R=1}^{N} \nu_R [\, \mu_{RS} J(R^R) - \lambda_{RS} K(R^R) \,]$$

we have

$$E = \sum_{S=1}^{N} \nu_S \text{tr} h_S R^S = \sum_{S=1}^{N} \nu_S \text{tr} C^{S\dagger} h_S C^S \quad (14.4)$$

14.3 The variational method

In order to generate the analogue of the Hartree–Fock equation we must now require this function of the C^S to be stationary with respect to variations in the

14.3. THE VARIATIONAL METHOD

$C^{S\dagger}$ and C^S subject to orthonormality requirements similar to those obtained in the single-determinant case. These requirements are the orthonormality of all the orbitals involved:
$$C^\dagger S C = 1$$
which may be expressed in terms of the "shell" coefficient matrices C^S as
$$C^{S\dagger} S C^R = \delta_{SR} \tag{14.5}$$
where δ_{SR} is rather loose notation for a matrix (of appropriate dimension) which is the unit square matrix if $S = R$ and the zero rectangular matrix if $S \neq R$.

The first-order variation of eqn (14.4) is derived in a way which completely parallels the single-determinant case, and the result is an expression which has a doubled contribution from the electron-repulsion terms as the coefficient matrix occurs twice in these terms. Without going through the explicit algebra again the result is clearly
$$\delta E^{(S)*} = \mathrm{tr}\,\delta C^{S\dagger} h_S^F C^S = 0 \tag{14.6}$$
for variations in the matrix $C^{S\dagger}$, and a similar result for variations in C^S:
$$\delta E^{(S)} = \mathrm{tr}\, C^{S\dagger} \nu_S h_S^F \delta C^S = 0$$
where the "shell Hartree–Fock matrices" h_S^F are
$$h_S^F = \nu_S h + \nu_S \sum_{R=1}^{N} \nu_R \,[\, \mu_{RS} J(R^R) - \lambda_{RS} K(R^R) \,]$$

The factor of 1/2 has been cancelled by the appearance of each repulsion term twice in the variation.

The corresponding variations in the orthonormality condition generate
$$\delta C^{S\dagger} S C^R = 0 \tag{14.7}$$
which, again following the earlier derivation, is equivalent to
$$\mathrm{tr}\,\delta C^{S\dagger} S C^R \epsilon^{RS} = 0$$
where ϵ^{RS} is a matrix of Lagrange multipliers for each pair of shells.

Combining the energy variation and the orthonormality constraint in the usual way and insisting on the vanishing of the resulting expression for arbitrary variations $\delta C^{S\dagger}$ and δC^S generates twice as many equations as there are shells; one for each C^S and one for each $C^{S\dagger}$:
$$h_S^F C^S = \sum_{R=1}^{N} S C^R \epsilon^{RS} \tag{14.8}$$
$$C^{S\dagger} h_S^F = \sum_{R=1}^{N} \epsilon^{SR} C^{S\dagger} S \tag{14.9}$$

Taking the conjugate of the second of these two equations shows that it is the same as the first if

$$\epsilon^{SR} = \epsilon^{RS\dagger}$$

or, what amounts to the same thing, if the matrix ϵ formed from the submatrices ϵ^{RS} is Hermitian. This full $m \times m$ matrix may be formed as follows:

$$\boldsymbol{\epsilon} = \begin{pmatrix} \epsilon^{11} & \epsilon^{12} & \ldots & \epsilon^{1N} \\ \epsilon^{21} & \epsilon^{22} & \ldots & \epsilon^{2N} \\ \epsilon^{31} & \epsilon^{32} & \ldots & \epsilon^{3N} \\ \ldots & \ldots & \ldots & \ldots \\ \ldots & \ldots & \ldots & \ldots \\ \ldots & \ldots & \ldots & \ldots \\ \epsilon^{N1} & \epsilon^{N2} & \ldots & \epsilon^{NN} \end{pmatrix}$$

if there are N shells. This structure for the ϵ matrix presupposes that the columns of the coefficient matrix C have been ordered so that those belonging to the same shell are adjacent, but this can always be arranged:

$$\boldsymbol{C} = (\overbrace{\boldsymbol{c}^1, \boldsymbol{c}^2, \boldsymbol{c}^3}^{\boldsymbol{C}^1}, \overbrace{\boldsymbol{c}^4, \boldsymbol{c}^5}^{\boldsymbol{C}^2}, \overbrace{\boldsymbol{c}^6, \ldots,}^{\boldsymbol{C}^3} \ldots, \overbrace{\ldots, \boldsymbol{c}^n}^{\boldsymbol{C}^N})$$

The net result is that we have a "Hartree–Fock-like" equation *for each shell* plus the requirement that the matrix ϵ be Hermitian:

$$\boldsymbol{h}_S^F \boldsymbol{C}^S = \sum_{R=1}^N \boldsymbol{S} \boldsymbol{C}^R \boldsymbol{\epsilon}^{RS}$$

If we take a typical pair of these equations, say the equation involving $\boldsymbol{h}_S^F$ and $\boldsymbol{h}_R^F$ and multiply the first by $\boldsymbol{C}^{R\dagger}$ and the second by $\boldsymbol{C}^{S\dagger}$, we can use the fact that

$$\boldsymbol{C}^{S\dagger} \boldsymbol{S} \boldsymbol{C}^R = \boldsymbol{\delta}_{SR}$$

for all S and R to obtain

$$\boldsymbol{C}^{R\dagger} \boldsymbol{h}_S^F \boldsymbol{C}^S = \boldsymbol{\epsilon}^{RS} \qquad (14.10)$$
$$\boldsymbol{C}^{S\dagger} \boldsymbol{h}_R^F \boldsymbol{C}^R = \boldsymbol{\epsilon}^{SR} \qquad (14.11)$$

However the matrix ϵ is required to be Hermitian, so that these two expressions may be combined to give a set of equations which are equivalent to the Hermiticity of ϵ:

$$\boldsymbol{C}^{S\dagger} \left(\boldsymbol{h}_S^F - \boldsymbol{h}_R^F \right) \boldsymbol{C}^R = \boldsymbol{0} \qquad (14.12)$$

This last equation may be given a neater appearance by noting that, in our definition of the matrices $\boldsymbol{h}_S^F$, the shell occupation number appears as a simple

14.3. THE VARIATIONAL METHOD

factor. If we extract this factor and quote it explicitly *without changing the notation* for the h_S^F we have:

$$C^{S\dagger}\left(\nu_S h_S^F - \nu_R h_R^F\right) C^R = 0$$

Equation (14.11) may also be used to remove the explicit appearance of the ϵ matrices in the shell Hartree–Fock equations which then become:

$$h_S^F C^S = \sum_{R=1}^{N} S C^R \left(C^{R\dagger} h_R^F C^S\right)$$

subject to eqn (14.12).

Rearranging the last equation slightly, the equations to be solved become

$$h_S^F C^S - \sum_{R=1}^{N} S C^R \epsilon = 0$$
$$C^{S\dagger}\left(\nu_S h_S^F - \nu_R h_R^F\right) C^R = 0$$

for $S, R = 1, N$.

This a a set of *simultaneous* equations for the variationally optimum C^S subject to the constraints originally arising from the orthonormality requirement.

In anticipation of the use of projection operators to help manipulate these equations into a tractable form, we write the equations to be solved in terms of the "shell R-matrices" and use an *orthonormal* basis to avoid algebraic complications arising from the overlap matrix S: in this case we have, in an orthonormal basis (without change of notation),

$$h_S^F C^S - \sum_{R=1}^{N} C^R \epsilon = 0$$
$$C^{S\dagger}\left(\nu_S h_S^F - \nu_R h_R^F\right) C^R = 0$$

which may be written in terms of shell R-matrices as[1]

$$h_S^F R^S - R^S h_S^F = 0$$
$$R^S (\nu_S h_S^F - \nu_R h_R^F) R^R = 0$$
$$R^S R^R = \delta_{RS} R^S$$

---
[1] This equivalence is proved in Chapter 26.

As noted in Appendix 14.A to this chapter, the shell $\boldsymbol{R}$-matrices, when supplemented by the "empty shell" $\boldsymbol{R}^Z$ matrix associated with the subspace not occupied at all, form a resolution of the m-dimensional identity:

$$\boldsymbol{R}^Z + \sum_{S=1}^{N} \boldsymbol{R}^S = 1$$

so that the first set of equations may be multiplied from the left by $\boldsymbol{R}^Z$ to give the equivalent set:

$$\boldsymbol{R}^Z \boldsymbol{h}_S^F \boldsymbol{R}^S = 0$$

for all S.

14.4 A single "Hartree–Fock" operator

In order to get these equations into a form suitable for computational implementation we need to do two things which were not required of the single-determinant theory:

- Incorporate the constraints into the equation determining the $\boldsymbol{C}^S$ so that the equation may be solved *unconditionally*. This was not necessary in the former cases because there was only one $\boldsymbol{h}^F$ matrix and so the constraint

$$\boldsymbol{C}^\dagger \left(\nu \boldsymbol{h}^F - \nu \boldsymbol{h}^F \right) \boldsymbol{C} = 0$$

was trivially satisfied.

- Attempt to obtain a single equation for *all* the $\boldsymbol{C}^S$. In this way we may be sure of the mutual orthogonality of all the orbitals which has been assumed throughout the derivation.

That is, we seek a single "Fock" matrix ($\bar{\boldsymbol{h}}$, say) for which the equation

$$\bar{\boldsymbol{h}} \bar{\boldsymbol{U}} = \bar{\boldsymbol{U}} \boldsymbol{\epsilon}$$

generates all the orbitals that solve the equations

$$\boldsymbol{R}^Z \boldsymbol{h}_S^F \boldsymbol{R}^S = 0 \tag{14.13}$$

$$\boldsymbol{R}^S (\nu_S \boldsymbol{h}_S^F - \nu_R \boldsymbol{h}_R^F) \boldsymbol{R}^R = 0 \tag{14.14}$$

where $R \neq S$ and, of course, $R, S \neq Z$.

14.4. A SINGLE "HARTREE–FOCK" OPERATOR

The matrix $\bar{U}$ will have the structure:

$$\bar{U} = (\overbrace{c^1,c^2,c^3}^{C^1},\overbrace{c^4,c^5}^{C^2},\overbrace{c^6,\ldots,}^{C^3}\ldots,\overbrace{\ldots,c^n}^{C^N},\overbrace{c^{n+1},\ldots,c^m}^{C^Z})$$

where each of the **R**-matrices is given by

$$R^S = C^S C^{S\dagger}$$

including the empty shell Z.

As noted at the end of the last section, we shall work in an orthogonal basis since the equations can always be transformed to a working non-orthogonal basis, in a similar manner to the single-determinant case, at the end of the calculation.

The generation of a single Hartree–Fock-like matrix for *all* the MO coefficients is not so much a *derivation* as an examination of the possibilities and using the properties

$$R^R R^S = \delta_{RS}$$

of the **R**-matrices.

For example, taking

$$h = \sum_{T \neq Z} a_T R^Z h_T^F R^T$$

and forming

$$R^Z h R^S = a_S R^Z h R^S$$

for all S shows the equivalence of this condition to those of eqn (14.13).

Similarly taking

$$h = \sum_T \sum_{U>T} b_{TU} R^T (\nu_T h_T^F - \nu_U h_U^F) R^U \tag{14.15}$$

$T,U \neq Z$ and forming

$$R^S h R^R = b_{SR} R^S (\nu_S h_S^F - \nu_R h_R^F) R^R \tag{14.16}$$

is equivalent to eqn (14.14). What is more these two sums do not interfere with each other because of the mutually exclusive properties of the projectors R^R, so that

$$h = \sum_{T \neq Z} a_T R^Z h_T^F R^T + \sum_T \sum_{U>T} b_{TU} R^T (\nu_T h_T^F - \nu_U h_U^F) R^U$$

$(T, U \neq Z)$ combines the conditions, eqns (14.13) and (14.14).

However, eqn (14.15), when inserted into the left-hand-side of eqn (14.16), because $R \neq S$, may have terms like

$$\boldsymbol{R}^T \boldsymbol{d}_T \boldsymbol{R}^T$$

added to it *for all* T since

$$\boldsymbol{R}^S \left(\boldsymbol{R}^T \boldsymbol{d}_T \boldsymbol{R}^T \right) \boldsymbol{R}^R = \boldsymbol{\delta}_{ST} \boldsymbol{d}_T \boldsymbol{\delta}_{TR}$$

where, since S and R are *different*, at least one of $\boldsymbol{\delta}_{ST}$ and $\boldsymbol{\delta}_{TR}$ must be zero. This is true *whatever the matrices* $\boldsymbol{d}_T$ *are*.

Thus, the most general form of the Hartree–Fock-like matrix equivalent to the variational conditions eqns (14.13) and (14.14) is

$$\begin{aligned} \bar{\boldsymbol{h}} &= \boldsymbol{R}^Z \boldsymbol{d}_Z \boldsymbol{R}^Z + \sum_{T \neq Z} \boldsymbol{R}^T \boldsymbol{d}_T \boldsymbol{R}^T + \sum_{T \neq Z} a_T \boldsymbol{R}^Z \boldsymbol{h}_T^F \boldsymbol{R}^T \\ &+ \sum_{T \neq Z} \sum_{U < T} b_{TU} \boldsymbol{R}^T (\nu_T \boldsymbol{h}_T^F - \nu_U \boldsymbol{h}_U^F) \boldsymbol{R}^U \end{aligned} \quad (14.17)$$

There are several points to be made immediately about this effective Hartree–Fock matrix, the "McWeenyan":[2]

- There is no *essential* difference between the a_T and the b_{TU} since, by noting that ν_Z is zero, the summation involving the b_{TU} may be extended to include the empty shell without changing its form since

$$\boldsymbol{R}^T (\nu_T \boldsymbol{h}_T^F - \nu_Z \boldsymbol{h}_Z^F) \boldsymbol{R}^Z = \nu_T \boldsymbol{R}^T \boldsymbol{h}_T^F \boldsymbol{R}^Z$$

  The difference is retained to emphasise the role of the empty shell.

- It contains a very large amount of freedom in its definition; the *numbers* a_T and b_{TU} are *arbitrary*. The values of these numbers determine the *eigenvalues* of the associated matrix equation

$$\bar{\boldsymbol{h}} \bar{\boldsymbol{U}} = \bar{\boldsymbol{U}} \boldsymbol{\epsilon}$$

  but not the all-important matrix of MO coefficients $\bar{\boldsymbol{U}}$.

- Similarly, there is one arbitrary (Hermitian) matrix per shell (the $\boldsymbol{d}_T$) which may be chosen in many ways. Again, these choices affect $\boldsymbol{\epsilon}$ but not $\bar{\boldsymbol{U}}$.

---

[2] It is my privilege to try to bring this term into general use, it has been used for many years in Sheffield (since McWeeny's departure, of course) as this formulation was first presented in a paper by Roy McWeeny in 1964: " A Self-Consistent generalisation of Hückel theory" (in *Molecular orbitals in physics, chemistry and biology*, Academic Press). Like "Fockian" for the Hartree–Fock matrix, it is a convenient term for a complicated object.

14.5. NON-ORTHOGONAL BASIS

- How can we attach a *physical interpretation* to the eigenvalues of the matrix equation if they are to be changed at will? The interpretation of the many-shell effective Hartree–Fock matrix is deferred until Chapter 23.

- Recall that, in the closed-shell Hartree–Fock equation,

$$h^F C = C\epsilon$$

(in an orthogonal basis) the notation disguised the fact that the equation is *non-linear* because C is involved in the definition of h^F. Comparing this relatively simple case with our expression for $\bar{h}$, we see that the associated equation is *extremely* non-linear since each R is quadratic in the coefficients in the corresponding C and these R-matrices appear both explicitly and hidden in the definitions of the h_T^F via the J and K matrices. It looks as if each term in $\bar{h}$ depends on the *sixth* power of some MO coefficients.

We therefore expect some technical difficulties with, for example, convergence of any iterative process; it may prove possible to attenuate the highly non-linear character of the equations by suitable choices of the arbitrary parameters in the definition.

14.5 Non-orthogonal basis

The general form of $\bar{h}$ is a sum of terms like

$$\bar{R}\bar{h}^F \bar{R}$$

where bars have been put over all matrices to emphasise that they are expressed in terms of an *orthogonal basis*.

We have seen earlier that, if the basis ϕ suffers a linear transformation

$$\bar{\phi} = \phi V$$

which may be such to generate an orthogonal basis $\bar{\phi}$, then the energy integral matrices undergo the induced transformation

$$\bar{h}^F = V^\dagger h^F V$$

and the density matrices actually undergo the contragredient transformation:

$$R = V^\dagger \bar{R} V$$

If $V = S^{-1/2}$ which is symmetric ($V^\dagger = V$) we have each of the above terms, when formed over the non-orthogonal basis and transformed, looking like

$$\bar{h}^F = S^{-\frac{1}{2}} h^F S^{-\frac{1}{2}}$$
$$\bar{R} = S^{\frac{1}{2}} R S^{\frac{1}{2}}$$

so that the terms in the single effective matrix undergo the transformation

$$\bar{R}\bar{h}^F \bar{R} = S^{\frac{1}{2}} R S^{\frac{1}{2}} S^{-\frac{1}{2}} h^F S^{-\frac{1}{2}} S^{\frac{1}{2}} R S^{\frac{1}{2}} = S^{\frac{1}{2}} R h^F R S^{\frac{1}{2}}$$

Thus the relationship between the single effective Hartree–Fock matrix in the non-orthogonal[3] (h) and orthogonal ($\bar{h}$) bases is simply

$$\bar{h} = S^{\frac{1}{2}} h S^{\frac{1}{2}} \tag{14.18}$$

This contrasts with the usual relationship between Hartree–Fock matrices in the two bases, since the occurrence of the R-matrices means that the *inverse* of the usual transformation is needed.

However, the same numerical procedure may be used if the following recipe is used in the implementation:

1. Form the individual Fock matrices, h^F using integrals over the non-orthogonal basis.
2. Form the single effective matrix (h, say) with these components and the R-matrices over the non-orthogonal basis.
3. Use the overlap matrix to form

$$ShS$$

4. Proceed as usual using the normal orthogonalisation matrix $S^{-1/2}$ since

$$S^{-\frac{1}{2}} ShS S^{-\frac{1}{2}} = S^{\frac{1}{2}} h S^{\frac{1}{2}}$$

14.6 Choice of the arbitrary matrices

Before any discussions of possible applications of this general theory and any implementations of those applications we must obviously say something about the fact that *the general equation which determines the optimum MOs associated with eqn (14.1) contains matrices which are arbitrary*. What does this mean

---
[3]This is temporary notation, h is usually used for the *one-electron Hamiltonian matrix*.

14.6. CHOICE OF THE ARBITRARY MATRICES

from a *scientific* point of view and, technically, how are we to cope with the consequences when we come to *program* these equations? It is worth noting straight away that it is only the *projections* of these arbitrary matrices which occur in the theory so that the optimum MOs still have a measure of "control" over the arbitrariness.

If we recall the definition of the canonical MOs of the UHF or closed-shell model as those MOs which *diagonalise* the Hartree–Fock matrix:

$$\boldsymbol{c}^{i\dagger}\boldsymbol{h}^F \boldsymbol{c}^j = \delta_{ij}\epsilon_i$$

where the ϵ_i have a well-defined meaning as the *orbital energies* of the MOs, we can begin an investigation of the meaning of the arbitrary matrices and coefficients in the generalised case.

The columns of coefficients which *diagonalise* $\bar{\boldsymbol{h}}$ are the $\boldsymbol{c}^i$ etc. which are grouped into shells, and the idempotent and exclusive properties of the shell $\boldsymbol{R}$-matrices mean that

$$\boldsymbol{R}^S \boldsymbol{c}^i = \begin{cases} \boldsymbol{c}^i & \text{if } i \in S \\ 0 & \text{if } i \notin S \end{cases} \qquad (14.19)$$

So, it is straightforward to see that

$$\boldsymbol{c}^{i\dagger}\bar{\boldsymbol{h}}\boldsymbol{c}^i = \boldsymbol{c}^{i\dagger}\boldsymbol{d}_S \boldsymbol{c}^i$$

if $\boldsymbol{c}^i$ belongs to shell S and zero otherwise. Thus, diagonalisation of $\bar{\boldsymbol{h}}$ implies the simultaneous diagonalisation of the whole set of Hermitian matrices $\boldsymbol{d}_S$ and, what is more, the SCF orbitals are *canonical* with respect to the shell $\boldsymbol{d}_S$ matrices. We can therefore choose the $\boldsymbol{d}_S$ in such a way as to make the diagonalised values *meaningful*; in particular, the choice

$$\boldsymbol{d}_S = \boldsymbol{h}^F_S$$

is particularly appealing, since the SCF coefficients for the orbitals associated with shell S diagonalise the associated shell Hartree–Fock matrix and provide a useful connection, by way of an analogue of Koopmans' Theorem, with experimental ionisation energies.

During the iterative SCF process, when the MO coefficients do *not* diagonalise $\bar{\boldsymbol{h}}$, i.e. do not diagonalise $\boldsymbol{d}_S$, the corrections made to each column of MO coefficients comes from *inter-shell* mixing in exactly the same way as our analysis showed that it was occupied/virtual mixing in the simple MO model which was responsible for corrections to the occupied MOs. In the present case, however, since there are several occupied shells as well as an empty shell, there are correspondingly more opportunities for changes in the MO coefficients.

In these cases convergence is often a problem and we can sometimes assist the convergence by "shifting" the energies of the MOs of different shells with respect to each other in the same way as we emphasised the occupied/virtual gap in simple SCF theory. Adding a multiple of the unit matrix to each h_S^F will generate a "level shifter" for each *shell* of orbitals, and optimum choice of these numerical factors often improves the convergence properties of the associated SCF procedure; that is, taking

$$d_S = h_S^F + \alpha_S \mathbf{1}$$

where α_S is a "level shifter" for shell S, will shift the "MO energies" of each shell.

The off-diagonal elements of the matrix $\bar{h}$ are, of course, zero at self-consistency but during the iterations of the SCF process a current set of shell coefficient matrices is available and it is again clear, because of the properties (14.19) of the shell $\mathbf{R}$-matrices, that[4]

$$c^{i\dagger} \bar{h} c^j = b_{TU} c^{i\dagger} (\nu_T h_T^F - \nu_U h_U^F) c^j$$

when $i \in T$ and $j \in U$. This gives the coefficients b_{TU} an interpretation and potential use as "damp factors" which are an extended analogue of the damp factors discussed in simple SCF theory.

An informal overall interpretation of the reduction to diagonal form of the single effective Fock-like matrix is now clear:

- The self-consistent MO coefficients bring all the shell Hartree–Fock matrices to diagonal form in a way which is expected from consideration of the simpler, "one shell" case.

- Simultaneously all the inter-shell mixing terms are eliminated.

So that, at self-consistency, there is a set of unmixed shell Hartree–Fock problems to be solved whose physical interpretation is an obvious generalisation of the simpler cases. This point is taken up in detail in Chapter 23.

14.7 Implementation: stacks of matrices

It is clear from the theory developed in this chapter that there are no serious problems to be faced in thinking about implementing this generalised form of the SCF equations; the problems are going to be ones of *access to and organisation*

---

[4] Even when the MOs are not self-consistent they are orthonormal, so that a given set of coefficients and projectors have the same mutual projection properties.

14.7. IMPLEMENTATION: STACKS OF MATRICES

of matrix elements. So, even before having a specific case to code, we can begin some system design for this class of problem.

Assuming that the various overlap and energy integrals have been generated in the same way as before we shall be involved in performing what are now routine tasks:

- Generation of various J and K matrices, either from a file of repulsion integrals or by direct computation.
- Formation of density matrices from current MO coefficient matrices.
- Transformation to an orthogonal basis using, typically, $S^{-1/2}$.
- Diagonalisation of the transformed matrix.
- Convergence testing.

But we shall now be dealing with *many more* matrices; we shall need a density matrix for each shell, a Hartree–Fock matrix for each shell *etc.*

Clearly we can simply set up the number of matrices needed in each individual case, but it is, as usual, better to plan ahead and generate software for the general case; a few routines which will deal with an arbitrary number of matrices of each type. The matrix storage method which we have been using is ideally suited to this extension. Matrices are currently stored in a singly subscripted array in columns so that, for example, element A_{ij} of a $m \times m$ matrix is kept at A(m*(j-1) + i) with the first element, A_{11} at A(1), of course. This fills up the first m^2 elements of the array A. If there is enough space allocated there is no reason why we should not simply continue this process and store another $m \times m$ (say) matrix from A(m*m +1) to A(2*m*m) *and so on.*

> We therefore make a slight extension to our matrix storage algorithm to say that the $(i,j)th$ element of the Kth matrix of a given type is stored at
>
> $$A(K\_offset + m*(j-1) + i)$$
>
> where K_offset = (K-1)*m*m.

This enables us to store an arbitrary number of R-matrices (say) in one place and simplifies access considerably.

The simplest approach is by example; here is a WEB for the calculation of a set of orbital R-matrices plus an R^Z for an empty shell if one is needed. The index to variables is omitted as it is quite trivial.

14.7.1 rmcz

General "MO" functional R matrix former. Forms all n R matrices of the form $R=TT^T$ where T is a column of U (the full matrix of MO coefficients). Then the last $n - nr$ are summed to form the empty shell R matrix i.e. $R(nr + 1, *)$ contains R^Z.

NAME rmcz

Forms a stack of orbital density matrices plus an "empty shell" matrix.

SYNOPSIS

```
subroutine rmcz(U,R,n,nr)
double precision U(ARB), R(ARB);
integer n, nr;
```

DESCRIPTION

The matrix stack R is formed from the columns of U, the first nr are as they stand, the last one is the sum of the last $n - nr$ formed in this way.

ARGUMENTS

U Input: The matrix of MO coefficients.

R Output: The stack of density matrices.

n Input: the dimension of U.

nr Input: the number of "occupied" R matrices.

DIAGNOSTICS

None

"rmcz.f" 14.7.1 ≡

 subroutine $rmcz(U, R, n, nr)$

  ⟨ rmcz interface declarations 14.7.1.1 ⟩

 {

 ⟨ rmcz local declarations and initialisation 14.7.1.2 ⟩

14.7. IMPLEMENTATION: STACKS OF MATRICES

⟨ Form all n R matrices 14.7.1.3 ⟩
⟨ Check for an empty shell and form its R matrix 14.7.1.4 ⟩
return;
}

⟨ rmcz interface declarations 14.7.1.1 ⟩ ≡
    implicit double precision $(a - h,\ o - z)$;
    double precision $R(ARB),\ U(ARB)$;
    integer $n,\ nr$;

This code is used in section 14.7.1.

⟨ rmcz local declarations and initialisation 14.7.1.2 ⟩ ≡
    integer $j\_offset,\ z\_offset$;
    integer $i,\ j,\ k,\ ik,\ ki,\ kj,\ nrp1,\ nrp2,\ nn$;

    $nn = n * n$;
    $nrp1 = nr + 1$;

This code is used in section 14.7.1.

⟨ Form all n R matrices 14.7.1.3 ⟩ ≡
   do $j = 1$, n;
     {
     $j\_offset = (j-1) * nn$;
     do $i = 1$, n;
       {
       $ij = (j-1) * n + i$;
       do $k = 1$, i;
         {
         $ik = (k-1) * n + i$;
         $ki = (i-1) * n + k$;
         $kj = (j-1) * n + k$;
         $R(j\_offset + ik) = U(ij) * U(kj)$;
         $R(j\_offset + ki) = R(j\_offset + ik)$;
       } /* end of k loop */
     } /* end of i loop */
   } /* end of j loop */

This code is used in section 14.7.1.

⟨ Check for an empty shell and form its R matrix 14.7.1.4 ⟩ ≡
   if $(nrp1 \geq n)$
     return;

   $z\_offset = nr * nn$;
   $nrp2 = nr + 2$;

   /* Just in case the empty shell is only one orbital: */

   if $(nrp2 > n)$
     return;

   do $j = nrp2$, n;
     {
     $j\_offset = (j-1) * nn$;
     do $ik = 1$, nn;
       $R(z\_offset + ik) = R(z\_offset + ik) + R(j\_offset + ik)$;
     }

This code is used in section 14.7.1.

The code to form the effective single Hartree–Fock matrix is correspondingly straightforward. Here is a **WEB** to do the job.

14.7. IMPLEMENTATION: STACKS OF MATRICES

14.7.2 heffz

Routine to form the single effective Hamiltonian matrix from the "orbital" Hamiltonians: the McWeenyan. There are nr "occupied" orbital Hamiltonians and nr corresponding $\boldsymbol{R}$ matrices; the "empty" $\boldsymbol{R}$ matrix $\boldsymbol{R}^Z$ is in $R(nr+1, *)$ The routine will work if $nr = n$ or $nr + 1 = n$ as well as the cases of "several unoccupied orbitals".

NAME heffz
    Forms the McWeenyan from stacks of orbital Hamiltonians and density matrices.

SYNOPSIS

```
subroutine heffz(Hi,R,S,n,nr,ndocc,Hbar);
dimension S(ARB), Hbar(ARB), Hi(ARB), R(ARB);
integer n, nr, ndocc;
```

DESCRIPTION
    The individual projections are formed in the most obvious and straightforward way and summed, taking due account of the "level shifters" and "damp factors".

ARGUMENTS

    Hi Input: the nr orbital hamiltonians are in Hi.

    R Input: the $nr + 1$ density (projection) matrices are in R.

    S Input: contains the overlap matrix.

    n Input: the number of basis orbitals.

    nr Input: the number of occupied orbitals.

    alpha Input: level shifters.

    Hbar Output: the (orthogonalised) McWeenyan.

DIAGNOSTICS
    None

"rmcw.f" 14.7.1 ≡
   @m ARB *   /* In f77 "ARB" can be "*" */

subroutine *heffz*(*Hi*, *R*, *S*, *n*, *nr*, *ndocc*, *alpha*, *Hbar*)
⟨ heffz interface declarations 14.7.1.1 ⟩
{
⟨ heffz local declarations 14.7.1.2 ⟩
⟨ heffz initialisation 14.7.1.3 ⟩
⟨ Form the contributions from the occupied orbitals 14.7.1.4 ⟩
⟨ Form the "empty orbital" contributions 14.7.1.5 ⟩
⟨ Finally, add the "redundant" level shifter terms 14.7.1.6 ⟩
⟨ Make the McWeenyan ready for orthogonalisation 14.7.1.7 ⟩
return; }

⟨ heffz interface declarations 14.7.1.1 ⟩ ≡
  implicit double precision $(a - h, o - z)$;
  double precision $S(ARB)$, $Hbar(ARB)$, $Hi(ARB)$, $R(ARB)$;
  double precision $alpha(ARB)$;
  integer n, nr;

This code is used in section 14.7.1.

This is awful practice, illustrating an accident just waiting to happen; the work areas here have explicit declarations and the damp factors are just as bad. These work areas should be combined into a single area passed through the interface. These declarations disobey the rule that there should be no "magic numbers" in the code. No matter how cosmetically attractive the codes, if they contain bad design faults, they will fail sooner or later.

⟨ heffz local declarations 14.7.1.2 ⟩ ≡
  double precision $D(2500)$, $C(2500)$, $E(2500)$;
  double precision $b(50, 50)$;
  integer i, j, ij, nn, ir, jr, ir_offset, jr_offset;
  integer $nrp1$, ik, ik_offset;
  double precision *zero*, *one*;
  data *zero*/$0.0d00$/, *one*/$1.0 \cdot 10^{00}$D/;

This code is used in section 14.7.1.

14.7. IMPLEMENTATION: STACKS OF MATRICES

This is the simplest possible choice of the $b(i, j)$; they must be *antisymmetric*.

⟨ heffz initialisation 14.7.1.3 ⟩ ≡
```
    /* Set up the simplest possible set of b(i, j) */
    do i = 1, n;
      {
        do j = 1, i;
          {
            b(i, j) = -one;
            b(j, i) = one;
          }
        b(i, i) = one;
      }
    nt = n * (n + 1)/2;
    nn = n * n;
        /* Initialise Hbar to zero, the contributions are to added. */
    do ij = 1, nn;
      Hbar(ij) = zero;
```

This code is used in section 14.7.1.

⟨ Form the contributions from the occupied orbitals 14.7.1.4 ⟩ ≡
```
    /* loop over the contributions from the 'occupied' orbitals */
    do ir = 1, nr;
      {
      ir_offset = (ir − 1) * nn;
      do jr = 1, ir;
        {
        jr_offset = (jr − 1) * nn;
        if (ir ≡ jr)
          break;
        do i = 1, nn;
          {
          D(i) = R(ir_offset + i);
          C(i) = Hi(ir_offset + i) − Hi(jr_offset + i);
          }
        call gmprd(D, C, E, n, n, n);
        do i = 1, nn;
          D(i) = R(jr_offset + i);
        call gmprd(E, D, C, n, n, n);    /* put C transpose in D; */
        call gmtra(C, D, n, n);
        do i = 1, nn;
          Hbar(i) = Hbar(i) + b(ir, jr) * (C(i) + D(i));
        }
      }
```

This code is used in section 14.7.1.

14.7. IMPLEMENTATION: STACKS OF MATRICES

⟨ Form the "empty orbital" contributions 14.7.1.5 ⟩ ≡
```
    /* First, check if there is at least one "unoccupied" MO */
    nrp1 = nr + 1;
    nr_offset = nr * nn;
    if (nrp1 < n)
    {
      do ik = 1, nr;
        {
        ik_offset = (ik - 1) * nn;
        do i = 1, nn;
            {
            D(i) = R(ik_offset + i);
            C(i) = Hi(ik_offset + i);
            }
        call gmprd(D, C, E, n, n, n);
        do i = 1, nn;
            D(i) = R(nr_offset + i);
        call gmprd(E, D, C, n, n, n);
        call gmtra(C, D, n, n);
        do i = 1, nn;
            Hbar(i) = Hbar(i) + b(ik, ik) * (C(i) - D(i));
        }
    }
```

This code is used in section 14.7.1.

⟨ Finally, add the "redundant" level shifter terms 14.7.1.6 ⟩ ≡
  /* Now add the "redundant" terms with level shifters */
  do $ir = 1,\ nr$;
    {
    $ir\_offset = (ir - 1) * nn$;
    $kp = nr$;   /* a pointer to an *alpha*, a level shifter */
    do $k = 1,\ nn$;
      {
      $C(k) = R(ir\_offset + k)$;
      $D(k) = Hi(ir\_offset + k)$;
      }
    do $k = 1,\ n$;
      {
      $kk = n * (k - 1) + k$;
      $D(kk) = D(kk) + alpha(kp)$;
      }
    call $gmprd(C,\ D,\ E,\ n,\ n,\ n)$;
    call $gmprd(E,\ C,\ D,\ n,\ n,\ n)$;
    do $k = 1,\ nn$;
      $Hbar(k) = Hbar(k) + D(k)$;
    }

This code is used in section 14.7.1.

Multiply the composite Hamiltonian fore and aft by the overlap matrix so that the usual transformation with $S^{1/2}$ will generate the correctly orthogonalised form.

⟨ Make the McWeenyan ready for orthogonalisation 14.7.1.7 ⟩ ≡
  call $gtprd(S,\ Hbar,\ E,\ n,\ n,\ n)$;
  call $gmprd(E,\ S,\ Hbar,\ n,\ n,\ n)$;

This code is used in section 14.7.1.

14.7.3 INDEX

The index is ommitted; it simply lists usage of subscripting variables.

Appendix 14.A

Projection operators and SCF

The R matrices have been used so far simply as convenient matrices to represent the total electron density. These matrices have a much more fundamental theoretical importance than this; in a very direct way they are a representation of a single-determinant wavefunction in a given basis. The SCF equations and many of their properties may be written in terms of the R-matrices.

Contents

14.A.1 Introduction: optimum single determinant 433
14.A.2 Alternative SCF conditions 435
14.A.3 R matrices as projection operators 436

14.A.1 Introduction: optimum single determinant

The equation which the optimum orbitals of the best single-determinant wavefunction must satisfy is

$$h^F C = C\epsilon$$

where, for simplicity, an orthogonal set of basis functions has been used to avoid the appearance of the overlap matrix S.

We have met the matrix R given by

$$R = CC^\dagger$$

mainly in its capacity as a numerical description of the total electron density due to the single-determinant wavefunction; if the basis functions are φ then this density is just

$$\rho(\vec{x}) = \varphi(\vec{x})\boldsymbol{R}\varphi^\dagger(\vec{x})$$

But the $\boldsymbol{R}$-matrix has some very interesting and useful formal, algebraic properties in addition to this physical interpretation.

The orbital coefficients are normalised and orthogonal:

$$\boldsymbol{C}^\dagger \boldsymbol{C} = 1$$

so that the action of $\boldsymbol{R}$ on $\boldsymbol{C}$ is

$$\boldsymbol{R}\boldsymbol{C} = (\boldsymbol{C}\boldsymbol{C}^\dagger)\boldsymbol{C} = \boldsymbol{C}(\boldsymbol{C}^\dagger\boldsymbol{C}) = \boldsymbol{C}$$

and

$$\boldsymbol{R}^2 = \boldsymbol{R}\boldsymbol{R} = \boldsymbol{R}$$

by the same construction.

These properties of $\boldsymbol{R}$ are similar to those of the unit matrix and the difference between $\boldsymbol{R}$ and $\boldsymbol{1}$ emerges only when one considers the action of $\boldsymbol{R}$ on matrices other than $\boldsymbol{C}$ from which it is constructed.

If we solve the larger matrix problem

$$\boldsymbol{h}^F \boldsymbol{U} = \boldsymbol{U}\boldsymbol{\epsilon}$$

where $\boldsymbol{U}$ is square, we may write

$$\boldsymbol{U} = (\boldsymbol{C}, \boldsymbol{V})$$

and ask what is the nature of

$$\boldsymbol{R}_V = \boldsymbol{V}\boldsymbol{V}^\dagger$$

Now all the columns of $\boldsymbol{U}$ are normalised and orthogonal so that[5]

$$\boldsymbol{C}^\dagger \boldsymbol{C} = \boldsymbol{V}^\dagger \boldsymbol{V} = 1$$

and

$$\boldsymbol{V}^\dagger \boldsymbol{C} = \boldsymbol{C}^\dagger \boldsymbol{V} = 0$$

which means that

$$\boldsymbol{R}_V^2 = \boldsymbol{R}_V$$

---

[5] Slightly loose notation here, the *dimension* of the zero or unit matrices may be different in some cases; it is n in the original case and $(m - n)$ in the other.

in addition to the similar property of R, and
$$RR_V = R_V R = 0$$
and
$$R + R_V = 1$$
These two matrices provide a resolution of the identity matrix 1 and are idempotent and mutually exclusive.

We noted earlier that the matrix eigenvalue equation which determines the solutions of the Hartree–Fock equation is equivalent to the condition
$$V^\dagger h^F C = C^\dagger h^F V = 0$$
multiplying from the appropriate directions by V or C this equation may be expressed in terms of the R matrices:
$$R_V h^F R = R h^F R_V = 0$$
or
$$(1 - R) h^F R = R h^F (1 - R) = 0$$
as a condition of the solution of the original equation.

14.A.2 Alternative SCF conditions

Again taking
$$h^F C = C\epsilon$$
as the equation determining C, we may take its Hermitian conjugate to form
$$C^\dagger h^F = \epsilon C^\dagger$$
because both h^F and ϵ are Hermitian. Multiplying the first equation from the right by $C^\dagger$, and the second from the left by C both equations then have identical right-hand-sides so that
$$h^F CC^\dagger = CC^\dagger h^F$$
or
$$h^F R - R h^F = 0$$
as a condition of solution of the SCF equations.

Conversely, if
$$h^F R = h^F CC^\dagger = CC^\dagger h^F = R h^F$$

multiplying from the right by C gives

$$h^F(CC^\dagger)C = C(C^\dagger h^F C)$$

which, because $RC = C$, becomes

$$h^F C = C\epsilon$$

where, as above,

$$\epsilon = C^\dagger h^F C$$

so that the equations

$$\begin{aligned} h^F C &= C\epsilon \\ h^F R - R h^F &= 0 \\ (1-R) h^F R = R h^F (1-R) &= 0 \end{aligned}$$

are all equivalent subject only to the constraint that

$$R^2 = R$$

which itself implies that

$$(1-R)^2 = (1-R)$$

The first two of these formulations may be made the basis of practical methods of obtaining the optimum U and hence C, but the first is by far the most convenient to use.

14.A.3 R matrices as projection operators

So far we have met *two* R matrices but it is clear that, from a $m \times m$ unitary matrix U, we may actually form (at most) m independent R matrices from the m columns (c^i, say) of U:

$$R^i = c^i c^{i\dagger}$$

($i = 1, m$). Since all the c^i are normalised and orthogonal,

$$c^{i\dagger} c = \delta^{ij}$$

(where δ^{ij} is an obvious matrix analogue of the scalar Kronecker delta), the set of R^i is a maximal resolution of the identity for this m-dimensional vector space:

$$\sum_{i=1}^{m} R^i = 1$$

14.A.3. R MATRICES AS PROJECTION OPERATORS

and they are, therefore, idempotent

$$\left(\boldsymbol{R}^i\right)^2 = \boldsymbol{R}^i$$

and mutually exclusive

$$\boldsymbol{R}^i \boldsymbol{R}^j = \boldsymbol{\delta}^{ij}$$

Now, these are just the properties of a set of *projection operators* of which the three projections onto three mutually perpendicular unit vectors in ordinary three-space are the prototypes. If the three unit vectors are $(\vec{i}, \vec{j}, \vec{k})$, then the corresponding projectors may be written as the dyadics

$$\hat{p}^i = \vec{i}\vec{i}\cdot \ , \hat{p}^j = \vec{j}\vec{j}\cdot \ , \hat{p}^k = \vec{k}\vec{k}\cdot$$

where the "dot" means scalar product.

These operators have the same formal properties as the $\boldsymbol{R}^i$, they form a resolution of the identity and they are idempotent and mutually exclusive. But they have a very familiar physical interpretation; they *resolve* a given vector in three-space into its *components*, for if,

$$\vec{x} = x_1 \vec{i} + x_2 \vec{j} + x_3 \vec{k}$$

then, for example,

$$\hat{p}^j \vec{x} = \vec{j}\vec{j} \cdot \vec{x} = \vec{j} x_2 = x_2 \vec{j}$$

so that

$$1 \times \vec{x} = \vec{x} = \hat{p}^i \vec{x} + \hat{p}^j \vec{x} + \hat{p}^k \vec{x} = (\hat{p}^i + \hat{p}^j + \hat{p}^k)\vec{x}$$

the resolution of the vector into its components in an orthonormal basis illustrating graphically what is meant by a "resolution of the identity".

All these abstract properties are carried over, together with their interpretation, onto the $\boldsymbol{R}^i$; each $\boldsymbol{R}^i$, when multiplying an arbitrary vector in the m-dimensional expansion space gives the *component* of that vector in the "direction" of the corresponding c^i. Further, in an obvious extension of the idea of projection onto a direction, sums of the $\hat{p}$-operators (*e.g.* $\hat{p}^i + \hat{p}^k$) are just projections onto *planes* in three-space. The generalisation here is that disjoint sums of projection operators resolve a given vector into components which lie in (orthogonal) *sub-spaces* of the overall vector space.

In the case of the single-determinant wavefunction, our two projection operators $\boldsymbol{R}$ and $\boldsymbol{R}_V$ will act on any vector of the m-dimensional overall space and produce the part which lies in the space spanned by the orbitals occupied in the single determinant and the remainder which lies in the complementary space because

$$\boldsymbol{R} = \sum_{i=1}^n \boldsymbol{R}^i$$

This fact gives an interesting algebraic interpretation to the process of solving the matrix Hartree–Fock equations particularly in view of the invariance property of the determinant.

Given an m-dimensional vector space of basis functions, the act of solving the Hartree–Fock equations is to induce a linear transformation of the vectors in that space to divide it into two sub-spaces:

1. A space of occupied orbitals which is defined by the projection operator $\boldsymbol{R}$.
2. The orthogonal complement which completes the original space and is defined by the projector
$$\boldsymbol{R}_V = (\boldsymbol{1} - \boldsymbol{R})$$

The fact that unitary transformations amongst the occupied orbitals do not change the projection matrix $\boldsymbol{R}$ but only its *composition* in terms of individual orbital projectors is nicely reflected in the fact that such transformations leave the single-determinant wavefunction unchanged, so that in a very real and physically transparent sense, the matrix $\boldsymbol{R}$ *represents* the determinantal wavefunction Φ. A knowledge of $\boldsymbol{R}$ (within a given basis approximation) is a knowledge of Φ.

This fact is most forcefully illustrated in the second of the three equivalent forms of the Hartree–Fock equation:

$$\boldsymbol{h}^F \boldsymbol{R} - \boldsymbol{R} \boldsymbol{h}^F = 0$$

where

$$\boldsymbol{h}^F = \boldsymbol{h} + \boldsymbol{G}(\boldsymbol{R})$$

and the orbital coefficient matrix $\boldsymbol{C}$ is not mentioned at all in the equations to be solved.

Since the $\boldsymbol{R}$-matrices are projection operators, they must necessarily induce transformations from a larger (bigger dimension) space into a smaller space. The analogue is a projection from three-space onto a line or a plane. This process involves *loss of information* since one or more components of a given vector is *lost* on projection; it is evidently possible for several vectors to have the same projection onto a given subspace. This elementary fact, of course, means that the act of projection *cannot be reversed* or, the algebraic statement of the same thing, projection operators cannot be inverted, they are necessarily represented by *singular* matrices.

This fact is extremely useful in the development of many-shell SCF theories as we see in this chapter, but it is also necessary to proceed with caution

14.A.3. R MATRICES AS PROJECTION OPERATORS

to ensure that an equation does not have its domain of validity inadvertently reduced by the action of projection operators.

The action of projection operators on matrices representing (Hermitian) operators clearly involves projection operators *twice* because of the way in which an operator matrix transforms under a change of basis; it is trivial to show that

$$A = B = 1A1 = 1B1$$

if and only if

$$R^i A R^j = R^i B R^j$$

for all i and j. In order to be sure that two matrices are equal in a situation where projection operators are in use it is necessary to be sure that all projections of the two matrices are equal; since a projection operator is singular it may not be "cancelled" from a matrix equation.

Chapter 15

Spin-restricted open shell

The most useful and familiar application of the theory developed in the last chapter is to the case of a set of "singly occupied" orbitals (usually with parallel spin) "outside" a set of "doubly occupied" orbitals; the familiar picture of an open-shell system which has spin-polarisation specifically excluded in order that all the spin properties of the system be concentrated into the explicit open-shell structure.

Contents

15.1	Introduction		441
15.2	The ROHF model		442
15.3	Implementation		444
15.4	A WEB for spin-restricted open shell		445
	15.4.1	grhf	448
	15.4.2	grhfR	456
	15.4.3	grhfGR	459
	15.4.4	Hone	462
	15.4.5	Hhalf	464
	15.4.6	addprd	467
	15.4.7	INDEX	467

15.1 Introduction

Although the *optimum* single-determinant wavefunction has all the formal advantages of a fully variational solution of the Hartree–Fock equations and has many conceptual and numerical advantages, perhaps the model of open-shell systems which has the most familiar "feel" is the one which has a set of closed-shell (doubly occupied) spatial orbitals and a separate set of singly occupied orbitals.

This model of open-shell systems has a number of qualitative conceptual features which seem more attractive than the spin-polarised orbitals of the optimum single-determinant (GUHF or DODS) wavefunction:

- The single determinant of a set of ν parallel-spin singly occupied orbitals outside a closed shell is a *spin eigenfunction*;

$$\hat{S}^2 \Phi = S(S+1)\Phi$$

and

$$\hat{S}_z \Phi = M_S \Phi$$

where $S = M_S = \nu/2$

- In many cases the wavefunction of lower multiplicity, $M_S < S$, which may be a linear combination of determinants, still generates an energy expression of the form required.

- The picture of singly-occupied MOs outside a closed shell is the conventional picture of open-shell systems.

Naturally, this model is a *restricted* or *constrained* model of molecular electronic structure (Restricted Open-shell Hartree–Fock, ROHF, say), and so may be expected to have an energy lying above that of the best (GUHF) single determinant and may well have some quantitative difficulties (the ROHF model of the ^2P, $1s^2 2p$ state of the lithium atom, for example will have no Fermi contact nuclear-spin electron-spin coupling since the unpaired electron density is zero at the nucleus).

15.2 The ROHF model

This model consists of just *two occupied shells*; the closed shell with occupation number 2, the open shell with occupation number 1 and the empty shell. The general MO energy functional eqn (14.1), becomes, under the circumstances:

- Two shells: Shell 1 with MOs labelled i and j and occupation number of each MO $\nu_1 = 2$, Shell 2 with MOs labelled k and ℓ and occupation number $\nu_2 = 1$.

15.2. THE ROHF MODEL

- Spin integration carried through $(d\tau \to dV, \chi \to \psi)$ generating zeroes for some exchange integrals.
- The multiplying factors for each intra- and inter-shell repulsions (λ_{RS}, μ_{RS}) written in explicitly. $\lambda_{11} = \lambda_{22} = \lambda_{12} = 1$, $\mu_{11} = \mu_{12} = 1/2$, $\mu_{22} = 1$
- The sums over R and S in eqn (14.1) being replaced by explicit expressions in terms of shells 1 and 2.

$$\begin{aligned}
E &= \nu_1 \left\{ \sum_{i \in 1} \left[\int \psi_i^* \hat{h} \psi_i dV \right] + \frac{1}{2} \sum_{i \in 1} \sum_{j \in 1} \nu_1 \left[(\psi_i \psi_i, \psi_j \psi_j) - \frac{1}{2}(\psi_i \psi_j, \psi_i \psi_j) \right] \right\} \\
&+ \nu_2 \left\{ \sum_{k \in 2} \left[\int \psi_k^* \hat{h} \psi_k dV \right] + \frac{1}{2} \sum_{k \in 1} \sum_{\ell \in 1} \nu_2 \left[(\psi_k \psi_k, \psi_\ell \psi_\ell) - (\psi_k \psi_\ell, \psi_k \psi_\ell) \right] \right\} \\
&+ \nu_1 \nu_2 \left\{ \sum_{i \in 1} \sum_{k \in 2} \left[(\psi_i \psi_i, \psi_k \psi_k) - \frac{1}{2}(\psi_i \psi_k, \psi_i \psi_k) \right] \right\}
\end{aligned} \quad (15.1)$$

Using the familiar relationships between the MO repulsion integrals and the J and K matrices

$$\sum_{i \in 1} \sum_{j \in 1} \nu_1 \left[(\psi_i \psi_i, \psi_j \psi_j) - \frac{1}{2}(\psi_i \psi_j, \psi_i \psi_j) \right]$$
$$= \text{tr} J(\nu_1 R^1) R^1 - \frac{1}{2} \text{tr} K(\nu_1 R^1) R^1$$

and

$$\nu_1 \nu_2 \sum_{i \in 1} \sum_{k \in 2} \left[(\psi_i \psi_i, \psi_k \psi_k) - \frac{1}{2}(\psi_i \psi_k, \psi_i \psi_k) \right]$$
$$= \text{tr} J(\nu_1 R^1) \nu_2 R^2 - \text{tr} K(\nu_1 R^1) \nu_2 R^2$$

etc., this energy expression takes the form

$$\begin{aligned}
E &= \nu_1 \text{tr} \left[h + \frac{1}{2} \left(J(\nu_1 R^1 + \nu_2 R^2) - \frac{1}{2} K(\nu_1 R^1 + \nu_2 R^2) \right) \right] R^1 \\
&+ \nu_2 \text{tr} \left[h + \frac{1}{2} \left(J(\nu_1 R^1 + \nu_2 R^2) - K(\frac{1}{2}\nu_1 R^1 + \nu_2 R^2) \right) \right] R^2
\end{aligned}$$

The equations therefore become particularly simple, with just two Fock matrices and we may now insert the values $(\nu_1 = 2, \nu_2 = 1)$[1]

$$h_1^F = h + J(2R^1 + R^2) - K(R^1 + \frac{1}{2}R^2) \quad (15.2)$$

$$h_2^F = h + J(2R^1 + R^2) - K(R^1 + R^2) \quad (15.3)$$

---

[1]Notice the infelicity of notation here $\nu_1 = 2$ and $\nu_2 = 1$ but we are used to thinking of the *closed* shell as the *first* shell (lower in energy) so the clash between the *subscripts* of ν and the *values* of ν must be borne stoically.

note that

$$h_2^F = h_1^F - \frac{1}{2}K(R^2)$$

With these expressions for the Fock matrices, the single Hartree–Fock matrix eqn (14.17) is just:

$$\begin{aligned}\bar{h} &= R^Z d_Z R^Z + R^1 d_1 R^1 + R^2 d_2 R^2 + a_1 R^Z h_1^F R^1 \\ &+ a_2 R^Z h_2^F R^2 + b_{12} R^1 \left(2h_1^F - h_2^F\right) R^2 \end{aligned} \quad (15.4)$$

which *looks* more complex than eqn (14.17) merely by virtue of having the sums of two terms written out in full. We can now proceed to implementation of the ROHF case, noting that we have, in any actual *implementation*, to face up to the *choice* of the matrices d_1, d_2 and d_Z which were *arbitrary* in the general theory.

15.3 Implementation

In spite of the apparent complexity of the matrix expressions for the equation determining the optimum MOs of the ROHF case, there is very little in the way of new *ideas* involved in the implementation of the solution of this equation.

As in the closed-shell or DODS case, we have the basic structure:

1. Set up the Fock matrices (two this time, not just one) from the one- and two-electron basis function integrals.

2. Form the single effective Fock matrix (this is a new step).

3. Transform the matrix to an orthogonalised basis and diagonalise it.

4. Transform the MO coefficients back to the original basis, form the density matrices, test for convergence and iterate to self-consistency.

The new tasks are contained in the first two of these steps.

Referring back to eqn (15.2) and eqn (15.3) we see that, although there are several J and K matrices involved there are actually only two *separate* basic quantities; if we write, temporarily,

$$A = J(2R^1 + R^2) - K(R^1 + \frac{1}{2}R^2)$$

we see that eqn (15.2) is just

$$h_1^F = h + A$$

15.4. A WEB FOR SPIN-RESTRICTED OPEN SHELL

and eqn (15.3) is
$$h_2^F = h_1^F - \frac{1}{2}K(R^2)$$

Thus any program which is to set up h_1^F and h_2^F has to form the matrices A and $K(R^2)$ which can be conveniently done in *one pass* of the file of repulsion integrals.[2]

In order to effect the transformation of the single Fock matrix between the non-orthogonal and orthogonal bases, we simply have to pass the overlap matrix S through the interface to the iterative process since we need to form ShS before transformation with $S^{-1/2}$. Equivalently, $S^{+1/2}$ could be passed to the iterative procedure.

All the other new steps are simply matrix multiplications.

15.4 A WEB for spin-restricted open shell

Since the WEBs are rather space-consuming, only the weave/LATEX output is given here in the hope that the reader may now begin to make sense of the "woven" form of the program as it is often called. This WEB is rather large so some of the more "routine" parts are omitted:

- A slightly modified matrix multiplication routine (addprd).
- A standard Gauss–Jordan matrix inversion code (minv).
- The basic primitives for adding a repulsion integral into the correct (up to) eight places in the J and K matrices (JofR, KofR).

Where this occurs weave inserts *"Sections skipped"*... as a reminder.

In this example a slightly different flavour is given to the WEB by moving the "declarations" sections further back in the text and making the major modules (subroutines) a set of minor modules whose names are simply phrases or sentences describing what they do. Thus, the major modules hardly contain any explicit code. It is left to the reader's taste to choose how far to go with "WEBising" the code. The relative advantages and disadvantages are obvious enough, the structure of the main routine subroutine grhf and of the core of the problem (<One iteration of SCF cycle>) can be seen at a glance each *on a single page* but the index of modules must be used to find the individual pieces of code.

---
[2]Or, of course, equivalently, using only one direct computation of each repulsion integral per SCF iteration.

The implementation given below is capable of performing the SCF calculation for an arbitrary number of shells and is not restricted to the "two-shell" case. A decision about which case to use is taken on the basis of the number of shells and their occupations which is supplied as part of the interface. In order to avoid much duplication which would be involved if an implementation of the Multi-Configuration method (described in Chapter 22) were to be given, much of the code below is a little too general for the open-shell spin-restricted MO model but could be used, basically unchanged, in, for example, the GVB method.

A rather cursory manual page is given here followed by a **WEB** for the implementation.

15.4. A WEB FOR SPIN-RESTRICTED OPEN SHELL

NAME grhf

SYNOPSIS

    subroutine grhf(H,V,nbasis,number_of_shells,norbs,noclst,
               nfile,nu,x,y,alpha,b,en,HF,R,eps,
               Work,Hbar,U, initial_scf,irite);

DESCRIPTION
Solves the spin-restricted many-shell "Hartree-Fock" problem.

ARGUMENTS H Input: One-electron Hamiltonian.

  V Input: Orthogonalistion Matrix.

  nbasis Input: Number of basis functions.

  number_of_shells Input: self explanatory

  norbs Input: `norbs(i)`, `i=1,number_of_shells` are the number of MOs in each shell.

  noclst Input: `noclst(j,i)`, `i = 1,norbs(j)`, `j= 1,number_of_shells` are the sequence numbers of the orbitals in each shell.

  nfile Input: Logical file for repulsion integrals.

  nu Input: $nu(i)$, $i = 1$, $number\_of\_shells$ are the shell occupation numbers.

  x,y Input: Symmetric matrices (**number_of_shells**×**number_of_shells**) of coupling coefficients.

  alpha Input `alpha(i)`, `i=1,number_of_shells` are "level shifters" for the shells.

  b Input: an (**number_of_shells**×**number_of_shells**) anti-symmetric matrix of "damp factors"

  HF Workspace: Large enough for a stack of **number_of_shells** Fock matrices.

  R Workspace: Must be large enough for a stack of **number_of_shells** density matrices.

  Work Workspace: Large enough for a stack of **number_of_shells** matrices.

  Hbar Workspace: The McWeenyan.

  eps Output: Eigenvalues of McWeenyan.

  U Input/Output: An initial set of MO coefficients if `initial_scf = NO`. On output final MO coefficients.

DIAGNOSTICS
Routine output of Energy and convergence information as calculation proceeds. Message on convergence failure.

15.4.1   grhf

This is an implementation of the General Energy Functional minimisation of Chapter 14. The code here has been very highly "WEBised" to illustrate one possible approach to coding; the major modules are simply lists of minor modules whose names are sentences or phrases describing what they do.

§.   @I "defns.hweb"   *Section(s) skipped...*

"grhf.f" 15.4.1 ≡

> **subroutine** *grhf* (H, V, *nbasis*, *number_of_shells*, *norbs*, *noclst*, *nfile*, *nu*, x, y, *alpha*, *b*, *en*, *HF*, *R*, *eps*, *Work*, *Hbar*, *U*, *initial_scf*, *irite*);
> **implicit double precision** ($a - h$, $o - z$);
> ⟨Interface argument declarations 15.4.1.15⟩
> {
> ⟨grhf local declarations 15.4.1.16⟩
> $n = nbasis$;
> $nn = n * n$;
> ⟨Choose which effective Fock matrix 15.4.1.1⟩
> ⟨Invert the orthogonalisation matrix 15.4.1.3⟩
> ⟨Compute total number of occupied orbitals 15.4.1.2⟩
> ⟨Calculate the R matrices if the input U is a sensible one 15.4.1.4⟩
> *iteration* = 0;
> **repeat**
> {
> ⟨One iteration of SCF cycle 15.4.1.5⟩
> }
> **until** (*icon* ≡ 0)   /∗ convergence ∗/
>   **return**;
> }

15.4. A WEB FOR SPIN-RESTRICTED OPEN SHELL

First, decide which form of the single effective Fock-like matrix is to be used; both are due to McWeeny and are known (at least in Sheffield) as McWeenyans.

⟨ Choose which effective Fock matrix 15.4.1.1 ⟩ ≡

    /* decide path i.e. which form of "Fock Operator" */
    $iw1 = \mathbf{int}(nu(1) + 0.1)$;
    $iw2 = \mathbf{int}(nu(2) + 0.1)$;
    $path = GENERAL$;
    if $((number\_of\_shells \equiv 2) \wedge (iw1 \equiv 2) \wedge (iw2 \equiv 1))$
        $path = HALF\_CLOSED$;

This code is used in section 15.4.1.

We only have the shell occupation numbers in the arguments, get the total.

⟨ Compute total number of occupied orbitals 15.4.1.2 ⟩ ≡
    $total\_occupied\_orbitals = 0$;
    do $i = 1$, $number\_of\_shells$;
        $total\_occupied\_orbitals = total\_occupied\_orbitals + norbs(i)$;

This code is used in section 15.4.1.

We need the orthogonalisation matrix *and* its inverse; the orthogonaliser comes in via the interface, so invert a copy and add it to the end of the original.

⟨ Invert the orthogonalisation matrix 15.4.1.3 ⟩ ≡
    do $i = 1$, nn;
        $V(nn + i) = V(i)$;   /* a copy to invert */
        call $minv(V(nn + 1), n, det, lwork, mwork)$;   /* V**-1 in V(nn+1) */

This code is used in section 15.4.1.

If, by chance, the incoming matrix is a guess at the MO coefficients, not just an orthogonaliser, form the corresponding R matrices.

⟨ Calculate the R matrices if the input U is a sensible one 15.4.1.4 ⟩ ≡
    if $(initial\_scf \neq YES)$
        call $grhfR(U, R, n, noclst, norbs, number\_of\_shells)$;

This code is used in section 15.4.1.

Here is the "inner" code for one complete SCF cycle. It is further broken down into smaller modules as appropriate.

⟨ One iteration of SCF cycle 15.4.1.5 ⟩ ≡
   $icon = 0$;
   $iteration = iteration + 1$;

   $en = zero$;   /∗ accumulator for total energy ∗/

   if $((initial\_scf) \equiv NO \land (iteration \equiv 1))$

   {

   ⟨ Start first cycle from one-electron Hamiltonian 15.4.1.6 ⟩

   }

   else

   {

     /∗ This is the normal SCF cycle procedure ∗/

   ⟨ Put one-electron Hamiltonian into each Fock matrix 15.4.1.7 ⟩

   ⟨ Use the repulsion integrals to form the Fock matrices 15.4.1.8 ⟩

   ⟨ Scale the Fock matrices by occupation number 15.4.1.14 ⟩

   ⟨ Choose the appropriate form of the effective Fock matrix 15.4.1.9 ⟩

   ⟨ Transpose the inverse-orthogonaliser for convenience 15.4.1.10 ⟩

   }

   ⟨ Transform and diagonalise the single Fock matrix 15.4.1.11 ⟩

   ⟨ Transform back to the original basis and form the new R 15.4.1.12 ⟩

   ⟨ Check for convergence and prepare to re-cycle 15.4.1.13 ⟩

This code is used in section 15.4.1.

If no guess at the R matrices is available, set the effective Hamiltonian to the one-electron Hamiltonian to start the process off.

⟨ Start first cycle from one-electron Hamiltonian 15.4.1.6 ⟩ ≡
   do $i = 1$, nn;
     {
     $Hbar(i) = H(i)$;
     $R(i) = V(i)$;
     }

This code is used in section 15.4.1.5.

15.4. A WEB FOR SPIN-RESTRICTED OPEN SHELL

Each shell Fock matrix must contain the one-electron Hamiltonian, so put it there. Here is a good place to start accumulating the total electronic energy in en. The one-electron contribution is entered here; note the occupation number is used to scale R.

⟨ Put one-electron Hamiltonian into each Fock matrix 15.4.1.7 ⟩ ≡
```
do shell = 1, number_of_shells;
   {
   shell_offset = (shell − 1) * nn;
   do i = 1, nn;
      {
      ioffset = shell_offset + i;
      HF(ioffset) = H(i);      /* put H into each 'HF' */
      en = en + nu(shell) * HF(i) * R(ioffset);
      }
   }
```
This code is used in section 15.4.1.5.

Now that each Fock matrix contains the one-electron Hamiltonian, use $grhfGR$ to add in the relevant electron-repulsion contributions. Also set up the "virtual" Fock matrix.

⟨ Use the repulsion integrals to form the Fock matrices 15.4.1.8 ⟩ ≡

```
virtual_offset = number_of_shells * nn;
do i = 1, nn;
   HF(virtual_offset + i) = zero;
```
call $grhfGR(R, HF, n, nfile, nu, x, y, number\_of\_shells, Work)$;

This code is used in section 15.4.1.5.

If the calculation is the simplest case, a closed shell plus one shell of parallel-spin singly occupied orbitals, use McWeeny's original form of the effective Fock matrix; otherwise use the general case. The general case will also work for the special case but convergence is poorer.

⟨ Choose the appropriate form of the effective Fock matrix 15.4.1.9 ⟩ ≡
    if ($path \equiv HALF\_CLOSED$)
        call $Hhalf(HF, R, n, Work, Work(nn + 1), Hbar)$;
            /* "half-closed case" use RMcW form */
    else
        call $Hone(HF, R, n, alpha, b, nu, number\_of\_shells, Work,$
        $Work(nn + 1), Hbar)$;
            /* general case; display stationary conditions */

This code is used in section 15.4.1.5.

It is convenient to transpose the transformation matrix in order to use *gtprd* and *gmprd* to transform the effective Fock matrix to the orthogonal basis.

⟨ Transpose the inverse-orthogonaliser for convenience 15.4.1.10 ⟩ ≡
    do $i = 1, n$;
    {
      do $j = 1, n$;
      {
      $ij = (j - 1) * n + i$;
      $ji = (i - 1) * n + j$;
      $R(ij) = V(nn + ji)$;   /* R has V**-1 transposed */
      }
    }

This code is used in section 15.4.1.5.

15.4. A WEB FOR SPIN-RESTRICTED OPEN SHELL

Use the inverse-orthogonaliser to transform the single Fock matrix to the orthogonalised basis, diagonalise it and save the eigenvalues. Notice that, if inter-shell level shifters have been used, this is the place to remove them from the eigenvalues if the eigenvalues are to be interpreted as orbital energies.

⟨ Transform and diagonalise the single Fock matrix 15.4.1.11 ⟩ ≡
    call *gtprd*(*R*, *Hbar*, *HF*, *n*, *n*, *n*);   /∗ HF now work space ∗/

    call *gmprd*(*HF*, *R*, *Hbar*, *n*, *n*, *n*);

    /∗ this is the single Hamiltonian in *Hbar* - orthogonalised ∗/

    call *eigen*(*Hbar*, *Work*, *n*, 0);

    do $i = 1$, n;
    {
    $ii = (i-1)*n + i$;
    $eps(i) = Hbar(ii)$;
    } /∗REMOVE THE SHIFTERS when they are passed back to the caller!! ∗/

This code is used in section 15.4.1.5.

The R matrices must always be in the original non-orthogonal basis because they must be compatible with the basis over which the repulsion integrals are available, so transform the eigenvectors back to this basis and form the new R matrices.

⟨ Transform back to the original basis and form the new R 15.4.1.12 ⟩ ≡
    call *gmprd*(*V*, *Work*, *U*, *n*, *n*, *n*);   /∗ HF now contains new U ∗/

    call *grhfR*(*U*, *R*, *n*, *noclst*, *norbs*, *number_of_shells*);

This code is used in section 15.4.1.5.

CHAPTER 15. SPIN-RESTRICTED OPEN SHELL

Compare the new eigenvector matrix with the saved one from the previous cycle and, if covergence has not been achieved, set things up for the next cycle. If the cycle limit is hit, quit with a message.

⟨ Check for convergence and prepare to re-cycle 15.4.1.13 ⟩ ≡
   do $i = 1,\ n$;
   {
     do $j = 1,\ total\_occupied\_orbitals$;
     {
       $ij = (j-1)*n + i$;
       $test = \mathbf{dabs}(U(ij) - U(nn+ij))$;
       $U(nn+ij) = U(ij)$;   /* replace old U by new U */
       if $(test > crit)$
          $icon = icon + 1$;
     }
   }
   if $(irite > 0)$
     write $(irite,\ 200)\ icon,\ en,\ iteration$; 200 **format** $(I4,$
          '␣␣Elements␣of␣T␣Changing␣', /, '␣␣Energy␣is␣', $E20.10$,
          '␣␣Iteration␣␣', $I4$) ;
   if $(iteration > max\_iteration)$
   {
     write $(ERROR\_OUTPUT\_UNIT,\ 201)$;
     return;
   } 201 **format** ('␣CONVERGENCE␣FAILURE:␣␣STOPPING␣') ;

This code is used in section 15.4.1.5.

15.4. A WEB FOR SPIN-RESTRICTED OPEN SHELL

We have factored out the shell occupation numbers from the numerical factors in the energy expression ("coupling coefficients"), so multiply the individual shell Fock matrices by these occupation numbers before using the coupling coefficients. This is also a convenient place to finish the energy accumulation.

⟨ Scale the Fock matrices by occupation number 15.4.1.14 ⟩ ≡
    do $shell = 1$, $number\_of\_shells$;
    {
      $shell\_offset = (shell - 1) * nn$;
      do $i = 1$, nn;
      {
        $ioffset = shell\_offset + i$;
        $HF(ioffset) = nu(shell) * HF(ioffset)$;
        $en = en + HF(ioffset) * R(ioffset)$;
      }
    }
    $en = half * en$;   /* all energy terms added */

This code is used in section 15.4.1.5.

⟨ Interface argument declarations 15.4.1.15 ⟩ ≡
    double precision $H(*)$, $U(*)$, $HF(*)$, $R(*)$, $Work(*)$, $Hbar(*)$, $eps(*)$;
    double precision $V(*)$;
    double precision $nu(number\_of\_shells)$, $alpha(number\_of\_shells)$;
    double precision $x(number\_of\_shells, number\_of\_shells)$;
    double precision $y(number\_of\_shells, number\_of\_shells)$;
    double precision $b(nbasis, nbasis)$;
    double precision en;
    integer $nbasis$, $nfile$, $irite$, $initial\_scf$;
    integer $number\_of\_shells$, $norbs(nbasis)$, $noclst(nbasis, nbasis)$;

This code is used in section 15.4.1.

These are declarations of working space which are local to *grhf* and some required initialisations.

⟨ grhf local declarations 15.4.1.16 ⟩ ≡
    integer *lwork*(100), *mwork*(100);   /* work space for minv */
    integer *irite, nbasis, nfile, path, iw1, iw2*;
    integer *initial_scf, iteration, max_iteration*;
    integer *shell_offset, ioffset, virtual_offset, shell*;
    integer *total_occupied_orbitals*;
    data *zero, half, crit*/0.0 · 10^{00}D, 0.5 · 10^{00}D, 1.0 · 10^{-06}D/;
    data *max_iteration*/100/;

This code is used in section 15.4.1.

15.4.2 grhfR

forms the grhf 'R' matrices from the single MO matrix U.

"grhf.f" 15.4.2 ≡

    subroutine *grhfR*(U, R, n, *noclst, norbs, number_of_shells*);
    implicit double precision ($a - h$, $o - z$);
    dimension $R(*)$, $U(*)$;
    integer n, *number_of_shells*;
    integer *noclst*(n, *), *norbs*(*); { **integer** *shell, nshell, offset_ij, offset,*
        offset_ji;
    data *zero*/0.0 · 10^{0}D/;

    ⟨ Form the virtual shell from unoccupied orbitals 15.4.2.1 ⟩

    nshell = *number_of_shells* + 1;   /* include the virtual shell */
    nn = $n * n$;

    do *shell* = 1, *nshell*;

    {

    ⟨ Form the R matrix for each shell using noclst 15.4.2.2 ⟩

    }

    return;

    }

15.4. A WEB FOR SPIN-RESTRICTED OPEN SHELL

Use the information about the orbital structure of the shells to form the list of orbitals in the virtual shell. This could be done once and for all outside *grhfR* if the orbital shell structure is preserved throughout.

⟨Form the virtual shell from unoccupied orbitals 15.4.2.1⟩ ≡
  $nz = number\_of\_shells + 1;$

  /* First make the virtual shell all the orbitals */
  do $i = 1, n;$
    $noclst(nz, i) = i;$

    /* Now, subtract each of the occupied orbitals in each shell from the full list; the remainder is the virtual shell */
  do $shell = 1, number\_of\_shells;$
    {
    $ntop = norbs(shell);$
    do $kd = 1, ntop;$
      {
      $k = noclst(shell, kd);$
      $noclst(nz, k) = 0;$
      }
    }

    /* Now reorder this random list and count the number of orbitals in the virtual shell */

  $iz = 0;$
  do $i = 1, n;$
    {
    $kz = noclst(nz, i);$
    if $(kz \neq 0)$
      {
      $iz = iz + 1;$
      $noclst(nz, iz) = kz;$
      }
    }
  $norbs(nz) = iz;$

This code is used in section 15.4.2.

The information about which orbitals belong to which shell is in *noclst* and so the R matrix for each shell is formed from the orbital coefficients and this information. All the R matrices are stored as contiguous arrays in R, each one is nn long; *offset* keeps track of where each one is to be put.

⟨ Form the R matrix for each shell using noclst 15.4.2.2 ⟩ ≡
    offset = (*shell* − 1) ∗ *nn*;
    do $i = 1$, n;
    {
      do $j = 1$, i;
        {
        $ij = n \ast (j - 1) + i$;
        $ji = n \ast (i - 1) + j$;
        sum = *zero*;
        ntop = *norbs*(*shell*);
        do $kd = 1$, *ntop*;
          {
          $k = noclst(shell, kd)$;
          $kk = n \ast (k - 1)$;
          $ik = kk + i$;
          $jk = kk + j$;
          $sum = sum + U(ik) \ast U(jk)$;
        }
        offset_ij = *offset* + ij;
        $R(\textit{offset\_ij}) = sum$;
        offset_ji = *offset* + ji;
        $R(\textit{offset\_ji}) = sum$;
      }
    }

This code is used in section 15.4.2.

15.4. A WEB FOR SPIN-RESTRICTED OPEN SHELL

15.4.3 grhfGR

Adds the G(R)s needed for grhf to the stack of copies of the one-electron Hamiltonian by collecting together all the R matrices contributing to one Fock matrix and using the simple linear result

$$\sum_i a_i \mathrm{G}(\mathrm{R}^i) = \mathrm{G}\left(\sum_i a_i \mathrm{R}^i\right)$$

for any linear combination of G matrices.

"grhf.f" 15.4.3 ≡

    subroutine *grhfGR*(R, G, n, *nfile*, *nu*, x, y, *number_of_shells*, *Rwork*);

    ⟨grhfGR interface declarations 15.4.3.1⟩

    {

    ⟨grhfGR local declarations 15.4.3.2⟩

      /∗ Loop over all shell G matrices ∗/

    for (*shellh* = 1; *shellh* ≤ *number_of_shells*; *shellh* = *shellh* + 1)

    {

    ⟨Set up the Combinations of R matrices 15.4.3.3⟩

    ⟨Obtain the repulsion integrals and form the Gs 15.4.3.4⟩

    }

    ⟨Now symmetrise all the Gs 15.4.3.5⟩

    return;

    }

⟨grhfGR interface declarations 15.4.3.1⟩ ≡
    implicit double precision ($a - h$, $o - z$);
    double precision $G(*)$, $R(*)$, $nu(number\_of\_shells)$;
    double precision $x(number\_of\_shells, number\_of\_shells)$;
    double precision $y(number\_of\_shells, number\_of\_shells)$;
    double precision $Rwork(*)$;

This code is used in section 15.4.3.

⟨ grhfGR local declarations 15.4.3.2 ⟩ ≡
    double precision *zero*;
    integer *shellh, shellr, hoffset, roffset, nn, n, nfile*;
    integer *pointer*;
    integer *getint*;

    @m $locg(i,j)$ $(((j) - 1) * n + (i))$ /* macro for subscripting */

    data $zero/0.0 \cdot 10^{00}D/$;

    $nn = n * n$;

This code is used in section 15.4.3.

⟨ Set up the Combinations of R matrices 15.4.3.3 ⟩ ≡
    $hoffset = (shellh - 1) * nn$;
    do $m = 1, \ nn$;
      {
      $Rwork(m) = zero$;   /* initialise the "net" R matrix */
      $Rwork(nn + m) = zero$;   /* and the one for K(R) */
      }
    for ($shellr = 1$; $shellr \leq number\_of\_shells$; $shellr = shellr + 1$)
      {
      $roffset = (shellr - 1) * nn$;
      do $m = 1, \ nn$;
        {
        $Rwork(m) = Rwork(m) + nu(shellr) * x(shellr, shellh) * R(roffset + m)$;
        $Rwork(nn + m) = Rwork(nn + m) + nu(shellr) * y(shellr,$
            $shellh) * R(roffset + m)$);
        }
      }

This code is used in section 15.4.3.

15.4. A WEB FOR SPIN-RESTRICTED OPEN SHELL

⟨ Obtain the repulsion integrals and form the Gs 15.4.3.4 ⟩ ≡
    rewind $nfile$;
    $pointer = 0$;   /∗ initialise the file ∗/
    while $(getint(nfile,\ i,\ j,\ k,\ l,\ mu,\ va,\ pointer) \neq END\_OF\_FILE)$
    {
      call $JofR(Rwork,\ G(hoffset+1),\ n,\ i,\ j,\ k,\ l,\ va)$;
      call $KofR(Rwork(nn+1),\ G(hoffset+1),\ n,\ i,\ j,\ k,\ l,\ va)$;
    }

This code is used in section 15.4.3.

⟨ Now symmetrise all the Gs 15.4.3.5 ⟩ ≡
    do $shellh = 1,\ number\_of\_shells$;
    {
      $hoffset = (shellh - 1) * nn$;
      do $mi = 1,\ n$;
        {
        do $mj = 1,\ mi$;
          {
          $ij = locg(mi,\ mj)$;
          $ji = locg(mj,\ mi)$;
          if $(mi \equiv mj)$
            next;
          $G(hoffset + ji) = G(hoffset + ij)$;
          }
        }
    }

§. @I "jandk.web" *Section(s) skipped...*

15.4.4 Hone

Given a "stack" of individual shell Fock matrices in H and a stack of shell R matrices in R, forms the single effective Hamiltonian for the combined set of shells: the general McWeenyan.

"grhf.f" 15.4.4 ≡

    subroutine $Hone(H, R, n, alpha, b, nu, number\_of\_shells, Work1, Work2, Hbar)$;
    ⟨ Hone interface declarations 15.4.4.1 ⟩
    {
    ⟨ Hone local declarations 15.4.4.2 ⟩
    ⟨ Hone initialisations 15.4.4.3 ⟩
    ⟨ Add the off-diagonal projections 15.4.4.4 ⟩
    ⟨ Now the diagonal terms including level shifters 15.4.4.5 ⟩
    return;
    }

⟨ Hone interface declarations 15.4.4.1 ⟩ ≡
    implicit double precision $(a - h, o - z)$;
    double precision $H(*), R(*), Work1(*), Work2(*), Hbar(*)$;
    double precision $alpha(number\_of\_shells)$;   /* shifters; */
    double precision $nu(number\_of\_shells)$;   /* occupations */
    double precision $b(n, n)$;   /* dampers */
    integer $n, number\_of\_shells$;

This code is used in section 15.4.4.

⟨ Hone local declarations 15.4.4.2 ⟩ ≡
    double precision $damp\_factor, shifter, half$;
    double precision $correct$;   /* used to divide by nu(shell1) */
    integer $shell1, shell2, nshell, number\_of\_shells$;
    integer $shell1\_offset, shell2\_offset$;
    data $one, zero, half/1.0 \cdot 10^{00}D, 0.0 \cdot 10^{00}D, 0.5 \cdot 10^{00}D/$;

This code is used in section 15.4.4.

15.4. A WEB FOR SPIN-RESTRICTED OPEN SHELL 463

⟨ Hone initialisations 15.4.4.3 ⟩ ≡
   $nshell = number\_of\_shells + 1;$ /∗ allow for virtual shell ∗/
   $nn = n * n;$
   do $i = 1,\ nn;$
      $Hbar(i) = zero;$ /∗ initialise Hbar since addprd cannot! ∗/

This code is used in section 15.4.4.

⟨ Add the off-diagonal projections 15.4.4.4 ⟩ ≡
   do $shell1 = 1,\ nshell;$
   {
   $shell1\_offset = (shell1 - 1) * nn;$
   do $shell2 = 1,\ nshell;$
     {
     if $(shell1 \equiv shell2)$
       next;
     $damp\_factor = b(shell1,\ shell2);$
     $shell2\_offset = (shell2 - 1) * nn;$
     do $i = 1,\ nn;$
       $Work1(i) = (H(shell2\_offset + i) - H(shell1\_offset + i)) * damp\_factor;$
     call $gmprd(R(shell1\_offset + 1),\ Work1,\ Work2,\ n,\ n,\ n);$
     call $addprd(Work2,\ R(shell2\_offset + 1),\ Hbar,\ n,\ n,\ n);$
     }
   }

This code is used in section 15.4.4.

⟨ Now the diagonal terms including level shifters 15.4.4.5 ⟩ ≡

    do *shell1* = 1, *number_of_shells*;
      {
    shell1_offset = (*shell1* − 1) * *nn*;
    shifter = *alpha*(*shell1*);   /* correct = one/nu(shell1) */
    correct = $1.0 \cdot 10^{00}$D;
    do $i = 1$, *nn*;
      Work1(*i*) = *correct* * *H*(*shell1_offset* + *i*);
    do $i = 1$, *n*;
      {
    ii = (*i* − 1) * *n* + *i*;   /* add a diagonal level-shifter */
    Work1(*ii*) = *Work1*(*ii*) + *shifter*;
      }
    call *gmprd*(*R*(*shell1_offset* + 1), *Work1*, *Work2*, *n*, *n*, *n*);
    call *addprd*(*Work2*, *R*(*shell1_offset* + 1), *Hbar*, *n*, *n*, *n*);
      }
    do $i = 1$, *nn*;
      Hbar(*i*) = *half* * *Hbar*(*i*);   /* get the eigenvalues right */

This code is used in section 15.4.4.

15.4.5 Hhalf

Special form for half-closed case of the effective many-shell Hamiltonian. Given a "stack" of individual shell Fock matrices in H and a stack of shell R matrices in R, forms the single effective Hamiltonian for the combined set of shells: the original McWeenyan.

"grhf.f" 15.4.5 ≡
    subroutine *Hhalf*(*HF*, *R*, *n*, *Work1*, *Work2*, *Hbar*);

    ⟨ Hhalf interface declarations 15.4.5.1 ⟩

    {

    ⟨ Hhalf local declarations 15.4.5.2 ⟩

    ⟨ Hhalf initialisation 15.4.5.3 ⟩

    ⟨ Form (R1+R2)(2H1-H2)(R1+R2) 15.4.5.4 ⟩

    ⟨ Add in (R1+RZ)(H1)(R1+RZ) + (R2+RZ)(H2)(R2+RZ) 15.4.5.5 ⟩

    return;

    }

15.4. A WEB FOR SPIN-RESTRICTED OPEN SHELL

⟨ Hhalf interface declarations 15.4.5.1 ⟩ ≡
    double precision $HF(*)$, $R(*)$, $Work1(*)$, $Work2(*)$, $Hbar(*)$;
    integer n;

This code is used in section 15.4.5.

⟨ Hhalf local declarations 15.4.5.2 ⟩ ≡
    double precision $half$, $zero$;
    integer $shell$, $offset$, $offset2$, $zoffset$;
    data $half$, $zero/0.5 \cdot 10^{00}$D, $0.0 \cdot 10^{00}$D/;

This code is used in section 15.4.5.

⟨ Hhalf initialisation 15.4.5.3 ⟩ ≡
    $nn = n * n$;
    $offset2 = nn$;
    $zoffset = nn + nn$;
    do $i = 1$, nn;
        $Hbar(i) = zero$;   /∗ initialise Hbar ∗/

This code is used in section 15.4.5.

⟨ Form (R1+R2)(2H1-H2)(R1+R2) 15.4.5.4 ⟩ ≡
    do $i = 1,\ nn$;
       $Work1(i) = R(i) + R(offset2 + i)$;   /∗ R1 + R2 ∗/
    call $gmprd(Work1,\ HF,\ Work2,\ n,\ n,\ n)$;
    call $addprd(Work2,\ Work1,\ Hbar,\ n,\ n,\ n)$;
       /∗ (R1+R2)(2H1)(R1+R2) ∗/
    call $gmprd(Work1,\ HF(offset2 + 1),\ Work2,\ n,\ n,\ n)$;
    do $i = 1,\ nn$;
       $Work1(i) = -Work1(i)$;   /∗ to get -H2 ∗/
    call $addprd(Work2,\ Work1,\ Hbar,\ n,\ n,\ n)$;
       /∗ (R1+R2)(-H2)(R1+R2) ∗/

    /∗ so far Hbar = (R1+R2)(2H1-H2)(R1+R2) ∗/

    do $i = 1,\ nn$;
       $HF(i) = half * HF(i)$;   /∗ correct for occupation number of 2 ∗/

This code is used in section 15.4.5.

⟨ Add in (R1+RZ)(H1)(R1+RZ) + (R2+RZ)(H2)(R2+RZ) 15.4.5.5 ⟩ ≡
    do $shell = 1,\ 2$;
      {
      $offset = (shell - 1) * nn$;
      do $i = 1,\ nn$;
         $Work1(i) = R(offset + i) + R(zoffset + i)$;
      call $gmprd(Work1,\ HF(offset + 1),\ Work2,\ n,\ n,\ n)$;
      call $addprd(Work2,\ Work1,\ Hbar,\ n,\ n,\ n)$;
      }

    /∗ Hbar = as before + (R1+RZ)(H1)(R1+RZ) + (R2+RZ)(H2)(R2+RZ) ∗/

    do $i = 1,\ nn$;
       $Hbar(i) = half * Hbar(i)$;   /∗ get eigenvalues right ∗/

This code is used in section 15.4.5.

15.4. A WEB FOR SPIN-RESTRICTED OPEN SHELL

15.4.6 addprd

Matrix multiplier which **adds** $A \times B$ into existing R i.e. there is no initialisation of R, otherwise this routine is identical to *gmprd*

§. @I "addprd.web" *Section(s) skipped...*

§. @I "minv.web" *Section(s) skipped...*

15.4.7 INDEX

A: 15.4.6.

addprd: 15.4.6, 15.4.4.4, 15.4.4.5, 15.4.5.4, 15.4.5.5.

alpha: 15.4.1, 15.4.1.9, 15.4.1.15, 15.4.4, 15.4.4.1, 15.4.4.5.

B: 15.4.6.

b: 15.4.1.15, 15.4.4.1.

correct: 15.4.4.2, 15.4.4.5.

crit: 15.4.1.13, 15.4.1.16.

dabs: 15.4.1.13.

damp_factor: 15.4.4.2, 15.4.4.4.

det: 15.4.1.3.

eigen: 15.4.1.11.

en: 15.4.1, 15.4.1.5, 15.4.1.7, 15.4.1.13, 15.4.1.14, 15.4.1.15.

END_OF_FILE: 15.4.3.4.

eps: 15.4.1, 15.4.1.11, 15.4.1.15.

ERROR_OUTPUT_UNIT: 15.4.1.13.

G: 15.4.3.1.

GENERAL: 15.4.1.1.

getint: 15.4.3.2, 15.4.3.4.

gmprd: 15.4.1.10, 15.4.1.11, 15.4.1.12, 15.4.4.4, 15.4.4.5, 15.4.5.4, 15.4.5.5, 15.4.6.

grhf: 15.4.1, 15.4.1.16.

grhfGR: 15.4.1.8, 15.4.3.

grhfR: 15.4.1.4, 15.4.1.12, 15.4.2, 15.4.2.1.

gtprd: 15.4.1.10, 15.4.1.11.

H: 15.4.1.15, 15.4.4.1.

half: 15.4.1.14, 15.4.1.16, 15.4.4.2, 15.4.4.5, 15.4.5.2, 15.4.5.4, 15.4.5.5.

HALF_CLOSED: 15.4.1.1, 15.4.1.9.

Hbar: 15.4.1, 15.4.1.6, 15.4.1.9, 15.4.1.11, 15.4.1.15, 15.4.4, 15.4.4.1, 15.4.4.3, 15.4.4.4, 15.4.4.5, 15.4.5, 15.4.5.1, 15.4.5.3, 15.4.5.4, 15.4.5.5.

HF: 15.4.1, 15.4.1.7, 15.4.1.8, 15.4.1.9, 15.4.1.11, 15.4.1.14, 15.4.1.15, 15.4.5, 15.4.5.1, 15.4.5.4, 15.4.5.5.

Hhalf: 15.4.1.9, 15.4.5.

hoffset: 15.4.3.2, 15.4.3.3, 15.4.3.4, 15.4.3.5.

Hone: 15.4.1.9, 15.4.4.

icon: 15.4.1, 15.4.1.5, 15.4.1.13.

ii: 15.4.1.11, 15.4.4.5.

ij: 15.4.1.10, 15.4.1.13, 15.4.2.2, 15.4.3.5.

ik: 15.4.2.2.

initial_scf: 15.4.1, 15.4.1.4, 15.4.1.5, 15.4.1.15, 15.4.1.16.

int: 15.4.1.1.

ioffset: 15.4.1.7, 15.4.1.14, 15.4.1.16.

irite: 15.4.1, 15.4.1.13, 15.4.1.15, 15.4.1.16.

iteration: 15.4.1, 15.4.1.5, 15.4.1.13, 15.4.1.16.

iw1: 15.4.1.1, 15.4.1.16.

iw2: 15.4.1.1, 15.4.1.16.

iz: 15.4.2.1.

ji: 15.4.1.10, 15.4.2.2, 15.4.3.5.

jk: 15.4.2.2.

JofR: 15.4.3.4.

kd: 15.4.2.1, 15.4.2.2.

kk: 15.4.2.2.

KofR: 15.4.3.4.

kz: 15.4.2.1.

l: 15.4.6.

locg: 15.4.3.2, 15.4.3.5.

lwork: 15.4.1.3, 15.4.1.16.

m: 15.4.6.

max_iteration: 15.4.1.13, 15.4.1.16.

mi: 15.4.3.5.

minv: 15.4.1.3.

mj: 15.4.3.5.

mu: 15.4.3.4.

mwork: 15.4.1.3, 15.4.1.16.

n: 15.4.2, 15.4.3.2, 15.4.4.1, 15.4.5.1, 15.4.6.

nbasis: 15.4.1, 15.4.1.15, 15.4.1.16.

nfile: 15.4.1, 15.4.1.8, 15.4.1.15, 15.4.1.16, 15.4.3, 15.4.3.2, 15.4.3.4.

nn: 15.4.1, 15.4.1.3, 15.4.1.6, 15.4.1.7, 15.4.1.8, 15.4.1.9, 15.4.1.10, 15.4.1.13, 15.4.1.14, 15.4.2, 15.4.2.2, 15.4.3.2, 15.4.3.3, 15.4.3.4, 15.4.3.5, 15.4.4.3, 15.4.4.4, 15.4.4.5, 15.4.5.3, 15.4.5.4, 15.4.5.5.

NO: 15.4.1, 15.4.1.5.

15.4. A WEB FOR SPIN-RESTRICTED OPEN SHELL 469

noclst: 15.4.1, 15.4.1.4, 15.4.1.12, 15.4.1.15, 15.4.2, 15.4.2.1, 15.4.2.2.

norbs: 15.4.1, 15.4.1.2, 15.4.1.4, 15.4.1.12, 15.4.1.15, 15.4.2, 15.4.2.1, 15.4.2.2.

nshell: 15.4.2, 15.4.4.2, 15.4.4.3, 15.4.4.4.

ntop: 15.4.2.1, 15.4.2.2.

nu: 15.4.1, 15.4.1.1, 15.4.1.7, 15.4.1.8, 15.4.1.9, 15.4.1.14, 15.4.1.15, 15.4.3, 15.4.3.1, 15.4.3.3, 15.4.4, 15.4.4.1.

number_of_shells: 15.4.1, 15.4.1.1, 15.4.1.2, 15.4.1.4, 15.4.1.7, 15.4.1.8, 15.4.1.9, 15.4.1.12, 15.4.1.14, 15.4.1.15, 15.4.2, 15.4.2.1, 15.4.3, 15.4.3.1, 15.4.3.3, 15.4.3.5, 15.4.4, 15.4.4.1, 15.4.4.2, 15.4.4.3, 15.4.4.5.

nz: 15.4.2.1.

offset: 15.4.2, 15.4.2.2, 15.4.5.2, 15.4.5.5.

offset_ij: 15.4.2, 15.4.2.2.

offset_ji: 15.4.2, 15.4.2.2.

offset2: 15.4.5.2, 15.4.5.3, 15.4.5.4.

one: 15.4.4.2.

path: 15.4.1.1, 15.4.1.9, 15.4.1.16.

pointer: 15.4.3.2, 15.4.3.4.

R: 15.4.1.15, 15.4.2, 15.4.3.1, 15.4.4.1, 15.4.5.1, 15.4.6.

roffset: 15.4.3.2, 15.4.3.3.

Rwork: 15.4.3, 15.4.3.1, 15.4.3.3, 15.4.3.4.

shell: 15.4.1.7, 15.4.1.14, 15.4.1.16, 15.4.2, 15.4.2.1, 15.4.2.2, 15.4.5.2, 15.4.5.5.

shell_offset: 15.4.1.7, 15.4.1.14, 15.4.1.16.

shellh: 15.4.3, 15.4.3.2, 15.4.3.3, 15.4.3.5.

shellr: 15.4.3.2, 15.4.3.3.

shell1: 15.4.4.2, 15.4.4.4, 15.4.4.5.

shell1_offset: 15.4.4.2, 15.4.4.4, 15.4.4.5.

shell2: 15.4.4.2, 15.4.4.4.

shell2_offset: 15.4.4.2, 15.4.4.4.

shifter: 15.4.4.2, 15.4.4.5.

subroutine: 15.4.1.

sum: 15.4.2.2.

test: 15.4.1.13.

total_occupied_orbitals: 15.4.1.2, 15.4.1.13, 15.4.1.16.

U: 15.4.1.15, 15.4.2.

V: 15.4.1.15.

va: 15.4.3.4.

virtual_offset: 15.4.1.8, 15.4.1.16.

Work: 15.4.1, 15.4.1.8, 15.4.1.9, 15.4.1.11, 15.4.1.12, 15.4.1.15.

Work1: 15.4.4, 15.4.4.1, 15.4.4.4, 15.4.4.5, 15.4.5, 15.4.5.1, 15.4.5.4, 15.4.5.5.

Work2: 15.4.4, 15.4.4.1, 15.4.4.4, 15.4.4.5, 15.4.5, 15.4.5.1, 15.4.5.4, 15.4.5.5.

x: 15.4.1.15, 15.4.3.1.

y: 15.4.1.15, 15.4.3.1.

YES: 15.4.1.4.

zero: 15.4.1.5, 15.4.1.8, 15.4.1.16, 15.4.2, 15.4.2.2, 15.4.3.2, 15.4.3.3, 15.4.4.2, 15.4.4.3, 15.4.5.2, 15.4.5.3.

zoffset: 15.4.5.2, 15.4.5.3, 15.4.5.5.

15.4. A WEB FOR SPIN-RESTRICTED OPEN SHELL

⟨ Add in (R1+RZ)(H1)(R1+RZ) + (R2+RZ)(H2)(R2+RZ) 15.4.5.5 ⟩ Used in section 15.4.5.

⟨ Add the off-diagonal projections 15.4.4.4 ⟩ Used in section 15.4.4.

⟨ Calculate the R matrices if the input U is a sensible one 15.4.1.4 ⟩ Used in section 15.4.1.

⟨ Check for convergence and prepare to re-cycle 15.4.1.13 ⟩ Used in section 15.4.1.5.

⟨ Choose the appropriate form of the effective Fock matrix 15.4.1.9 ⟩ Used in section 15.4.1.5.

⟨ Choose which effective Fock matrix 15.4.1.1 ⟩ Used in section 15.4.1.

⟨ Compute total number of occupied orbitals 15.4.1.2 ⟩ Used in section 15.4.1.

⟨ Form (R1+R2)(2H1-H2)(R1+R2) 15.4.5.4 ⟩ Used in section 15.4.5.

⟨ Form the R matrix for each shell using noclst 15.4.2.2 ⟩ Used in section 15.4.2.

⟨ Form the virtual shell from unoccupied orbitals 15.4.2.1 ⟩ Used in section 15.4.2.

⟨ Hhalf initialisation 15.4.5.3 ⟩ Used in section 15.4.5.

⟨ Hhalf interface declarations 15.4.5.1 ⟩ Used in section 15.4.5.

⟨ Hhalf local declarations 15.4.5.2 ⟩ Used in section 15.4.5.

⟨ Hone initialisations 15.4.4.3 ⟩ Used in section 15.4.4.

⟨ Hone interface declarations 15.4.4.1 ⟩ Used in section 15.4.4.

⟨ Hone local declarations 15.4.4.2 ⟩ Used in section 15.4.4.

⟨ Interface argument declarations 15.4.1.15 ⟩ Used in section 15.4.1.

⟨ Invert the orthogonalisation matrix 15.4.1.3 ⟩ Used in section 15.4.1.

⟨ Now symmetrise all the Gs 15.4.3.5 ⟩ Used in section 15.4.3.

⟨ Now the diagonal terms including level shifters 15.4.4.5 ⟩ Used in section 15.4.4.

⟨ Obtain the repulsion integrals and form the Gs 15.4.3.4 ⟩ Used in section 15.4.3.

⟨ One iteration of SCF cycle 15.4.1.5 ⟩ Used in section 15.4.1.

⟨ Put one-electron Hamiltonian into each Fock matrix 15.4.1.7 ⟩   Used in section 15.4.1.5.

⟨ Scale the Fock matrices by occupation number 15.4.1.14 ⟩   Used in section 15.4.1.5.

⟨ Set up the Combinations of R matrices 15.4.3.3 ⟩   Used in section 15.4.3.

⟨ Start first cycle from one-electron Hamiltonian 15.4.1.6 ⟩   Used in section 15.4.1.5.

⟨ Transform and diagonalise the single Fock matrix 15.4.1.11 ⟩   Used in section 15.4.1.5.

⟨ Transform back to the original basis and form the new R 15.4.1.12 ⟩   Used in section 15.4.1.5.

⟨ Transpose the inverse-orthogonaliser for convenience 15.4.1.10 ⟩   Used in section 15.4.1.5.

⟨ Use the repulsion integrals to form the Fock matrices 15.4.1.8 ⟩   Used in section 15.4.1.5.

⟨ grhf local declarations 15.4.1.16 ⟩   Used in section 15.4.1.

⟨ grhfGR interface declarations 15.4.3.1 ⟩   Used in section 15.4.3.

⟨ grhfGR local declarations 15.4.3.2 ⟩   Used in section 15.4.3.

Chapter 16

Banana skins: unexpected disasters

After our earlier discussion of virtual orbitals with its surprising and, perhaps, unpalatable conclusions, a few more examples are given here of artifacts of the LCAO SCF method; some surprising and some less so.

Contents

16.1	Symmetry restrictions	474
16.2	Anions	475
16.3	Aufbau exceptions	477
16.4	Summary	478

The use of the linear expansion method has some drawbacks which although very obvious when pointed out, are not initially obvious. All these drawbacks are different aspects of the same fundamental fact: when we make the transition from the variational optimisation of the Hartree–Fock energy *functional* to the optimisation of the parametric energy *function* we implicitly *assume* that the solutions to the variational problem *exist*. That is, unless we have chosen our parametrised method unusually ineptly, it will generate a well-behaved energy *function* of the parameters. Of course, any reasonably well-behaved *function* of a set of parameters that models the energy of a set of interacting particles is guaranteed to have a minimum even if the parameters used do not cover the whole space spanned by the original *functional*.

That is, we have to rely on physical and chemical intuition in our choice of parametrisation to try to ensure that our solution of the SCF (finite-basis approximated) "Hartree–Fock" equations are, indeed, realistic and acceptable approximations to the solutions of the (differential) true Hartree–Fock equations. Our parameters are the expansion coefficients, so that these matters all hinge on choice of basis function by which the parameters are multiplied; we must ensure that there are sufficient basis functions available to locate a minimum in the energy *function* which reflects both the *existence* and *form* of the putative solution of the differential Hartree–Fock equations that we seek.[1]

The possibilities of pitfalls are both quantitative *and qualitative*:

- The choice of basis functions may be inadequate to reflect the detailed form of the SCF orbitals.

- Basis functions of some particular type or symmetry may not be present in the basis.

- The action of the variation principle may be unwittingly limited by our choice of basis function.

The case of "virtual orbitals" has been discussed earlier in a different context since these orbitals are not even formally available from the differential Hartree–Fock equation.

16.1 Symmetry restrictions

Perhaps the most familiar (non-trivial) example of the last of the effects listed above is a minimal basis calculation of the MOs of a highly symmetrical system. It is a familiar fact that the Hückel MOs of the π-system of benzene also solve the SCF equations when both are expressed in terms of a minimal basis of $2p_\pi$ carbon AOs. The solid-state analogy of this Hückel/SCF case is the fact that the plane-wave orbitals of a non-interacting electron gas solve the Hartree–Fock equations for the system for exactly the same reason:

> In both of these systems there is only one orbital of each symmetry type so that the symmetry orbitals formed from the basis functions *are* the non-interacting *and* the Hartree–Fock orbitals; there is no freedom for the variation principle to act to optimise the form of the orbitals. In order for there to be genuine variational freedom in the calculation there must be *more than one* basis function of each (occupied) symmetry type.

---

[1] As we shall see, we are in something of a "deadly embrace" here since we have no independent way of verifying the existence and form of the sought solution *except* by trial and error using the very expansion method under test.

16.2. ANIONS

In the case of the plane waves this is trivially obvious because they do not even have to be combined; they *are* the symmetry orbitals. In the benzene case the AOs have to be combined to form symmetry orbitals to reveal the nature of the limitation but, if an SCF calculation is performed, the mere diagonalisation of the Hartree–Fock matrix does this task.

In the case of the π-system of benzene the solution to the difficulty is obvious; using more than one $2p_\pi$ AO per carbon atom produces the corresponding number of symmetry orbitals of each type and the variation principle can act to produce MOs which are genuinely influenced by electron repulsion. The solution in the plane-wave case is less obvious since there is nothing analogous to the orbital exponent of an AO in the form of simple plane waves.

So, although these type of limited-basis expansions do formally solve the Hartree–Fock equations, it is better, perhaps, to refer to them as merely *satisfying* the SCF equations rather than *solving* them; to try and distinguish between actively solving the equations and the rather passive fact of satisfying the equations by dint of some accidental fact of symmetry. The Hückel and SCF orbitals of pyridine, for example, are different even in a minimal basis since pyridine has a lower symmetry than benzene.

16.2 Anions

Many molecules of acute chemical interest are charged; in particular many species containing transition metal atoms are anions. Sometimes these anions are closed-shell, sometimes open-shell. There exists a formal proof that the solutions of the (differential, GUHF) Hartree–Fock equations actually exist for neutral systems and cations. The proof apparently cannot be extended to anions: all that can be proved is that a molecule with $n+1$ electrons is stable if the net nuclear charge sum is $n+\epsilon$ where ϵ may be small but *non-zero*. This means that the existence of the solutions of the Hartree–Fock equations for anions is contingent on the particular case; in some cases the solutions will exist, in other cases not. Clearly, it is extremely unlikely that the Hartree–Fock equations for multiply charged anions will exist.

However, as we have already seen much earlier, if we perform an SCF calculation at minimal-basis level on the dianion of the water molecule, we obtain smooth convergence to a stable self-consistent ground state. We even took the "testbench" program to the limit and performed calculations on $(H_2O)^{4-}$. The conflict between the formal proof and actual practice is complete; but the resolution is obvious. If we recall the discussion about virtual orbitals it is clear that the existence of these "approximations to non-existent functions" are, once again, artifacts of the use of a limited basis of expansion functions. The minimal basis of functions is a set of (approximate) atomic orbitals of the ground states of the oxygen and hydrogen atoms; adapted to a description of electron density close to the nuclei. If these functions are used to expand the MOs of an anion which, presumably, contains very weakly bound or unbound electrons the resulting MOs will be totally inadequate and will simply be an artifact of the parametrised expansion method; a minimum in the energy function which has no connection with any minima in the energy functional. We should be warned about this by the fact that the total energy is higher than that of the neutral molecule and, sometimes, the orbital energies are positive; all indications that the system is "trying" to lose electrons but the basis is too inflexible to allow this to happen. If we augment the basis with some suitably diffuse functions the highest anion orbitals will take advantage of this to either expand to a more realistic diffuse form or, more usually, to try to simulate a free electron.

Often the *combination* of some symmetry constraints and a limited basis conspire to produce artifacts. If we perform a closed-shell calculation on the H^- anion we find, as usual for most closed-shell calculations on small systems, a smooth convergence to a self-consistent ground state with energy higher than that of the H atom *even using a basis including diffuse functions*. This result is a little surprising in view of what we have just said, but not if we realise that we are *imposing* the condition that the lowest state *must* be a doubly occupied orbital. With this constraint, notwithstanding the availability of diffuse functions to simulate ionisation, the choice open to the variation principle is between H^+ (no electrons, impossible) and H^- (two electrons with identical spatial distributions, possible but unrealistic); naturally the "solution" found is the bogus[2] ground state of H^-.

If we remove the closed-shell constraint and simply use the same basis to compute the best single-determinant wavefunction without regard for any other constraints we do obtain a DODS-type solution which has one tightly bound electron and one very loosely bound electron. If we go on extending the basis, we quickly see that the limit of the solution is a hydrogen *atom* and a free electron which is the SCF solution of the ground state of one proton and two electrons in this context.

Sometimes more interesting cases arise, particularly in the case of systems with degenerate ground states. The conventional restricted open-shell single-determinant solution for the oxygen atom, cation or anion (configuration $1s^2 2s^2 2p^n$) has the same radial function for *all* the n $2p$ electrons. In particular the anion O^- has just one radial function which must be optimised to do duty for the description of all five $2p$ electrons. Now, whatever the quality of the basis set in an SCF calculation, the variation principle is presented with an interesting dilemma: it must either have a single $2p$ radial orbital which is a compromise between that of the atom and that of a more diffuse orbital or have no occupation of a $2p$ AO at all, the choice is between O^{4+} and O^-. Of course the

---

[2] In fact H^- does exist and its wavefunction may be computed by more advanced methods but there is no stable *Hartree–Fock* wavefunction for H^-.

16.3. AUFBAU EXCEPTIONS

solution generates the compromise solution which has no correspondence with, for example, the numerical solution of the HF equation for the system.

If a UHF single determinant is used where each $2p$ electron can have its own radial distribution the dilemma is removed and a solution is generated which, within the confines of the actual basis used, is an oxygen atom and a diffuse electron.

These are all simple examples and provide pathological cases to illustrate the more general point:

> When dealing with anions, always be alert to the possibility that a combination of symmetry constraints or basis set limitations will generate "SCF approximations to non-existent HF solutions".

16.3 Aufbau exceptions

We have seen earlier that the validity of the aufbau principle is dependent on the optimisation of the *unconstrained* (GUHF) single-determinant model of electronic structure. The most familiar constraint on the single-determinant model occurs in the theory of open-shell systems, particularly symmetrical open-shell systems.

The familiar model of the electronic structure of atoms has the orbitals grouped into angular-momentum "shells" in which there is the ever-increasing number of $(2\ell+1)$ AOs with the same orbital energy and radial function, as the orbital angular momentum value, ℓ, increases. What is more, the concept of a partially filled shell ensures that there are often atomic structures which have an implicit HOMO-LUMO gap of zero, thereby automatically ensuring that there is a possible violation of the aufbau principle. The simplest case is the lithium atom in which the restricted-model LUMO is the $2s\beta$ AO, degenerate with the occupied $2s\alpha$ AO.

In some small radicals containing the oxygen atom the aufbau principle is violated by the ROHF method in such a way that the conventional picture of a radical (closed shells with an outer open shell) is not even true. In both the OH and OOH radicals the ROHF model gives an electronic structure in which the singly-occupied MO is not the HOMO; it is an "inner" MO which is singly-occupied. Removing the spin constraint and performing a GUHF calculation removes the problem and restores the aufbau principle.

One interpretation of these results is the fact that the ground state of (*e.g.*) the OH$^+$ molecule is a triplet (two unpaired electrons in symmetry-related π

MOs) and, if the HOMO of the parent molecule were singly-occupied, ionisation of the molecule would produce a closed-shell singlet for OH^+ which is clearly wrong.

The most celebrated and venerable problem associated with the aufbau principle is the case of the $nd/(n + 1)s$ electronic structures of the transition-metal atoms. There is no satisfactory way *within the conventional spherical SCF model* to explain why (*e.g.*) the $4s$ AO of the nickel atom is occupied before the $3d$ shell is complete in view of their relative (HF or SCF) orbital energies. If a GUHF calculation is carried through on these atoms what emerges is the correct electron configuration without any necessity of *imposing* that configuration on the calculation. Allowing each AO to have its own choice of radial form is sufficient to ensure that, if there are $(m + 2)$ electrons available in this region, there are m AOs of predominantly d type occupied and 2 AOs of mainly s-type lying immediately above.[3] What is wrong is that the "strain" involved in constraining all the AOs of the d-shell to be the same causes the failure of the aufbau principle.

Of course in the last case and many of the examples quoted here the single-determinant wavefunction does not have the spatial or spin symmetry of the atom or molecule. This is probably just an indication of the limitations of the HF model in coping with these situations. What *is* true is that the HF model will give its best energetic results and its most unambiguous physical interpretation if no constraints are placed on the action of the variation principle other than those implicit in the length and quality of the basis-set expansion.

16.4 Summary

There are theoretical limitations of the Hartree–Fock *model* of electronic structure and, where these are known, one simply has to use a more advanced *model* of those structures. However, many of the more familiar failings of practical SCF calculations can be avoided by a careful examination of the implicit or explicit *constraints* placed on the SCF model by choice of basis or presumed symmetries.

---

[3]In those cases where the configuration is $d^{(m+1)}s^1$ this emerges correctly from the GUHF calculation.

Chapter 17

Molecular symmetry

When a molecules has one or more elements of symmetry in its geometrical structure there are several possible ways in which this symmetry can be used in electronic structure calculations. The principle ways in which symmetry is used are introduced in this chapter and details of one way of reducing the labour involved in the calculation of the molecular integrals are presented.

Contents

17.1	Introduction	**479**
17.2	Symmetry and the HF method	**480**
17.3	Permutational symmetry of the basis	**483**
17.4	Implementation	**488**
17.4.1	Two-electron integrals	488
17.4.2	One-electron integrals	492
17.4.3	Numerical values	501
17.4.4	Conditions on the array perms	502
17.5	Permutation symmetry: summary	**504**

17.1 Introduction

It often happens in the performance of molecular electronic structure calculations that the system for which a calculation is being planned has some elements of symmetry; a rotation axis, a mirror plane, a centre of symmetry etc. It is therefore natural to ask if advantage can be taken of this symmetry to speed up the calculations.

In this and the following chapters we will investigate the use of molecular symmetry in the context of the computation of molecular electronic structure. There are three distinct areas of applicability of molecular symmetry with a chapter devoted to each area with some inevitable overlap:

- The calculation of molecular integrals; if the integrals are required for *any* type of model of the molecular electronic structure (not just SCF) then certain savings due to molecular symmetry may be made, particularly in the time-consuming calculation of the repulsion integrals.[1]

- The diagonalisation of the Fock matrix during the iterative process of SCF (or MCSCF) convergence. This step is clearly not applicable to non-SCF models.

- The use of the *one-electron* symmetry of the Fock matrix to speed up the processing of *two-electron* repulsion integrals. Again, this method is only available in SCF models.

In this chapter the first of these three points is addressed. However, by way of general introduction to the use of molecular symmetry, it is interesting to consider the SCF case in qualitative terms.

17.2 Symmetry and the HF method

There are three rather self-evident ways in which symmetry should help a calculation of molecular electronic structure (at least, at the LCAOSCF level):

1. Atoms which are related by the symmetry operations will, presumably, have the same set of basis functions centred on them. In fact, one usually places the same basis functions on *all* atoms of the same atomic number whatever their chemical environment.

2. The one-electron Hamiltonian and, by implication, the Hartree–Fock matrices should have the symmetry of the molecule; matrix elements between basis functions which are related by the symmetry operations of the molecule might reasonably be expected to be identical (apart from a phase factor, usually just a sign). This, of course, presumes that item 1 above holds.

---
[1]These methods are based on an idea in the first generally-available set of *ab initio* programs: the POLYATOM system. The particular methods used were developed by P. D. Dacre (*Chem. Phys. Lett.*, **7**, 47, (1970)) and extended by M. Elder (*Int. J. Quant. Chem.*, **VII**,75 (1973)).

17.2. SYMMETRY AND THE HF METHOD

3. Again, assuming 1 and 2 hold then it necessarily follows and might well be expected on general intuitive grounds, that the coefficients of basis functions in a given MO which are related by symmetry operations will be of identical modulus. That is, the electron density due to occupancy of a given MO will be the same at symmetry-equivalent points in the molecule (neglecting the effects due to degenerate MOs for the moment).

However intuitively self-evident these three points appear to be, unfortunately only the first one is true and it is only true because we decide to make it so by our *choice* of basis functions. This is an entirely sensible choice and will be assumed in what follows; it would be simply perverse to place different basis functions on symmetry-related atoms.

To see that the last point is not necessarily true it is only necessary to state its implications a little more carefully. While it is, perhaps, not unreasonable to expect that the *total* electron density at symmetry-equivalent points in a molecule to be the same, point 3 above goes very much further than this. What it says is that in an n-electron system the total electron density at symmetry-equivalent points is the same *because* it is composed of n separate electron densities, *each of which is separately* identical at symmetry-equivalent points in the molecule. This is surely *a priori* unreasonable; to ask that the separate electron densities *compensate* for each others' distributions to give a symmetrical electron distribution is fair but to ask each electron *individually* to have a symmetrical distribution is a little restrictive. In practice, it is found that even this *weak and reasonable* constraint is not obeyed in every case.

In most practical computational work, however, even in the *absence* of any symmetry constraints, the molecular orbital coefficients of typical stable molecules *do* turn out to be symmetry-adapted. It is worth a brief recapitulation of the Hartree–Fock method to see why this should be so.

The symmetries associated with a particular Schrödinger equation would emerge as the integral invariants of the Lagrangian, of which the Schrödinger equation is the Euler–Lagrange condition for an extremising solution. In principle the symmetries of the Hartree–Fock equation are determined in a similar way. In practice, the symmetries of a Schrödinger equation are found by looking for operators which commute with the Hamiltonian and similar conditions are simply *assumed* to hold for the Hartree–Fock equation.

The usual "electrostatic" Hamiltonian of quantum chemistry has its symmetry determined by the nuclear configuration. If

$$\hat{H} = \sum_{i=1}^{n} \hat{h}(i) + \sum_{i>j=1}^{n} \frac{1}{r_{ij}}$$

where

$$\hat{h}(i) = -\frac{1}{2}\nabla^2 - \sum_{A=1}^{N} \frac{Z_A}{r_{iA}}$$

then the electron-repulsion operators and the kinetic energy operators all depend on *differences* between the coordinates of identical particles and so have very high symmetry (full spherical symmetry). The nuclear-attraction operators involve *differences* between the coordinates of the electrons and the nuclei. Thus, since all point groups are sub-groups of the full spherical group, the symmetry of the Hamiltonian is just that of the nuclear geometry.

Now the Hartree–Fock "Hamiltonian" is of the general form

$$\hat{h}^F = \hat{h}(i) + \hat{J}(i) - \hat{K}(i)$$

where, for example,

$$\hat{J}(1) = \int \frac{\rho(2)}{r_{12}} dV_2$$

and

$$\rho = \sum_{i=1}^{n} \psi_i \psi_i^*$$

This situation is quite different from that of the Schrödinger equation, since the Hartree–Fock operator contains a term which depends, not on coordinate *differences*, but on a coordinate itself; the operator $\hat{J}$ depends on the bare *coordinates* of an electron, not on the difference between this electron and other electrons. It is not at all clear *what* consequences this simple fact has for the symmetry of the canonical Hartree–Fock orbitals. What is, however, certainly clear is that there is no *requirement* that the canonical Hartree–Fock orbitals have the symmetry of the nuclear framework, or, indeed of any operator which commutes with the "parent" Hamiltonian of the system under study. This fact is, of course, completely familiar from our calculations of the UHF orbitals of the water cation; they do not have the "spin symmetry" of the parent Hamiltonian. The above result has much wider consequences; in general there is no reason why the Hartree–Fock orbitals should have any of the symmetries of the parent Hamiltonian.

If the orbitals ψ_i *do* have the symmetry of the molecular framework, then so do the operators $\hat{J}$ and $\hat{K}$. In this case there must be some *self-consistent symmetries* which will, typically, be generated if an initial guess at the ψ_i produces orbitals which do have the molecular symmetry. These self-consistent symmetries will, of course, tend to *persist* throughout the calculation and the final self-consistent MOs will then be self-consistent in two senses: energetically and symmetrically. But we cannot rely on these orbitals generating the lowest state of the system concerned.

As we noted above, *in fact* calculations on stable molecules usually *do* generate canonical Hartree–Fock orbitals which have the symmetry of the molecular framework. It seems likely that this is a contingent, numerical fact rather than one of principle. If the one-electron terms in the Hartree–Fock Hamiltonian are so large as to *dominate* the form of the molecular orbitals, these MOs will,

17.3. PERMUTATIONAL SYMMETRY OF THE BASIS

presumably, take up a distribution which optimises the energy due to the dominant terms in the Hamiltonian, and the electron-repulsion terms involving the potentially symmetry-breaking effects are too weak to change this situation. In molecules, radicals or (particularly) anions which have very weakly bound electrons, one would expect to see the effect of broken symmetry.

Thus, in deciding to incorporate the effects of molecular symmetry into an implementation of the LCAO method (for example), we must be aware of the possible pitfalls in this decision. In any case, however, the use of the same basis on symmetry-equivalent atoms seems quite innocent and we can always attempt to use this piece of information to reduce the redundant computation of molecular integrals; particularly the time- and storage-consuming repulsion integrals.

We now turn to the use of molecular symmetry in the *general* calculation of molecular integrals.

17.3 Permutational symmetry of the basis

The algorithm we designed in earlier chapters for the storage and retrieval of the electron-repulsion integrals (implemented as `putint` and `getint`) only required that all the (non-zero) integrals be present, together with their labels in standard order, on an external file. There was no requirement that those integrals in the file be in any particular *order* in the file. This design is particularly flexible since a file generated by calls to `putint` will always be capable of being processed by `getint` even if there are integrals:

1. Missing; when some approximation schemes are being investigated, or

2. In arbitrary order; when it is computationally convenient to calculate groups of integrals together for some reason.

The same kind of remarks apply if `getint` is implemented to *calculate* (rather than read) the repulsion integrals (so-called direct methods).

The use of molecular symmetry demands that one create a file of integrals in random order, so that symmetry-related integrals may be found and generated *together* independent of the order in which the basis functions are placed.

If we consider *all* the symmetry operations which are associated with a particular molecular geometry, these operations form a point group and all these operations have the property of *permuting* atoms in identical environments in the molecule. However, if a set of identical Cartesian basis functions is placed on each symmetry-equivalent atom then, in addition to the *permutation* of symmetry-equivalent basis functions, some of the symmetry operations will send these basis functions into *linear combinations* of themselves (it is only necessary to think of the action of a three- or five-fold rotation on a set of p basis functions to see this).

It proves possible to use a very elegant algorithm to make use of the *permutational* symmetries (including a possible change of sign) but rather more difficult to make effective use of linear combinations of basis functions. Formally, this restricts the use of symmetry to those molecules which have an Abelian point-group symmetry; symmetry groups with one-dimensional representations. In fact, due to the fact that the dimension of the representations which are actually *realised* in a particular case depend both on the nature of the point group *and* the particular basis functions used, it is possible to use more symmetry by a slight adaptation of the permutational algorithm.

The algorithm itself is, perhaps, best understood by its application to a simple example and then the essential features may be abstracted for general use. We shall consider the water molecule using the minimal basis of Chapter 5 with the numbering system:

$1s_O$	$2s_O$	$2p_{xO}$	$2p_{yO}$	$2p_{zO}$	$1s_{H1}$	$1s_{H2}$
ϕ_1	ϕ_2	ϕ_3	ϕ_4	ϕ_5	ϕ_6	ϕ_7
1	2	3	4	5	6	7

and consider the electron-repulsion integral

$$(77,21) = \int \phi_7(\vec{r_1})\phi_7(\vec{r_1})\frac{1}{r_{12}}\phi_2(\vec{r_1})\phi_1(\vec{r_1})dV_1 dV_2$$

and how it is affected by one of the symmetry elements of the H$_2$O molecule; the reflection in the plane perpendicular to the molecular plane bisecting the H-O-H angle (the xz-plane in the usual convention).

Under this operation the basis functions are simply permuted; in fact, only basis functions ϕ_6 and ϕ_7 are actually *permuted*, the other functions are sent into themselves except, possibly, for a change of sign:

$$\begin{pmatrix}\phi_1\\\phi_2\\\phi_3\\\phi_4\\\phi_5\\\phi_6\\\phi_7\end{pmatrix} \rightarrow \begin{pmatrix}\phi_1\\\phi_2\\\phi_3\\-\phi_4\\\phi_5\\\phi_7\\\phi_6\end{pmatrix}$$

17.3. PERMUTATIONAL SYMMETRY OF THE BASIS

The symmetry operations of the point group have no effect on the electron-repulsion *operator* $1/r_{12}$ (since it is invariant with respect to the full three-dimensional rotation group), and so the only effect of symmetry operations on the repulsion integrals are those due to the effect of the operations on the basis functions in the integrand.

The symmetry operation sends the integral $(77, 21)$ into $(66, 21)$ in which the labels are already in canonical order, and so do not need rearranging. This means that these two integrals *must be identical* as, of course, is obvious from the geometry of the molecule.

If we consider the repulsion integral $(76, 42)$, then similar considerations show that it is transformed into itself by the reflection operation *but with a sign change*; orbital ϕ_4 changes sign but the others do not. This means that this particular integral *must be zero* and need not be computed. Some of the electron-repulsion integrals are simply sent into themselves by the transformation so that no useful information is obtained for them.[2]

The symmetry group of the water molecule is actually C_{2v} in which there are four symmetry operations and the effect of these operations on the basis functions can be conveniently collected into a 7×4 array.

E	C_2^z	σ^{xz}	σ^{yz}
1	1	1	1
2	2	2	2
3	-3	3	-3
4	-4	-4	4
5	5	5	5
6	7	7	6
7	6	6	7

---

[2]See, however, later applications of the symmetry algorithm.

Here the standard group-representation theory symbols have been used for the rotation and two reflection operations.

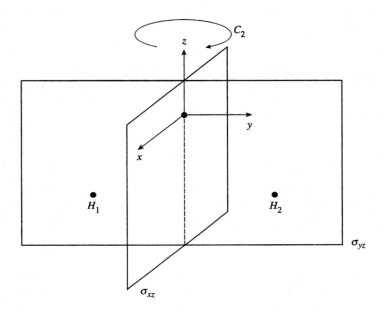

Now suppose we apply all four of these permutations to our original electron-repulsion integral (77,21). There are four symmetry-equivalent integrals generated which, in the order of the above table are:

$$(77,21) \quad (66,21) \quad (66,21) \quad (77,21)$$

This does not give us any new information, it merely confirms that the two integrals $(77,21)$ and $(66,21)$ must be the same. In fact the identity operation E can never give any new information and can always be omitted.

If, however, we were to consider a more general example (CH_4, for example with symmetry group T_d —24 operations— and a minimal basis of nine functions) we *would* find new equivalences and zeroes among the transformed integrals.

What is clear is that this simple technique can be made the basis of a very powerful method of reducing redundant computations in molecular structure calculations. The main difficulty is not the generation of equivalences amongst

17.3. PERMUTATIONAL SYMMETRY OF THE BASIS

the integrals but making sure that *repetitions* are discarded. As we have found in applying the four transformations to the integral $(77, 21)$ the procedure generates a list of equivalences containing repetitions and, of course, if we were to look at $(66, 21)$ by the same technique we would generate the whole list (with repetitions) *again*; the two *distinct* integrals which are equal would be generated four times each!

These two types of repetition:

1. Due to different symmetry operations generating the same result.
2. Due to application of symmetry operations to integrals that are equivalent.

are dealt with by different techniques:

1. In this case, it is necessary to keep a list of distinct labels and only add to the list if the new labels are different from the ones already present. In the simple case above only *one copy* of $(77, 21)$ or $(66, 21)$ is kept.

2. If we arrange to generate the labels in the usual canonical "reverse dictionary" order

$$\begin{aligned} i &\geq j \quad k \geq \ell \\ (ij) &\geq (k\ell) \\ (ij) &= i(i-1) + j \quad (k\ell) = k(k-1) + \ell \end{aligned}$$

then, as we progress through these labels and apply the transformations to them, if we encounter an integral which is *earlier* than the current one in the canonical order, *this whole list* must have been generated when that earlier integral was being processed and the whole list may be discarded. In the example considered above, the integral $(66, 21)$ would be encountered first and the list comprising $(66, 21)$ and $(77, 21)$ would be generated. Then, when the integral $(77, 21)$ is encountered later, it will be discovered that, amongst the equivalent integrals is $(66, 21)$ which must have already been processed and so the current copy of the list comprising $(77, 21)$ and $(66, 21)$ must be discarded if repetitions are to be avoided.

The list of integrals which are symmetry-related to a given one (the one to be actually *computed*, say) must also contain the *sign* of the transformed integral so that the correct relative signs may be assigned to the integral values.

One final touch extends the algorithm to those operations which induce linear combinations amongst the basis functions. If a basis function is sent into a linear combination of basis functions by an operation of the symmetry group and we have a knowledge of those linear combination coefficients, then there is no problem of *principle* in simply using this knowledge to compute the new integral as a linear combination of known, symmetry-related integrals. However, this is impractical since the integrals are computed and stored in a buffer which is written out to an external medium when it becomes full. Thus to *find* the integrals to which a given integral is related by linear combinations of the basis is a non-trivial task involving searching an external file (in general). This searching would always prove more time-consuming than computing the required integral. If the integrals are computed as they are needed (the "direct" method) then the integrals for the linear combinations would no longer be available or not yet have been computed and the book-keeping problems would be even worse. In these cases therefore, where the effect of a symmetry operation is to send a given basis function into a linear combination of members of the basis, the array of permutations simply has a suitable marker inserted at that point to say that our algorithm is no use in this case and this integral *must* be computed. The marker has to be an arbitrary integer not equal to 1 or -1: zero will do nicely.

17.4 Implementation

17.4.1 Two-electron integrals

The only data required for the implementation of the algorithm implied by the discussion of the last section is the number of symmetry operations and the array of permutations. In order for the algorithm to work successfully we must arrange to generate the integral labels in the canonical order; that is the way in which the loops over i, j, k and l were arranged in genint.[3] If this is assumed to be done, then for the current set of labels i, j, k, and l, the following steps must be performed:

1. Put the primary labels i j k l and a zero marker to say that this integral has to be computed as the first items in a list.

2. Apply the transformations in the permutation array to generate it jt kt lt (say) in each case

3. If one of it jt kt lt is zero, then mark this integral as one which must be computed.

4. If it jt kt lt are the same as i j k l, go on to the next permutation.

---
[3]Using next_label of Appendix 5.D.

17.4. IMPLEMENTATION

5. If it jt kt lt are the same as i j k l with an odd number of sign changes, discard the whole list (since all members refer to zero integrals) and go on to the next set of primary labels.

6. If it jt kt lt, when placed in canonical order, are prior to i j k l, discard the whole list and go on to the next set of primary labels.

7. If it jt kt lt, when placed in canonical order, are different from i j k l and do not already appear in the list, place them and the sign of their product in the list and go on to the next permutation.

Obviously, the change to **genint** (to produce **gensym**, say) which is necessary is to replace the direct use of **generi** by a call to a procedure which will generate a list of symmetry-equivalent labels and then the processing of this list using **generi** with the appropriate equivalencing.

Here is the relevant part of **genint**[4] with the code to deal with the list once it has been generated by **lister** which is given the task of implementing the symmetry algorithm and generating the correct list in list.

---

[4] A code fragment, not a **WEB**.

```
    define(MAX_PERMS,24)      # allow for up to Td (or Oh) symmetry
    integer perms(*);                    # Orbital permutations
    integer mmax;                        # maximum number of operations
    integer list(eval(MAX_PERMS*5));  # space for list of labels
    integer lister;           # symmetry function generator

    define(locl, ($2-1)*mmax + $1)   # macro for subscripting list
#
    rewind nfile;  pointer = 0;           # initialisation for putint
    last = NO;                            # ditto

    i = 1; j = 1; k = 1; l = 0;           # initialise next_label

    while(next_label(i,j,k,l,nbfns) == YES)
              {
              if ( l == nbfns ) last = YES;         # last integral

# lister has to find out if there are any integrals related to
# the current one; returns YES if there are

              if(lister(i,j,k,l,perms,n,mmax,list,nlist) == YES)
                {
# Go through the list and either compute or equivalence
                for (kk = 1; kk <= nlist; kk = kk + 1)
                    {
                    it = list(locl(kk,1));  jt = list(locl(kk,2));
                    kt = list(locl(kk,3));  lt = list(locl(kk,4));
                    itag = list(locl(kk,5));
# This "switch" decides which of the 3 possibilities to use
#   for each member of the list
                    switch (itag)
                        {
                        case -1: { val = - calc; break;}
                        case  0: {
                                calc = generi (i,j,k,l,0,eta,ngmx,nfirst,
                                    nlast,ntype,nr,NO_OF_TYPES)
                                  val = calc; break;
                                 }
                        case  1: { val = calc; break;}
                        default: STOP # can't happen!
                        }
# Store each integral
                    call putint(nfile,it,jt,kt,lt,itag,val,pointer,NO);

                    }   # end of kk loop
                }       # end of if(lister)
```

17.4. IMPLEMENTATION

The list of labels related by symmetry to the current canonical set is returned by lister in list, where each entry in list has five members; four basis function labels and a sign.

If these changes alone were to be made to genints and the file generated were to be submitted to scf it would quickly be discovered that the file is not in the correct format; scf would attempt to read beyond the end of the file nfile. The reason is clear enough; in the above code, calls to putint always have NO as the indicator for the last integral in the file. A file is only capable of being successfully processed by getint if the last call to putint has YES for this entry. In the case of the original genint the solution was simple; the last integral (nn, nn) is always non-zero and so when this last integral is computed it should be offered to putint with the argument YES to successfully complete the file. When symmetry is used the situation is very different; the last integral (nn, nn) will certainly be non-zero but it will not necessarily be actually *computed* as the last integral; it may well have been set equal to a symmetry-equivalent integral earlier in the calculation. This is precisely what happens with our H_2O example; the integral $(77, 77)$ is the same as $(66, 66)$ and will be put onto the file when $(66, 66)$ was computed.

There is no general way of knowing which integral will be physically the last one to be output to the file. Given this impasse there are two ways of patching the situation:

1. When the loops over the integral labels are complete and it is known that all the integrals must have been computed, backspace the file by a record and rewrite that record with the correct information.

2. Issue a "dummy" call to putint with a zero integral (with arbitrary labels), telling putint that this is the last. Something like:

    call putint(nfile,1,1,1,1,0,zero,pointer,YES);

   will suffice

Neither of these methods is very elegant, but the second is very easy and so will be adopted. Notice that this means that the file will contain *two* integrals $(11, 11)$ and one should convince oneself that this will not affect the use of getint or of scf at all *provided that one of them is zero*!

Before giving an implementation (or even a manual-page specification) of **integer function lister**, it is possible to save effort by noting that the above analysis for the electron-repulsion integrals is capable of being repeated almost word-for-word for the one-electron (overlap, kinetic energy and nuclear attraction) integrals.

17.4.2 One-electron integrals

In the case of the electron-repulsion integrals, we noted that the electron-repulsion *operator* ($1/r_{12}$) was "spherically symmetric" and so it is only the permutation (transformation) properties of the basis functions which mattered in using molecular symmetry. The situation is similar in the case of the one-electron integrals:

- The overlap integrals involve no operator so the basis functions are the only things to be considered.

- The kinetic energy integrals involve the operator ∇^2 which, like the electron-repulsion operator has much higher symmetry than the molecular point group so, again, only the lower symmetry of the permutations of the basis functions is involved.

- The nuclear-attraction integrals contain the nuclear attraction terms from the electrostatic Hamiltonian, which, of course(!), has the symmetry of the nuclear framework and so is left invariant by exactly the same permutation operations which we are considering. Again, we may consider only the symmetry properties of the permutations of the basis functions.

The upshot is that the one-electron integrals may be treated in exactly the same way as the repulsion integrals with the obvious simplification that only two labels need to be considered; the steps in the algorithm are exactly the same if the labels are generated in canonical order ($i \geq j$).

Thus, the one-electron calculation part of **subroutine genint** may be modified to use symmetry in an almost identical way to the earlier two-electron code:

17.4. IMPLEMENTATION

```
#
#     One-electron integrals
#
    for ( i = 1; i <= n; i = i + 1)
      {
      for( j = 1; j <= i; j = j + 1)
        {
        if(lister(i,j,0,0,perms,n,mmax,list,nlist) == YES)
          {
# Go through the list and either compute or equivalence the
#    integrals (with sign)
            for (kk = 1; kk <= nlist; kk = kk + 1)
              {
              it = list(locl(kk,1));  jt = list(locl(kk,2));
              itag = list(locl(kk,5));
              ij = (jt-1)*n + it;   ji = (it-1)*n+jt;
# This "switch" decides which of the 3 possibilities to use
#    for each member of the list
              switch (itag)
                {
                case -1: { valH = - calc; valS = -ovltot; break;}
                case  0: {
                          calc = genoei (i,j,eta,ngmx,nfirst,nlast,ntype,
                             nr,NO_OF_TYPES,vlist,noc,ncmx,ovltot,kintot)
                          valH = calc; valS = ovltot; break;
                         }
                case  1: { valH = calc; valS = ovltot; break}
                default: STOP # can't happen!
                }
              H(ij) = valH;   H(ji) = valH;
              S(ij) = valS;   S(ji) = valS;
              } # end of kk loop
            }   # end of if (lister
        }       # end of j loop
      }         # end of i loop
#
```

The difference between this requirement of **lister** and the electron-repulsion equivalent is simply the zero values for "k" and "l" in the calling sequence. Thus, **lister** must be taught to recognise that these zeroes signal its use as a one-electron list generator.

Here is a specification for **lister** and an implementation. The implementation is annotated and reference to the steps in the algorithm given above should make its performance clear.

17.4.2.1 lister

Generate a set of labels equivalent to the supplied one by using the permutations in *perms*.

NAME lister
   Generate a list of symmetry-equivalent one- or two-electron labels from information about basis function permutations.

SYNOPSIS

   integer function lister(i,j,k,l,perms,n,mmax,list,nlist)
   integer i, j, k, l, n, mmax, nlist;
   integer perms(ARB), list(ARB);

DESCRIPTION
   `lister` takes a set of four integers i, j, k, l which label an electron-repulsion integral and uses the permutational information about the n basis functions contained in `perms` to generate a `list` of symmetry-equivalent sets of integral labels. If k is zero, `lister` assumes that i and j must refer to a one-electron integral and performs the corresponding task for the two labels.

ARGUMENTS

   i,j,k,l Input: The labels (in canonical order) of a repulsion integral or, if k = 0, of a one-electron integral.

   n Input: The number of basis functions.

   mmax Input: The number of symmetry operations.

   perms Input: The array of permutations, n by mmax stored by columns in a singly subscripted way; the (i,j) element is accessed by `perms((j-1)*n + i)`

   nlist Output: The number of items in the output `list`.

   list Output: The list of one- or two-electron labels equivalent to i, j, k, l; i, j, k, l is the first member of the `list`. `list` is an array of size mmax by 5 stored by single subscripting and the (ij) element is accessed by `list((j-1)*mmax+i)`.

RETURNS
   YES if it has found some integrals to process, NO if not.

SEE ALSO
   gensym

DIAGNOSTICS
   None, it is the user's responsibility to generate a correct **perms** array.

17.4. IMPLEMENTATION

"lister.f" 17.4.2.1 ≡

  @m *MAX_PERMS* 24
  @m $locp(i,j)$ $((j-1)*n+i)$ /* macro to access *perms* */

  @m $locl(i,j)$ $((j-1)*mmax+i)$ /* and one to access *list* */

    integer function $lister(i, j, k, l, perms, n, mmax, list, n\_canon)$
       ⟨lister interface declarations 17.4.2.1.7⟩
    {
    ⟨lister local declarations 17.4.2.1.8⟩
    ⟨Initialise the list 17.4.2.1.2⟩
      /* Check that there is something to do */
    if $(mmax \neq 0)$
    {
      /* If there are symmetry operations, loop through them */
    for $(m = 1;\ m \leq mmax;\ m = m + 1)$
    {
    ⟨Form the current permuted list item 17.4.2.1.1⟩
    ⟨Put the permuted list item into canonical form 17.4.2.1.3⟩
    ⟨Check to see if current canonical list item is needed 17.4.2.1.4⟩
    ⟨Put the saved labels in the sub-list 17.4.2.1.5⟩
    ⟨Discard the repetitions 17.4.2.1.6⟩
    }
    }
    return $(dolist)$;
    }

⟨ Form the current permuted list item 17.4.2.1.1 ⟩ ≡
    /* Get the labels of the related integrals and the characteristic canonical number and sign */
$it = perms(locp(i, m))$;
$jt = perms(locp(j, m))$;
if $(k \neq 0)$
  {
  $kt = perms(locp(k, m))$;
  $lt = perms(locp(l, m))$;
  }
$t\_sign = it * jt$;
if $(k \neq 0)$
  $t\_sign = it * jt * kt * lt$;
isign $= 1$;   /* Check if *perms* contains a zero, if so the integral must be computed */
if $(t\_sign \equiv 0)$
  next;
if $(t\_sign < 0)$
  isign $= -1$;

This code is used in section 17.4.2.1.

17.4. IMPLEMENTATION

⟨ Initialise the list 17.4.2.1.2 ⟩ ≡
    $canon = (i * (i - 1))/2 + j$;
    if $(k \neq 0)$
        $canon = term4\,(i,\ j,\ k,\ l)$ /* canonical (ijkl) */
        /* Set up a sub-list of integrals generated by the symmetry operations
          from this integral, started by the integral itself */

    $list(locl(1,\ 1)) = i$;
    $list(locl(1,\ 2)) = j$;
    if $(k \neq 0)$
      {
      $list(locl(1,\ 3)) = k$;
      $list(locl(1,\ 4)) = l$;
      }
    $list(locl(1,\ 5)) = 0$;
        /* this zero means this integral must be computed, if it survives */
    $dolist = YES$;
    $s\_canon(1) = canon$;   /* s_canon is a saved list of canonicals */
    $n\_canon = 1$;   /* The above is the first */
    sign $= i * j$;
    if $(k \neq 0)$
        sign $= i * j * k * l$;

This code is used in section 17.4.2.1.

⟨ Put the permuted list item into canonical form 17.4.2.1.3 ⟩ ≡
/∗ The permuted labels will not, in general, be in standard order *order*
orders them enablng the transformed canonical index to be computed.
The signs can now be discarded, having been used. ∗/
$it = \mathbf{abs}(it);$
$jt = \mathbf{abs}(jt);$
if $(it < jt)$
  {
  $id = it;$
  $it = jt;$
  $jt = id;$
  }
$t\_canon = (it * (it - 1))/2 + jt;$
if $(k \neq 0)$
  {
  call $order(it, jt, kt, lt);$
  $t\_canon = term4(it, jt, kt, lt);$
  }

This code is used in section 17.4.2.1.

⟨ Check to see if current canonical list item is needed 17.4.2.1.4 ⟩ ≡
/∗ Now see if the whole list may be omitted. If the transformed
canonical number is less than the original canonical number, then this
whole list will have been generated before, when *t_canon* was first
met, so discard it. ∗/
if $(t\_canon < canon)$
  return $(NO);$

/∗ If the transformed canonical is the same as the original and the sign
is different, it must be zero as must the whole list ∗/
if $(t\_canon \equiv canon)$
  {
  if $(\mathbf{sign} \neq t\_sign)$
    return $(NO);$
  next;
  }

This code is used in section 17.4.2.1.

17.4. IMPLEMENTATION 499

⟨ Put the saved labels in the sub-list 17.4.2.1.5 ⟩ ≡
  /* At last, a permuted set of labels worth saving, put them in the
  sub-list and keep their canonical index */
  $ntemp = n\_canon + 1$;
  $list(locl(ntemp, 1)) = it$;
  $list(locl(ntemp, 2)) = jt$;
  if $(k \neq 0)$
    {
    $list(locl(ntemp, 3)) = kt$;
    $list(locl(ntemp, 4)) = lt$;
    }
  $list(locl(ntemp, 5)) = \boldsymbol{isign}$;

This code is used in section 17.4.2.1.

⟨ Discard the repetitions 17.4.2.1.6 ⟩ ≡
  /* Now check for repetitions in the list and skip them */
  $id = NO$;
  for $(kk = 1;\ kk \leq n\_canon;\ kk = kk + 1)$
    if $(t\_canon \equiv s\_canon(kk))$
      $id = YES$;
  if $(id \equiv YES)$
    next;
  $n\_canon = n\_canon + 1$;
  $s\_canon(n\_canon) = t\_canon$;

This code is used in section 17.4.2.1.

⟨ lister interface declarations 17.4.2.1.7 ⟩ ≡
  integer $i,\ j,\ k,\ l,\ n,\ mmax,\ n\_canon$;
  integer $perms(*),\ list(*)$;

This code is used in section 17.4.2.1.

⟨ lister local declarations 17.4.2.1.8 ⟩ ≡
  integer *canon*, *t_canon*, *s_canon*(*MAX_PERMS*);
  integer *term4*;   /∗ function to generate canonical (ijkl) ∗/
  integer *dolist*, **sign**, *t_sign*, **isign**, *ntemp*, *m*, *id*;
  integer *it*, *jt*, *kt*, *lt*;

This code is used in section 17.4.2.1.

17.4.2.2 term4

Evaluates the unique integer which puts i, j, k, l in canonical order.
"lister.f" 17.4.2.2 ≡

  integer function *term4* (i, j, k, l)
    integer i, j, k, l;
  {
  integer ij, kl;
  $ij = (i \ast (i-1))/2 + j$;
  $kl = (k \ast (k-1))/2 + l$;
  $term4 = (ij \ast (ij-1))/2 + kl$;
  return;
  }

17.4.2.3 INDEX

abs: 17.4.2.1.3.

canon: 17.4.2.1.2,   17.4.2.1.4,
   17.4.2.1.8.

dolist: 17.4.2.1, 17.4.2.1.2, 17.4.2.1.8.

i: 17.4.2.1.7, 17.4.2.2.

id: 17.4.2.1.3, 17.4.2.1.6, 17.4.2.1.8.

ij: 17.4.2.2.

isign: 17.4.2.1.1,   17.4.2.1.5,
   17.4.2.1.8.

it: 17.4.2.1.1, 17.4.2.1.3, 17.4.2.1.5,
   17.4.2.1.8.

j: 17.4.2.1.7, 17.4.2.2.

jt: 17.4.2.1.1, 17.4.2.1.3, 17.4.2.1.5,
   17.4.2.1.8.

k: 17.4.2.1.7, 17.4.2.2.

kk: 17.4.2.1.6.

17.4. IMPLEMENTATION

kl: 17.4.2.2.

kt: 17.4.2.1.1, 17.4.2.1.3, 17.4.2.1.5, 17.4.2.1.8.

l: 17.4.2.1.7, 17.4.2.2.

list: 17.4.2.1, 17.4.2.1.2, 17.4.2.1.5, 17.4.2.1.7.

lister: 17.4.2.1.

locl: 17.4.2.1, 17.4.2.1.2, 17.4.2.1.5.

locp: 17.4.2.1, 17.4.2.1.1.

lt: 17.4.2.1.1, 17.4.2.1.3, 17.4.2.1.5, 17.4.2.1.8.

m: 17.4.2.1.8.

MAX_PERMS: 17.4.2.1, 17.4.2.1.8.

mmax: 17.4.2.1, 17.4.2.1.7.

n: 17.4.2.1.7.

n_canon: 17.4.2.1, 17.4.2.1.2, 17.4.2.1.5, 17.4.2.1.6, 17.4.2.1.7.

NO: 17.4.2.1.4, 17.4.2.1.6.

ntemp: 17.4.2.1.5, 17.4.2.1.8.

order: 17.4.2.1.3.

perms: 17.4.2.1, 17.4.2.1.1, 17.4.2.1.7.

s_canon: 17.4.2.1.2, 17.4.2.1.6, 17.4.2.1.8.

sign: 17.4.2.1.2, 17.4.2.1.4, 17.4.2.1.8.

t_canon: 17.4.2.1.3, 17.4.2.1.4, 17.4.2.1.6, 17.4.2.1.8.

t_sign: 17.4.2.1.1, 17.4.2.1.4, 17.4.2.1.8.

term4: 17.4.2.1.2, 17.4.2.1.3, 17.4.2.1.8, 17.4.2.2.

YES: 17.4.2.1.2, 17.4.2.1.6.

The only new support which is needed by lister is integer function term4 which computes the canonical number for a particular set of repulsion integral labels. term4 is deemed to be too small to merit a manual entry!

17.4.3 Numerical values

To give some idea of the orders of magnitude of the possible savings obtained by using the permutational symmetry of the basis functions here is a small sample of data obtained by running lister for the electron-repulsion integrals associated with some typical basis sets of symmetrical molecules.[5] In the table

---

[5] The manual preparation of permutation matrices for basis functions is extremely tedious and, of course, error-prone. The matrices used here were generated by Mike Elder's original program which is available from the ftp site.

Examples of numbers of repulsion integrals

Molecule (Basis)	Group	n	Total	Stored	Distinct
CH_4 (MIN)	T_d	9	1035	840	80
CH_4 (DZV)	T_d	17	11781	10328	745
CH_4 (DZVP)	T_d	38	274911	274457	17381
CH_4 (DZVP10)	T_d	39	304590	287452	14655
$PtCl_4$ (MIN)	D_{4h}	25	52975	8601	1153
$PtCl_4$ (MIN10)	D_{4h}	26	61776	5911	717
UF_6 (DZV)	O_h	66	2445366	605843	33585
UF_6 (DZV10)	O_h	67	2595781	550588	24284

(MIN) means a "minimal" basis (the AOs occupied in the ground states of the component atoms of the molecule), (DZV) means "double zeta valence" (two sets of AOs for the valence shell) and (DZVP) means (DZV) plus one set of AOs of the next higher angular momentum type on each atom. (MIN10), (DZV10) and (DZVP10) mean that the set of d functions is the "monomial set" which has d_{x^2}, d_{y^2} and d_{z^2} in place of the cubic harmonic set $d_{x^2-y^2}$ and $d_{3z^2-r^2}$. n is the total number of basis functions. In the case of heavy atoms (Pt and U) only the outer valence shell is used, the inner shells are neglected to avoid having huge numbers in the tables.

One is immediately struck by the very large reduction in the numbers of integrals which have to be computed and stored. The difference between the 4th and 5th columns is the number of integrals which must be zero by symmetry. The last column is the number of integrals which must be actually computed.

It is also noticeable that, in spite of the monomial sets of d functions being larger than the conventional cubic harmonic sets, there are in these examples actually *fewer* integrals to compute. This is because the monomial set has better *permutation* properties than the smaller set which has preferable formal properties (linear independence, transformations properties etc.).

17.4.4 Conditions on the array perms

The algorithm which we have used will clearly always work if the operations on the basis functions form a *group*, in particular if they form the point group of the molecule or one of its sub-groups. It is also clear that the identity operation need not be included as a column of **perms** since the natural generation of the labels in canonical order ensures that the identity is always implicitly present.

17.4. IMPLEMENTATION

Equally obviously, since `lister` discards repetitions as it processes the labels it does not matter if any column of `perms` is duplicated (or triplicated) except that time will be wasted on redundant computations.

It is, therefore, natural to ask if there are any formal criteria by which we can be assured that the list generated by `lister` will be valid.

A little thought shows that, in the case of the highly symmetrical kinetic energy and electron-repulsion operators it is the symmetry of our choice of basis functions which is the determining factor in the overall symmetry, not that of the operator; only in the case of nuclear attraction is the symmetry of the operator a consideration. In these two cases, therefore, it is possible to add "operations" to the symmetry operations which are not symmetries of the molecule. For example, the C_{2v} symmetry of the water molecule and the associated array `perms` given in the table above cannot know that *all* the $2p$ basis functions on the oxygen atom are the same apart from orientation in space; none of the operations of C_{2v} send $2p_x$ into $2p_y$ since this is a symmetry of the oxygen *atom* not of the water molecule. Of course, one may wish to take advantage of this fact to use, for example, a more diffuse $2p$ function to describe the out-of-plane electron density than the two in-plane $2p$ functions. But, in general, the basis functions within a "shell" will be equivalent (in atomic terms, they will have the same radial function).[6]

We can invent such "local symmetry operations" by using the fact that `perms` may have zeroes: originally introduced to circumvent the problem of linear combinations. In the case of H_2O the original `perms` array may be augmented to include the local symmetries of the oxygen atom. Since *real* basis functions are being used, this means C_4 rotations, not C_∞. Here is the augmented array, with appropriate zeroes.

E	C_2^z	σ^{xz}	σ^{yz}	C_4^z	C_4^{z3}
1	1	1	1	1	1
2	2	2	2	2	2
3	-3	3	-3	-4	4
4	-4	-4	4	3	-3
5	5	5	5	5	5
6	7	7	6	0	0
7	6	6	7	0	0

(C_4^{z2} is, of course, the same as C_2^z and so is omitted)

---

[6] Many methods of rapid integral evaluation make this constraint "hard-wired" into the code to enable some potential duplications of calculation to be avoided.

17.5 Permutation symmetry: summary

These techniques can make considerable savings[7] in the calculation of the molecular integrals which is often the most time-consuming step in a molecular electronic structure calculation.

However:

- there is a small processing overhead involved and, more important,
- there is not much saving in file storage space.

Certain integrals which are zero by molecular symmetry are detected and are neither *computed* nor stored (without the use of `lister` they would have been computed and discarded). All the remaining non-zero integrals will be stored.

As we shall see in a later chapter, if we make the further constraint on the SCF method that the molecular orbitals are *symmetry-adapted*, then the above technique can be extended in such a way that only the symmetry-distinct integrals need be computed *and stored*. What is more, the restriction to *permutational symmetries* can be removed so that the full molecular point group symmetry can be used, which may involve operations which induce linear combinations among the basis functions.

We therefore turn to some preliminaries which are involved in the application of spatial symmetry constraints on the HF model.

---

[7] It must be noted here that this particular saving and a good deal more could be achieved by the much simpler expedient of arranging for a special "one-centre" electron-repulsion integral function to be used as appropriate; but local symmetry is not always atomic symmetry; the C_2 fragment in ethane, for example, could be treated as locally D_{4h}.

Chapter 18

Symmetry orbital transformations

Independently of the use of symmetry to facilitate the calculation of energy integrals in the LCAO approximation, it is always possible to reduce the dimension of the matrix equations involved in an effective-Hamiltonian model of molecular electronic structure by the use of any molecular symmetry. The prototype of these approximations is the LCAOSCF model and in this chapter we look at the effects of using symmetry orbitals on the implementation of this model.

Contents

18.1	Introduction	505
18.2	Symmetry-adapted basis	509
18.3	Generation of symmetry orbitals	512
18.4	Conclusions	514

18.1 Introduction

In the introduction to the last chapter, it was noted that the variationally optimum molecular orbitals do not, in general, have the symmetry of the nuclear framework (or the spin operators), but that, as a matter of empirical experience, they are very often found to have these symmetries when the molecules are of conventional structure.

It would, therefore, be pedantic to insist that the Hartree–Fock equations *always* be solved in their full generality when, in most cases, considerable savings can be effected by *constraining* the molecular orbitals to have certain molecular symmetries. Most routine uses of LCAOSCF theory can safely assume that the optimum MOs will have the symmetry of the molecular framework; the dominant nuclear-attraction term in the Hamiltonian. The situation is similar to the use of the closed-shell constraint in place of a full GUHF solution for molecules of conventional known closed-shell structure. In any case, the numerical techniques used to impose spatial symmetry constraints on the MOs are very similar to techniques used for a variety of other purposes in LCAOSCF theory—the transformation of the Fock matrix to a new basis—so we include a discussion of the techniques here.

The main computational attraction of the SCF method is that it reduces to a *one-electron* matrix problem; the manipulations are those of matrices of the size of the number of basis functions. This means, for example, that any *orbital transformations* which may have to be done involve transformations of matrix representations of *one*-electron operators; no time-consuming transformations of the four-index electron-repulsion integrals have to be done explicitly. The electron-repulsion terms are all contained in the Hartree–Fock matrix via G or, at worst, J and K separately. This intuitively reasonable conclusion is proved in Appendix 19.A to the next chapter.

There are several occasions where one might like to transform the basis over which the SCF calculation is being performed:

1. It may prove advantageous to use a particular basis to improve the convergence properties of the iterative procedure: a particular method of orthogonalisation might be appropriate for atoms and a different one for molecules. We have seen already that it is *essential* to transform the basis to an orthogonal set in order to carry through the calculation at all.

2. The use of the "Cartesian Monomials" for the 6 d and 10 f basis functions is not always appropriate; it is often preferable to work explicitly in terms of the cubic harmonic (5) d functions or (7) f functions.

3. As we are discussing here, it may be useful to work in terms of *symmetry-adapted* basis functions which are linear combinations of the raw basis functions in fixed amounts determined by the representations of the molecular point group; these combinations are discussed in the next section.

If we recall the implementation of the SCF method in, for example, Chapter 12, the matrix which generated an orthogonal basis from the original set of basis functions was supplied as data to the **integer function scf**. That is, the matrix V (say) (such that the basis $\bar{\phi}$ is orthonormal):

$$\bar{\phi} = \phi V$$

18.1. INTRODUCTION

was generated from the original basis-function-overlap matrix S.

Now suppose that we wish to implement *more than one* of the above procedures; let us say transformation from the Cartesian monomials to the cubic harmonics, followed by orthogonalisation. Let the matrix which generates the cubic harmonics from the monomials be H, then

$$\phi_H = \phi H$$

in an obvious notation.[1] Since there are *fewer* cubic harmonics than monomials, the matrix H is rectangular.

The overlap matrix in the new basis is:

$$S_H = H^T S H$$

(where H^T is the transpose of H) and is, of course, square and of smaller dimension than the original overlap matrix.

Using the symmetric orthogonalisation method outlined earlier, the matrix which orthogonalises this (smaller) basis is:

$$V = (S_H)^{-\frac{1}{2}}$$

say.

Thus the relationship of the final, symmetrically orthogonalised, cubic harmonic basis to the original, non-orthogonal monomial basis is

$$\bar{\phi} = \phi H V = \phi V_H$$

(say), where V_H is rectangular.

There is, therefore, no reason at all why *this matrix* should not be used in place of the original V matrix (which simply orthogonalised the monomial basis as data to scf). More precisely we should say that, provided **integer function scf** is taught to handle *rectangular* matrices, we can continue as before except that we are working in a different basis. However, no changes have to be made to either the one-electron integrals or to the electron-repulsion integrals; they do not have to be transformed to the new basis any more than they had to be transformed to an orthogonal basis in the original implementation; all the effects of the transformation to the new, working, basis are contained in the *one-electron* transformation of the Fock matrix.

Now to the problem in hand; the use of symmetry-adapted linear combinations of the basis functions in the SCF procedure. The method is now obvious,

---

[1] The matrix H is rather sparse, it only contains entries to change the d_{xx}, d_{yy}, d_{zz} to d_{z^2} and $d_{x^2-y^2}$ and similar transformations of the f functions; s and p functions are left unchanged.

it is simply one more additional step in the chain of transformations from the "raw" basis to the actual "working" basis. Deferring the actual *technique* of calculation of the elements of the transformation between the raw basis and the symmetry-adapted basis until later in this chapter, we assume that the matrix U effects this transformation. The matrix is actually unitary or, more commonly in the real case, orthogonal so that, unlike the monomial/harmonic transformation above, it does not disturb the orthogonality properties of the basis and it is, of course, square.

Orthogonalisation methods do not necessarily preserve the symmetry properties of the basis, however, and so it is always safer to form symmetry-adapted orbitals *before* the orthogonalisation step and allow the orthogonalisation method to clean up any loose ends.

Let us assume then that we have a given raw basis and wish to transform it into a working basis which is in the form of an orthogonalised, symmetry-adapted linear combination of cubic harmonic functions. The steps are clear:

$$\begin{aligned}\phi_H &= \phi H \\ \phi_U &= \phi_H U = \phi H U \\ \bar{\phi} &= \phi_U V\end{aligned}$$

where, this time,

$$V = S_U^{-1/2} = \left(U^T H^T S H U\right)^{\frac{1}{2}}$$

and the final transformation (again rectangular) is:

$$\bar{\phi} = \phi H U V = \phi V_U$$

(say). No other changes are required to `integer function scf`.

The general conclusion is that *by the generation of the appropriate single transformation matrix* we can work in any basis of our choice which is linearly related to the raw basis over which the molecular integrals are computed. No transformations of integrals are involved and there is no additional overhead in the iterative procedure. Clearly, the GUHF procedure may be specialised to DODS by an analogous method; DODS is just GUHF in which the Fock matrix consists of two diagonal blocks[2] each of the size of the spatial basis and the two off-diagonal ($\alpha - -\beta$ mixing) blocks set to zero.

Having seen how to *implement* the use of a symmetry-adapted basis we should now turn to its formal properties and the way in which the use of a symmetry-adapted basis can speed up the SCF process.

---

[2] As we noted in Appendix 12.B.

18.2 Symmetry-adapted basis

If a basis of functions is chosen on the reasonable grounds that symmetry-equivalent atoms in the molecule have identical basis functions centred on them, then this basis will carry a representation of the molecular point group; any operation of the point group ($\hat{\mathcal{G}}^i$) will send the basis functions into linear combinations of themselves without the generation of any functions outside the basis.

$$\hat{\mathcal{G}}^i \boldsymbol{\phi} = \boldsymbol{\phi} \boldsymbol{G}^i$$

where

$$G^i_{kj} = \int \phi_k \hat{\mathcal{G}}^i \phi_j dV$$

If the matrix $\boldsymbol{G}^i$ is *diagonalised*, then the eigenvectors define a set of functions which are simply sent into themselves (multiplied by a numerical factor, an eigenvalue) by the operation $\hat{\mathcal{G}}^i$. That is, the functions which reduce the matrix $\boldsymbol{G}^i$ which represents the symmetry operator $\hat{\mathcal{G}}^i$ are *symmetry-adapted*.

If we restrict attention to Abelian point groups for simplicity, then the operations of the point groups *commute* and, therefore, so do their matrix representations. Two (or more) matrices which commute

$$\boldsymbol{AB} = \boldsymbol{BA}$$

can be *simultaneously* diagonalised. That is, there is a set of vectors $\boldsymbol{c}^i$, say, for which *both* of

$$\boldsymbol{A}\boldsymbol{a}^i = a_i \boldsymbol{a}^i$$

and

$$\boldsymbol{B}\boldsymbol{a}^i = b_i \boldsymbol{a}^i$$

can be made to hold for some sets of eigenvalues a_i and b_i.

For, if the first of the two equations holds, then

$$\boldsymbol{BA}\boldsymbol{a}^i = \boldsymbol{B}a_i\boldsymbol{a}^i = a_i\boldsymbol{B}\boldsymbol{a}^i = a_i b_i \boldsymbol{a}^i$$

with a similar equation for the eigenvalues of $\boldsymbol{B}$.

If the matrix representation $\boldsymbol{G}^i$, of a symmetry operation $\hat{\mathcal{G}}^i$ is used to transform a matrix representation, $\boldsymbol{h}$, of an operator $\hat{h}$ which has the symmetry of that operation, then the matrix should be *invariant* with respect to that transformation i.e. the operator matrix will be *unchanged* by the transformation:

$$\boldsymbol{G}^{i\dagger}\boldsymbol{h}\boldsymbol{G}^i = \boldsymbol{G}^{iT}\boldsymbol{h}\boldsymbol{G}^i = \boldsymbol{h}$$

(if the matrix $\boldsymbol{G}^i$ is real-unitary i.e. orthogonal). That is

$$\boldsymbol{G}^i \boldsymbol{h} = \boldsymbol{h} \boldsymbol{G}^i$$

the matrices of the operator and symmetry operations *commute* and (for Abelian groups) the whole set of symmetry operation matrices and symmetrical operator matrices commute and can therefore have a common set of eigenvectors.

Furthermore, it is an elementary exercise in use of the Hermitian properties of the matrices h and G^i to show that if the matrix

$$U = (u^1, u^2, \ldots)$$

diagonalises G^i:

$$G^i U = U g$$

(where g is the diagonal matrix of eigenvalues) then the matrix

$$U^T h U$$

has a particularly "structured" form.

For, if we choose two of the eigenvectors of G and multiply them by h and G in turn we have

$$G^i h u^k = h G^i u^k = g_k h u^k$$
$$G^i h u^\ell = h G^i u^\ell = g_\ell h u^\ell$$

Multiplying the first of these by $u^{\ell T}$ and the second by u^{kT} gives

$$u^{\ell T} G^i h u^k = g_k u^{\ell T} h u^k$$
$$u^{kT} G^i h u^\ell = g_\ell u^{kT} h u^\ell$$

Now the matrix h is symmetrical (or Hermitian, in general) and so the two matrix products on the left-hand-sides are identical. The products on the right-hand-sides are also identical but this is not quite so obvious, because the product of two symmetrical (Hermitian) matrices is not necessarily symmetrical (Hermitian).

$$(AB)^T = B^T A^T$$

However, if the matrices commute the product is symmetrical (Hermitian). Thus using these equalities and subtracting the two equations we have

$$(g_k - g_\ell) u^{\ell T} h u^k = 0$$

meaning that, *unless the two matrices u^ℓ and u^k have identical eigenvalues of the symmetry operator G^i*, the element of the representation of the operator h in the basis of functions which diagonalises G^i is zero.

18.2. SYMMETRY-ADAPTED BASIS

Thus the matrix representation of the operator $\hat{h}$ in the basis of eigenvectors of any of the symmetry operators GF^i is "sparse". The transformation of basis induced by the diagonalisation of any G^i:

$$\phi_U = \phi U$$

means that the matrix

$$h_U = U^T h U$$

only has non-zero elements between functions ϕ_{Uk} and $\phi_{U\ell}$ if these functions transform under the action of the symmetry operation $\hat{\mathcal{G}}^i$ in an identical manner.

These conclusions apply to the whole set of symmetry operations which form the point group of the molecule, so that it is possible to take advantage of the symmetry of the molecule to make the transformed matrix h_U as sparse as possible. Further, by suitable numbering of the members of the basis ϕ_U, the matrix h_U can be brought into a form in which it has (square) blocks of non-zero elements in a diagonal form *and zeroes everywhere else* with obvious advantages in the diagonalisation procedure. If a square matrix of dimension m can be reduced to a set of smaller square matrices of dimension m_1, m_2 etc. where

$$\sum m_i = m$$

then the time to diagonalise such a matrix becomes proportional to

$$\sum (m_i)^k$$

rather than to m^k for some k which depends on the diagonalisation algorithm.

All of the above hinges on the assumption that h is a representation of an operator $\hat{h}$ which is *invariant with respect to the operations of the molecular point group*.[3] Obviously the one-electron Hamiltonian and the unit operator satisfy this condition and so the matrices h and S of the LCAO method can be "symmetry blocked" in this way. We have seen that, in general, the matrix representation of the Hartree–Fock operator h^F will *not* satisfy this condition, so that it is a *constraint* on the LCAO method to make the assumption that the h^F matrix can be treated in this way. As we have seen, this constraint consists of generating "self-consistent" symmetries which in certain critical cases may prevent us from obtaining the lowest-energy determinant.

However, it is often the case that the savings are so great and the broken-symmetry solutions are so rare that it is worth implementing this symmetry-blocking procedure.

---

[3] And, of course, that the basis functions have (at least) the symmetry of the molecular point group.

18.3 Generation of symmetry orbitals

Although the derivation above of the symmetry-blocking procedure used the fact that all the operations of the point group commute with the matrix $\boldsymbol{h}$ *and with each other*, this latter restriction is not necessary and the result holds for arbitrary point groups. The method of generation of a symmetry-adapted basis does not, in fact, depend on the diagonalisation of the matrix representations of the symmetry operations but on a technique using the group projection operators which are defined in terms of the standard irreducible matrix representations of the point groups. The full derivation of these projection operators would take us some way from our aims here so the result is simply quoted with an explanation.[4]

Most texts on the use of group representation theory in physical science list the character tables for the commonly occurring point groups and the better ones will list the full standard irreducible representation matrices. If one has the full representation matrices it is possible to use these to effect a complete solution of the problem of the formation of symmetry-adapted basis from a given set of basis functions.

If the operations of the point group are $\mathcal{G}^i$ the standard method of naming the irreducible representation matrices is the *functional* one:

> The matrix which provides a representation of the operation $\mathcal{G}^i$ of type α (A_1, E_2, etc.) is written as
>
> $$\boldsymbol{D}_\alpha(\mathcal{G}^i)$$

The "matrices" $\boldsymbol{D}_\alpha(\mathcal{G}^i)$ are one-dimensional for A and B representations, two-dimensional for E representations and three-dimensional for T representations. In the cases usually considered in computational work the matrices are orthogonal (or, at worst, unitary) so that the distinction between the co- and contragredient matrices need not be made.

A symmetry-adapted function corresponding to a pure symmetry type may be formed ("projected") from an arbitrary function by the operator

$$\rho_{k\ell}^{(\alpha)} = \sum_{i=1}^{g} \left(\boldsymbol{D}_\alpha(\mathcal{G}^i)\right)_{k\ell} \mathcal{G}^i$$

The formulae are particularly easy to use when the symmetry group is Abelian and all the representation matrices are one-dimensional.

Here are the irreducible representation matrices for the group C_{3v} referred to standard axes.

---

[4]The clearest introduction to these techniques is to be found in Roy McWeeny's book *Symmetry* (Pergamon, 1963) which is, unfortunately, out of print.

18.3. GENERATION OF SYMMETRY ORBITALS

Irreducible representation matrices for C_{3v}

Matrix Element	E	C^3	$(C^3)^2$	σ^1	σ^3	σ^2
$\left(\boldsymbol{D}^{A_1}\right)_{11}$	1	1	1	1	1	1
$\left(\boldsymbol{D}^{A_2}\right)_{11}$	1	1	1	-1	-1	-1
$\left(\boldsymbol{D}^{E}\right)_{11}$	1	$-c$	$-c$	1	$-c$	$-c$
$\left(\boldsymbol{D}^{E}\right)_{12}$	0	$-s$	s	0	$-s$	s
$\left(\boldsymbol{D}^{E}\right)_{21}$	0	s	$-s$	0	$-s$	s
$\left(\boldsymbol{D}^{E}\right)_{22}$	1	$-c$	$-c$	-1	c	c

($c = cos(2\pi/6) = \sqrt{3}/2$, $s = sin(2\pi/6) = 1/2$)

The projection operators each have the form:

$$\begin{aligned}\rho^\alpha_{k\ell} &= \left(\boldsymbol{D}^E\right)_{k\ell} E + \left(\boldsymbol{D}^{C^3}\right)_{k\ell} C^3 + \left(\boldsymbol{D}^{(C^3)^2}\right)_{k\ell} (C^3)^2 \\ &+ \left(\boldsymbol{D}^{\sigma^1}\right)_{k\ell} \sigma^1 + \left(\boldsymbol{D}^{\sigma^3}\right)_{k\ell} \sigma^3 + \left(\boldsymbol{D}^{\sigma^2}\right)_{k\ell} \sigma^2\end{aligned}$$

where α is one of A_1, A_2 or E.

These operators can be used to derive the symmetry-adapted functions equivalent to an atomic basis. For example, a single s-type function on each of three atoms at the corners of an equilateral triangle (ϕ_1, ϕ_2 and ϕ_3) generates three symmetry-adapted functions which may be obtained by application of the above six operators to any one of the three functions. It is elementary to verify that

$$\begin{aligned}\rho^{A_1}_{11}\phi_1 &= 2(\phi_1 + \phi_2 + \phi_3) \rightarrow \frac{1}{\sqrt{3}}(\phi_1 + \phi_2 + \phi_3) \\ \rho^{A_2}_{11}\phi_1 &= 0 \\ \rho^{E}_{11}\phi_1 &= 2\phi_1 - \phi_2 - \phi_3 \rightarrow \frac{1}{\sqrt{6}}(2\phi_1 - \phi_2 - \phi_3) \\ \rho^{E}_{12}\phi_1 &= 0 \\ \rho^{E}_{21}\phi_1 &= 0 \\ \rho^{E}_{22}\phi_1 &= \frac{\sqrt{3}}{2}(2\phi_2 - 2\phi_3) \rightarrow \frac{1}{\sqrt{2}}(\phi_2 - \phi_3)\end{aligned}$$

It is conventional to "normalise" these symmetry-adapted functions assuming orthonormality[5] of the basis functions. Of course, in the case of the two E-type

---
[5]This "normalisation" is simply to ensure that the transformation matrices are orthogonal (or unitary).

functions, they are not unique; using the ρ operators on ϕ_2 would generate an equivalent but different set.

The effect of these projection operators can be summarised by the definition of a set of three (there can be no more than three linearly-independent combinations of ϕ_1, ϕ_2, ϕ_3) symmetry-adapted functions ("symmetry orbitals") $(\lambda_1, \lambda_2, \lambda_3)$ say given by:

$$(\lambda_1, \lambda_2, \lambda_3) = (\phi_1, \phi_2, \phi_3) \begin{pmatrix} \frac{1}{\sqrt{3}} & \frac{2}{\sqrt{6}} & 0 \\ \frac{1}{\sqrt{3}} & \frac{1}{\sqrt{6}} & \frac{1}{\sqrt{2}} \\ \frac{1}{\sqrt{3}} & -\frac{1}{\sqrt{6}} & -\frac{1}{\sqrt{2}} \end{pmatrix}$$

This case is analogous to the plane-wave case; there is only one (or zero) orbital of any symmetry type, so that there would be no variation problem to solve. The generalisation of this kind of procedure is the generation of so-called Bloch orbitals in solid-state theory, where translational symmetry will generate one symmetry orbital from any orbital in the unit cell and the variation problem is solved in terms of orbitals of the same symmetry type generated from different orbitals in the same unit cell.

18.4 Conclusions

The solution of the matrix HF equations can be facilitated by the use of a basis of symmetry-orbitals obtained from the basis functions by the projection operators of the point group. When symmetry-orbitals are used the degrees of freedom available to the variation principle are exposed as the number of different orbitals of each symmetry type. The matrix which defines the symmetry orbitals can be combined with the orthogonalisation matrix in the SCF procedure so that, once formed, this transformation adds nothing to the implementation of the process. Techniques and codes for the diagonalisation of blocked matrices have already been met in Appendix 12.B.

Chapter 19

A symmetry-adapted SCF method

The two main areas of the application of symmetry to the calculation of molecular electronic structure— calculation of integrals and symmetry blocking— can now be combined to generate an efficient way of solving the matrix SCF equations.

Contents

19.1	Introduction		**515**
19.2	Permutations only		**518**
	19.2.1	scount	521
	19.2.2	symG	525
19.3	Full implementation; linear combinations		**528**
	19.3.1	Symmetry and "shells" of functions	533
19.4	Summary		**533**

19.1 Introduction

If we accept the idea that most routine uses of the SCF method will generate molecular orbitals which *do* have the symmetry of the nuclear framework of the molecule, then there are considerable advantages to be gained by using this restriction in the actual implementation of the SCF method.

There are two key ideas which enable these savings to be made:

- The transformation properties of the electron-repulsion integrals are not explicitly needed in an SCF calculation since they appear in particular linear combinations which have one-electron transformation properties (within the matrices J and K).

- Since the matrices J and K (or the particular combination of them in current use) are kept in main memory, it is possible to keep and use the effects of linear combinations of the basis functions during the calculation.

Recall that the main reason why it was not possible to use point-group transformations which induce linear transformations among the basis functions in the calculation of the repulsion integrals was the fact that the repulsion integrals are usually stored on an external medium and so the integrals required might be anywhere in the file; giving a very time-consuming searching overhead which would quickly swamp any processing saving.

Appendix 19.A to this chapter gives the necessary description of the behaviour of the electron interaction matrices when transformation of the basis is performed; it is only necessary to consider the special cases of transformations of the basis induced by point-group symmetry operations. First, let us recall some obvious properties of these transformations.

- A symmetry operation necessarily *permutes* identical atoms which have identical molecular environments and these atoms will necessarily have the same basis functions centred on them.

- Therefore a symmetry operation can, at most, induce a transformation of the basis functions on one atom to linear combinations of the identical basis functions on another symmetry-equivalent atom (including, of course, the possibility of linear combinations of the basis functions on that atom itself).

- In most cases of the use of standard basis functions (cubic harmonic Cartesians), the basis functions on any atom will have much higher symmetry (typically T_d) than the molecular point group therefore it will be the rule rather than the exception for symmetry operations to induce such linear combinations.

It was this last point which enabled us to use the "local symmetry" of the $2p$ functions on the oxygen atom of the water molecule to the "global" C_{2v} symmetry.

We simply need, therefore, to investigate the effect of certain small "blocked" linear transformations of the basis; those involving linear combinations of functions on the same atom. Taking the general four-centre case, assume that the matrix

$$(ab, cd)$$

19.1. INTRODUCTION

consists of all the electron-repulsion integrals formed from the basis functions on atomic centres a, b, c, and d in the molecule, and that the basis functions on these centres are ϕ_a, ϕ_b, ϕ_c and ϕ_d, respectively. A symmetry operation ($\hat{G}$, say) sends the basis functions ϕ_a (say) into a linear combination of the set $\phi_{a'}$ where the atom a' is one of those which are symmetry-equivalent to atom a and similarly for the other atomic bases.

$$\hat{G}\phi_a = \phi_{a'} \boldsymbol{V}^{aa'}$$

where the matrix $\boldsymbol{V}^{aa'}$ is of the dimension of the basis on atoms a and a' and is, typically, a sparse unitary matrix although this is not necessary to the derivation (a C_3^z operation might send a $2p_x$ basis function on atom a into a linear combination of $2p_x$ and $2p_y$ on atom a').

This same symmetry operation will send the other three atomic bases into some linear combinations of the bases on equivalent atoms, so that the effect on the set of electron-repulsion integrals is the transformation

$$(ab,cd) \to \left(\boldsymbol{V}^{aa'} \otimes \boldsymbol{V}^{bb'}\right)(a'b',c'd')\left(\boldsymbol{V}^{cc'} \otimes \boldsymbol{V}^{dd'}\right)$$

using the results of Appendix 19.A applied to this special case.

There will be a transformation of this type induced by each of the operations in the point group; a $\boldsymbol{V}^{aa'}$ etc. for each $\hat{G}$.

The contributions of each of these sets of repulsion integrals to the Coulomb matrix will be (in the notation of Appendix 19.A):

$$\Delta \boldsymbol{\mathcal{J}}^{ab^T} = (ab,cd)\,\boldsymbol{\mathcal{R}}^{cd}$$

where $\Delta \boldsymbol{\mathcal{J}}^{ab}$ is the row of contributions of the integrals to the Coulomb matrix elements referring to the basis functions on atoms a and b, and $\boldsymbol{\mathcal{R}}^{cd}$ is the column of density matrix elements referring to the basis functions on the atoms c and d.

Similarly, the contributions from the integrals and density matrix elements to the symmetry-equivalent sets (under the operation $\hat{G}$) are

$$\Delta \boldsymbol{\mathcal{J}}^{a'b'^T} = (a'b',c'd')\,\boldsymbol{\mathcal{R}}^{c'd'}$$

But, since the matrix of repulsion integrals and the density matrices are related by the linear transformations $\boldsymbol{V}^{aa'}$ etc., we may use the result of Appendix 19.A to show

$$\Delta \boldsymbol{\mathcal{J}}^{a'b'^T} = \left(\boldsymbol{V}^{aa'} \otimes \boldsymbol{V}^{bb'}\right)(a'b',c'd')\left(\boldsymbol{V}^{cc'} \otimes \boldsymbol{V}^{dd'}\right)\left(\boldsymbol{V}^{cc'} \otimes \boldsymbol{C}^{dd'}\right)^{-1}\boldsymbol{\mathcal{R}}^{cd}$$

That is,

$$\Delta \mathcal{J}^{a'b'} = \Delta \mathcal{J}^{ab} \left(V^{aa'} \otimes V^{bb'} \right)$$

so that a knowledge of the contributions to the Coulomb matrix from one set of repulsion integrals and a density matrix is sufficient to determine the contributions from sets of symmetry-related integrals *because of the one-electron* nature of the resulting transformation.

Therefore, provided that we know all the integrals arising from basis functions on all symmetry-distinct sets of atoms in a molecule, we can obtain the full Coulomb matrix if we also know the density matrix. Considerations similar to the ones in Appendix 19.A show that the same theorem applies to the exchange matrix $\mathcal{K}$. It must be emphasised that these ideas are *restricted* to those cases where the computational problem can be reduced to the diagonalisation of an effective one-electron matrix, unlike the methods of integral calculation based on the permutational symmetry of the basis which are applicable to *any* method since all that these latter methods do is produce a file of integrals in a non-standard order which, if the method of processing the electron-repulsion integrals is well-designed, is immaterial.

Although these proofs are completely elementary, as is the method of implementing them, it is sometimes difficult to appreciate the simplicity of the implementation without convincing oneself by an explicit small case. Let us consider the implementation in two stages:

1. The simplest case of all; when the basis is simply *permuted* by the symmetry transformations of the molecular point group (the case considered earlier in treating the integral computation problem). The matrices V are unit matrices; the centres are permuted (and so the basis is permuted, but no further mixing occurs).

2. The *full* case; when the symmetry operations of the point group induce linear transformations amongst the basis functions. The matrices V are non-trivial orthogonal matrices. The centres are permuted *and* the basis functions are mixed by the transformations.

19.2 Permutations only

Let us consider an explicit example; a "molecule" consisting of three hydrogen atoms in an equilateral triangular configuration each having a single $1s$-type basis function. Considering the one-electron operator, it is easy to see that of the six possible matrix elements, only two are distinct in the symmetry group C_{3v}. These may be taken to be h_{11} and h_{12}. The action of the six symmetry operations of C_{3v} on these three basis functions and the distinct one-electron matrix elements is particularly simple:

19.2. PERMUTATIONS ONLY

Orbital and one-electron transformations for $1s$ basis for triangular H_3

E	C^3	$(C^3)^2$	σ^1	σ^3	σ^2
1	2	3	1	3	2
2	3	1	3	2	1
3	1	2	2	1	3
h_{11}	h_{22}	h_{33}	h_{11}	h_{33}	h_{22}
h_{12}	h_{23}	h_{13}	h_{13}	h_{23}	h_{12}

Now, suppose we apply the transformations to the distinct one-electron integrals and add up the results to generate a one-electron matrix. The result is a one-electron matrix with the correct symmetry but the elements are all *doubled*; a matrix of $2h_{ij}$. It is easy to see why this doubling has occurred; the result of applying the symmetry operations to the distinct elements does not *result in a disjoint set of new elements*; repetitions occur as they do in the list of transformed orbitals. Of course, these repetitions must occur *with the molecular symmetry*; if h_{11} occurs twice, so must h_{22} and h_{33}. In our case both h_{11} and h_{12} occur twice but this need not be so; in general there will be a characteristic number of repetitions of each of the symmetry-distinct matrix elements. Thus the method is clear, we must

1. Apply the operations of the point group to the symmetry-distinct objects and sum the results for each set of basis-function indices.

2. Divide each of the symmetry-related objects by the number of times that its symmetry-distinct "parent" is *sent into itself* by the operations of the group.

Clearly, this whole process can be made much cleaner by performing the division *when the integrals are computed* and the symmetrisation is then just the first of the above two steps.

If we now consider the two-electron repulsion matrix J, during an SCF calculation we shall typically have to handle the full density matrix R and a file of symmetry-distinct repulsion integrals. Again, it is easy to record, during the computation of these distinct integrals, the number of times each integral is sent into itself by the point-group operations.

It is much more tedious to follow through the calculation by hand but the *result* is the same.

If the "partial" Coulomb matrix J, formed from the full R matrix and the symmetry-distinct repulsion integrals, each divided by the number of times it is sent into itself by the operations of the point group, is subject to the operations of the point group and the results summed for each index pair, the result is the correct full Coulomb matrix J.

This result has some important ramifications if we wish to incorporate it into our SCF implementation since, in order to obtain the correct electron-repulsion matrices, the matrix describing the effects of the symmetry operations must be available to a new **integer function scf** as well as to the integral generation programs.

However, the symmetry part of the integral calculation programs can be much simpler; it is no longer necessary to keep a (non-repeating) list of symmetry-equivalent integrals during the calculation all that is necessary is to *count* the number of times a given integral is sent into itself by the operations of the point group and to ensure that a symmetry-distinct set is generated. Here is a suitable procedure, **integer function scount**.

19.2.1 scount

A **function** to count the number of times an integral is sent into itself by the permutations in *perms*.

NAME scount
> Use **perms** to compute how many times a given integral is sent into itself by the operations of a point group which only permutes the basis functions.

SYNOPSIS
>
>   integer function scount(i,j,k,l,perms,n,mmax)
>   integer i, j, k, l, n, mmax
>   integer perms(ARB)

DESCRIPTION
> **scount** takes a set of four integers i, j, k, l which label an electron-repulsion integral and uses the permutational information about the n basis functions contained in **perms** to generate a an integer which is the number of times the integral $(ij, k\ell)$ is sent into itself by the transformations summarised in **perms**.

ARGUMENTS

> **i,j,k,l** Input: The labels (in canonical order) of a repulsion integral.
>
> **n** Input: The number of basis functions.
>
> **mmax** Input: The number of symmetry operations.
>
> **perms** Input: The array of permutations, n by mmax stored by columns in a singly subscripted way; the (i, j) element is accessed by perms((j-1)*n + i).

RETURNS
> The required number or zero if the current set of labels should not be included in the electron-repulsion file.

SEE ALSO
> gensym, lister

DIAGNOSTICS
> None, it is the user's responsibility to generate a correct **perms** array. Notice that, unlike **lister**, the information in **perms** must not be redundant and must represent a group.

"scount.f" 19.2.1 ≡

    integer function $scount(i, j, k, l, perms, n, mmax)$
       ⟨ scount interface declarations 19.2.1.5 ⟩
    {
    ⟨ scount local declarations 19.2.1.6 ⟩
    ⟨ Initialise the count and form canonical number 19.2.1.1 ⟩
       /* Loop over all the permutations in $perms$ */
    for $(m = 1;\ m \leq mmax;\ m = m + 1)$
    {
    ⟨ Apply the mth permutation to i, j, k, l 19.2.1.3 ⟩
    ⟨ Put the transformed labels in canonical order 19.2.1.2 ⟩
    ⟨ Check for zero and increment the count 19.2.1.4 ⟩
    }
    return (num);
    }

⟨ Initialise the count and form canonical number 19.2.1.1 ⟩ ≡
   $num = 0$;
   $canon = term4\,(i,\ j,\ k,\ l)$;   /* canonical (ijkl) */
   sign $= i * j * k * l$;

This code is used in section 19.2.1.

19.2. PERMUTATIONS ONLY

⟨ Put the transformed labels in canonical order 19.2.1.2 ⟩ ≡
/* The permuted labels will not, in general, be in standard order "order" puts them so and then the transformed canonical index can be computed. */
call $order(it, jt, kt, lt)$;
$t\_canon = term4(it, jt, kt, lt)$;
/* If the new label is earlier than the original, return */
if $(t\_canon < canon)$
    return (0);

This code is used in section 19.2.1.

⟨ Apply the mth permutation to i, j, k, l 19.2.1.3 ⟩ ≡
/* Get the labels of the related integrals and the characteristic canonical number and sign */

$it = perms(locp(i, m))$;
$jt = perms(locp(j, m))$;
$kt = perms(locp(k, m))$;
$lt = perms(locp(l, m))$;
$t\_sign = it * jt * kt * lt$;

if $(t\_sign \equiv 0)$
    $STOP$; /* perms contains a zero, */

This code is used in section 19.2.1.

⟨ Check for zero and increment the count 19.2.1.4 ⟩ ≡
/* If the transformed canonical is the same as the original and the sign is different, it must be zero */

if $(t\_canon \equiv canon)$
{
if $(sign \neq t\_sign)$
    return (0);
$num = num + 1$;
}

This code is used in section 19.2.1.

⟨ scount interface declarations 19.2.1.5 ⟩ ≡
    integer i, j, k, l, num;
    integer $perms(*)$;   /* Orbital permutations */
    integer n, $mmax$;   /* maximum number of operations */

This code is used in section 19.2.1.

⟨ scount local declarations 19.2.1.6 ⟩ ≡
    integer $term4$;   /* function to generate canonical (ijkl) */
    integer $canon$, t_canon;
    integer *sign*, t_sign, m;
    integer it, jt, kt, lt;

This code is used in section 19.2.1.

There has to be an associated change to **subroutine scfGR** which is called by **integer function scf** to use the file generated by a **subroutine gensym** which makes use of **scount**.

The most important difference between the data supplied to **lister** and to **scount** is that the columns of **perms** must form a group and must not be redundant. This is simply because **scount** does *not* maintain a list and therefore cannot discard repetitions. It is trivial to make the changes to **scount** (analogous to those in **lister**) which will make the function work in an identical way for the one-electron quantities.

The changes to the code of **scfGR** which must be made to make the integral processing compatible with **scount** is easily made in the form of a *subroutine* which has to be called when the processing of the repulsion integral file is finished. It would therefore be convenient to tack onto the end of the repulsion integral file a copy of **perms** which was used to generate that file and this copy can be picked up and used by the new **scf**.

19.2.2 symG

Use the "partial" HF repulsion matrix in H and the permutations in *perms* to form the full repulsion matrix.

On entry, H contains the partial G(R) matrix, on exit H has the full G(R) matrix. G is workspace.

NAME symG

Use **perms** to transform the partial electron repulsion matrix in H into the full SCF repulsion matrix.

SYNOPSIS

```
subroutine symG(H,G,m,perms,pmax);
integer m, pmax, perms(*);
double precision H(*), G(*);
```

DESCRIPTION

symG takes the "skeletal" electron-repulsion G matrix which has been formed using a list of scaled, symmetry-distinct repulsion integrals (using only *permutational* symmetry) and uses the permutational symmetry information in **perms** to compute the full G matrix.

ARGUMENTS

H Input/Output: On entry, H contains the partial G matrix; on exit it contains the full matrix.

m Input: The number of basis functions.

perms Input: The array of permutations, **m** by **pmax** stored by columns in a singly subscripted way; the (i, j) element is accessed by perms((j-1)*m + i).

pmax Input: The number of symmetry operations.

RETURNS

The full electron-repulsion matrix.

SEE ALSO

scount, scfGR

DIAGNOSTICS

None, it is the user's responsibility to generate a correct **perms** array.

"symg.f" 19.2.1 ≡

 @m $locp(i,j)$ $((j-1)*m+i)$ /* Usual subscripting macro */

  subroutine $symG(H, G, m, perms, pmax)$

   /* symG; routine to use the "permutations only" Dacre algorithm to form the full G(R) from the partial one */

  ⟨symG interface declarations 19.2.1.4⟩

{

⟨symG local declarations 19.2.1.5⟩

⟨Take a working copy of H 19.2.1.1⟩

 /* Loop over symmetry operations; exclude the identity ($p = 1$) since H already has this. */

for $(p = 2;\ p \leq pmax;\ p = p + 1)$

{

 /* Loop over the matrix columns */

for $(i = 1;\ i \leq m;\ i = i + 1)$

{

⟨Get the orbital permutation of orbital i and its sign 19.2.1.2⟩

 /* Loop over the matrix rows */

for $(j = 1;\ j \leq i;\ j = j + 1)$

 {

⟨Now the orbital permutation of orbital j and its sign 19.2.1.3⟩

$ij = (j-1)*m + i;$
$ijt = (jt-1)*m + it;$
$jit = (it-1)*m + jt;$

 /* Now add in the symmetry-related term with sign product */

$H(ijt) = H(ijt) + signi * signj * G(ij);$
$H(jit) = H(ijt);$

 }

}

}

return;

}

19.2. PERMUTATIONS ONLY

⟨ Take a working copy of H 19.2.1.1 ⟩ ≡
    for $(i = 1;\ i \leq m * m;\ i = i + 1)$
        $G(i) = H(i);$

This code is used in section 19.2.1.

⟨ Get the orbital permutation of orbital i and its sign 19.2.1.2 ⟩ ≡
    $it = perms(locp(i,\ p));$
    $signi = one;$
    if $(it < 0)$
        $signi = -one;$
    $it = \boldsymbol{abs}(it);$

This code is used in section 19.2.1.

⟨ Now the orbital permutation of orbital j and its sign 19.2.1.3 ⟩ ≡
    $jt = perms(locp(j,\ p));$
    $signj = one;$
    if $(jt < 0)$
        $signj = -one;$
    $jt = \boldsymbol{abs}(jt);$

This code is used in section 19.2.1.

⟨ symG interface declarations 19.2.1.4 ⟩ ≡
    integer m, $pmax$, $perms(*);$
    double precision $H(*)$, $G(*);$

This code is used in section 19.2.1.

⟨ symG local declarations 19.2.1.5 ⟩ ≡
    integer i, j, it, jt, p, ij, ijt, $jit;$
    double precision one, $signi$, $signj;$
    data $one/1.0 \cdot 10^{00}\text{D}/;$

This code is used in section 19.2.1.

This routine must be called (ideally from **scfGR** in order to minimise changes to **integer function scf**) after the file of repulsion integrals has been processed.

19.3 Full implementation; linear combinations

Having seen how the implementation goes for the simplest possible case, we can proceed to the full case when symmetry operations induce linear combinations of basis functions on symmetry-equivalent centres.

Again, it is useful to consider a simple example; let us retain the set of three planar hydrogen atoms but make this system more realistic by placing a nitrogen atom above the plane to generate the ammonia molecule. Using a minimal basis of AOs on the atoms of this molecule includes a pair of coplanar $2p$ AOs which are sent into linear combinations of each other by the operations of the point group C_{3v}. The coordinate system is the conventional one; the nitrogen atom is at the origin and the plane of the hydrogen atoms is at a negative value of z and parallel to the x, y plane. Thus it is the $2p_x$ and $2p_y$ AOs which form linear combinations under the action of the operations of C_{3v}.

The basis functions are numbered ϕ_1 to ϕ_8 in the order $1s_N$, $2s_N$, $2p_{xN}$, $2p_{yN}$, $2p_{zN}$, $1s_{H1}$, $1s_{H2}$, $1s_{H3}$.

Thus, in the old style of **perms**, a matrix of permutation-only transformations would look like:

Basis "permutations" for the minimal basis of NH_3

E	C^3	$(C^3)^2$	σ^1	σ^3	σ^2
1	1	1	1	1	1
2	2	2	2	2	2
3	0	0	3	0	0
4	0	0	-4	0	0
5	5	5	5	5	5
6	7	8	6	8	7
7	8	6	8	7	6
8	6	7	7	6	8

A set of transformations which only uses *one* of the symmetry planes of the molecule to give useful information and, incidentally, shows the contrast between molecular and global-basis-function symmetry.

There are two organisational problems to be overcome here:

19.3. FULL IMPLEMENTATION; LINEAR COMBINATIONS

1. How to code the information which replaces the zeroes in the table; the linear combinations.

2. How to find out what (if anything) replaces the "number of times an integral is sent into itself" by the operations of the point group since the zeroes in the table preclude the simple usage in the permutations-only case. Most of the ϕ_i are not sent into themselves *at all* by the operations of the group.

The first of these is solved easily enough; it is simply a question of using a code in **perms** (now a misnomer since it includes transformations other than **perms**) which does not arise in the permutations-only usage. Using an integer which is not zero or plus/minus one of the basis function numbers is adequate. Here, to draw attention to these codes we start them at 1001. Then, each of these codes "points" to an array of linear combination information. These ideas are made abundantly obvious by application to the example, here is a table to replace **perms**:

Coded basis transformations for the minimal basis of NH$_3$

E	C^3	$(C^3)^2$	σ^1	σ^3	σ^2
1	1	1	1	1	1
2	2	2	2	2	2
3	1001	1002	3	1003	1004
4	1005	1006	-4	1007	1008
5	5	5	5	5	5
6	7	8	6	8	7
7	8	6	8	7	6
8	6	7	7	6	8

The codes 1001 to 1008 are easily collected as follows:

Array of linear combinations for NH$_3$

Code	Coefficient	Orbital	Coefficient	Orbital
1001	$-1/2$ (1)	ϕ_3	$-\sqrt{3}/2$ (2)	ϕ_4
1002	$-1/2$ (1)	ϕ_3	$\sqrt{3}/2$ (3)	ϕ_4
1003	$-1/2$ (1)	ϕ_3	$\sqrt{3}/2$ (3)	ϕ_4
1004	$-1/2$ (1)	ϕ_3	$-\sqrt{3}/2$ (2)	ϕ_4
1005	$-1/2$ (1)	ϕ_3	$\sqrt{3}/2$ (3)	ϕ_4
1006	$-1/2$ (1)	ϕ_3	$-\sqrt{3}/2$ (2)	ϕ_4
1007	$\sqrt{3}/2$ (3)	ϕ_3	$1/2$ (4)	ϕ_4
1008	$-\sqrt{3}/2$ (2)	ϕ_3	$1/2$ (4)	ϕ_4

It is readily seen that the number of *distinct* coefficients is very small since (of course, because of the group property) the same set of linear combinations occurs again and again. In fact, there are only *four* in this case and they are indicated by the numbers in parenthesis in the above table. For completeness, they are given in the table below.

$$\begin{array}{cc}
\multicolumn{2}{c}{\text{Distinct coefficients of the linear combinations}} \\
1 & -1/2 \\
2 & -\sqrt{3}/2 \\
3 & \sqrt{3}/2 \\
4 & 1/2
\end{array}$$

Thus, the rather daunting table of linear combinations can be *coded* in a very compact form; each linear combination requires (in general):

1. The number of terms in that linear combination.

2. A list of *pairs* of integers giving the number of a linear combination coefficient in a list of such and the number of the basis function by which this coefficient is multiplied.

We must now turn to the problem of how to *use* this information in an actual implementation; what (if anything) replaces the number characteristic of each distinct integral in the permutations-only case?

To establish this theorem it is really a question of stating the result and convincing oneself by experiment and generalisation that it is actually true. It is a very simple result and depends on two things:

1. Identical atomic centres and therefore identical *sets* of basis functions are permuted by the symmetry operations.

2. The symmetry operations induce orthogonal (in general, unitary) transformations among the basis functions so that any one basis function "appears in full" in the transformed set; the sum of the squares of the columns *and* rows of the transformation matrices are unity.

We must therefore use the transformation properties of the atomic centres as well as the properties of the basis functions. In our NH_3 example the transformation properties of the centres are identical to those of the four $1s$ functions centred on them whose transformations are given in the above table.

19.3. FULL IMPLEMENTATION; LINEAR COMBINATIONS

Atomic centre transformations for NH_3

E	C^3	$(C^3)^2$	σ^1	σ^3	σ^2
1	1	1	1	1	1
2	3	4	2	4	3
3	4	2	4	3	2
4	2	3	3	2	4

Now consider the application of the same techniques as we have used before to the labels of the centres and the basis functions; using a, b, c and d to label the centres and the usual i, j, k and ℓ for the basis function labels.

- We first form the canonical index $((ab)(cd))$ from the atomic centre labels.
- Then apply the permutations induced by the symmetry operations (summarised in the above table) and form the canonical index $((a'b')(c'd'))$ for each permutation; there are three possibilities:

  1.
  $$((a'b')(c'd')) > ((ab)(cd))$$

  2.
  $$((a'b')(c'd')) = ((ab)(cd))$$

  3.
  $$((a'b')(c'd')) < ((ab)(cd))$$

  If case 3 arises for *all* the operations the integrals associated with this centre will be computed elsewhere. If case 2 occurs there is the analogue of an integral being sent into itself and this case must be examined further by applying the transformations to the basis functions.

- We now form the canonical index of an integral involving a basis function on each of the centres; let $((ij)(k\ell))$ be such an index, it is only necessary that one of ϕ_i, ϕ_j, ϕ_k and ϕ_ℓ be on each of the centres a, b, c and d, not that they be on the respective centres. Now apply the transformations contained in the coded linear transformation array to this set of basis-function labels; there are, this time, four possibilities:

  1.
  $$((i'j')(k'\ell')) > ((ij)(k\ell))$$

2.
$$((i'j')(k'\ell')) = ((ij)(k\ell))$$

3.
$$((i'j')(k'\ell')) < ((ij)(k\ell))$$

4. $((i'j')(k'\ell'))$ contains one of the labels of a coded linear transformation (one greater than 1000, in our example).

Again, if case 3 occurs the integral will be computed elsewhere. The interesting cases are cases 2 and 4 since they are the ones which either send an integral into itself or *partly* into itself.

- The result is that an integral is retained in the file if *none* of the operations give one the results:

1.
$$((a'b')(c'd')) < ((ab)(cd))$$

2. Both of
$$((i'j')(k'\ell')) = ((ij)(k\ell))$$
and
$$((i'j')(k'\ell')) < ((ij)(k\ell))$$

and the integral is divided by the number of times it is sent into itself either partially or wholly, i.e. the number of times the result

$$((a'b')(c'd')) = ((ab)(cd))$$

plus either
$$((i'j')(k'\ell')) = ((ij)(k\ell))$$

or $((i'j')(k'\ell'))$ contains one of the labels of a coded linear transformation (one greater than 1000, in our example).

These ideas can be implemented in much the same way as the "permutations only" method; the situation is a little more involved because of the necessity of storing a permutation matrix for the atomic centres *and* a set of simple symmetry projection operator coefficients.

19.3.1 Symmetry and "shells" of functions

In the routine calculation of electron-repulsion integrals there are considerable savings to be made by utilising the trivial fact that there may be many Gaussian functions on the same centre with the same exponents. Indeed, it is possible to arrange for the expansion of, for example, the ns and np functions in terms of the same Gaussian radial functions, taking up the differences between the two types of function in the contraction coefficients. In this situation, we can see from the considerations of Chapter 8 that the geometrical factors of many of the integrals will be shared and need be computed only once.

It is appropriate here to use the symmetry of the "shells" of basis functions as these groupings are called and this development has been pioneered by M. Dupuis and H. F. King[1] using the methods of Dacre and Elder.[2]

19.4 Summary

Molecular point-group symmetry may be used to its maximum extent in the symmetry-constrained SCF model of molecular electronic structure by combining the savings available in integral calculations with the fact that the HF matrix is a *one*-electron matrix and has, therefore, the transformation properties of a one-electron matrix. Naturally, this type of usage precludes the investigation of "symmetry-breaking" single-determinant wavefunctions but, if these situations are suspected, one can always fall back on the unconstrained method.

---

[1] *Int. J. Quant. Chem.*, **XI**, 613 (1977).
[2] Referred to in Chapter 17.

Appendix 19.A

Kronecker product notation

Contents

    19.A.1 Basis transformations 534
    19.A.2 Basis-product transformations 535
    19.A.3 Density matrix transformations 536
    19.A.4 Transformations in the HF matrix 537
    19.A.5 Practice 539

19.A.1 Basis transformations

A new set of basis functions ϕ' may be defined by a linear transformation of the original basis ϕ by

$$\phi' = \phi V$$

where V is the $m \times m$ transformation matrix. Any one-electron operator, $\hat{A}$, say, ($\hat{A}$ might be $\hat{h}$ or 1) then generates a matrix representation in the old basis

$$A = \int \phi^T \hat{A} \phi \, dV$$

and in the transformed basis

$$A' = \int \phi'^T \hat{A} \phi' \, dV$$

where, of course,

$$A' = V^T A V$$

19.A.2 Basis-product transformations

Although the relationship between the transformed two-electron integrals and the original set is easy to express in terms of summations over products of the transformation matrix elements and the original integrals:

$$(i'j', k'\ell') = \sum_{r=1}^{m}\sum_{s=1}^{m}\sum_{t=1}^{m}\sum_{u=1}^{m} V_{ri} V_{sj} (rs, tu) V_{tk} V_{u\ell}$$

the manipulations of the summations often prove confusing.

It is often more convenient to use a basis of *products* of the basis functions and express the repulsion integrals as a matrix in this $m \times m$-dimensional space. We define

$$\boldsymbol{\Phi} = (\phi_1\phi_1, \phi_1\phi_2, \phi_1\phi_3, \phi_1\phi_4, \phi_1\phi_m, \ldots, \phi_2\phi_1, \ldots \phi_m\phi_m,)$$

as a row of all possible products of the ϕ_i. Notice that the *order* of the product suffixes is *dictionary* order, which is the opposite convention from that used in our implementations of matrix storage:

$$(ij) = (i-1) * m + j$$

This is unfortunate but this ordering is used in the standard mathematical conventions on the Kronecker products of matrices which will be used in the following theory.

The electron repulsion integrals can then be written as

$$\boldsymbol{\Gamma} = \int \boldsymbol{\Phi}^T \frac{1}{r_{12}} \boldsymbol{\Phi} dV_1 dV_2$$

a square matrix of dimension m^2 with typical element

$$\Gamma_{(ij),(k\ell)} = (ij, k\ell)$$

The transformation matrix of the row matrix $\boldsymbol{\Phi}$ of *products* of basis functions is:

$$\Phi'_{ij} = \phi'_i \phi'_j = \sum_{r=1}^{m}\sum_{s=1}^{m} \phi_r \phi_s V_{ri} V_{sj}$$

The *Kronecker product* of two $m \times m$ matrices $\boldsymbol{A}$ and $\boldsymbol{B}$ is defined to be the $m^2 \times m^2$ matrix $\boldsymbol{A} \otimes \boldsymbol{B}$ whose elements are given by

$$\left(\boldsymbol{A} \otimes \boldsymbol{B}\right)_{(ik),(j\ell)} = A_{ij} B_{k\ell}$$

so that, in terms of the Kronecker product of $\boldsymbol{V}$ with itself the relationship of the row of transformed products of basis functions to the original products is:

$$\boldsymbol{\Phi}' = \boldsymbol{\Phi} \boldsymbol{V} \otimes \boldsymbol{V}$$

This notation then means, in complete analogy with the transformation of one-electron integrals which involve the *basis functions themselves*, that

$$\mathbf{\Gamma}' = (\mathbf{V} \otimes \mathbf{V})^T \mathbf{\Gamma} (\mathbf{V} \otimes \mathbf{V})$$

Defining

$$\mathcal{V} = \mathbf{V} \otimes \mathbf{V}$$

makes the analogy more striking:

$$\mathbf{\Gamma}' = \mathcal{V}^T \mathbf{\Gamma} \mathcal{V}$$

19.A.3 Density matrix transformations

If a set of molecular orbitals $\boldsymbol{\psi}$ is determined by the variational procedure in terms of the basis $\boldsymbol{\phi}$ then

$$\boldsymbol{\psi} = \boldsymbol{\phi}\mathbf{C} = \boldsymbol{\phi}'(\mathbf{V})^{-1}\mathbf{C}$$

where the matrix of expansion coefficients $\mathbf{C}$ is $m \times n$.

The electron density ρ due to occupation of these n MOs is

$$\begin{aligned}\rho = \boldsymbol{\psi}\boldsymbol{\psi}^T &= \boldsymbol{\phi}\mathbf{C}\mathbf{C}^T\boldsymbol{\phi}^T = \boldsymbol{\phi}\mathbf{R}\boldsymbol{\phi}^T \\ &= \boldsymbol{\phi}'\mathbf{V}^{-1}\mathbf{C}\mathbf{C}^T(\mathbf{V}^{-1})^T\boldsymbol{\phi}'^T = \boldsymbol{\phi}'\mathbf{R}'\boldsymbol{\phi}'^T\end{aligned}$$

The equation connecting ρ and $\mathbf{R}$ may also be written in terms of the matrices $\boldsymbol{\Phi}$ and $\boldsymbol{\Phi}'$ of basis-function products;

$$\rho = (\phi_1\phi_1, \phi_1\phi_2, \phi_1\phi_3, \phi_1\phi_4, \phi_1\phi_m, \ldots, \phi_2\phi_1, \ldots \phi_m\phi_m,) \begin{pmatrix} R_{11} \\ R_{12} \\ R_{13} \\ R_{14} \\ \ldots \\ R_{1m} \\ \ldots \\ R_{mm} \end{pmatrix}$$

which may be written as

$$\rho = \boldsymbol{\Phi}\mathcal{R}$$

(say), where $\mathcal{R}$ is a *column* of the elements of $\mathbf{R}$ in the appropriate order.

$$\mathcal{R}_{(rs)} = R_{rs}$$

The one-electron integrals cannot be written in a simple way in terms of the basis-function products because of the essential *operator* nature of parts

19.A.4. TRANSFORMATIONS IN THE HF MATRIX

of $\hat{h}$; however, for simple *multiplicative* "operators", A (no hat signifying no derivatives in A) the result is

$$\mathcal{A} = \int A\mathbf{\Phi}dV = (\mathcal{A}_{(11)}, \mathcal{A}_{(12)}, \mathcal{A}_{(13)}, \ldots, \mathcal{A}_{(mm)})$$
$$= (A_{11}, A_{12}, A_{13}, \ldots, A_{mm})$$

and this notation may be taken over for the general operator $\hat{A}$ in spite of the fact that we may **not** write, in the general case:

$$\mathcal{A} = \int \hat{A}\mathbf{\Phi}dV$$

Notice that matrices of operators are *rows* in this convention while the density matrix is a *column*.

Transformations of the basis now induce transformations of the operator and density matrices in terms of the Kronecker product of the transformation matrix with itself:

$$\mathcal{A}' = \mathcal{A}\mathcal{V}$$

and

$$\mathcal{R}' = \mathcal{B}\mathcal{R}$$

where, temporarily,

$$B = (V)^{-1}$$

and

$$\mathcal{B} = B \otimes B$$

has been used to avoid too complex an expression.

Any mean value, which is the trace of a product of an operator matrix (A, say) and a density matrix in the usual orbital basis representation becomes a simple scalar product in the basis-product representation:

$$<\hat{A}> = \text{tr}\mathbf{AR} = \mathcal{A}\mathcal{R}$$

and is easily seen to be invariant against linear transformations of the basis-products

$$<\hat{A}> = \mathcal{A}'\mathcal{R}' = \mathcal{A}\mathcal{V}\mathcal{V}^{-1}\mathcal{R} = \mathcal{A}\mathcal{R}$$

19.A.4 Transformations in the HF matrix

It is then easy to show that the "Coulomb matrix" J, appearing in the HF method, has elements given by:

$$J_{rs} = \sum_{t=1}^{m}\sum_{u=1}^{m}(rs,tu)R_{tu} = \sum_{tu=1}^{m^2}\Gamma_{(rs),(tu)}\mathcal{R}_{tu}$$

and a *row* matrix $\mathcal{J}$, related in the by now obvious way to $\boldsymbol{J}$ may be defined so that

$$\mathcal{J}^T = \boldsymbol{\Gamma}\mathcal{R}$$

Now, if the orbital basis is subject to a linear transformation both the matrices $\mathcal{R}$ and $\boldsymbol{\Gamma}$ undergo the transformations outlined above, so that the matrix $\mathcal{J}$ suffers the combined transformation given by

$$\begin{aligned}\mathcal{J}'^T &= \boldsymbol{\Gamma}'\mathcal{R}' \\ &= \mathcal{V}^T\boldsymbol{\Gamma}\mathcal{V}\mathcal{V}^{-1}\mathcal{R}\end{aligned}$$

since

$$(\boldsymbol{V})^{-1} \otimes (\boldsymbol{V})^{-1} = \mathcal{V}^{-1}$$

which is a special case of the general rule for Kronecker products that

$$(\boldsymbol{A} \otimes \boldsymbol{B})(\boldsymbol{C} \otimes \boldsymbol{D}) = (\boldsymbol{AC}) \otimes (\boldsymbol{BD})$$

Thus,

$$\mathcal{J}'^T = \boldsymbol{\Gamma}'\mathcal{R}' = \mathcal{V}^T\boldsymbol{\Gamma}\mathcal{R} = \mathcal{V}^T\mathcal{J}^T$$

or, much more directly,

$$\mathcal{J}' = \mathcal{J}\mathcal{V}$$

which is simply confirmation of what would be expected intuitively: the Coulomb matrix transforms under any linear transformation of the basis by the same rule as a one-electron operator. Indeed, if it did not, it would be impossible to transform the matrix $\boldsymbol{h}^F$ since it is composed of a sum of one-electron integrals and the Coulomb and exchange matrices.

The definition of the exchange matrix precludes its calculation as a simple matrix product involving $\boldsymbol{\Gamma}$. The $\boldsymbol{K}$ matrix involves the same $\boldsymbol{R}$ matrix but with different electron-repulsion integrals:

$$K_{rs} = \sum_{t=1}^{m}\sum_{u=1}^{m} R_{tu}(ru, ts)$$

If we define a new square matrix of repulsion integrals by

$$\Gamma^e_{(ij)(k\ell)} = [ij, k\ell] = (i\ell, kj)$$

then, under a transformation of basis

$$\boldsymbol{\Gamma}^{e'} = \mathcal{V}^T\boldsymbol{\Gamma}^e\mathcal{V}$$

and the whole derivation may be repeated to show that the exchange matrix undergoes the same "one-electron" transformation as the Coulomb Matrix.

$$\mathcal{K}' = \mathcal{K}\mathcal{V}$$

The fact that the two matrices $\mathcal{J}$ and $\mathcal{K}$ were calculated from different orderings of the repulsion integrals does not affect the transformation properties, of course. There is no reason why the relationship between the electron-repulsion integrals and the Hartree–Fock matrix for a particular case should not be incorporated into the definition of Γ so that

$$\Gamma_{(rs),(tu)} = a(rs,tu) - b(ru,ts)$$

and so

$$\mathcal{G} = a\mathcal{J} - b\mathcal{K} = \Gamma\mathcal{R}$$

The upshot is that all the components of the Hartree–Fock matrix undergo the same transformation when the basis is subject to a linear transformation. This result can be expressed as

$$h^{F'} = V^T h^F V$$

or

$$\mathcal{H}^{F'} = \mathcal{H}^F \mathcal{V}$$

19.A.5 Practice

The relationships between the one-electron transformations involving the orbital basis and the two-electron transformations involving the basis-products are useful for formal purposes but would certainly never be implemented as a practical way of solving the SCF equations as they stand since they involve $m^2 \times m^2$ matrices which contain much redundant information. However, techniques of storage compression, analogous to use of the permutational symmetries of the repulsion integral labels, can be used to enable what is known as the "supermatrix" formulation of the SCF equations to be implemented in an economical way.

In particular, the fact that many of the operations in a supermatrix form become simple scalar products has practical advantages under certain hardware conditions.

Chapter 20

Linear multi-determinant methods

The most straightforward way to include electron correlation into a description of molecular electronic structure is to use a (variationally optimised) linear combination of antisymmetric products of orbitals. If these orbitals are the solutions of the Hartree–Fock matrix equations an energy below the best HF energy is guaranteed and an improved description of the electrons' distribution and their mutual repulsions will be available. If the orbitals are approximate atomic orbitals, a similar formalism generates the valence bond model of electronic structure.

Contents

20.1	Correlation and the Hartree–Fock model	541
20.2	The configuration interaction method	543
20.3	The valence bond method	544
20.4	Restricted CI	545
20.5	Symmetry-restricted CI	550
20.6	More general CI	553
20.7	Nesbet's method for large matrices	554
20.8	"Direct" CI	560
	20.8.1 A small example; paired excitations	562
	20.8.2 A perturbation/CI method	564
20.9	Conclusions	565

20.1 Correlation and the Hartree–Fock model

In the normal (probability theory) use of the term, two probability distributions are not *correlated* if their joint (combined) probability distribution is just the simple *product* of the individual probability distributions. In the case of the Hartree–Fock model of electron distributions the probability distribution for *pairs* of electrons is a product "corrected" by an exchange term. The two-particle density function cannot be obtained from the one-particle density function; the one-particle density *matrix* is needed which depends on *two* sets of spatial variables. In a word, the two-particle density matrix is a (2×2) *determinant* of one-particle density *matrices* for each electron:

$$\rho_2(\vec{x}_1, \vec{x}_2; \vec{x}_1', \vec{x}_2') = \rho_1(\vec{x}_1; \vec{x}_1')\rho_1(\vec{x}_2; \vec{x}_2') - \rho_1(\vec{x}_1, \vec{x}_2')\rho_1(\vec{x}_2; \vec{x}_1')$$

so that the two-particle density *function* is just

$$\rho_2(\vec{x}_1, \vec{x}_2) = \rho(\vec{x}_1)\rho(\vec{x}_2) - \rho_1(\vec{x}_1, \vec{x}_2)\rho_1(\vec{x}_2; \vec{x}_1)$$

Thus, strictly speaking, there is some partial electron correlation in the HF model.

Putting $\vec{x}_1 = \vec{x}_2$ in the above we see that

$$\rho_2(\vec{x}_1, \vec{x}_1) = 0$$

independently of the value of $\vec{x}_1$, in contrast to the result expected if neither one-particle density were zero. This result arises because the term coming from antisymmetry cancels the simple product (uncorrelated) term. Remembering that we are using *space–spin* variables $\vec{x}$, it is clear that, in general, this result holds *only for electrons of the same spin* since only then can $\vec{x}_1 = \vec{x}_2$. The *spatial* requirement $\vec{r}_1 = \vec{r}_2$ is not sufficient to make the two-particle density zero. We can conclude then

> The Hartree–Fock model, by dint of its antisymmetry requirement, provides some partial correlation of the motions of electrons of like spin while electrons of unlike spin are completely uncorrelated.

Thus, for example, in the HF description of the helium atom as two electrons of opposite spin with the same spatial distribution, the two electrons may occupy the same point in space notwithstanding their mutual repulsion. Attempts at the physical interpretation of these phenomena are important in the density-functional description of electronic structure which we discuss in Chapter 33.

If we wish to attempt to describe the "genuine" electron correlation brought about by the mutual repulsions,[1] we must use a linear combination of determinants or some antisymmetric function which is equivalent to a linear combination of determinants. The simplest approach is to use, directly, a linear combination of determinants of one-electron functions.

---

[1] Rather than the Pauli principle.

20.2 The configuration interaction method

Going back to Chapter 1, we express an approximation to the wavefunction for $2n$ electrons[2] as

$$\Psi = \sum_{K=0}^{N} D_K \Phi_K \qquad (20.1)$$

where each determinant Φ_K is composed of n doubly-occupied orbitals as the simplest example. Specifically, let us assume that an SCF calculation has been performed using a basis of m spatial functions which has generated m spatial MOs ($2m$ spin-orbitals), the lowest $2n$ of which are occupied to form the leading determinant Φ_0. The other N determinants are formed by occupying a selection of the $2m$ spin-orbitals. With $2n$ occupied from a total of $2m$ this process can generate as many as

$$N = \left(\begin{array}{c} 2m \\ 2n \end{array} \right)$$

independent determinants. For example, a minimal basis calculation on the benzene molecule has $m = 36$, $n = 21$ giving an absolutely huge number of possible determinants. Of course millions of these determinants make microscopic or zero contribution to the sum in eqn (20.1) because:

- They are of the wrong spin symmetry; the ground state of the molecule is closed-shell so that only *spin singlets* contribute to the expansion in eqn (20.1).

- Even if spin singlets, they are "quantitatively unsuitable" because they contain orbital occupancies which are too high in energy to make any meaningful contribution to the ground-state wavefunction.

Nevertheless, expansions like eqn (20.1), unless restricted in some way, generate expansions of completely unmanageable length.

Using the algebraic method (expansion of the one-electron functions in terms of basis functions), it is clear that if eqn (20.1) is used including (*e.g.*) all possible singlet functions,[3] the resulting expansion will include all possible determinants of all possible orbitals and so will represent the *best possible* wavefunction obtainable using the given basis. This natural limit provides a benchmark for all other calculations using the given basis and is called Full Configuration Interaction (FCI) and generates the so-called basis set limit for the given basis. However, other than giving useful information about the relative quality of other, more approximate, models of electronic structure, this full expansion is of little practical utility. The *scientific* problem in using the CI model is to get the best possible approximate wavefunction with the minimum expansion length in eqn (20.1).

---
[2]Closed-shell, for simplicity.
[3]That is, not just the determinants of doubly occupied MOs assumed for simplicity above.

20.3 The valence bond method

The simplest form of the VB model follows the Heitler–London method for the structure of the H_2 molecule; an approximate wavefunction is written as a (antisymmetrised) product of electron-pair functions. Each electron-pair function is a product of a (symmetric) spatial function and a (antisymmetric) singlet spin function. The spatial function is chosen to model the chemist's intuition about the structure of the electron pair; usually a simple symmetrised product of (hybrid) atomic orbitals. The resulting total approximate wavefunction will not be a determinant but, since it is a function which is antisymmetric with respect to exchange of electrons' coordinates, it must be capable of being expanded as a linear combination of determinants. If polar structures are added to the VB model the same *general* result must obtain; any antisymmetric function may be expressed as a linear combination of determinants of the space/spin functions:

$$\Psi = \sum_{K=0}^{N} C_K \Phi'_K \qquad (20.2)$$

where the determinants Φ'_K have been distinguished by a prime to avoid confusion between the determinants in the CI expansion eqn (20.1) and those in the "VB expansion" eqn (20.2).

The formal analogy between eqns (20.1) and (20.2) masks the enormous difference between both the scientific content of the two models and, equally important, the ease of actual implementation of the two approximations.

> The orbitals in the CI expansion eqn (20.1) are *orthogonal MOs* while the orbitals in the VB expansion eqn (20.2) are *non-orthogonal AOs*.

The physical and chemical appeal of the VB model hinges on this non-orthogonality which is an enormous barrier to *routine use* of this model.

If we recall the relative complexity of the expressions for integrals involving determinants of orthogonal and those for non-orthogonal orbitals in Appendix 2.A, it is easy to appreciate the difficulties associated with the implementation of the VB model. Of course, it is possible to transform the AOs to an orthogonalised set and so generate a VB expansion in terms of an expansion of determinants of *orthogonalised* AOs, but this expansion is at least as long as any CI expansion and, perhaps more important, loses the essential physical appeal of the limited number of chemical "structures" of the VB model. In terms of an orthogonalised set of AOs many unfamiliar highly charged polar structures appear in the description of a basically covalent electronic structure. If only

20.4. RESTRICTED CI

"covalent" structures are retained in the VB model it is easy to show that the model will not describe any binding.[4]

The fact that the orbitals in a CI expansion are orthogonal does mean that the transformation of the energy integrals from the raw AO basis in which they are computed to the MO basis must be performed. But this transformation is done only once to generate those integrals which are needed.

Considerations like these mean that the VB model, although it has seen a considerable resurgence of interest over the past few years with several key breakthroughs being made, is still predominantly perceived as a "specialist" area with applications confined to small numbers of electrons. The so-called Generalised Valence Bond (GVB)/indexGVB method which is a self-consistent development of the Heitler–London model is developed in Chapter 22 and, for the moment, this is as far as we shall go with the details of the more classical VB model until we have been able to develop the tools necessary to continue in Chapter 21.

20.4 Restricted CI

Making the central assumption that the best possible single-determinant function formed from a given basis is also the best possible starting point from which to make a multi-determinant expansion[5] we take the first term in the expansion eqn (20.1) to be Φ_0, the HF wavefunction so that eqn (20.1) becomes

$$\Psi = D_0 \Phi_0 + \sum_{K=1}^{N} D_K \Phi_K \quad (20.3)$$

and construct the other Φ_K from the orbitals occupied in Φ_0 and the virtual orbitals formed as a byproduct of the SCF process.

We have noted already in Chapter 10 that the virtual orbitals have the desirable property of being orthogonal to the occupied orbitals but may well not have any particularly useful physical properties. This means that the corrections to the energy and electron distribution to be obtained by adding determinants containing these virtual orbitals may not be very "compact"; the expansion length may well have to be long to achieve an acceptable improvement.

The determinants used in the expansion may be usefully classified according to the number of *replacements* of orbitals in Φ_0 by virtual orbitals; we may call

---
[4]This result is due to the fact that the so-called "exchange energy" which is responsible for the binding in the VB model between non-orthogonal orbitals includes a (binding) one-electron contribution which is absent in the orthogonal-orbital "exchange energy": see Appendix 20.A.

[5]This assumption, although plausible is not always true.

them "excitations" by analogy with the excitation of individual electrons from orbitals occupied in the ground state of the molecule into unoccupied orbitals.[6] So, we may have

- Single excitations; one orbital occupied in Φ_0 is replaced by one virtual MO.

- Double excitations; two occupied MOs are replaced by two virtuals. There are several classes of these, *e.g.* a doubly occupied MO may be replaced by a doubly occupied virtual or by two singly occupied virtuals forming an overall singlet etc.

- Triple and higher excitations.

There is a certain "natural break" at double excitations because of the form of the energy integrals involving determinants of orthogonal orbitals; Slater's rules mean that there are no non-zero matrix elements between the ground state and determinants with more than two excitations.

Let us look briefly at the interpretation and associated energy integrals of these three classes.

Single Excitations.

The addition of two determinants which differ by a single row (or column) generates a new single determinant with a new row (column) which is the same linear combination of the rows (columns) which differ as the original combination of determinants, *e.g.*:

$$a \begin{vmatrix} A_{11} & A_{12} \\ A_{21} & A_{22} \end{vmatrix} + b \begin{vmatrix} A_{11} & B_{12} \\ A_{21} & B_{22} \end{vmatrix} = \begin{vmatrix} A_{11} & aA_{12} + bB_{12} \\ A_{21} & aA_{22} + bB_{22} \end{vmatrix}$$

Thus the addition of a set of single excitations to a given single determinant has the effect of changing the forms of the component orbitals in the original. Therefore the solution of the variational problem associated with the approximation

$$\Psi = D_0 \Phi_0 + \sum_{i,a} D_i^a \Phi_i^a$$

where the Φ_i^a are determinants of *single* excitations $\chi_i \longrightarrow \chi_a$, say (where χ_i is occupied in Φ_0 and χ_a is not), generates a single determinant with the best possible (variationally determined) component orbitals. The notation of the

---

[6]That is not to say that these determinants are approximations to the structure of real molecular excited states, it is just a useful mnemonic device.

20.4. RESTRICTED CI

coefficients D has been changed to match the more descriptive notation of the Φ. The summation ranges over all occupied χ_i and all virtual χ_a.

However, if the component orbitals of Φ_0 are the SCF orbitals in the given basis they *already are* the best possible orbitals of a single determinant. Thus there can be no *improvement* of an SCF single determinant by the addition of single-excitation determinants. If we recall the form of the linear variation problem given in Chapter 1, it is clear that this result implies that all integrals of the form

$$\int \Phi_i^a \hat{H} \Phi_0 d\tau$$

must be zero. This result is easily verified by the use of the relevant Slater Rule from Appendix 2.A. Indeed, one way of *defining* the optimum SCF orbitals is by the condition

$$\int \Phi_i^a \hat{H} \Phi_0 d\tau = 0$$

for all i and a.

As we shall see shortly, this result does not mean that single excitations should always be excluded from a CI expansion, what it does mean is that their role is to change the form of the orbitals and that there is no advantage in this if they are used in the CI expansion *alone*. Similar, but more involved, considerations apply to triple and higher excitations.

Double excitations.

It is easy to convince oneself empirically that a linear combination of two determinants which differ in more than one row (or column) *cannot* be written as a single determinant.[7] Therefore, double excitations are the first terms to introduce correlation to the wavefunction; such a wavefunction is said to "contain" electron correlation or to correlate the electrons which it describes.

Not all double excitation determinants are singlets and therefore, if the target ground state is a singlet, we shall be doing redundant work if we simply set up the "doubly excited" wavefunction and attempt to determine the linear coefficients in the expansion:

$$\Psi = D_0 \Phi_0 + \sum_{i,j;a,b} D_{ij}^{ab} \Phi_{ij}^{ab}$$

but, of course, we will not be *in error*; the irrelevant coefficients will turn out to be zero by spin integration.

---
[7]This is why we could not obtain the best single determinant wavefunction by adding small amounts of other determinants in the derivation of the Hartree–Fock equations in Chapter 2.

The simple notation which we have used for single-excitation wavefunctions has been extended in an obvious way for double excitations:

$$\Phi_{ij}^{ab}$$

is the determinant obtained by removing the occupied MOs ψ_i and ψ_j from Φ_0 and replacing them by the MOs ψ_a and ψ_b.

The expansion is redundant because the determinants are antisymmetric:

$$\Phi_{ij}^{ab} = -\Phi_{ji}^{ab} = \Phi_{ji}^{ba} = -\Phi_{ij}^{ba}$$

so that

$$D_{ij}^{ab} = -D_{ji}^{ab} = D_{ji}^{ba} = -D_{ij}^{ba}$$

thus the summations may be restricted to $i < j$ and $a < b$ to give a non-redundant (but still not spin-eigenfunction) expansion:

$$\Psi = D_0 \Phi_0 + \sum_{i=1}^{2n} \sum_{j=1}^{i} \sum_{a=2n+1}^{2m} \sum_{b=2n+1}^{a} D_{ij}^{ab} \Phi_{ij}^{ab} \tag{20.4}$$

This is simply a standard linear variation problem to which we have the solutions from Chapter 1; a single matrix equation involving the matrix of $\hat{H}$ the full many-electron Hamiltonian operator and the overlap matrix:

$$\boldsymbol{HD} = E\boldsymbol{SD}$$

if we require only the lowest eigenvalue E.

The matrix elements required are of the types:

$$\int \Phi_0 \quad \hat{H} \quad \Phi_0 d\tau = E_0$$

$$\int \Phi_0 \quad \hat{H} \quad \Phi_{ij}^{ab} d\tau$$

$$\int \Phi_{ij}^{ab} \quad \hat{H} \quad \Phi_{i'j'}^{a'b'} d\tau$$

and the corresponding overlap integrals. In fact the expansion functions are *orthonormal* since all the off-diagonal elements involve an integration over at least one pair of orthogonal orbitals.

There are a number of conventions used in setting up these CI variational problems which reflect the physics of the situation more clearly than the standard notation used here:

- The dominant contribution to the diagonal terms of the matrix $\boldsymbol{H}$ is E_0, the energy of the SCF ground-state. It is usual to subtract this from these diagonal terms so that the lowest eigenvalue of the matrix is the correlation energy directly.

20.4. RESTRICTED CI

- Because the coefficient D_0 is by far the largest one, it is often set to unity in the expansion. This means that the resulting CI expansion is not normalised but, because the overlaps between the SCF function and all the excited determinants are zero, the overlap between this un-normalised CI expansion and the SCF determinant is unity:

$$\int \Phi_0 \Psi d\tau = 1$$

so-called intermediate normalisation. Once all the CI coefficients have been obtained, the fact that the expansion functions are orthonormal may be used to obtain a true normalised function by dividing by the square root of the sum of the squares of the coefficients (including 1.0 for Φ_0).

- The CI matrix H is *partitioned* to reflect the relative sizes of the effects and to enable perturbation methods to be used if the matrix proves to be very large.

Thus writing

$$\Psi = \Phi_0 + \sum_{i=1}^{2n} \sum_{j=1}^{i} \sum_{a=2n+1}^{2m} \sum_{b=2n+1}^{a} D_{ij}^{ab} \Phi_{ij}^{ab}$$

and using

$$\hat{H} - E_0$$

in place of $\hat{H}$ the CI matrix equation becomes

$$\begin{pmatrix} 0 & H_{12} \\ H_{21} & H_{22} \end{pmatrix} \begin{pmatrix} 1 \\ D \end{pmatrix} = E_{corr} \begin{pmatrix} 1 \\ D \end{pmatrix}$$

where, of course

$$H_{12} = H_{21}^T$$

and $E_{corr} = E - E_0$.

What remains to be done is:

- Develop a unique numbering system for the elements of the matrices.
- Use Slater's rules to obtain the values of the elements of H in terms of MO integrals
- Actually *solve* the above equation.

Now the above matrix system may be very large so the partitioning scheme may be used to obtain an equation of reduced dimension which may be solved *iteratively*. Writing out the partitioned matrix equation as two separate equations we have

$$\begin{aligned} E_{corr} &= H_{12} D \\ H_{21} + H_{22} D &= E_{corr} D \end{aligned}$$

The second of these may be rearranged to

$$(\mathbf{H}_{22} - E_{corr}\mathbf{1})\,\mathbf{D} = -\mathbf{H}_{21}$$

where the matrix $\mathbf{H}_{22}$ is square, and the matrix on the left is non-singular and so may be inverted to give:

$$\mathbf{D} = -\left(\mathbf{H}_{22} - E_{corr}\mathbf{1}\right)^{-1}\mathbf{H}_{21}$$

which, when substituted into the first of the two partitioned equations, gives a final, working, expression:

$$E_{corr} = -\mathbf{H}_{12}\left(\mathbf{H}_{22} - E_{corr}\mathbf{1}\right)^{-1}\mathbf{H}_{21} \quad (20.5)$$

an equation which may be solved iteratively for E_{corr} if we can make a reasonable starting approximation.

We can always make the most reasonable of all estimates of the correlation energy; it is small with respect to E_0, so why not start with $E_{corr} = 0$ and obtain a first approximation of

$$E_{corr} \approx -\mathbf{H}_{12}\mathbf{H}_{22}^{-1}\mathbf{H}_{21}$$

and proceed from there?

Although very simple-looking the method of Double-excitation CI (DCI) involves some computational tasks which are very demanding:

- The energy integrals must be transformed to the orthogonal MO basis; in particular the repulsion integrals over this basis must be computed. This is taken up in Chapter 29.

- Huge numbers of matrix elements must be formed, indexed and stored.

- A very large matrix must be repeatedly inverted in the iterative solution of the final expression for E_{corr}.

20.5 Symmetry-restricted CI

Although we said that our DCI expansion was to be a linear combination of *determinants*, nothing in the scheme above is actually dependent on this choice; the whole scheme would be formally identical if the Φs were any suitable antisymmetric (orthonormal) functions. Indeed, the method we have outlined does not depend on the Φs being doubly excited either; it is just a general iterative method of solving the linear variational problem by a partitioning technique. It

20.5. SYMMETRY-RESTRICTED CI

is also obvious that the same partitioning technique could be used if the starting function were a (fixed) linear combination of determinants; in this case the partitioned matrices would have a "top-left" block of dimension greater than one.

This elementary fact enables us to reduce the size of the CI matrices by choosing linear combinations of the double excitations that have a particular spin symmetry. In particular, since we are normally concerned with ground-state *singlets*, we can use linear combinations of double-excitation determinants that are singlet spin functions. The only difference which this will make to our method is the detailed form of the H_{KL} in the matrix $\boldsymbol{H}$ (the singlets are still orthonormal).

The choice of singlets (or triplets) arising from the various doubly excited determinants is quite straightforward to do if a little tedious and can be classified according to the various equalities amongst i and j, a and b of Φ_{ij}^{ab}. The easiest type is the one where one excites a pair of electrons from the same spatial occupied MO into the same virtual: clearly the determinant $\Phi_{ii}^{a\bar{a}}$ is already a singlet since it is a determinant of doubly-occupied orbitals.[8] Here we have introduced a new notation which is "intermediate" between that used for spatial orbitals and the spin-orbital notation:

- i is used to mean a spatial orbital occupied by an electron of α spin, i.e. it is χ_r whose product form is

$$\psi_i(\vec{x}) \equiv \chi_r(\vec{x}) \quad \text{(say)} \quad = \psi_i(\vec{r}) \times \alpha(s)$$

- $\bar{i}$ is used to mean the same spatial orbital occupied by an electron of β spin:

$$\psi_{\bar{i}}(\vec{x}) \equiv \chi_s(\vec{x}) \quad \text{(say)} \quad = \psi_i(\vec{r}) \times \beta(s)$$

Since, in the closed-shell-singlet Φ_0 one is normally dealing with spatial orbitals, this notation, although a little clumsy and, actually, strictly ambiguous[9] concentrates attention on those spatial orbitals and avoids the "doubling" of the MOs by the two spin factors.

The determinants involved in the DCI expansion obviously contain lots of orbitals in common (all except i, j, a and b, in fact) and so it is useful to use a contracted notation which only involves the "active" orbitals, we write

$$\Phi_{ij}^{ab} = ||ijab|| = \begin{vmatrix} \psi_i(\vec{x}_1) & \psi_i(\vec{x}_2) & \psi_i(\vec{x}_3) & \psi_i(\vec{x}_4) \\ \psi_j(\vec{x}_1) & \psi_j(\vec{x}_2) & \psi_j(\vec{x}_3) & \psi_j(\vec{x}_4) \\ \psi_a(\vec{x}_1) & \psi_a(\vec{x}_2) & \psi_a(\vec{x}_3) & \psi_a(\vec{x}_4) \\ \psi_b(\vec{x}_1) & \psi_b(\vec{x}_2) & \psi_b(\vec{x}_3) & \psi_b(\vec{x}_4) \end{vmatrix}$$

---
[8]This type of double excitation is made the basis of a class of MCSCF methods in Chapter 22.
[9]Note that, in the first of the two definitions given above ψ_i is used with two *different* kinds of argument! There is no justification for this type of duplication of notation except that it is in use.

where each of i, j, a, b may have "bars" and the normalisation factor is understood.

With this notation the singlets arising from the DCI determinants are classified according to equalities amongst the spatial MOs in i, j, a, b and we use a similar notation ($^1\Phi_{ij}^{ab}$) for the singlets formed from the determinants Φ_{ij}^{ab}:

- $i, \bar{i}, a, \bar{a}$
$$^1\Phi_{i\bar{i}}^{a\bar{a}} = \Phi_{i\bar{i}}^{a\bar{a}} = ||i\bar{i}a\bar{a}||$$

- $i, \bar{i}, a, \bar{b}$
$$^1\Phi_{i\bar{i}}^{a\bar{b}} = \frac{1}{\sqrt{2}} \left(||i\bar{i}a\bar{b}|| - ||i\bar{i}\bar{a}b|| \right)$$

- $i \neq j, a = b$
$$^1\Phi_{i\bar{j}}^{a\bar{a}} = \frac{1}{\sqrt{2}} \left(||i\bar{j}a\bar{a}|| - ||\bar{i}j\bar{a}a|| \right)$$

- $i \neq j, a \neq b$
In this case there are *two* linearly independent singlet functions and there are several ways to choose two orthonormal functions. Here is one way:

$$^1_A\Phi_{ij}^{ab} = \frac{1}{2} \left(||ij\bar{a}\bar{b}|| + ||\bar{i}\bar{j}ab|| - ||\bar{i}ja\bar{b}|| - ||i\bar{j}\bar{a}b|| \right)$$

and

$$^1_B\Phi_{ij}^{ab} = \frac{1}{2\sqrt{3}} \left[2 \left(||ij\bar{a}\bar{b}|| + ||\bar{i}\bar{j}ab|| \right) \right.$$
$$\left. - \left(||i\bar{j}a\bar{b}|| + ||\bar{i}j\bar{a}b|| + ||\bar{i}ja\bar{b}|| + ||i\bar{j}\bar{a}b|| \right) \right]$$

where the labels A and B are simply to distinguish the two singlets.

Notice that the superscripts on the Φs do not strictly correspond to the actual allocation of spin factors in the determinants; that is, for example $^1_B\Phi_{ij}^{ab}$ does not have MO ϕ_j with β spin in every term but the notation (which is more appropriate for single determinants) has been retained as a reminder of the *number* of α and β spin factors in each term.

All of these singlets are orthonormal and the evaluation of the matrix elements H_{KL} is a simple exercise in the use of Slater's rules of Appendix 2.A. The results of this exercise are collected in Appendix 20.B to this chapter. Naturally, the *triplets* arising from double excitations may be generated and treated in a completely analogous way.

20.6 More general CI

With the small exception of the use of the partitioned, iterative method for solving the linear variation problem, what we have described so far is the most straightforward method of performing CI calculations; what might be called "aboriginal CI", setting up the linear variation problem and solving it. This method is only realistic for expansions of small or moderate length; for the improvement of the HF wavefunction and energy by the addition of a large part of the correlation energy. These methods cannot be used to perform calculations of great accuracy where the maximum amount of correlation energy is required.

The techniques that we have outlined above very quickly become completely inadequate for CI expansions involving multiple excitations using a quantitatively satisfactory basis for even quite small molecules by chemical standards (20 electrons, say). There are three *kinds* of problem:

- Generating the form of the multiply excited functions in a standard and unique way.
- Transforming the basis-function (repulsion) integrals to the orthonormal MO basis.
- Diagonalising the resulting huge matrices.

These problems have been the object of a huge amount of research over the past several years and each one has been solved by one or more specialised techniques:

- The method of second quantization is ideally suited to the manipulation of determinants of orthonormal orbitals, and the definition of "excitation" operators composed of products of "creation" and "annihilation" operators has proved extremely productive in this field. The group properties of these operators has generated the "unitary group" approach to large-scale CI. The CI method can be thought of as the natural end of the algebraic approximation; unlike the HF model, there is no differential equation to which the CI method is an approximation. In the CI approach the concept of "electron" is really replaced by the concept of "occupied orbital" and, in the case of orthogonal orbitals, second quantization is just a set of techniques for manipulating the algebra of occupied and unoccupied orbitals.[10]

- "Direct" or "MO-integral-driven" techniques are used to overcome problems of storage of very large matrices.

---
[10] The "field operators" of the more basic quantum field theories which involve creation and annihilation of an *electron* rather than of an *occupied orbital* are not used in the CI method.

- A variety of specialised techniques are now used to obtain the lowest root of very large and sparse matrices. It is not necessary ever to construct the whole of the CI matrix.

These methods are a whole study in themselves, even presenting the formal tools required for their efficient development would require far too much space and the implementation is similarly specialised.

However, the general "feel" of modern CI methods may be given by a discussion of one of the main techniques now in use for the diagonalisation of large matrices.

20.7 Nesbet's method for large matrices

In an admirably compact note in 1965[11] Nesbet used a very simple idea to generate a powerful method for the calculation of the individual eigenvalues and eigenvectors of very large matrices, which is particularly well-behaved if the lowest root is required. The method is iterative and uses the following technique:

1. Estimate the value of an eigenvalue/eigenvector pair; typically for the CI case this will be the energy of the "root function"; the single-determinant HF function and an eigenvector consisting of unity in the first position and zeroes everywhere else.

2. Form the set of linear equations resulting from the multiplication of *one* row of the CI Hamiltonian matrix and the current approximate CI eigenvector.

3. Solve these equations exactly using the current (approximate) eigenvalue to obtain a new set of approximate CI coefficients (a new eigenvector).

4. Use the simplest (second-order) perturbation theory to update the variational expression for the Energy (numerator and normalisation denominator).

5. Repeat this procedure for *all* rows of the CI Hamiltonian matrix.

6. Repeat the *whole* procedure (from 2) until the CI eigenvector coefficients have converged to some pre-set tolerance.

---

[11] R. K. Nesbet *J. Chem. Phys.*, **43**, 311 (1965).

20.7. NESBET'S METHOD FOR LARGE MATRICES

This methods works extremely well, the mathematical derivation is straightforward and a fragment of code to perform the calculation is shorter than its description.

If we attack the full eigenvalue problem with overlap

$$\boldsymbol{Hc} = E\boldsymbol{Sc} \tag{20.6}$$

for a single eigenvalue/eigenvector pair, a single one of the N (say) secular equations which compose this matrix equation is

$$\sum_{i=1}^{N} H_{\mu i} c_i = E \sum_{i=1}^{N} S_{\mu i} c_i$$

for $\mu = 1, 2, \ldots N$ or

$$\sum_{i=1}^{N} H_{\mu i} c_i - E \sum_{i=1}^{N} S_{\mu i} c_i = 0 \tag{20.7}$$

for each μ. Now suppose that E and the coefficients c_i are *not* the *solution* of eqn (20.6), but are *given* numbers (obviously in the application here an *approximation* to the eigenvalue/eigenvector pair). Since all the quantities in eqn (20.7), the $H_{\mu i}$ and the $S_{\mu i}$, the c_i and E are now known, the equation will not hold *unless* E and the c_i happen to be the solution.

So, if we pick just one of the coefficients (c_μ, say) and allow it to be corrected (by Δc_μ, say) we can actually restore the consistency of eqn (20.7). Changing c_μ to $c_\mu + \Delta c_\mu$ in eqn (20.7) we have

$$\sum_{i=1}^{N} H_{\mu i} c_i - E \sum_{i=1}^{N} S_{\mu i} c_i + (H_{\mu\mu} - E S_{\mu\mu})\Delta = 0$$

or

$$\Delta c_\mu = \frac{\sigma_\mu}{(E S_{\mu\mu} - H_{\mu\mu})} \tag{20.8}$$

where

$$\sigma_\mu = \sum_{i=1}^{N} H_{\mu i} c_i - E \sum_{i=1}^{N} S_{\mu i} c_i$$

which gives an exact solution of eqn (20.7) containing a (presumably) improved value of the μth coefficient $c_\mu + \Delta c_\mu$.

Since the energy associated with any given approximate eigenvector is (to second order)

$$E = \frac{\sum_{i,j=1}^{N} H_{ij} c_i c_j}{\sum_{i,j=1}^{N} S_{ij} c_i c_j} = \frac{N}{D}$$

say, we have at this level of approximation

$$\Delta E = \frac{\sigma_\mu \Delta c_\mu}{D + \Delta D} \quad (20.9)$$

$$\Delta D = \left(2\sum_{i=1}^{N} S_{\mu i} c_i + S_{\mu\mu}\Delta c_\mu\right)\Delta c_\mu$$

This gives an improved estimate of the μth coefficient and a corresponding improved estimate of the eigenvalue E. The process can then be repeated for each $\mu = 1, \ldots N$ and the whole process repeated until satisfactory convergence of the coefficients c_μ is achieved.

In practice this method "converges for all eigenvalues of some nontrivial matrices and at least the lowest eigenvalue of all matrices tried"[12] and is particularly suitable for CI calculations where one normally has a very good approximation to the lowest eigenvalue/eigenvector pair namely the energy of the reference function and the vector

$$(1, 0, 0, 0, \ldots 0)$$

Of course, in the case of CI calculations the overlap matrix between the configuration functions is the unit matrix

$$S_{\mu i} = \delta_{\mu i}$$

and the calculation is a little simpler:

$$\sigma_\mu = \sum_{i=1}^{N} H_{\mu i} c_i - E c_\mu$$

$$\Delta E = \frac{\sigma_\mu \Delta c_\mu}{D + \Delta D}$$

$$\Delta D = (2c_\mu + \Delta c_\mu)\Delta c_\mu$$

In this important case the implementation is very simple and requires no comment:

---

[12] In Nesbet's modest phrase.

20.7. NESBET'S METHOD FOR LARGE MATRICES

```
    c(1) = ONE;
    D = ONE;
    E = HCI(1,1);
    iter = 0;
    nlow = 2;
    repeat
     {
      kount = 0;
      do mu = nlow,n;
        {
         sigma = ZERO;
         do i = 1, n;
            {
             sigma = sigma + HCI(mu,i)*c(i);
            }
         sigma = sigma - E*c(mu);
         delC = sigma/(E-HCI(mu,mu));
         delD = (TWO*c(mu) + delC)*delC;
         delE = sigma*delC/(D + delD);
         E = E + delE;
         D = D + delD;
         c(mu) = c(mu) + delC;
         if ( abs(delC) > CRIT) kount = kount + 1
        }
      write(6,*) " Iteration ",iter, " Energy = ", E;
      iter = iter + 1;
      nlow = 1;
     }
    until ( (kount == 0) | (iter == 20))
```

558 CHAPTER 20. LINEAR MULTI-DETERMINANT METHODS

This piece of code, as it stands, requires the presence in main random-access memory of the full CI matrix (HCI) and space for one eigenvector (c). Now, as we have seen above the algorithm only actually requires *one* row of HCI to be available at one time, so that, strictly speaking, we could get away with allocating storage for one row of HCI and reading one row at a time from disk storage. This is, indeed one strategy for the use of the algorithm, but the convergence of CI expansions in general is so poor that very large *vectors* (of the order of millions, at least) are routinely used. This would mean that either the storage and manipulation of just one row of HCI would be a large demand on storage or that the whole algorithm would become I/O bound with transfers between disk and memory or both.

In fact, although Nesbet did not make this point in his paper, the solution to this type of problem can be extracted from the way in which the matrix he used in the paper as an example was formed. In 1965, the largest matrix which could be stored (at IBM's Research Centre) was 250 × 250 and, rather than supply the elements of such a huge matrix for verification of the algorithm in the paper, the matrix was specified by a *rule* for generating the elements.

- The off-diagonal elements were all unity.
- The diagonals were the (positive) odd numbers starting at HCI(1,1) = 1.

Here is the algorithm. modified to calculate the lowest eigenvalue/eigenvector pair of the test matrix:

20.7. NESBET'S METHOD FOR LARGE MATRICES

```
# Initialize the eigenvector, the eigenvalue and the Normalisation
  n = 250;
  c(1) = ONE;   D = ONE;   E = ONE;
# HCI(1,1) is 2*1-1 = 1 for the test matrix
  iter = 0;   nlow = 2;
# now use the algorithm
  repeat
    {
    kount = 0;
    do mu = nlow,n;
      {
      Hmumu = 2*mu -1;
      sigma = ZERO;
      do i = 1, n;
        {
# Form the elements of the matrix from the rule
        if ( mu == i) Hmui = 2*i-1;
        else Hmui = ONE;

        sigma = sigma + Hmui*c(i);
        }
      sigma = sigma - E*c(mu);
      delC = sigma/(E-Hmumu);
      delD = (TWO*c(mu) + delC)*delC;
      delE = sigma*delC/(D + delD);
      E = E + delE;
      D = D + delD;
      c(mu) = c(mu) + delC;
      if ( abs(delC) > CRIT) kount = kount + 1
      }
    write(6,*) " Iteration ",iter, " Energy = ", E;
    iter = iter + 1;
    nlow = 1;
    }
  until ( (kount == 0) | (iter == 20))
```

This is *exactly* analogous to the situation in CI calculations:

> The CI matrix is, typically, sparse and the non-zero elements are formed from two or three simple *rules*. Thus, Nesbet's algorithm can be implemented *without ever explicitly forming the CI matrix*. Each individual element is formed as required from the one- and two-electron integrals and Slater's rules or some equivalent.

20.8 "Direct" CI

Of course, in using the idea of generating the CI matrix elements from their defining rules we shall immediately come across the "integral access problem"; each (MO) electron integral will appear many times in many elements of the CI matrix, just as each (AO) electron-repulsion integral appears many times in many elements of $G(R)$ in an SCF calculation. We solve this problem in exactly the same way:

> The rules for forming the required matrix elements from the various energy integrals are "inverted"; that is formulated so that we answer the question
> "I have a particular (MO) integral to hand, where does it appear in the target matrix being formed?"
> Rather than answering the question
> "I need to form a particular element of the target matrix, where are all the integrals I need?"

In the case of the CI matrix it is really the elements of σ which are formed from the one- and two-electron integrals and the current elements of the CI vector c, the CI matrix is never explicitly formed, one forms the product

$$\boldsymbol{\sigma} = \boldsymbol{Hc}$$
$$\sigma_\mu = \sum_{\nu=1}^{N} H_{\mu\nu} c_\nu$$

and the elements of the CI matrix are just linear combinations of the one- and two-electron integrals:

$$H_{\mu\nu} = \sum_{i,j} C_{ij}^{\mu\nu} h_{ij}^F + \sum_{i,j,k,\ell} A_{ijk\ell}^{\mu\nu}(ij,k\ell)$$

$$\sigma_\mu = \sum_{\nu=1}^{N} \left(\sum_{i,j=1}^{m} C_{ij}^{\mu\nu} h_{ij}^F + \sum_{i,j,k,\ell=1}^{m} A_{ijk\ell}^{\mu\nu}(ij,k\ell) \right) c_\nu$$

20.8. "DIRECT" CI

for some choice of coefficients $C_{ij}^{\mu\nu}$, $A_{ijk\ell}^{\mu\nu}$ and a CI expansion of length N and m MOs. The vector $\boldsymbol{\sigma}$ may be formed *directly*[13] from the MO integrals. The key points which make this approach feasible are:

- The Nesbet algorithm (or one of a number of similar methods)
- The fact that the vast majority of the C and A coefficients ("coupling coefficients" as they are called) are zero; the CI matrix is sparse and each element involves only a few non-zero As.

The coupling coefficients for a particular choice of CI expansion are obviously fixed by the expressions for the CI matrix elements. For example the expressions in Appendix 20.B contain all the information to generate the coupling coefficients for Doubly-excited CI for singlet states generated from a closed-shell determinant. Equally obviously, the expressions of Slater's Rules in Appendix 2.A imply the coupling coefficients for CI using determinants.

Even a casual examination of these two sets of formulae from the two appendices shows some important differences between (for example) the two cases for a DCI expansion with a closed shell single determinant as starting function:

- There are *more* determinants than there are singlet functions. In other words some part of the *determinant* DCI will be redundant in the sense that the coefficients generated by the variational process could have been anticipated by algebraic spin eigenfunction considerations.
- The single-determinant expressions of Appendix 2.A are a good deal simpler in form than the corresponding expressions for the spin-eigenfunction case of Appendix 20.B. In fact, for determinants, all the non-zero $C_{ij}^{\mu\nu}$, $A_{ijk\ell}^{\mu\nu}$ have the value ± 1 while the spin-eigenfunction values contain characteristic surds from the normalisations of linear combinations of determinants.

Now, there is a decision to take. Typically, the determinant CI expansion is about three times longer than the spin-eigenfunction expansion for equivalent wavefunctions, but the complexity of the implementation for the determinant expansion is considerably less than for the spin-eigenfunction case. There is a good deal of evidence now available to suggest that the *simplicity* of the determinant expansion is the deciding factor.[14]

---

[13] It is, perhaps, not out of place to note here that the term "direct" is used rather loosely in quantum chemistry implementations. What it tends to mean is just "quicker or more convenient than before". Thus, as we have seen in an SCF context "direct" means "computing the repulsion integrals as they are needed" while in the current CI context "direct" means "forming from a stored list of integrals". Thus, the traditional SCF procedure is "direct" in the CI sense.

[14] N. C. Handy (see, for example, *Chem. Phys. Lett.*, **74**, 280 (1980)) has been most active in pursuing this point.

20.8.1 A small example; paired excitations

In Chapter 22 we outline a method of obtaining the optimum MOs and expansion coefficients for a particularly simple DCI expansion; the one with which we started this chapter the set of all determinants of doubly occupied MOs. This small model (all paired excitations) provides a working example for the techniques discussed here without the tedium and complexity of the full direct DCI case.

The approximate wavefunction is

$$\begin{aligned}\Psi &= D_0\Phi_0 + \sum_{K=1}^{n(m-n)} D_K\Phi_K \\ &= D_0\Phi_0 + \sum_{i=1}^{n}\sum_{a=n+1}^{m} D_{i\bar{i}}^{a\bar{a}}\Phi_{i\bar{i}}^{a\bar{a}} \\ &= D_0\Phi_0 + \sum_{i=1}^{n}\sum_{a=n+1}^{m} D_i^a\Phi_i^a\end{aligned}$$

(say) where one set of subscripts has been omitted in this (temporary) notation for simplicity. Thus each determinant Φ_i^a has an orbital which is doubly occupied in Φ_0 (ψ_i) replaced by a doubly occupied virtual orbital (ψ_a). There are n doubly occupied MOs in Φ_0 and m MOs in total (occupied plus virtual).

The diagonal terms in the CI matrix are all of a similar form being the energy of a set of doubly occupied MOs. They are most conveniently expressed as the energy of Φ_0 minus the energy of the removed MO plus the energy of the added MO plus the repulsion "corrections":

$$E_i^a = E_0 + 2\left(\epsilon_a - \epsilon_i - 2(\psi_a\psi_a, \psi_i\psi_i) + (\psi_a\psi_i, \psi_a\psi_i)\right) + (\psi_a\psi_a, \psi_a\psi_a) + (\psi_i\psi_i, \psi_i\psi_i)$$

The off-diagonal terms are all of the same type, having two orbital differences in every case:

$$\int \Phi_0\hat{H}\Phi_i^a dV = \int \Phi_i^a\hat{H}\Phi_0 dV = (\psi_i\psi_a, \psi_i\psi_a)$$

$$\int \Phi_{i'}^a\hat{H}\Phi_i^a dV = \int \Phi_i^a\hat{H}\Phi_{i'}^a dV = (\psi_{i'}\psi_i, \psi_{i'}\psi_i)$$

$$\int \Phi_i^{a'}\hat{H}\Phi_i^a dV = \int \Phi_i^a\hat{H}\Phi_i^{a'} dV = (\psi_{a'}\psi_a, \psi_{a'}\psi_a)$$

The coupling constants here all having the values ± 1, and there are very few of them. They will not appear in the implementation since they all involve multiplication by unity, only the *logic* will appear.

The mapping from (i, a) to K is also very simple:

$$K = (i-1) \times (m-n) + (a-n) + 1$$

20.8. "DIRECT" CI

gives a unique association between the paired excitations and the first $n(m-n)+1$ integers.

If we have to hand the MO orbital energies (the diagonal representation of the Hartree–Fock matrix in the MO basis) and a stored list of MO repulsion integrals,[15] the direct solution of this CI problem is straightforward. Here is part of an implementation which depends on the repulsion integrals being stored with the usual labelling convention; the contributions to the σ vector from the exchange integrals which arise from the *off-diagonal* elements of the CI matrix are of three types:

- $(\psi_i\psi_j, \psi_i\psi_j)$, both MOs ψ_i and ψ_j are occupied in Φ_0 then:

$$\sigma(K) = \sigma(K) + (\psi_i\psi_j, \psi_i\psi_j) \times c(L)$$
$$\sigma(L) = \sigma(L) + (\psi_i\psi_j, \psi_i\psi_j) \times c(K)$$

since

$$H_{KL} = H_{LK} = (\psi_i\psi_j, \psi_i\psi_j)$$

where

$$K = (i-1) \times (m-n) + (a-n) + 1$$
$$L = (j-1) \times (m-n) + (a-n) + 1$$

for all values of the "unoccupied MO" ψ_a.

- Similarly, the integral $(\psi_a\psi_b, \psi_a\psi_b)$ will appear in the CI matrix H in all positions H_{KL} where

$$K = (i-1) \times (m-n) + (a-n) + 1$$
$$L = (i-1) \times (m-n) + (b-n) + 1$$

for all ψ_i i.e. this integral, multiplied by c_K or by c_L will make a contribution to the positions σ_L and σ_K.

- The exchange integrals involving one "occupied" MO (ψ_i) and one "virtual" MO (ψ_a) make contributions to just one element of the vector σ with

$$K = (i-1) \times (m-n) + (a-n) + 1$$
$$\sigma(K) = \sigma(L) + (\psi_i\psi_j, \psi_i\psi_j) \times c(K)$$

---

[15] As we can see from the matrix element expressions, only the "Coulomb" and "exchange" integrals are actually needed so that a full transformation would not be done for this case. But we are using this simplified case to *illustrate* the general case.

The contributions to $\boldsymbol{\sigma}$ from the diagonal elements of the CI matrix are similarly easily formed from the Coulomb and exchange integrals and the MO orbital energies, so that it is particularly easy to see how the whole calculation of $\boldsymbol{\sigma}$ might be accomplished along the lines:

```
rewind nfile; pointer = 0;

while (getint(nfile,i,j,k,l,mu,val,pointer) != END_OF_FILE
  {
  if ( ( i == j ) && ( k == l ) )
      {
       process the Coulomb integral contributions to
         sigma
      }
  if ( ( i == k ) && ( j == l ) )
      {
       process the exchange integral contributions to
         sigma
      }
  }
Complete the diagonals using the MO energies
Use the Nesbet algorithm to use the current sigma to
generate a new approximate set of CI coefficients.
```

20.8.2 A perturbation/CI method

With the availability of implementations of the direct CI method it has become relatively routine to perform CI calculations with huge expansion lengths; millions or even billions of determinants. Calculations of this type are scarcely capable of physical, much less chemical, interpretation. They are performed in order to obtain very accurate molecular energies and do enable the effect of electron correlation to be studied quantitatively. With *very* long expansion lengths, which necessarily contain terms with *very* small contributions to the energy, it becomes imperative to be able to choose which determinants to include in the expansion, even which *types* of determinant to include. Perturbation theory provides an excellent tool for this choice and an extraordinarily effective method for the calculation of molecular energies to very high accuracy can be summarised as follows:

1. Starting with some reference function (perhaps the HF determinant) use perturbation theory to estimate the coefficients in the CI expansion.

2. Using some criterion on these approximate coefficients, set up and solve the direct CI problem.

3. Lower the criterion used to include excited functions.

4. Repeat from 2 until further lengthening of the expansion has no effect (obviously to some energy criterion).

The use of this technique has led to the largest-ever CI calculation:[16] over 10^{15} determinants.

20.9 Conclusions

The method of configuration interaction can be made to give the most accurate wavefunctions ever calculated for many-electron systems. The results are strictly variational and therefore provide true upper bounds to exact energies.[17]

However, the expansion of the wavefunction in terms of billions of determinants is of little conceptual or interpretational value. There are two possible views on this latter fact:

- Why *should* the mutual interactions and motions of a large number of strongly interacting particles be susceptible to interpretation by some simple conceptual scheme? This might be termed the "mechanical view".

- Notwithstanding the mechanical view, it simply is the case that the distributions and energies of systems of dozens, even hundreds of electrons *are* capable of being brought into a relatively simple conceptual scheme involving the environment-insensitive properties of electronic sub-structures. This is the "chemical view"

The creative tension between these two general philosophies has been part of what has driven quantum chemistry forward over the past decades. The question for the mechanical view is "can these very accurate calculations be given an interpretation?" and for the chemical view "can the conceptual model be made to yield accurate numbers?"

---

[16] To date, late 1997.
[17] Given the usual Born–Oppenheimer, electrostatic Hamiltonian.

Appendix 20.A

The "orthogonal VB" model

The simple Heitler–London valence bond model for an electron-pair bond formed by the combination of two (hybrid atomic) orbitals has the wavefunction

$$\Psi(\vec{x}_1, \vec{x}_2) = N\left[\phi_1(\vec{r}_1)\phi_1(\vec{r}_1) + \phi_1(\vec{r}_1)\phi_1(\vec{r}_1)\right] \times \left[\alpha(s_1)\beta(s_2) - \beta(s_2)\alpha(s_1)\right]$$
(20.A.10)

where the normalisation constant N is given by

$$N = \frac{1}{\sqrt{2(1+S^2)}} \times \frac{1}{\sqrt{2}}$$

and S is the overlap integral between the two orbitals.

The energy expression for this wavefunction, using the Hamiltonian

$$\hat{H} = \hat{h}(\vec{r}_1) + \hat{h}(\vec{r}_2) + \frac{1}{r_{12}}$$

is

$$E = \frac{1}{1+S^2}\left[h_{11} + h_{22} + 2Sh_{12} + (22,11) + (21,21)\right] + V_R$$
(20.A.11)

The integrals h_{ij} are

$$h_{ij} = \int \phi_i \hat{h} \phi_j dV,$$

the repulsion between the parent atoms[18] at an inter-nuclear distance R is V_R (say) and the repulsion integrals are given their usual notation.

In this expression:

---

[18] The residual atom not counting the electron involved in the bond pair.

- The quantities V_R, $(22,11)$ and $(21,21)$ are *necessarily positive*.

- The diagonal one-electron terms h_{ii} are the energies of the orbitals in their "parent" atoms plus a (negative) term due to the attraction of an electron in ϕ_1 by atom 2 and vice versa.

- The one-electron integral h_{12} is negative representing the attraction between a charge distribution $\phi_1 \phi_2$ and the two atoms plus a small positive kinetic energy component.

Now, if the orbitals ϕ_1 and ϕ_2 are orthogonal $S = 0$ and the largest negative contribution to the *binding* energy vanishes and the energy expression eqn (20.A.11) becomes:

$$E = h_{11} + h_{22} + (22,11) + (21,21) + V_R$$

(retaining the same notation for the orthogonal ϕ_i). We may write

$$h_{ii} = E_i + V_i$$

where E_i is the energy of ϕ_i on its own atom and V_i is the (negative) interaction between an electron in ϕ_i and the other atom. Thus the energy for a pair of orthogonal orbitals becomes

$$E = E_1 + E_2 + (V_1 + V_2) + [(22,11) + 21,21) + V_R]$$

If this expression is to describe a bonding situation then

$$E < E_1 + E_2$$

but only the small terms V_1 and V_2 *are negative*. The three terms in the final bracket are all *positive*. In practice it is found that these positive terms dominate and the use of the HL model with orthogonal orbitals does not lead to a bound system.

If the orbitals are not orthogonal then the large negative term $2Sh_{12}$ is retained and the HL model does generate a bound system whose energy is lower than that of the separate atoms.

The two terms $2Sh_{12}$ and $(21,21)$ in eqn (20.A.11) both arise from the exchange of electrons to ensure electron antisymmetry and, in the context of VB theory, both are sometimes grouped together as the "exchange term" notwithstanding the more familiar (MO) use of "exchange integral" for $(21,21)$ alone.

Appendix 20.B

DCI matrix elements

Here is a collection of the matrix elements

$$H_{KL} = \int {}^1\Phi_K(\hat{H} - E_0)\, {}^1\Phi_L d\tau = <K|H|L> \quad \text{(say)}$$

for the doubly excited singlet functions of Chapter 20. K and L here are simply dummy subscripts, the full notation for each singlet is given below. The relevant functions Φ_K are:

$${}^1\Phi_{i\bar{i}}^{a\bar{a}} = \Phi_{i\bar{i}}^{a\bar{a}} = ||i\bar{i}a\bar{a}||$$

$${}^1\Phi_{i\bar{i}}^{a\bar{b}} = \frac{1}{\sqrt{2}}\left(||i\bar{i}a\bar{b}|| - ||i\bar{i}\bar{a}b||\right)$$

$${}^1\Phi_{i\bar{j}}^{a\bar{a}} = \frac{1}{\sqrt{2}}\left(||i\bar{j}a\bar{a}|| - ||\bar{i}ja\bar{a}||\right)$$

$${}^1_A\Phi_{i\bar{j}}^{a\bar{b}} = \frac{1}{2}\left(||i\bar{j}a\bar{b}|| + ||\bar{i}j\bar{a}b|| - ||\bar{i}ja\bar{b}|| - ||i\bar{j}\bar{a}b||\right)$$

$${}^1_B\Phi_{i\bar{j}}^{a\bar{b}} = \frac{1}{2\sqrt{3}}\left[2\left(||ij\bar{a}\bar{b}|| + ||\bar{i}\bar{j}ab||\right) - \left(||i\bar{j}a\bar{b}|| + ||\bar{i}j\bar{a}b|| + ||\bar{i}ja\bar{b}|| + ||i\bar{j}\bar{a}b||\right)\right]$$

defined in Chapter 20.

Obviously, the notation for each singlet function contains the symbol ${}^1\Phi$, which is omitted from the expressions below; i.e. ${}^1_B\Phi_{ij}^{a\bar{b}}$ is written simply as $|i\bar{j}a\bar{b}(B)>$.

Using this contracted notation the results for the distinct DCI matrix elements are:

$$<i\bar{i}a\bar{a}|H|k\bar{k}c\bar{c}> = 2\delta_{ik}\delta_{ac}\left[h^F_{aa} - h^F_{ii} - 2(ii,aa) + (ia,ai)\right]$$
$$+\ \delta_{ik}(ac,ac) + \delta_{ac}(ki,ki)$$

$$<i\bar{i}a\bar{a}|H|k\bar{\ell}c\bar{c}> = \sqrt{2}\delta_{ac}\{\delta_{ik}\left[h^F_{\ell i} + 2(\ell i,aa) - (\ell a,ai)\right]$$
$$-\ (ki,\ell i)\}$$

$$<i\bar{i}a\bar{a}|\hat{H}|k\bar{k}c\bar{d}> = \sqrt{2}\delta_{ik}\{\delta_{ac}\left[h^F_{ad} - 2(ii,ad) + (id,ai)\right]$$
$$+\ (ac,ad)\}$$

$$<i\bar{i}a\bar{a}|\hat{H}|k\bar{\ell}c\bar{d}(1)> = \delta_{ik}\delta_{ac}\left[2(\ell i,ad) - (\ell d,ai)\right]$$

$$<i\bar{i}a\bar{a}|\hat{H}|k\bar{\ell}c\bar{d}(2)> = -\sqrt{3}\delta_{ik}\delta_{ac}(\ell d,ai)$$

$$<i\bar{j}a\bar{a}|\hat{H}|k\bar{\ell}c\bar{c}> = 2\delta_{ik}\delta_{j\ell}\delta_{ac}h^F_{aa}$$
$$-\ \delta_{ik}\delta_{ac}\left[h^F_{\ell j} + 2(\ell j,aa) - (\ell a,aj)\right]$$
$$-\ \delta_{j\ell}\delta_{ac}\left[h^F_{ki} + 2(ki,aa) - (ka,ai)\right]$$
$$+\ \delta_{ik}\delta_{j\ell}(ac,ac) + \delta_{ac}\left[(ki,\ell j) + (kj,\ell i)\right]$$

$$<i\bar{j}a\bar{a}|\hat{H}|k\bar{\ell}c\bar{d}> = \delta_{ik}\delta_{ac}\left[2(ij,ad) - (id,aj)\right]$$

$$<i\bar{j}a\bar{a}|\hat{H}|k\bar{\ell}c\bar{d}(1)> = \frac{1}{\sqrt{2}}\{2\delta_{ik}\delta_{j\ell}\delta_{ac}h^F_{ad}$$
$$-\ \delta_{ik}\delta_{ac}\left[2(\ell l,ad) - (\ell d,al)\right]$$
$$-\ \delta_{j\ell}\delta_{ac}\left[2(ki,ad) - (kd,ai)\right]$$
$$+\ 2\delta_{ik}\delta_{j\ell}(ac,ad)\}$$

$$<i\bar{j}a\bar{a}|\hat{H}|k\bar{\ell}c\bar{d}(2)> = \sqrt{\frac{3}{2}}\delta_{ac}\left[\delta_{ik}(\ell d,al) - \delta_{j\ell}(kd,ai)\right]$$

$$<i\bar{i}a\bar{b}|\hat{H}|k\bar{k}c\bar{d}> = -2\delta_{ik}\delta_{ac}\delta_{bd}h^F_{ii}$$
$$+\ \delta_{ik}\delta_{ac}\left[h^F_{bd} - 2(ii,bd) + (id,bi)\right]$$
$$+\ \delta_{ik}\delta_{bd}\left[h^F_{ac} - 2(ii,ac) + (ic,ai)\right]$$
$$+\ \delta_{ac}\delta_{bd}(ki,ki) + \delta_{ik}\left[(ac,bd) + (ad,bc)\right]$$

$$<i\bar{i}a\bar{b}|\hat{H}|k\bar{\ell}c\bar{d}(1)> = \frac{1}{\sqrt{2}}\{2\delta_{ik}\delta_{ac}\delta_{bd}h^F_{\ell i} + \delta_{ik}\delta_{ac}\left[2(\ell i,bd) - (\ell d,bi)\right]$$
$$+\ \delta_{ik}\delta_{bd}\left[2(\ell i,ac) - (\ell c,ai)\right] - 2\delta_{ac}\delta_{bd}(ki,\ell i)\}$$

$$<i\bar{i}a\bar{b}|\hat{H}|k\bar{\ell}c\bar{d}(2)> = -\sqrt{\frac{3}{2}}\delta_{ik}\left[\delta_{ac}(\ell d,bi) - \delta_{bd}(\ell c,ai)\right]$$

APPENDIX 20.B. DCI MATRIX ELEMENTS

$$
\begin{aligned}
<i\bar{j}a\bar{b}(1)|\hat{H}|k\bar{\ell}c\bar{d}(1)> \;=\;& \delta_{ik}\delta_{j\ell}\left[\delta_{ac}h^F_{bd}+\delta_{bd}h^F_{ac}\right] \\
-\;& \delta_{ac}\delta_{bd}\left[\delta_{ik}h^F_{\ell j}+\delta_{j\ell}h^F_{ki}\right] \\
-\;& \delta_{ik}\left\{\delta_{ac}\left[(\ell j,bd)-(1/2)(\ell d,bj)\right]\right. \\
+\;& \left.\delta_{bd}\left[(\ell j,ac)-(1/2)(\ell c,aj)\right]\right\} \\
-\;& \delta_{j\ell}\left\{\delta_{ac}\left[(ki,bd)-(1/2)(kd,bi)\right]\right. \\
+\;& \left.\delta_{bd}\left[(ki,ac)-(1/2)(kc,ai)\right]\right\} \\
+\;& \delta_{ik}\delta_{j\ell}\left[(ac,bd)+(ad,bc)\right] \\
+\;& \delta_{ac}\delta_{bd}\left[(ki,\ell j)+(kj,\ell i)\right] \\
<i\bar{j}a\bar{b}(1)|\hat{H}|k\bar{\ell}c\bar{d}(2)> \;=\;& \sqrt{\tfrac{3}{2}}\,[\delta_{ik}\delta_{ac}(\ell d,bj)-\delta_{ik}\delta_{bd}(\ell c,al) \\
-\;& \delta_{j\ell}\delta_{ac}(kd,bi)+\delta_{j\ell}\delta_{bd}(kc,ai)] \\
<i\bar{j}a\bar{b}(2)|\hat{H}|k\bar{\ell}c\bar{d}(2)> \;=\;& <i\bar{j}a\bar{b}(1)|\hat{H}|k\bar{\ell}c\bar{d}(1)> +\delta_{ik}\delta_{ac}(\ell d,bj) \\
+\;& \delta_{ik}\delta_{bd}(\ell c,aj)+\delta_{j\ell}\delta_{ac}(kd,bi)+\delta_{j\ell}\delta_{bd}(kc,ai) \\
-\;& 2\,[\delta_{ik}\delta_{j\ell}(ad,bc) \\
+\;& \delta_{ac}\delta_{bd}(kj,\ell i)]
\end{aligned}
$$

Chapter 21

The valence bond model

Although it is possible to obtain good estimates of the electron correlation energy by either density functional or configuration interaction methods both of these methods (for different reasons) suffer from the same defect; it is not possible to obtain a clear physical and chemical interpretation of the results of the calculation. The valence bond model, in principle and in practice, puts the physical interpretation as a top priority but pays a price in the complexity of its implementation.

Contents

21.1	**Non-orthogonality in expansions**	**571**
21.2	**Spins and spin functions**	**573**
21.3	**Spin eigenfunctions and permutations**	**576**
21.3.1	Spin eigenfunctions	577
21.3.2	Permutations of spin variables	579
21.3.3	Spin eigenfunctions in the VB model	580
21.4	**Spin-free VB theory**	**581**
21.4.1	The simplest model	581
21.4.2	Generalisation	583
21.5	**Summary**	**585**

21.1 Non-orthogonality in expansions

The practical strength of the CI method for the computation of electronic structures which include electron correlation depends on the fact that the set of orbitals used to construct the n-electron terms in the expansion of the wavefunction (determinants or symmetry-adapted combinations of determinants, usually) are *orthogonal*. This orthogonality makes the evaluation of the matrix elements of the molecular Hamiltonian between the expansion functions trivial; the vast majority of them are zero and the non-zero ones are a combination of just a few energy integrals over the orthogonal basis functions (occupied and virtual MOs).

The optimum (HF) single determinant (or short linear combination of determinants) "reference function" is completely dominant in these CI expansions; it provides the vast majority of the energy and its expansion coefficient is always close to unity. Thus, the overlap of the reference function and the resulting CI expansion is usually very close to unity.

These facts can be illustrated by using a simple (spatial) vector analogy which throws into sharp focus some shortcomings of the CI method. Let us assume that the true wavefunction is a vector in three-space and the HF determinant (or other reference function) is another vector in three-space. The angle between these two vectors will be very small since they are both normalised to unit length and their mutual overlap is large. The determinants which can be used to correct the reference function are orthogonal to it and, therefore, almost orthogonal to the target wavefunction, having, therefore, very little component in the "direction" of the target function.

In contrast, if we look at the simple VB function for the hydrogen molecule (Heitler–London plus the two equivalent polar structures), these three functions are not orthogonal to each other (or to the MO single determinant). For a typical value of the $1s_A/1s_B$ overlap these two-electron functions overlap by about 0.7 for the Heitler–London/polar terms and about 0.4 between the two polar structures. In our analogy this translates into the expansion of a target vector by three vectors which are at mutual angles of about 45 and 60 degrees; clustering around the direction of the target vector.

In general, this type of situation is the aim of the VB model:

By using a set of overlapping (hybrid) atomic orbitals to construct antisymmetric non-orthogonal functions which are interpretable as chemical structures, the VB model aims to:

- Reduce the length of the linear expansion in terms of chemical structures.
- Generate an expansion which is capable of being given a physical and chemical interpretation while still recovering a large part of the "post Hartree–Fock" (correlation) energy.

Of course, since the VB model is *multi-structure* (that is, equivalently, multi-determinant) *by definition*, correlation energy is not computed separately and, if an explicit evaluation of the correlation energy is required, a HF calculation using the same orbital basis must be done to provide a baseline.

21.2 Spins and spin functions

We have seen in Chapter 1 that, although we commonly say that we are finding approximate solutions to the many-electron Schrödinger equation, we are in fact, finding approximations to the solutions of the two *simultaneous* equations

$$\hat{H}\Psi(\vec{x}_1,\ldots,\vec{x}_n) = E\Psi(\vec{x}_1,\ldots,\vec{x}_n)$$
$$\hat{A}\Psi(\vec{x}_1,\ldots,\vec{x}_n) = \Psi(\vec{x}_1,\ldots,\vec{x}_n) \tag{21.1}$$

Where $\hat{H}$ is the Hamiltonian and $\hat{A}$ is the antisymmetriser.

Since our chosen molecular Hamiltonian is always independent of electron *spin*, and so commutes with any operator which depends only on electron spin, we can without loss of generality add two more equations to our list:

$$\hat{S}^2\Psi(\vec{x}_1,\ldots,\vec{x}_n) = S(S+1)\Psi(\vec{x}_1,\ldots,\vec{x}_n)$$
$$\hat{S}_z\Psi(\vec{x}_1,\ldots,\vec{x}_n) = M\Psi(\vec{x}_1,\ldots,\vec{x}_n) \tag{21.2}$$

Any function of a set of space-spin variables must be an eigenfunction of total electron spin and one (Cartesian) component of total spin.

In fact what we actually do, of course, is to regard all the equations except the Schrödinger equation as *constraints* on that equation and simply "solve" them by a suitable choice of variational form, and variationally optimise the resulting expression *subject to* the trial functions being constrained to be antisymmetric and eigenfunctions of the spin operators.[1] The whole apparatus for dealing with electron spin in the case of a spin-free Hamiltonian has a rather redundant air about it.

However, it is the Pauli principle which prevents us from simply ignoring the existence of electron spin altogether. The trial wavefunction must be antisymmetric with respect to the exchange of the coordinates (space-spin) of any two particles. Without this constraint the solutions of the many-electron Schrödinger equation would be wrong; there are many more solutions of the Schrödinger equation than there are *antisymmetric* solutions of that equation. Electron spin, at this level, simply ensures that the spatial part of the wavefunction behaves properly when the electrons' coordinates are permuted. Thus, notwithstanding the manipulational convenience of the use of spin "functions" it would be attractive to be able to deal explicitly only with a *spatial* trial function and solve a *spatial* variational problem.

However the valence bond model seems inextricably bound up with the idea of electron spin; indeed the chemical bond is often qualitatively described as

---
[1] Note that, in spite of the symmetrical form in which the constraints have been presented here, antisymmetry is a *requirement*, while spin eigenfunctions are *not* required and will emerge from the calculation (or not) when it is performed.

"due to" the pairing of electron spins. What is certainly true is that the largest contributions to a description of electron-pair bonds are functions which have a singlet spin function for each bond; what is not so obvious is that the bonding is *due* to this spin pairing. Ideas of electron spin have become "entangled" with the Pauli principle and antisymmetry.

What is true is that, *for a given type* of spin function, the requirement of antisymmetry *imposes* a particular *form* of permutation "symmetry" on the spatial form of any acceptable wavefunction which is to be variationally optimised, and this *form* must be retained during the variation procedure. That is, for a given form of spin function the *spatial part* of the wavefunction must behave in a predetermined way when the *spatial* coordinates of electrons are permuted.

To see that this is true it is only necessary to write out each full permutation operator in the definition of the antisymmetrising operator explicitly as a "product" of a spatial permutation and a spin permutation:

$$\hat{P}(\vec{x}) = \hat{P}_s(s)\hat{P}_r(\vec{r})$$

to give

$$\hat{A} = \frac{1}{\sqrt{N!}} \sum_{\hat{P}} (-1)^p \hat{P}_s \hat{P}_r \qquad (21.3)$$

Now, taking an arbitrary "primitive" product of a spatial function $\Omega(\vec{r}_1 \ldots \vec{r}_n)$ and a spin function $\Theta(s_1 \ldots s_n)$ and generating its antisymmetric component (Ψ, say) by application of $\hat{A}$ we have

$$\begin{aligned}\Psi(\vec{x}_1 \ldots \vec{x}_n) &= \hat{A}\Omega\Theta \qquad (21.4)\\ &= \frac{1}{\sqrt{N!}} \sum_{\hat{P}} (-1)^p \left(\hat{P}_r \Omega\right)\left(\hat{P}_s \Theta\right)\end{aligned}$$

showing clearly that the antisymmetric part of any space-spin product may be written as a *sum* of products of separate spatial and spin factors.

What is more any integral involving this function and any *spinless* operator (like the usual molecular Hamiltonian) may be written entirely in terms of integrals over the spatial functions since the spin integrations are always trivial.

What makes this result especially valuable in the context of a VB approach to molecular electronic structure[2] is the fact that the expansion eqn (21.4) takes a particularly attractive form if the functions Θ are chosen to be *spin eigenfunctions* which will be a *requirement* of the solutions of any spinless Schrödinger equation or approximation to such a solution.

---
[2] And to the technology of the CI method.

21.2. SPINS AND SPIN FUNCTIONS

The simplest case is familiar enough; the (un-normalised) Heitler–London (covalent) function for the homopolar bond involving two AOs ϕ_1 and ϕ_2 is

$$\Psi = \Psi^r \Theta^s = (\phi_1(\vec{r}_1)\phi_2(\vec{r}_2) + \phi_1(\vec{r}_2)\phi_2(\vec{r}_1)) \times \frac{1}{\sqrt{2}}(\alpha(s_1)\beta(s_2) - \alpha(s_2)\beta(s_1))$$

Here, the space-spin function is a one-term product of a spatial part and a spin factor. The overall function is antisymmetric, the spin factor is antisymmetric with respect to exchange of spin coordinates thus the spatial part must be symmetric with respect to exchange of spatial coordinates. Having used this formalism to ensure that the total wavefunction is antisymmetric, we are finished with spin completely. This result is clearly independent of the *specific* form of the HL trial function and:

> So long as we retain that part of the spatial factor Ψ_r in the wavefunction *which has the correct behaviour with respect to exchange of electronic coordinates*, we can use this spatial factor as the total wavefunction in any energy calculation involving a spinless Hamiltonian.

$$\int \Psi \hat{H} \Psi d\tau = \int \Psi_r \hat{H} \Psi_r dV$$

This is the germ of the whole approach; we use only spatial functions for which we *know* permutations of the *spatial* coordinates will generate an antisymmetric total wavefunction if their partner spin permutations were applied to the associated spin factor in the total wavefunction. The rest is technique: how to carry out this programme.

In using the determinantal method, the explicit use of a spin "coordinate" and spin-orbitals for each electron has provided the simplest mathematically elementary technique for ensuring that this permutational symmetry is obeyed and retained. No new algebraic or analytical tools were required; the definition of spin "integration" is all that is needed. But the price which had to be paid for this *mathematical* simplicity was the entanglement of space and spin variables. It would be physically and chemically attractive to be able to *remove* any explicit appearance of electron spin from the valence bond model and concentrate on the main *physical* features of the model which are:

- Modelling the real electronic structure of a molecule as a linear combination of a series of "chemical structures"[3] each of which has a physical interpretation and makes a strong connection with the well-tried ideas of empirical chemistry.

---

[3]Note that we are using the term "structure" here in two different senses in dangerously close proximity; "electronic structure" is used here as we have used the term elsewhere meaning the energetics and distributions of electrons, on the other hand "chemical structure" is a higher-level term meaning the way in which atoms are joined together in a molecule.

- Ensuring that the mathematical description of these chemical structures is consistent with the requirements of the Pauli principle and generates a chemical structure which has the observed electron-spin multiplicity (overwhelmingly singlet for most molecules).

- Removing the unnecessary emphasis on "spin" and "spin-pairing" from valence bond theory in favour of the idea of bond formation between *spatial* orbitals.

Historically, because of the difficulties of evaluating the various spatial integrals arising in the development of the VB method, emphasis was placed on the "spin" part of the space/spin separation and the VB model was presented as a "spin-Hamiltonian" theory, with the spatial integrals being absorbed into the theory as disposable parameters in much the same way as the theory of electron spin resonance is rationalised as a purely spin theory with the coupling parameters being determined either experimentally or by spatial integrals. That is to say, for entirely practical reasons, the opposite point of view was taken to the one being taken here.

However, this mode of presentation led to the idea that chemical bonding could be interpreted as the actual interaction between electron *spins* with the coupling constants being tens of kilocalories per mole while the actual couplings between electron spins are only thousandths of a kilocalorie (as measured in EPR experiments). The *formal framework* of this presentation was given an unfortunate literal interpretation which tended to discredit the whole VB model. When it became possible to actually evaluate the spatial integrals involved in VB theory, it was then possible to discard the explicit "spin" dependence and concentrate attention on the associated "permutation" dependence of the spatial factors in the VB model.

With the idea of the actual chemical structure of a molecule being a linear combination of idealised chemical structures, our analogy at the start of this chapter becomes a real model, quantifying the idea of the structure of a molecule being a "resonance hybrid" between classical structures. The CI model offers no such attractive visualisation.

21.3 Spin eigenfunctions and permutations

There is a variety of methods for dealing with permutations of coordinates in VB theory; all lead to equivalent results but some are more "visually" attractive than others, and some require a good deal of the formal apparatus of the theory of the so-called symmetric group[4] (the group of all permutations of n identical

---

[4]The theory of the symmetric group would involve us in a considerable detour. Some results will be used but the general theory will not be developed even though it has considerable importance for both the VB and CI methods.

21.3. SPIN EIGENFUNCTIONS AND PERMUTATIONS

objects). At the risk of a complete loss of rigour, a jumble of different techniques will be used to try to give a general flavour of the ideas involved.

21.3.1 Spin eigenfunctions

First, it is essential to obtain some of the properties of the spin functions, spin eigenfunctions and the application of permutations of spin variables to spin functions. Clearly, for n electrons there are 2^n possible spin functions; a product of n individual electron spin functions each of which may be α or β. These simple product functions may by classified by the numbers of α factors and the number of β factors. The n-electron spin operator $\hat{S}_z$ given by

$$\hat{S}_z = \sum_{i=1}^{n} \hat{s}_z(s_i)$$

has these simple products as eigenfunctions. If Θ is a product of n_α factors of type α and n_β (where $n_\beta = n - n_\alpha$) β factors, then

$$\hat{S}_z \Theta = \frac{1}{2}(n_\alpha - n_\beta)\Theta = M\Theta$$

(say) independently of *which* of the n electrons have α or β spin factor.

So there is just one such product with $n_\alpha = n$ with spin eigenvalue $M = n/2$ and just one with spin eigenvalue $M = -n/2$ when all the electrons have β spin. If all electrons but one have α spin there are n ways of realising this situation and so there are n independent spin products with spin eigenvalue $M = n/2 - 1$. Similarly, if two α factors are replaced by βs we get $n(n-1)/2$ products with spin eigenvalue $M = n/2 - 2$ and so on.

In general, a product containing n_α spin α factors and n_β spin β factors has spin eigenvalue

$$M = \frac{1}{2}(n_\alpha - n_\beta)$$

and there are a total of

$$\binom{n}{n_\beta} = \binom{n}{n_\alpha}$$

linearly independent spin products. It is intuitively obvious that all these products must sum to 2^n:

$$\sum_{n_\beta=0}^{n_\beta=n} \binom{n}{n_\beta} = 2^n$$

These simple products, although eigenfunctions of $\hat{S}_z$ by construction are not eigenfunctions of the total spin operator

$$\hat{S}^2 = \sum_{i,j=1}^{n} \hat{s}(s_i) \cdot \hat{s}(s_j) = \sum_{i=1}^{n} \hat{s}^2(s_i) + \sum_{i \neq j=1}^{n} \hat{s}(s_i) \cdot \hat{s}(s_j)$$

but a set of simultaneous eigenfunctions of $\hat{S}^2$ and $\hat{S}_z$ may be formed from linear combinations of the products. For the moment we concentrate on the *number* of members in each set of simultaneous eigenfunctions which may be obtained by elementary methods:

- For a set of n electrons there is only *one* function with $M = n/2$ so that this must be an eigenfunction of $\hat{S}^2$ (there being no scope for linear combinations). This can by verified by using the full form of $\hat{S}^2$.

- There must, therefore, be a further $2M$ functions amongst the 2^n which have the same value of S (the eigenvalue of $\hat{S}^2$ is $S(S+1)$) as this function since, for a given S, there are $2S+1$ functions which differ by their M-value. Of these one must have $M = n/2 - 1$.

- But there are n with this value of M as we have seen above. Thus the remaining $(n-1)$ functions *must* have $S = M = n/2 - 1$, there being no further functions belonging to $M = S$.

- Going one step further into the $n(n-1)/2$ functions with two β factors, n of these are spoken for; one is associated with the $M = n/2 - 2$ component of the set with $S = n/2$ and another with the $M = n/2 - 2$ component of each of the previous $S = n/2 - 1$ functions. This leaves

$$\frac{n(n-1)}{2} - n = \frac{n(n-3)}{2}$$

functions which must have $S = n/2 - 2$.

- This process can be continued until all the functions are exhausted and the *number* of functions within each simultaneous spin-eigenfunction set has been obtained.

Notice that, although language like "one of these functions" has been used, it is obvious that more circumspect terminology should have been used since the functions are *not* already eigenfunctions of $\hat{S}^2$. What should have been said was something like "one of the possible linearly independent combinations of these functions" or "one of the degrees of freedom amongst the current set of functions". But since we have been concerned with the *numbers* of functions rather than the functions themselves the correct result has been obtained since they are linearly independent. In fact, using standard spin-algebra methods the actual functions may be generated in a process which uses exactly the same steps as the outline above.

Extrapolating to the general case, the number of simultaneous eigenfunctions of $\hat{S}^2$ and $\hat{S}_z$ for a system of n electrons (with a given S) is just the *difference*

21.3. SPIN EIGENFUNCTIONS AND PERMUTATIONS

between the numbers of functions for $M = S$ and for $M = S + 1$. This number is always written f_S^n:

$$\begin{aligned} f_S^n &= \binom{n}{n_\alpha} - \binom{n}{n_\alpha + 1} \\ &= \binom{n}{\frac{n}{2} + S} - \binom{n}{\frac{n}{2} + S + 1} \end{aligned} \qquad (21.5)$$

where the binomial coefficients are now written in terms of S rather than n_β or n_α to agree with the notation f_S^n.

21.3.2 Permutations of spin variables

We have seen in eqn (21.4) that the action of the antisymmetrising operator, when factored into space and spin terms, ensures that any many-particle wavefunction generated by antisymmetrising a single primitive product of separate space and spin factors can be written as a *sum* of products of space and spin factors. In general, this sum contains $n!$ terms since there are $n!$ terms in the antisymmetriser $\hat{A}$. This general sum can be reduced by some simple considerations involving spin eigenfunctions.

Both of the spin operators $\hat{S}^2$ and $\hat{S}_z$ are totally symmetric with respect to any exchanges of spin coordinates so that each of them commutes with all the $n!$ permutations of n identical coordinates contained in $\hat{A}$:

$$\begin{aligned} \hat{P}\hat{S}_z f(s_1, s_2 \ldots, s_n) &= \hat{S}_z \hat{P} f(s_1, s_2 \ldots, s_n) \\ \hat{P}\hat{S}^2 f(s_1, s_2 \ldots, s_n) &= \hat{S}^2 \hat{P} f(s_1, s_2 \ldots, s_n) \end{aligned}$$

Since $\hat{S}_z$ and $\hat{S}^2$ commute with each other, they, together with the $n!$ permutation operators, form a set of commuting operators which may have a set of mutual eigenfunctions. So that, if we have a set of simultaneous eigenfunctions only of $\hat{S}_z$ and $\hat{S}^2$, these eigenfunctions are, at most, sent into linear combinations of each other by the operations of the permutation operators. In particular this means that, if the spin factor in the primitive product eqn (21.4) is a *spin eigenfunction* then the action of the antisymmetriser is constrained to generate only other spin eigenfunctions[5] *with the same values* of S and M. That is, the sum in eqn (21.4) contains, at most, only f_S^n (different) terms not the $n!$ which appears in the formal sum. Of course, the spin eigenfunctions generated by the action of the antisymmetriser may:

- Contain repetitions.
- Be non-orthogonal.

---
[5]From now on "spin eigenfunction" means eigenfunction of both $\hat{S}_z$ and $\hat{S}^2$.

However, if the terms are collected so that the sum is expressed as a sum of spatial factors multiplied by distinct orthogonal spin eigenfunctions:

$$\Psi = \sum_{K=1}^{f_S^n} F_K \Theta_K \qquad (21.6)$$

(say),[6] then the mean value of any (spinless) operator like the molecular Hamiltonian is just

$$\int \Psi \hat{H} \Psi d\tau = f_S^n \int dV F_K \hat{H} F_K$$

$$\int \Psi \Psi d\tau = f_S^n \int dV F_K F_K$$

$$E = \frac{\int dV F_K \hat{H} F_K}{\int dV F_K F_K} \qquad (21.7)$$

for *any* of the F_K because of the orthonormality of the spin eigenfunctions and the identity of the f_S^n spatial integrals.

21.3.3 Spin eigenfunctions in the VB model

Several separate strands of material are beginning to come together:

- The fact of a spin-free molecular Hamiltonian really requires trial functions to be spin eigenfunctions.

- The VB method is traditionally associated with the use of spin eigenfunctions.

- Spin eigenfunctions have certain optimum properties from the point of view of permutations and antisymmetrisation.

The sets of spin eigenfunctions with common values of S and M may have their properties changed by suitable linear combinations amongst the set. For example, there are intuitive or formal reasons for choosing particular linear combinations; we may require orthogonality or chemical familiarity for the members of such a set and there are various ways of systematically choosing linear transformations amongst the sets to achieve certain aims.[7]

---

[6]Note that these expansions of the model wavefunction have the *look* of a linear variational expansion, but they cannot be "generalised" by the introduction of variational parameters because all the terms must appear with *fixed weights* to ensure that the wavefunction is *antisymmetric*. They cannot be used with variational coefficients any more than the individual products can in the expansion of a determinant of spin-orbitals.

[7]A very complete treatment of the definition of various types of spin eigenfunction and the relationships amongst the various choices can be found in *Spin eigenfunctions* by R. Pauncz (Plenum, 1979).

21.4 Spin-free VB theory

We have seen in eqn (21.7) that, in order to compute the energy of a wavefunction of the form of eqn (21.6):

$$\Psi = \sum_{K=1}^{f_S^n} F_K \Theta_K$$

we only need to know the explicit form of *one* of the F_K which are obtained as the spatial partners when a total wavefunction Ψ is expanded in terms of a set of spin-eigenfunctions. If we have a method for the construction of a function of the correct spatial permutational symmetry containing variational parameters we have a working VB method with the desirable property of concentrating attention on the real (spatial) nature of the Valence Bond model. What are the optimum hybrid orbitals and in what chemical structures are they involved?

21.4.1 The simplest model

What is involved here is to *choose* a spin function Θ_K (or functions) and *reverse* the interpretation of equation (21.6); that is, generate an associated F_K. The choice of the spin function may be the easy part, for a closed-shell saturated organic molecule we would normally seek a description of the molecule in terms of a set of localised electron-pair bonds formed from individually localised hybrid "atomic orbitals". In spin terms this is just a product of $n/2$ Heitler–London spin factors

$$\Theta_K = \prod_{k=1,3,\ldots}^{n-1} \frac{1}{\sqrt{2}} (\alpha(s_k)\beta(s_{k+1}) - \alpha(s_{k+1})\beta(s_k))$$

for n electrons. The question now is

> How do we construct an F_K associated with this Θ_K and what variational parameters should it contain?

We can obviously set up an heuristic primitive for the spatial part using the very idea which was used to generate Θ_K. If each electron pair is a localised bond then the spatial part should be capable of being generated from a *product* of spatial orbitals similar to the primitive product used in the Heitler–London model of the electron-pair bond. Clearly, if these individual orbitals are formed from some set of basis functions, then the matrix which defines the orbitals in terms of the basis functions forms a familiar variational problem

$$\boldsymbol{\lambda} = \boldsymbol{\phi}\boldsymbol{A} \tag{21.8}$$

(say) where $\boldsymbol{A}$ is to be determined.

So we may use a primitive spatial product (Ω, say) of the form

$$\Omega(\vec{r}_1, \vec{r}_2, \ldots, \vec{r}_n) = \lambda_1(\vec{r}_1)\lambda_2(\vec{r}_2)\ldots\lambda_n(\vec{r}_n) = \prod_{k=1}^{n} \lambda_k(\vec{r}_k) \qquad (21.9)$$

The generalisation of the HL model applied to eqn (21.9) must therefore be the generalisation of the formation of the full spatial HL function from the primitive product: the function F_K must simply be a sum of terms which can be generated from Ω by *permutations* of the spatial variables of the electrons. That is

$$F_K(\vec{r}_1, \vec{r}_2, \ldots, \vec{r}_n) = \sum_{\hat{P}_r} D_K(\hat{P}_r)\hat{P}_r\Omega(\vec{r}_1, \vec{r}_2, \ldots, \vec{r}_n)$$

where the $D_K(\hat{P})$ are numbers characteristic of each permutation $\hat{P}$ which are *fixed by the nature of the chosen* Θ_K (hence their subscript K). Once these coefficients are known for a particular Θ_K then the *form* of the associated spatial function F_K is fixed and the energy of this function may be computed from equation eqn (21.7); any parameters in the F_K may be optimised using the variation theorem in the usual way.

The remaining problem is to evaluate the permutation-related combination coefficients $D_K(\hat{P})$. The way in which these coefficients are fixed by the spin function Θ_K may be seen from eqn (21.3):

$$\hat{A} = \frac{1}{\sqrt{N!}} \sum_{\hat{P}} (-1)^p \hat{P}_s \hat{P}_r$$

When a spin/space permutation is applied to a function of space and spin the Pauli principle requires that the *product* of the coefficient of the spatial part multiplied by the coefficient of the spin part must be $(-1)^p$, where p is the parity of the permutation. So, if we know the coefficients which occur in the effect of a spin permutation on the chosen Θ_K, the coefficients of the associated spatial part are fixed. But, as we have seen, the effect of permutations of spin on a spin eigenfunction is simply to send it into a linear combination of other spin eigenfunctions with the same value of S and M; the spin functions form a basis for the irreducible representations of the symmetric group of permutations of n objects.

The formal group-theoretical result is that the representation matrices of the two spatial and spin permutations ($\hat{P}_r$ and $\hat{P}_s$) are linked by the Pauli principle through

$$\boldsymbol{D}_S(\hat{P}_r) = (-1)^p \boldsymbol{D}_S(\hat{P}_s^{-1})$$

where S is the spin-state involved. Since the set of Θ_K of common (S, M) values can be chosen to be orthogonal (we have assumed this in obtaining eqn

21.4. SPIN-FREE VB THEORY

(21.7)) the expression is even simpler in terms of the matrices of an orthogonal representation:
$$\boldsymbol{D}_S(\hat{P}_r) = (-1)^p \boldsymbol{D}_S(\hat{P}_s) \qquad (21.10)$$

With these representation matrices we can set up the projection operators which will project, from an arbitrary spatial function, the component by which Θ_K must be multiplied in order to form a total wavefunction which

1. Satisfies the Pauli principle.
2. Is a spin eigenfunction.
3. Expresses the model of molecular electronic structure contained in the original choice of Θ_K and Ω: a set of localised bond functions formed as linear combinations of pairs of localised "AOs".

In fact, as is familiar from the use of point groups to generate symmetry orbitals (Chapter 18), it is only necessary to know the diagonal elements of the representation matrices to form a projection operator which will generate a suitable function, in our case the projector

$$\rho^S = \frac{f_S^n}{n!} \sum_{\hat{P}} (-1)^p \left(\boldsymbol{D}_S(\hat{P})\right)_{11} \hat{P} \qquad (21.11)$$

is sufficient, which only uses the $(1,1)$ element (taking our K as 1 for convenience) of the representation matrix.[8] These representation matrices are readily available and are easily computed, they are the spin "integrals"

$$\left(\boldsymbol{D}_S(\hat{P})\right)_{11} = \int ds \Theta_1 \hat{P}_s \Theta_1 = \int ds \Theta_1 \Theta_{1'}$$

where $\Theta_{1'}$ is the spin eigenfunction of the same (S, M) after the permutation of spin variables in Θ_1. No actual integration is involved, of course, and these numbers may be evaluated extremely rapidly by graphical or combinatorial techniques known since the very start of VB theory.[9]

21.4.2 Generalisation

The method developed above has been for a single spatial "configuration" (orbital pairing scheme) using a single spin eigenfunction which generates the spatial permutation symmetry ensuring the effects of the Pauli principle are fully taken into consideration. As it stands, the only possible room for optimisation

---

[8] The subscript "r" has been dropped from the permutation operator for simplicity.
[9] G. Rumer *Göttinger Nachr.*, **3**, 337 (1932) and see R. McWeeny and B T Sutcliffe *Methods of Molecular Quantum Mechanics*, (Academic Press, 1969).

of the spatial function is in the form of the orbitals in the spatial configuration, the matrix $\mathbf{A}$ in eqn (21.8).

There are two possible ways of generalising the method which may be used separately or together:

1. We could use several spatial configurations while retaining the same spin function. This corresponds to several alternative "bonding schemes" in a given model. For example

$$\begin{aligned}\Omega_1 &= \lambda_1(\vec{r}_1)\lambda_2(\vec{r}_2)\lambda_3(\vec{r}_3)\lambda_4(\vec{r}_4) \\ \Omega_2 &= \lambda_1(\vec{r}_1)\lambda_4(\vec{r}_2)\lambda_2(\vec{r}_3)\lambda_3(\vec{r}_4)\end{aligned}$$

each used with the single spin function

$$\Theta = \frac{1}{\sqrt{2}}(\alpha(s_1)\beta(s_2) - \alpha(s_2)\beta(s_1)) \times \frac{1}{\sqrt{2}}(\alpha(s_3)\beta(s_4) - \alpha(s_4)\beta(s_3))$$

would generate F_1 corresponding to a "bond" between λ_1, λ_2 and a bond between λ_3, λ_4, and F_2 corresponding to a "bond" between λ_1, λ_4 and a bond between λ_2, λ_3. A situation which might be appropriate to describe the situation in the π-system of cyclobutadiene.

A suitable wavefunction for the system might then be

$$\Psi = (C_1 F_1 + C_2 F_2)\Theta \qquad (21.12)$$

with the coefficients C_K determined by the linear variation method[10] of Chapter 1.

2. We might wish to retain the same spatial configuration and use more than one spin function of the same (S, M), that is, retain Ω_1 from above but use Θ and Θ' given by

$$\begin{aligned}\Theta' &= \frac{1}{\sqrt{12}}\{[\alpha(s_1)\beta(s_2) + \alpha(s_2)\beta(s_1)][\alpha(s_3)\beta(s_4) + \alpha(s_4)\beta(s_3)] \\ &\quad -2\alpha(s_1)\alpha(s_2)\beta(s_3)\beta(s_4) - 2\beta(s_1)\beta(s_2)\alpha(s_3)\alpha(s_4)\}\end{aligned}$$

corresponding to the electrons in λ_1, λ_2 forming a triplet and those in λ_3, λ_4 generating a triplet and the two triplets combined to form an overall singlet.

In this case the linear variational problem would be generated by the ratio of singlet spin functions, determining D and D' in

$$\Psi = D F_1 \Theta + D' F_1' \Theta'$$

---

[10]This approach is used in a spin-free context by F. A. Matsen in the paper which introduced the idea of spin-free methods: "Spin-Free Quantum Chemistry" in *Advances in Quantum Chemistry*, **1**, 59, (1964). Matsen uses a rather austere algebraic method, dispensing with the spin factors from the very start.

The form of the spatial function generated from the single spatial configuration Ω_1 is different in the two cases but contains no additional variational parameters so that one is still using a single spatial configuration.[11]

Each of these two *types* of approach may be combined with the optimisation of the form of the spatial orbitals λ_i in a technique reminiscent of MCSCF (see Chapter 22). Finally, of course, one could combine the two linear variation schemes to optimise both the spin function and the spatial partners. In this context it is useful to know when to stop, that is, when the set of spin functions and/or spatial configurations exhausts the capacity of the orbital basis and become linearly dependent. This corresponds to the limit set by a CI calculation which uses all possible determinants (with common spin eigenfunctions) which can be generated from a given set of molecular orbitals. Again, group-theoretical or even simpler graphical rules are available for the determination of such "completeness".

Once these "spin-free" variational problems have been set up the evaluation of the spatial integrals involving the molecular Hamiltonian and the spatial functions F_K must be carried through. This, of course, involves the generation of the orbital products contained in the expansion of the F_K and the breakdown of the integrals over these products into products of overlap, one- and two-electron integrals. This is formally identical to the evaluation of the matrix elements over determinants of non-orthogonal orbitals, since these also involve using permutations to generate a set of orbital products and the subsequent evaluation and summation of the results of these primitive product integrals.

21.5 Summary

The VB method is capable of expressing the electronic structure of a molecule in an extremely compact and intuitively appealing manner. Technically the VB model generates a multi-configuration expansion of the wavefunction of very modest length but involving non-orthogonal spatial functions. The unavoidable necessity of evaluating matrix elements for the linear variation problem involving non-orthogonal products means that some equivalent of the evaluation of matrix elements of determinants of non-orthogonal orbitals must be involved. We have seen in Appendix 2.A that this must place an upper limit on the number of electrons which can be treated. Currently, the VB model can be used for between 10 and 14 electrons. Naturally, the VB model can be combined with a closed-shell "core" and/or effective core potentials to limit the full VB model to the

---

[11] This approach has led to a revival of interest in and application of VB theory in recent years; a summary of methods and applications may be found in "Modern Valence Bond Theory" by D. L. Cooper, J. Gerratt and M. Raimondi in Ab Initio *methods in quantum chemistry* ed. K. P. Lawley, Wiley, 1987. Here, unlike Matsen's algebraic approach, traditional spin-eigenfunction techniques are employed.

chemically interesting electrons which considerably extends the scope of the method.

Implementation of the VB model involves a series of specialised techniques which are dependent on permutation and group algebras and will not be attempted here.

Chapter 22

Doubly-occupied MCSCF

The general theory outlined in Chapter 14 has been used in the RO-HF model. It can also provide the basis for a whole class of multi-determinant models of molecular electronic structure which are constructed from sets of doubly occupied orbitals. Two such models are outlined in this chapter: Paired-Electron MCSCF (PEMCSCF) and the General Valence Bond (GVB). These "pair expansion" theories are based on the idea of natural orbitals which diagonalise the one-electron density matrix.

Contents

22.1	Introduction: natural orbitals	**588**
22.2	Paired-excitation MCSCF	**590**
22.2.1	The orbital equation	593
22.2.2	The CI equation	595
22.3	Implementation	**595**
22.4	Partial Paired-Excitations; GVB	**596**
22.4.1	Simple valence-bond model	597
22.4.2	Pair natural orbitals	598
22.5	Details of GVB	**599**
22.5.1	Strong orthogonality	599
22.5.2	The energy expression	600
22.5.3	The orbital Fock matrices	602
22.5.4	The pair expansion coefficients	603
22.6	Implementation	**604**

22.1 Introduction: natural orbitals

The idea of natural orbitals has been mentioned briefly in Chapter 13 where it was said that it was useful to define a set of orbitals (ψ_k, say) for which the one-electron spatial density ($P(\vec{r})$) is given by

$$P(\vec{r}) = \sum_{k=1}^{m} n_k |\psi(\vec{r})|^2 \tag{22.1}$$

or, analogously, a set of spin-orbitals (χ_k, say) which generate the one-electron space/spin density by

$$\rho(\vec{x}) = \sum_{k=1}^{m} n_k |\chi(\vec{x})|^2 \tag{22.2}$$

where, of course, the two sets of n_k are different in the two cases.

In the single-determinant case, the MOs are just such a set where, for example, in the closed-shell case the n_k of eqn (22.1) are all 2 and, in the DODS case, the n_k of eqn (22.2) are all 1. There is nothing new here since our choice of a single-determinant of *equally-occupied* MOs has pre-empted the definition of natural orbitals. But what about the multi-determinant[1] case?

If an approximate, n-electron, closed-shell wavefunction is written

$$\Psi = \sum_{K=1}^{N} C_K \Phi_K \tag{22.3}$$

where the Φ_K are determinants then, for example,

$$P(\vec{r}) = \sum_{K,L=1}^{N} C_K C_L^* \int \Phi_K(\vec{x}, \vec{x}_2 \ldots \vec{x}_n) \Phi_L^*(\vec{x}, \vec{x}_2 \ldots \vec{x}_n) ds d\tau_2 \ldots d\tau_n \tag{22.4}$$

Each of the integrals on the right of this expression must reduce to a sum of numerical multiples of *products* of the orbitals of which the determinant is composed:

$$\int \Phi_K(\vec{x}, \vec{x}_2 \ldots \vec{x}_n) \Phi_L^*(\vec{x}, \vec{x}_2 \ldots \vec{x}_n) ds d\tau_2 \ldots d\tau_n = \boldsymbol{\psi}(\vec{r}) \boldsymbol{P}^{KL} \boldsymbol{\psi}^*(\vec{r})$$

where the matrices $\boldsymbol{P}^{KL}$ are two-determinant generalisations of the MO density matrix

$$\boldsymbol{P} = 2\boldsymbol{R}$$

for the closed-shell case.

---
[1] Or a more general form of wavefunction which is equivalent to a linear combination of determinants.

22.1. INTRODUCTION: NATURAL ORBITALS

By construction, all these matrices are Hermitian (symmetric in the real case) and so, therefore, is the matrix

$$\boldsymbol{P} = \sum_{K,L=1}^{N} C_K C_L^* \boldsymbol{P}^{KL}$$

which generates the required one-electron spatial density function $P(\vec{r})$:

$$P(\vec{r}) = \boldsymbol{\psi}(\vec{r})\boldsymbol{P}\boldsymbol{\psi}^*(\vec{r})$$

This is, therefore, a quadratic form in the ψ_k which, like any other, may be diagonalised

$$\boldsymbol{U}^\dagger \boldsymbol{P} \boldsymbol{U} = \boldsymbol{n}$$

(where $\boldsymbol{n}$ is diagonal) to generate a set of eigenvectors ($\boldsymbol{U}$) which define a new set of "MOs" ($\boldsymbol{\gamma}$) as a linear transformation of the $\boldsymbol{\psi}$:

$$\boldsymbol{\gamma} = \boldsymbol{\psi}\boldsymbol{U}$$

to yield a diagonal expansion of the density $P(\vec{r})$:

$$P(\vec{r}) = \sum_{k=1}^{m} n_k |\gamma_k(\vec{r})|^2 \tag{22.5}$$

Since the $\boldsymbol{\psi}$ form an orthonormal set, so do the $\boldsymbol{\gamma}$ and so, integrating the above expression for $P(\vec{r})$ gives

$$\int P(\vec{r}) dV = \sum_{k=1}^{m} n_k \int |\gamma_k(\vec{r})|^2 dV = \sum_{k=1}^{m} n_k = n$$

The n_k are now a set of *occupation numbers* (≤ 2, since the γ_k are *spatial* orbitals) which sum to the total number of electrons, and the natural orbitals γ_k are a unique set of spatial functions (for a given expansion Ψ) which have some interesting properties, obvious and not-so-obvious.

- The *magnitudes* of the n_k tell us something about the relationship between the general expansion eqn (22.3) and, for example, the best single determinant. If such an expansion has as its leading term the closed-shell SCF determinant, the way in which the largest $n/2$ values of n_k differ from 2 and the remaining ones differ from zero is a good indication of the quality of the single-determinant approximation.

- The intuition that, if the γ_k are put in order of decreasing n_k, and these most heavily-occupied orbitals are used in an expression like eqn (22.3) the length of the expansion may be reduced without loss of quality, proves to be borne out.

590   CHAPTER 22. DOUBLY-OCCUPIED MCSCF

If the *number of one-electron orbitals* can be reduced in an expansion like eqn (22.3), the number of possible determinants is considerably reduced and the occupation numbers of the natural orbitals would seem to be an eminently suitable criterion for discarding orbitals from a given set.

The fact is that if a given function of n (vector) variables is to be approximated by a sum of products (or determinants if the function is antisymmetric) of functions of a single (vector) variable then this approximation is at its numerical best[2] for a given length of expansion (N in eqn (22.3)) when the single-variable functions are the natural orbitals defined above. Conversely, for a given least-squares accuracy, the number of terms in the expansion eqn (22.3) is at its minimum when the single-particle orbitals of which the determinants are composed are natural orbitals.

This is all extremely encouraging except for one very obvious and painful fact:

> The natural orbitals (NOs) are only available *after* a multi-determinant calculation has been performed. One could *verify* that the above theory is true, but apparently, not use it in order to simplify the calculations.

In order to generate the NOs γ one needs the matrix U which is obtained by diagonalising P which involves a knowledge of the coefficients C_{KL} which are only available *after* the full multi-determinant calculation has been carried through.

However, the important point here is the *theoretical* one; our knowledge of the desirable features of the NOs will enable us to adopt better *strategies* for the generation of multi-determinant models of molecular electronic structure.

Let us summarise this discussion of NOs with one simple fact: the shortest expansion of a closed-shell multi-determinant wavefunction can be written in terms of a set of orbitals which are doubly occupied in the expansion of the electron density.

We can now go on to consider specific multi-determinant models of molecular electronic structure.

22.2 Paired-excitation MCSCF

We begin by considering a multi-determinant wavefunction of the form eqn (22.3) and try to find approximations to the orbitals which will *validate* our particular choice of:

---

[2]In a least-squares sense.

22.2. PAIRED-EXCITATION MCSCF

1. Number and construction of determinantal functions Φ_K.

2. Detailed form (in the LCAO sense) of the one-electron orbitals of which the determinants are composed.

There are both theoretical and practical considerations involved in the strategy.

Slater's rules for the evaluation of the energy integral between two determinants of *orthonormal* orbitals ensure that:

- The integral involving three (or more) orbital differences between the determinants is zero.

- The integral involving two differences is just a pair of electron-repulsion integrals ("Coulomb minus exchange").

- Determinants differing by two (or more) orbitals are orthogonal.

The considerations of the last section indicate that the NOs will provide the theoretical background for a choice of orbital set.

If, therefore, we set up a wavefunction which is a linear combination of determinants composed of doubly occupied spatial orbitals only:

$$\Psi = C_0\Phi_0 + \sum_{i,a} C_{tu}\Phi_t^u \qquad (22.6)$$

where Φ_0 is some basic determinant (the Hartree–Fock solution, say), and the determinants Φ_t^u are formed from Φ_0 by removing a *pair* of electrons from (spatial) orbital ψ_t (which is, therefore, doubly occupied in Φ_0, n of them) and placing them into ψ_u (which is not occupied in Φ_0, $(m-n)$ of them).

If the orbitals are the (occupied and virtual) solutions of the LCAO Hartree–Fock equations, it is easy to see that the NOs of such a wavefunction are the same as these SCF orbitals since all the "off-diagonal" integrals are zero

$$\int \Phi_K(\vec{x},\vec{x}_2\ldots\vec{x}_n)\Phi_L^*(\vec{x},\vec{x}_2\ldots\vec{x}_n)dsd\tau_2\ldots d\tau_n = \delta_{KL}\boldsymbol{\psi}(\vec{r})\boldsymbol{P}^{KK}\boldsymbol{\psi}^*(\vec{r})$$

and the $\boldsymbol{P}^{KK}$ are diagonal; only the *occupation numbers* of the SCF orbitals are changed by the use of the paired-excitation model.

We therefore take this paired-excitation model and allow *both* the LCAO forms of the orbitals *and* the coefficients C_{tu} (a restricted class of C_K) to be determined by the variational process (the Paired-Excitation Multi-Configuration or PEMCSCF model). Of course, because of the practical points outlined above, it is anticipated that the energy expression will involve only the matrices h, J and K.

The energy expression for the wavefunction given by this sum is particularly straightforward if a little lengthy. We assume that the coefficients (C_0, C_t^u) are normalised:

$$|C_0|^2 + \sum_{t=1}^{n} \sum_{u=(n+1)}^{m} |C_t^u|^2 = 1$$

since the determinants are orthogonal if the MOs are orthogonal. Therefore

$$\begin{aligned} E &= |C_0|^2 E_0 + \sum_{t=1}^{n} \sum_{u=(n+1)}^{m} |C_t^u|^2 E_{tu} \\ &+ C_0 \sum_{t=1}^{n} \sum_{u=(n+1)}^{m} C_t^u \int \Phi_0 \hat{H} \Phi_t^u dV \\ &+ \sum_{t=1}^{n} \sum_{u=(n+1)}^{m} \sum_{t' \neq} \sum_{u' \neq u} C_t^u C_{t'}^{u'} \int \Phi_t^u \hat{H} \Phi_{t'}^{u'} dV \end{aligned}$$

The terms in this expression are of three types:

1. Φ_0 is the "ground-state" determinant, with energy given by the familiar expression

$$E_0 = 2 \operatorname{tr} h R + \operatorname{tr} G(R) R$$

where the usual basis-expansion has been assumed:

$$\boldsymbol{\psi} = \boldsymbol{\phi} U$$

and, if C is the first m columns of U,

$$R = CC^{\dagger}$$

If each column of U is called separately c^i, then

$$R^i = c^i c^{i\dagger}$$

and the above expression becomes

$$\int \Phi_0 \hat{H} \Phi_0 dV = E_0 = 2 \sum_{t=1}^{n} \operatorname{tr} h R^t + \sum_{t=1}^{n} \sum_{t'=1}^{n} \operatorname{tr} G(R^{t'}) R^t \qquad (22.7)$$

2. Each Φ_t^u is a determinant so that its energy is given by an expression identical in *form* to that for E_0, but the "R" involves a summation over the columns c^i which are occupied in Φ_t^u.

$$\int \Phi_t^u \hat{H} \Phi_t^u dV = E_{tu} = 2 \sum_{k \in \Phi_t^u} \operatorname{tr} h R^k + \sum_{k \in \Phi_t^u} \sum_{k' \in \Phi_t^u}^{n} \operatorname{tr} G(R^{k'}) R^k$$

22.2. PAIRED-EXCITATION MCSCF

For the manipulations involved later, the expression is most conveniently expressed in terms of E_0 and the *differences* which arise when one $\boldsymbol{R}^t$ is replaced by an $\boldsymbol{R}^u$, corresponding to removal of orbital ψ_u from Φ_0 and addition of orbital ψ_t to Φ_0 (doubly occupied in each case).

$$\begin{aligned}
E_t^u &= E_0 - 2\mathrm{tr}\boldsymbol{h}\boldsymbol{R}^t + 2\mathrm{tr}\boldsymbol{h}\boldsymbol{R}^u \\
&- 2\sum_{t'=1}^{n}\mathrm{tr}\boldsymbol{G}(\boldsymbol{R}^{t'})\boldsymbol{R}^t + 2\sum_{t'=1}^{n}\mathrm{tr}\boldsymbol{G}(\boldsymbol{R}^{t'})\boldsymbol{R}^u \\
&+ 2\mathrm{tr}\boldsymbol{G}(\boldsymbol{R}^u)\boldsymbol{R}^t + \mathrm{tr}\boldsymbol{G}(\boldsymbol{R}^u)\boldsymbol{R}^u + \mathrm{tr}\boldsymbol{G}(\boldsymbol{R}^t)\boldsymbol{R}^t
\end{aligned}$$

3. The inter-determinant terms always involve at least *two* orbital differences which are the two spin-occupations of the same spatial orbital:

$$\begin{aligned}
\int \Phi_0 \hat{H} \Phi_t^u dV &= \mathrm{tr}\boldsymbol{K}(\boldsymbol{R}^t)\boldsymbol{R}^u \\
\int \Phi_t^u \hat{H} \Phi_t^{u'} dV &= \mathrm{tr}\boldsymbol{K}(\boldsymbol{R}^u)\boldsymbol{R}^{u'} \\
\int \Phi_t^u \hat{H} \Phi_{t'}^u dV &= \mathrm{tr}\boldsymbol{K}(\boldsymbol{R}^t)\boldsymbol{R}^{t'}
\end{aligned}$$

with all the remaining terms zero because they involve determinants differing in *four spin*-orbitals (remember each of ψ_t and ψ_u is doubly occupied).

The total energy expression is then

$$\begin{aligned}
E &= C_0^2 E_0 + \sum_{t=1}^{n}\sum_{u=(n+1)}^{m} C_{tu}^2 E_{tu} \\
&+ 2C_0 \sum_{t=1}^{n}\sum_{u=(n+1)}^{m} C_{tu}\mathrm{tr}\boldsymbol{K}(\boldsymbol{R}^t)\boldsymbol{R}^u \\
&+ \sum_{t=1}^{n}\sum_{u=(n+1)}^{m} C_{tu} \sum_{t'=1}^{n} C_{t'u}\mathrm{tr}\boldsymbol{K}(\boldsymbol{R}^t)\boldsymbol{R}^{t'} \\
&+ \sum_{t=1}^{n}\sum_{u=(n+1)}^{m} C_{tu} \sum_{u'=(n+1)}^{m} C_{tu'}\mathrm{tr}\boldsymbol{K}(\boldsymbol{R}^u)\boldsymbol{R}^{u'} \quad (22.8)
\end{aligned}$$

This expression is exactly of the anticipated form; it involves only two-electron contributions involving the matrices $\boldsymbol{J}$ and $\boldsymbol{K}$; in fact only $\boldsymbol{G} = 2\boldsymbol{J} - \boldsymbol{K}$ and $\boldsymbol{K}$ occur independently.

22.2.1 The orbital equation

In comparing the expression with the general functional there are some points to note:

- There are no "shells"; each MO has its own occupation number which was one of the factors involved in our definition of a shell.
- There is no "empty shell"; all the MOs are occupied to some extent.
- The occupation numbers of the MOs do not come out of the theory quite so cleanly as in the ROHF case; they do not, in fact, appear as numerical factors in an effective Fock matrix.

Since each MO is doubly occupied in each determinant, the occupation numbers of the MOs are simply given in terms of the squares of the expansion coefficients C_0 and C_{tu}:

$$\nu_t = 2(|C_0|^2 + \sum_{u=(n+1)}^{m} |C_t^u|^2)$$

$$\nu_u = 2\sum_{t=1}^{n} |C_t^u|^2$$

However the analysis of Chapter 14 can still be carried through in its full generality to yield a Fock matrix for each *orbital*. The analogue of the "shell" structure is, in this case, the two different *forms* for the two types of MO; occupied or unoccupied in Φ_0:

$$\begin{aligned}
\boldsymbol{h}_t^F &= (1-A_t)\boldsymbol{h} + 2A_t\boldsymbol{G}(\boldsymbol{R}^t) + \sum_{t'}(1-A_t-A_{t'})\boldsymbol{G}(\boldsymbol{R}^{t'}) \\
&+ 2\sum_u [(B_u - |C_t^u|^2)\boldsymbol{G}(\boldsymbol{R}^u) + C_0 C_t^u \boldsymbol{K}(\boldsymbol{R}^u)] \\
&+ \sum_{t'\neq t} A_{tt'}\boldsymbol{K}(\boldsymbol{R}^{t'}) \qquad (22.9) \\
\boldsymbol{h}_u^F &= B_u\left[\boldsymbol{h} + 2\boldsymbol{G}(\boldsymbol{R}^u) + 2\sum_t \boldsymbol{G}(\boldsymbol{R}^t)\right] \\
&+ \sum_t \left[C_0 C_t^u \boldsymbol{K}(\boldsymbol{R}^t) - |C_t^u|^2 \boldsymbol{G}(\boldsymbol{R}^t)\right] \\
&+ \sum_{u'\neq u} B_{uu'}\boldsymbol{K}(\boldsymbol{R}^{u'}) \qquad (22.10)
\end{aligned}$$

where

$$A_{tt'} = \sum_{u=(n+1)}^{m} C_t^u C_{t'}^u \quad \text{with} \quad A_{tt} = A_t$$

$$B_{uu'} = \sum_{t=1}^{n} C_t^u C_t^{u'} \quad \text{with} \quad B_{uu} = B_u$$

22.3. IMPLEMENTATION

These expressions can now be simply inserted into the equation for the single effective Fock-like matrix which generates the associated matrix eigenvalue equation to be solved iteratively for the MO coefficients c^k.

However, in order to set up the matrices we need to know the CI expansion coefficients C_0 and C_t^u which must be generated for a particular set of known c^k.

22.2.2 The CI equation

The linear variation method introduced in Chapter 1 is the most straightforward way of generating the coefficients C_0 and C_t^u since, during the calculation of the matrices h_t^F and h_u^F we must evaluate all the G and K matrices which are required to evaluate all the elements of the matrix H with elements

$$\int \Phi_t^u \hat{H} \Phi_{t'}^{u'} dV$$

etc.

All that is necessary is to set up this matrix and solve the (orthogonal) matrix eigenvalue problem

$$HD = ED$$

for the lowest eigenvalue (E) and the associated lowest eigenvector D whose elements are the required coefficients (C_0, C_t^u).

This one-step process is added to each step of the iterative procedure for the generation of the SCF orbitals. The resulting orbitals and CI expansion coefficients are then self-consistent in a more general sense than the MO coefficients of the single-determinant model. The MO coefficients are the optimum set which can be generated (from a given choice of basis) *for a CI expansion of the chosen type*. The combined MO and CI coefficients should then *validate* the chosen model as far as the constraints of the CI expansion allow.

The matrix H is $n(m-n) \times n(m-n)$ which may become quite large if a flexible basis set is used. The maximum of this dimension occurs when $m = 2n$. However, unlike the solution of the *orbital* coefficient problem, where we need *all* the coefficients, in this case we only need the coefficients associated with the lowest eigenvalue. There are special methods for generating single eigenvectors of large matrices, some of which do not even require all of H to be in main memory at once.

22.3 Implementation

Clearly, we shall be dealing with stacks of matrices by the methods we have already met; there are now stacks of Rs, Gs, Ks and h^Fs to be formed and manipulated.

A new feature is the generation of $\boldsymbol{H}$ from these stacks of matrices.

22.4 Partial Paired-Excitations; GVB

Some of the general results of the PEMCSCF calculation could be predicted in advance by thinking about the general features of the CI part of the calculation. In particular, we can estimate on the basis of perturbation theory that some of the Φ_t^u will make very small contributions to Ψ.

A perturbation analysis of the CI problem shows that the energy differences $(E_0 - E_{tu})$ occur in the denominator of an approximation to C_{tu}, and so the highly excited Φ_t^u will make little contribution to the MCSCF wavefunction. Such Φ_t^u occur when very tightly bound ("core") electrons are removed or when electrons are placed in virtual orbitals which are associated with high E_{tu} or both, of course.

This general result can be incorporated into the PEMCSCF method simply by truncating the CI expansion either *a priori* or on the basis of calculated energy differences $(E_0 - E_{tu})$. The only difference that this choice would make to our implementation would be the possible appearance of an "empty shell" of unused virtual MOs, which would be an $\boldsymbol{R}^Z$ in the earlier notation, and there would be associated changes to the form of the final effective Fock matrix.

However, the main objection to MCSCF methods of this sort is, perhaps, that they are too "coarse" in the sense of not being sufficiently tailored to any physical or chemical knowledge we might have about the *overall* electronic structure. For example

- We may not wish to improve the theoretical description of certain parts of a molecule; it may well be enough that these sub-structures are merely sources of potential for the interesting molecular sub-structures. Atomic inner shells spring to mind immediately; these electrons are tightly bound to the nuclei and a single-determinant model is usually sufficient.

- Certain parts of the molecular electronic structure may be weakly bound (in the chemical sense of inter-atomic bonding) while others may be strongly bound structures in a conventional bonding scheme.

In short, it would be more sensible to be able to use different explicit models for different electronic sub-structures of a molecule.

22.4. PARTIAL PAIRED-EXCITATIONS; GVB

In particular it is well-known that the closed-shell single-determinant model of electronic structure is inappropriate for the description of bond-breaking situations in molecules; the very fact that one insists on a set of *doubly occupied* MOs in this model means that a bond will dissociate heterolytically into closed-shell ions rather than homolytically into open-shell fragments.

22.4.1 Simple valence-bond model

The most intuitively familiar model of molecular electronic structure which gives a satisfactory qualitative and quantitative picture of both stable bonds and bond-breaking is the famous Heitler–London non-polar covalent model. A chemical bond between a pair of (hybrid atomic) orbitals λ_1 and λ_2 is described by the two-electron wavefunction

$$\Psi_{HL} = N[\lambda_1(\vec{r}_1)\lambda_2(\vec{r}_2) + \lambda_1(\vec{r}_2)\lambda_2(\vec{r}_1)] \times [\alpha(s_1)\beta(s_2) - \alpha(s_2)\beta(s_1)]$$

where N is a normalising factor.

The key point in this model is that the orbitals λ_i are *non-orthogonal*; the inter-atomic *overlap* is essential for the description of bonding at short internuclear distances, while the symmetric nature of the spatial factor must be present to ensure correct dissociation behaviour.

If we introduce two new orbitals, ψ_1 and ψ_2, so that the λ_i are defined by the sum and difference of these new orbitals:

$$\begin{aligned} \lambda_1 &= c_1\psi_1 + c_2\psi_2 \\ \lambda_2 &= c_1\psi_1 - c_2\psi_2 \end{aligned} \qquad (22.11)$$

where we may, without loss of generality, assume that

$$c_1^2 + c_2^2 = 1$$

then it is easy to show that

$$\Psi_{HL} = 2N[c_1^2\psi_1(\vec{r}_1)\psi_1(\vec{r}_2) - c_2^2\psi_2(\vec{r}_1)\psi_2(\vec{r}_2)] \times [\alpha(s_1)\beta(s_2) - \alpha(s_2)\beta(s_1)]$$

Now, the overlap between the λ_i is

$$\int \lambda_1\lambda_2 dV = c_1^2 \int \psi_1\psi_1 dV - c_2^2 \int \psi_2\psi_2 dV \qquad (22.12)$$

independently of the overlap

$$\int \psi_1\psi_2 dV$$

between ψ_1 and ψ_2 which cancels from the expansion. This means that we may choose the ψ_i to be *orthogonal* and maintain the λ_i non-orthogonal.

However, eqn (22.12) contains only *doubly occupied* orbitals; it is, therefore a candidate for a simplified MCSCF treatment of the type we have been discussing in this chapter; the c_i^2 playing the role of CI coefficients and the ψ_i being capable of LCAO expansion in the usual way with both types of degree of freedom being determined variationally.

Before continuing, let us look at this result from another angle.

22.4.2 Pair natural orbitals

It is one of the most basic assumptions of chemistry that a complex molecular electronic structure can be thought of as a series of environment-insensitive substructures with a large degree of autonomy. In particular for many molecules we think of the total structure as composed of *pairs* of electrons (inner shells, lone pairs, bond pairs) and this idea can be translated into a quantum-mechanical model in which each separate pair (or group) has its own wavefunction. If this is so, then most of the analysis which we have used for the total electronic structure can be taken over unchanged in a description of the separate pairs.

It is tempting, therefore, to see the simple result of the last section as the simplest case of the *natural orbital* expansion of a bonding pair of electrons:

$$\Psi_R(\vec{x}_1, \vec{x}_2) = \left[\sum_{k=1}^{N} C_k^R \gamma_k^R(\vec{r}_1) \gamma_k^R(\vec{r}_2)\right] \times \frac{1}{\sqrt{2}}[\alpha(s_1)\beta(s_2) - \alpha(s_2)\beta(s_1)] \quad (22.13)$$

where the γ_k^R are natural orbitals for the expansion of the bonding pair R, and to write the total wavefunction as the (antisymmetrised) product of a set of such pair functions:

$$\Psi = \hat{A} \prod_R \Psi_R$$

We can combine these two strands of argument:

1. The Heitler–London model.

2. The pair-NO extension of the simple Heitler–London model.

to generate the Generalised Valence-Bond (GVB) model of molecular electronic structure; each pair of electrons has its wavefunction expanded as a sum of doubly occupied orbitals in which both expansion coefficients and orbitals are variationally determined.

Notice that if just *one* doubly occupied orbital is used to expand the wavefunction of each electron pair, we simply regenerate the closed-shell MO model.

But the closed-shell model does *not* generate Heitler–London (localised) descriptions of the electronic structure in our experience. This is simply because we have pre-empted the form of the individual MOs in the single determinant by insisting on *canonical* form for those MOs. A localised set of doubly occupied MOs may be generated from this canonical set by a linear transformation which does not change the total wavefunction, but these functions will not *diagonalise* the HF matrix when self-consistent.

This simple fact ought to make us alert to the possibility that the optimum wavefunction of our chosen form might *not* be localised but might simply be a CI-type wavefunction using disjoint sets of *delocalised* MOs. Just as the SCF MOs of a single-determinant wavefunction may be transformed to a more chemically appealing localised set, it may be necessary to transform the expansion of the pair functions, which were introduced on the basis of an expected localised description, into a localised form.

22.5 Details of GVB

22.5.1 Strong orthogonality

It has already been noted that the doubly occupied approximations to the pair NOs may be taken to be orthogonal to each other; in order to obtain a manageable total wavefunction which is the product of these expansions, it is necessary to insist that the expansion functions of one pair are orthogonal to those of the other pairs of electrons in the system. This is, of course, a constraint but, if the assumption of semi-autonomous groups of electrons is realistic, then it is not an onerous constraint.

In general, the so-called strong orthogonality constraint on two separate-pair wavefunctions[3] is

$$\int \Psi_R(\vec{x}_1, \vec{x}_2) \Psi_S(\vec{x}_1, \vec{x}_2) d\tau_2 = 0 \qquad (22.14)$$

that is, integration over the coordinates of just *one* of the electrons (and, because of antisymmetry, it may be *any* one) is sufficient to give zero; the *function* of $\vec{x}_2$ generated by the above integration is *identically zero* everywhere in space.

It is easy to see that if the functions which are used to expand Ψ_R are orthogonal to those used to expand Ψ_S then eqn (22.14) is satisfied.

---

[3] Or separate wavefunctions for any number of electrons in general, as we shall see in discussing pseudopotentials in Chapter 24.

22.5.2 The energy expression

The GVB wavefunction is not a single determinant, so we cannot use our routine Slater's-rule formulae for the evaluation of the total energy integral; however, if we compute the energy of a single pair of electrons described by the simplest two-term expansion, the general result is intuitively clear.

Using eqn (22.13) with the approximate NOs written as ψ_i rather than γ_i and dropping the superscript R in the case of just one pair we have

$$\Psi(\vec{x}_1, \vec{x}_2) = \left[\sum_{k=1}^{N} C_k \psi_k(\vec{r}_1)\psi_k(\vec{r}_2)\right] \times \frac{1}{\sqrt{2}}[\alpha(s_1)\beta(s_2) - \alpha(s_2)\beta(s_1)]$$

The two-term special case is

$$\begin{aligned}\Psi(\vec{x}_1, \vec{x}_2) &= [C_1\psi_1(\vec{r}_1)\psi_1(\vec{r}_2) + C_2\psi_2(\vec{r}_1)\psi_2(\vec{r}_2)] \\ &\times \frac{1}{\sqrt{2}}[\alpha(s_1)\beta(s_2) - \alpha(s_2)\beta(s_1)]\end{aligned}$$

and the energy expression is just

$$E = \int \Psi(\vec{x}_1, \vec{x}_2)(\hat{h}(\vec{r}_1) + \hat{h}(\vec{r}_2) + \frac{1}{r_{12}})\Psi(\vec{x}_1, \vec{x}_2) d\tau_1 d\tau_2$$

if the ψ_k are normalised and orthogonal and

$$C_1^2 + C_2^2 = 1$$

As usual, the electronic Hamiltonian does not include spin so, since the singlet spin function for two electrons is a separate factor, we may integrate over spin to give

$$\begin{aligned}E &= \int [C_1\psi_1(\vec{r}_1)\psi_1(\vec{r}_2) + C_2\psi_2(\vec{r}_1)\psi_2(\vec{r}_2)] \\ &\times (\hat{h}(\vec{r}_1) + \hat{h}(\vec{r}_2) + \frac{1}{r_{12}}) \\ &\times [C_1\psi_1(\vec{r}_1)\psi_1(\vec{r}_2) + C_2\psi_2(\vec{r}_1)\psi_2(\vec{r}_2)] dV_1 dV_2\end{aligned}$$

which may be evaluated explicitly by separating the integral into products for the one-electron terms and retaining the full integral for the repulsion terms. The result is

$$\begin{aligned}E &= 2C_1^2 h_{11} + C_1^2(\psi_1\psi_1, \psi_1\psi_1) + 2C_2^2 h_{22} \\ &+ C_2^2(\psi_2\psi_2, \psi_2\psi_2) + 2C_1 C_2(\psi_1\psi_2, \psi_1\psi_2)\end{aligned}$$

in the familiar notation. Two things are worth noting:

22.5. DETAILS OF GVB

1. The "self-interaction" term has not appeared and so is not corrected for by the "diagonal" exchange term. If we had used the (MO) single term standard expression we might have expected

$$2h_{11} + 2(\psi_1\psi_1, \psi_1\psi_1) - (\psi_1\psi_1, \psi_1\psi_1)$$

with its explicit cancellation of the self-interaction.

2. The singlet spatial function is *symmetric* and so the exchange integral $(\psi_1\psi_2, \psi_1\psi_2)$ appears with *positive* sign.

Making the usual "LCAO" expansion of each ψ_k in terms of a (spatial) basis ϕ:

$$(\psi_1, \psi_2) = \phi(c^1, c^2)$$

and defining "orbital" R-matrices

$$R^k = c^k c^{k\dagger}$$

in the usual way, the energy expression may be translated into "AO" form:

$$\begin{aligned} E &= 2C_1^2 \text{tr} h R^1 + C_1^2 \text{tr} J(R^1) R^1 \\ &+ 2C_2^2 \text{tr} h R^2 + C_2^2 \text{tr} J(R^2) R^2 \\ &+ 2C_1 C_2 \text{tr} K(R^1) R^2 \end{aligned}$$

where the J terms could be replaced by $G = 2J - K$ and the two R matrices may be permuted in the last term without changing the value of the expression.

The generalisation to an expansion of arbitrary length is clearly

$$E = \sum_k \left\{ 2C_k^2 \text{tr} h R^k + C_k^2 \text{tr} J(R^k) R^k + 2C_k \sum_{j \neq k} C_j \text{tr} K(R^j) R^k \right\}$$

Obviously, the total energy of several interacting pairs of electrons described in this way is just the sum of expressions like the above *plus* the interaction terms. The interaction terms are easily seen to be of "MO" type since they all involve (single spin-paired) doubly occupied orbitals; a typical interaction term between pairs R and S, say, is

$$V_{RS} = \sum_{k \in R} \sum_{j \in S} C_k^2 C_j^2 (2\text{tr} J(R^k) R^j - \text{tr} K(R^k) R^j)$$

Here, the notation is beginning to show the strain a little; clearly the coefficients C_k and the R-matrices R^k should ideally have another symbol to indicate with which pair they are associated, but such notation is inclined to be rather fussy.

The set-inclusion notation has been used rather loosely; $k \in R$ means "ψ_k and all its associated quantities belong to pair R".

We may now collect all this information together in a compact form as

$$E = \sum_S \sum_{k \in S} \nu_k \text{tr} \boldsymbol{h}_k \boldsymbol{R}^k \tag{22.15}$$

where

$$\nu_k = 2C_k^2 \qquad k \in S$$
$$\boldsymbol{h}_k = \boldsymbol{h} + \frac{1}{2}\boldsymbol{G}_k$$

and

$$\boldsymbol{G}_k = \boldsymbol{J}(\boldsymbol{R}^k) + \sum_{j \in S \neq k} \frac{C_j}{C_k} \boldsymbol{K}(\boldsymbol{R}^j) + \sum_{R \neq S} \sum_{i \in R} \nu_i [2\boldsymbol{J}(\boldsymbol{R}^i) - \boldsymbol{K}(\boldsymbol{R}^i)]$$

Using the redundancy associated with the self-interaction and

$$\boldsymbol{G} = 2\boldsymbol{J} - \boldsymbol{K}$$

the expression for $\boldsymbol{G}_k$ may be written

$$\boldsymbol{G}_k = \boldsymbol{G}(\boldsymbol{R}^k) + \sum_{j \in S \neq k} \frac{C_j}{C_k} \boldsymbol{K}(\boldsymbol{R}^j) + \sum_{R \neq S} \sum_{i \in R} \nu_i \boldsymbol{G}(\boldsymbol{R}^i) \tag{22.16}$$

which shows the relationship to the familiar closed-shell expression when all the C_j are zero and C_k is unity; all the $\boldsymbol{G}_k$ are then identical.

Notice that the expression is general enough to have different pairs described by different expansion lengths and, in particular, we may choose to have some pairs described by a one-term expansion. It may well be appropriate to use this simple "MO" description for inner shells, lone pairs and, generally, the less vital parts of the molecular electronic structure.

22.5.3 The orbital Fock matrices

Having taken the trouble to put the energy of the GVB wavefunction into the form eqn (22.15), and being familiar with the variational method, it is clear that the self-consistent sets of *orbital* coefficients are determined by the equations derived in Chapter 14 and the individual orbital Fock matrices are obtained from the $\boldsymbol{h}_k$ of eqn (22.15) simply by doubling the multiplier of $\boldsymbol{G}_k$, so that

$$\boldsymbol{h}_k^F = \boldsymbol{h} + \boldsymbol{G}_k$$

22.5. DETAILS OF GVB

where $\boldsymbol{G}_k$ is given by eqn (22.16).

The machinery for assembling these matrices into the single effective Fock matrix is, by now, familiar and runs exactly parallel to the ROHF and PEMC-SCF cases.

Note that even if *all* the electrons in a system are to be described by pair functions with more than a single doubly occupied term, there may still be an empty shell, unlike the PEMCSCF case, since we may use more basis functions than the sum of the required doubly occupied orbitals.

22.5.4 The pair expansion coefficients

In order to set up the $\boldsymbol{h}_k^F$ we need the pair expansion coefficients C_k for each pair of electrons in exactly the same way as the PEMCSCF case. However, in this case the expansion lengths are much shorter but there are more of them. If we choose to work with the simplest, Heitler–London, case of two terms per pair there would be a 2×2 CI problem to solve for the lowest eigenvalue of each pair; the solutions here can be simply written down. Longer expansion lengths will require the lowest eigenvector of a small CI problem for each pair.

The matrix elements of the CI matrix are as simple as the ones involved in PEMCSCF for the same reason; only doubly occupied orbitals are involved and, in fact, each pair is described by a local PEMCSCF wavefunction. What is required for each pair is the energy of each doubly occupied orbital ψ_k in the field of the other pairs. This is obviously obtainable from our earlier expressions by the omission of the "interaction" terms between different orbitals of the same pair. These interaction terms are the off-diagonal elements of the CI matrix.

More precisely, if the CI matrix for a particular pair S is $\boldsymbol{H}$ then

$$\begin{aligned} H_{kk} &= 2\mathrm{tr}\boldsymbol{h}\boldsymbol{R}^k + \mathrm{tr}\boldsymbol{G}(\boldsymbol{R}^k)\boldsymbol{R}^k \\ &+ \sum_{R \neq S}\sum_{i \in R} \nu_i \mathrm{tr}\boldsymbol{G}(\boldsymbol{R}^i)\boldsymbol{R}^k \end{aligned}$$

and

$$H_{kj} = \mathrm{tr}\boldsymbol{K}(\boldsymbol{R}^k)\boldsymbol{R}^j$$

where $k, j \in S$.

This matrix is assembled from the matrices already prepared for the Fock matrices.

22.6 Implementation

There are no new problems of principle here; the framework of Chapter 14 is used. What is new is simply the detailed *form* of the individual orbital Fock matrices and the fact that there are several simultaneous CI problems. It is clear that we need all the R^k, all the $G(R^k)$ and all the $K(R^k)$ for the h_k^F and the elements of H so the *same* codes which we developed for PEMCSCF for the generation of these matrices can be used in the GVB program. The two cases only differ in the way in which the orbital Fock matrices are assembled. Techniques which are very similar to those used in the many-shell MO method of Chapter 15 are used to assemble the various J and K matrices and the elements of the CI matrices are all formed from these matrices.

There are available at the ftp site WEBs for both PEMCSCF and GVB.

Chapter 23

Interpreting the McWeenyan

The full form of the single effective Fock matrix for all the orbitals of an energy functional can be interpreted to shed some light on the nature of the SCF process and the way in which a stationary condition in the energy functional is determined by the optimum orbitals.

Contents

23.1	Introduction		605
23.2	Stationary points		607
23.3	Many shells		608
23.4	Summary		610

23.1 Introduction

The effective Hartree–Fock matrix equation for a many-shell system has been derived in Chapter 14 and used in several applications; open shells and some MCSCF models. So far, it has been seen simply as the formally correct equation to generate SCF orbitals for these many-shell structures *without any interpretation*. In particular, the fact that the effective Hartree–Fock matrix (the "McWeenyan") contains many *arbitrary* parameters has not been addressed, nor has the practical problem of the actual grounds for the *choice* of values for these parameters been systematised. In looking at this problem we must bear two points in mind:

1. The variational method which was used to generate *all* the SCF equations only ensures a *stationary* point in the energy functional; we have not looked at the second variation to test for a (local) minimum.

2. The occurrence of any "empty shell" of virtual MOs is an artifact of the linear expansion approximation in the sense that the "empty space" is simply the residuum when the occupied spaces have been chosen by the variation process.

We said in Chapter 14 that, in the single-determinant UHF or closed-shell SCF models, it is not necessary to incorporate the "shell constraints"

$$\boldsymbol{R}^K \left(\nu_K \boldsymbol{h}_K^F - \nu_L \boldsymbol{h}_L^F \right) \boldsymbol{R}^L = 0$$

into the matrix Hartree–Fock equation since there is just *one* Hartree–Fock matrix for all the orbitals and so these constraints are trivially satisfied;

$$\nu_L = \nu_K \quad \boldsymbol{h}_L^F = \boldsymbol{h}_K^F$$

for all K and L.

This is not quite true since, when an expansion method is used, there is always an "empty shell". If there were no empty shell (*i.e.* if the number of basis functions were not greater than the number of MOs) there would be no linear variational problem to solve.

If we therefore apply the "many-shell" formalism to the closed-shell HF problem we obtain the effective Fock matrix

$$\bar{\boldsymbol{h}} = a\boldsymbol{R}\boldsymbol{d}\boldsymbol{R} + b\boldsymbol{R}\boldsymbol{h}^F(\boldsymbol{1} - \boldsymbol{R}) - b(\boldsymbol{1} - \boldsymbol{R})\boldsymbol{h}^F\boldsymbol{R}$$

where the expression is considerably simplified since:

- There is only one occupied $\boldsymbol{R}$ matrix and so the empty-shell $\boldsymbol{R}$ matrix is just $\boldsymbol{1} - \boldsymbol{R}$.

- There is only one $\boldsymbol{h}^F$.

- There are only two arbitrary parameters a and b.

- There is only one arbitrary matrix $\boldsymbol{d}$.

We can use this "degenerate form" of the SCF problem to cast some light on the interpretation of the single effective Fock matrix.

23.2 Stationary points

There are two general possibilities with the above Fock matrix:

1. Set $a \neq 0$ and $b = 0$ and make a choice of $\boldsymbol{d}$.
2. Set $a = 0$ and $b \neq 0$.

If we make the first choice and choose
$$\boldsymbol{d} = \boldsymbol{h}^F$$
this generates the usual Fock matrix for the closed-shell case and when the equation
$$\bar{\boldsymbol{h}}\boldsymbol{C} = \boldsymbol{C}\boldsymbol{\epsilon}$$
is solved self-consistently, the fact that the matrix $\boldsymbol{\epsilon}$ is *diagonal* and the mutual orthogonality of the occupied and virtual orbitals ensures that
$$\boldsymbol{R}\boldsymbol{h}^F(1-\boldsymbol{R}) = (1-\boldsymbol{R})\boldsymbol{h}^F\boldsymbol{R} = 0$$
so that the second term in $\bar{\boldsymbol{h}}$ vanishes automatically when the coefficients are self-consistent.

Conversely, if we make the second choice setting $a = 0$ and *were able*, by some numerical procedure, to find a set of MO coefficients $\boldsymbol{C}$ which generated a matrix $\boldsymbol{R}$ such that
$$\boldsymbol{R}\boldsymbol{h}^F(1-\boldsymbol{R}) = 0$$
then, of course, such a $\boldsymbol{C}$ would solve
$$\bar{\boldsymbol{h}}\boldsymbol{C} = \boldsymbol{C}\boldsymbol{\epsilon}$$
and the two choices would seem to be equivalent and the full definition of the matrix $\bar{\boldsymbol{h}}$ redundant.

This conclusion is too facile for at least two reasons:

1. The practical: while choice 1 can be implemented by solving an iterative eigenvalue problem, choice 2 clearly cannot; since, with $a = 0$, as self-consistency is approached
$$\bar{\boldsymbol{h}} \longrightarrow 0$$
which is not useful if one is solving
$$\bar{\boldsymbol{h}}\boldsymbol{C} = \boldsymbol{C}\boldsymbol{\epsilon}$$

2. The condition
$$Rh^F(1 - R) = 0$$
simply ensures that the MO coefficients which generate R represent a stationary point of the energy functional.

Without the direct involvement of h^F in $\bar{h}$, we have no way of choosing *which* stationary point we might be at. Again, from a practical point of view, it is not the vanishing of the off-diagonal part of h^F which is crucial to the generation of the optimum orbitals, but the *diagonalisation* of ϵ and the use of the *aufbau* principle which gives us some orientation in the whole SCF process.

We can see, therefore, that in a certain sense it is good luck that the UHF or closed-shell SCF process can be solved uniquely without the involvement of the additional "off-diagonal" terms in the Fock matrix. If there is only *one* occupied "shell" then the "empty shell" is uniquely determined. Expressing the stationary condition in terms of the occupied-empty projection of the Fock matrix is formally correct but not useful.

If we picture the SCF process as it converges to a stationary (self-consistent) set of MOs using $\bar{h}$, what is happening numerically is that

$$\bar{h} \longrightarrow h^F$$
$$Rh^F(1 - R) \longrightarrow 0$$

simultaneously because of the properties of the unique matrix h^F.

23.3 Many shells

With this analysis of the simple one-shell case the interpretation of the many-shell case is much more straightforward.

Just as in the closed-shell case, the formal requirement for the existence of a stationary point in the energy functional is the vanishing of the "off-diagonal" projections of the differences between the shell Fock matrices. But equally, this is neither a practical procedure for the actual computation of the optimum MO coefficients determining that stationary point nor is it sufficient to *identify* the particular stationary point found. In particular, we are normally looking for the ground state.

However, unlike the closed-shell case, the diagonalisation of the individual Fock matrices is *not* sufficient to find a stationary point since, now that there are *several* Fock matrices, simultaneous diagonalisation of them does not necessarily generate a set of MO coefficients which are immune to improvement by

23.3. MANY SHELLS

mixing with each other. The diagonalisation of, for example, two Fock matrices generates two sets of MO coefficients which cannot all be orthogonal let alone non-mixing.

If, therefore, we arrange to diagonalise all the shell Fock matrices while *simultaneously* reducing the off-diagonal projections to zero the resulting MO coefficients will be optimum and self-consistent. By choosing the "arbitrary" matrices d_T to depend linearly on the shell Fock matrices this is exactly what is accomplished by the *self-consistent* diagonalisation of $\bar{h}$ since

- The projected parts of the relevant shell Fock matrices generate the required shell MOs.
- The off-diagonal projections must be reduced to zero before subsequent diagonalisations will *reproduce* the same $\bar{h}$ to ensure self-consistency.

It is also now clear why the coefficients by which the off-diagonal projections are multiplied are arbitrary; if these matrices are to be reduced to zero by the SCF iterative process it does not matter (formally) by how large or small a factor they are scaled. Naturally, it is of some *practical* importance to choose sensible multipliers since one would not want to swamp the individual Fock matrices which are "driving" $\bar{h}$ to a self-consistent solution.

The reason for the fact that the d_T may be arbitrary is also clear but, if anything, even more remote from the practical details of ground-state calculations. If the vanishing of the off-diagonal projections is the sole formal requirement for a stationary point in the energy functional then the matrices d_T are free to be used to determine *which* stationary point is to be located. The formal determination of the stationary point determines only the shell *sub-spaces*, not the individual orbitals which span that sub-space, just as in the one-shell case the wavefunction is unchanged by linear transformations amongst the MOs of each shell. Since diagonalisation of $\bar{h}$ implies diagonalisation of the d_T, choosing a set of d_T will do two things:

1. By being diagonalised, the d_T determine the particular set of MOs which span the shell sub-space; these orbitals might be called "canonical with respect to these d_T" by analogy with *the* canonical set in the UHF or closed-shell cases.

2. The initial choice of the d_T determines *which* stationary point is found by choosing the initial R matrices.[1] For example, choosing

$$d_T = -h_T^F$$

would give an unphysical initial occupancy of the highest-energy MOs if the usual *aufbau* occupancy were built into the program.

---
[1]This statement can never be made precise because of the way the iterative process works but (*e.g.*) often initial symmetry choices are "locked into" the iterative process.

It is clear both from practical, computational experience and the considerations of this chapter that the choice of the as bs and ds may be used to manipulate both the convergence properties of the SCF process and the stationary point to be located.

23.4 Summary

The single effective matrix Hartree–Fock equation for systems composed of several shells of electrons combines *all* the requirements of a self-consistent turning point in the corresponding energy functional; it ensures that:

- Each shell has an optimised invariant "shell space" which may be spanned by a variety of linearly independent canonical MOs depending on the choice of the matrices $\boldsymbol{d}_K$.
- The inter-shell elements of the Hartree–Fock matrix are simultaneously reduced to zero.

The freedom of choice in the form of the matrices $\boldsymbol{d}_K$ and in the scalar parameters may be utilised to affect the rate of convergence of the iterative SCF procedure and the particular self-consistent "state" to which convergence is required.

Chapter 24

Core potentials

There are two kinds of barrier to the extension of ab initio *methods of calculation of molecular electronic structure to molecules containing heavy atoms. The number of basis functions required to describe the inner shell electronic sub-structures causes the calculation to become prohibitively large and the effects of relativity begin to be noticeable, which causes the assumptions of a "spin-only" Hamiltonian and the implicit Russell–Saunders coupling scheme to break down. Both of these effects are particularly frustrating since we are prone to regard the inner shells as basically unchanged on molecule formation and the effects of relativity on the chemical properties of any molecule to be indirect. In this chapter we examine the possibility of circumventing the problem of atomic inner shells.*

Contents

24.1	Introduction	612
24.2	Simple orthogonalization	614
24.3	Transforming the Hartree–Fock equation	615
24.4	The pseudopotential	618
24.4.1	Frozen cores	620
24.5	Arbitrariness in the pseudo-orbital	621
24.6	Modelling atomic pseudopotentials	624
24.7	Modelling atomic core potentials	626
24.8	Several valence electrons	629
24.9	Atomic cores in molecules	633
24.10	Summary	634

24.1 Introduction

Examination of the results of any calculation of the electronic structure of any molecule shows that, where the molecule actually contains atoms with "inner shells" of electrons, the inner shells are essentially undisturbed by molecule formation; there are always low-lying "MOs" which are well-separated in energy from the next "band" of MOs and which are composed almost exclusively of the inner-shell AOs or basis functions of the component atoms of the molecule.

To take a concrete example, any MO calculation of the electronic structure of the ethene (ethylene) molecule will generate two lowest MOs (almost degenerate) which are just the in-phase and out-of-phase linear combinations of the basis functions used to describe the $1s$ shells of the carbon atoms. The fact that they occur as *molecular* orbitals rather than remaining actually unchanged as *atomic* orbitals is simply an artifact of the symmetry of ethene; the MOs are computed as symmetric or antisymmetric with respect to the operations of the point group which in this case includes reflection in a plane perpendicular to the C—C axis. If a calculation is carried through on the isoelectronic methanal (formaldehyde) molecule the oxygen $1s$ AO and the carbon $1s$ AO survive the calculation almost unscathed as the lowest "MOs".

These results are totally expected; they reflect the concept of core-valence separation: the most obvious of the ideas of the environment-insensitivity of groups of electrons in molecular structure.

If Gaussian functions are used as a basis for these calculations then it is a matter of common experience that, because of the very large contribution to the electronic energy of these rapidly moving electrons close to the nuclei, a much larger contraction length has to be used for the core basis functions than for the valence basis. A factor of two is typical for the ratio of core primitives to valence primitives.

This latter fact exacerbates the problem; for variational reasons we are forced to use a long contraction to give an adequate description of sub-structures of electrons which are substantially indifferent to their environment.

It would obviously be convenient to be able to incorporate the fact that the inner shells of electrons are not affected by molecule formation into our computational technique *at the outset*. In a word it would be extremely useful to be able to *simulate* the effect of the core electrons on the valence shells.

The difficulties associated with such a project are quickly apparent if we consider the simplest case; the lithium atom. A crude way of simulating the two $1s$ electrons of lithium is simply to assume that they *completely screen* the outer $2s$ valence electron from the core to generate a net effective nuclear

24.1. INTRODUCTION

charge of $Z_{eff} = 3 - 2 = 1$. This apparently gives us a one-electron Schrödinger equation to solve for the valence electron.

$$\left(-\frac{1}{2}\nabla^2 - \frac{Z_{eff}}{r}\right)\chi = E\chi$$

But this equation is the Schrödinger equation for the *hydrogen* atom ($Z_{eff} = 1$) and its lowest solution is the $1s$ AO of a hydrogenic atom. The required "valence" orbital is the *second* solution of the equation. This second solution is, of course, *orthogonal* to the lowest solution; what is, in fact, required is the lowest solution of this Schrödinger equation which is orthogonal to the wavefunctions of the simulated electrons.

The fact that we require the *second* solution of the effective Schrödinger equation is not due to the Schrödinger equation itself but due to the fact that we have *implicitly* made some rather gross *model* assumptions about the electronic structure of atoms in setting up this simple case.

The idea that the ground-state electronic structure of the lithium atom is $1s^22s$ contains two (unstated) assumptions:

- The idea that the wave function in the lithium atom must satisfy the *Pauli principle*. If it were not for the Pauli principle, we would describe the electronic structure of all the atoms as $1s^Z$ and there would be no problem; all we would ever require of the Schrödinger equation for atoms would be its lowest solution, something like the $1s$ AO. If, however, the Pauli exclusion principle is to operate we require a new solution for (at least) every third atom in the periodic table.

- The electronic structure is being described by a special case of the single-determinant model; the spin-eigenfunction-restricted, spherical-approximation open-shell Hartree–Fock model.

The requirement of orthogonality to inner orbitals is a guarantee that a valence orbital would satisfy the requirements of the Pauli principle *whether or not the inner orbitals are explicitly present in the calculation.*

The effects of the use of the *restricted* HF model are associated with the idea of "frozen cores" and will become clear shortly.

There are, therefore, two requirements to be met if we are to simulate the effect of any core electrons and avoid explicitly computing the wave functions for these invariant sub-structures:

- The potential generated by the frozen cores must be present in the calculation; there has to be some way of simulating the interactions (Coulomb repulsions and exchange interactions) between the core electrons and the valence shells.

- The "Pauli repulsion" must also be simulated; the problem that the valence shell orbitals are *not* simply the *lowest* solutions of the effective Schrödinger (or Hartree–Fock) equation in the presence of the electrostatic core potential must be addressed.

24.2 Simple orthogonalization

From what was said in the last section it is obvious that, provided a means can be found to generate the electrostatic potential due to the core electrons, the use of a set of basis orbitals which are *orthogonal* to the (omitted) core orbitals, enables calculations to be carried through which are only limited by the quality of the description of the core potential.

Unfortunately, the way to make the valence orbitals orthogonal to the core orbitals is to form new valence orbitals which are linear combinations of the original valence orbitals *and* the core orbitals. In the case of the lithium atom, the valence basis functions used to describe the $2s$ AO would each have to contain linear combinations of the *core* basis functions to ensure core/valence orthogonality (this is one role of the node in the hydrogenic $2s$ AO; to ensure orthogonality to the $1s$ AO).

This defeats any *computational* advantages of using only a valence basis; all the molecular *integrals* over the entire (core plus valence) basis must be computed. The only (small) saving would be a reduction in the dimension of the eigenvalue problem to be solved to the size of the new valence basis. In the language which we have developed for the description of basis sets this means very long contractions for the valence basis functions: the elimination of the core basis is merely formal; all the practical difficulties remain.

It is not too difficult to see that this simple example contains *all* the difficulties associated with the full problem of the replacement of core electrons by some "effective" method. The potential due to filled shells of electrons might be expected to be not *too* difficult to simulate; after all:

- Classically, the potential due to a spherical charge distribution acts like a point charge at external points and the core distribution falls away very sharply with distance from the nucleus.

- The electrostatic potential from a charge distribution is generally a smoother function than the distribution itself.

There are some counter-points, however:

24.3. TRANSFORMING THE HARTREE–FOCK EQUATION

- The valence orbitals are not, in fact, "external" to the cores; each function vanishes only at infinity. There is definite *penetration* of the core region by the valence electrons.

- The quantum-mechanical energy expression does not correspond to the classical electrostatic model; there are *exchange* terms in the energy expression which might prove difficult to simulate.

- And, perhaps less vital, there is always the question of the importance of the polarisation of the cores by the valence electrons. For example a DODS calculation of the electronic structure of the lithium atom yields a split core structure ($1s1s'$) the effects of which are *automatically* excluded by a core potential.

While it might prove possible to transform a given Schrödinger (or Hartree–Fock) equation into an equivalent effective equation for the valence-electrons only; any *exact* transformation cannot possibly be a net saving because the total electronic energetics and distribution must still be governed by the resulting equation; the equations governing the behaviour of groups of charged particles cannot be solved by prestidigitation.

What *can* be done, however, is to make the exact transformation and use the results as a guide for the development of *models* of the valence electronic structure which *do* yield net savings by removing any explicit treatment of the core electrons *and* do not involve the covert use of the core basis functions.

24.3 Transforming the Hartree–Fock equation

The most straightforward way to approach the problem is, perhaps, to work within the model of molecular electronic structure which is the most familiar and which transparently satisfies the Pauli principle: the single-determinant model.

Suppose we have a set of $n+1$ electrons occupying $n+1$ orbitals χ_i to form a determinant Φ:

$$\Phi = \frac{1}{\sqrt{n+1}} \begin{vmatrix} \chi_1(\vec{x}_1) & \chi_1(\vec{x}_2) & \cdots & \chi_1(\vec{x}_{n+1}) \\ \chi_2(\vec{x}_1) & \chi_2(\vec{x}_2) & \cdots & \chi_2(\vec{x}_{n+1}) \\ \chi_3(\vec{x}_1) & \chi_3(\vec{x}_2) & \cdots & \chi_3(\vec{x}_{n+1}) \\ \cdots & \cdots & \cdots & \cdots \\ \cdots & \cdots & \cdots & \cdots \\ \cdots & \cdots & \cdots & \cdots \\ \chi_{n+1}(\vec{x}_1) & \chi_{n+1}(\vec{x}_2) & \cdots & \chi_{n+1}(\vec{x}_{n+1}) \end{vmatrix}$$

and we wish to develop an equation for the "highest" orbital χ_{n+1} which is equivalent to the full Hartree–Fock equation in the sense of generating an orbital with the same energy but which is the lowest solution of the equivalent equation.

Now it is well-known from the theory of self-adjoint differential equations that the lowest solution of a "Schrödinger-like" equation has *no nodes* and so it is quite impossible that the solution we seek will be *identical* to the $(n+1)$th solution of the HF equation.

If we use the UHF model, all the χ_i ($i=1, n+1$) are solutions of the HF equation
$$\hat{h}^F \chi_i = \epsilon_i \chi_i$$
and, as such, are an orthonormal set. These orbitals are not *required* to be orthonormal, and our approach will be to remove the condition that the highest orbital χ_{n+1} be orthogonal to the others in anticipation of the nodeless property of the equivalent orbital. The single-determinant Φ is unchanged by linear transformations amongst the orbitals χ_i, so we leave the first n orbitals unchanged and define a new orbital χ (without a subscript) by the condition that it is related to χ_{n+1} by Schmidt orthogonalisation against the other χ_i:
$$\chi_{n+1} = \chi + \sum_{i=1}^{n} c_i \chi_i$$
Since only one orbital is involved we may use the simple two-orbital Schmidt method which gives:
$$c_i = -\int d\tau \chi_i^* \chi$$
The normalisation is achieved by
$$\chi_{n+1} = N(\chi + \sum_{i=1}^{n} c_i \chi_i)$$
where
$$N = \left(1 - \sum_{i=1}^{n} c_i^2\right)^{\frac{1}{2}}$$
Then the determinant with χ_{n+1} replaced by χ is identical to the original determinant of HF solutions.
$$\hat{h}^F \chi_i = \epsilon_i \chi_i$$
for $(i = 1, n+1)$. In particular
$$\hat{h}^F \chi_{n+1} = \epsilon_{n+1} \chi_{n+1}$$
so that
$$\hat{h}^F (\chi + \sum_{i=1}^{n} c_i \chi_i) = \epsilon_{n+1}(\chi + \sum_{i=1}^{n} c_i \chi_i)$$

24.3. TRANSFORMING THE HARTREE–FOCK EQUATION

where the normalisation constant cancels. The fact that the χ_i are eigenfunctions of $\hat{h}^F$ gives

$$\hat{h}^F \chi + \sum_{i=1}^n c_i \epsilon_i \chi_i = \epsilon_{n+1} \chi + \sum_{i=1}^n c_i \epsilon_{n+1} \chi_i$$

which may be rearranged to

$$\hat{h}^F \chi + \sum_{i=1}^n -c_i(\epsilon_{n+1} - \epsilon_i)\chi_i = \epsilon_{n+1}\chi$$

which is a "Hartree–Fock-like" equation for χ containing a correction term involving the χ_i. We would like to bring this equation into the form

$$(\hat{h}^F + V)\chi = \epsilon\chi$$

(dropping the subscript on ϵ_{n+1} to match the unsubscripted χ) and there are two ways to achieve this transformation; both of them depend on knowing the form of the orbitals χ_i:

1. The coefficients c_i are given explicitly by the overlap integrals

$$-c_i = \int d\tau \chi_i^*(\vec{x})\chi(\vec{x}) = \int d\tau \chi(\vec{x})\chi_i^*(\vec{x})$$

($d\tau$ is the volume element associated with the space-spin coordinates $\vec{x}$) so that we may generate an operator form for the terms in V by writing

$$-c_i\chi_i = \hat{P}_i\chi = \chi_i(\vec{x}) \int d\tau' \chi_i^*(\vec{x}')\chi(\vec{x}')$$

where the *projection operator* $\hat{P}_i$ needs a little more specification than the above symbolic definition;

> The action of $\hat{P}_i$ is to change the name of the variable of its operand (to $\vec{x}'$ in our case), multiply by the product of χ_i of the old variable and χ_i^* of the renamed variable, and integrate over this newly named variable.

This technique is the standard way of introducing the action of operators which, when operating on a function in their domain, generate results which do not simply depend on that function and its infinitesimal neighbourhood; so-called *non-local* operators.[1] We may now write

$$\left(\hat{h}^F + \sum_{i=1}^n (\epsilon - \epsilon_i)\hat{P}_i\right)\chi = \epsilon\chi$$

---
[1] The exchange operator of Hartree–Fock theory is another such non-local operator.

i.e.
$$V = \sum_{i=1}^{n}(\epsilon - \epsilon_i)\hat{P}_i$$
which is an equation of the required *form*; it is an equation for the "outer" electron only in which all the effects of the Pauli principle have been absorbed into V.

2. The second approach is mathematically more simple. In order to make the correction term appear as an operator acting on χ we simply multiply it by unity in the form of χ/χ:

$$\sum_{i=1}^{n} -c_i(\epsilon_{n+1} - \epsilon_i)\chi_i = \sum_{i=1}^{n} -c_i(\epsilon_{n+1} - \epsilon_i)\chi_i \times \frac{\chi}{\chi} = V\chi$$

where
$$V = \frac{1}{\chi}\sum_{i=1}^{n} -c_i(\epsilon_{n+1} - \epsilon_i)\chi_i$$

Notice that, since we are seeking the *lowest* solution of the resulting equation determining χ, there is no problem of dividing by zero since this χ is nodeless. It is also worth emphasising that this form of the potential is most useful because we are seeking its action on just *one* function (χ). If we were to require the action of the potential on other orbitals we would have to define a similar potential *for each orbital*.

These two procedures give expressions for V which are very different in appearance and have different mathematical properties and yet, because of the way that they have been constructed, they must be equivalent and be satisfied[2] by the same lowest solution χ.

The most obvious difference between the two forms of V is that the first represents V as a *non-local* operator while the second has just a multiplying factor. We shall see that these two forms are, to a large extent *complementary*; the local form is ideal for *visualization* of the operator (it can be calculated and plotted), while the non-local form enables us to study some of the formal properties of V.

24.4 The pseudopotential

Our equation for the orbital χ has been obtained by a transformation of the HF equation and so only has the range of validity of the single-determinant model of electronic structure. Thus far, we have written the equation as:

$$(\hat{h}^F + V)\chi = \epsilon\chi$$

---
[2] They can hardly be said to *generate* the same lowest solution χ, since one must know χ in order to set up the equations.

24.4. THE PSEUDOPOTENTIAL

where the effect of V is *entirely to replace the effects of the Pauli principle* in the original single-determinant model. We must not be *too* euphoric that the transformation has proved possible; all our theory so far applies to *one electron* "outside" a set of n electrons. The question of an analogous theory for *many* valence electrons outside a core has not yet been addressed. In our context here we hardly want to develop a theory whose only application is Rydberg spectra.

Because V is not a "genuine potential" in the usual sense of the word, it is now universally called a "pseudopotential". In view of the fact that it is possible to express the pseudopotential as a non-local operator, we will extend the notation to include the operator nature of V and give it the subscript PS to remind ourselves that it *is* the pseudopotential i.e.

$$V \to \hat{V}_{PS}$$

If we write our "HF" equation in a more explicit notation

$$\left(-\frac{1}{2}\nabla^2 + \sum_A V_{nuc}^A + \hat{J} - \hat{K} + \hat{V}_{PS}\right)\chi = \epsilon\chi$$

and call all the non-kinetic-energy terms "potentials" then we can see the relative role of $\hat{V}_{PS}$ in the scheme of things and begin to think about how we might *model* this equation in such a way that it is not simply tautologous; a restatement of the original $(n+1)$-particle HF equation.

Here

- V_{nuc}^A is the nuclear-attraction term for nucleus A:

$$V_{nuc}^A = -\frac{Z_A}{r - R_A}$$

and is a genuine (local) potential.

- $\hat{J}$ is the electrostatic (Coulomb) term representing the repulsions between the electron occupying χ and all the others in the molecule:

$$\hat{J} = \int d\tau_2 \frac{\rho(\vec{x}_2)}{|r_1 - r_2|}$$

and is again a genuine local potential.

- $\hat{K}$ is the exchange term from the single-determinant expression,

$$\hat{K}\chi(\vec{x}_1) = \sum_{i=1}^{n+1} \int d\tau_2 \frac{\chi_i(\vec{x}_1)\chi_i^*(\vec{x}_2)\chi(\vec{x}_2)}{|r_1 - r_2|}$$

the largest part of which is to cancel the "self-repulsion" wrongly included in $\hat{J}$ but containing a contribution due to the Pauli principle which generates the "Fermi hole"; a non-local term with no classical equivalent. Notice that the non-local exchange operator $\hat{K}$ is Pauli-principle dependent like the non-local pseudopotential.

- $\hat{V}_{PS}$ is our pseudopotential arising from the Pauli principle and capable of non-local formulation.

The moment we begin to think about the pseudopotential we begin to appreciate the size of the problem before us, both theoretical and computational. The expression for $\hat{V}_{PS}$ contains *all the orbitals* other than the one we are trying to compute; that is, in a molecule, all the *molecular* orbitals whose computation we are trying to avoid. Now each molecule will have its own $\hat{V}_{PS}$ since each molecule has its own electron distribution and its own shape etc. There is clearly no future in the endeavour of using pseudopotentials *unless* they can be chosen in a more modular way. That is, pseudopotentials will remain an interesting theoretical transformation unless they can be broken down along the same lines as all the other quantities in the electronic theory of molecules are broken down; along atomic lines. Just as we use a nuclear potential which is the sum of atomic terms and atomic basis functions, the most urgent need if we are to develop a practical pseudopotential method is to investigate *atomic* pseudopotentials and see if they are *additive and transferable*. That is, to decide if molecular pseudopotentials are capable of being written as

$$\hat{V}_{PS} = \sum_A \hat{V}_{PS}^A$$

for a set of atoms A?

Quite apart from the problem of generating a workable form for the pseudopotential there is the question of the *grouping together* of some of the "potential" terms in the equation for χ; it is obvious that the "net potential" experienced by an electron in a molecule, when expressed as the *sum* of all the potentials, might well show large cancellations to yield, for example, the net potential as a sum of "atomic potentials" whose forms might prove as easy or easier to approximate than the individual contributing terms.

24.4.1 Frozen cores

The theory developed thus far has been for "one electron outside an inner core". Of course, nothing in the general development actually *requires* the single electron to be the outer one but the expectation is that this will be the main useful case. Our derivation actually hinged on the fact that *all* the orbitals

24.5. ARBITRARINESS IN THE PSEUDO-ORBITAL

χ_i ($i = 1, n+1$) were solutions of the same HF equation. But, if this is the case, the inner core orbitals will be spin-polarised and there will be different HF potentials for different spins which rather defeats the idea of *the* core potential augmented by the pseudopotential.

In fact, for the special case of one electron outside a set of occupied inner orbitals, we can use frozen core orbitals in a formulation which is identical in form to the above theory. We assume that the HF equation has been solved for the inner orbitals:

$$\hat{h}^F \chi_i = \left[\hat{h} + \sum_{i=1}^{n} \left(\hat{J}_i - \hat{K}_i \right) \right] \chi_i = \epsilon_i \chi_i$$

Now we "freeze" these orbitals and introduce the outer orbital and, since there is no self-consistency requirement for the introduced single electron, the equation for the outer electron is just the same as the HF equation for the frozen core orbitals.

That is, if we do not allow the outer electron to disturb the inner electrons' distributions, *all* the orbitals are solutions of the *same* equation which is:

- The Hartree–Fock equation from the point of view of the n-electron system.
- The single-electron equation from the point of view of the outer electron in the presence of the frozen cores.

This result means that we can, in the frozen-core approximation, use the idea of "doubly occupied" orbitals in the usual sense and this formalism is identical to the one we have already developed.

24.5 Arbitrariness in the pseudo-orbital

Projection operators are dangerous objects to have in equations used to determine orbitals. They are capable of destroying functions, indeed their whole method of operation is to destroy those parts of a function which are not in the function space onto which they project. So, for example, if we have some operator $\hat{A}$ and a projection operator $\hat{P}$ then if

$$\hat{A}f = g$$

and

$$\hat{P}h = 0$$

it follows that
$$\hat{A}\hat{P}(f + \lambda h) = g$$
Multiplication of an operator on the right by a projection operator means that any linear combination of functions which are orthogonal to the space onto which $\hat{P}$ projects can be added to a function on which an operator acts without effect. Let us see how this bears on our equation for χ. Obviously we choose to work with the non-local form of $\hat{V}_{PS}$:

$$\hat{V}_{PS} = \sum_{i=1}^{n}(\epsilon - \epsilon_i)\hat{P}_i$$

where each $\hat{P}_i$ projects onto χ_i:

$$\hat{P}_i f = a_i \chi_i$$

$$a_i = \int d\tau \chi_i^* f$$

so that, in particular, for the set of orthonormal χ_i:

$$\hat{P}_i \left(\sum_{i=1}^{n} b_i \chi_i \right) = b_i \chi_i$$

Now, our pseudopotential equation for this choice of $\hat{V}_{PS}$ is

$$\left(\hat{h}^F + \sum_{i=1}^{n}(\epsilon - \epsilon_i)\hat{P}_i \right) \chi = \epsilon \chi$$

If we replace χ in this equation by $\chi + \chi_k$ where k lies between 1 and n, i.e. χ_k is one of the inner orbitals then, evaluating the terms in the equation:

$$\hat{h}^F(\chi + \chi_k) = \hat{h}^F \chi + \epsilon_k \chi_k$$
$$\left(\sum_{i=1}^{n}(\epsilon - \epsilon_i)\hat{P}_i \right)(\chi + \chi_k) = \left(\sum_{i=1}^{n}(\epsilon - \epsilon_i)\hat{P}_i \right)\chi + \epsilon_k \chi_k$$
$$\epsilon(\chi + \chi_k) = \epsilon\chi + \epsilon\chi_k$$

Combining these results shows that the function $\chi + \chi_k$ *also satisfies our equation*. What has happened is that the fact that the χ_i are *eigenfunctions* of $\hat{h}^F$ has conspired with the projection properties of the $\hat{P}_i$ to make sure that the additional terms exactly cancel.

With this result it is trivial to show that the addition of any *linear combination* of the functions χ_i to χ

$$\chi + \sum_{i=1}^{n} d_i \chi_i$$

24.5. ARBITRARINESS IN THE PSEUDO-ORBITAL

also solves the pseudopotential equation for χ. That is, our pseudopotential equation for χ has infinitely many solutions which differ only by the addition of linear combinations of the transformed-away functions! Or, to put a more optimistic face on things, there is additional freedom in the choice of χ which can be utilised to insist on χ having some properties other than just solving the pseudopotential equation.

This conclusion may be reinforced from another angle by examining the effect of the valence Hamiltonian on one of the core functions χ_k, say:

$$\left(\hat{h}^F + \sum_{i=1}^{n}(\epsilon - \epsilon_i)\hat{P}_i\right)\chi_k = \epsilon_k\chi_k + (\epsilon - \epsilon_k)\chi_k = \epsilon\chi_k$$

where we have used the fact that

$$\hat{h}^F\chi_k = \epsilon_k\chi_k$$

and the mutual exclusiveness of the core projectors:

$$\hat{P}_i\chi_k = \delta_{ik}\chi_k$$

That is, all the core orbitals χ_i, $i = 1, n$ are eigenfunctions of the *valence* Hamiltonian with the same valence eigenvalue. This $(n + 1)$-fold degeneracy of the valence eigenvalue means that arbitrary linear combinations of these functions also satisfy the valence equation as we found by another route.

It is also clear that having the inner orbitals present in the definition of the pseudopotential *and* in the definition of the orbital χ as arbitrary linear combinations is probably redundant. We have introduced the outer orbital χ as a nodeless function on two grounds:

- Anticipating the fact that χ will be the *lowest* solution of the final equation.

- Ensuring that the local form of the pseudopotential is well-behaved; it contains no division by zero at nodes in χ.

But within the area of nodeless functions there is much freedom. For example, it is possible for a function to have no nodes and yet be extremely "bumpy" that is, to have regions of large gradient which means, physically, large local kinetic energy distributions. One way to use the freedom in the choice of χ is to include that linear combination of the core functions which cancels out any oscillations in the core region and so minimises the kinetic energy of χ; to impose a condition of "maximum smoothness" on χ.

If we do this then *both* the form of χ and the local form of the pseudopotential are fixed since the latter contains χ. In this case it is now worth looking briefly at the physical interpretation of the pseudopotential and the orbital χ which are inextricably linked together in the local form. The orbital calculated by the solution of an effective Schrödinger equation is often called a **pseudo-orbital** to emphasise its origins and to remind us of its non-unique nature.

The pseudopotential/pseudo-orbital pair are linked and what is achieved by the formulation of the valence orbital problem is a replacement of the effect of the Pauli principle. The Pauli principle causes electrons (of like spin) to avoid each other independently of their mutual repulsion; it generates the so-call Fermi hole around a particular electron. Now as the valence electron penetrates the core space it must have a distribution which reflects this Fermi hole; it must avoid the phantom core electrons or they must avoid it.[3] So the pseudopotential/pseudo-orbital pair must reflect this fact and this is why they are linked. If we choose to make the pseudo-orbital smooth then the local form of the pseudopotential becomes oscillatory and vice versa, so that the imposition of pseudo-orbital smoothness may have some ramifications for the choice of a model potential to simulate the effect of the pseudopotential.

24.6 Modelling atomic pseudopotentials

The most pressing practical problem is the one outlined in an earlier section; can the pseudopotentials be expressed as a sum of atomic terms? This demands an *empirical* investigation and the local form of the pseudopotential operator is ideal for this type of work.

Let us take, for example, the sodium atom (electron configuration $1s^2 2s^2 2p^6 (n\ell m)^1$) in the ground state and those excited states which are the lowest of a given symmetry type, so that there are no problems with the variational solution of the HF equations $(n\ell) = 3s, \ 3p, \ 3d \ 4f \ldots)$. We can (with a given choice of basis functions) solve the HF equations and therefore plot the pseudopotential

$$V_{PS} = \frac{1}{\chi_{n\ell m}} \sum_{i="1s"}^{"2p"} -c_i(\epsilon_{n\ell} - \epsilon_i)\chi_i$$

where $\chi_{n\ell m}$ is the $(n\ell m)$ outer AO and the sum goes over the "core" set $(1s^2 2s^2 2p^6)$.

What we find is that:

- The pseudopotential is very dependent on the the ℓ-value of the outer AO; the form of pseudopotential is very different for the $3s$ and $3p$ AOs.

---

[3] Perhaps it is not too fanciful to think of the effect of the Pauli principle on the passage of a valence electron through the core region as generating a "Fermi tube"; a linked set of Fermi holes.

24.6. MODELLING ATOMIC PSEUDOPOTENTIALS

- But it is obviously independent of the m-value of the outer AO; the core is frozen to be spherically symmetrical.

- If we assume that the higher AOs can be approximated by STOs of reasonable exponent, it is evident that the pseudopotential is *not* very dependent on the principle quantum number n for a given series of AOs with the same ℓ-value. In other words speaking approximately, there is an "s-type pseudopotential" and a "p-type pseudopotential" for the sodium core.

- Subject to the above, if the outer AO has no "precursor" of the same ℓ-value, then the pseudopotential is obviously *zero* since, there being no core precursor, there are no non-zero overlaps between core and outer orbital so all the terms in the pseudopotential are separately zero.

The most obvious conclusion from this very simple examination is:

notwithstanding the possibility of writing the pseudopotential in a local form, any *model* of this atomic pseudopotential must have the capability to *decide* the symmetry type of the valence AO on which it has to operate.

Now, a decision about the symmetry type (ℓ-value) of an AO is, colloquially speaking, essentially a non-local type of act; the ℓ-value of a function cannot be deduced from its properties in an infinitesimal neighbourhood of a point. We are therefore forced to incorporate some form of non-local operator if we are to develop any sort of realistic modelling of the atomic pseudopotential.

An obvious candidate for the non-local part of the pseudopotential operator is the "spherical harmonic projection operator":

$$\hat{P}_{\ell_1 m_1} \chi(r, \theta, \phi) = R_{n\ell}(r) Y_{\ell_1}^{m_1}(\theta, \phi) \int d\Omega' Y_{\ell_1}^{m_1*}(\theta', \phi') \chi(r', \theta', \phi')$$

where the general AO χ on which the operator acts is assumed to have the "central field" form which is separable in spherical polar coordinates:

$$\chi(r, \theta, \phi) = R_{n\ell}(r) Y_\ell^m(\theta, \phi)$$

for which

$$d\Omega = d(cos\theta) d\phi$$

is the volume element with the usual limits.

$\hat{P}_{\ell_1 m_1}$, acting on such separable functions generates either zero (if $(n_1 \ell_1 m_1) \neq (n\ell m)$) or reproduces the AO (if $(n_1 \ell_1 m_1) = (n\ell m)$). We shall see shortly that this operator has more interesting and useful properties when acting on functions which are *not* separable in spherical polars.

Since the pseudopotential is independent of m-values we may combine all the $2\ell_1 + 1$ operators $(\hat{P}_{\ell_1 m_1}, m_1 = -\ell_1, \ldots, \ell_1)$ to form the operator $\hat{P}_{\ell_1}$ given by

$$\hat{P}_{\ell_1} = \sum_{m_1=-\ell}^{\ell} \hat{P}_{\ell_1 m_1}$$

whose action is to generate, from a given separable function, *any* component of a given ℓ-type. Again, the action of this operator on functions of three-space which are *not* separable in spherical polars centred on the nucleus in question will be interesting and useful later.

This evidence is by no means exhaustive or convincing in itself, but there is a vast body of such evidence to support the idea that the most convenient form for an atomic model potential which will simulate the effect of the true atomic pseudopotential is

$$V_{Model} = \sum_{\ell-values} (Model\ potential\ for\ an\ \ell value) \\ \times\ (Projector\ for\ that\ \ell value)$$

Or, in a more precise form

$$V_{Model} = \sum_{\ell=0}^{\ell_{max}} V_\ell \hat{P}_\ell$$

where:

- ℓ_{max} is the maximum value of ℓ occurring in the core AOs.

- V_ℓ is the potential for AOs of the type $\chi_{n\ell m}$ for all values of n and m. V_ℓ has not been given an operator "hat" because we specifically expect to use a *local form* for V_ℓ.

For the moment we can leave the development at that point since, rather than devote effort to the generation of appropriate model potentials V_ℓ, we will defer this task until we can decide if the V_ℓ can be generalised to include other terms in the core potential.

24.7 Modelling atomic core potentials

Having put some effort into modelling the effect of the Pauli principle on a single outer electron we should now return to the main problem of this chapter which is to model the *total* effect of a set of core electrons on an outer electron.

24.7. MODELLING ATOMIC CORE POTENTIALS

We said earlier that the Coulomb/exchange potential generated by a set of core electrons should not be too difficult to simulate. Let us also recall that, so far, we are working entirely within a single-determinant model.

If we begin to think about the Coulomb/exchange potential generated by a nucleus and a determinant of frozen core orbitals it is clear that we should, initially at least, give separate consideration to two quantities:

- The Coulomb potential of a spherically symmetrical nucleus plus a spherically symmetric set of n doubly-occupied atomic orbitals has the obvious asymptotic form
$$\frac{Z - 2n}{r}$$
for large r, and should present no serious problems since it can be expanded as a series in (say) inverse powers of r times Gaussian functions.

- The non-local exchange potential for a single orbital may be cast into local form by the same device which we used for the pseudopotential (multiplication by unity in the form χ/χ) and the same techniques used to examine its form and likely approximation methods.

$$\begin{aligned}\hat{K}\chi(\vec{x}_1) &= \sum_{i=1}^{n} \int d\tau_2 \frac{\chi_i(\vec{x}_1)\chi_i^*(\vec{x}_2)\chi(\vec{x}_2)}{|r_1 - r_2|} \\ &= \left(\sum_{i=1}^{n} \frac{\chi_i(\vec{x}_1)}{\chi(\vec{x}_1)} \int d\tau_2 \frac{\chi_i^*(\vec{x}_2)\chi(\vec{x}_2)}{|r_1 - r_2|}\right)\chi(\vec{x}_1)\end{aligned}$$

Since we are dealing explicitly with one electron outside a core, the self-energy correction which cancels part of the Coulomb and exchange terms has been removed from both to leave the "pure" Coulomb and exchange terms.

Again, it is important to note for the future that this local form of the exchange operator is only valid for one particular function (χ) and, in general, the non-local exchange operator can only be replaced by a *set* of local operators; one for each orbital explicitly included in the treatment. Also, of course, the local form is not valid at nodal points of χ.

This local form of our special exchange operator may be studied as a function of space for what one might hope to be typical functions χ. This time, however, since the non-local form has no obvious "cut-off" due to projection operators, the summation over atomic orbital types goes on indefinitely, in principle:

$$V_{Model}^{EXCH} = \sum_{\ell=0}^{\infty} V_\ell \hat{P}_\ell$$

Thus, we might reasonably expect that the *total* core potential (pseudopotential, Coulomb and exchange potentials) might well be reasonably modelled (for one electron outside a closed-shell atomic core) by an expression of the above type with suitable changed definitions of the individual V_ℓ to include pseudo, Coulomb and exchange terms:

$$V_{Model}^{Core} = \sum_{\ell=0}^{\infty} V_\ell \hat{P}_\ell$$

In practice, it is found that it is worthwhile retaining the distinction between those outer orbitals which do and those which do not have a core precursor of the same ℓ-value. So, using the fact that the angular momentum projection operators sum to unity

$$\sum_{\ell=0}^{\infty} \hat{P}_\ell = 1$$

we may reorganise the sum to give an explicit potential for those orbitals with no core precursor and the (smaller) *differences* between this potential and the angular-specific potentials for those orbitals which do have a core precursor:

$$V_{Model}^{Core} = V_L^{Core} + \sum_{\ell=0}^{L-1} \left(V_\ell^{Core} - V_L^{Core} \right) \hat{P}_\ell$$

where L is one greater than the maximum value of ℓ occurring amongst the core orbitals.

That is,

- For $\ell \geq L$, a valence orbital has no core precursor and only experiences the potential V_L^{Core}.

- For $\ell < L$, the potential experienced by a valence orbital is V_L^{Core} plus a number of (small) corrections due to the pseudo and exchange potentials of the core precursors.

Recall that this development is for a single valence electron outside an atomic core. There are, therefore, two major steps needed if the theory is to be useful for molecular calculations:

- Methods for the expansion of the V_L^{Core} and the $(V_\ell^{Core} - V_L^{Core})$ in terms of computationally tractable functions.

- Extension of the theory in three directions:

    1. To include several electrons outside an atomic core.

2. To cope with several atomic cores in the same molecule.
3. To generalise the method beyond the single-determinant model.

We will devote the next chapter to a sketch of the theories and results for the first of these problems since it is an essentially computational problem. The second, a trio of theoretical problems, will be addressed in the next section.

24.8 Several valence electrons

The moment we contemplate more than one valence electron outside an atomic core the central assumption of our theory so far is immediately invalidated. The frozen-core and valence electrons *no longer satisfy the same equation* even at the single-determinant level. The valence electron HF equation does not take the particularly simple one-electron form, which is the same as the HF equation for the isolated core, because of valence-valence electron repulsion.

In the case of two or more valence electrons we have to make a choice which is absent from the single-electron case; we must choose a *model* for the electronic structure. We have to decide if we shall use a single determinant for the pseudo-wavefunction (using an obvious generalisation of the term pseudo-orbital) or a more accurate model containing electron correlation.[4] Obviously the detailed form of the pseudopotential and of the pseudo-wavefunction will depend on this choice of model and the development will become too complex to be useful. Let us make the opposite choice: look at the formal equations independent of model and see if there are some *general* decisions to be made which will enable us to use the theory developed so far for a single electron.

We take a variational approach so that there is no question of requiring an exact solution of the Schrödinger equation for reference. Let Ψ' be a variational trial function for the valence electrons of a many-electron system and let $\hat{h}$ be the valence many-electron Hamiltonian. We seek a minimum in the mean value of $\hat{H}$ with respect to such (normalised) trial functions *together with* the constraint that Ψ' be orthogonal to the wavefunction of a subset of the electrons (the core). We will then recast the equation into a pseudopotential form and examine this form with a view to modelling the pseudopotential.

It is not immediately obvious what the meaning of this orthogonality constraint might be since orthogonality between two functions which are functions of different *numbers* of coordinates is not defined in general. However, if we assume that the core wavefunction is a single determinant (or, in general that

---

[4]Strictly, we can make a series of such choices for the core electrons but we are *modelling* the effect of these electrons and so an over-sophisticated treatment of this group it not appropriate.

the core wavefunction is constructed from linear combinations of determinants) then the condition known as *strong orthogonality* ensures that our constraint holds; that is that

$$\int \Psi'(\vec{x}_1, \vec{x}_2, \vec{x}_3, \ldots, \vec{x}_n)\chi_i(\vec{x}_1)d\tau_1 = 0$$

for $i = 1, n$. The integration over the coordinates of just *one* of the n coordinates contained in Ψ' generates a function of $n-1$ coordinates which is required by strong orthogonality to be *identically zero* for all values of these coordinates. This is to be true for all the orbitals χ_i used in the construction of the core wavefunction.

This condition can be expressed in terms of a core projection operator $\hat{P}$ of identical structure to our earlier Hartree–Fock core projector:

$$\hat{P} = \sum_{i=1}^{m_c} \hat{P}_i$$

where m_c is the number of core orbitals (greater than or equal to the number of core electrons, of course) and each $\hat{P}_i$ is given by

$$\hat{P}_i f = \chi_i(\vec{x}) \int d\tau' \chi_i^*(\vec{x}')f(\vec{x}')$$

for any function f of the coordinates of a single electron.

In terms of $\hat{P}$ the condition of orthogonality to the core may be written formally as

$$\hat{P}\Psi' = 0$$

so let us incorporate the orthogonality into the variational principle in such a way that we may vary a trial function *without* any constraint. We put

$$\Psi' = (\hat{1} - \hat{P})\Psi$$

and vary Ψ subject only to the normalisation condition

$$\int (\hat{1} - \hat{P})^*\Psi^*(\hat{1} - \hat{P})\Psi d\tau = 1$$

which gives a familiar variational problem:

$$\delta \frac{\int (\hat{1} - \hat{P})^*\Psi^*\hat{H}(\hat{1} - \hat{P})\Psi d\tau}{\int (\hat{1} - \hat{P})^*\Psi^*(\hat{1} - \hat{P})\Psi d\tau} = 0$$

Since $\hat{P}$ and therefore $(\hat{1} - \hat{P})$ is Hermitian, the numerator of this expression may be rewritten

$$\int \Psi^*(\hat{1} - \hat{P})^*\hat{H}(\hat{1} - \hat{P})\Psi d\tau$$

24.8. SEVERAL VALENCE ELECTRONS

and so, using the variational calculus from Chapter 2 we have the Euler–Lagrange equation

$$(\hat{1} - \hat{P})\hat{H}(\hat{1} - \hat{P})\Psi = E(\hat{1} - \hat{P})\Psi$$

for the valence wavefunction.

This equation can be rearranged and the terms regrouped to give

$$(\hat{H} - \hat{H}\hat{P} - \hat{P}\hat{H} + \hat{P}\hat{H}\hat{P} + E\hat{P})\Psi = E\Psi$$

which is obviously capable of being rearranged to

$$(\hat{H} + \hat{V}_{PS})\Psi = E\Psi$$

by defining

$$\hat{V}_{PS} = -\hat{H}\hat{P} - \hat{P}\hat{H} + \hat{P}\hat{H}\hat{P} + E\hat{P}$$

This pseudopotential has some points in common with our earlier one-electron case:

- It is non-local; it contains the core projection operator $\hat{P}$.
- It depends on the valence energy eigenvalue; this time the many-electron energy E rather than the one-electron orbital energy ϵ.

But it has a number of differences from the one-electron case:

- It looks more complicated; it involves projections of the *many-electron* Hamiltonian $\hat{H}$.
- We have not explicitly specified the core wavefunction, we only indicated the m_c basis functions from which it must be constructed. This means the core state does not have to satisfy the valence equation with eigenvalue E; indeed we can get no information about the core state from this equation, only that it is formed from the m_c functions used to define $\hat{P}$.

The first of these points is the most important and requires some comment.

If the valence system were just *two* electrons then $\hat{H}$ would be just

$$\hat{H} = \hat{h}(\vec{x}_1) + \hat{h}(\vec{x}_1) + \frac{1}{|r_1 - r_2|}$$

which, when put into the expression for the pseudopotential, would generate terms like

$$\hat{P}\frac{1}{|r_1 - r_2|}$$

$$\frac{1}{|r_1 - r_2|}\hat{P}$$

$$\hat{P}\frac{1}{|r_1 - r_2|}\hat{P}$$

in addition to projections of the one-electron Hamiltonian $\hat{h}$ for the two electrons. Now these formal expressions have a clear physical interpretation in terms of the more complicated action of the Pauli principle when two valence electrons are involved in interaction with a core.

The core projector has its largest effect in the core region and so, when two electrons both penetrate the core region, the effect of the Pauli principle is the generation of the explicit Fermi hole in their mutual interaction (if the two have the same spin). Thus, in simulating the Fermi hole due to the phantom core electrons around each electron, the pseudopotential changes the law of interaction between the electrons from the simple Coulomb law to some screened effective law *which changes from point to point in space*. It is not too difficult to verify that, in the case of more than two valence electrons, the action of the projectors on the electron-repulsion operators is to generate *three-body* and higher terms in the effective Hamiltonian.

From a practical point of view these last facts are very bad news indeed; we have seen earlier in some detail the problems associated with the calculation of the large numbers of electron-repulsion integrals when a molecular wavefunction is expanded in terms of a basis set. If this calculation is to be complicated by a space-dependent law of interaction and three-body interactions then it will become prohibitively expensive in computing resources.

We will take this exact development no further and simply use the fact that we have proved the *existence* of a pseudopotential transformation to underpin the use of physical intuition to suggest a form for a suitable *model* potential. The criteria we use will be a combination of

- Experience with the one-electron case: the one-electron model potentials work.

- Computational practicality: there are no new integrals of overwhelming complexity.

- Physical intuition and experience with the relative sizes of terms in the pseudopotential: we shall neglect the effect of the core projector on the electron-repulsion operators.

Basically therefore we make the central approximation that the effects of the Pauli principle due to the existence of an atomic core can be simulated by a model potential which is simply an addition to the valence *one-electron* Hamiltonian; we take for n valence electrons:

$$\hat{H} = \sum_{i=1}^{n} \hat{h}^{eff}(\vec{x}_i) + \sum_{i,j\neq i=1}^{n} \frac{1}{|r_i - r_j|}$$

where
$$\hat{h}^{eff} = \hat{h} + V^{Core}$$
in which $\hat{h}$ is the standard valence one-electron Hamiltonian and the entire core potential (Coulomb, exchange and pseudopotentials) is simulated by the model core potential V^{Core} whose notation has been simplified a little by dropping the explicit *Model* subscript.

24.9 Atomic cores in molecules

In molecules, of course, we may well be concerned with several spatially separated atomic cores which can be simulated on the separate atoms by the above methods. The way to proceed is obvious but is the "obvious" method valid?

In the case when there is no appreciable overlap between atomic cores there is every reason to believe that the simulation of the Pauli principle in the region of each atomic core would be approximated to a high degree of accuracy by the simple sum of the separate-atom core potentials. This statement can be made more precise by saying that, if the core-core overlap is negligible, then the product of the corresponding core projectors will be zero, which is equivalent to saying that the electrons of one core have no *Pauli-principle driven* interaction with those of a remote core. In this case the molecular case is just the same as the above many-electron atomic case with the model potential replaced by a sum of model potentials for each atomic core:

$$\hat{h}^{eff} = \hat{h} + \sum_{atoms\ A} V_A^{Core}$$

We are left with two outstanding problems:

1. What if the atomic cores *do* overlap appreciably?

2. What replaces the simple
$$\sum \frac{Z_A Z_B}{|r_A - r_B|}$$
for the core repulsion energy in the expression for the total energy?

The first point may always be resolved empirically by "reducing the size of the core". For example, if it is found that the core of an iron atom, when considered to be $1s^2 2s^2 2p^6 3s^2 3p^6$, overlaps with other atomic cores in particular molecules, then one can always include, say, the $n = 3$ orbitals in the valence shell and consider the core to be $1s^2 2s^2 2p^6$ at the expense of a larger valence calculation.

Although the core repulsion energy is a constant for a given configuration of the nuclei in a molecule, its value is important if one is using the total energy to optimise the geometry of that molecule, so that there is scope here for the development of core-core repulsion energy expressions that will answer the second point.

24.10 Summary

From the point of view of the computation of molecular electronic structure this chapter may be summarised as follows:

- It has proved to be always possible to transform a Schrödinger equation for a many-electron system into an effective equation for one sub-structure of the electronic structure described by that Schrödinger equation; the valence electrons.

- The pseudopotential and pseudo-wavefunction resulting from this transformation have some interesting properties which enable the effect of the Pauli principle to be simulated by a pseudopotential.

- The combination of this pseudopotential with the Coulomb and exchange potential to form a model potential for each atomic core is possible, and a sensible form for this combined atomic core potential is the "semi-local" form:

$$V_{Model}^{Core} = \sum_{\ell=0}^{L} V_{\ell}^{Core} \hat{P}_{\ell}$$

which, under some circumstances, may have the terms in the summation redefined to emphasise the ℓ-independence of a major component of the core potential to:

$$V_{Model}^{Core} = V_{L}^{Core} + \sum_{\ell=0}^{L-1} \left(V_{\ell}^{Core} - V_{L}^{Core}\right) \hat{P}_{\ell}$$

where L is one greater than the maximum value of ℓ occurring amongst the core orbitals.

- For molecular calculations the combination of these atomic core potentials in the obvious way should provide a background for the performance of valence-only molecular calculations.

The remaining practical problem is the generation of the explicit forms of the potentials

$$V_{L}^{Core}$$

24.10. SUMMARY

and
$$\left(V_\ell^{Core} - V_L^{Core}\right)$$
which will be sketched in the next chapter, along with formulae for the one-electron integrals which arise when these potentials *and* the atomic angular projectors are used in molecular calculations.

If we are able to carry through this plan then, since we have explicitly excluded the effect of the pseudopotential transformation on the valence-valence electron repulsions, then the effects of restricting the calculation to valence electrons only will be:

1. A reduction in the size of the basis from the all-electron basis to a valence-only basis.

2. The replacement of the valence-electron one-electron Hamiltonian matrix by the matrix representation of an effective one-electron Hamiltonian containing the core potential. This result is true *whatever the complexity of the valence wavefunction*. Although our initial investigations were based on the single-determinant model, the extensions and model approximations we have made enable the final model to be represented simply by a change in one-electron Hamiltonian which can be made quite independently of the nature of the valence wavefunction.

This transformation of the problem has some very obvious advantages, principally the reduction in basis size and all the practical and theoretical advantages which follow from this reduction. But, as we shall see when we look at calculations of the electronic structure of molecules containing heavy atoms, there are additional benefits which are not quite so obvious.

Chapter 25

Practical core potentials

The last chapter used pseudopotential theory to obtain a plausible form for model potentials to simulate atomic cores. In this chapter we address the practical implications of these model potentials. There are some parallels with the derivations of Chapter 8 which means that some aspects of the derivation can be skipped.

Contents

25.1	Introduction	637
25.2	Forms for the core potentials	638
25.3	Core potential integrals	641
25.3.1	The local potential integrals	642
25.3.2	The non-local potential integrals	649
25.4	Implementation	651

25.1 Introduction

We have decided to use the expression

$$V_{Model}^{Core} = \sum_{\ell=0}^{L} V_{\ell}^{Core} \hat{P}_{\ell}$$

or

$$V_{Model}^{Core} = V_{L}^{Core} + \sum_{\ell=0}^{L-1} \left(V_{\ell}^{Core} - V_{L}^{Core} \right) \hat{P}_{\ell}$$

for the potential due to a frozen, closed-shell core of each atomic centre of a molecule. What remains to do is:

- Choose a *general form* for quantities like V_ℓ^{Core} on theoretical and practical grounds.

- Obtain the values of the parameters in this general form for the particular atoms.

- Derive and implement the expressions for the energy integrals which arise when these forms of the core potential are used together with the expansion method for the electronic wave function.

This chapter is devoted to giving an outline of each of these tasks. Fortunately, we can simply *refer* to existing tabulations of core potential parameters.

25.2 Forms for the core potentials

The over-riding factor which governs the choice of the mathematical form of anything which appears in the molecular electronic Hamiltonian is

"Will the energy integrals which arise from this above form be tractable?"

In particular one needs to ask if the integrals

$$\int \phi_i^* V_\ell^{Core} \hat{P}_\ell \phi_j dV$$

will be computationally tractable when the basis functions ϕ_i are Gaussian functions.

We have seen earlier that it is possible to expand almost any smooth function which goes to zero at infinity in terms of Gaussian functions, so that the natural first choice for an expansion of the core potential is a linear combination of Gaussians. We have seen how to generate the explicit numerical forms of the pseudo, Coulomb and exchange potential available from atomic calculations so that we may use both these forms and the Gaussian expansion method to guide our choice.

> Notice now that we are proposing to model the *whole core potential* at once; from now on we do not make any attempt to fit the pseudopotential, the Coulomb potential or the exchange potential separately. We assume that these three components may all be collected together and fitted by a suitably flexible single expansion method.

25.2. FORMS FOR THE CORE POTENTIALS

Obviously, the core potential due to a closed-shell, spherically symmetrical electron distribution ought to be spherically symmetrical *for a particular type of valence basis function*; the angular dependence of the *total* core potential being entirely in the non-local angular projectors. Thus the individual terms V_ℓ^{Core} or $(V_\ell^{Core} - V_L^{Core})$ might reasonably be expected to have the form

$$V_\ell^{Core} = \sum_k f_{k,\ell}(r)\exp(-\zeta_{k,\ell}r^2)$$

for a particular atomic core, where r is the radial distance measured from the nucleus of that core, and the $f_{k,\ell}(r)$ are some functions to be found and the $\zeta_{k,\ell}$ are parameters to be optimised for each atomic core.

In practice the $f_{k,\ell}(r)$ are chosen to be simple monomials :

$$f_{k,\ell}(r) = d_{k,\ell} r^{n_{k,\ell}}$$

to give, for a particular angular component of the core potential,

$$V_\ell^{Core} = \sum_k d_{k,\ell} r^{n_{k,\ell}} \exp(-\zeta_{k,\ell}r)$$

leaving the parameters of the fit to be:

1. The expansion length; the number of terms in each summation.
2. The values of $d_{k,\ell}$, $n_{k,\ell}$ and $\zeta_{k,\ell}$ for each angular component of each atomic core potential.

There are now a number of important decisions to be made[1] which affect the way in which the core potentials will be used and, more importantly, the interpretation of the results of calculations made with the potentials. The most important decision is

> "Which theoretical model of an atomic core and what accuracy of atomic core wavefunctions shall we use as a 'standard' in setting up our core potentials?"

This choice has several important ramifications:

1. It is probably prudent to use a single-determinant model of the atomic cores since all our introductory theory assumed this model. However since we are now *modelling* the core potential not transforming an exact equation for the electronic structure, there is no *a priori* reason to stay with the HF model.

---
[1] Assuming that this fit is adequate for the core potentials, of course.

2. Within the HF model shall we fit to a core function as accurate as possible? That is should we fit to a core function which, for all practical purposes, could never be actually used in a molecular calculation? For example a clear candidate for a core function would be accurate *numerical* HF functions. Or, should we fit to a core function which *could* be used in an all-electron calculation but which we wish to avoid in order to effect computational savings?

3. Should we use a *relativistic* core function as our standard for fitting? In general, should we use a Hamiltonian for the core electrons which is more accurate than (or, at least, different from) the valence-electron Hamiltonian to which the core potential is to be added? This has similar consequences to the point above; should we always retain the possibility of performing a full all-electron calculation which would provide a standard for the valence-only calculation?

All three of these individual items point towards one central decision which must be made explicit if we are to use the core potential method in a scientific manner:

Do we wish to treat the core potential method of calculation as an *approximation* to some reference calculation which could, realistically, be carried through, at least for a representative selection of molecules, or do we wish to model the atomic cores in such a way that our calculations could not realistically be thought of as approximations since the basis functions and/or the Hamiltonian are not available for the full all-electron calculation.

If we use the second alternative we are obviously moving into unknown territory since we shall have no *ab initio* calculations for comparison purposes; we shall have to rely on comparisons with *experiment* with all that that implies for what we hope to be a basically non-empirical method.

In fact, we have no real choice for several reasons:

- There are, in fact, very few all-electron bases available for atoms from the bottom of the periodic table where the use of core potentials is most valuable.

- It is a frustratingly familiar fact that, in all-electron calculations it is necessary to use a more accurate basis for the core electrons than for the chemically interesting valence electrons. This is, basically, because the region close to a nucleus is very "energy-rich" and, unless this region is "saturated" with basis functions to provide a good description of the core

25.3. CORE POTENTIAL INTEGRALS

orbitals, the phenomenon of "practical variational collapse" occurs. This is the phenomenon of the nominal core basis not being sufficient to give a good description of the core orbitals and so forcibly using some of the nominal valence basis at the expense of the description of the valence electronic structure. Clearly this last is an entirely computational effect and, while quite distinct from the "variational collapse" caused by omission of the Pauli principle, has some affinity to it.

Thus, an acceptable basis for atoms of the first row of the periodic table often has twice as many basis functions for a $1s$ core orbital as for a $2s$ in the valence region. For the third and fourth rows of the periodic table it is scarcely practical to attempt to saturate this extremely energy-rich region with an enormous number of basis functions.

- It is not all obvious that relativistic effects are important at the valence level *per se*. What is known to be important is the effect *on the valence electronic structure* of relativistic effects in the core.

So, attempting to carry through a full *all-electron* relativistic calculation of the electronic structure of, for example, Ziese's salt simply because it contains a platinum atom might well turn out to be a quixotic adventure. The uncertainties in the basis functions, apprehensions about variational collapse and doubt about the area of applicability of the Hamiltonian used would cast the interpretation of the whole work into difficulties.

We therefore, once and for all, take the decision to use model potentials which are fitted to the results of calculations of the electronic structure of atomic cores, which are as accurate as it is possible to make them both numerically and in terms of correctness of the atomic Hamiltonian.

There are a number of methods in the literature for generating fits to core potentials associated with core wavefunctions of various types.

25.3 Core potential integrals

Assuming now the existence of suitable parametrised expansions of the core potentials of the forms outlined in the last section we need to go ahead and obtain expressions for the integrals

$$\int \phi_i^* V_\ell^{Core} \hat{P}_\ell \phi_j dV$$

where the ϕ_i are contracted Gaussian functions; that is we need the basic integrals over the Gaussian primitives η_i:

$$\int \eta_i^* V_\ell^{Core} \hat{P}_\ell \eta_j dV$$

Having been through the analysis of the nuclear-attraction integrals and the electron-repulsion integrals over a basis of Gaussian primitives, we expect that, except for the angular projectors, the integrals can be done since they involve integrals over products of *three* Gaussians which is, in some sense, intermediate in complexity between the nuclear-attraction integrals and the repulsion integrals. The new problem is the effect, on a Cartesian GTF on a given centre, of an angular projector which is, in general, expressed in terms of spherical harmonics *centred on a different nucleus*. Fortunately the machinery has been developed in other contexts and is ready for us to use.

What we must do in order to use both the potential from a given centre and the angular projector on that centre effectively is to:

- Express each GTF as a combination of functions centred on the site of the angular projector,

- Express each of these functions on the site of the projector as a combination of spherical harmonics,

- Use the fact that the angular projector has particularly simple properties when acting on spherical harmonics:

$$\hat{P}_\ell Y_{\ell''}^m(\theta,\phi) = \delta_{\ell,\ell'} Y_\ell^m(\theta,\phi)$$

to remove the projector from the integrand,

- Perform the remaining integration.

The procedure is complicated a little by the fact that the usual GTFs are written in Cartesian form:

$$\eta_i = N_i x^{n_i} y^{\ell_i} z^{m_i} \exp(-\alpha_i r^2)$$

but the angular projectors are adapted to spherical polar coordinates. The relationship between the two systems is, however, straightforward.

In fact, because the expansion of the V_ℓ^{Core} is based on the site of the angular projector, we must do the first of these actions even when no projector is present; that is even for V_L^{Core}, so we will start there, calling the centre on which the core potential is based C.

25.3.1 The local potential integrals

The simplest new integral arises if we use the second form of the core potential which has a purely local (angular projector-independent) term V_L^{Core} which we have chosen to expand in terms of monomials times Gaussians, terms like:

$$d_{k,\ell} r^{n_{k,\ell}} \exp(\zeta_{k,\ell} r^2)$$

25.3. CORE POTENTIAL INTEGRALS

where the $n_{k,\ell}$ are generally negative integers. Thus the integral to be evaluated has the general form:

$$\int \eta_A(\vec{r}_A) \frac{\exp(-\zeta r_C^2)}{r_C^n} \eta_B(\vec{r}_B) dV$$

where, in general, the two GTF primitives (η_A and η_B) will be on different centres (A and B, say), both of which may be different from C (the potential-generating centre) and the notation η_A and η_B has been used to replace the more conventional η_i and η_j to emphasise this point. The techniques and manipulations required to evaluate this integral are substantially identical to those used in the more involved non-local case but less labyrinthine, and so, because of the lower complexity, this integral will be used to exemplify the method of attack so that the more complex case can be seen as an extension of the techniques used here.

The exponential part of a GTF on centre A may be written in terms of an exponential on centre C using Pythagoras' Theorem

$$r_A^2 = r_C^2 + |\vec{CA}|^2 + 2\vec{CA} \cdot \vec{r}_C$$

where

$\vec{r}_C$ is the position vector of a point with respect to centre C, $|\vec{r}_C| = r_C$
$\vec{r}_A$ is the position vector of a point with respect to centre A, $|\vec{r}_A| = r_A$
$\vec{CA} = \vec{C} - \vec{A}$ and $\vec{C}, \vec{A}$ are the position vectors of the two centres

So that

$$\exp(-\alpha r_A^2) = \exp(-\alpha r_C^2) \times \exp(-\alpha|\vec{CA}|^2) \times \exp-(2\alpha \vec{CA} \cdot \vec{r}_C)$$

the first term on the right here is just a (spherical) GTF on centre C, the second is a constant and the problem lies in the expansion of the third term which contains the exponential of the Cartesians *themselves* not their squares.

Fortunately, the expansion of the third term is a well-known theorem in the theory of Bessel functions; if we write

$$2\alpha \vec{CA} \cdot \vec{r}_C = kr_C cos(\theta)$$

where θ is the angle between the two vectors and $k = 2\alpha|\vec{CA}|$, we may use

$$\exp(kr_C cos(\theta)) = 4\pi \sum_{\lambda=0}^{\infty} M_\lambda(kr_C) P_\lambda(\theta)$$

where the P_λ are the Legendre polynomials and the M_λ are the modified spherical Bessel functions of the first kind. This expression may be reduced to a more familiar form by the expansion of the P_λ in terms of the spherical polar angles of the two vectors $\vec{CA}$ and $\vec{r}_C$, since θ is the angle between these vectors:

$$P_\lambda = \sum_{\mu=-\lambda}^{\lambda} Y_\lambda^\mu(\theta_{AC}, \phi_{AC}) Y_\lambda^\mu(\theta_C, \phi_C)$$

where the $Y_\ell^m(\theta, \phi)$ are the familiar spherical harmonics.

We can, of course, perform this transformation for the exponential factor of the other GTF primitive on another centre (B, say) and collect the two results together by defining
$$\vec{k} = 2(\alpha_A \vec{CA} + \alpha_B \vec{CB})$$
to give

$$\exp(\vec{k} \cdot \vec{r}_C) = 4\pi \sum_{\lambda=0}^{\infty} \sum_{\mu=-\lambda}^{\lambda} M_\lambda(kr_C) Y_\lambda^\mu(\theta_k, \phi_k) Y_\lambda^\mu(\theta_C, \phi_C)$$

where the explicit form of the Bessel functions can be obtained from the defining relationship

$$M_\lambda(x) = x^\lambda \left(\frac{1}{x} \frac{d}{dx}\right)^\lambda \frac{sinh(x)}{x}$$

Notice that the spherical harmonic $Y_\lambda^\mu(\theta_k, \phi_k)$ is, for a particular pair of GTF primitives, *a constant* (r_k, θ_k, ϕ_k are the spherical polar coordinates associated with the vector $\vec{k}$) and will remain to be evaluated after the integration over space has been completed. Notice, also, the uncomfortable-looking sum to infinity involved in the final expansion. In fact, because of the way in which the infinite sum is involved in angular integrals with other angular functions, the sum always terminates in the sense of only generating a finite number of non-zero integrals.

It is clear now, referring back to the derivations of the nuclear-attraction and electron-repulsion integrals, how the evaluation of the integral must go; taking the primitives as

$$\eta_A(\vec{r}_A) = N_A x_A^{n_A} y_A^{\ell_A} z_A^{m_A} \exp(-\alpha_A r_A^2)$$

and
$$\eta_B(\vec{r}_B) = N_B x_B^{n_B} y_B^{\ell_B} z_B^{m_B} \exp(-\alpha_B r_B^2)$$

the necessary steps will be:

1. Refer the Cartesian monomials of the primitives centred on A and B to a Cartesian set on the potential-generating centre C *e.g.*

$$x_A = x_C + \vec{CA}_x$$

25.3. CORE POTENTIAL INTEGRALS

etc. This will generate a sum of products of powers of the components of the two vectors $\vec{CA}$ and $\vec{CB}$ multiplied by powers of the Cartesians (x_C, y_C, z_C) in the usual binomial distribution.

$$x_A^{n_A} x_B^{n_B} = \sum_{a=0}^{n_A} \sum_{b=0}^{n_B} \binom{n_A}{a} \binom{n_B}{b} CA_x^{n_A-a} CB_x^{n_B-b} x_C^{a+b}$$

with similar expressions for the y and z components.

In terms of the "f_j" coefficients introduced in the reduction of the nuclear-attraction integrals this product is

$$x_A^{n_A} x_B^{n_B} = \sum_{j=0}^{n_A+n_B} f_j(n_A, n_B, CA_x, CB_x) x_C^j$$

2. Use a spherical polar system on centre C to re-express the Cartesians on C in a form adapted to the action of the angular projectors:

$$\bar{x} = \frac{x_C}{r_C} = sin\theta_C cos\phi_C$$
$$\bar{y} = \frac{y_C}{r_C} = sin\theta_C sin\phi_C$$
$$\bar{z} = \frac{x_C}{r_C} = cos\theta_C$$

using the "bar" notation for brevity.

3. Use the above substitution to separate the integral into:

- "Angular" parts involving the products of the spherical harmonics (from the projectors) and the angular factors of the Cartesian products (the terms involving $\bar{x}, \bar{y}$ and $\bar{z}$). For example, the above expansion of the product $x_A^{n_A} x_B^{n_B}$ may be written

$$x_A^{n_A} x_B^{n_B} = \sum_{a=0}^{n_A} \sum_{b=0}^{n_B} \binom{n_A}{a} \binom{n_B}{b} CA_x^{n_A-a} CB_x^{n_B-b} \bar{x}^{a+b} r_C^{a+b}$$

with the factor in $\bar{x}$ contributing to the angular integral. Similar terms in the other Cartesian components ($\bar{y}$ and $\bar{z}$) generate other contributions.

- "Radial" parts involving the product of the Bessel function and the radial factor of the Cartesian products times the radial exponential factor. In the example used above, a radial contribution of r_C^{a+b} comes from the expansion of $x_A^{n_A} x_B^{n_B}$.

The final step involves two integrals of a type which we have not met before in our earlier integral evaluations.

The angular integrals.

The angular integral is a sum of terms like

$$\int d\Omega \bar{x}^I \bar{y}^J \bar{z}^K Y_\lambda^\mu(\Omega)$$

(inheriting the notation λ, μ from the Bessel function) which occur in groups, multiplied by the value of the same spherical harmonic evaluated at $\Omega_k = (\theta_k, \phi_k)$ as follows:

$$\boldsymbol{A}_\lambda^{IJK} = \sum_{\mu=-\lambda}^{\lambda} Y_\lambda^\mu(\Omega_k) \int d\Omega \bar{x}^I \bar{y}^J \bar{z}^K Y_\lambda^\mu(\Omega)$$

Now since the product of powers of $\bar{x}$ etc. is merely compact notation for a product of powers of trigonometric functions, we can expand the real spherical harmonics in terms of $\bar{x}, \bar{y}, \bar{z}$ to generate a sum of integrals which are entirely products of powers of these angular variables, and use the fact that

$$\int d\Omega \bar{x}^i \bar{y}^j \bar{z}^k = 4\pi \frac{(i-1)!!(j-1)!!(k-1)!!}{(i+j+k+1)!!}$$

if i, j and k are *all* even and zero otherwise.

Thus, apart from the book-keeping problem of keeping track of all the expansions we have used, the angular integration is solved. Further, the angular integration places an upper limit on the troublesome-looking sum to infinity on λ. If we were to evaluate the angular integral in the "opposite sense" of expanding the products of powers of $\bar{x}, \bar{y}, \bar{z}$ in terms of the real spherical harmonics, the maximum order of such polynomials would be the sum of the exponents of the $\bar{x}, \bar{y}, \bar{z}$, so that all the integrals arising from the infinite summation over λ which have values greater than this maximum would be zero by the *orthogonality* of the spherical harmonics.

This result also places an upper limit on λ in the *overall* summation since each radial integral involving a λ is *multiplied* by an angular integral involving the same λ.[2]

The radial integrals.

What remains is the evaluation of the "radial" integrals which contain products of powers of r_C, Gaussian exponentials and the Bessel functions. A typical one

---

[2] This result is very typical of methods of integral evaluation by separation in polar coordinates; the expansion of an *operator* in terms of products of spherical harmonics has an infinite number of terms but the forms of the *orbitals* involved in any integrand ensures that the expansion cuts off after a finite number of terms. The atomic electron-repulsion integrals expanded as a sum of F and G Slater–Condon parameters is perhaps the prototype.

25.3. CORE POTENTIAL INTEGRALS

of these integrals is:

$$\boldsymbol{R}_\lambda^N(k,\alpha) = \int_0^\infty dr\, r^N \exp(-\alpha r^2) M_\lambda(kr)$$

There are several ways to approach integrals of this kind and we just outline one of them, the strategy is:

- Expand the Bessel function in terms of the hyperbolic functions *sinh* and *cosh*:

$$M_\lambda(x) = (-1)^{\lambda+1} \sum_{i=1}^{\lambda} \frac{\alpha_{i\lambda}}{x^i} \cosh(x) + (-1)^\lambda \sum_{i=1}^{\lambda+1} \frac{\beta_{i\lambda}}{x^i} \sinh(x)$$

where the expansion coefficients $(\alpha_{i\lambda}, \beta_{i\lambda})$ may be obtained from the explicit forms of the Bessel functions or from the recursion relationships connecting them.

- This gives integrals like

$$\int_0^\infty dr\, r^n \exp(-\alpha r^2) \cosh(kr)$$

to be evaluated (plus, of course, the same expression with *cosh* replaced by *sinh*). Now,

$$\cosh(kr) = \frac{\exp(kr) + \exp(-kr)}{2}$$

generating integrals like

$$\int_0^\infty dr\, r^n \exp(-\alpha r^2) \exp(kr) = \int_0^\infty dr\, r^n \exp(-\alpha r^2 + kr)$$

- Making the substitution

$$x = r + \frac{k}{2\alpha} = x + a$$

(say) reduces this integral to

$$\exp(a^2) \int_{-a}^\infty dx\, (x+a)^n \exp(-\alpha x^2)$$

which may be integrated by parts to give a recursion relationship which terminates involving one or other of the three forms:

$$erf(x) = \frac{2}{\sqrt{\pi}} \int_0^x \exp(-t^2) dt$$

$$D(x) = \exp(-x^2) \int_0^x \exp(-t^2) dt$$

$$H(x) = \exp(-x^2) \int_0^x \exp(-t^2) erf(t) dt$$

which are, respectively, the error function (familiar as $F_0(x)$ in the nuclear-repulsion integral), the Dawson function and the hybrid Dawson-error function. These functions are well studied and there are series expansions for all of them optimised for various ranges of x.

This outline means that we have the techniques available to evaluate the energy integrals over GTF basis functions for the local part of the core potential. Collecting all these fragments together involves a considerable amount of tedious algebra which we shall not give.

If we choose a definite explicit form for the core potentials of the core of atom C (which has N_C core electrons, say)

$$r_C^2 \left[V_L^{Core}(r_C) - \frac{N_C}{r_C} \right] = \sum_k d_{k,L} r^{n_{k,L}} \exp(-\zeta_{k,L} r_C)$$

$$r_C^2 \left[V_\ell^{Core}(r_C) - V_L^{Core}(r_C) \right] = \sum_k d_{k,\ell} r^{n_{k,\ell}} \exp(-\zeta_{k,\ell} r_C)$$

then all the potential arising from core C is expressed in terms of the Gaussian expansion and all the integrals which arise from the first expansion are of the type outlined in this section. Using some obvious contractions of notation for both the Gaussian primitives and the form of the potential, the general result is:

$$\int \eta_A(\vec{r}_A) \left[\exp(-\zeta r_C^2) r_C^{n-2} \right] \eta_B(\vec{r}_B) dV$$

$$= N_{ABC} \sum_{a=0}^{n_A} \sum_{b=0}^{\ell_A} \sum_{c=0}^{m_A} \sum_{d=0}^{n_B} \sum_{e=0}^{\ell_B} \sum_{f=0}^{m_B}$$

$$\left\{ \binom{n_A}{a} \binom{\ell_A}{b} \binom{m_A}{c} CA_x^{n_A-a} CA_y^{\ell_A-b} CA_z^{m_A-c} \right.$$

$$\times \binom{n_B}{d} \binom{\ell_B}{e} \binom{m_B}{f} CB_x^{n_B-d} CB_y^{\ell_B-e} CB_z^{m_B-f}$$

$$\times \left. \sum_{\lambda=0}^{\infty} \boldsymbol{A}_\lambda^{a+d,b+e,c+f} \boldsymbol{R}_\lambda^{a+b+c+d+e+f+n}(k,\alpha) \right\} \quad (25.1)$$

where, as above

$$k = |\vec{k}| = |2(\alpha_A \vec{CA} + \alpha_B \vec{CB})|$$
$$\alpha = \alpha_A + \alpha_b + \zeta$$

and N_{ABC} is a product of constants: the normalisation constants of the Gaussian primitives and the familiar factor from the product of two Gaussians:

$$N_{ABC} = 4\pi N_A N_B \exp(-\alpha_A |\vec{CA}|^2 - \alpha_B |\vec{CB}|^2)$$

25.3. CORE POTENTIAL INTEGRALS

The summation over λ is only formally infinite since the angular integrals vanish for
$$\lambda > \lambda_{max} = a + b + c + d + e + f$$
and the orthogonality of the spherical harmonics means that $\lambda - \lambda_{max}$ must be even so that the whole expression is finite.

25.3.2 The non-local potential integrals

The second of the new integrals involves the same derivations as given above *plus* the complication of the existence of the angular projectors which we now have to write out in full, abandoning the simple symbolic terminology $\hat{P}_\ell$. The angular projector is a non-local integral operator whose kernel is a sum of products of spherical harmonics:
$$\hat{P}_\ell = \sum_{m=-\ell}^{m=\ell} \hat{P}_\ell^m$$
and the action of a typical $\hat{P}_\ell^m$ centred on nucleus C on an arbitrary GTF $\eta_j(\vec{r}_B)$ on another centre (B, say) is:
$$\hat{P}_\ell^m \eta_j(\vec{r}_B) = Y_\ell^m(\Omega_C) \int d\Omega'_C Y_\ell^m(\Omega'_C) \eta_j(\vec{r}_B)$$
where, as before, the volume element has been written out more explicitly as
$$dV = r_C^2 dr_C [d(cos\theta_C) d\phi_C] = r_C^2 dr_C d\Omega_C$$
using Ω_C as a compact notation for the angular variables and $d\Omega_C$ as the associated volume element.

Thus, the second type of integral, arising from the non-local terms in the core potential, has the form:
$$\int dr_C \left\{ r_C^2 \left(\int d\Omega_C Y_\ell^m(\Omega_C) \eta_A(\vec{r}_A) \right) \frac{\exp(-\zeta r_C^2)}{r_C^n} \left(\int d\Omega_C Y_\ell^m(\Omega_C) \eta_B(\vec{r}_B) \right) \right\}$$
where the use of brackets means that the primed variable Ω'_C and Ω_C may be replaced by Ω_C. Notice that the two GTF primitives are bracketed *separately* with a spherical harmonic so that the formula
$$\exp(\vec{k}_A \cdot \vec{r}_C cos(\theta)) = 4\pi \sum_{\lambda=0}^{\infty} M_\lambda(k_A r_C) P_\lambda(\theta)$$
must be used *for each primitive separately* (i.e. for
$$\begin{aligned} \vec{k}_A &= -2\alpha_A \vec{CA} \\ \vec{k}_B &= -2\alpha_B \vec{CB} \end{aligned}$$

it cannot be collapsed into a single expression with a single $\vec{k} = -2(\alpha_A \vec{CA} + \alpha_B \vec{CB})$ as it was in the local case).

Fortunately, these complications which arise from the presence of the angular projection operators are not too great and we have already met all the techniques required to deal with them in our discussion of the local case. The appearance of additional Bessel function and spherical harmonic *factors* in the integrands simply means another layer or layers of expansions to be performed when reducing the potential integral to more basic integrals to be evaluated. We shall therefore give this more complicated case a more cursory treatment, emphasising only the *differences* between the simpler local case and the current case.

In fact, the two cases are surprisingly similar and the differences can be concentrated into a new, more general, form of each of the two main integrals; the angular and radial terms.

The angular integrals.

The angular integral contains an additional spherical harmonic factor ($Y_\ell^\mu(\Omega)$) from the projector (orthogonality of the $Y_\ell^m(\Omega)$ means that only the μ value occurs) and reduces to

$$\int d\Omega \bar{x}^I \bar{y}^J \bar{z}^K Y_\lambda^\mu(\Omega) Y_\ell^\mu(\Omega)$$

which can be treated by exactly the same methods as the local case; expanding *both* spherical harmonics as products of $(\bar{x}, \bar{y}, \bar{z})$ and using the same integral formula as before. Also, in the same way, it is a property of *these* integrals which causes the summation over λ to truncate.

We can therefore use a very similar notation for these angular integrals; the difference is in the number of subscripts on A which arise from the differing number of spherical harmonics occurring in the local and non-local case:

$$\boldsymbol{A}_{\lambda\ell m}^{IJK} = \sum_{\mu=-\lambda}^{\lambda} Y_\lambda^\mu(\Omega_k) \int d\Omega \bar{x}^I \bar{y}^J \bar{z}^K Y_\lambda^\mu(\Omega) Y_\ell^\mu(\Omega)$$

Of course, since there are *two* occurrences of these angular integrals in the non-local potential integral, the summation will contain *two* of these factors. With some algebra, these integrals reduce to integrals of products of $(\bar{x}, \bar{y}, \bar{z})$.

The radial integrals.

The non-local core-potential radial integral has an additional Bessel function factor in the integrand due to the separate expansion of each primitive GTF about centre C, and takes the general form

$$\boldsymbol{R}_{\lambda\bar{\lambda}}^N(k_A, k_B, \alpha) = \int_0^\infty dr\, r^N \exp(-\alpha r^2) M_\lambda(k_A r) M_{\bar{\lambda}}(k_B r)$$

again, as in the case of the angular integral, retaining the main symbol $\boldsymbol{R}$ for the integral and distinguishing the local and non-local cases by the number of subscripts.

The complications here over and above those of the local case are ones of book-keeping rather than substance; there are now *two* Bessel functions to expand in terms of the hyperbolic functions $sinh$ and $cosh$ which simply generates more of the basic integrals ($erf(x)$, $D(x)$ and $H(x)$) which we met in discussing the local case. Taking care to note the similarity of the notation which has been adopted for the two *types* of radial and angular integrals, the general result for a typical term in the non-local core-potential integral is:

$$\int \left(\hat{P}_\ell \eta_A(\vec{r}_A)\right) \left[\frac{\exp(-\zeta r_C^2)}{r_C^{n-2}}\right] \left(\hat{P}_\ell \eta_B(\vec{r}_B)\right) dV$$

$$= 4\pi N_{ABC} \sum_{a=0}^{n_A} \sum_{b=0}^{\ell_A} \sum_{c=0}^{m_A} \sum_{d=0}^{n_B} \sum_{e=0}^{\ell_B} \sum_{f=0}^{m_B}$$

$$\left\{ \binom{n_A}{a} \binom{\ell_A}{b} \binom{m_A}{c} CA_x^{n_A-a} CA_y^{\ell_A-b} CA_z^{m_A-c} \right.$$

$$\times \binom{n_B}{d} \binom{\ell_B}{e} \binom{m_B}{f} CB_x^{n_B-d} CB_y^{\ell_B-e} CB_z^{m_B-f}$$

$$\times \left[\sum_{\lambda=0}^\infty \sum_{\bar{\lambda}=0}^\infty \boldsymbol{A}_\lambda^{a+d,b+e,c+f} \boldsymbol{R}_{\lambda\bar{\lambda}}^{a+b+c+d+e+f+n'}(k_A, k_B, \alpha) \right.$$

$$\left.\left. \sum_{m=-\ell}^\ell \boldsymbol{A}_{\lambda\ell m}^{abc} \boldsymbol{A}_{\bar{\lambda}\ell m}^{def} \right] \right\} \qquad (25.2)$$

The summations over λ and $\bar{\lambda}$ are both finite for the same reasons as in the similar summation in the local expression.

These sketch derivations are based on the derivations published by L. E. McMurchie and E. R. Davidson (*J. Comp. Phys.*, **44**, 289 (1981)) although the notation has been changed for some of the auxiliary integrals. Very similar derivations have been given almost simultaneously by M. Kolar (*Comp. Phys. Comm.*, **23**, 275 (1981)) using notation which is different again.

25.4 Implementation

In any discussion of the implementation of the complex expressions for the pseudopotential integrals we come across the same problems encountered in

Chapter 8; the complexity of the expressions make the code very boring to look at and very long. The same decision has been taken for these integrals as was taken for the main molecular integrals in that chapter; their coding is available as a WEB available by ftp from the site listed at the end of the text.

Chapter 26

SCF perturbation theory

The effect of small changes in the molecular Hamiltonian due to interactions of the molecule with external fields, solvents and other molecules are often too small to be included directly in the SCF process. Perturbation theory provides a suitable tool for the investigation of these effects since they can be computed directly rather than as small differences between large numbers. If we insist on retaining the single-determinant form of the wavefunction while allowing the orbitals to change and retain electron interaction self-consistency, the self-consistent perturbation theory is obtained.

Contents

26.1		Introduction	**654**
26.2		Two forms for the HF equations	**654**
	26.2.1	Introduction	654
	26.2.2	An equation for the MO density matrix	655
	26.2.3	Recovering the SCF equations	656
26.3		Self-consistent perturbation theory	**657**
	26.3.1	Introduction	657
	26.3.2	Choice of perturbation method	658
26.4		The method	**659**
	26.4.1	Single-determinant	659
	26.4.2	Orthonormality constraints	660
	26.4.3	The first-order equation	662
	26.4.4	Self-consistency	663
	26.4.5	The energy expressions	664
26.5		Conclusions	**667**
	26.5.1	Implementation	668

CHAPTER 26. SCF PERTURBATION THEORY

26.1 Introduction

In everything we have done so far we have been concentrating on the calculation of the wavefunction associated with the "usual" molecular Hamiltonian (the non-relativistic, Born–Oppenheimer, electrostatic Hamiltonian) using the variation method. However, many of the most interesting properties of molecules arise from *interactions* of the molecule with various electric and magnetic fields as well as with other molecules.

Of course, one could simply add the relevant terms to the usual Hamiltonian and repeat the variational calculation to obtain the desired effects by subtraction. But this brute-force approach is inappropriate for two kinds of reason:

- The effects sought may be very small compared to the total energies involved; the subtraction may be an attempt to compute a very small number by differencing two very large numbers.

- More important, it is often desirable to find out the *form* of the dependence of the energy on a particular parameter (field strength etc.). Particular properties are often defined by their behaviour with respect to particular (e.g.) *powers* of field strengths and a single number lumping together all the effects of a field may not generate the required information.

Perturbation theory is a tool specially developed for these purposes: it deals directly with *differences* in energy due to changes in the Hamiltonian and has the analytic power to separate various orders of dependence of the energy on these changes.

There are many forms of perturbation theory but for our purposes they may be broken into two broad classes:

1. Methods treating the whole wavefunction.

2. Methods treating the orbitals of which the wavefunction is composed.

In this chapter we consider the most widely used method of the second type; the so-called self-consistent perturbation theory. Before developing and implementing the theory it is necessary to put the SCF equation into a new form.

26.2 Two forms for the HF equations

26.2.1 Introduction

We have already met the "charge and bond order" or "density" matrix R as a notational convenience in the theory of the SCF method and as a summary

26.2. TWO FORMS FOR THE HF EQUATIONS

of the charge distribution associated with a single-determinant wavefunction. In looking at the role of $\boldsymbol{R}$ in summarising the charge distribution, we noted that the $\boldsymbol{R}$ matrix is invariant against transformations amongst the (equally) occupied MOs. That is to say, the invariance properties of the $\boldsymbol{R}$ matrix are similar to those of the single-determinant wavefunction itself, unlike the MOs which are not invariants of the theory.

In fact, the equations of the SCF method may be cast into a form which emphasises this fundamental importance of the invariant $\boldsymbol{R}$ matrix. In this section it is shown that the "orbital" form of the SCF equations is completely equivalent to an equation for the $\boldsymbol{R}$ matrix and vice versa. This form is particularly suitable for the application of perturbation theories which, typically, involve *sums* over orbitals not individual orbitals.

26.2.2 An equation for the MO density matrix

The SCF equation for the MO coefficients $\boldsymbol{C}$ in any given basis is

$$\boldsymbol{h}^F \boldsymbol{C} = \boldsymbol{S} \boldsymbol{C} \boldsymbol{\epsilon}$$

with the definition of $\boldsymbol{R}$ as

$$\boldsymbol{R} = \boldsymbol{C} \boldsymbol{C}^\dagger = \sum_{i=1}^{n} \boldsymbol{c}_i \boldsymbol{c}_i^\dagger$$

Multiplication of the SCF equation from the right by $\boldsymbol{C}^\dagger$ generates

$$\boldsymbol{h}^F \boldsymbol{R} = \boldsymbol{S} \boldsymbol{C} \boldsymbol{\epsilon} \boldsymbol{C}^\dagger$$

that is,

$$\boldsymbol{S}^{-1} \boldsymbol{h}^F \boldsymbol{R} = \boldsymbol{C} \boldsymbol{\epsilon} \boldsymbol{C}^\dagger$$

since $\boldsymbol{S}$ must be non-singular for the SCF calculation to be valid.

The Hermitian conjugate of the SCF equation is:

$$\boldsymbol{C}^\dagger \boldsymbol{h}^F = \boldsymbol{\epsilon} \boldsymbol{C}^\dagger \boldsymbol{S}$$

since $\boldsymbol{h}^F$, $\boldsymbol{\epsilon}$ and $\boldsymbol{S}$ are all Hermitian.

Multiplying this equation from the left by $\boldsymbol{C}$ and from the right by the inverse of $\boldsymbol{S}$ we have

$$\boldsymbol{R} \boldsymbol{h}^F \boldsymbol{S}^{-1} = \boldsymbol{C} \boldsymbol{\epsilon} \boldsymbol{C}^\dagger$$

and equating the two separate expressions for $\boldsymbol{C} \boldsymbol{\epsilon} \boldsymbol{C}^\dagger$ gives

$$\boldsymbol{R} \boldsymbol{h}^F \boldsymbol{S}^{-1} = \boldsymbol{S}^{-1} \boldsymbol{h}^F \boldsymbol{R}$$

which rearranges to the required equation for R:

$$h^F RS - SRh^F = 0$$

The orthonormality condition on the MOs:

$$C^\dagger SC = 1$$

translates into a condition on the R matrix:

$$RSR = R$$

26.2.3 Recovering the SCF equations

In order to be sure that the original SCF equation is completely equivalent to the equation involving the density matrix it is useful to derive the SCF equation for C from the equation for R.

Starting with

$$h^F RS - SRh^F = 0$$
$$C^\dagger SC = 1$$

and multiplying the first equation from the right by R we have

$$h^F RSR - SRh^F R = 0$$

which becomes, using the second equation,

$$h^F R - SRh^F R = 0$$

Noting that

$$R = CC^\dagger$$

gives

$$h^F CC^\dagger - SCC^\dagger h^F CC^\dagger = 0$$

which may be rearranged to

$$\left(h^F C - SCC^\dagger h^F C\right) C^\dagger = 0$$

and, assuming that $C^\dagger$ is not zero,

$$h^F C - SC\left(C^\dagger h^F C\right) = 0$$

Using the notation

$$\lambda = C^\dagger h^F C$$

26.3. SELF-CONSISTENT PERTURBATION THEORY

the above equation becomes
$$h^F C - SC\lambda = 0$$
which is just the matrix SCF equation.

In particular, if we use the invariance of the matrix R to linear transformations:
$$R = CC^\dagger = CUU^\dagger C^\dagger$$
for unitary U, we may choose U to *diagonalise* λ:
$$U^\dagger \lambda U = \epsilon$$
$$\lambda = U\epsilon U^\dagger$$
to give the SCF equation its usual canonical form:
$$h^F C = SCU\epsilon U^\dagger$$
or
$$h^F(CU) = S(CU)\epsilon$$
$$h^F \bar{C} = S\bar{C}\epsilon$$
$$\bar{C} = CU$$

So that the equation for the density matrix implies and is implied by the SCF equation.

26.3 Self-consistent perturbation theory

26.3.1 Introduction

The conceptual attractiveness and simplicity of implementation of the single-determinant model is such that it is worthwhile asking if the model may be retained in treating the theory of the interactions of molecules with fields and with other molecules; that is in the theory of molecules suffering changes in environment.

There are two aspects to any theory of electric and magnetic properties:

- Getting the *numbers* right; the quantitative calculation of the effects.
- Identifying the phenomena; being able to say correctly how a given physical quantity depends on powers of the components of an electromagnetic field.

658   CHAPTER 26. SCF PERTURBATION THEORY

Perturbation theory is an excellent choice for use here since, usually, the sizes of the energy terms involved are very small compared with the total energy of the molecule and perturbation theory computes these small energy differences *directly* not by subtraction.

At least equally importantly perturbation theory, with its separation of energy changes explicitly into dependence on powers of chosen parameters, enables separate conceptual effects to be identified and evaluated.

The "global" nature of the variation method which was so powerful in the calculation of the ground-state wavefunction is far too coarse for these small effects. One would have to perform variational calculations both with and without the presence of the perturbing term and obtain the required results by differencing or by some form of numerical differentiation; a notoriously inaccurate procedure.

26.3.2 Choice of perturbation method

Let us assume that we have computed a single-determinant wavefunction for a molecule and we wish to investigate its properties in the presence of some small additional term in the Hamiltonian. We have used the LCAO method; that is we have divided the finite-dimensional space into an occupied space and a virtual space each spanned by (say) the canonical MOs which diagonalise the HF matrix.

There are two kinds of approach open to us if we want to use the same basis to investigate the new situation:

- Use the determinants formed by occupying arbitrary selections of the occupied and virtual orbitals as a basis for perturbation theory. That is, allow the perturbation to mix the single-determinant ground state with various "excited" single determinants to give a multi-determinant wavefunction in the presence of the perturbation.

- Retain the single-determinant *form* of the wavefunction and allow the perturbation to change the detailed structure of the optimum orbitals.

The choice has already been pre-empted by the title of this chapter; let us investigate the possibility of the change in the *orbitals* in the single determinant while retaining the all-important *self-consistency* of the orbitals necessary because of electron repulsion.

26.4 The method

26.4.1 Single-determinant

We shall use the density matrix form of the equations so that we do not have to think about the arbitrariness of the choice of orbital basis in the occupied and virtual spaces. Recall the equations satisfied by the self-consistent $\boldsymbol{R}$-matrix:

$$\boldsymbol{h}^F \boldsymbol{RS} - \boldsymbol{SR}\boldsymbol{h}^F = 0$$

subject to the orthonormality condition:

$$\boldsymbol{RSR} = \boldsymbol{R}$$

It is convenient to choose to work in an *orthonormal* basis to reduce the algebra; there is no loss of generality here since the results of any single determinant are invariant against linear transformations of the basis. Thus we choose

$$\boldsymbol{S} = \boldsymbol{1}$$

and so

$$\boldsymbol{h}_0^F \boldsymbol{R}_0 - \boldsymbol{R}_0 \boldsymbol{h}_0^F = 0 \qquad (26.1)$$
$$\boldsymbol{R}_0 \boldsymbol{R}_0 = \boldsymbol{R}_0^2 = \boldsymbol{R}_0 \qquad (26.2)$$

where

$$\boldsymbol{h}_0^F = \boldsymbol{h}_0 + \boldsymbol{G}(\boldsymbol{R}_0)$$

and all the original quantities have been given a zero subscript to indicate their unperturbed nature.

We now investigate the effect of changing the *one-electron part* of the Hamiltonian by a small parametrised amount, equivalent to adding the term $\lambda \hat{h}_1$ to the one-electron Hamiltonian $\hat{h}_0$:

$$\boldsymbol{h}_1 = \int \boldsymbol{\phi}^\dagger \hat{h}_1 \boldsymbol{\phi} dV$$

and assume a corresponding power series development in the matrix $\boldsymbol{R}$:

$$\boldsymbol{h} = \boldsymbol{h}_0 + \lambda \boldsymbol{h}_1 \qquad (26.3)$$
$$\boldsymbol{R} = \boldsymbol{R}_0 + \lambda \boldsymbol{R}_1 + \lambda^2 \boldsymbol{R}_2 + \ldots \qquad (26.4)$$
$$\boldsymbol{G} = \boldsymbol{G}(\boldsymbol{R}_0 + \lambda \boldsymbol{R}_1 + \lambda^2 \boldsymbol{R}_2 + \ldots) \qquad (26.5)$$

and, for convenience, we use the notation

$$\boldsymbol{h}_i^F = \boldsymbol{h}_i + \boldsymbol{G}(\boldsymbol{R}_i)$$

for the two relevant cases $i = 0, 1$.

Inserting these expressions into the new density matrix equations

$$h^F R - R h^F = 0$$

subject to

$$R^2 = R$$

generates a familiar set of linked equations; one pair for each order in λ.

The first-order pair are:

$$h_1^F R_0 + h_0^F R_1 - R_0 h_1^F - R_1 h_0^F = 0 \qquad (26.6)$$
$$R_0 R_1 + R_1 R_0 = R_1 \qquad (26.7)$$

We may also easily recover expressions for the various energy corrections by substitution in the single-determinant energy expression:

$$E = \operatorname{tr}\left(h + \frac{1}{2}G(R)\right) R$$

We obtain

$$\begin{aligned} E_0 &= \operatorname{tr}\left(h_0 + \frac{1}{2}G(R_0)\right) R_0 \\ E_1 &= \operatorname{tr} h_1 R_0 + \operatorname{tr} h_0^F R_1 \\ E_2 &= \operatorname{tr} h_1^F R_1 - \frac{1}{2}\operatorname{tr} G(R_1) R_1 + \operatorname{tr} h_0 R_2 \end{aligned}$$

where use has been made of the identity

$$\operatorname{tr} G(A) B = \operatorname{tr} G(B) A$$

The third expression for the second-order energy apparently requires the second-order density matrix. In fact, as we shall see, when the perturbation equations are solved, the expressions can be cast into a much more compact form involving only the *first-order* density matrix R_1.

Let us start by deriving an expression for the all-important matrix R_1.

26.4.2 Orthonormality constraints

Looking at the second of these equations we can extract some conditions on the form of R_1. First let us define a "virtual" density matrix R_v as the complement of R_0:

$$R_v = 1 - R_0 = \sum_{a=n+1}^{m} c_a c_a^\dagger$$

26.4. THE METHOD

and notice that these two matrices have the properties

$$\boldsymbol{R}_0^2 = \boldsymbol{R}_0 \tag{26.8}$$
$$\boldsymbol{R}_v^2 = \boldsymbol{R}_v \tag{26.9}$$
$$\boldsymbol{R}_0\boldsymbol{R}_v = \boldsymbol{R}_v\boldsymbol{R}_0 = 0 \tag{26.10}$$

Multiplying eqn (26.7) from the left and right by each of these pairs of operators and using the above properties it is easy to see that:

$$\boldsymbol{R}_0\boldsymbol{R}_1\boldsymbol{R}_0 = 0$$
$$\boldsymbol{R}_v\boldsymbol{R}_1\boldsymbol{R}_v = 0$$
$$\boldsymbol{R}_0\boldsymbol{R}_1\boldsymbol{R}_v \neq 0$$
$$\boldsymbol{R}_v\boldsymbol{R}_1\boldsymbol{R}_0 \neq 0$$

These conditions severely restrict the possible form of $\boldsymbol{R}_1$ and enable eqn (26.7) to be solved.

It is, however, necessary to return to a particular basis in the occupied and virtual spaces since the simple division of the total space into just *two* subspaces does not provide the full use of all the degrees of freedom available; it is not possible to calculate the first (or any other) order correction to the density matrix simply by adding a multiple of the virtual density matrix.

If we use the most obvious basis — the canonical MOs — then the above relationships translate easily into a condition on the canonical MOs if we remember the orthogonality relationships among the MOs, which are conveniently expressed as:

$$\boldsymbol{R}_0\boldsymbol{c}_i = \boldsymbol{c}_i$$
$$\boldsymbol{R}_v\boldsymbol{c}_a = \boldsymbol{c}_a$$
$$\boldsymbol{R}_0\boldsymbol{c}_a = 0$$
$$\boldsymbol{R}_v\boldsymbol{c}_i = 0$$

where i is understood to label the occupied MOs and a the virtual MOs ($1 \leq i \leq n$; $n+1 \leq a \leq m$). Because of the rather severe constraints on its form specified above, the first-order correction to the density matrix can have contributions only from "cross products" of the MOs:

$$\boldsymbol{R}_1 = \sum_{i=1}^{n} \sum_{a=n+1}^{m} \left(D_{ia}\boldsymbol{c}_i\boldsymbol{c}_a^\dagger + D_{ai}\boldsymbol{c}_a\boldsymbol{c}_i^\dagger \right)$$

where the D_{ia} and D_{ai} are numerical coefficients to be found. The density matrix is *Hermitian* $\boldsymbol{R}_1^\dagger = \boldsymbol{R}_1$ so that

$$D_{ia} = D_{ai}^*$$

and so
$$R_1 = \sum_{i=1}^{n} \sum_{a=n+1}^{m} \left(D_{ia} c_i c_a^\dagger + D_{ia}^* c_a c_i^\dagger \right)$$

This is as far as we can go using only the information in the first-order orthogonality constraint; we must now use this form of R_1 to solve the perturbation equations.

26.4.3 The first-order equation

Recall the first-order equation eqn (26.7) is:
$$h_1^F R_0 + h_0^F R_1 - R_0 h_1^F - R_1 h_0^F = 0$$
i.e.
$$h_0^F R_1 - R_1 h_0^F = R_0 h_1^F - h_1^F R_0$$
and that
$$h_0^F c_k = \epsilon^k c_k$$
$$c_k^\dagger h_0^F = \epsilon^k c_k^\dagger$$
for $1 \leq k \leq m$ and we know the form of R_1 from the above considerations.

If we insert the form of R_1 into the first-order equation we obtain, for the left-hand-side:
$$h_0^F R_1 - R_1 h_0^F = \sum_{i,a} \left(D_{ia}(\epsilon_i - \epsilon_a) c_i c_a^\dagger + D_{ia}^*(\epsilon_a - \epsilon_i) c_a c_i^\dagger \right)$$

which, if we select a particular term by multiplying from the left by $c_j^\dagger$ (the Hermitian transpose of one of the *occupied* MOs) and from the right by c_b (one of the *virtual* MOs) gives:
$$c_j^\dagger \left(h_0^F R_1 - R_1 h_0^F \right) c_b = D_{jb}(\epsilon_j - \epsilon_b)$$

where we have used the fact that these vectors are orthonormal solutions of the zeroth-order (unperturbed) equation. The right-hand-side of the equation is simply
$$c_j^\dagger h_1^F c_b$$
Thus, for all choices of j and b we have
$$D_{jb} = \frac{c_j^\dagger h_1^F c_b}{(\epsilon_j - \epsilon_b)}$$
for the expansion coefficients of the first-order correction to the density matrix.

It remains to evaluate the numerators $c_j^\dagger h_1^F c_b$.

26.4.4 Self-consistency

It should be noted at this point that *nothing we have said yet involves the self-consistency requirement*. All the manipulations so far are independent of the detailed nature of the original one-electron operator. Of course, in anticipation of the final application, the notation $\boldsymbol{h}^F$ has been used and we defined the matrix $\boldsymbol{h}^F$ in the usual way *but none of the properties* of the matrix $\boldsymbol{h}^F$ have been used.

Now that we have to evaluate the numerators in the expressions for the D_{ia} we shall have reason to use these particular properties, and it will be these properties which will make the results *self consistent* rather than being the results of any single determinant.

If we use the definition of $\boldsymbol{h}_1^F$ to write out one of the numerators explicitly we obtain

$$\begin{aligned}\boldsymbol{c}_j^\dagger \boldsymbol{h}_1^F \boldsymbol{c}_b &= \boldsymbol{c}_j^\dagger \left(\boldsymbol{h}_1 + \boldsymbol{G}(\boldsymbol{R}_1)\right) \boldsymbol{c}_b \\ &= \boldsymbol{c}_j^\dagger \boldsymbol{h}_1 \boldsymbol{c}_b + \sum_{i,a} \left(D_{ia} \boldsymbol{c}_j^\dagger \boldsymbol{G}(\boldsymbol{c}_i \boldsymbol{c}_a^\dagger) \boldsymbol{c}_b + D_{ia}^* \boldsymbol{c}_j^\dagger \boldsymbol{G}(\boldsymbol{c}_a \boldsymbol{c}_i^\dagger) \boldsymbol{c}_b\right)\end{aligned}$$

The terms involving $\boldsymbol{G}$ are just electron repulsion integrals over the *molecular orbitals* which we may write, for example,

$$\boldsymbol{c}_j^\dagger \boldsymbol{G}(\boldsymbol{c}_i \boldsymbol{c}_a^\dagger) \boldsymbol{c}_b = [ai, jb]$$

where $[ai, jb]$ involves the usual two electron repulsion integrals with coefficients depending on the SCF method used; if the method is UHF then

$$[ai, jb] = (ai, jb) - (ab, ji)$$

and the integrals are over spin-orbitals. If the method is closed shell then the first term will be multiplied by a factor of 2 and the integrals are over spatial orbitals. Collecting these terms in the new notation gives:

$$\boldsymbol{c}_j^\dagger \boldsymbol{h}_1^F \boldsymbol{c}_b = \boldsymbol{c}_j^\dagger \boldsymbol{h}_1 \boldsymbol{c}_b + \sum_{i,a} \left(D_{ia}[ai, jb] + D_{ia}^*[ia, jb]\right)$$

so that the expression for the coefficients becomes:

$$D_{jb} = \frac{1}{(\epsilon_j - \epsilon_b)} \left[\boldsymbol{c}_j^\dagger \boldsymbol{h}_1 \boldsymbol{c}_b + \sum_{i,a} \left(D_{ia}[ai, jb] + D_{ia}^*[ia, jb]\right)\right]$$

exposing clearly the self-consistency requirement. The value of any *one* of the coefficients (D_{jb}) which multiplies a contribution to $\boldsymbol{R}_1$ depends on all the others (D_{ia}).

However the coefficients only appear *linearly* in the above set of equations so that the solution to the problem is simply a matrix inversion; no iteration is needed to satisfy the self-consistency requirement. It is easy to see that the size of this matrix inversion may become quite large if a large basis is used; typically one is inverting a matrix of dimension $n(m-n)$ for m basis functions and n occupied orbitals.

In particular, if we assume that the coefficients D_{ia} are *real* so that $D_{ia} = D_{ai}$, then we can write the equations for the coefficients as a single matrix equation by reinterpreting the *pairs* of orbital labels (i, a etc.) as a single index (ia say), ranging from 1 to $n(m-n)$ when n of the MOs chosen from m are occupied in the ground state.

$$MD = x$$
$$\sum_{jb=1}^{n(m-n)} M_{ia,jb} D_{jb} = x_{ia}$$

where

$$\begin{aligned} M_{ia,jb} &= \delta_{ij}\delta_{ab}(\epsilon_i - \epsilon_a) - \nu(ia,jb) \\ &\quad + (ib,ja) + (ij,ba) \end{aligned} \quad (26.11)$$
$$x_{ia} = c_i^\dagger h_1 c_a \quad (26.12)$$

Here, the factor ν has been included to cover both the closed-shell case ($\nu = 2$, when the integrals would be over *spatial* orbitals) and the UHF case ($\nu = 1$, the integrals are over *spin-orbitals*).

26.4.5 The energy expressions

The expressions for the first- and second-order energies have been given earlier simply by extracting the terms of each order from the total energy expression. We shall now see that the form of the self-consistent first-order density matrix and its properties enable these expressions to be simplified considerably.

The first-order energy[1] is

$$E_1 = \text{tr} h_1 R_0 + \text{tr} h_0^F R_1$$

Now, if R_1 is simply a weighted sum of products of the unperturbed MOs:

$$R_1 = \sum_{i=1}^{n} \sum_{a=n+1}^{m} \left(D_{ia} c_i c_a^\dagger + D_{ia}^* c_a c_i^\dagger \right)$$

---

[1] This is for the UHF optimum single determinant, the closed-shell case would be expressed in a basis of spatial functions of half the dimension and multiplied by a factor of 2 overall with a corresponding change in the definition of the elements of G. The same is true for the second-order energy expression considered shortly.

26.4. THE METHOD

then so is $h_0^F R_1$:

$$h_0^F R_1 = \sum_{i=1}^{n} \sum_{a=n+1}^{m} \left(\epsilon_i D_{ia} c_i c_a^\dagger + \epsilon_a D_{ia}^* c_a c_i^\dagger \right)$$

because the c_p are all eigenvectors of h_0^F. The trace of this product is then the sum of multiples of terms like

$$\mathrm{tr}\, c_p c_q^\dagger = \mathrm{tr}\, c_q^\dagger c_p = 0$$

since $p \neq q$ in every case. Thus

$$\mathrm{tr}\, h_0^F R_1 = 0$$

and the first-order energy is simply

$$E_1 = \mathrm{tr}\, h_1 R_0$$

which is a result similar to that of ordinary perturbation theory; the *first-order* energy is obtained without performing any perturbation calculation, self-consistent or otherwise. It is determined by the perturbation matrix and the unperturbed density matrix.

The expression for the most widely used term (the second-order term) in the perturbed energy expression was given earlier:

$$E_2 = \mathrm{tr}\, h_1^F R_1 - \mathrm{tr}\, \tfrac{1}{2} G(R_1) R_1 + \mathrm{tr}\, h_0 R_2$$

and this expression may be reduced to a much simpler form by using two facts:

- The expansion of the second-order matrix R_2 in terms of products of the zeroth-order eigenvectors.

$$R_2 = \sum_{p,q} C_{pq} c_p c_q^\dagger$$

(using p and q as indices since we do not yet know the range of the summations) then, because the c_p are solutions of the unperturbed equation,

$$h_0^F R_2 = \sum_{p,q} C_{pq} \left(h_0^F c_p \right) c_q^\dagger = \sum_{p,q} \epsilon_p C_{pq} c_p c_q^\dagger$$

and

$$\mathrm{tr}\, c_p c_q^\dagger = \mathrm{tr}\, c_q^\dagger c_p = \delta_{pq}$$

we have

$$\mathrm{tr}\, h_0^F R_2 = \sum_{p} \epsilon_p C_{pp}$$

again without a knowledge of the range of the summation over p.

- The second-order "orthonormality conditions"

$$R_0 R_2 + R_1 R_1 + R_2 R_0 = R_2$$

Multiplying this equation from each side by R_0 and, using $R_0^2 = R_0$, gives the identity

$$R_0 R_1^2 R_0 = R_0 R_2 R_0$$

and we have already the expansion for R_1 in terms of the products of MO coefficients:

$$R_1 = \sum_{i=1}^{n} \sum_{a=n+1}^{m} \left(D_{ia} c_i c_a^\dagger + D_{ia}^* c_a c_i^\dagger \right)$$

which means that the "diagonal" terms required to evaluate the trace expression for the energy in the second-order expansion are

$$\begin{aligned} C_{ii} &= -\sum_a |D_{ia}|^2 \\ C_{aa} &= +\sum_i |D_{ia}|^2 \end{aligned}$$

where the notation i and a have their usual meanings in the summations: i labels an occupied MO and a a virtual one.

The combination of these two pieces of information yields

$$\mathrm{tr}\, h_0^F R_2 = -\sum_{i,a} (\epsilon_i - \epsilon_a)|D_{ia}|^2 = -\sum_{i,a} \frac{|h_1^F|_{ia}^2}{(\epsilon_i - \epsilon_a)}$$

where the minus sign arises from the expression of C_{ii} in terms of the D_{ia} given above.

Now

$$\begin{aligned} \mathrm{tr}\, h_1^F R_1 &= \sum_{i,a} \left\{ D_{ia}(\mathrm{tr}\, h_1^F c_i c_a^\dagger) + D_{ia}^*(\mathrm{tr}\, h_1^F c_a c_i^\dagger) \right\} \\ &= \sum_{i,a} \left\{ D_{ia}(h_1^F)_{ai} + D_{ia}^*(h_1^F)_{ia}^* \right\} \\ &= 2 \sum_{i,a} D_{ia}(h_1^F)_{ai} \\ &= 2 \sum_{i,a} \frac{|h_1^F|_{ia}^2}{\epsilon_i - \epsilon_a} \end{aligned}$$

if the energy is real.

26.5. CONCLUSIONS

Thus
$$\mathrm{tr}\boldsymbol{h}_0^F \boldsymbol{R}_2 = -\frac{1}{2}\mathrm{tr}\boldsymbol{h}_1^F \boldsymbol{R}_1$$

Thus E_2 reduces to a very simple expression involving only $\boldsymbol{R}_1$ and the perturbation matrix:

$$E_2 = \frac{1}{2}\mathrm{tr}\boldsymbol{h}_1^F \boldsymbol{R}_1 - \frac{1}{2}\mathrm{tr}\boldsymbol{G}(\boldsymbol{R}_1\boldsymbol{R}_1 = \frac{1}{2}\mathrm{tr}\boldsymbol{h}_1\boldsymbol{R}_1$$

with its straightforward interpretation as the interaction energy between the perturbing operator and the (first-order) perturbed density.

Summarising the self-consistent energy expressions through second order we have:

$$E_0 = \frac{1}{2}\left(\mathrm{tr}\boldsymbol{h}_0\boldsymbol{R}_0 + \mathrm{tr}\boldsymbol{h}_0^F \boldsymbol{R}_0\right) \tag{26.13}$$

$$E_1 = \mathrm{tr}\boldsymbol{h}_1\boldsymbol{R}_0 \tag{26.14}$$

$$E_2 = \frac{1}{2}\mathrm{tr}\boldsymbol{h}_1\boldsymbol{R}_1 \tag{26.15}$$

There is a very close analogy between these results and the results for conventional perturbation theory given in Appendix 10.A.

26.5 Conclusions

If we wish to study the effect of (small) changes in the one-electron Hamiltonian on the results of an SCF wavefunction by:

- retaining the single-determinant form of the total wavefunction,
- applying a perturbation $\lambda \hat{h}_1$ to an existing self-consistent set of MOs which form the determinant,

then, in order to calculate the energy due to the perturbation which is self-consistent to *second order* in the perturbation parameter λ, it is only necessary to compute the coefficients D_{ia} which determine the expansion of the *first-order* perturbed matrix $\boldsymbol{R}_1$.

This energy includes the "direct" effect of the perturbing term in the Hamiltonian and the "indirect" effect of changes in energy due to electron reorganisations induced by the perturbation.

26.5.1 Implementation

At first sight, all that is necessary to implement the self-consistent perturbation method is some book-keeping (getting the elements of the matrix M in the right place) and a matrix inversion program. However, in deciding to use a *linear* matrix technique we are working in a basis of *molecular orbitals* and the electron-repulsion integrals contained in the elements of the matrix M are over these molecular orbitals. Thus before we can implement the SCF perturbation method we need to think about the problem of *transforming* the electron-repulsion integrals from the form in which they are computed (over basis functions) into the form in which they are required (over molecular orbitals).

Of course, the MOs are just linear combinations of the basis functions so there is no problem of *principle* in performing this transformation; the problem is one of *practice*. There are about $m^4/8$ repulsion integrals over a basis of m functions and about the same number of repulsion integrals over the MOs (occupied and virtual). It looks, at first sight, as if the problem of generating the repulsion integrals over the MOs involves a calculation depending on the *eighth power* of the number of basis functions; an overwhelming task.

In mitigation, it must be said that for the special case in hand we do not require *all* $m^4/8$ of the MO integrals; only about m^2 of them appear in the expressions above. But any calculation involving a dependence on m^6 still appears rather daunting.

There are two ways forward:

- "Brute force"; simply go ahead and generate the MO repulsion integrals by the most obvious method. This will work provided the size of the problem is not too great; in particular it proves acceptable if all the MO integrals can be kept in main memory at once.

- Specialised techniques; look into the problem of repulsion-integral transformation more carefully and try to optimise the method. This will be most appropriate for large calculations.

Let us therefore defer any implementation of this method until we have considered the general problem of the transformation of electron-repulsion integrals induced by a change of basis; an issue which will be taken up in Chapter 29.

Chapter 27

Time-dependent perturbations: RPA

The reaction of a many-electron system to a time-dependent perturbation is almost as important as its adjustment to a static time-independent environment change. The theory of the first-order response of a system to an oscillating field follows very similar lines to the perturbation theory of the last chapter.

Contents

27.1	Introduction	669
27.2	Time-dependent Hartree–Fock theory	670
27.3	Oscillatory time-dependent perturbations	671
27.4	Self consistency	675
27.5	Implementation	676

27.1 Introduction

In the general spirit of the previous chapter, it is interesting to look at the case of a perturbation of the single-determinant wavefunction which is time-dependent and *which does not disturb the single-determinant form* of the wavefunction.

Before we do so it is worth-while to establish some conventions and terminology in this area. The obvious name for a model of electronic structure which has a time-dependent Hamiltonian and consists of a single determinant of orbitals and remains a single determinant at all times is the Time-Dependent Hartree–Fock (TDHF) model, and this is the terminology which will be used here. However, there is, particularly in the theoretical physics literature, another related usage. Because the use of perturbation theory is so much their stock-in-trade, many theoretical physicists use the term "time-dependent Hartree–Fock" to mean the *first-order* (in the sense of perturbation theory) *approximation* to what we will call the time-dependent Hartree–Fock model.

Much more enigmatically, the first-order time-dependent self-consistent field approximation is also widely called the Random Phase Approximation (RPA). This terminology is entrenched and so, although the name "time-dependent self-consistent perturbation theory" is more descriptive and preferable to both "time-dependent Hartree–Fock" and "RPA" all are used more or less interchangeably. The evolution of the concept and phrase "random phase approximation" is sketched in Appendix 27.A to this chapter.

27.2 Time-dependent Hartree–Fock theory

The problem of a perturbation approximation to the solutions of the time-dependent Hartree–Fock equations presuppose a knowledge of what those equations are. We do not, as yet, know these equations but we can quickly derive them from a variation principle in a way analogous to the matrix Hartree–Fock equations which are now familiar.

The time-independent variation method used in the earlier chapters for a single determinant Φ:

$$\delta E[\Phi] = \delta \frac{\int \Phi^* \hat{H} \Phi d\tau}{\int \Phi^* \Phi d\tau} = 0$$

may be recast to give a more direct physical meaning

$$\delta \int \left(\Phi^* \hat{H} \Phi - \Phi^* \hat{E} \Phi \right) d\tau = 0$$

where the energy operator ($\hat{E}$) is just a constant in the time-independent case.

If now Φ is a *time-dependent* single determinant, the arguments of Chapter 1 are enough to convince that the corresponding time-dependent variation principle is obtained by replacing the constant energy operator by

$$\hat{E} = i \frac{\partial}{\partial t}$$

to give

$$\delta \int \left(\Phi^* \hat{H} \Phi - \Phi^* (i \frac{\partial \Phi}{\partial t}) \right) d\tau dt = 0$$

27.3. OSCILLATORY TIME-DEPENDENT PERTURBATIONS

If the MOs of which Φ are composed are expanded in the usual way in terms of a *fixed* (time-independent) set of basis functions ϕ then the time-dependence is confined to the expansion coefficients C in

$$\psi = \phi C(t)$$

where the time-dependence has been stated explicitly for emphasis.

This choice of basis means that the time-dependence of the various energy integrals only arises from time-dependence of the operators in $\hat{H}$ which, in practice, means the *one-electron* operators, since the electron-repulsion operators in $\hat{H}$ will be the same as in the time-independent case.

With our knowledge of the invariance properties of the single-determinant wavefunction we may, for simplicity of notation, assume an orthogonal basis and quickly arrive at the conditions on the time-dependent coefficients:

$$\int \delta C^\dagger \left(h + G + i\frac{\partial}{\partial t} \right) C \, dt = 0$$

$$\int C^\dagger \left(h + G + i\frac{\partial}{\partial t} \right) \delta C \, dt = 0$$

which, for arbitrary variations in C are equivalent to the equation

$$h^F C = i\frac{\partial C}{\partial t}$$

This equation may be brought into a "density matrix" form by using the usual

$$R = CC^\dagger$$

and noting that

$$\frac{\partial R}{\partial t} = \frac{\partial C}{\partial t} C^\dagger + C \frac{\partial C^\dagger}{\partial t}$$

to give the time-dependent Hartree–Fock equation:

$$h^F R - R h^F = i\frac{\partial R}{\partial t}$$

which we wish to solve by a self-consistent perturbation method.

27.3 Oscillatory time-dependent perturbations

The most usual way in which a molecule is subject to a time-dependent perturbation is by interaction with electromagnetic radiation. This radiation may interact with the molecule in two main ways:

- It may induce oscillations in the electronic distribution of the molecule or
- If the frequency of the oscillations has certain critical values, it may induce electronic transitions in the molecule.

A theory of the interaction of radiation with matter ought to be able to cope with both these phenomena.

In fact, if we develop a theory for the effect of a time-dependent perturbation oscillating at a single frequency we can always use a sum of such terms to give the Fourier synthesis of an arbitrary time-dependent perturbation. Let us try to develop such a theory for the self-consistent (i.e. including changes in electron repulsion) corrections to the energies and orbitals of a single-determinant wavefunction.

In this section we follow the notation and techniques of the time-independent case closely; the perturbation will be taken to be one Fourier component

$$h_1 = \frac{1}{2}\left(fe^{-i\omega t} + f^\dagger e^{+i\omega t}\right)$$

where the matrix f is the (time-independent) representation of some operator $\hat{f}$ describing the details of the perturbation. It is now necessary to be a little more careful of Hermitian conjugacy since we are dealing with complex perturbations.

We now make the assumption that the angular frequency ω is not close to one of the transition energies ($\Delta E = h\nu = \hbar\omega$) of the system, that is, if we shake the electron distribution with frequency ω, it will meekly respond by vibrating with the same frequency (at least in the order of our calculation); there will be no discontinuous large change in the electron distribution.

Thus, the *first-order* response of the system to this applied perturbation ($\boldsymbol{R}_1$) is

$$\boldsymbol{R}_1 = \frac{1}{2}\left(re^{-i\omega t} + r^\dagger e^{+i\omega t}\right)$$

Applying these corrections to the TDHF equation

$$h^F \boldsymbol{R} - \boldsymbol{R} h^F = i\frac{\partial \boldsymbol{R}}{\partial t}$$

and requiring self-consistency to first order in the $\boldsymbol{R}$-matrix gives a pair of equations which are very similar to those of the time-independent case. Each of these two equations is complex and may be split into a pair of equations which are complex conjugates of each other by separately equating the terms involving $e^{\pm i\omega t}$ which is equivalent to equating real and imaginary parts of the complex equations.

27.3. OSCILLATORY TIME-DEPENDENT PERTURBATIONS

The first-order equation to be satisfied is similar to the time-independent case but, of course, has a non-zero right-hand-side arising from the time-dependent term in the TDHF equation. Here is eqn (26.7), slightly rearranged

$$h_0^F R_1 - R_1 h_0^F + h_1^F R_0 - R_0 h_1^F = 0$$

The term to replace the zero right-hand-side for the $\exp(-i\omega t)$ component is

$$i\frac{\partial r e^{-i\omega t}}{\partial t} = \omega r e^{-i\omega t}$$

and so, dividing through by $e^{-i\omega t}$ which multiplies all the terms, we have

$$h_0^F r - r h_0^F + (f + G(r)) R_0 - R_0 (f + G(r)) = \omega r$$

subject, as usual, to the "orthonormality constraint" on r equivalent to the time-independent constraint on R_1. The $e^{i\omega t}$ term generates an equation which turns out to be the conjugate of the above.

The only non-zero contributions to r are those of the "off-diagonal" type:

$$R_0 r R_v$$
$$R_v r R_0$$

where the virtual projector is

$$R_v = 1 - R_0$$

as before.

Again writing the expansion of the first-order correction in terms of the eigenvectors of the unperturbed Hartree–Fock matrix:

$$r = x + y$$

where

$$x = \sum_{i,a} X_{ia} c_a c_i^\dagger$$
$$y = \sum_{i,a} Y_{ia} c_i c_a^\dagger$$

where, as before, i is an occupied MO and a is a virtual MO. The coefficients X_{ia} and Y_{ia} are analogous to the coefficients D_{ia} of the time-independent case but *without* the complex conjugacy relationship.

Making the substitution $r = x+y$ generates a pair of simultaneous equations for x and y which, in turn, generate a set of linear equations for the expansion coefficients X_{ia} and Y_{ia}.

$$h_0^F x - x h_0^F + R_v (f + G(x+y)) R_0 = \omega x$$
$$h_0^F y - y h_0^F + R_0 (f + G(x+y)) R_v = \omega y$$

where the properties of $\boldsymbol{R}_0$ and $\boldsymbol{R}_v$ have been used.

The solution of these *two* equations parallels the solution of the single time-independent case exactly:

1. Rearrange the equation for $\boldsymbol{x}$, say, to

$$\boldsymbol{h}_0^F \boldsymbol{x} - \boldsymbol{x} \boldsymbol{h}_0^F = \omega \boldsymbol{x} - \boldsymbol{R}_v \left(\boldsymbol{f} + \boldsymbol{G}(\boldsymbol{x} + \boldsymbol{y}) \right) \boldsymbol{R}_0$$

2. Make the substitution

$$\boldsymbol{x} = \sum_{i,a} X_{ia} \boldsymbol{c}_a \boldsymbol{c}_i^\dagger$$

for $\boldsymbol{x}$ (or $\boldsymbol{y}$ in the other case).

3. Multiply each of the equations from the left by the Hermitian conjugate of an virtual MO ($\boldsymbol{c}_b^\dagger$, say) and from the right by an occupied MO ($\boldsymbol{c}_j$, say), and use the orthonormality of the MOs to give

$$X_{jb} = -\frac{\Delta_{bj}}{(\epsilon_b - \epsilon_j - \omega)}$$

where Δ_{bj} (a scalar) is just the analogue of the numerator in the D_{jb} of the time-independent case:

$$\Delta_{bj} = \boldsymbol{c}_b^\dagger \left(\boldsymbol{f} + \boldsymbol{G}(\boldsymbol{x} + \boldsymbol{y}) \right) \boldsymbol{c}_j$$

4. Thus

$$\boldsymbol{x} = -\sum_{i=1}^{n} \sum_{a=(n+1)}^{m} X_{ia} \boldsymbol{c}_a \boldsymbol{c}_i^\dagger$$

A similar analysis gives

$$Y_{jb} = -\frac{\Delta_{jb}}{(\epsilon_j - \epsilon_b + \omega)}$$

and

$$\boldsymbol{y} = -\sum_{i=1}^{n} \sum_{a=(n+1)}^{m} Y_{ia} \boldsymbol{c}_i \boldsymbol{c}_a^\dagger$$

Again, note that the self-consistency requirement lies in the fact that each of the X_{ia} and Y_{ia} coefficients involves all the others because of the occurrence of the electron-repulsion correction $\boldsymbol{G}(\boldsymbol{x} + \boldsymbol{y})$ and because $\boldsymbol{x}$, $\boldsymbol{y}$ involve all the coefficients.

27.4 Self consistency

The development is so similar to the time-independent case that it is sufficient to sketch the imposition of self-consistency and the solution of the equations for X_{ia} and Y_{ia} which reduce to a (large) matrix inversion problem.

1. Expand out x and y in terms of the unperturbed MOs c_k.

2. Note that the resulting expansion of the electron repulsion terms in $G(x+y)$ simply involves terms like

$$c_j^\dagger G(c_i c_a^\dagger) c_b = [ai, jb]$$

as before where the integrals may be over the spin-orbitals (UHF) or spatial orbitals (closed-shell) with appropriate factors inherited from the definition of G in each case.

3. Collect terms to generate the linear equations.

The self-consistency requirement links the equations for the two sets of coefficients: any one of the X_{ia}, for example, depends on all the X_{jb} and all the Y_{jb} (in principle).

It is useful to display the coefficients which express this mutual dependency in the self-consistency equations by using $A_{ia,jb}$ for a coefficient which relates the *intra*-dependence of the X_{ia} on the other X_{jb} (or Y_{ia} on the Y_{jb}, since the dependence is identical apart from sign), and the notation $B_{ia,jb}$ for the *inter*-dependence of, for example, X_{ia} on the Y_{jb} (and *vice versa*). That is, we write

$$\begin{pmatrix} f \\ f^\dagger \end{pmatrix} + \begin{pmatrix} A & B \\ B & A \end{pmatrix} \begin{pmatrix} X \\ Y \end{pmatrix} = \omega \begin{pmatrix} 1 & 0 \\ 0 & -1 \end{pmatrix} \begin{pmatrix} X \\ Y \end{pmatrix} \qquad (27.1)$$

The fact that ω and $-\omega$ both appear in the time-dependent perturbation has been absorbed into the "blocked" matrix on the right-hand-side.

The elements of the matrices (which are all of dimension $n(m-n)$, the product of the dimensions of the occupied and virtual spaces) are:

- X and Y are *columns* of the coefficients X_{ia} and Y_{ia} which are to be found and which determine the first-order correction R_1.

- f and $f^\dagger$ are the matrix representatives of the spatial part of the time-dependent oscillating perturbation, written as a *column* matrix by combining the indices i and a into ia as for X and Y. Notice that we have been trapped by our own notation here; the matrix representation of one part of the spatial perturbation $\hat{f}^\dagger$ has been denoted by $\boldsymbol{f}^\dagger$ which usually means the Hermitian conjugate of the representation of $\hat{f}$ *not* the representation of the Hermitian conjugate[1] *operator*. This infelicity must be borne in mind when performing actual calculations.

- $\boldsymbol{A}$ and $\boldsymbol{B}$ are *square* matrices labelled by the two combined indices ia, jb (say) where

$$A_{ia,jb} = A_{jb,ia} = \delta_{ij}\epsilon_{ii} - \delta_{ab}\epsilon_{aa} + (ia,jb) - (ij,ab)$$
$$B_{ia,jb} = B_{jb,ia} = +(ia,jb) - (ib,ja)$$

in a very similar form to the time-independent case of the previous chapter. Again, it is worth emphasising that the matrix representations are over the unperturbed *molecular orbital* basis, presuming that a transformation of the *basis function* repulsion integrals has been performed.

ω is, of course, the frequency of the perturbation.

27.5 Implementation

Rewriting the equation temporarily as

$$M \begin{pmatrix} X \\ Y \end{pmatrix} = \begin{pmatrix} f \\ f^\dagger \end{pmatrix}$$

where

$$M = \begin{pmatrix} A - \omega \mathbf{1} & B \\ B & A + \omega \mathbf{1} \end{pmatrix} \begin{pmatrix} X \\ Y \end{pmatrix}$$

shows that there is a matrix inversion problem to solve.

The results we have obtained are valid for a single frequency and, if we wish to be able to perform time-dependent perturbation theory for a *general* perturbation of the form

$$\hat{h}_1 = \hat{f}(\vec{r})\hat{g}(t)$$

we shall require the Fourier (frequency) components of $\hat{g}(t)$ and the results of the perturbation calculation for $\hat{f}(\vec{x})$ at these frequencies. This may well turn out to be a daunting task. However, ω appears in a particularly simple way in the problem.

The central assumption behind the derivations in this chapter has been that the frequency of the time-dependent perturbation was such that the system responded to the perturbation by vibrating *with that frequency*. Consideration of

---
[1] Which would be a *row* matrix.

27.5. IMPLEMENTATION

the effect of perturbations for which this assumption is not true will lead to a much more convenient form for the time-dependent SCF perturbation expressions, which eliminates the need to solve separate equations at each frequency and, more surprisingly, for each separate spatial form of the perturbation $\hat{f}$.

Any discussion of implementation is therefore deferred until these problems are solved. Notice that the implementation has also been put back because of the problem of transformation of electron-repulsion integrals; we must soon clear up the mountain of debt which is accumulating.

Appendix 27.A

"Random phase approximation"

The name "random phase approximation" comes from a classical mechanical treatment of the collective properties of the electron gas by Bohm and Pines[2] in the early 1950s.

Bohm and Pines transformed the electron-repulsion terms in the classical many-electron Hamiltonian into its Fourier components. This Fourier transform has the effect of interpreting the familiar Coulomb repulsion potential term as a series of momentum-transfers between the states of the electrons. They showed how an important series of results could be obtained by the assumption that the terms in this Fourier transform which depended on a non-zero phase difference in the **k**-vector could be neglected. The idea behind this approximation is that these terms, having random phases, have a zero mean value and contribute only to random fluctuations in the electron plasma which are negligible under the circumstances of their study, or, what amounts to the same thing, the momentum transfers could be replaced by their ensemble average.

More important from the point of view of the connection with *self-consistent* perturbation theory, the approximation, when translated back into ordinary Coulomb repulsion language, has the physical interpretation of retaining only the electron repulsion *in the mean*, which is just the classical equivalent of the physics of the single-determinant model. Thus the approximation of neglecting these terms in a (classical) time-dependent treatment has the same physical interpretation as the single-determinant model.

This connection was made explicitly by Nozieres and Pines in 1958[3]. who showed that, in a quantum-mechanical treatment, the RPA is equivalent to the

---
[2]D. Bohm and D. Pines *Phys. Rev.*, **82**, 625 (1951) and **85**, 338 (1952) is the second in the series of papers and contains a discussion of the random phase approximation.

[3]P. Nozieres and D. Pines *Nuovo Cimento*, [X]**9**, 470 (1958)

addition of electron interaction to an independent-particle single determinant by superposition of determinants containing single excitations from the independent particle determinant. This is one way of defining the Hartree–Fock model; inclusion of all possible single excitations from a given determinant is just a single determinant of optimum orbitals for a given Hamiltonian.

Finally, Ehrenreich and Cohen[4] showed that the perturbed self-consistent-field model of a many-electron system in interaction with an electromagnetic field is equivalent to the RPA of Bohm and Pines.

The RPA, together with a simpler variant, the so-called Tamm–Dancoff Approximation, is now used in all forms of many-body theories (nuclear, solid state and molecular) in a wide variety of notations and formulations. It is an interesting exercise to consult these publications and contrast the notation and techniques used in this area with the ones used in this work. In this respect the book *The nuclear many-body problem* by P. Ring and P. Schuck (Springer-Verlag, 1980) might prove a useful starting point.

---

[4]H. Ehrenreich and M. H. Cohen *Phys. Rev.*, **115**, 786 (1959).

Appendix 27.B

Time-dependent variation principle

There has been a great deal of discussion over the years about the correct form for the quantum-mechanical time-dependent variation principle. The difficulties have arisen, in the main, from confusions about the role of variational principles in quantum (and other) theories. There are two points of view which are not often made explicit:

- The variation principle is more fundamental than the differential equation to which it is "equivalent"; the differential equation is simply the Euler–Lagrange equation generated by application of the variation principle within the boundaries of applicability.

- The variation principle is a mathematical technique for the generation of approximations to a differential equation and is therefore less fundamental than that differential equation.

Most of the literature takes the latter view, albeit unconsciously and implicitly.

However, the physical law contained in the Schrödinger equation is not just contained in the (partial) differential equation, it is contained in the differential equation in the interior of the relevant region of configuration space *and* the conditions at the boundaries of that region. The variation principle generates *both* of these conditions. Thus, there are very strong grounds for making the law of nature be contained in the variation principle.[5]

---

[5]That is not to say that the variation principle does not continue to be an extremely powerful tool for generating whole families of approximations to differential equations, simply that *in this context of an interpreted physical theory* the law of nature is the primary focus.

Thus we have provided justification elsewhere[6] for the view that the dynamical law of Schrödinger's mechanics is the equality of the mean values of the Hamiltonian density and the energy density:

$$\int \Psi^* \hat{H} \Psi d\tau dt = \int \Psi^* \hat{E} \Psi d\tau dt$$

which generates the variation principle

$$\int \delta\Psi^* \hat{H} \Psi d\tau dt - \int \delta\Psi^* \hat{E} \Psi d\tau dt = \int \delta\Psi^*(\hat{H}\Psi - \hat{E}\Psi)d\tau dt = 0 \quad (27.\text{B}.2)$$

It is simple to show that, if $\hat{H}$ is independent of time,

$$\int \Psi^* \hat{E} \Psi d\tau dt = constant = E$$

(say) and the usual variation principle follows.

If $\hat{H}$ is dependent on time the full form must be used as we have done in the last chapter.

It has been remarked earlier that the fact that we use a linear expansion method enables us to get away with very simple and incomplete expressions of the variation principle. The Hamiltonian and energy densities are dependent (in general) on the *gradients* of Ψ (and Ψ^*) as well as their values, and it is these gradient-dependent terms which, when integrated by parts, generate the boundary conditions on Ψ which we routinely satisfy not explicitly but simply by choosing our expansion (basis) functions with an appropriate physical interpretation in mind. To generate the *differential* equations *and* the boundary conditions we must, of course, use the fact that the integrals in the theory depend on Ψ, Ψ^*, $\nabla\Psi$ (or $\nabla^2\Psi$), $\nabla\Psi^*$, $\partial\Psi/\partial t$, $\partial\Psi^*/\partial t$ etc. When this is done it is clear that the general time-dependent variation principle is just eqn (27.B.2), requiring no special considerations.

---

[6]D. B. Cook *Schrödinger's Mechanics*, (World Scientific, 1988).

Chapter 28

Transitions and stability

In both the time-dependent and time-independent SCF perturbation theories the equations determining the effect of the perturbation look as if they can generate a finite effect with no applied perturbation. These cases of "infinitesimal perturbations" are genuine ones which have important ramifications for the stabilities of the single-determinant states of a many-electron system.

Contents

28.1	Introduction	**683**
28.2	Transitions	**684**
28.3	The transition frequencies	**685**
28.4	Finite perturbations; oscillations	**686**
28.5	Stability; the time-independent case	**688**
28.6	Implementation	**689**

28.1 Introduction

Both types of SCF perturbation theory generate similar-looking expressions for the first-order corrections to the density matrix:

$$\boldsymbol{R}_1 = \sum_{i,a}(X_{ia}\boldsymbol{c}_a\boldsymbol{c}_i^\dagger + \ldots)$$

where

$$X_{ia} = -\frac{\Delta_{ai}}{(\epsilon_a - \epsilon_i - \omega)}$$

and the numerator Δ_{ai} (or $(h_1)_{ai}$ in the time-independent case) of the expression for the expansion coefficients X_{ia} (or D_{ia} in the other case) contains the matrix representation of the actual perturbation operator *plus* an electron-repulsion term depending on the perturbed density:

$$\Delta_{ai} = c_a^\dagger \left(f + G(R_1) \right) c_i$$

Now what is the interpretation of the fact that *if f is vanishingly small* these coefficients may still be non-zero? Obviously the case we have discussed presents one kind of answer: it is f (i.e. $\hat{f}$) which is *generating* R_1 so, in the absence of an $\hat{f}$, we should have R_1 zero.

The time-dependent case is the easiest to interpret because it is absolutely basic to the quantum theory that there *are* quantum effects where very large changes in electron distribution are caused by time-dependent perturbations of arbitrarily small magnitude; indeed, one of these processes — the photo-electric effect — was among the key experiments in demolishing the classical model at the electronic level.

For certain characteristic frequencies *electronic transitions* are caused by the application of a perturbation (of the appropriate type) of arbitrarily small magnitude. Clearly this is a process of considerable intrinsic interest and, as we shall see, throws light on the interpretation of the time-independent case and, equally important, provides a technique for the solution of the *general* time-dependent case.

28.2 Transitions

The equations to be solved for the time-dependent SCF first-order perturbation "correction" are, in the absence of a perturbation:

$$\begin{pmatrix} A & B \\ B & A \end{pmatrix} \begin{pmatrix} X \\ Y \end{pmatrix} = \omega \begin{pmatrix} 1 & 0 \\ 0 & -1 \end{pmatrix} \begin{pmatrix} X \\ Y \end{pmatrix} \qquad (28.1)$$

which may be written in a more compact, recognisable, form as

$$\Omega U = \omega \Delta U \qquad (28.2)$$

with obvious notation

$$\Omega = \begin{pmatrix} A & B \\ B & A \end{pmatrix}$$
$$U = \begin{pmatrix} X \\ Y \end{pmatrix}$$
$$\Delta = \begin{pmatrix} 1 & 0 \\ 0 & -1 \end{pmatrix}$$

28.3. THE TRANSITION FREQUENCIES

for the collected matrices.

Unlike the case *with* the perturbation matrix, this is an *eigenvalue problem* not a matrix inversion problem. That is, the above equation only has solutions for *certain* values of ω; ω is not *input* to the equations but is *determined by them*. Since the matrices A, B etc. are of dimension $n(m-n)$, there are therefore $n(m-n)$ values of ω^1 and corresponding sets of values of X, Y.

These characteristic ω-values are those frequencies at which even an infinitesimally small perturbation will cause changes in the electron density in the molecule. As we shall shortly see, the magnitude of the changes in the density becomes arbitrarily large at these frequencies. In view of what was said at the end of the last chapter *and* bearing in mind the fact that the diagonal terms of the matrix whose eigenvalues are being determined are *orbital energy differences*, it is clear that theses frequencies are nothing more than (approximations to) those which will induce electronic transitions[2] in the system.

28.3 The transition frequencies

When eqn(28.1) is expanded out it generates a pair of simultaneous matrix equations for X and Y:

$$AX + BY = \omega X$$
$$BX + AY = -\omega Y$$

Addition and subtraction of these two equations generates a pair of equivalent equations for the sum and difference of X and Y which have the essential dimension $(n(m-n))$ of the problem:

$$(A+B)(X+Y) = \omega(X-Y) \qquad (28.3)$$
$$(A-B)(X-Y) = \omega(X+Y) \qquad (28.4)$$

The second of these may be inverted to yield

$$(X-Y) = \omega(A-B)^{-1}(X+Y) \qquad (28.5)$$

which may be inserted in the other equation to generate an eigenvalue equation for $(X+Y)$:

$$(A+B)(X+Y) = \omega^2(A-B)^{-1}(X+Y) \qquad (28.6)$$

---

[1] The *total* matrix equation is of twice this dimension but it is easy to see that, if ω is a solution, so is $-\omega$ associated with X and Y *interchanged*

[2] In his pioneering work on these matters Thouless (D. J. Thouless *The quantum mechanics of many-body systems*, Academic Press, 1961) calls these frequencies "free vibrations" of the system which occur about the Hartree–Fock solution without, however, saying why a vibrating distribution of charge would not lose its energy by radiation.

where the "overlap" term in this new eigenvalue equation is $(A-B)^{-1}$ so that, by analogy with the solution of the SCF equations with overlap,

$$(A-B)^{\frac{1}{2}}$$

will be needed. The values of ω are *both* the roots of the eigenvalues ω^2. If $(X+Y)$ is obtained from this equation, the associated $(X-Y)$ may be generated from eqn (28.5).

This completes the solution of the problem for the evaluation of the transition frequencies; the (eigenvalue, coefficient) pairs can be distinguished by a superscript (q), say, so that the first-order time-dependent perturbation theory approximation to the transition frequencies[3] are the $\omega^{(q)}$, and the composition of the transition in terms of excitations between molecular orbitals occupied in the SCF single determinant and the virtuals of that SCF calculation are given by the elements of $X^{(q)}$ and $Y^{(q)}$.

These matrices, $X^{(q)}$ and $Y^{(q)}$, together with the $\omega^{(q)}$, form a solution of eqn (28.2) with

$$U = \begin{pmatrix} X^{(1)} & X^{(2)} & \ldots & X^{n(m-n)} & Y^{(1)} & Y^{(2)} & \ldots & Y^{n(m-n)} \\ Y^{(1)} & Y^{(2)} & \ldots & Y^{n(m-n)} & X^{(1)} & X^{(2)} & \ldots & X^{n(m-n)} \end{pmatrix}$$

(recall the relationship between the two solutions $\pm\omega^{(p)}$), and the eigenvalues may be collected into the diagonal matrix $\boldsymbol{\omega}$:

$$\boldsymbol{\omega} = \begin{pmatrix} \omega^{(1)} & 0 & \ldots & 0 \\ 0 & \omega^{(2)} & \ldots & 0 \\ \ldots & \ldots & \ldots & \ldots \\ 0 & 0 & \ldots & -\omega^{n(m-n)} \end{pmatrix}$$

so that

$$U^\dagger \Omega U = \boldsymbol{\omega}$$

and we may find any *function* of the matrix Ω (in particular its inverse) by the methods used for finding $S^{-1/2}$ in the SCF process.

28.4 Finite perturbations; oscillations

In the presence of a finite perturbation of the type originally outlined in the last chapter, eqn (28.1) becomes

$$\begin{pmatrix} f \\ f^\dagger \end{pmatrix} + \begin{pmatrix} A & B \\ B & A \end{pmatrix} \begin{pmatrix} X' \\ Y' \end{pmatrix} = \omega \begin{pmatrix} 1 & 0 \\ 0 & -1 \end{pmatrix} \begin{pmatrix} X' \\ Y' \end{pmatrix} \quad (28.7)$$

---
[3] Actually, angular frequencies $\omega = 2\pi\nu$.

28.4. FINITE PERTURBATIONS; OSCILLATIONS

which may be rearranged to[4]

$$(\mathbf{\Omega} - \omega \mathbf{\Delta}) \mathbf{U}' = \mathbf{f} \tag{28.8}$$

where

$$\mathbf{U}' = \begin{pmatrix} \mathbf{X}' \\ \mathbf{Y}' \end{pmatrix}$$

and the prime has been added to the coefficient matrices to emphasise that they differ from the case of zero $\mathbf{f}$.

The formal solution to this matrix problem for a given ω, which is now *given in the form of the perturbation* not generated by the solution, is simply

$$\mathbf{U}' = (\mathbf{\Omega} - \omega \mathbf{\Delta})^{-1} \mathbf{f} \tag{28.9}$$

which requires the inverse matrix for every ω.

However, the inverse of $\mathbf{\Delta}$ is trivial because it is diagonal and we can form the inverse of $\mathbf{\Omega}$ from the results of the last section. In fact, because the matrix $\mathbf{U}$ is unitary it diagonalises $\mathbf{\Omega}$ and does not disturb the diagonal form of $\mathbf{\Delta}$, so that the $\mathbf{U}$ of the last section is such that

$$\mathbf{U}^\dagger (\mathbf{\Omega} - \omega \mathbf{\Delta}) \mathbf{U} = \mathbf{d}$$

where $\mathbf{d}$ is *diagonal* with elements

$$d_{ij} = \delta_{ij}(\omega \pm \omega^{(i)})$$

with the choice of sign determined by the two occurrences of the eigenvalues with opposite sign.

Thus

$$(\mathbf{\Omega} - \omega \mathbf{\Delta})^{-1} = \mathbf{U} \mathbf{d}^{-1} \mathbf{U}^\dagger \tag{28.10}$$

which is an *explicit function* of ω.

This is the full solution of the time-dependent self-consistent perturbation problem since this inverse, when multiplied by the matrix elements of the spatial form of the perturbation, generates the expansion coefficients of the (oscillating) first-order correction to the density matrix.

It is important to note that the *mathematics* of this derivation reveals a surprising *physical interpretation* of the result.

---

[4] This is a slight liberty with notation, the matrix $\mathbf{f}$ has been used with two slightly different meanings here.

The inverse matrix eqn (28.10), which determines the effects of the time-dependent perturbation through eqn (28.9), is entirely fixed by the natural transition frequencies of the unperturbed system and their composition. This matrix is characteristic of the molecule itself and determines the response of the system to any periodic time-dependent perturbation and therefore, by implication, any Fourier synthesis of time-dependent perturbations. For obvious reasons this important matrix is called the linear[5] response matrix.

The explicit form of the response matrix may be written down in terms of the $\omega^{(i)}$ and associated coefficient matrices. If we use the relationship between the solutions for $\pm\omega^{(i)}$, the summation can be limited to *positive* $\omega^{(i)}$:

$$(\Omega - \omega\Delta)^{-1} = \sum_{i=1}^{n(m-n)} \left\{ \frac{1}{\omega - \omega^{(i)}} \begin{pmatrix} X^{(i)} \\ Y^{(i)} \end{pmatrix} \begin{pmatrix} X^{(i)\dagger} & Y^{(i)\dagger} \end{pmatrix} \right.$$
$$\left. - \frac{1}{\omega + \omega^{(i)}} \begin{pmatrix} Y^{(i)} \\ X^{(i)} \end{pmatrix} \begin{pmatrix} Y^{(i)\dagger} & X^{(i)\dagger} \end{pmatrix} \right\}$$

which shows clearly that the effect of a finite perturbation at one of the transition frequencies will cause arbitrarily large changes in the electron density.

28.5 Stability; the time-independent case

The equations for time-independent self-consistent perturbation theory are of very similar form to those time-dependent ones we have just studied, but the interpretation of the case of infinitesimal perturbations is, if anything, more problematical. What is the meaning of the fact that there are non-trivial solutions of the time-independent equations for arbitrarily small perturbations?

The clue lies in the idea of the *transitions* of the analogous time-dependent case; there are solutions of the ordinary time-independent SCF equations which are *unstable* or *metastable* in the following senses:

- The equations may have been *constrained* in some way so that the self-consistent solution obtained is not an approximation to true minimum in the Hartree–Fock energy functional.

- The solution may exist at a turning point in the energy functional which is a saddle point or point of inflection on that surface; there are other solutions which are lower in energy which are true (local) minima.

---

[5] Because of the first-order derivation.

28.6. IMPLEMENTATION

In each of these types of case an infinitesimal perturbation of the electron density may be enough to cause the "transition" to a more stable solution. In the first case the removal of the constraint may or may not effect the transition since, even with the constraint removed, the solution may be of the second type, exhibiting a "self-consistent symmetry".

With this interpretation it is clearly useful to use the SCF perturbation equations to check that any solution of the SCF equations is, indeed, of the type we seek. Conversely, it is often useful to look for "excited" solutions of the SCF equations.

Using the notation already established in eqn (26.11), the second-order energy expression when there is no perturbation is

$$E_2 = \boldsymbol{D}^\dagger \boldsymbol{M} \boldsymbol{D}$$

and we may study the nature of the turning points in the SCF energy functional (which has $E_1 = 0$, defining such a turning point) by examining the *sign* of E_2. Multiplying the above equation from the left by $\boldsymbol{D}$ and using $\boldsymbol{DD}^\dagger = 1$ we have an equivalent eigenvalue equation:

$$\boldsymbol{MD} = E_2 \boldsymbol{D} \qquad (28.11)$$

where, as before in the time-dependent case, the numbers E_2 and coefficients $\boldsymbol{D}$ are now *determined* by the equation and tell us, by their sign, the nature of the single-determinant wavefunction composed of the self-consistent molecular orbitals used in setting up the matrix $\boldsymbol{M}$:

$$M_{ia,jb} = \delta_{ij}\delta_{ab}(\epsilon_i - \epsilon_a) - f(ia, jb) + (ib, ja) + (ij, ba)$$

with f one or two depending on the SCF case.

Diagonalising the matrix $\boldsymbol{M}$ and checking for positive eigenvalues will, therefore, tell us whether or not our SCF solution is stable. If the eigenvalue is negative we may use the information in the associated coefficient matrix $\boldsymbol{D}$ to correct the current $\boldsymbol{R}$ matrix and use this new $\boldsymbol{R}$ in the iterative process to find a stable solution.[6]

28.6 Implementation

There are no new requirements for the implementation of the time-independent or time-dependent SCF equations except the generation of the electron-repulsion integrals over the molecular orbitals. We now turn to this integral transformation problem.

---

[6]The most comprehensive study of the stabilities of HF wavefunctions has been carried out by H. Fukutome and he has reviewed the whole area in *Int. J. Quant. Chem.*, **XX**, 955 (1981).

Chapter 29

Two-electron transformations

In order to implement the perturbed SCF method and any other, more advanced, models of molecular electronic structure, we must be able to compute the one- and two-electron integrals over different sets of orbitals. These transformations are simple in principle but the one involving the repulsion integrals is sufficiently demanding of resources to be worth some examination.

Contents

29.1	Orbital transformations	**691**
29.2	Strategy	**693**
29.3	Transformation without sorting	**695**
	29.3.1 tran2e	696
	29.3.2 cdgvc	703
	29.3.3 INDEX	704
29.4	Transformations with sorting	**705**
29.5	Summary	**708**

29.1   Orbital transformations

If a set of orbitals ψ is defined as linear combinations of an existing set of functions ϕ by

$$\psi = \phi C$$

then any integrals involving the energy operators in the Hamiltonian suffer an induced transformation:

$$h' = \int \boldsymbol{\psi}^\dagger \hat{h} \boldsymbol{\psi} dV = \boldsymbol{C}^\dagger \boldsymbol{h} \boldsymbol{C}$$

where

$$\boldsymbol{h} = \int \boldsymbol{\phi}^\dagger \hat{h} \boldsymbol{\phi} dV$$

is the matrix of integrals over the original basis.

A similar transformation is induced in the electron-repulsion integrals which may be written in Kronecker-product form or, more commonly, simply written out in full:

$$(\psi_i \psi_j, \psi_k \psi_\ell) = \sum_{r=1}^{m} \sum_{s=1}^{m} \sum_{t=1}^{m} \sum_{u=1}^{m} C_{ri}^* C_{sj} C_{tk}^* C_{u\ell} (\phi_r \phi_s, \phi_t \phi_u)$$

where the *charge-cloud* notation has been used:

$$(\psi_i \psi_j, \psi_k \psi_\ell) = \int \psi_i^*(\vec{r}_1) \psi_j(\vec{r}_1) \frac{1}{r_{12}} \psi_k^*(\vec{r}_2) \psi_\ell(\vec{r}_2) dV_1 dV_2$$

and the complex-conjugate notation is merely for completeness, normally both our basis functions and transformation coefficients are *real*.

We have met some of the components of the some of the transformed repulsion integrals earlier. If $\boldsymbol{C}$ is the MO coefficient matrix, then the "diagonal" repulsion integrals ($i = j$, and $k = \ell$) over the MOs are just:

$$\begin{aligned}
(\psi_i \psi_i, \psi_k \psi_k) &= \sum_{r=1}^{m} \sum_{s=1}^{m} \sum_{t=1}^{m} \sum_{u=1}^{m} C_{ri} C_{si} C_{tk} C_{uk} (\phi_r \phi_s, \phi_t \phi_u) \\
&= \sum_{r=1}^{m} \sum_{s=1}^{m} \sum_{t=1}^{m} \sum_{u=1}^{m} R_{rs}^i \left[R_{tu}^k (\phi_r \phi_s, \phi_t \phi_u) \right] \\
&= \mathrm{tr} \boldsymbol{R}^i \boldsymbol{J}(\boldsymbol{R}^k)
\end{aligned}$$

where

$$R_{rs}^i = C_{ri} C_{si}$$

Indeed, it is this convenient fact which enables all the SCF methods outlined so far to be implemented compactly; the underlying two-electron integral transformation (four-index multiplications) has been contained into the formation of $\boldsymbol{J}$ and $\boldsymbol{K}$ matrices and some one-electron (two-index matrix multiplications) transformations. However, if we use any method which demands the existence of the MO-based repulsion integrals $(\psi_i \psi_j, \psi_k \psi_\ell)$ with either $i \neq j$ or $k \neq \ell$ (or both) then we cannot use this approach since these integrals cannot be written in terms of R matrices involving, as they do, products of *different* columns of C.

There are no *theoretical* considerations involved here, the matter is entirely a *technical* one; what is the most effective way of generating this very large set of numbers ($\approx m^4/8$) each of which may involve all $\approx m^4/8$ of the original integrals?

29.2 Strategy

The general dilemma involved in transforming the repulsion integrals to a new basis can be appreciated by looking briefly at the one-electron case. Written out explicitly the transformation is

$$h'_{ij} = \sum_{r=1}^{m} \sum_{s=1}^{m} C_{ri} C_{sj} h_{rs}$$

For the sake of generality, let us assume that the matrix C is rectangular, defining m' (say) new functions in terms of the original m basis functions, of course $m' \leq m$. There are about m'^2 integrals to form, each of which involves all m^2 original integrals if there are no zeroes in C. That is, it looks as if about $m^2 m'^2$ multiplications[1] are required to effect the transformation. But during the transformation, many products are evaluated many times; if we are willing to *store* some of the intermediate products we can reduce the number of multiplications.

In particular, if we form

$$b = hC$$

which involves about m^2 multiplications, and then use b to form h':

$$h' = C^\dagger b$$

requiring about m'^2 multiplications, we have cut the dependence on m and m' from $\approx m^2 m'^2$ to $\approx (m^2 + m')^2$ *at the expense of storing* $\approx m^2$ intermediate products. This is typical of all computing applications; there is some balance to be struck between computing and storage. We have already experienced this fact more acutely in a decision to use "conventional" or "direct" methods of treating the basis-function repulsion integrals in SCF processes.

The situation with the transformation of repulsion integrals is much more acute; we can scarcely afford to increase the amount of work 256-fold when we

---
[1] The matrices are symmetrical so only $(m(m+1))/2$ are needed but each term involves two multiplications.

double the size of the basis. We must decide on the basis of available storage (and time to access that storage) how many intermediate products to save and how many to recompute.

It is easy to see what the minimum number of multiplications can be for the two-electron transformation simply by writing the four-index multiplication as four separate summations, saving the intermediate products in each case:

$$(\psi_i\phi_s, \phi_t\phi_u) = \sum_{r=1}^{m} C_{ri}^*(\phi_r\phi_s, \phi_t\phi_u)$$

$$(\psi_i\psi_j, \phi_t\phi_u) = \sum_{s=1}^{m} C_{sj}(\psi_i\phi_s, \phi_t\phi_u)$$

$$(\psi_i\psi_j, \psi_k\phi_u) = \sum_{t=1}^{m} C_{tk}^*(\psi_i\psi_j, \phi_t\phi_u)$$

$$(\psi_i\psi_j, \psi_k\psi_\ell) = \sum_{u=1}^{m} C_{tu\ell}(\psi_i\psi_j, \psi_k\phi_u)$$

The first summation involves about m multiplications for all $m'm^3$ results with similar expressions for the other three, giving a total of about

$$m'm^4 + m'^2 m^3 + m'^3 m^2 + m'^4 m$$

multiplications which has its maximum when $m' = m$, a total of $4m^5$ multiplications. The main result being that, with unlimited *storage*, the calculation cannot be made to involve less than an m^5 dependence on basis size if all the transformed repulsion integrals are required. Notice that, at any one time, the most straightforward implementations of these single-index transformations would require only *two* sets of integrals to be available; there is never any need to keep the integrals from any step other than the last one.

It is also possible to separate the four-index transformation into two two-index transformations:

$$(\psi_i\psi_j, \phi_t\phi_u) = \sum_{r=1}^{m}\sum_{s=1}^{m} C_{ri}^* C_{sj}(\phi_r\phi_s, \phi_t\phi_u)$$

$$(\psi_i\psi_j, \psi_k\psi_\ell) = \sum_{t=1}^{m}\sum_{u=1}^{m} C_{tk}^* C_{u\ell}(\psi_i\psi_j, \phi_t\phi_u)$$

where each transformation involves m'^2 terms being formed by the use of $2m^3$ products, giving the same $4m^5$ maximum dependence on m and the same requirement to save a set of intermediate products. Thus, the possibilities arise of performing the total transformation as:

- Four single-index transformations

- Two two-index transformations
- Two single-index and one two-index transformation

whichever is most appropriate for the technology available.

If there is space in main working storage for two complete sets of repulsion integrals then these expressions can be coded as they stand and the problem is relatively simple.

If this is not possible, then two strategies are possible:

1. The *whole* transformation can be done for one, two or three of the indices by reading the AO integral file once and the file then has to be read repeatedly to complete the remaining transformations.
2. The transformation can be done for a *subset* of i and j (say) for which all $(\psi_i \psi_j, \phi_t \phi_u)$ will fit into working store and repeat this partial transformation be reading these sets into main store in turn.

29.3 Transformation without sorting

The first of the above methods is easy to implement, we do the full transformation over three of the indices and repeat this work for the fourth sequentially.

The WEB is given below, the scheme is

1. Transform the first index $r \to i$ and store all $\approx m^3$ intermediate integrals for a given first index. These intermediate integrals are stored as a set of m symmetric matrices, with only the upper half of each matrix stored since storage space is at a premium:

$$V(tu, s) = (\boldsymbol{V})^{(s)}_{tu} = (\boldsymbol{V})^{(s)}_{ut} = \sum_{r=1}^{m} C_{ri}(\phi_r \phi_s, \phi_t \phi_u)$$

where $tu = t * (t-1)/2 + u$, if $t \geq u$.

2. Perform a "one-electron" transformation on these matrices to transform $t, u \to k, l$:

$$\left(\boldsymbol{C}^\dagger \boldsymbol{V}^s \boldsymbol{C}\right)_{kl} = (\psi_i \phi_s, \psi_k \psi_\ell)$$

3. Perform the final transformation of $s \to j$ by summation over the transformed matrices:

$$(\psi_i \psi_j, \psi_k \psi_\ell) = \sum_{s=1}^{m} \left(\boldsymbol{C}^\dagger \boldsymbol{V}^s \boldsymbol{C}\right)_{kl} C_{sj}$$

These steps must all be done m times, involving reading the file of AO integrals m times.

29.3.1 tran2e

Two-electron (four-index) transformation routine. Does the full transformation and assumes that there is room in main memory for $nbasis^2 * (nbasis + 1)/2$ intermediate results in V.

The file ($nfile1$) of "AO" repulsion integrals contains the integrals ($getint$-legible) in random order (as they have been computed) and is read $nbasis$ times; once for each transformed index.

Integrals over the new basis generated by C are output into $nfile2$ in the usual standard $putint$ format.

NAME
   tran2e

SYNOPSIS

```
subroutine tran2e(nfile1,nfile2,nbasis,ntt,C,W,RS,V);
implicit double precision (a-h,o-z);
double precision C(nbasis,nbasis),V(ntt,nbasis);
double precision W(*);
integer RS(*);
integer nbasis,ntt,nfile1,nfile2;
```

DESCRIPTION
Transforms the repulsion integrals residing on *nfile1* to the ones defined by the transformation matrix C and puts them onto file *nfile2*.

ARGUMENTS

nfile1 Input: The logical number of the (**getint**-legible) input repulsion integrals.

nfile2 Input: The logical number of the (**putint**-generated) output repulsion integrals.

nbasis Input: The number of basis functions.

ntt Input: $nbasis \times (nbasis + 1)/2$.

C Input: $nbasis \times nbasis$ matrix of transformation coefficients (columns define new orbitals).

V, W, RS Workspace. RS is an integer indexing array.

DIAGNOSTICS
None

698   CHAPTER 29. TWO-ELECTRON TRANSFORMATIONS

§. **@I** "tran2e.defns" *Section(s) skipped...*
   0

"tran2e.f" 29.3.1 ≡

{
  ⟨ tran2e local declarations 29.3.1.1 ⟩
  ⟨ Initialise RS and output file 29.3.1.2 ⟩
      /* Start outer loop on first "MO" index i */
  do $i = 1$, *nbasis*;
  {
    ⟨ Initialise input file and Zeroise V 29.3.1.3 ⟩
        /* For each i, read the "AO" integrals and transform r. */
    ⟨ Transform r to i 29.3.1.4 ⟩
        /* We now have the integrals (is, tu) */
        /* Next, use these integrals to transform t and u by normal matrix multiplication */
    ⟨ Transform t and u to k and l 29.3.1.5 ⟩
        /* This gives us the integrals (is, kl), so, complete the job by transforming s */
    ⟨ Transform s to j 29.3.1.6 ⟩
    ⟨ Pack the Output in putint form 29.3.1.7 ⟩
  }   /* End of main loop on i, finished */
  return;
}

29.3. TRANSFORMATION WITHOUT SORTING

⟨ tran2e local declarations 29.3.1.1 ⟩ ≡
    integer *point1*, *point2*, *getint*;   /∗ File control integers ∗/
    integer i, j, k, l;   /∗ labels for output MO integrals ∗/
    integer r, s, t, u;   /∗ labels for input AO integrals ∗/
    integer *rtt*, *rsx*, *tux*, *maxkl*, *kl*, *lmax*;   /∗ working temporaries ∗/
    double precision *val*;
    double precision *zero*;
    data *zero* $/0.0 \cdot 10^{00}$D$/$;

This code is used in section 29.3.1.

⟨ Initialise RS and output file 29.3.1.2 ⟩ ≡
    /∗ Initialise output file ∗/
    rewind *nfile2*;
    point2 = 0;

    /∗ Set up indexing array ∗/
    do $r = 1$, *nbasis* + 1;
      $RS(r) = (r \ast (r - 1))/2$;
    /∗ Note that $RS(n + 1) = $ (n∗(n+1))/2 ∗/

This code is used in section 29.3.1.

⟨ Initialise input file and Zeroise V 29.3.1.3 ⟩ ≡
    /∗ Initialise file of AO integrals for each i ∗/
    rewind *nfile1*;
    point1 = 0;

    /∗ Initialise V with zeroes for each i ∗/
    do $r = 1$, *nbasis*;
    {
    do *rtt* = 1, *ntt*;
      $V(rtt, r) = zero$;
    }

This code is used in section 29.3.1.

Module to do the work for each of the **four** possible contributions from inequalities amongst the i, j, k, l. Notice that there are indeed (at most) **four not eight** because of the fact that we are only forming a non-redundant list with respect to the two last indices. So, if there is any contribution from interchanges due to inequalities between the **last two** indices, they must not be included.

The Macro *USE* therefore begins with a redundant **if**; this is just belt and braces (belt and suspenders in the USA).

⟨Transform r to i 29.3.1.4⟩ ≡

```
@m  USE(r, s, t, u) if ((t) ≥ (u))
      {
      indtu = RS((t)) + (u);
      V(indtu, (s)) = V(indtu, (s)) + C((r), i) * val;
      }

    /* getint reads the AO integrals which are the usual non-redundant set
     */
    while (getint(nfile1, r, s, t, u, mu, val, point1) ≠ END_OF_FILE)
      {
      USE(r, s, t, u);
      if (r ≠ s)
        USE(s, r, t, u);
      rsx = (r * (r − 1))/2 + s;
      tux = (t * (t − 1))/2 + u;
      if (rsx ≠ tux)
        {
        USE(t, u, r, s);
        if (t ≠ u)
          {
          USE(u, t, r, s);
          }
        }
      }
```

/* This completes the reading of the AO integrals and the transformation of the AO index r to the MO index i */

This code is used in section 29.3.1.

29.3. TRANSFORMATION WITHOUT SORTING   701

⟨ Transform t and u to k and l 29.3.1.5 ⟩ ≡

/* The transformation of the last two indices t and u to k and l is just a straightforward "one-electron" (two-index) transformation; two matrix multiplications for each s and i */

do $s = 1$, $nbasis$;
  call $CdgVC(nbasis, V(1, s), C, W)$;

This code is used in section 29.3.1.

⟨ Transform s to j 29.3.1.6 ⟩ ≡
  $maxkl = RS(i+1)$;   /* This is to generate $(i*(i+1))/2$, the upper limit of the non-redundant set of indices */

do $kl = 1$, $maxkl$;
{
  do $s = 1$, $nbasis$;
    $W(s) = V(kl, s)$;

  /* W is workspace to enable the transformed integrals to be put back into V The transformation is just a matrix multiplication */

  do $j = 1$, i;
  {
    $V(kl, j) = zero$;
    do $s = 1$, $nbasis$;
      $V(kl, j) = V(kl, j) + C(s, j) * W(s)$;
  }
}

This code is used in section 29.3.1.

Now, for each i, reorganise the transformed integrals in V to the standard form of packed intgrals and use *putint* to fire them out to the file *nfile2* which is then readable by *getint* */

⟨ Pack the Output in putint form 29.3.1.7 ⟩ ≡

/* For each i the set of integrals in V is (ij, kl) in the usual standard order of indices i, j, k, l. So, setting up the usual order of loops on j, k and l will enable the labels and values to be identified and packed for output to *putint* */

$id = 0$;   /* a dummy for packing */
$last = NO$;   /* for *putint*, means that the integral is not the last */
do $j = 1, i$;
  {
  do $k = 1, i$;
    {
    /* Get the upper limit, *lmax*, for l right */

    $lmax = k$;
    if $(i \equiv k)$
      $lmax = j$;
    do $l = 1, lmax$;
      {
      $kl = RS(k) + l$;
      $val = V(kl, j)$;

      /* If you want to leave out small integrals from the output file, *nfile2*, this is the place to do it: **if** $(val < crit)$ **next**; would work OK */

      if $(l \equiv nbasis)$
        $last = YES$;   /* Signals last integral */
      call *putint*$(nfile2, i, j, k, l, id, val, point2, last)$;

      /* *putint* outputs the MO integrals to file *nfile2*. It keeps track of things with *point2* and has to know when to close the file, hence *last* has to be NO or YES */
      }
    }
  }

This code is used in section 29.3.1.

29.3. TRANSFORMATION WITHOUT SORTING

29.3.2 cdgvc

One-electron transformation.

"tran2e.f" 29.3.2 ≡
```
subroutine CdgVC(n, A, B, R);
implicit double precision (a − h, o − z);
double precision A(*), B(*), R(*);
integer n;
{
integer i, j, k, ijr, ik, kj, ki;
double precision zero;
data zero/0.0 · 10⁰⁰D/;
```
$$\text{data } zero/0.0 \cdot 10^{00}\text{D}/;$$
```
   /* Form R = AB in full storage mode */
   do i = 1, n;
     {
     do j = 1, n;
       {
       ijr = n * (j − 1) + i;
       R(ijr) = zero;
       do k = 1, n;
         {
         ik = (i * (i − 1))/2 + k;
         if (k > i)
            ik = (k * (k − 1))/2 + i;
         kj = n * (j − 1) + k;
         R(ijr) = R(ijr) + A(ik) * B(kj);
         }
       }
     }
```
/* Now form $A = B^\dagger R$ which must be symmetric */
```
   do i = 1, n;
     {
     do j = 1, i;
       {
       ijr = (i * (i − 1))/2 + j;
       A(ijr) = zero;
       do k = 1, n;
         {
         ki = n * (i − 1) + k;
         kj = n * (j − 1) + k;
         A(ijr) = A(ijr) + B(ki) * R(kj);
         }
```

```
            }
        }
        return;
    }
```

29.3.3 INDEX

A: 29.3.2.

B: 29.3.2.

$CdgVC$: 29.3.1.5, 29.3.2.

$crit$: 29.3.1.7.

END_OF_FILE: 29.3.1.4.

$getint$: 29.3.1, 29.3.1.1, 29.3.1.4, 29.3.1.7.

i: 29.3.1.1, 29.3.2, 29.3.1.

id: 29.3.1.7.

ijr: 29.3.2.

ik: 29.3.2.

$indtu$: 29.3.1.4.

j: 29.3.1.1, 29.3.2.

k: 29.3.1.1, 29.3.2.

ki: 29.3.2.

kj: 29.3.2.

kl: 29.3.1.1, 29.3.1.6, 29.3.1.7.

l: 29.3.1.1.

$last$: 29.3.1.7.

$lmax$: 29.3.1.1, 29.3.1.7.

$maxkl$: 29.3.1.1, 29.3.1.6.

mu: 29.3.1.4.

n: 29.3.2.

$nbasis$: 29.3.1, 29.3.1.2, 29.3.1.3, 29.3.1.5, 29.3.1.6, 29.3.1.7.

$nfile1$: 29.3.1, 29.3.1.3, 29.3.1.4.

$nfile2$: 29.3.1, 29.3.1.2, 29.3.1.7.

NO: 29.3.1.7.

ntt: 29.3.1.3.

$point1$: 29.3.1.1, 29.3.1.3, 29.3.1.4.

$point2$: 29.3.1.1, 29.3.1.2, 29.3.1.7.

$putint$: 29.3.1, 29.3.1.7.

R: 29.3.2.

r: 29.3.1.1.

RS: 29.3.1, 29.3.1.2, 29.3.1.4, 29.3.1.6, 29.3.1.7.

rsx: 29.3.1.1, 29.3.1.4.

rtt: <u>29.3.1.1</u>, 29.3.1.3.

s: <u>29.3.1.1</u>.

t: <u>29.3.1.1</u>.

tux: <u>29.3.1.1</u>, 29.3.1.4.

u: <u>29.3.1.1</u>.

USE: 29.3.1.4.

val: <u>29.3.1.1</u>, 29.3.1.4, 29.3.1.7.

YES: 29.3.1.7.

zero: <u>29.3.1.1</u>, 29.3.1.3, 29.3.1.6, 29.3.2.

29.4 Transformations with sorting

Each single-index transformation which transforms just one of the orbitals in the definition of the original integrals:

$$(\psi_i \phi_s, \phi_t \phi_u) = \sum_{r=1}^{m} C_{ri}^* (\phi_r \phi_s, \phi_t \phi_u)$$

can, by defining an appropriate matrix of repulsion integrals, be seen to be a single matrix multiplication. The problem is, of course, that the elements of the "matrices" of original and transformed integrals:

1. Are not all capable of being stored in main storage.

2. May occur in random order since they have been computed by methods which may group together integrals over similar basis functions and zeroes will not be stored.

3. Are not all present; only the permutationally distinct integrals are ever stored.

However, if we are willing to pay the price of reading the original integrals and *sorting* and possibly *duplicating* many of them, the returns in speed of execution of the subsequent actual numerical computing will more than offset this overhead.

The technique is most easily appreciated if we take the simplest possible case. Suppose we wish to perform the single-index transformation given above followed by the transformation

$$(\psi_i \psi_j, \phi_t \phi_u) = \sum_{s=1}^{m} C_{sj} (\psi_i \phi_s, \phi_t \phi_u)$$

of the second index *for a fixed pair* ϕ_t, ϕ_u. To do this we need:

1. *All* the integrals
$$(\phi_r\phi_s, \phi_t\phi_u)$$
with r and s taking all possible values and t, u with their fixed chosen values.

2. To arrange these integrals into a matrix form and duplicate the ones with $r \neq s$ into appropriate positions in this matrix.

3. The matrix C, of course.

4. A matrix multiplication code to perform the actual numerical work.

For a *fixed* pair t, u the storage required is simply that of an $m \times m$ matrix which we are assuming is always available.

If all this is done and the matrix of integrals for the pair t, u is, say, V^{tu} with elements
$$V^{tu}_{rs} = (\phi_r\phi_s, \phi_t\phi_u)$$
then the transformation is effected by the matrix multiplications
$$C^\dagger V^{tu} C = V^{ij}$$
(say). Clearly, from what we have said before, the two matrix multiplications would, in practice, be done *separately* to minimise the duplication of numerical work.

Obviously, having formed these integrals they are a part of the data required to perform a complete transformation of all four indices.

All this is elementary but it does not take us any further forward since, as it stands, this approach would involve reading all the original integrals for each pair t, u. What we must do is find a way of performing the sort and matrix formation *all at once* for all the pairs t, u. This must involve some kind of use of additional storage since there simply will not be room for the sorted integrals in main storage.

The trick is to use *virtual storage*; to use a paged storage technique in much the same way as operating systems manage huge storage requests. The method is to separate the sorting and duplication from the matrix formation step:

- Assume that unlimited storage is available and that one "page" of this unlimited storage is available for each pair t, u of indices.

- Read the original integrals, sort them into the relevant t, u sets, duplicating them as appropriate and store them in the relevant page associated with each t, u pair.

29.4. TRANSFORMATIONS WITH SORTING

- As the pages become full, write them out to the unlimited storage and reinitialise the pages. This generates sorted "buckets" of integrals but the integrals themselves are still in random order within each "bucket", reflecting the way in which they occurred in the original file.

- When the whole input has been processed, discard it and discard the pages of main storage used in the sort.

- For each required pair t, u read back *all* the integrals from the relevant bucket and structure them in the required matrix fashion.

- Perform the transformations by matrix multiplication.

- Repeat the last two steps until complete.

- Discard the sorted original integrals.

This will generate a list of new integrals

$$(\psi_i\psi_j, \phi_t\phi_u)$$

which may be put through an identical process to generate the fully transformed set by sorting into sets of fixed i, j etc.

The method is extremely simple in principle and is well-adapted to performing *partial* transformations; when only *some* of the integrals are required over the new functions ψ_i.

There are some technical considerations:

- If enough main storage is available, it is possible to sort into a *range* of t, u during the sort rather than having a separate bucket for each pair t, u.

- The main storage, which is used as pages during the sorting stage, will be reused as storage for the matrix form of the sorted integrals, so it is necessary to use a large main storage "buffer" and partition this buffer to reflect the stages in the calculation, rather than assigning specific storage for each element of the process.

- There has to be a simple algorithm for labelling the pages in each bucket in order to read the correct sets of integrals.

- It is not usually necessary to emit the intermediate integrals in standard, labelled, form (`putint`/`getint` form) since they are only used internally.

- The final list of transformed integrals must be emitted in standard form if they are to be used by other programs.

29.5 Summary

The generation of molecular energy integrals required for any model of molecular electronic structure for which the total energy functional cannot be brought into the form of Chapter 14 requires a numerical procedure which is dependent on m^5 at best. There are strategies for performing these calculations; the choice of method depends on the number of original basis functions and the number of integrals over the new set of orbitals. The implementations given in this chapter have assumed the existence of a *file* of repulsion integrals over the original basis; only trivial modifications are necessary to make the codes *calculate* the original integrals.

Appendix 29.A

A bit of fun: MP2

Here is a "brute-force" simple method for the computation of the simplest estimate of the correlation energy of a many-electron system. The transformation methods of this chapter could be used to generate the required repulsion integrals much more efficiently!

29.A.1 Derivation

The simplest model of molecular electronic structure which includes electron correlation results from the application of perturbation theory to the Hartree–Fock single determinant.

Taking the sum of the n (identical, apart from labelling) Hartree–Fock (UHF) Hamiltonians as the unperturbed Hamiltonian and the usual full Hamiltonian as the target we have

$$\hat{H} = \hat{H}_0^F + \lambda(\hat{H} - \hat{H}_0^F)$$

where

$$\hat{H}_0^F = \sum_{i=1}^{n} \hat{h}^F(i)$$

where the individual $\hat{h}^F$ only differ in the label of the electron and, for the moment we may write them as:

$$\hat{h}^F(i) = \hat{h}(i) + \hat{G}(i)$$

using the (temporary) notation $\hat{G}$ for the electron interaction (Coulomb plus exchange) operator. The full Hamiltonian is the usual one of Chapter 1:

$$\hat{H} = \sum_{i=1}^{n} \hat{h}(i) + \sum_{i>j=1}^{n} \frac{1}{r_{ij}}$$

Thus, the partitioning of the full Hamiltonian becomes

$$\hat{H} = \hat{H}^F + \hat{V}$$

where

$$\hat{V} = \left(\frac{1}{2}\sum_{i,j\neq i}\frac{1}{r_{ij}}\right) - \sum_i \hat{G}(i)$$

and λ has been given the value unity. In the above equation the more symmetrical expression involving full summations over i and j (excluding $i=j$) has been used with double summation being corrected by the factor of $1/2$.

The solutions of the one-particle Hartree–Fock equation

$$\hat{h}^F \chi_k = \epsilon_k \chi_k$$

are assumed to be known.

The solutions of the zeroth-order equation

$$\hat{H}_0^F \Phi_0^K = E_0^K \Phi_0^K$$

are, of course,[2] *determinants* of the MOs χ_k, but the unperturbed *energies* are the corresponding sums of the HF MO energies, ϵ_k which are not the *Hartree–Fock* energies of those determinants. If we denote the *lowest-energy* solution by Φ_0 then the associated energy (E_0, say) is the sum of the lowest n eigenvalues of the one-particle HF equation:

$$\begin{aligned} E_0 &= \sum_i \epsilon_i \\ E_{HF} &= \sum_i \epsilon_i - \frac{1}{2}\sum_{i,j}[(ii,jj) - (ij,ij)] \end{aligned}$$

The Hartree–Fock energies are obtained by subtracting repulsion terms[3] from the sum of the orbital energies.

However, since the perturbing Hamiltonian consists of only electron-repulsion terms, we find the first-order perturbation energy exactly corrects for the omission:

$$E_1 = \int \Phi_0^* \hat{V} \Phi_0 dV$$

This is because, by Slater's rules of Appendix 2.A

$$\int \Phi_0^* \hat{V} \Phi_0 dV = \frac{1}{2}\int \Phi_0^* \left(\sum_{i,j\neq i}\frac{1}{r_{ij}}\right)\Phi_0 dV$$

---

[2] As we saw in Chapter 1.
[3] The form of the summand makes it immaterial whether or not we include the term with $i=j$ since it is zero.

29.A.1. DERIVATION

$$-\int \Phi_0^* \left(\sum_i \hat{G}(i)\right) \Phi_0 dV$$

$$= \left(\frac{1}{2}\sum_{i,j}[(ii,jj)-(ij,ij)]\right)$$

$$- \left(\sum_{i,j}[(ii,jj)-(ij,ij)]\right)$$

$$= -\left(\frac{1}{2}\sum_{i,j}[(ii,jj)-(ij,ij)]\right)$$

so that
$$E_{HF} = E_0 + E_1$$

The effects of electron correlation are therefore seen first in the *second-order* perturbed energy:

$$E_2 = \sum_K \frac{\left(\int \Phi_0 \hat{V} \Phi_0^K dV\right)^2}{E_0 - E_0^K}$$

where the Φ_0^K are the other solutions of the zeroth-order equation and the E_0^K are their energies.

$$\hat{H}_0 \Phi_0^K = E_0^K \Phi_0^K \tag{29.A.1}$$

What remains is to find the form of these E_0^K and Φ_0^K.

The solutions of eqn (29.A.1) are all determinants composed of orbitals which are the solutions of the one-electron HF equation. But there are a huge number of such determinants which may be classified by the number of differences between Φ_0^K and the lowest-energy determinant Φ_0 which enables us to use Slater's rules to decide which ones have non-zero values of the numerator of E_2.

1. Single "excitations"; Φ_0^K is obtained from Φ_0 by replacing just one of the orbitals by an orbital not occupied in Φ_0. In the notation of Chapter 22:

$$\Phi_0^K \equiv \Phi_i^a$$

where χ_i is occupied in Φ_0 and χ_a is not.

2. Double "excitations":
$$\Phi_0^K \equiv \Phi_{ij}^{ab}$$

3. Triple and higher "excitations":
$$\Phi_0^K \equiv \Phi_{ijk\cdots}^{abc\cdots}$$

The integrals

$$\int \Phi_0 \hat{V} \Phi_0^K dV$$

are zero for case 1 because use of Slater's rules reduces them to the off-diagonal component of the matrix ϵ of orbital energies which is zero. The similar ones for case 3 are zero because Slater's rules make all integrals with more than two excitations vanish. This leaves only the double excitations to contribute to the second-order sum:

$$\begin{aligned} E_2 &= \sum_K \frac{\left(\int \Phi_0 \hat{V} \Phi_0^K dV\right)^2}{E_0 - E_0^K} \\ &= \sum_{i,j=1}^n \sum_{a,b=n+1}^m \frac{\left(\int \Phi_0 \hat{V} \Phi_{ij}^{ab} dV\right)^2}{E_0 - E_{ij}^{ab}} \end{aligned}$$

In this expression the energies are just sums of orbital energies so that the energy denominators are simply orbital energy differences. The numerators are the squares of related pairs of MO repulsion integrals (by Slater's rules).

$$\begin{aligned} E_0 - E_{ij}^{ab} &= \epsilon_i + \epsilon_j - \epsilon_a - \epsilon_b \\ \int \Phi_0 \hat{V} \Phi_{ij}^{ab} dV &= (ia,jb) - (ib,ja) \end{aligned}$$

All that is required to calculate the second-order energy are the HF orbital energies and the MO repulsion integrals of the type

$$\begin{aligned} (ia,jb) &= \int d\tau_1 \int d\tau_2 \chi_i(\vec{x}_1)\chi_a(\vec{x}_1)\frac{1}{r_{12}}\chi_j(\vec{x}_2)\chi_b(\vec{x}_2) \\ (ib,ja) &= \int d\tau_1 \int d\tau_2 \chi_i(\vec{x}_1)\chi_b(\vec{x}_1)\frac{1}{r_{12}}\chi_j(\vec{x}_2)\chi_a(\vec{x}_2) \end{aligned}$$

where χ_i and χ_j are orbitals occupied in the ground state and χ_a, χ_b are virtual orbitals.

These latter integrals can be coded directly by the use of a simple macro, and, although inefficient, the program to compute the MP2 energy is very simple indeed.

29.A.2 Implementation

Here is the WEB/LaTeX output from the WEBtext.

29.A.2. IMPLEMENTATION

29.A.2.1 ump2

Compute the UHF Møller–Plesset second-order energy correction. This is the simplest "pick and shovel" method. The **function** simply reads through the file of repulsion integrals and expands them to a spin-orbital basis and adds each basis-function integral multiplied by the appropriate product of MO coefficients into each MO integral. No intermediate products are formed so that many contributions are duplicated. The integral file is read only once.

The function exists mainly to illustrate the use of macros in WEB.

Arguments:

eps The MO orbital energies; they are needed for the energy denominators.

U The MO coefficients for the transformation.

n The number of spin-orbital basis functions (the matrix U is $n \times n$).

nelec The number of electrons in the system; since this is DODS this is the number of occupied MOs.

nfile The logical number of a file containing the spatial basis function repulsion integrals (the basis here is $n/2$).

V Workspace for the temporary accomodation of the MO repulsion integrals; minimum storage space required is $maxab \times maxpq$ where

$$maxpq = (nelec * (nelec + 1))/2 \text{ and}$$
$$maxab = ((n - nelec) * (n - nelec + 1))/2.$$

"ump2.f" 29.A.2.1 ≡

@m *YES* 1
@m *NO* 0
@m *END_OF_FILE* −1
/∗ matrix subscripting for U ∗/
@m $loc(i,j)$ $(j-1)*n+i$
/∗ subscripting for V ∗/
@m $vloc(i,j)$ $(j-1)*maxpq+i$

/∗ *UMP2SUM* is a macro which, when given the labels i, j, k, l and value *va* of a repulsion integral adds its contribution to the MO Coulomb/Exchange term in the numerator of the UMP2 sum ∗/

@m $UMP2SUM(i,j,k,l)$
$V(vloc(pq,\ ab)) = V(vloc(pq,\ ab)) + U(loc(i,\ p))*U(loc(k,\ q))*(U(loc(j,\ a))*U(loc(l,\ b)) - U(loc(j,\ b))*U(loc(l,\ a)))*va$

double precision function $ump2\,(eps,\ U,\ n,\ nelec,\ nfile,\ V)$;

⟨ump2 interface declarations 29.A.2.1.5⟩

{

⟨ump2 local declarations 29.A.2.1.6⟩

⟨ump2 Initialisation 29.A.2.1.7⟩

⟨Read the file and form the MO integrals 29.A.2.1.1⟩

⟨Now use the MO integrals to form the UMP2 sum 29.A.2.1.4⟩

return $(-sum)$;

}

29.A.2. IMPLEMENTATION

spin is an integer which is used to distinguish the four possible spin cases which a given repulsion integral over the spatial orbital basis will generate over the spin-orbital basis. $spin = 1, 4$ corresponds to $\alpha - \alpha, \beta - \beta, \beta - \alpha, \alpha - \beta$.

⟨ Read the file and form the MO integrals 29.A.2.1.1 ⟩ ≡
   while $(getint(nfile, is, js, ks, ls, mu, va, pointer) \neq END\_OF\_FILE)$
   {
   do $spin = 1, 4$;
   { ⟨ Get spin-orbital integrals from spatial integrals 29.A.2.1.2 ⟩
   ⟨ Add in the spin-orbital contributions to each MO integral 29.A.2.1.3 ⟩
   }
   }

This code is used in section 29.A.2.1.

⟨ Get spin-orbital integrals from spatial integrals 29.A.2.1.2 ⟩ ≡
   $ijs = is * (is - 1)/2 + js;$
   $kls = ks * (ks - 1)/2 + ls;$
   $skip = NO;$
   switch $(spin)$
    {
   case 1:
     {
     $i = is;\ j = js;\ k = ks;\ l = ls;$
     break;
     }
   case 2:
     {
     $i = is + halfn;\ j = js + halfn;\ k = ks + halfn;\ l = ls + halfn;$
     break;
     }
   case 3:
     {
     $i = is + halfn;\ j = js + halfn;\ k = ks;\ l = ls;$
     break;
     }
   case 4:
     {
     if $(ijs \equiv kls)$
       $skip = YES;$
     $i = is;\ j = js;\ k = ks + halfn;\ l = ls + halfn;$
     call $order(i, j, k, l);$
     break;
     }
    }   /∗ end of "switch" ∗/
   if $(skip \equiv YES)$
    next;
   $ij = i * (i - 1)/2 + j;$
   $kl = k * (k - 1)/2 + l;$

This code is used in section 29.A.2.1.1.

29.A.2. IMPLEMENTATION

⟨ Add in the spin-orbital contributions to each MO integral 29.A.2.1.3 ⟩ ≡
/* loop over p, q, a, b to add contributions from (ij,kl) to all (pq ab) (pq ab) is stored in $V(pq, ab)$ where pq and ab are symmetric one-dimensional subscripts for the pairs */

```
p = 1;
q = 1;
a = lumo;
b = lumo − 1;
while (next_mp2_label(p, q, a, b, n, nelec) ≡ YES)
  {
  if ((p ≡ q) ∨ (a ≡ b))
    next;
  pq = (p ∗ (p − 1))/2 + q;    /∗ sequence for (pq ab) ∗/
  aa = a − nelec;
  bb = b − nelec;
  ab = (bb ∗ (bb − 1))/2 + aa;  /∗ sequence for (pq ab) ∗/
  UMP2SUM (i, j, k, l);
  if (i ≠ j)
    UMP2SUM (j, i, k, l);
  if (k ≠ l)
    {
    UMP2SUM (i, j, l, k);
    if (i ≠ j)
      UMP2SUM (j, i, l, k);
    }
  if (ij ≠ kl)
    {
    UMP2SUM (k, l, i, j);
    if (i ≠ j)
      UMP2SUM (k, l, j, i);
    if (k ≠ l)
      {
      UMP2SUM (l, k, i, j);
      if (i ≠ j)
        UMP2SUM (l, k, j, i);
      }
    }
  }
```

This code is used in section 29.A.2.1.1.

⟨ Now use the MO integrals to form the UMP2 sum 29.A.2.1.4 ⟩ ≡
    $sum = zero$;
    $p = 1$;
    $q = 1$;
    $a = lumo$;
    $b = lumo - 1$;
    while ($next\_mp2\_label(p,\ q,\ a,\ b,\ n,\ nelec) \equiv YES$)
      {
      if (($p \equiv q$) ∨ ($a \equiv b$))
        next;
      $pq = (p * (p - 1))/2 + q$;
      $dpq = -eps(p) - eps(q)$;
      $aa = a - nelec$;
      $bb = b - nelec$;
      $ab = (bb * (bb - 1))/2 + aa$;
      $dab = dpq + eps(a) + eps(b)$;
      $dab = one/dab$;
      $term = V(vloc(pq,\ ab)) * dab * V(vloc(pq,\ ab))$;
      $sum = sum + term$;
      } /* no factor of 0.25 is needed because there are 4 equal (pq ab) by permutation of p,q and a,b; the pairs p, q and a, b are necessarily different */

This code is used in section 29.A.2.1.

⟨ ump2 interface declarations 29.A.2.1.5 ⟩ ≡
    implicit double precision ($a - h,\ o - z$);
    double precision $U(*)$, $eps(*)$, $V(*)$;
    integer n, $nfile$, $nelec$;

This code is used in section 29.A.2.1.

29.A.2. IMPLEMENTATION

⟨ ump2 local declarations 29.A.2.1.6 ⟩ ≡
   integer $halfn$, $lumo$, a, b, p, q, ab, pq, aa, bb, $maxpq$, $maxab$;
   double precision va, $zero$, one, sum, $term$, dab, dpq;
   integer i, j, k, l, mu, $pointer$, ij, kl;
   integer is, js, ks, ls, $spin$;
   integer $getint$;
   data $zero/0.0 \cdot 10^{00}\text{D}/$, $one/1.0 \cdot 10^{00}\text{D}/$;

This code is used in section 29.A.2.1.

⟨ ump2 Initialisation 29.A.2.1.7 ⟩ ≡
   /* n is UHF (spin-orbital) basis size */
   $halfn = n/2$;   /* spatial basis for integrals */
   $lumo = nelec + 1$;   /* lowest unoccupied MO */
   rewind $nfile$;
   $pointer = 0$;

   /* maxima of pq and ab */
   $maxpq = (nelec * (nelec + 1))/2$;
   $maxab = ((n - nelec) * (n - nelec + 1))/2$;
   do $i = 1$, $maxpq$;
   {
   do $j = 1$, $maxab$;
      $V(vloc(i, j)) = zero$;   /* initialise the (pq ab)s */
   }

This code is used in section 29.A.2.1.

29.A.2.2 next_mp2_label

Given a set of integral labels i, j, k, l which are such that i and j are two of the occupied set $(1, nelec)$ and k, l are two of the virtual set $(nelec + 1, n)$, this procedure generates the next valid label for the UMP2 calculation. It simulates the loops:

do $i = 1$, $nelec$; **do** $j = 1$, i; **do** $k = lumo$, n; **do** $l = l$, n;

where $lumo = nelec + 1$.

It must be initiated by setting $1 = 1$; $j = 1$; $k = lumo$; $l = lumo - 1$.

Notice that, like the loops it simulates it does not exclude $i \equiv j$ or $k \equiv l$. Notice also that it is not a candidate for much structure; the use of WEB is unnecessary here.

"ump2.f" 29.A.2.2 ≡

```
integer function next_mp2_label(i, j, k, l, n, nelec);
integer i, j, k, l, n, nelec;

{
integer lumo;

next_mp2_label = YES;
lumo = nelec + 1;
ld = l - k + 1;
kk = n - k + 1;
if (ld < kk)
    ld = ld + 1;
else
    {
    ld = 1;
    if (k < n)
        k = k + 1;
    else
        {
        k = lumo;
        if (j < i)
            j = j + 1;
        else
            {
            j = 1;
            if (i < nelec)
                i = i + 1;
            else
                {
```

29.A.2. IMPLEMENTATION

$$next\_mp2\_label = NO;$$
$$\}$$
$$\}$$
$$\}$$
$$\}$$
$$l = ld + k - 1;$$

return;

}

29.A.2.3 INDEX

a: <u>29.A.2.1.6</u>.

aa: 29.A.2.1.3, 29.A.2.1.4, <u>29.A.2.1.6</u>.

ab: 29.A.2.1, 29.A.2.1.3, 29.A.2.1.4, <u>29.A.2.1.6</u>.

b: <u>29.A.2.1.6</u>.

bb: 29.A.2.1.3, 29.A.2.1.4, <u>29.A.2.1.6</u>.

dab: 29.A.2.1.4, <u>29.A.2.1.6</u>.

dpq: 29.A.2.1.4, <u>29.A.2.1.6</u>.

END_OF_FILE: 29.A.2.1, 29.A.2.1.1.

eps: 29.A.2.1, 29.A.2.1.4, <u>29.A.2.1.5</u>.

$getint$: 29.A.2.1.1, <u>29.A.2.1.6</u>.

$halfn$: 29.A.2.1.2, <u>29.A.2.1.6</u>, 29.A.2.1.7.

i: <u>29.A.2.1.6</u>, <u>29.A.2.2</u>.

ij: 29.A.2.1.2, 29.A.2.1.3, <u>29.A.2.1.6</u>.

ijs: 29.A.2.1.2.

is: 29.A.2.1.1, 29.A.2.1.2, <u>29.A.2.1.6</u>.

j: <u>29.A.2.1.6</u>, <u>29.A.2.2</u>.

js: 29.A.2.1.1, 29.A.2.1.2, <u>29.A.2.1.6</u>.

k: <u>29.A.2.1.6</u>, <u>29.A.2.2</u>.

kk: 29.A.2.2.

kl: 29.A.2.1.2, 29.A.2.1.3, <u>29.A.2.1.6</u>.

kls: 29.A.2.1.2.

ks: 29.A.2.1.1, 29.A.2.1.2, <u>29.A.2.1.6</u>.

l: <u>29.A.2.1.6</u>, <u>29.A.2.2</u>.

ld: 29.A.2.2.

loc: 29.A.2.1.

ls: 29.A.2.1.1, 29.A.2.1.2, <u>29.A.2.1.6</u>.

$lumo$: 29.A.2.1.3, 29.A.2.1.4, <u>29.A.2.1.6</u>, 29.A.2.1.7, <u>29.A.2.2</u>.

$maxab$: 29.A.2.1, <u>29.A.2.1.6</u>, 29.A.2.1.7.

maxpq: 29.A.2.1, 29.*A*.2.1.6, 29.A.2.1.7.

mu: 29.A.2.1.1, 29.*A*.2.1.6.

n: 29.*A*.2.1.5, 29.*A*.2.2.

nelec: 29.A.2.1, 29.A.2.1.3, 29.A.2.1.4, 29.*A*.2.1.5, 29.A.2.1.7, 29.*A*.2.2.

next_mp2_label: 29.A.2.1.3, 29.A.2.1.4, 29.*A*.2.2.

nfile: 29.A.2.1, 29.A.2.1.1, 29.*A*.2.1.5, 29.A.2.1.7.

NO: 29.A.2.1, 29.A.2.1.2, 29.A.2.2.

one: 29.A.2.1.4, 29.*A*.2.1.6.

order: 29.A.2.1.2.

p: 29.*A*.2.1.6.

pointer: 29.A.2.1.1, 29.*A*.2.1.6, 29.A.2.1.7.

pq: 29.A.2.1, 29.A.2.1.3, 29.A.2.1.4, 29.*A*.2.1.6.

q: 29.*A*.2.1.6.

skip: 29.A.2.1.2.

spin: 29.A.2.1.1, 29.A.2.1.2, 29.*A*.2.1.6.

sum: 29.A.2.1, 29.A.2.1.4, 29.*A*.2.1.6.

term: 29.A.2.1.4, 29.*A*.2.1.6.

U: 29.*A*.2.1.5.

ump2: 29.*A*.2.1.

UMP2SUM: 29.A.2.1, 29.A.2.1.3.

V: 29.*A*.2.1.5.

va: 29.A.2.1, 29.A.2.1.1, 29.*A*.2.1.6.

vloc: 29.A.2.1, 29.A.2.1.4, 29.A.2.1.7.

YES: 29.A.2.1, 29.A.2.1.2, 29.A.2.1.3, 29.A.2.1.4, 29.A.2.2.

zero: 29.A.2.1.4, 29.*A*.2.1.6, 29.A.2.1.7.

Chapter 30

Geometry optimisation: derivatives

The most common single use for the calculation of molecular electronic structure is the computation of the minimum-energy conformation of a molecule. Naturally, this involves finding the values of the geometrical parameters (bond distance, inter-bond angles, dihedral angles etc.) for the molecule at this energy minimum: that is the values of these parameters where the derivatives of the total energy with respect to these parameters are zero. Clearly this can be achieved by some kind of "assisted brute force" method: systematic searches for a minimum along a grid of parameter values. But the most reliable and, ultimately, most efficient way is to actually compute the derivatives of the energy and extrapolate along the energy hypersurface to find and confirm a minimum.

Contents

30.1	Introduction	**724**
30.2	Derivatives and perturbation theory	**725**
30.3	Derivatives of variational solutions	**727**
	30.3.1 Constrained variational optimisations	728
30.4	Parameter-dependent basis functions	**729**
30.5	The derivative of the SCF energy	**730**
30.6	Derivatives of molecular integrals	**734**
30.7	Derivatives of non-variational energies	**735**
30.8	Higher derivatives	**737**
30.9	Summary	**737**

30.1 Introduction

Throughout this work we have been concentrating on the problem of the calculation of the electronic structure of atoms, molecules and molecular entities as a problem which is worthwhile in its own right; the distributions and energies of the *electrons* have been our main area of investigation. The fact that one of the results of the calculation is the total energy of the molecule has not been put to any use. However, the ability to compute the energy of a set of electrons in the field of an arbitrary collection of *nuclei* in an arbitrary relative disposition means that one can compute the *minimum* energy of a set of electrons in the field of these nuclei with respect to the relative positions of those nuclei merely by a sufficiently exhaustive set of calculations.

This minimum is, of course, of considerable chemical interest; it is the "natural" geometry of the isolated molecule.[1] Since it is a central intuition of chemistry that the shape of many molecules is relatively "environment insensitive", this piece of information is of enormous value, and this computational technique holds out the hope of being able to *compute* the shapes of experimentally inaccessible species: transients, transition states, etc. in addition to the very useful checks such calculations provide on the validity of our model and numerical approximations.

Of course, there may well be *many* minima in the variation of the energy as a function of the geometrical parameters and one would normally be interested in locating the *global* minimum as the one of chemical interest. Thus, the simple "grid search" method of making (systematic) changes in the parameters and choosing a minimum might prove to be error-prone as well as time-consuming since every point on the hypersurface requires a full molecular wavefunction calculation. There have, historically, been two kinds of approach to this problem:

- "Numerical" methods; the simple grid search is replaced by a more systematic method of constructing the approximations to the energy gradients from the energy computations at the grid points (using more or less complicated differencing methods on the grid-point energies). If enough points are used, a reasonable approximation to the vector of gradients may be formed and even the matrix of second derivatives (the "Hessian") may be approximated. The problem of finding a minimum (or maximum) in a function of which only the *values* (not the derivatives) are available is a very well-researched one and many related standard methods are known.

- "Derivative methods"; the values *and* derivatives of the molecular energy are computed analytically and standard methods of analysis are available for the location of a (local) minimum.

---

[1] At absolute zero of temperature and isolated in outer space.

30.2. DERIVATIVES AND PERTURBATION THEORY

The problems associated with the latter method can be seen clearly enough when we think about the methods we have used to generate approximate solutions of the Hartree–Fock equation, for example:

- The *one-electron Hamiltonian* depends on the nuclear positions as parameters.

- The *atomic orbitals* (or basis functions) are *centred* on the nuclei and so any integrals involving these basis functions will depend on the nuclear positions parametrically.

We can attempt to see these two aspects of the problem separately, at least initially. The first of these is the more general case and applies to the Schrödinger equation associated with any parameter-dependent Hamiltonian. The second is, of course, dependent on the *approximation* of the use of the linear (basis-set) expansion method and using atom-dependent functions.

30.2 Derivatives and perturbation theory

Our approach to perturbation theory has been a little utilitarian so far; it has been used as an aid to SCF convergence and in the self-consistent perturbation method. It is clear that the variation of a Hamiltonian with respect to any parameters that this Hamiltonian may depend on *contains* perturbation theory as a special (linear) case.

The perturbed Hamiltonian

$$\hat{H}(\lambda) = \hat{H}_0 + \lambda \hat{H}_1$$

is the simplest possible case of the variation of a Hamiltonian with a parameter λ where

$$\frac{\partial \hat{H}}{\partial \lambda} = \hat{H}_1$$

extracts the linear dependence of $\hat{H}$ on λ and

$$\frac{\partial^n \hat{H}}{\partial \lambda^n} \equiv 0$$

for $n \geq 2$.

In the general case where the Hamiltonian depends on a parameter in a non-linear way (which is the case when the parameter is a nuclear coordinate) partial derivatives of *all orders* may well be non-zero. But, for the moment, we may consider the first-order derivative of the Schrödinger equation:

$$\frac{\partial}{\partial \alpha}\left\{\hat{H}\Psi = E\Psi\right\} \equiv \left\{\left(\frac{\partial \hat{H}}{\partial \alpha}\right)\Psi + \hat{H}\left(\frac{\partial \Psi}{\partial \alpha}\right) = \left(\frac{\partial E}{\partial \alpha}\right) + E\left(\frac{\partial \Psi}{\partial \alpha}\right)\right\}$$
$$\equiv \hat{H}^\alpha \Psi + \hat{H}\Psi^\alpha = E^\alpha \Psi + E\Psi^\alpha$$

(say) where the dependence of the equation on just one of the possible parameters has been picked out for emphasis.

Just as in perturbation theory, it makes sense to arrange for the "equilibrium" Schrödinger equation to be the one for which the parameters may be chosen to be zero giving an "unperturbed" equation, the solutions of which are assumed to be known:

$$\hat{H}_0 \Psi_0 = E_0 \Psi_0$$

and, having evaluated the derivative equation we can seek *its* "equilibrium solution" when all the parameters take their equilibrium (zero) values:

$$\hat{H}_0^\alpha \Psi + \hat{H}_0 \Psi_0^\alpha = E_0^\alpha \Psi_0 + E_0 \Psi_0^\alpha$$

yielding an expression for the partial derivative of the energy with respect to the parameter α "at equilibrium" :

$$E_0^\alpha = \left(\hat{H}_0 - E_0\right)\Psi_0^\alpha + \hat{H}_0^\alpha \Psi_0$$

This may be multiplied from the left by Ψ_0^* and integrated to yield the familiar expression

$$E_0^\alpha = \int \Psi_0^* \hat{H}_0^\alpha \Psi_0 dV$$

where the Hermiticity of $\hat{H}_0$ and the fact that Ψ_0 solves the zeroth-order equation, ensures that the integral involving Ψ_0^α vanishes.

What this simple result, and its extension to higher-order derivatives of the Schrödinger equation, means is that the gradient of the total energy may be obtained from the (assumed known) zeroth-order wavefunction and the derivative of the Hamiltonian (again assumed known): a result entirely analogous to the "$2n+1$" rule of perturbation theory:

> The nth derivative of the wavefunction determines the $(2n+1)$th derivative of the energy.

With $n = 0$ this gives the first derivative of the energy depending on the original wavefunction (the zeroth derivative). This looks optimistic; we should be able to determine the gradients of (*e.g.*) the LCAOSCF energy from a knowledge of the LCAO expansion coefficients and the basis functions. There are two reasons to mitigate this initial optimism:

- We do not have the *solutions* of the Hartree–Fock equations, just the variationally optimum set in a given basis.

- In addition to the fact that the Hamiltonian clearly depends on the positions of the nuclei parametrically,[2] the *basis functions* are usually *atom-centred* functions and so they depend parametrically on the nuclear coordinates.

Both of these points can be addressed simultaneously by considering the derivatives of the energy with respect to two *sets* of parameters: one set which is variational and one set which is not.

30.3 Derivatives of variational solutions

Suppose that, instead of an exact solution to the Schrödinger equation, we have some variational solution; a set of parameters, collectively called C, have been optimised (these parameters are often linear coefficients, hence the notation). In addition, suppose that the Hamiltonian depends parametrically on a set of parameters which, in anticipation of their geometric meaning, we may call X, that is

$$E = \frac{\int \Psi^*(C)\hat{H}(X)\Psi(C)dV}{\int \Psi^*(C)\Psi(C)dV} = W'(C;X) \tag{30.1}$$

(say); the dash on W is in anticipation of a further development.[3] If the parameters C are optimised then

$$\frac{\partial W'}{\partial C_i} = 0 \tag{30.2}$$

and it is *this* condition which replaces the earlier condition that the zeroth-order wavefunction solves the original Schrödinger equation.

Of course, the precise *values* of these optimising variational parameters will depend on the parameters X (nuclear coordinates) so that a fuller notation would show this as a functional dependence of the C on X: $C(X)$.

If we now ask for the derivative of the energy *function* with respect to one of the members of the set X, we have:

$$\frac{\partial E}{\partial X_a} = \frac{\partial W'}{\partial X_a} + \sum_i \left(\frac{\partial W'}{\partial C_i}\right)\left(\frac{\partial C_i}{\partial X_a}\right) = \frac{\partial W'}{\partial X_a} \tag{30.3}$$

---
[2]This is the central assumption of the Born–Oppenheimer model of Chapter 1.
[3]This and subsequent sections use notation which follows closely that of P. Pulay (*Advances in Chemical Physics*, Wiley Interscience 1987) which has become the *de facto* standard in this area; however, we use X in place of Pulay's R for the geometric parameters because of the possibility of collision with the use of R for the "charge and bond-order" matrix.

since the variational parameters are optimised and the associated derivatives are all zero. Thus the variational analogue of the $2n+1$ rule is the fact that the first-order derivatives of the variational parameters are not required in order to evaluate the first-order derivatives of the energy.

Notice that the next set of energy derivatives are also of considerable chemical interest; the second derivatives of the energy with respect to nuclear positions are the (harmonic) *force constants* which determine the vibration frequencies of the molecule. To obtain these numbers it *is* necessary to obtain the (first-order) derivatives of the variational parameters with respect to the nuclear coordinates (with $n = 1$, $2n + 1 = 3$).

30.3.1 Constrained variational optimisations

Implicit in the above considerations is the idea that both sets of parameters occurring in the energy function are *independent* and, while this is clearly true for the geometrical parameters,[4] it is certainly *not* true for the variational parameters if they are of the usual linear-expansion type. These coefficients are constrained by the orthogonality requirements on the wavefunction. In our derivations of the SCF equations we allowed for this redundancy among the parameters by using the method of Lagrange multipliers; we shall do the same here but use the previous experience to set the problem up in an advantageous way.

Any constraints on the variational parameters will, at least indirectly, depend on the geometrical parameters so that we write these constraints as equations

$$f_m(\boldsymbol{C}, \boldsymbol{X}) = 0 \tag{30.4}$$

where the dependence on $\boldsymbol{R}$ might be indirect via $\boldsymbol{C}(\boldsymbol{X})$.

These constraints may be incorporated into a "Lagrangian function"[5]

$$W(\boldsymbol{C}, \boldsymbol{\lambda}, \boldsymbol{X}) = W'(\boldsymbol{C}, \boldsymbol{X}) - \sum_m f_m(\boldsymbol{C}, \boldsymbol{X})\lambda_m \tag{30.5}$$

so that the variational extremisation of $W(\boldsymbol{C}, \boldsymbol{\lambda}, \boldsymbol{X})$ generates the two conditions:

$$\left(\frac{\partial W}{\partial C_i}\right) = 0$$
$$\left(\frac{\partial W}{\partial \lambda_m}\right) = f_m = 0$$

---
[4]But not if one wants to *constrain* the geometry in some way: to maintain symmetry, for example.
[5]This terminology, as noted earlier, is a little unfortunate since it collides with the Lagrangian of classical mechanics; however, it is common to use the term "Lagrangian" to denote any functional (or function) from which equations may be derived by variational techniques.

30.4. PARAMETER-DEPENDENT BASIS FUNCTIONS

which determine the C_i and f_m.

When these optimum values are used we have, for the energy of the system,

$$E = W(\boldsymbol{C}(\boldsymbol{X}), \boldsymbol{\lambda}, \boldsymbol{X})$$

emphasising again the dependence of the variational parameters on the geometrical parameters. The derivative of this expression with respect to one of the geometrical parameters is

$$\begin{aligned}\frac{\partial E}{\partial X_a} &= \frac{\partial W}{\partial X_a} + \sum_i \left(\frac{\partial W}{\partial C_i}\right)\left(\frac{\partial C_i}{\partial X_a}\right) + \sum_m \left(\frac{\partial W}{\partial \lambda_m}\right)\left(\frac{\partial \lambda_m}{\partial X_a}\right) \\ &= \frac{\partial W}{\partial X_a}\end{aligned}$$

since, from the constrained variational equations,

$$\left(\frac{\partial W}{\partial C_i}\right) = \left(\frac{\partial W'}{\partial C_i}\right) = 0$$

$$\left(\frac{\partial W}{\partial \lambda_m}\right) = f_m = 0$$

30.4 Parameter-dependent basis functions

So far, the variational parameters C_i have not been identified with any particular form of appearance in a trial wavefunction but the very notation suggests the familiar *linear* expansion parameters. If these linear expansion parameters are for the *orbitals* out of which the total wavefunction is to be constructed (by the determinantal method of Chapter 1, for example), then the question of possible dependence of the *expansion functions* on molecular geometric parameters naturally arises. There are at least three possibilities in general use:

1. "Atomic orbitals": that is atom-centred functions. These clearly depend on nuclear geometry since, when a nucleus moves, its basis functions move with it.

2. Plane waves: used historically in the theory of the solid state, these functions are being used increasingly in molecular theories in conjunction with the density functional method discussed in Chapter 32. These functions are not dependent on the positions of the nuclei and offer considerable simplifications in gradient calculations.

3. "Floating functions": these functions are, typically, s-type functions whose positions are allowed to be optimised by the variation method. Although not actually centred on nuclei, the optimum positions of these functions are clearly dependent on the positions of the nuclei.

The use of basis functions which depend on the geometrical parameters shows itself in the final energy function as *molecular integrals* (overlap, one-electron and electron-repulsion) depending on these parameters; so that, for example

$$\frac{\partial(ij,kl)}{\partial X_a} \neq 0$$

for cases 1 and 3 above.

This dependence of the basis functions on nuclear geometry is a major factor in the calculation of the derivatives of the energy associated with *self-consistent* orbital-based theories (SCF, MCSCF) to which we now turn.

30.5 The derivative of the SCF energy

The finite-basis-expansion form of the Hartree–Fock equations —the SCF equations— are, from the point of view of the current development, a set of constrained variational equations in which:

- The one-electron Hamiltonian term in the Hartree–Fock Hamiltonian depends on the nuclear coordinates,

- The optimum expansion coefficients depend on the nuclear coordinates,

- The basis functions ("AOs") depend on the nuclear coordinates and

- The Coulomb and exchange operators in the Hartree–Fock Hamiltonian depend on the expansion coefficients and through them on the nuclear coordinates.

The constraints are on the expansion coefficients since the MOs must be orthogonal and (usually, but not necessarily) normalised. If n MOs are expanded in terms of m basis functions, the number of degrees of freedom is $n(m-n)$ not mn. In our derivation of the SCF equations we worked in terms of the *matrices* rather than the individual coefficients using δC, for example, to induce a first-order change in energy δE rather than

$$\frac{\partial E}{\partial C_{ij}}$$

and then incorporated the constraints *via* the Lagrange multiplier matrix ϵ.

In this section, since the presence of constraints is a matter of some importance, we use the "UHF" (DODS) model which is the optimum single determinant *without any "model" constraints*. The formalism is identical to the more

30.5. THE DERIVATIVE OF THE SCF ENERGY

common closed-shell model, the only difference being that, in the UHF case, R is over the spin-orbital basis so that "$P = R$" while, in the closed-shell case R is over the spatial-orbital basis: "$P = 2R$".

As was noted at the time this is entirely equivalent to using the *unconstrained* minimisation of the "Lagrangian":

$$W = \text{tr}\, C^\dagger h^{ad} C - \text{tr}\left(C^\dagger S C - 1\right)\epsilon \tag{30.6}$$

where the "additive Hamiltonian" h^{ad} is given[6] by

$$h^{ad} = \frac{1}{2}\left(h^F + h\right)$$

Requiring a variational extremum:

$$\frac{\partial W}{\partial C_{ij}} = 0$$

for all C_{ij}, generates the matrix SCF equations, but in explicit component form.

In the spirit of previous sections, we write eqn(30.6) as

$$W = W' - \text{tr}\left(C^\dagger S C - 1\right)\epsilon \tag{30.7}$$

where

$$W' = \text{tr}\, C^\dagger h^{ad} C = \frac{1}{2}\text{tr}\, C^\dagger \left\{h + [h + G(CC^\dagger)]\right\} C$$

Then, *at a variational extremum* where the partial derivatives of W' with respect to the C_{ij} are all zero, the derivative of this Lagrangian with respect to a geometrical (or any other non-variational) parameter involves only the derivatives of the energy integrals with respect to this parameter:

$$W^a = \frac{\partial W}{\partial X_a} = \frac{\partial W'}{\partial X_a} - \frac{\partial}{\partial X_a}\left\{\text{tr}\left(C^\dagger S C - 1\right)\epsilon\right\} \tag{30.8}$$

there being no terms involving

$$\frac{\partial C_{ij}}{\partial X_a}$$

since these are all multiplied by the variationally zero derivatives.

The resulting expression is just

$$W^a = \text{tr}\, C^\dagger \left\{h^a + \frac{1}{2} G^a(CC^\dagger)\right\} C - \text{tr}\, C^\dagger S^a C \epsilon$$

---

[6] As usual, we use h^F to mean the matrix formed from the one-electron integrals and the Coulomb and exchange matrices whatever the coefficient matrix C; we do not distinguish the *self-consistent* matrix by specific notation.

which is more conveniently written in terms of the $\boldsymbol{R}$ matrix.

$$\boldsymbol{R} = \boldsymbol{CC}^\dagger$$

$$W^a = \text{tr}\left(\boldsymbol{h}^a + \frac{1}{2}\boldsymbol{G}^a(\boldsymbol{R})\right)\boldsymbol{R} - \text{tr}\boldsymbol{C}^\dagger\boldsymbol{S}^a\boldsymbol{C}\boldsymbol{\epsilon} \tag{30.9}$$

In this latter equation the molecular-integral derivatives have been written in shorthand form:

$$(\boldsymbol{h}^a)_{ij} = \frac{\partial}{\partial X_a}(\boldsymbol{h})_{ij} = \frac{\partial h_{ij}}{\partial X_a}$$
$$(\boldsymbol{S}^a)_{ij} = \frac{\partial}{\partial X_a}(\boldsymbol{S})_{ij} = \frac{\partial S_{ij}}{\partial X_a}$$

and a particularly contracted notation has been used for the derivative of the electron-repulsion matrix $\boldsymbol{G}$. In general, since the matrix $\boldsymbol{G}$ depends on $\boldsymbol{R}$, derivatives would involve several terms but the variational condition removes several of these and the notation

$$\boldsymbol{G}^a(\boldsymbol{R})$$

is used to mean

> Formation of the electron-repulsion matrix in the usual way from the elements of the matrix $\boldsymbol{R}$ but using the *derivatives* of the electron-repulsion integrals in place of the integrals themselves.

i.e.
$$G^a_{ij} = \sum_{r,s} R_{rs}[(ij,rs)^a - (ir,js)^a]$$

with
$$(ij,kl)^a = \frac{\partial}{\partial X_a}(ij,kl)$$

In the case of the overlap integrals and the repulsion integrals it is only the basis functions which depend on the geometrical parameters, but in the one-electron integrals the basis functions *and the operator* $\hat{h}$ depend on molecular geometry.

Notice that, in line with the "$2n+1$" rule, eqn (30.9) does *not* involve the derivatives of the MO coefficients and so, at a particular self-consistent point, the derivatives of the total SCF energy may be obtained from a knowledge of the SCF matrix $\boldsymbol{C}$ and the derivatives of the molecular integrals which have to be computed just once *and not stored*. Note also that each original molecular integral, e.g. $(ij,k\ell)$ generates *several* derivative integrals $(ij,k\ell)^a$ depending on the number of geometrical parameters on which the particular integral depends (if ϕ_i, ϕ_j, ϕ_k, ϕ_ℓ are on four different centres there may be as many as twelve

30.5. THE DERIVATIVE OF THE SCF ENERGY

derivative integrals $(ij, k\ell)^a$, one for each X_a) Each derivative may be computed by the "direct" method of forming the traces of products of matrices formed by the familiar techniques of SCF theory, computing the derivative integrals as they are required.

The values and signs of these derivatives now enable a systematic search to be made for a minimum in the SCF energy with respect to molecular geometry.

The extension of the above result to any *variational* approximation to the solution of the Schrödinger equation involving optimisation of orbitals is surprisingly straightforward. In the MCSCF model of molecular electronic structure the model wavefunction is written as a linear combination of antisymmetric terms (usually determinants) constructed from a set of MOs which are themselves variationally optimised as linear combinations of some basis functions:

$$\Psi = \sum_{K=1}^{M} D_K \Phi_K$$
$$\Phi_K = det\{\psi_{K_1}, \psi_{K_2}, \psi_{K_3}, \ldots, \psi_{K_n}\}$$
$$\psi_{K_i} = \boldsymbol{\phi} \boldsymbol{c}_i^K$$

extending the SCFMO notation in an obvious way. All the linear variational parameters are optimised in the MCSCF procedure so that derivatives of the resulting energy with respect to the D_K and the elements of $\boldsymbol{c}_1^K$ are *all zero*.

The only difference between the derivative of the MCSCF energy and the SCF energy eqn (30.9) is simply that which reflects the difference in the original energy expression in terms of the basis-function integrals. The (idempotent) *R*-matrix of SCFMO theory determines the energy expression of the SCFMO model completely. But in the MCSCF model *both* the one-electron *and* two-electron density matrices are required and, of course, in this many-determinant case, the one-electron density does *not* fix the two-electron density. However, when the energy expression in terms of the molecular integrals over the basis functions is known, its derivatives are simply given by that same expression with the energy *integrals* replaced by their derivatives in exactly the same way as we have found in the SCFMO case.

The expression for the one- and two-particle density matrices for linear combinations of determinants are, essentially, restatements of Slater's rules in the case of orthogonal MOs, which is, by far, the most common case, or, in the case of non-orthogonal MOs, Løwdin's rules. The detailed forms depend on the *structure* of the determinants in the sense of the number of MOs they have in common, exactly as in Slater's rules so that a general expression is cumbersome to quote and not much use in practice. For the present purposes we simply assume them to be known with elements

$$P_{ij}^1$$
$$P_{ij,k\ell}^2$$

respectively, so that the MCSCF energy expression is:

$$E = \sum_{i,j=1}^{m} P^1_{ji} h_{ij} + \sum_{i,j,k,\ell=1}^{m} P^2_{j\ell,ik}(ij,k\ell)$$

and the derivative is

$$E^a = \sum_{i,j=1}^{m} P^1_{ji} h^a_{ij} + \sum_{i,j,k,\ell=1}^{m} P^2_{j\ell,ik}(ij,k\ell)^a - \text{tr}\,\boldsymbol{C}^\dagger \boldsymbol{S}^a \boldsymbol{C}\boldsymbol{\epsilon} \qquad (30.10)$$

30.6 Derivatives of molecular integrals

There are two obvious main points to be noted in preparing to calculate the derivatives of the overlap, one-electron and electron-repulsion integrals:

1. The exponential form of all atom-centred basis functions (STOs or GTFs) ensure that the spatial derivatives of these basis functions are of a similar exponential form so that, at worst, we shall be involved in the calculation of molecular integrals involving, perhaps, new types of basis function involving higher powers of the Cartesian coordinates.[7]

2. The one-electron Hamiltonian only involves *differences* of geometrical parameters; the nuclear attraction terms involve the electron-nucleus distance *not* the absolute position of either particle. Likewise the molecular integrals' dependence on molecular geometry is only *via* inter-centre distance; the fact that the basis functions are atom-centred does not induce any dependence of the integrals on absolute position of the integrals.

The first of these points merely threatens a tedious coding problem similar to the ones encountered in Chapter 8 but the second is immediately and obviously useful.

For a molecule containing N nuclei, there are only $(3N - 6)$ internal coordinates, the remaining six coordinates are just motions of the molecule *as a whole* (translations and rotations), so that there are only $(3N - 6)$ useful derivatives to be found for the optimisation of the geometry of a molecule. Remarks of a similar nature apply to the evaluation of the derivatives of the molecular integrals. There are many dependencies amongst them due to the fact that only *differences* of coordinates are involved in the integrals; translational invariances are highlighted by expressing the basis functions and nuclear-attraction operators

---

[7]This merely means that our formulae of Chapter 7 will still be suitable for the evaluation of derivative integrals, but the implementation of Chapter 8 will be limited to derivatives of s, p and d GTFs only.

30.7. DERIVATIVES OF NON-VARIATIONAL ENERGIES

in terms of Cartesian coordinates and rotational invariances can be emphasised by using polar coordinates.

In practice, the translational invariance is *very* much easier to use than the rotational property, because a rotation of coordinates sends any non-s-type basis function into a *linear combination* of other basis functions within the same shell and this leads to book-keeping problems which are not easy to handle. Translational invariance simply generates one "free" integral from a given group of related integrals and so is quite straightforward to implement.

30.7 Derivatives of non-variational energies

The evaluation of the derivatives of the energy of both the exact solutions of the Schrödinger equation and variational approximations to it are considerably simplified by the fact that, either the associated wavefunction solves the Schrödinger equation, or the derivatives of the energy with respect to the variational parameters are all zero. If we have an approximate solution to the Schrödinger equation which is *not* variational then the situation is, in principle, much more complicated; there are more terms to evaluate in the derivative expression. The problem is that many of the most useful and common approximations which are more accurate than the SCF HF wavefunction are exactly of this type: perturbation expansions in terms of the SCF MOs, in particular, Møller–Plesset (many-body) perturbation expansions.

The non-vanishing of the energy derivatives with respect to the linear expansion coefficients means that the full expression for the derivative of the energy, eqn (30.3) must be used, which demands a knowledge of many more derivatives in view of the large number of linear parameters involved in expansion of the orbitals. However, in studies of the evaluations of the derivatives of the second-order Møller–Plesset energy an important technical result was obtained by Handy and Schaefer[8] which drastically reduces the work involved in this (and other) models of electronic structure.

If the linear parameters (in general, the non-geometrical parameters in the wavefunction) are not variationally optimised, then an infinitesimal perturbation will cause them to change. We have seen already in Chapter 26 that, using perturbation theory for the orbitals of a single determinant we can generate a set of equations for the response of the linear parameters, a change in a Hamiltonian, such as the movement of a nucleus. These "response equations" will give (to some chosen order) the required changes in the coefficients and hence the derivatives needed. In the case of the Møller–Plesset model we have the situation where the response equations for the linear parameters are obtained

---

[8] N. C. Handy and H. F. Schaefer III, *J. Chem. Phys.*, **81**, 5031 (1984).

from a *simpler* model of the electronic structure; the required linear parameters are the LCAO coefficients from the SCF model which underlies the MP2 (say) model. Unlike (say) the CI model, the MP2 method does not have any dependence on the coefficients of the many-electron determinants which are implicitly involved.

The original expression for the geometrical derivatives

$$\frac{\partial E}{\partial X_a} = \frac{\partial W}{\partial X_a} + \sum_i \left(\frac{\partial W}{\partial C_i}\right)\left(\frac{\partial C_i}{\partial X_a}\right) \tag{30.11}$$

must be augmented by a response equation for the reaction of the non-variational parameters to the geometrical changes which, in the notation of this chapter, is:

$$\sum_j \frac{\partial^2 w}{\partial C_i \partial C_j}\frac{\partial C_j}{\partial X_a} = -\frac{\partial^2 w}{\partial C_i \partial X_a} \tag{30.12}$$

for each geometrical parameter X_a. In eqn (30.12) the symbol w has been used in place of W in eqn (30.11) to emphasise the fact that the response equations are derived from the SCF energy function(al) while the full derivative expression must be obtained from the MP2 energy function(al). This equation may be conveniently written in matrix notation as

$$\boldsymbol{w}\boldsymbol{C}^a = \boldsymbol{b}^a$$

at the cost of doing some violence to the consistency of our subscript and superscript notation since here

$$(\boldsymbol{w})_{ij} = \frac{\partial^2 w}{\partial C_i \partial C_j}$$
$$(\boldsymbol{b}^a)_i = \frac{\partial^2 w}{\partial C_i \partial X_a}$$

while

$$(\boldsymbol{C}^a)_i = \frac{\partial C_i}{\partial X_a}$$

The formal solution of this equation (since the Hessian $\boldsymbol{w}$ is not singular) is

$$\boldsymbol{C}^a = -\boldsymbol{w}^{-1}\boldsymbol{b}^a = -\boldsymbol{z}\boldsymbol{b}^a$$

(say) which may be inserted (in component form) into eqn (30.11) to yield

$$\frac{\partial E}{\partial X_a} = \frac{\partial W}{\partial X_a} - \sum_i \left(\frac{\partial W}{\partial C_i}\right)\left(\sum_j z_{ij}\frac{\partial^2 w}{\partial C_j \partial X_a}\right)$$

30.8. HIGHER DERIVATIVES

and the order of the summations over i and j on the right may be interchanged to define a new matrix Y (say):

$$\frac{\partial E}{\partial X_a} = \frac{\partial W}{\partial X_a} - \sum_j \left[\sum_j z_{ij} \left(\frac{\partial W}{\partial C_i} \right) \right] \left(\frac{\partial^2 w}{\partial C_j \partial X_a} \right)$$

$$= W^a - Y^T b^a$$

where the (column) matrix Y is obtained *once and for all perturbations* by the solution of a single response equation

$$w^T Y = W$$

The (column) matrix W is composed of the derivatives of the MP2 energy function with respect to the linear coefficients:

$$W_i = W_i = \frac{\partial W}{\partial C_i}$$

30.8 Higher derivatives

The evaluation of higher derivatives of the energy as a function of the molecular geometry is possible by extending the techniques developed here. However, for the second and higher derivatives the response equations must be solved since, of course, the second and higher derivatives of the energy with respect to the variational parameters are never zero. The perturbation theory outlined in Chapter 26 is used for this purpose and much of the theory of derivatives of the energy has very close parallels in perturbation theory. The second derivatives of the energy, which are required to obtain the (harmonic) force constants and hence the harmonic vibration frequencies and normal modes of vibration, require the solution of the first-order response equations which, for SCF theories, are the equations obtained in Chapter 26.

30.9 Summary

The derivatives of the total energy required to obtain the equilibrium geometries of molecules, vibration frequencies and other spectroscopic properties can be obtained. Additional implementations are required for:

- Derivatives of the molecular integrals involving the basis functions.
- Derivatives of the energy with respect to the linear ("variational") parameters contained in the wavefunction.

- Solution of the first-order perturbed SCF equations.
- Numerical methods of function minimisation well adapted to the optimisation of molecular geometry.

The last of these has not been addressed here.

Chapter 31

The Semi-empirical approach

The energy integrals appearing in the algebraic approximation all have straightforward physical interpretations and these interpretations may be used to make estimates of their relative sizes and importance in a calculation. The physical interpretation may also be used to correlate some integrals with experimentally accessible data directly or may be used as a basis of approximations based on very simple models of the underlying charge distributions. In this chapter we make an overview of the general philosophy of this approach which leads inexorably to the idea of "calibrating" a given model of molecular electronic structure against experimental data.

Contents

31.1	Introduction	739
31.2	Use of Coulomb's law	741
31.3	Atomic data	742
31.4	Simulation or calibration?	744
31.5	General conclusions	745

31.1 Introduction

The methods we have been outlining are to be evaluated by two general criteria:

- Agreement with experimental measurements where the experimental results are available.

- Explanation of phenomena in terms of a set of concepts generated by a combination of intuition and theoretical analysis.

Clearly, the second of these is of little value unless the first is satisfied within some well-defined and well-understood hierarchy of approximation; we do not wish to have explanations of the strength of a chemical bond using theories which are not capable of giving a good quantitative calculation of bond energies, for example.

The models of molecular electronic structure we have been using fall into a more-or-less strict hierarchy:

1. The potential-energy terms in the Hamiltonian are only those due to Coulomb's law. We exclude magnetic and relativistic effects completely.
2. The Born–Oppenheimer (fixed-nucleus) model is assumed throughout.
3. The algebraic approximation is the key numerical approximation to make the whole project feasible.
4. The use of only atom-centred basis functions is based on our intuitions about the likely distribution of electrons in molecules.
5. The number and type of basis functions has to be chosen as a compromise between accuracy and convenience.
6. Core potentials are often used both for reasons of economy and to avoid difficulties with the description of core electrons.

These points are raised explicitly here to emphasise the fact that what we call *ab initio* calculations contain a good deal of intuitive input, and it is not an exaggeration to say that the *calculation* might be *ab initio* but the *theory* is not.

The natural question to ask is "can we extend our use of intuitive information and physical interpretation to *numerical* approximations within the calculation?" If the answer is "yes" then there are two possible ways in which we might go forward which are not mutually exclusive:

- We can use the physical interpretation of the energy integrals appearing in the algebraic approximation to estimate their relative sizes and to make numerical estimates of their values.
- The values (or functional forms) of the energy integrals may be used as a method of *forcing* a particular model of molecular electronic structure to agree with experiment. That is, we can *calibrate* a particular model against experiment for some chosen property.

31.2. USE OF COULOMB'S LAW

Each of these approaches has its own advantages and disadvantages and this will become particularly clear when we discuss the implementation of density functional methods in Chapter 33.

31.2 Use of Coulomb's law

The Hamiltonian operator which we have used throughout contains only kinetic energy and potential energy terms which all involve Coulomb's law of electrostatics. The integrals involving the Coulomb interaction "operators"[1] and the various orbitals and basis functions are capable of being given a very simple interpretation as the interaction of a charge density $\rho_{ij}(\vec{r}) = \phi_i(\vec{r})\phi_j(\vec{r})$ (say) with:

- A point charge; these are the nuclear-attraction integrals

$$\int dV \phi_i(\vec{r})\phi_j(\vec{r}) \frac{1}{|r - R_A|} = \int dV \rho_{ij}(\vec{r}) \frac{1}{|r - R_A|} \qquad (31.1)$$

- Another charge density $\rho_{k\ell}(\vec{r}) = \phi_k(\vec{r})\phi_\ell(\vec{r})$; these are the electron-repulsion integrals

$$\int dV_1 \int dV_2 \phi_i(\vec{r}_1)\phi_j(\vec{r}_1) \frac{1}{|r_1 - r_2|} \phi_k(\vec{r}_2)\phi_\ell(\vec{r}_2)$$
$$= \int dV_1 dV_2 \rho_{ij}(\vec{r}_1) \frac{1}{|r_1 - r_2|} \rho_{k\ell}(\vec{r}_2) \qquad (31.2)$$

Both of these integrals have obvious asymptotic forms. If the nucleus A in eqn (31.1) is very remote from the charge distribution ρ_{ij}[2] then we may replace $1/(r - R_A)$ by some mean value: the distance from the centroid of the distribution to the nucleus (R, say, a constant independent of r). In this case

$$\int dV \phi_i(\vec{r})\phi_j(\vec{r}) \frac{1}{|r - R_A|} = \int dV \rho_{ij}(\vec{r}) \frac{1}{|r - R_A|}$$
$$= \frac{1}{R} \int dV \rho_{ij}(\vec{r}) = \frac{S_{ij}}{R}$$

where S_{ij} is the overlap integral between the two functions ϕ_i and ϕ_j and is a measure of the "amount of charge" contained in the charge distribution ρ_{ij}.

Obviously, if $i = j$, $S_{ii} = 1$ and the approximate expression for the electron-nucleus attraction integral become the simple point-charge formula $1/R$.

---

[1] These operators are all simply multiplicative.
[2] Of course, each charge distribution is a product of basis functions which cover all of three-space but what is meant here is the "majority" of the charge distribution.

In a completely analogous way, if the two charge distributions in eqn (31.2) are remote from each other the electron-repulsion integral has the asymptotic form

$$\int dV_1 \int dV_2 \phi_i(\vec{r}_1)\phi_j(\vec{r}_1)\frac{1}{|r_1 - r_2|}\phi_k(\vec{r}_2)\phi_\ell(\vec{r}_2)$$
$$= \frac{1}{R}\int dV_1 dV_2 \rho_{ij}(\vec{r}_1)\rho_{k\ell}(\vec{r}_2) = \frac{S_{ij}S_{k\ell}}{R}$$

Again, if $i = j$ and $k = \ell$ the electron-repulsion integral takes on the simple point-charge asymptotic form of $1/R$, where this time R is the distance between the centroids of the two distributions. Since the basis functions are always atom-centred functions, the centroid of the "diagonal" charge distributions (ρ_{ii}) are the relevant atoms and so the distances R are actually *inter-atomic* distances.

The other extreme for the "amount of electron" in a charge distribution is zero when the two basis functions are orthogonal ($S_{ij} = 0$). In this case the asymptotic form for both types of integral is zero.

It is natural to ask:

1. What is the range of numerical applicability of the asymptotic forms?
2. Can the size of the overlap integral be made into a systematic criterion for the generation of these integrals?

These matters have been discussed in detail elsewhere.[3] Obviously, these simple forms are not valid for $R = 0$; the repulsion between two electrons occupying orbitals on the same atom is not infinite. But quantities like this, which are essentially *atomic* in character, may be obtained by another route; they can be inferred from a study of atomic spectra. This is another approach to the estimation of (parts of) energy integrals and which particularly justifies the name "semi-empirical" for the technique.

31.3 Atomic data

The analysis of atomic spectral data has traditionally been carried through in terms of an orbital model; the separation of the spectral lines has involved angular integrals related to Clebsch–Gordan coefficients which are determined by symmetry and energy-dependent "radial integrals". These radial integrals and angular coefficients are precisely what are involved in the one-centre repulsion

---
[3]See, for example, D. B. Cook *Structures and Approximations for Electrons in Molecules*, (Ellis Horwood, 1978).

31.3. ATOMIC DATA

integrals. Thus we may obtain, directly from experiment, the values of AO repulsion integrals of the previous section for $R = 0$. For example, let us say that for two functions ϕ_i ϕ_j, which are both the same actual AO but may be on different atomic centres, the integral

$$I = \int dV_1 \int dV_2 \phi_i(\vec{r}_1)\phi_i(\vec{r}_1) \frac{1}{|r_1 - r_2|} \phi_j(\vec{r}_2)\phi_j(\vec{r}_2)$$

will have the asymptotic form of $1/R$ for large R and a value of A for $R = 0$ taken from atomic spectral data. Then we know its value at $R = \infty$ and at $R = 0$ and we can combine these two known values in the formula

$$I \approx \frac{1}{R+a}$$

(where $a = 1/A$) and hope that this might be acceptable at intermediate values of R.

In a similar spirit we can use the fact that the molecular *one*-electron Hamiltonian may be partitioned:

$$\hat{h} = \left(-\frac{1}{2}\nabla^2 + V_A\right) + \sum_{S \neq A} V_S = \hat{h}^A + \sum_{S \neq A} V_S \qquad (31.3)$$

to reflect the atomic nature of some of the one-electron integrals. Here V_A is the potential due to atomic core[4] A and the V_S are the potentials due to other atomic cores. If ϕ_i is an orbital centred on atom A then the above partitioning of the Hamiltonian means that the integral

$$\int dV \phi_i \hat{h} \phi_i = \int dV \phi_i \hat{h}^A \phi_i + \sum_{S \neq A} \int dV \phi_i V_S \phi_i$$

where the first term on the right is the energy of an electron occupying ϕ_i on a *separate* atom A and the remaining terms are "corrections" for the presence of the other atoms. This first term is (approximately) the ionisation energy of orbital ϕ_i in atom A and may be taken from experiment. Equally, it should be an approximation to (minus) the electron affinity of orbital ϕ_i on atom A again available experimentally. The correction terms may be approximated by the asymptotic nuclear attraction formula or some parametric generalisation of this.

In this way various semi-empirical approximation schemes may be developed in which a combination of experimental data and simple formulae may be used in place of the full calculations outlined in, for example, Chapter 8.

However, there is a more radical approach to the whole problem; one which uses the underlying SCF method and the energy integrals simply as a convenient framework for reproducing experimental results.

---

[4]This may simply be the attraction of a nucleus or it may also include repulsions from core electrons not explicitly considered in the calculation (*cf.* Chapter 24).

31.4 Simulation or calibration?

The methods outlined in the previous section aim to *approximate* the various energy integrals appearing in a conventional calculation and therefore if they are used in, say, an SCF calculation the result can only be as good as the full calculation in which the integrals are not approximated; the aim is simply to *simulate* a full calculation. If this is possible it means that calculations on large molecules are possible which still retain the *interpretation* of the full calculation, including the inadequacies of the underlying model of molecular electronic structure.

Suppose, however, that the interpretation of the result of a calculation is less important than its closeness to some empirical data; suppose that the calculations are being performed to probe some experimentally inaccessible energies, for example. In this case the whole approach of a graduated series of models, each with its own strengths and limitations, might be inappropriate. It is more important that the numbers be right than that the physical validity of the model be correct. We might, in cases like this, simply abandon the distinction between "model" and "numerical" approximations and treat the whole calculation as an interpolation and (perhaps) extrapolation device for experimental results.

This approach has a long history in quantum physics and chemistry, being most systematically carried through for the calculation of the energetics of transformations involving molecules composed of the first nine atoms of the periodic table (H – F). The LCAOMOSCF model is used as a vehicle for the project and those energy integrals which are retained[5] in the calculation are given functional forms which are both physically reasonable and contain parameters. The parameters in these forms are then optimised for a particular property (usually geometry optimisation and heat of formation) by choosing a "representative" set of molecules and fitting the results of the calculation to experimentally known values of that property by some least-squares procedure. The fact that the LCAOMOSCF method is a realistic but not exact model for the electronic structure means that, for example, the fitted parameters must make the *numerical values* of the energy integrals compensate for the theoretical failings of the model. Thus effects of electron correlation are accounted for, not by using a multi-determinant wavefunction but by adjustments in the values of the integrals appearing in the implementation of the single-determinant model. Using this approach, we cannot "improve" the description of the electronic structure by using the orbitals of the optimum single determinant in, for example, a configuration interaction scheme because the effects of CI have already been included in the parametrisation. Naturally, the detailed form of the molecular orbitals obtained by this procedure will not have a physical interpretation since this has been abandoned in using an empirical fitting method.

---

[5]The vast majority of the repulsion integrals are neglected in a systematic way based on the overlap ideas we have met above.

31.5. GENERAL CONCLUSIONS

We shall see that the method of Kohn and Sham in density functional theory actually provides a sound theoretical base[6] for this method which has been used over the years simply as a numerical convenience. The density functional method uses a set of "fictional" molecular orbitals which do not themselves have any physical interpretation and whose only property is to generate an electron density which is exact. The whole of the experimental calibration procedure is thrown into the generation of a potential (the exchange/correlation potential) which can, in principle, be *universal*: that is, not dependent on the particular molecule under study. The huge number of parameters required in earlier semi-empirical methods (some for every atom) is replaced by choice of a *form* for this potential and a few universal parameters (up to a dozen).

31.5 General conclusions

The possibility of using the general form of the LCAOMOSCF equations as a framework for the generation of an interpolation scheme for molecular energies has been widely investigated and used over the past few decades. Provided that only minimal physical interpretation is to be given to results of such calculations they provide a very useful tool for the generation of (*e.g.*) molecular shapes.

The most powerful and general of these semi-empirical methods is density functional theory which holds out the possibility of the systematic and cumulative development of empirically fitted exchange/correlation potentials which are independent of the particular molecule (and it atomic constituents) under study. However, whatever the numerical successes of this or any other semi-empirical method, the problems of explanation and interpretation of the phenomena remain just as acute. The tension here is between accuracy of computed results and physical interpretation: similar to the conclusions of Chapter 20.

---

[6]See Chapter 32.

Chapter 32

Density functional theory

In solid-state physics, theorists are dealing with many-electron systems where "many" mean billions not just dozens as in molecular theories. This means that methods based on the electron density are much more widely used and much more intuitively appealing. Their constant efforts to develop such methods have been rewarded by a series of amazing theorems showing that it is possible to obtain the exact electron density without having recourse to the wavefunction. Naturally, these results have been taken up with some enthusiasm by workers in the field of molecular electronic structure. In this chapter the celebrated Kohn–Hohenberg–Sham approach is developed and its close relationship to the Hartree–Fock model is used to indicate how it can be implemented. The very different "intuitions" of chemists and physicists about electronic structure generates some tensions in the interpretation of the results of these theories.

Contents

32.1	Introduction		**748**
32.2	Hohenberg and Kohn's proofs		**750**
	32.2.1	The first theorem; existence	750
	32.2.2	Kohn–Hohenberg's second theorem: variation	753
32.3	Kohn–Sham equations: introduction		**755**
32.4	Kohn–Sham equations		**757**
32.5	Non-local operators in orbital theories		**760**

32.1 Introduction

We have seen earlier that the Hartree–Fock model of molecular electronic structure is basically a *density-matrix* theory. The physical interpretation and even the formalism of the theory may be expressed in terms of the basic invariant of the theory either expressed as a function of six spatial and two spin variables or as an orbital expansion in terms of some set of MOs[1] $\{\chi_i\}$:

$$\rho_1(\vec{x};\vec{x}') = \boldsymbol{\varphi}(\vec{x})\boldsymbol{R}\boldsymbol{\varphi}^\dagger(\vec{x}')$$

The density matrix associated with a single-determinant function Φ is defined as

$$\rho_1(\vec{x};\vec{x}') = n\int \Phi(\vec{x},\vec{x}_2,\ldots,\vec{x}_n)\Phi^*(\vec{x}',\vec{x}_2,\ldots,\vec{x}_n)d\tau_2 d\tau_3\ldots d\tau_n$$

which is clearly independent of linear transformations amongst the individual orbitals comprising the determinant since that determinant itself is unchanged by such transformations. The two-particle density matrix associated with the determinant Φ is

$$\rho_2(\vec{x}_1,\vec{x}_2;\vec{x}'_1,\vec{x}'_2) = n(n-1)/2 \int \Phi(\vec{x}_1,\vec{x}_2,\ldots,\vec{x}_N)\Phi^*(\vec{x}'_1,\vec{x}'_2,\ldots,\vec{x}_N)d\tau_3\ldots d\tau_n$$

which may be written entirely in terms of ρ_1

$$\rho_2(\vec{x}_1,\vec{x}_2;\vec{x}'_1,\vec{x}'_2) = \rho_1(\vec{x}_1;\vec{x}'_1)\rho_1(\vec{x}_2;\vec{x}'_2) - \rho_1(\vec{x}_1;\vec{x}'_2)\rho_1(\vec{x}_2;\vec{x}'_1)$$

The Hartree–Fock equations are, in density matrix notation:

$$\hat{h}^F\rho_1 - \rho_1\hat{h}^F = 0$$
$$\rho_1\rho_1 = \int \rho(\vec{x};\vec{x}'')\rho(\vec{x}'';\vec{x})d\tau'' = \rho_1$$
$$\int \rho_1(\vec{x};\vec{x})d\tau = n$$

Leading one to think that, at least for the Hartree–Fock model, the orbitals and single-determinant wavefunction are not needed; why not solve the equations for the density matrix directly? As usual in these matters it is not the solution of the variational problem which causes the difficulties but the incorporation of the *constraints* on the acceptable solutions.

While it is obvious that any wavefunction will generate a density matrix, the opposite is by no means obvious; indeed it is false. Not every function of six variables has the correct pedigree; not every putative density matrix has the property of being capable of being generated from an n-electron *antisymmetric*

---
[1] Until explicitly stated to the contrary we shall work in *spin-orbitals* in this chapter; hence the use of χ_i, φ_k and $\vec{x}$ rather than ψ_i, ϕ_k and $\vec{r}$.

32.1. INTRODUCTION

wavefunction.[2] So that, if one attempts to solve the above equation directly for the density matrix one quickly goes below the observed energy, for example.

However, it is the density matrix, or more usually, the "diagonal component" of the density matrix—the electron density:

$$\rho(\vec{x}) = \rho_1(\vec{x}; \vec{x})$$

which carries most of the burden of physical interpretation of atomic and molecular properties. Now the electron density would seem to be a quantity "below" the one-particle density matrix in its relationship to the wavefunction and, from what we have observed so far, it would seem extremely unlikely that one could use *this* quantity as the basic function in a variationally based method. For example, one cannot obtain the conventional kinetic energy expression from the density. The exchange energy which is characteristic of the single-determinant model involves the full one-particle density *matrix*; using the methods we have introduced thus far one may only obtain the mean values of *local* (multiplicative) operators from $\rho(\vec{x})$.

In spite of these apparently pessimistic ideas it has been proved that

- The electron density alone characterises the potential energy experienced by the electrons.

- It *is* possible to use the electron density directly as the fundamental quantity in a variational method *and* what is more, the proof holds out the promise of being able to use the density in methods which are, in principle, more accurate than the Hartree–Fock method.

These proofs are, as we might expect, *existence* proofs rather than *constructive* proofs so that the task of developing methods which rely on these proofs is still largely a matter of experience and trial and error. The penalty to be paid for these very powerful-looking results are that the energy quantities turn out to be *functionals* of the electron density rather than the *functions* with which we are familiar.

The first of these results is not surprising, indeed when quoted carefully enough it seems obvious; one would expect that a given one-particle density might only arise from a unique external potential. In the simplest possible case for molecules where the external potential is just a series of point charges (nuclei) then we know that, at least for the ground state, there is a maximum (even a cusp) in the electron density at each nucleus. Thus even the *topological* properties of the density are enough to locate the nuclei and hence determine the Hamiltonian which itself determines the wavefunction. It is worth emphasising

---
[2] As usual, it is mainly the Pauli principle which thwarts our schemes.

here that it is only the *ground state* which has this i cusp property; many excited states have *nodes* at the nuclei. Even the magnitude of the charge on a nucleus may be inferred from the density and the limit of its gradient as the cusp is approached. If $\bar{\rho}(r)$ is the mean "radial" electron density (the average of $\rho(\vec{r})$ over all angles around that point) then the limit as the position of a nucleus is approached ($\vec{r} \longrightarrow \vec{R}_A$, for nucleus A) of the gradient of this average density fixes the charge (Z_A) on nucleus A:

$$\lim_{\vec{r} \longrightarrow \vec{R}_A} \left[\left(\frac{\partial}{\partial r} + 2Z_A \right) \bar{\rho}(r) \right] = 0$$

The second point the *variational* use of the density is not so conceptually straightforward.

32.2 Hohenberg and Kohn's proofs

Hohenberg and Kohn were able to give two very simple existence theorems which underpin the whole of current density functional theory.

32.2.1 The first theorem; existence

This proof is by the indirect method of proving that a given assumption generates a contradiction and hence establishing the opposite of that assumption.

Let us assume that a Hamiltonian $\hat{H}$ for the motion of a system of electrons is composed of a sum of the kinetic energy operator ($\hat{T}$), the "external potential" expression ($\hat{V}$) (i.e. this is the potential in which the electrons move in addition to their own mutual repulsions) and the inter-electronic repulsions ($\hat{G}$):

$$\hat{H} = \hat{T} + \hat{V} + \hat{G}$$

and let the (non-degenerate ground-state) solution of the associated Schrödinger equation be Ψ:

$$\hat{H}\Psi = E\Psi$$

The electron density associated with this solution is, of course,

$$\rho(\vec{x}) = n \int |\Psi(\vec{x}, \vec{x}_2, \ldots \vec{x}_N)|^2 d\tau_2 \ldots d\tau_n$$

The result to be proved is that the potential expression for a typical electron; $v(\vec{x}_i)$ in:

$$\hat{V} = \sum_{i=1}^{n} v(\vec{x}_i)$$

32.2. HOHENBERG AND KOHN'S PROOFS

is uniquely determined by the one-particle density $\rho(\vec{x})$.

Let us assume the opposite; that there are *two* possible one-particle potentials which generate the same density $\rho(\vec{x})$. Let them be $v(\vec{x})$ above and $v'(\vec{x})$, say. The second potential will be associated with a different Hamiltonian:

$$\hat{H}' = \hat{T} + \hat{V}' + \hat{G}$$

(say) and different solution of the associated Schrödinger equation:

$$\hat{H}'\Psi' = E'\Psi'$$

But the density associated with Ψ' is the same as that associated with Ψ:

$$\rho(\vec{x}) = n \int |\Psi'(\vec{x}, \vec{x}_2, \ldots \vec{x}_N)|^2 d\tau_2 \ldots d\tau_n$$

Except for the trivial case when $v(\vec{x})$ and $v'(\vec{x})$ differ by a constant, the solutions Ψ and Ψ' will be *different*.

Now suppose we make use of the variation theorem and evaluate the variational expressions:

$$W = \int \Psi'^* \hat{H} \Psi' d\tau > E$$

and

$$W' = \int \Psi^* \hat{H}' \Psi d\tau > E'$$

Where we have assumed both wavefunctions normalised.

These two expressions both differ by an integral over the density (common to both, by assumption) and the difference between the two one-particle external potentials:

$$E < W = \int \Psi'^* \hat{H} \Psi' d\tau = \int \Psi'^* (\hat{H}' + (\hat{V} - \hat{V}'))\Psi' d\tau$$

The first of these integrals is, of course, E' while the second (remember the external potential is just a local, multiplicative, operator) is

$$\int (v'(\vec{x}) - v(\vec{x}))\rho(\vec{x})d\tau$$

so that we have

$$E < E' + \int (v'(\vec{x}) - v(\vec{x}))\rho(\vec{x})d\tau$$

Repeating this breakdown of the other variational expression (for W' we obtain the analogous result

$$E' < E + \int (v(\vec{x}) - v'(\vec{x}))\rho(\vec{x})d\tau$$

i.e.
$$E' < E - \int (v'(\vec{x}) - v(\vec{x}))\rho(\vec{x})d\tau$$

Adding this to the expression for E above leads to:
$$E + E' < E + E'$$

which is the required contradiction. Notice that these are *strictly less than* conditions since the two wavefunctions Ψ and Ψ' are assumed *different*.

Notice that this proof hinges on the *multiplicative* nature of the external potential $v(\vec{x})$. In fact the proof can be extended to a wider class of external potential operator.

This result establishes the first, more straightforward, of the two results of the previous section.

The result can be taken further than the simple "topological fact". Since the density $\rho(\vec{x})$ uniquely determines the external potential and the external potential determines the Hamiltonian which, in turn, determines the ground-state wavefunction, the ground-state wavefunction must be determined uniquely by the ground-state density. That is, the ground-state wavefunction must be a *functional* of the ground-state density.

$$\Psi(\vec{x}_1, \ldots, \vec{x}_n) = f[\rho(\vec{x})]$$

Similarly the ground-state total energy, which is determined by the Hamiltonian and the ground-state wavefunction must be a functional of the ground-state density:

$$E = W[\rho(\vec{x})]$$

Clearly, if these results hold then the sum of the kinetic energy and electron-repulsion energies must be a functional of the density:

$$\int \Psi^*(\hat{T} + \hat{G})\Psi d\tau = F[\rho(\vec{x})]$$

and we know the relationship between the (multiplicative) operator $v(\vec{x})$ and the density from above:

$$\int \Psi^*\hat{V}\Psi d\tau = \int v(\vec{x})\rho(\vec{x})d\tau$$

So that the total ground state energy can be written as a functional of the ground-state density:

$$E = W[\rho(\vec{x})] = F[\rho(\vec{x})] + \int v(\vec{x})\rho(\vec{x})d\tau$$

32.2. HOHENBERG AND KOHN'S PROOFS

Notice that we have said nothing at all about $\hat{V}$ except that it can be expressed as a sum of one-electron potentials $v(\vec{x})$ so that this functional must be a *universal* functional, independent of the form of the external potential.

Of course we know expressions for *parts* of this functional since it must, for example, contain the familiar mean Coulomb repulsion energy of the total electron density:

$$\int\int \rho(\vec{x}_1)\frac{1}{r_{12}}\rho(\vec{x}_2) d\tau_1 d\tau_2$$

We should note for the future that this Coulomb expression contains the electronic self-interaction energies which are (in HF theory) cancelled by the exchange terms. We therefore write

$$F[\rho(\vec{x})] = G[\rho(\vec{x})] + \int\int \rho(\vec{x}_1)\frac{1}{r_{12}}\rho(\vec{x}_2) d\tau_1 d\tau_2$$

which concentrates everything that is unknown about the functional into $G[\rho(\vec{x})]$, giving

$$E = G[\rho(\vec{x})] + \int v(\vec{x})\rho(\vec{x}) d\tau + \int\int \rho(\vec{x}_1)\frac{1}{r_{12}}\rho(\vec{x}_2) d\tau_1 d\tau_2$$

where the form of the functional $G[\rho(\vec{x})]$ is sought whose numerical values are given by:

$$G[\rho(\vec{x})] = \int \Psi^*\hat{T}\Psi d\tau + \int \Psi^*\hat{U}\Psi d\tau - \int\int \rho(\vec{x}_1)\frac{1}{r_{12}}\rho(\vec{x}_2) d\tau_1 d\tau_2$$

That is, what is required is nothing less than the form of a functional $G[\rho(\vec{x})]$ which generates, *from the one-electron density* $\rho(\vec{x})$, the values of the *kinetic* energy ($T[\rho(\vec{x})]$, say) *and* the non-Coulomb part of the electron-repulsion energy ($E_{XC}[\rho(\vec{x})]$, say); the *exchange-correlation* energy.

$$G[\rho(\vec{x})] = T[\rho(\vec{x})] + E_{XC}[\rho(\vec{x})]$$

Deferring any detailed consideration of the obvious difficulties *of principle* as well as of practice which must be associated with this approach we will press on optimistically. However, one word of caution is, perhaps, in order. Our proofs are, as we emphasised at the outset, *existence* proofs and the mere fact that the electron density fixes the external potential should not cause *too* much celebration since, after all, we *know* what the external potential in molecules is and this is of no help at all in solving the resulting Schrödinger equation.

32.2.2 Kohn–Hohenberg's second theorem: variation

The generation of any *practical* method for the computation of molecular electron distributions and their energies usually depends on the existence of a variational principle which may, by way of a parametric method, be made to yield a computationally accessible technique. The generation of such a variational principle for the density is now straightforward since the first, existence, theorem enables us to use familiar "wavefunction methods".

A given electron density function $\tilde{\rho}(\vec{x})$ (say) is known from above to determine an external potential ($\tilde{v}(\vec{x})$, say) in which the electrons move and hence both the Hamiltonian ($\hat{\tilde{H}}$, say) and the solution ($\tilde{\Psi}$, say) of the associated Schrödinger equation:

$$\hat{\tilde{H}}\tilde{\Psi} = \tilde{E}\tilde{\Psi}$$

If, however we use this density function to evaluate the energy of a system of electrons with a *different, known* external potential ($v(\vec{x})$, say) which is associated with a known Hamiltonian ($\hat{H}$), then this energy is known in terms of the conventional wavefunction expectation value *and* as a functional of $\tilde{\rho}(\vec{x})$:

$$\tilde{E} = \int \tilde{\Psi}^* \hat{H} \tilde{\Psi} d\tau = W[\tilde{\rho}(\vec{x})]$$

where, for convenience we have assumed that:

- $\tilde{\Psi}$ is normalised
- $\tilde{\rho}(\vec{x})$ is n-representable and integrates to the number of electrons in the system

two conditions which are basically equivalent here.

Obviously, by the conventional Schrödinger variation condition

$$\tilde{E} \geq E = \int \Psi^* \hat{H} \Psi d\tau$$

where E is the eigenvalue of the target Schrödinger equation

$$\hat{H}\Psi = E\Psi$$

But this energy eigenvalue is guaranteed to be the same functional of the true density:

$$E = \int \Psi^* \hat{H} \Psi d\tau = W[\rho(\vec{x})]$$

that is

$$W[\tilde{\rho}(\vec{x})] \geq W[\rho(\vec{x})]$$

for *any* $\tilde{\rho}(\vec{x}) \neq \rho(\vec{x})$ which establishes the required variational condition subject, of course, to the n-representability of $\tilde{\rho}(\vec{x})$, in particular

$$\int \tilde{\rho}(\vec{x}) d\tau = n$$

32.3 Kohn–Sham equations: introduction

There are two very obvious conceptual difficulties which will have a bearing on the generation of electron densities directly:

- The kinetic energy: how can we possibly tell how fast electrons are moving merely from a knowledge of their distributions in space?
- The Pauli principle: what is the role of the antisymmetry requirement in the form of an electron density function?

Taking the second problem first, it is very well-known that not all functions of the coordinates of a single particle (space and spin) can be generated by partial integrations of the squared modulus of an antisymmetric function of the coordinates of n electrons; not all candidates for the role of an electron density are n-representable.[3] There is a class of functions which *always* generate an n-representable function; the so-called natural orbital expansion of an n-particle wavefunction's single particle density matrix and hence the diagonal component, the density function:

$$\rho(\vec{x}; \vec{x}') = \sum_{i=1}^{M} n_i \chi_i(\vec{x}) \chi_i^*(\vec{x}')$$

is n-representable if

$$\sum_{i=1}^{M} n_i = n$$

and $M \geq n$ and M may well be infinite.

In this expansion the natural orbitals are *orthonormal*:

$$\int \chi_i^*(\vec{x}) \chi_j(\vec{x}) d\tau = 0$$

In particular the special case of a single determinant of orbitals for which all the $n_i, i = 1, n$ are unity is n-representable.

If we make an assumption of this sort for the *form* of the electron density function then the calculation of the kinetic energy is straightforward, it is:

$$T'[\rho(\vec{x})] = \sum_{i=1}^{n} \int \chi_i^*(\vec{x}) \left(\frac{1}{2} \nabla^2 \right) \chi_i^*(\vec{x}) d\tau$$

(because of the orthonormality of the χ_i, there are no "cross terms").[4]

---
[3] Any expansion of the same *form* as the natural orbital expansion but where the functions are not orthonormal is not n-representable, for example.
[4] The notation T' rather than T has been used in anticipation of later developments.

This is exactly the same as the expression for the kinetic energy of a single determinant of MOs χ_i. If, however, the expansion is restricted to only n terms the *exact* kinetic energy of a system of *interacting* electrons cannot be expressed in this way; the best that we can hope for is to be able to choose this restricted expansion in such a way that, perhaps, the *majority* of the kinetic energy may be computed in this way. Because, if it can, we have an expansion for the electron density of a flexible form (identical to the familiar single-determinant MO expression) from which we can easily compute:

- The density itself
- The majority of the kinetic energy
- The interaction between the density and any "external" sources of potential (nuclei, in a molecule)
- The interaction of the density with itself *via* the Coulomb law

leaving just:

- The exchange energy
- The correlation energy
- A contribution to the kinetic energy

to be computed in order to provide us with an exact result for the many-electron system. That is, we have isolated the "essence" of the (unknown) energy functional into a functional which has to compute the sum of three small contributions to the total energy. Indeed, if we omit the last two of these energy terms we are simply left with the MO model itself.

The importance of this simple result is that an implementation of this density-functional method only involves the generation of an "exchange-correlation-residual-kinetic-energy" functional, and, if this can be found, the generation of a set of *one-electron* functions must involve a technology of implementation very similar to our SCF technique. In particular, using a similar heuristic justification as we have used for the *form* of the HF MOs, the linear expansion method is available for the expansion of the "natural orbitals" which generate the density.

Before taking this theory any further let us be sure that we are not making any unwarranted assumptions. Underlying all the above is the assumption that there actually *is* an n-term expansion of the exact density of natural orbital form. That is, in more physical terms, there is an "independent-electron-type" density which is exactly the same as the exact electron density. Worded in

32.4. KOHN–SHAM EQUATIONS

this way, the assumption about the n-term "natural orbital" expansion of the density function seems more remote than it actually is. What is required is that the expansion of a function of ordinary three-dimensional space can be done in terms of a set of other functions of ordinary three-dimensional space *of arbitrary form*; we are not proposing an expansion of an unknown function in terms of some known functions, all that we require is an *existence* theorem. Phrased in this way it seems extremely unlikely that such an expansion would not be possible. If this is true we can go ahead any try to find approximate, empirical forms for the missing energy terms.

There is one very obvious pitfall in all that we have said so far; the Coulomb energy of repulsion of a density "with itself" includes the interaction of each electron with all the others *plus* the interaction of that electron *with itself*. This self-interaction energy is clearly spurious; in the MO method it is automatically cancelled by the diagonal exchange terms

$$J_{ii} - K_{ii} = 0$$

but, in *separating* the calculation of the Coulomb energy from the exchange energy as we have we *exclude* this automatic cancellation and must therefore take steps to include it explicitly, if possible.

In other words, the existence of the so-called Self-Interaction-Correction (SIC) makes the method reminiscent of the Hartree (product) wavefunction rather than the Hartree–Fock (determinant) case.

As we have seen in Chapter 31, a central theme in work on semi-empirical MO theories is that these methods are *not* approximations to the full Hartree–Fock model but are a framework for the generation of (in principle) exact energetic results by *calibrating* the model against experiment *via* the "integral approximations". Such models have been developed to be a highly efficient method of computing equilibrium molecular geometries by treating the molecular integrals retained in the calculation as entirely disposable parameters which are only required to reproduce the geometries of a standard set of molecules chosen to "span" the set of molecular types met in the likely usage of the system. Now what is purely optimism on the part of the semi-empirical models is underpinned by the existence proof of Kohn and Hohenberg in the case of DFT but what they have in common is the *idea* that a technology which is no more demanding than the linear-expansion SCFMO model can be made to yield exact results.

32.4 Kohn–Sham equations

If the density is determined by a set of single-electron functions then what we require in order to compute the density is an equation which determines these functions. Generally speaking, such an equation and its boundary conditions will be generated from a variation principle, and the Kohn–Hohenberg proof is based on the ordinary Schrödinger variation principle so the prospects are optimistic.

We start with the combination of the second Hohenberg-Kohn theorem and the conclusions of the last section:

- The total energy functional is a minimum with respect to variations in trial n-representable density functions:

$$W[\tilde{\rho}(\vec{x})] \geq E$$
$$\int \rho(\vec{x}) d\tau = n$$

- The trial density is written in the restricted natural-orbital or single-determinant form:

$$\tilde{\rho}(\vec{x}) = \sum_{i=1}^{n} \tilde{\chi}_i(\vec{x}) \tilde{\chi}_i^*(\vec{x})$$

With these two assumptions variations in the trial density $\tilde{\rho}(\vec{x})$ reduce to variations in the "orbitals" $\tilde{\chi}_i(\vec{x})$ and their conjugates $\tilde{\chi}_i^*(\vec{x})$.

We therefore write out the energy functional in terms of the orbitals as far as this is possible; as before, this means the energy of a given choice of $\tilde{\rho}(\vec{x})$ may be given in terms of the $\tilde{\chi}_i(\vec{x})$ except for the "exchange-correlation-residual-kinetic-energy" functional, which from now on in line with established terminology we call the "exchange-correlation" functional.

$$\tilde{E} = W[\tilde{\rho}(\vec{x})] = T'[\tilde{\rho}(\vec{x})] + V[\tilde{\rho}(\vec{x})] + J[\tilde{\rho}(\vec{x})] + E_{XC}[\tilde{\rho}(\vec{x})] \qquad (32.1)$$

Where the notation E_{XC}, used earlier for the true exchange-correlation functional, has had its meaning changed to include the residual (non-MO-form) kinetic energy component. In this expression in addition to the above expression for T':

$$V[\tilde{\rho}(\vec{x})] = \int v(\vec{x}) \tilde{\rho}(\vec{x}) d\tau$$
$$J[\tilde{\rho}(\vec{x})] = \int\int \tilde{\rho}(\vec{x}_1) \frac{1}{r_{12}} \tilde{\rho}(\vec{x}_2) d\tau_1 d\tau_2$$

and, of course, the form of E_{XC} is unknown.

The orbital form of the trial density ensures that it is n-representable and so we seek a minimum in $W[\tilde{\rho}(\vec{x})]$ subject to this orthonormality constraint on

32.4. KOHN–SHAM EQUATIONS

the orbitals $\tilde{\chi}_i(\vec{x})$. Since the functionals T', V and J are all available explicitly in terms of the orbitals, the variational problem becomes identical to the Hartree–Fock variational problem set up and solved in Chapter 2 *except* for the problematic exchange-correlation functional E_{XC} which is not known explicitly as a functional of $\tilde{\rho}(\vec{x})$ or the orbitals $\tilde{\chi}_i(\vec{x})$. Thus we must simply *carry* the variation in E_{XC} induced by a variation in $\tilde{\rho}(\vec{x})$ into the differential equation for the optimum orbitals

The procedure is:

- A variation in the density $\tilde{\rho}(\vec{x})$ is associated (to first-order) with linearly independent variations in $\tilde{\chi}_i(\vec{x})$ and $\tilde{\chi}_i^*(\vec{x})$.

- Anticipating the result from the experience in Chapter 2, take the variation in $\tilde{\chi}_i^*(\vec{x})$ to generate an equation for the optimum $\chi_i(\vec{x})$ and the variation in $\tilde{\chi}_i(\vec{x})$ to generate an equation for $\chi_i^*(\vec{x})$.

- Generate the form of the variation for each of the separate functionals T', V and J involving $\tilde{\chi}_i(\vec{x})$.

- Add a multiple of the orthonormality constraints using the Lagrange method.

- Collect terms and set the multiplier of the variation in $\tilde{\chi}_i^*(\vec{x})$ to zero to generate the differential equation for the optimum $\chi_i(\vec{x})$.

- The same procedure for variations in $\tilde{\chi}_i(\vec{x})$ generates an equation for the optimum $\chi_i^*(\vec{x})$ which is merely the complex conjugate of the equation for optimum $\chi_i(\vec{x})$, *provided* the matrix of Lagrangian multipliers is Hermitian.

This is fine *except that* the functional E_{XC} is not known and, *a fortiori*, is not in the orbital form.

We simply *define* the variation in the functional E_{XC} induced by a variation in the density $\rho(\vec{x})$ as v_{XC}:

$$v_{XC} = \frac{\delta E_{XC}}{\delta \rho} \tag{32.2}$$

which is, in fact, the so-called *functional derivative* of $E_{XC}[\rho(\vec{x})]$ with respect to $\rho(\vec{x})$.

For our purposes this simply replaces the unknown E_{XC} in the energy expression with the unknown v_{XC}, which is an additional *potential*,[5] in the equation for the optimum $\chi_i(\vec{x})$. This potential contains the exchange "potential" as well as the correlation and residual kinetic energy and so must be non-local.

---
[5] Notwithstanding its kinetic energy component.

We therefore arrive at the Kohn–Sham equation for the optimum orbitals which define the *exact* one-electron density:

$$\hat{h}^{KS}\chi_i = \epsilon_i \chi_i \tag{32.3}$$

where we have taken the *canonical* set of orbitals which diagonalise the matrix (ϵ) of Lagrange multipliers.

Here

$$\hat{h}^{KS} = \frac{1}{2}\nabla^2 + \hat{v} + \hat{J} + \hat{v}_{XC} \tag{32.4}$$

The operator $\hat{v}$ is just the (multiplicative) nuclear potential, the Coulomb operator is familiar from HF theory:

$$\hat{J} = \int d\tau' \frac{\rho(\vec{x}')}{|\vec{r} - \vec{r}'|}$$

and $\hat{v_{XC}}$ must be obtained by a combination of heuristic arguments, modelling based on simple systems and calibration against known results.

32.5 Non-local operators in orbital theories

There have now been three occasions when, in dealing with the equations determining optimum orbitals, we have met the problem of non-local operators in these equations:

- The exchange operator in Hartree–Fock theory.
- The pseudopotential operator.
- The exchange-correlation operator

It is important to realise that this is not some inconvenience but is an *essential* part of the single-determinant (and related) methods.

Of course, all electrons are the same and any theory of molecular electronic structure which claims to have any validity must reflect this elementary fact whether it uses a single-determinant model or something more advanced. But in the single-determinant ("independent particle") model, while each electron is the same in the sense of charge, mass and laws of motion etc., each electron has its own, *different distribution*. If we are to have a method of calculating the distributions of the electrons in such a model (whether it be a "real" distribution as in HF theory or a "fictitious" one as in KS theory) we are trapped between the two conflicting requirements:

32.5. NON-LOCAL OPERATORS IN ORBITAL THEORIES

1. The indistinguishability of the electrons suggests that the orbitals should be determined by a *single* equation.

2. The fact that the individual electrons' distributions are different suggests that there should be a *separate* equation for each orbital.

In our discussion of the interpretation of the exchange operator the way in which this impasse is resolved was indicated.

The single equation which determines all the orbitals (electron distributions) of these models *must* contain an operator which *generates* each separate equation for each orbital when actually applied to that orbital.

This type of operator must necessarily be non-local since its function is (colloquially speaking) to recognise which orbital it is acting on and produce some modifications to its action accordingly. The act of recognition may involve:

- A permutation operator (the exchange case).
- A projection operator (the pseudopotential case).
- Some unknown action including, at least, the removal of self-interaction (the exchange-correlation case).

It is therefore clear that the occurrence of non-local operators is an *essential* part of independent-particle models of molecular electronic structure *if* we are to have a *single equation* which determines all the electron distributions.[6]

Naturally, we may wish to replace or approximate any non-local operator by one or more local approximations to it but, from what we have seen above, this can *never* be exact. At the very least we must expect to have to replace the single effective equation by a separate equation for each orbital. As we noted in Chapter 2, Hartree when using the simple *product* wavefunction, noticed that his equation would include self-interaction since there was no exchange operator to remove it, but he simply ignored the self-interaction on the grounds that it should not be there! We shall see later that attempts to remove the SIC from the KS model generates the same problem in a new form as it must; any approximation to the exchange-correlation potential which includes a local self-interaction-correction must necessarily generate a separate equation for each orbital. Now this result is much more discouraging in DFT than in HF theory since it leads back inexorably to an *orbital* rather than a *density* model of molecular electronic structure.

---

[6]This is made most explicit in the reduction of the sets of equations of the multi-shell SCF model to a single equation where sets of projection operators are required explicitly to effect the generation of the single effective equation.

Note that the considerations so far have been entirely in terms of a single determinant of *spin-orbitals* $\chi_i(\vec{x})$ and spin-basis-functions $\varphi_k(\vec{x})$. The whole derivation can be repeated for the closed-shell case simply by changing the names of the symbols: MOs to $\psi_i(\vec{r})$ and $\phi_k(\vec{r})$ with the associated change in occupation numbers from 1 to 2.

Chapter 33

Implementing the Kohn–Sham equations

There are several approaches to the problem of obtaining approximate solutions of the Kohn–Sham equations. The problem is to find an approximation to the exchange-correlation functional and to decide whether or not to use an entirely numerical method since the simplest approximate function(al)s involve numerical integration. These questions are still under active investigation and we shall outline one approach which fits most neatly into our HF implementation. Some considerations which have been used to guide the choice of energy functionals are given. The use of numerical integration for the specific case of molecular electron density functions calls for very specialised techniques which are outlined but not coded.

Contents

33.1	A precursor: The Hartree–Fock–Slater model	**764**
33.2	Implementation of the Kohn–Sham method	**766**
33.2.1	Constraints on exchange-correlation	768
33.3	The kinetic energy density	**771**
33.4	Gradients in the exchange-correlation energy	**772**
33.5	Numerical integration of densities	**773**
33.6	Summary	**776**

33.1 A precursor: The Hartree–Fock–Slater model

The idea that the electron density might serve as the fundamental object in the theory of many-electron systems has a venerable history starting with the Thomas–Fermi model which makes no reference to the wavefunction. Perhaps the best-known adaptation of these ideas in molecular theory is the use, by Slater, of the (known) expression for the exchange energy of the free-electron gas (the solid-state physicists' favourite model) as an *approximation* to the exchange energy in the Hartree–Fock equations. The expression for this exchange energy is both simple and general; it contains only the *electron density* not the individual orbitals. Thus, Slater proposed to simply replace the Hartree–Fock exchange operator by the operator which generates this exchange energy defining the Hartree–Fock–Slater (HFS) method.

There are two methods to derive this free-electron-gas exchange potential:

- From the application of the free-electron approximation in the Hartree–Fock exchange potential
- From application of the approximation in the total energy

Both of these methods yield expressions of the form

$$v_X(\vec{x}) = -\alpha \frac{3}{2} \left[\frac{3}{\pi} \rho(\vec{x}) \right]^{\frac{1}{3}} \qquad (33.1)$$

The first approach gives $\alpha = 1$ while the second gives $\alpha = 2/3$.[1]

It is not out of place to note here that the electron density for a free-electron gas is *constant*; what is important for the theory is not the *value* of this density but the way in which the exchange energy and exchange potential *depend* on that density. The electron density in any molecular system of interest is far from constant but the assumption is that the functional relationship between the density and the exchange potential might well be similar. The important thing here is the *form* of the expression; it depends simply on $\rho(\vec{x})$ and, since it is an approximation to the exchange potential for an electron gas moving in a potential, the value of α is taken to be a disposable parameter which is chosen to optimise the performance of the approximation in molecular situations (0.75 is a compromise value).

---

[1] Perhaps, this difference should set alarm bells ringing.

33.1. A PRECURSOR: THE HARTREE–FOCK–SLATER MODEL

In line with our general approach of using spin-orbitals wherever possible we should mention that, the electron density function generated from a single determinant will have the general form

$$\rho(\vec{x}) = \boldsymbol{\varphi}(\vec{x})\boldsymbol{R}\boldsymbol{\varphi}^T(\vec{x})$$

(for real spin-basis functions). If, as usual, the spin-basis functions are simply a set of spatial basis function "doubled" by multiplication by the two spin factors (α and β) then

$$\rho(\vec{x}) = P^\alpha(\vec{r})|\alpha|^2 + P^\beta(\vec{r})|\beta|^2$$

Where P^α and P^β are the *spatial* distributions of the electrons of each spin type.

Two cases may now be distinguished:

- The "spin-paired" closed-shell case for which the spatial electron density is

$$\rho(\vec{x}) \longrightarrow P(\vec{r}) = 2P^\alpha(\vec{r}) = 2P^\beta(\vec{r})$$

  This model is called the "Local Density Approximation" (LDA) since the associated exchange-correlation operator is, in fact, only an exchange operator and is certainly local since it is multiplicative.

- The "different densities for different spins" case (analogous to DODS in HF theory) for which the two spatial densities must be retained and used separately. This method will generate the closed-shell case if the system is, indeed, actually spin-paired. This is often called the "Local Spin Density Approximation" (LSD)[2] for obvious reasons.

Since the LSD obviously *includes* the LDA case when the system is spin-paired, we may use the acronym LSD for both; the defining characteristic is that the exchange-correlation energy involves a *local* exchange-correlation potential of the form

$$E_{XC} = \int d\tau \rho(\vec{x}) v_{XC}(\rho(\vec{x})$$

and this is now taken to be the *characteristic* of LSD; the multiplicative nature of v_{XC} in the exchange-correlation functional.

The HFS model is, of course, an example of LSD and was developed *before* the work of Kohn and Hohenberg so we have the interesting choice of point of view:

- Do we regard the HFS model as a *precursor* to the KS equation or

---

[2] The "A" for approximation has been dropped, perhaps, because of the requirement for all abbreviations to be TLAs (Three Letter Acronyms).

766   CHAPTER 33. IMPLEMENTING THE KOHN–SHAM EQUATIONS

- do we regard it as an *approximation* to the HF equation?

Clearly, since the term inserted is *exchange-only* it must be an approximation to the HF equations in the sense that the exact exchange is replaced by the free-electron model. But, it may also be regarded as the exchange-only precursor of the KS equation.

It is worth noting that the matrix elements of this operator (which is, of course, a multiplicative, *local* operator) involve quantities related to $\rho^{4/3}(\vec{x})$:

$$\int \chi_i^* \rho^{\frac{1}{3}} \chi_j d\tau$$

and the total exchange energy will be proportional to $\rho^{4/3}(\vec{x})$. We would anticipate problems with this type of integral when the density and MOs are expressed in terms of an "AO" basis; the fractional power of the basis functions makes analytical methods difficult. The usual approach is to evaluate these integrals *numerically*, a point to which we will return shortly.

33.2 Implementation of the Kohn–Sham method

The similarity between the Kohn–Sham (KS) and the (differential) Hartree–Fock equation is so great that is too tempting *not* to try to use the same linear expansion methods for its approximate (parametric) solution. We already have the mathematics and the software technology to evaluate all the terms in the matrix form of the Kohn–Sham equation.

Without repeating the details it is obvious that the matrix form of the KS equation obtained by making the usual finite basis expansion of the KS orbitals is

$$\boldsymbol{h}^{KS} \boldsymbol{C} = \boldsymbol{S} \boldsymbol{C} \boldsymbol{\epsilon} \tag{33.2}$$

where

$$\begin{aligned}
\boldsymbol{\chi} &= \boldsymbol{\varphi} \boldsymbol{C} \\
\boldsymbol{R} &= \boldsymbol{C} \boldsymbol{C}^\dagger \\
\boldsymbol{h}^{KS} &= \boldsymbol{h} + \boldsymbol{J} + \boldsymbol{v}_{XC} \\
h_{ij} &= \int \varphi_i^* (-\frac{1}{2} \nabla^2) \varphi_j d\tau + \int \varphi_i^* \hat{v} \varphi_j d\tau \\
J_{ij} &= \sum_{r,s=1}^{m} R_{rs}(ij, rs) \\
S_{ij} &= \int \varphi_i^* \varphi_j d\tau
\end{aligned}$$

33.2. IMPLEMENTATION OF THE KOHN–SHAM METHOD

with the completely new (and unknown) matrix representation of the exchange-correlation operator:

$$(v_{XC})_{ij} = \int \varphi_i^* \hat{v}_{XC} \varphi_j d\tau$$

In practical terms, all that this means is the replacement of the finite-expansion exchange matrix ($\boldsymbol{K}$) of the HF equation with the matrix $\boldsymbol{v}_{XC}$. If the elements of this matrix can be computed then all that needs to be changed in the implementation of the SCF method is the subroutine scfGR and our existing SCF code will perform DFT calculations *with no other changes*.

The problem with this elegant-looking plan is two-fold:

1. We do not know what the exchange-correlation potential is
2. When approximated or modelled, this potential is certain to be more complex than the simple Slater exchange potential ($X\alpha$, LDA, LSD) and therefore will generate basis-function energy integrals

$$\int \varphi_i^* v_{XC} \varphi_j d\tau$$

which are analytically intractable.

These points have a bearing on the design of an implementation of the KS method *as a whole* rather than on the specific point of technical integral evaluation. If the integrals involving the exchange-correlation potential have to be evaluated numerically then the following points spring to mind immediately:

- If we have the values of the basis-function products and the densities "to hand" during the numerical integration of *these* integrals, why not use these quantities to evaluate *all* the integrals in the KS method? Why use the analytical integrals at all?

- If the method is in principle exact we do not need to do perturbation theory or CI or any other "post HF" calculation, so the basis-function integrals are never needed *after* the KS calculation. This would suggest that the "direct" SCF method would be most appropriate here; computing all energy integrals on demand and not storing them.

The answers to questions like these can only be answered by looking, in practice, at the difficulties involved with the numerical evaluation by quadrature of the various energy integrals needed which we do in Chapter 34.

33.2.1 Constraints on exchange-correlation

We can say a good deal about the most elementary properties of an exchange-correlation functional by an examination of some of the integral constraints on the densities arising from many-electron wavefunctions. Basically, the normalisation constraints on the wavefunction and its associated one- and two-particle densities generate "normalisation conditions" on the conditional probability distributions which are involved in the definition of the exchange-correlation functional and these conditions place rather severe constraints on the form of any functional.

Since the electron-repulsion operator is simply multiplicative the total electron repulsion energy associated with a two-particle density $\rho_2(\vec{x}_1, \vec{x}_2)$ must be

$$\frac{1}{2} \int d\tau_1 \int d\tau_2 \frac{1}{r_{12}} \rho_2(\vec{x}_1, \vec{x}_2)$$

the factor 1/2 arises because, in symmetrical form, the electron repulsion operator is[3]

$$\sum_{i \geq j} \frac{1}{r_{ij}} = \frac{1}{2} \sum_{i,j} \frac{1}{r_{ij}}$$

But, of course, the two-electron density function is unknown. However, the largest contribution to this two-particle density must, presumably, be the *product* of two copies of the single-particle density $\rho(\vec{x})$ so that it would seem sensible to write

$$\rho_2(\vec{x}_1, \vec{x}_2) = \rho(\vec{x}_1)\rho(\vec{x}_2) \times [1.0 + f(\vec{x}_1, \vec{x}_2)]$$

where the effects of electron correlation are absorbed into the *correlation function* $f(\vec{x}_1, \vec{x}_2)$. This function incorporates all the non-classical effects in the repulsion energy *plus* the self-interaction correction which is both classical and part of the exchange energy as it is usually defined.

If we extract the dependence of $\rho(\vec{x}_1)$ from this expression for ρ_2 we have

$$\rho_2(\vec{x}_1, \vec{x}_2) = \rho(\vec{x}_1)[\rho(\vec{x}_2) + \rho(\vec{x}_2)f(\vec{x}_1, \vec{x}_2)]$$

and use the notation

$$\rho_{xc}(\vec{x}_1, \vec{x}_2) = \rho(\vec{x}_2)f(\vec{x}_1, \vec{x}_2) \tag{33.3}$$

for the so-called *exchange-correlation hole* we find that

$$\int \rho_2(\vec{x}_1, \vec{x}_2) d\tau_2 = \rho(\vec{x}_1) \int \rho(\vec{x}_2) d\tau_2 + \rho(\vec{x}_1) \int \rho_{xc}(\vec{x}_1, \vec{x}_2) d\tau_2$$

---

[3]Note here that there are two conventions about the normalisation of density functions; the one used here is the one in which the n-particle functions are divided by $n!$. The difference only appears for $n \geq 2$. If one uses the second convention then the expression for the n-particle *operator* must *not* be divided by $n!$

33.2. IMPLEMENTATION OF THE KOHN–SHAM METHOD

But the definition of the one- and two-particle densities gives

$$\int \rho_2(\vec{x}_1, \vec{x}_2) d\tau_2 = (n-1)\rho(\vec{x}_1)$$

$$\int \rho(\vec{x}_2) d\tau_2 = n$$

so that

$$\int \rho_{xc}(\vec{x}_1, \vec{x}_2) d\tau_2 = -1 \tag{33.4}$$

a surprising result since the right-hand-side might have been expected to be a function of $\vec{x}_1$. However, it is less surprising when we realise that the largest non-classical effect in the electron-repulsion energy is the self-interaction correction (SIC), which is precisely the removal of *one unit of charge* from the "classical" self-repulsion energy of a charge *distribution*. Thus the exchange-correlation hole must, at the very least, remove the SIC.

In view of this interpretation of eqn (33.4) we might expect that this normalisation condition is essentially due to the exchange part of the exchange-correlation hole and that it might be useful to divide this density into an "exchange hole" and a "correlation hole". Guided by the product form of the (real) two-particle density for the single-determinant model which involves the density function ρ and the density *matrix* ρ_1:

$$\rho_2(\vec{x}_1, \vec{x}_2) = \rho(\vec{x}_1)\rho(\vec{x}_2) - \rho_1(\vec{x}_1; \vec{x}_2)\rho_1(\vec{x}_2; \vec{x}_1)$$
$$\rho(\vec{x}_1) = \rho_1(\vec{x}_1; \vec{x}_1)$$

and the fact that, in this model, the exchange energy is

$$-\frac{1}{2} \int d\tau_1 \int d\tau_2 \frac{1}{r_{12}} \rho_1(\vec{x}_1; \vec{x}_2) \rho_1(\vec{x}_2; \vec{x}_1)$$

we can define an "exchange-only" hole by:

$$\rho_x(\vec{x}_1; \vec{x}_2) = -\frac{\rho_1(\vec{x}_1; \vec{x}_2)\rho_1(\vec{x}_2; \vec{x}_1)}{\rho(\vec{x}_1)} \tag{33.5}$$

In the same spirit as the normalisation of the exchange-correlation hole we see that

$$\int \rho_x(\vec{x}_1; \vec{x}_2) d\tau_2 = -\int \frac{\rho_1(\vec{x}_1; \vec{x}_2)\rho_1(\vec{x}_2; \vec{x}_1)}{\rho(\vec{x}_1)} d\tau_2$$
$$= \frac{1}{\rho(\vec{x}_1)} \int \rho_1(\vec{x}_1; \vec{x}_2)\rho_1(\vec{x}_2; \vec{x}_1) d\tau_2$$
$$= -1 \tag{33.6}$$

Since the integral over $d\tau_2$ is nothing more than the idempotency condition on the one-particle density matrix:

$$\int \rho_1(\vec{x}_1; \vec{x}_2)\rho_1(\vec{x}_2; \vec{x}_1) d\tau_2 = \rho_1(\vec{x}_1; \vec{x}_1) = \rho(\vec{x}_1)$$

This confirms our expectation about the content of the exchange hole; notice once more, that the integral has the same value whatever the value of $\vec{x}_1$. If the "correlation-only" hole is defined by default to be

$$\rho_c(\vec{x}_1, \vec{x}_2) = \rho_{xc}(\vec{x}_1, \vec{x}_2) - \rho_x(\vec{x}_1, \vec{x}_2) \tag{33.7}$$

then, necessarily, it is normalised to zero:

$$\int \rho_c(\vec{x}_1, \vec{x}_2) d\tau_2 = 0 \tag{33.8}$$

again, independently of the value of $\vec{x}_1$.

This result confirms the idea that the principle effect of exchange is to make the SIC while correlation involves a *reorganisation* of the electron density with no net charge; the analogy is:

- The exchange hole (*via* the SIC) is charge-like; for every position of an electron in a many-electron system the exchange hole scoops out one electron from around that position.

- The correlation hole is multipole-like; for each position of one electron any charge scooped out by the Coulomb repulsion near to that position must be compensated by a build-up of charge somewhere else to give no net charge.

If we now write down the expression for the total electron repulsion energy in terms of the densities and associated "holes" it has an attractive local-looking form:

$$\frac{1}{2} \int d\tau_1 \int d\tau_2 \frac{1}{r_{12}} \rho_2(\vec{x}_1, \vec{x}_2) =$$
$$\frac{1}{2} \int d\tau_1 \int d\tau_2 \rho(\vec{x}_1) \frac{1}{r_{12}} [\, \rho(\vec{x}_2) \quad + \quad \rho_x(\vec{x}_1, \vec{x}_2) + \rho_c(\vec{x}_1, \vec{x}_2)\,]$$

but the hole densities are not known; they must be *functionals* of the electron density ρ. These transformations simply provide a possible *model* for the generation of approximate exchange-correlation functionals by noting the fact that the required energy terms are *interpretable* as repulsion with a density of net positive[4] charge and, separately, a density with no net charge. These results both provide an intuitive model for estimates of the exchange and correlation potentials which are, in this notation:

$$v_X(\vec{x}) = \int d\tau' \frac{1}{|\vec{r} - \vec{r}'|} \rho_x(\vec{x}, \vec{x}')$$
$$v_C(\vec{x}) = \int d\tau' \frac{1}{|\vec{r} - \vec{r}'|} \rho_c(\vec{x}, \vec{x}')$$

---

[4]The one-particle density is a density of *electrons* of course so that the normalisation of the exchange hole to -1 means a net positive charge.

33.3 The kinetic energy density

There are two *major* problems associated with any attempt to generate the kinetic energy density from the particle density, one which is obvious and another, more fundamental one, which puts the whole program in jeopardy:

- Except for systems which only have a *single, real* occupied orbital (ψ, say) (one electron or two closed-shell electrons) the orbitals of even the single-determinant model cannot be recovered from the density function:

$$\rho_1(\vec{x}_1; \vec{x}_2) = n\psi(\vec{x})\psi(\vec{x}')$$
$$\rho(\vec{x}) = \rho_1(\vec{x}; \vec{x})$$
$$\psi(\vec{x}) = \frac{1}{\sqrt{n}} \rho(\vec{x})$$

- The variational method of generating the Schrödinger equation and the Hartree–Fock equation contains a possible ambiguity in the sense that the total kinetic energy of a single electron is given by *both* of

$$- \int \psi^* \left(\frac{1}{2}\nabla^2\right) \psi d\tau$$
$$\frac{1}{2} \int (\nabla \psi^*)(\nabla \psi) d\tau$$

the difference between the two *integrals* is a divergence integral which is zero for bound-state orbitals.[5]

Thus, some decision has to be taken about the very *meaning* of "kinetic energy *density*" before facing the essentially technical difficulty of approximating this density as a functional of the particle density. This conclusion applies just as forcefully to the Kohn–Sham method of approximating the dominant part of the kinetic energy by the single-determinant form:

$$-\sum_{i=1}^{n} \int \psi_i \left(\frac{1}{2}\nabla^2\right) \psi_i d\tau = \frac{1}{2} \sum_{i=1}^{n} \int (\nabla \psi_i)(\nabla \psi_i) d\tau$$

---

[5]Strictly, for systems with no electron net electron flux across the boundaries of the region concerned.

but it is certainly *not* true that

$$-\sum_{i=1}^{n} \psi_i \left(\frac{1}{2}\nabla^2\right) \psi_i = \frac{1}{2}\sum_{i=1}^{n}(\nabla\psi_i)(\nabla\psi_i)$$

for example the right-hand-side is always positive (desirable for a kinetic-energy density) while the left-hand-side may be positive or negative.[6]

The implicit choice for the single-determinant kinetic-energy density will, naturally, affect the form of the "exchange-correlation" functional since it contains the residual kinetic energy density which is different in the two cases. Fortunately, the residual kinetic energy does not seem to be a large correction.

The fact that the kinetic energy density for a (singly or doubly occupied) real orbital may be expressed *exactly* in terms of the electron density generates an alternative approach to modelling the kinetic energy density for *many-electron* systems. There are, of course, two possibilities for this kinetic energy density depending on which of the two orbital forms is preferred:

$$\left(\nabla \rho^{\frac{1}{2}}(\vec{x})\right)^2$$

or

$$\rho^{\frac{1}{2}}(\vec{x})\nabla^2\rho^{\frac{1}{2}}(\vec{x})$$

Both of these expressions are quite different[7] from the expression for the kinetic energy density of a free-electron gas:

$$\frac{3}{10}\left(3\pi^2\right)^{\frac{2}{3}}\left(\rho(\vec{x})\right)^{\frac{5}{3}}$$

so that the question of the *compatibility* between the use of the LSD model and the Kohn–Sham orbital form for the kinetic-energy density naturally arises.

33.4 Gradients in the exchange-correlation energy

The natural appearance of the gradient of the electron density in the kinetic energy density is not surprising in view of the familiar interpretation of the gradient as the momentum in quantum mechanics and the gradient expressions for kinetic energy in ordinary orbital models. Of course, the gradient of the (constant) electron density for the free-electron gas would simply generate a vanishing contribution. Let us now recall that

---

[6]Thus, unlike the electron density which is unique for a given state, the kinetic energy density may differ by a divergence without affecting the overall energy.

[7]To put the point no more strongly!

33.5. NUMERICAL INTEGRATION OF DENSITIES

- The "exchange-correlation" functional actually contains a kinetic-energy component.
- There are two expressions for the kinetic-energy density which differ by the divergence of a gradient.

Thus, it is entirely possible that the exchange-correlation energy (and potential) contain *gradients* of the electron density possibly in addition to the powers of the density which we have met already in the free-electron case.

One particularly successful conjecture was the use of a multiplicative *factor* in the exchange-correlation potential which is a function of a gradient-related variable. Naturally, the contributions of the gradients of the density to exchange and correlation may be investigated separately. Thus, writing, for example, the exchange energy as

$$E_X \approx \int \rho^{\frac{4}{3}}(\vec{x}) F(s) d\tau$$

where s is a variable related to the gradient of ρ extends the capability of the original LSD model considerably. For convenience, the variable s is "scaled" by the 4/3 power of the density in order to generate a dimensionless variable:

$$s = \frac{\nabla \rho}{\rho^{4/3}}$$

Investigation of these "gradient expansions" proceeds apace. Simply satisfying the asymptotic conditions of the potential generated by the "exchange hole" generates an exchange functional which represents a considerable improvement on the simple LSD.

33.5 Numerical integration of densities

The standard methods of numerical quadrature all require the value of the integrand at a set of standard points (x_i, say) and express the value of the integral as:

$$\int_a^b f dx \approx \sum_{i=1}^{N} w_i f(x_i)$$

where the w_i are weight-factors dependent on the nature of the quadrature method. This formula may be made exact for $f(x)$ which are reducible to polynomials of degree N and is approximate for higher polynomials or other functions. The extension to functions of several dimensions is obvious:

$$\int_{a_x}^{b_x} \int_{a_y}^{b_y} g(x,y) dx dy \approx \sum_{i=1}^{N_x} \sum_{j=1}^{N_y} w_i w_j g(x_i, y_j)$$

etc. There are specially developed sets of (x_i, w_i) for functions which contain $\exp(-\zeta x)$ or $\exp(-\alpha x^2)$.

Now these methods all work well for sufficiently "smooth" functions; functions with no sudden maxima or cusps etc.. It is extremely easy to see that the electron density in a molecule is *not* smooth in this sense. The density is extremely variable with very peaked cusps at the nuclei and falling away rapidly with distance from the nuclei. The function is very lumpy and markedly inhomogeneous with direction in space. It is easy to choose sets of numerical integration points which would give very different answers for any integral having the electron density in its integrand; a set which sampled only the region around each nucleus would perform differently from a set which sampled only the internuclear regions, for example. In short, to obtain accurate results for integrals which contain the exchange-correlation potential (and the basis functions) the method must be specifically tailored to this particular type of work.

One attractive idea is to partition the molecular density (and energy densities) into individual *atomic* densities and perform the integration outwards from each nucleus until the "boundary" of another atom is encountered or the density falls to zero (to some criterion) if that atom is on the exterior of the molecule. Using a system of spherical polar coordinates on each atom is attractive but the spherical boundaries would leave the familiar cusped regions uncovered by the integration procedure. The related method of (for example) bisecting each bond axis by a plane and performing the integration inside each polyhedron formed by the intersection of these planes (the so-called Voronoi cells) has proved successful and it certainly includes all the molecular space. Obviously the "outer" polyhedrons must have their outer planes fixed by some criterion which indicates when the value of the electron density has fallen to a negligible value.

This partitioning into mutually exclusive polyhedral regions of space may be simulated by using a set of additional multiplicative weight factors in the numerical expression. If we invent a suitable function $(p_i(\vec{r})$, say) which has the properties:

$$p_i(\vec{r}) = \begin{cases} 1 & \text{for } \vec{r} \text{ inside the } i\text{th polyhedron} \\ 0 & \text{for } \vec{r} \text{ outside the } i\text{th polyhedron} \end{cases}$$

then this function may be incorporated into the weight factors used in the integration over each polyhedron.

In fact, a closer look at the theory of numerical integration shows how this simple intuitive approach may be both justified and improved by choosing a particular form of the $p_i(\vec{r})$. Any integration over a (one-dimensional) region may, by a suitable transformation of variable, be transformed into an integration from 0 to 1:

$$\int_a^b f(x)dx = \frac{b-a}{\beta-\alpha}\int_\alpha^\beta g(y)dy$$

33.5. NUMERICAL INTEGRATION OF DENSITIES

where
$$(\beta - \alpha)x = (b - a)y + (a\alpha - b\beta)$$
that is
$$\int_a^b f(x)dx = (b - a) \int_0^1 g(y)dy$$
where $x = (b-a)y + a$ so that attention may be concentrated on integrals with end-points $(0,1)$. In more formal terms the Jacobian of the (one-dimensional) transformation of variable is
$$\frac{dx}{dy} = (b - a)$$

The Euler–Maclaurin numerical quadrature formula for such an integral (using $n + 1$ equally spaced values of y: $y_i = i/n$, for $i = 0, n$) is:

$$\begin{aligned}\int_0^1 g(y)dy &= \frac{1}{n}\left(\sum_{i=1}^{n-1} g(y_i) + \frac{1}{2}\left[g(0) + g(1)\right]\right) \\ &\quad - \frac{1}{12n^2}\left(g'(1) - g'(0)\right) + \frac{1}{720n^4}\left(g'''(1) - g'''(0)\right) \\ &\quad + (\text{ terms involving differences of higher-order derivatives } g^{(m)} \\ &\qquad (\text{say}) \text{ evaluated at the end-points })\end{aligned}$$

where the g', g'' etc. are the first, second etc. derivatives of the function g. This quadrature clearly becomes more and more accurate if the differences between the derivatives at the end-points

$$g^{(m)}(1) - g^{(m)}(0)$$

are zero; either because this *difference* is zero or because both terms are zero.

Now the Euler–Maclaurin numerical integration method is not the most accurate method available for a given number of function evaluations— the Gauss series of quadratures are usually the most efficient for a function which can be evaluated at an arbitrary point[8]— but, presumably, the general conclusion still holds; if those derivative differences vanish then the quadrature will be more accurate than if they are non-zero.

Thus the scheme based on an intuitive idea of what the electron density (and, presumably, the associated energy densities) are like may be improved if:

- The whole of three-space is divided up into polyhedral cells, each containing one atomic nucleus at its centre and the integration is performed "outwards" from each nucleus up to the boundaries of the cells.

---
[8]That is, not a tabulated function which is only available at pre-set points.

- The rigid division is modelled by a smoother cell-dividing function in order to be able to define derivatives at the edges of each cell.

- A transformation of variable within each cell is used so that the function of the transformed variable has zero derivatives (to some high order) at the boundaries of the cells.

Most numerical integration schemes for molecular electronic structure calculations now (implicitly or explicitly) use such schemes.

Many schemes use a change of variable whose Jacobian involves powers of $(1-y^2)$:

$$\frac{dx}{dy} \approx (1-y^2)^m$$

and the choice of m is made as a compromise between:

- Making the maximum number of derivatives at the end-points zero.
- Keeping the weighting function generated by the change of variable "realistic".

This is necessary since the number of vanishing derivatives is larger for larger m, but if m is *too* large, the resulting effective weighting function for the integration defeats the original object of being constant inside the atomic polyhedron and zero outside it. The larger is m the more "peaked" is the effective weight function leading to poor sampling of the space within an atomic polyhedron.

33.6 Summary

The most straightforward way to implement the solution of the Kohn–Sham equations using all the techniques we have used for the Hartree–Fock case is simply to replace the formation of the exchange part of the Hartree–Fock matrix by a matrix element of some chosen "exchange-correlation" potential.

The available exchange-correlation potentials have been generated by a combination of normalisation requirements and parametrisations of models of the electronic structure. The complicated dependence of the resulting matrix elements on the electron density (and therefore on the basis functions) means that these matrix elements must be computed by numerical integration techniques.

Families of numerical integration methods have been suggested which use the standard (Gaussian) numerical quadrature techniques *within* each of a set

33.6. SUMMARY

of mutually exclusive polyhedra formed by the planes which bisect[9] each bond emanating from a given atom. Integration proceeds outwards from the nucleus at the "centre" of each polyhedron with some suitable choice for the boundaries of the atoms on the periphery of the molecule.

No attempt is made here to give explicit codes for these numerical integration methods since they are still under active theoretical development and the codes are easy, if a little tedious, to generate.

---

[9]Naturally, the performance of the numerical method could be improved by dividing each bond at a point other than the mid-point. This choice could be made on the basis of the relative electronegativities of the atoms at each end of a bond, for example, so that the polyhedra are chosen to contain similar amounts of electron density.

Chapter 34

Semi-numerical methods

We have dismissed both totally numerical methods and the use of Slater-type orbitals as computationally intractable. There are quite separate reasons for this, both of them involving the evaluation of certain integrals required in the variational process. It is possible to use numerical methods which completely avoid the evaluation of these troublesome integrals. Methods which combine the advantages of numerical techniques and the linear expansion method are known and are becoming computationally tractable.

Contents

34.1	Non-variational expansions	**779**
34.2	The pseudospectral method	**782**
	34.2.1	Difficulties with the pseudospectral method	785
34.3	The discrete variational method	**786**
	34.3.1	The DVM and DFT	788

34.1 Non-variational expansions

All our use of the linear variational method has been based on what might be called the "integral approach"; in using the equations which result from the minimisation of the energy *functional* we have necessarily been involved in the manipulation of basis-function energy *integrals* and expansion coefficients. This fact has coloured what we have been *able* to do; we have excluded the use of intuitively attractive basis functions on the simple grounds that they generate energy integrals which are too difficult to evaluate. This is scientifically unfortunate since, as we have seen, once we are "inside" a given expansion method we cannot always tell if our results, while mathematically correct, are physically meaningful.

There is another approach which does not use the energy integrals at all. Instead of starting from the energy function after inserting the linear expansion approximation into the energy functional, we start from the *differential* equation which results from the functional minimisation of the energy expression and *then* apply the linear expansion method.

Taking the differential Hartree–Fock equation as our principle example, we have
$$\hat{h}^F \psi_i = \epsilon_i \psi_i$$
and we use
$$\psi_i = \boldsymbol{\phi} \boldsymbol{c}^i$$
to expand ψ_i in terms of a set of basis functions $\boldsymbol{\phi}$. As usual we generate the set of equations to solve by choosing a set of N "representative" points in real, physical (3D) space, and evaluate the function ψ_i and the quantity $\hat{h}^F \psi_i$ at each of these points through the values of the known basis functions at each of these points. This generates a set of N simultaneous equations for the m coefficients:
$$\hat{h}^F \psi_i(\vec{r}_k) = \epsilon_i \psi_i(\vec{r}_k)$$
i.e.
$$\sum_{j=1}^m c_j^i \left(\hat{h}^F \phi_j(\vec{r}_k) \right) = \epsilon_i \sum_{j=1}^m c_j^i \phi_j(\vec{r}_k)$$
for $k = 1, 2, \ldots, N$. Everything is known except the coefficients c_j^i and ϵ_i so that this system of linear equations may be solved for these unknowns. Clearly, if the number of coordinate points N is equal to the number of basis functions then we have a matrix eigenvalue problem similar in *form* to the matrix eigenvalue problem resulting from the linear variation method:
$$\boldsymbol{h}^F \boldsymbol{c}^i = \epsilon_i \boldsymbol{c}^i$$
but this time
$$h_{ik}^F = \left(\hat{h}^F \phi_j(\vec{r}_k) \right)$$
and, of course,
$$h_{ik}^F \neq h_{ki}^F$$
the matrix $\boldsymbol{h}^F$ is not symmetrical and so our existing matrix diagonalisation method is not appropriate.

There are no troublesome and numerous electron-repulsion integrals to evaluate, if the effect of $\hat{h}^F$ on each basis function is known, we may solve the

34.1. NON-VARIATIONAL EXPANSIONS

system. Of course $\hat{h}^F$ depends on the ψ_i (i.e. the coefficients c_j^i) in the usual way but this simply makes the solution of the equations *iterative* in exactly the same way as the usual linear SCF equations; we must solve a system of linear equations several times *but no integrals are involved*.

The real question is, of course, "can it be made to work?" What is immediately obvious is that a set of m points in space is not nearly enough to determine the m expansion coefficients; to take the simplest case, if we expand the MO of H_2 in terms of two $1s$ AOs and do not use two spatial coordinates which are interchanged by the symmetry operations of the molecule, we shall not get two equal coefficients. On the other hand, if we introduce a very large number of spatial coordinate points the system becomes formally (if not actually) over-determined and becomes a least-squares problem rather than an eigenvalue problem.

As we noted much earlier in Chapter 3, making the LCAO expansion independently of the linear variation method is usually invalid in the sense that the operation of an arbitrary differential operator ($\hat{h}$, say) on a linear expansion will normally generate a function which is, at least in part, *outside* the space spanned by the expansion functions:

$$\hat{h}\sum_{i=1}^{m} c_i \varphi_i \neq \sum_{i=1}^{m} d_i \varphi_i$$

for any choice of coefficients d_i. But the linear variational equivalent is always guaranteed:

$$\boldsymbol{hc} = \boldsymbol{d}$$

for any *matrix* $\boldsymbol{h}$ one cannot get "outside" the linear space by the action of such operators.

The key point seems to be to regard the linear expansion method as an *auxiliary* technique rather than the basic method as it is in the conventional approach. When the number of basis functions is the same as the number of grid points, the expansion technique is seen as a convenient way of obtaining *those parts of the effect of* $\hat{h}^F$ on the numerical function which are difficult to obtain by numerical methods. In practice, this means

- *Derivatives* (the effect of ∇^2), which are notoriously difficult to obtain accurately by numerical (differencing) methods.

- *Integral operators* like the Coulomb and exchange operators which involve (three-dimensional) integrals over the MOs which would be too inaccurate if evaluated using the grid of points used to solve the equations themselves.

So, just as one may use numerical integration techniques to obtain analytically intractable integrals in the usual algebraic expansion method; so one may use the converse, an expansion technique, to obtain numerically intractable terms in what is essentially a numerical method.

This is the basic approach in the so-called pseudospectral method.[1]

34.2 The pseudospectral method

Let the standard basis-function expansion of the MOs be given by the usual notation

$$\boldsymbol{\psi} = \boldsymbol{\phi} \boldsymbol{C} \tag{34.1}$$

and let us define a column matrix for each MO ψ_i and each basis function ϕ_k by the values of the (vector) variable on which each depends:

$$\boldsymbol{\psi}_i = \begin{pmatrix} \psi_i(\vec{r}_1) \\ \psi_i(\vec{r}_2) \\ \psi_i(\vec{r}_3) \\ \psi_i(\vec{r}_4) \\ \ldots \\ \ldots \\ \psi_i(\vec{r}_m) \end{pmatrix} \quad \boldsymbol{\phi}_k = \begin{pmatrix} \phi_k(\vec{r}_1) \\ \phi_k(\vec{r}_2) \\ \phi_k(\vec{r}_3) \\ \phi_k(\vec{r}_4) \\ \ldots \\ \ldots \\ \phi_k(\vec{r}_m) \end{pmatrix}$$

where $\vec{r}_j$ is the *value* of the coordinates of a particular point in space, *not* the coordinates of a particular electron as we have used earlier. The number, m, of such points has been deliberately chosen to be the same as the number of basis functions.

These columns matrices may be assembled side-by-side to form (square) matrices $\boldsymbol{\Psi}$ and $\boldsymbol{\Phi}$, say:

$$\boldsymbol{\Psi} = (\boldsymbol{\psi}_1, \boldsymbol{\psi}_2, \boldsymbol{\psi}_3, \ldots, \boldsymbol{\psi}_n)$$

and

$$\boldsymbol{\Phi} = (\boldsymbol{\phi}_1, \boldsymbol{\phi}_2, \boldsymbol{\phi}_3, \ldots, \boldsymbol{\phi}_m)$$

which are related by the coefficient matrix $\boldsymbol{C}$ exactly as in eqn (34.1):

$$\boldsymbol{\Psi} = \boldsymbol{\Phi} \boldsymbol{C}$$

However, since the matrix $\boldsymbol{\Phi}$ is *square*, this result can be inverted to give

$$\boldsymbol{C} = \boldsymbol{\Phi}^{-1} \boldsymbol{\Psi}$$

---

[1]This technique originated in the numerical solution of the diffusion equation and other areas of fluid mechanics where numerical derivatives are required. In many areas of physics the algebraic expansion method is seen historically as related to the "expansion in eigenfunctions" method and is known as the spectral method because of the common use of Fourier transform techniques.

34.2. THE PSEUDOSPECTRAL METHOD

or, written in terms of the individual MOs:

$$\boldsymbol{\psi}_i = \boldsymbol{\Phi} \boldsymbol{c}^i$$

and

$$\boldsymbol{c}^i = \boldsymbol{\Phi}^{-1} \boldsymbol{\psi}_i$$

Thus, we may look at the matrix of basis functions values (and its inverse) as an "active" transformation matrix rather than regarding the basis functions as simply "passive" elements in the process.

The matrices $\boldsymbol{\Phi}$ and $\boldsymbol{\Phi}^{-1}$ effect the transformation between two *representations* of the MOs:

- $\boldsymbol{\Phi}$, acting on the basis-expansion representation of an MO (the column $\boldsymbol{c}^i$ of expansion coefficients in the basis used) generates the physical-space representation of the MO (the column $\boldsymbol{\psi}_i$ of numerical values of the MO):

$$\boldsymbol{\Phi} \times \boldsymbol{c}^i = \boldsymbol{\psi}_i$$

The whole set may be collected as

$$\boldsymbol{\Phi} \times \boldsymbol{C} = \boldsymbol{\Psi}$$

- Conversely, the inverse of $\boldsymbol{\Phi}$, acting on the physical-space representation of an MO generates the basis-expansion representation of that MO:

$$\boldsymbol{\Phi}^{-1} \times \boldsymbol{\psi}_i = \boldsymbol{c}^i$$

or, together,

$$\boldsymbol{\Phi}^{-1} \times \boldsymbol{\Psi} = \boldsymbol{C}$$

A pair of relationships which are useful in manipulating the quantities involved in the use of the discrete variational method when used in conjunction with a basis-expansion approach. Any function which is a linear expansion of the basis functions and for which we have a physical space representation evaluated at our choice of coordinates may have its expansion coefficients determined by multiplication by the inverse of $\boldsymbol{\Phi}$.

Note, for emphasis that

$$\begin{aligned}
(\boldsymbol{\Psi})_{ij} &= \psi_j(\vec{r}_i) \\
(\boldsymbol{\Phi})_{k\ell} &= \phi_\ell(\vec{r}_k) \\
(\boldsymbol{C})_{ij} &= (\boldsymbol{c}^j)_i
\end{aligned}$$

Now, suppose that we want a *matrix* representation of the operator $\hat{A}$ (say) on the spatial function ψ_i:

$$\hat{A}\psi_i = \begin{pmatrix} \hat{A}\psi_i(\vec{r}_1) \\ \hat{A}\psi_i(\vec{r}_2) \\ \cdot \\ \cdot \\ \cdot \\ \hat{A}\psi_i(\vec{r}_{m-1}) \\ \hat{A}\psi_i(\vec{r}_m) \end{pmatrix}$$

We may, of course, insert a unit matrix in the form of $\boldsymbol{\Phi\Phi}^{-1}$ into this expression so that

$$\hat{A}\psi_i = \hat{A}\mathbf{1}\psi_i = \hat{A}\boldsymbol{\Phi\Phi}^{-1}\psi_i = \left(\boldsymbol{A\Phi}^{-1}\right)\psi_i$$

where

$$\tilde{\boldsymbol{A}} = \hat{A}\boldsymbol{\Phi} = \begin{pmatrix} \hat{A}\phi_1(\vec{r}_1) & \ldots & \hat{A}\phi_m(\vec{r}_1) \\ \hat{A}\phi_1(\vec{r}_2) & \ldots & \hat{A}\phi_m(\vec{r}_2) \\ \ldots & \ldots & \ldots \\ \ldots & \ldots & \ldots \\ \hat{A}\phi_1(\vec{r}_m) & \ldots & \hat{A}\phi_m(\vec{r}_m) \end{pmatrix}$$

i.e.

$$\tilde{A}_{ij} = \hat{A}\phi_j(\vec{r}_i)$$

So, we have

$$\hat{A}\psi_i = \boldsymbol{A}\psi_i \tag{34.2}$$

where

$$\boldsymbol{A} = \tilde{\boldsymbol{A}}\boldsymbol{\Phi}^{-1}$$

so that, for example, we may write with appropriate definition of operators

$$\hat{h}^F\psi_i = \boldsymbol{h}^F\psi_i$$

and note that this "Hartree–Fock" matrix is very different from out previous $\boldsymbol{h}^F$, being a matrix representation of the effect of an operator in ordinary, physical, space. This $\boldsymbol{h}^F$ is not symmetrical but we may still hope to solve the physical-space-grid equation

$$\boldsymbol{h}^F\psi_i = \epsilon_i\psi_i$$

The scheme would be to construct each term in the effect of the Hartree–Fock operator on a real-space representation of an MO, using either that spatial representation *or* the linear expansion representation, whichever is appropriate, easier or more tractable, and transform the expansion-method terms to the spatial representation by means of $\boldsymbol{\Phi}$ or its inverse and solve the resulting equation on a chosen grid of points.

Specifically, one would

34.2. THE PSEUDOSPECTRAL METHOD

- Use the expansion method to evaluate the "AO" kinetic energy expressions:
$$T = T_{ao}\Phi^{-1}$$
where
$$(T_{ao})_{ij} = \frac{1}{2}\nabla^2 \phi_j(\vec{r}_i)$$

- The nuclear-attraction operator is simply multiplicative and so is diagonal in the physical-space representation:
$$V = \begin{pmatrix} V(\vec{r}_1) & 0 & \cdots & 0 \\ 0 & V(\vec{r}_2) & \cdots & 0 \\ 0 & 0 & \cdots & V(\vec{r}_m) \end{pmatrix}$$

- The Coulomb operator involves integrals over the squares of the MOs and so is evaluated using the expansion method:
$$\hat{J}(\vec{r}) = \sum_i c^{i\dagger} J_{ao}(\vec{r}) c^i$$
where
$$(J_{ao})_{k\ell} = \int \frac{\phi_k(\vec{r}')\phi_\ell(\vec{r}')}{|\vec{r}-\vec{r}'|} dV'$$

- The effect of the exchange operator is evaluated in a similar way to that of the Coulomb operator but requiring the definition of a $\tilde{K}$ matrix since it is not diagonal; defining
$$\left(\tilde{K}\right)_{ij} = \sum_k \psi_k(\vec{r}_i) \left(c^k J_{ao}(\vec{r}_i)\right)$$
gives the physical-space operator K as
$$K = \tilde{K}\Phi^{-1}$$
completing the formation of
$$h^F = T + V + J - K$$
which may be diagonalised by the use of any algorithm for the solution of non-symmetric matrices.

34.2.1 Difficulties with the pseudospectral method

The central objection which anyone familiar with traditional LCAO expansion methods has to the pseudospectral (PS) method is that there does not seem to be nearly enough information in the value of a function at m points to determine m expansion coefficients, or, what amounts to the same thing, one can scarcely expect to obtain the values of 3D-space integrals numerically with only tens or even hundreds of points when one is using tens or hundreds of basis functions. This obvious objection turns out to be true, and extensions and enhancements of the PS method are all aimed at meeting this objection by:

- Using more basis functions specifically in order to be able to use more integration points and later discarding these ("aliasing" [2]) functions.

- Using more basis functions, hence more spatial points, and obtaining a least-squares method rather than an exactly determined method.

- More and more careful choice of the positioning of the integration points with respect to the atoms in the molecule (avoiding the nuclear positions and sampling the space uniformly etc.

The effect of these complications is the same in all numerically based methods (the discrete variational method later in this chapter and the implementation of density-functional methods of Chapter 33 are other cases); the originally simple concept becomes a complex and empirical technology demanding an enormous amount of specialised experience and knowledge to implement efficiently and usefully. That is not to say that these procedures are not extremely valuable; they are, but it is *not* realistic is to present here any kind of "generic" implementation as we have been able to do in the linear-variationally based methods.

34.3 The discrete variational method

Again starting from a set of simultaneous equations

$$\hat{h}^F \psi_i(\vec{r}_k) = \epsilon_i \psi_i(\vec{r}_k)$$

obtained by choosing points on a spatial grid, we can define an error function by

$$d_i(\vec{r}_k) = (\hat{h}^F - \epsilon_i)\psi_i(\vec{r}_k) \qquad (34.3)$$

which measures how far the value of ψ_i at grid point $\vec{r}_k$ is from satisfying the Hartree–Fock equation.

Exactly as in the algebraic method, we expand the ψ_i in terms of a set of basis functions ϕ_j:

$$\psi_i = \boldsymbol{\phi} \boldsymbol{c}^i$$

---
[2] The name comes from earlier hydrodynamical applications of the same technique.

34.3. THE DISCRETE VARIATIONAL METHOD

where, as usual, the values of the basis functions ϕ_j are easily evaluated at each grid point.

We now need, with the aid of some criterion or set of criteria, to *minimise* d_i for all values of $\vec{r}_k$ used on the grid. That is, we need to define a suitable *error functional* which measures the deviation of eqn (34.3) from zero and minimises it with respect to the quantities which are at our disposal; the expansion coefficients c_k.

With the familiar variational method in mind, let us define the quantities Δ_{ij} by:

$$\Delta_{ij} = \sum_{k=1}^{N} w(\vec{r}_k)\psi_i^*(\vec{r}_k)(\hat{h}^F - \epsilon_i)\psi_j(\vec{r}_k) \tag{34.4}$$

where some flexibility in the method has been built in by using $w(\vec{r}_k)$ as a *weight function* to retain the possibility of making certain $\vec{r}_k$ more important than others in the minimisation of the error.[3]

This definition is motivated by the fact that, simply by choosing the $w(\vec{r}_k)$ to be related to volume elements, we may make the Δ_{ij} into approximations of the corresponding integrals

$$\int \psi_i^*(\vec{r}_k)(\hat{h}^F - \epsilon_i)\psi_j(\vec{r}_k)dV$$

with the associated variational energy functional minimisation method in mind.

So, the procedure is simply to minimise the Δ_{ij} with respect to the expansion coefficients; we require

$$\frac{\partial \Delta_{ij}}{\partial c_\ell} = 0 \tag{34.5}$$

a procedure formally identical to the use of the linear variation method.

The result is the familiar secular eigenvalue problem:

$$\boldsymbol{h}^F \boldsymbol{c} = E \boldsymbol{S} \boldsymbol{c} \tag{34.6}$$

where the matrices are not integrals but components of our error functional Δ_{ij}:

$$h_{ij}^F = \sum_{k=1}^{N} w(\vec{r}_k)\psi_i^*(\vec{r}_k)\hat{h}^F \psi_j(\vec{r}_k) \tag{34.7}$$

$$S_{ij} = \sum_{k=1}^{N} w(\vec{r}_k)\psi_i^*(\vec{r}_k)\hat{h}^F \psi_j(\vec{r}_k) \tag{34.8}$$

---

[3]It may be more important that the valence region of space be fitted more accurately than the core region, for example.

Again we must note that in the limit of a large number of points on the grid and a particular choice of the weighting function $w(\vec{r}_k)$, this equation becomes identical to the linear variation method.

The advantage of the method lies in the freedom to choose:

- the number of points (N),
- the function $w(\vec{r}_k)$
- *and* the form of the expansion functions ϕ_j.

It is the last point that is crucial here; in order to evaluate the matrices $\boldsymbol{h}^F$ and $\boldsymbol{S}$ it is only necessary to be able to evaluate the values of the basis functions at any point in space and the values of their second derivatives, *not* to be able to evaluate *integrals* involving these quantities.

This means, of course, that we can use basis functions chosen on *scientific* grounds rather than have to compromise and use functions of convenience. In a word the Discrete Variational Method (DVM) (as this technique has come to be known) enables the use of Slater functions in molecular calculations. In fact, the method is essentially independent of the form of the basis functions since the evaluation of any likely function is computationally trivial.

Unlike the pseudospectral method, the DVM method does rely on the values of the functions on the grid to provide enough information to obtain meaningful values of the expansion coefficients, and so must use a large enough grid to go some way towards the variational method which is its natural limit. This necessarily means that the DVM will tend to be more time-consuming than the corresponding conventional variational method *if the integrals in the conventional method can be evaluated.* It follows, since the evaluation of the Δ_{ij} are relatively independent of the detailed form of the expansion functions, the DVM will be more time-consuming than any variational method with tractable basis functions since the evaluation of a many-dimensional integral (or Δ_{ij}) by numerical means will always be more time-consuming than the implementation of an analytical form for that integral.

34.3.1 The DVM and DFT

It is both the weakness and strength of numerical integration methods (the formation of weighted sums of values of integrands) that the difficulties of evaluation of those integrands and the time-consumption involved in forming those sums are, to a large degree, independent of the nature of the simple functions involved in the integrands. The bulk of the work lies in handling the large *numbers* of terms needed in the sums.

34.3. THE DISCRETE VARIATIONAL METHOD

Thus, up to a limit, the complexity of the integrand is not a major factor in evaluating an integral numerically and the same applies to the evaluation of the Δ_{ij} of the DVM. Now, the Hartree–Fock equations and the Kohn–Sham equations (particularly) involve some integrals which are of an analytically intractable form. The exchange term in the conventional HF equations is difficult to handle for all but Gaussian functions and the exchange/correlation/kinetic energy functional of the Kohn–Sham equations is, strictly, unknown and even known approximate forms generate integrals which are too complex to be evaluated even for Gaussian basis functions.

The denouement is obvious; the DVM may be used for basis functions of arbitrary[4] form and operators of arbitrary form. In particular, the DVM can be used with Slater functions and exchange/correlation functionals of any kind. We have seen that the Kohn–Sham equations may be implemented in a Gaussian-expansion form simply by deleting the HF exchange term and substituting an exchange/correlation functional evaluated numerically. The DVM enables the whole process to be done in a Slater basis. We must always be aware of the fact that the DVM is not variational, and only approaches a variational result for grid size for which the Δ_{ij} may be identified with the corresponding integrals.

---

[4] Arbitrary but physically reasonable, of course.

Chapter 35

Additional reading and other material

Here is a compilation of material related to this work: suggestions for additional reading and sources of program listings and software tools used.

35.1 Additional reading

The number of references has been purposely kept small in the hope that they will be actually consulted. However, there is an enormous amount of material available bearing on the theory of molecular electronic structure and applications of various models to specific cases. There is, on the other hand, a very much smaller literature on the actual implementation of the methods. Generally speaking, literature on the use of modern software tools in the generation and maintenance of quantum chemistry software is rather thin on the ground.

The principle difficulty in getting to grips with this vast literature is finding a way into it and being able, from the point of view of this work, to find the key books and papers in the field: the ones which present the models and methods rather than the applications to particular chemical systems. Of course, the literature does not necessarily split cleanly into "theory" and "applications" but the main point is clear.

35.1.1 Texts with a theoretical emphasis

The following three texts are quite different in "feel" from each other and present an interesting contrast and emphasis to that given in this work; all are required reading for an understanding of the theory of electronic structure.

- *Methods of molecular quantum mechanics*, (2nd Ed, Academic Press, 1989) by R. McWeeny uses similar notation to ours. The first edition of this work (by McWeeny and B. T. Sutcliffe, Academic press 1969) is still useful.

- *Modern quantum chemistry*, (Revised 1st Ed, McGraw-Hill, 1989) by A. Szabo and N. S. Ostlund has a slightly more computational and "applied" approach than McWeeny's book.

- *Energy density functional theory of many-electron systems*, (Kluwer, 1990) by E. S. Kryachko and E. V. Ludena is rather modestly titled; in fact this very large book contains discussions of many of the areas with which we have been involved (not just DFT) at a deeper theoretical level with many explicit examples. This is a valuable book with a comprehensive bibliography.

35.1.2 Texts describing applications

There are now a number of books which describe how to use the most popular *ab initio* software (GAUSSIAN): giving help and advice on various matters like choice of model and basis set etc.

- *Gaussian 94 user's reference*, (Gaussian Inc., 1996) is distributed with GAUSSIAN.

- *Exploring chemistry with electronic structure methods*, (2nd Ed, Gaussian Inc., 1996) by M. A. Frisch and J. B. Foresman is a simple and step-by-step introduction to *ab initio* calculations using the GAUSSIAN suite of programs.

- Ab initio *molecular orbital theory*, (Wiley-Interscience 1986) by W. J. Hehre, L. Radom, P. v.R. Schleyer and J. A. Pople is an admirably detailed account of the use of the GAUSSIAN series of programs with a huge amount of detail on methodology and lots of results, mainly for organic molecules.

- *Computational quantum chemistry*, (Wiley, 1988) by A. Hinchliffe is a small introductory handbook describing the use of several *ab initio* systems.

- *Modelling molecular structures*, (Wiley, 1996) by A. Hinchliffe is a general introductory text about the use of classical ("molecular mechanics") methods as well as both semi-empirical and *ab initio* quantum-mechanical methods of computing molecular structures.

35.1. ADDITIONAL READING

- *Handbook of computational chemistry*, (Wiley, 1985) by T. Clark is a rather older guide to the use of semi-empirical and *ab initio* software.

A different system which has a graphical display system built in is **spartan** by Wavefunction Inc. There are associated manuals and the following texts of general interest:

- *Practical strategies for electronic structure calculations*, (Wavefunction Inc., 1995) by W. J. Hehre.
- *Experiments in computational organic chemistry*, (Wavefunction Inc., 1993) by W. J. Hehre, L. D. Burke, A. J. Shusterman and W J Pietro.

35.1.3 Guide to the primary literature

H. F. Schaefer III has written a valuable guide to the primary literature with a list of the key papers in the field with a short commentary on each area; the reference is:

- *Quantum chemistry: the development of* ab initio *methods in molecular electronic structure theory*, (OUP, 1984)

This compact text provides a very convenient entry into all the major areas of molecular electronic structure theory with the exception of density functional theory which has come to such prominence since 1984.

35.1.4 Guides to software tools

Almost all the guides, manuals etc. for the software tools used in the production of this book and the programs which it contains are available free *via* the Internet. In particular the entire Linux operating system which includes

- All the Free Software Foundation (GNU) enhanced versions of all the Unix tools.
- **g77**, the GNU f77 compiler.
- **emacs** Richard Stallman's famous customisable editor.
- Donald Knuth's TeX, Leslie Lamport and friends' LaTeX, and all the **dvi** and PostScript support.

- Walter Tichy's RCS: the version control system.

(and much more) is available from the sites listed later or for the price of a CD.

There are a number of implementations of the WEB philosophy, with their own manuals and examples available *via* the Internet. The sites are listed in a later section.

The general philosophy of the WEB system is outlined by its creator (D. E. Knuth) and S. Levy in:

- *The CWEB system of structured documentation*, (Addison-Wesley, 1994)

emacs and its support is so central to modern software development and text processing that it has its own large manual; users will find

- *Learning GNU emacs*, (O'Reilly and Associates, 1991) by D. Cameron and B. Rosenblatt a very useful gentle introduction before getting to grips with the main manual.

35.2 Additional material by ftp

Some of the WEBs referred to in the text are too large or too uninteresting to merit inclusion in the text and they are available by anonymous ftp from an Internet site:

http://spider.shef.ac.uk/OUP/

The index and instructions are self-explanatory.

Also at this site are copies of some of the tools referred to and used in the text:

- The Literate Programming FAQ (Frequently-Asked Questions) which contains references to a variety of WEB systems and general information.

- John Krommes' FWEB system; the compressed tar archive contains the source for the system, texinfo files for online documentation and a comprehensive manual in texinfo-source form.

35.2. ADDITIONAL MATERIAL BY FTP

- A FAQ for FWEB.

- Ratfor (in Ratfor), instructions for bootstrapping the system and a Ratfor manual.

- The auctex major mode for emacs which takes much of the pain out of maintaining large LaTeX files.

As an experiment in which readers might like to take part and comment on, WEBs of a GAUSSIAN-like system which has been generated from the last public-domain GAUSSIAN (GAUSSIAN-80) is also available, together with a manual as LaTeX source and some sample data etc. The system is made available under the terms of the GNU public licence.[1] This suite of programs will perform most of the types of calculation discussed in the previous chapters, but is not meant as a rival to the much more efficient and highly developed current GAUSSIAN software. It is made available as a working system which illustrates both quantum-chemical implementations and the effect of trying to make an existing piece of software into a WEB *a posteriori*.

Other items of potential interest are added to this site from time to time; there may well be WEBs for semi-empirical SCF programs and for Molecular Mechanics ("MMn") methods in the fullness of time.

---

[1] Roughly speaking, the author retains copyright but it may be distributed freely and modified but no charge may be made for any software which incorporates it.

Index

ab initio, 28, 740
Abelian point group, 509
algebraic approximation, 23, 78, 740
α-spin function, 29
anions, 475
antisymmetrised product, see determinant
antisymmetriser, 16
antisymmetry, 9
   in VB, 574
AO, 97–99
   as basis function, 76, 220
   hybrid in VB, 572
approximations
   model, 5, 29
   numerical, 5, 740
arbitrariness
   in many-shell HF, 420
   in pseudo-orbital, 621
argument
   "dummy", 135
   of **subroutine**, 135
array storage
   single subscripted, 126
atom-centred functions, 221–222
atomic orbital, see AO
atomic units, 4, 24–26
   values in SI, 25
aufbau principle, 394–397
   effect of constraints, 397
   exceptions, 477
   $nd/(n+1)s$, 478

bar to indicate β spin, 30, 551
basis function, 219–220
   AO as natural choice, 220
   atom-centred functions, 221

contracted, 252
cubic harmonic, 502
desireable properties, 98, 219
for atoms, 221
for GUHF, 347
Gaussian (GTF), 233
monomial, 502
practical criteria, 232
primitive, 234
real, 149
spatial (ϕ_k), 30, 78
spin (φ_k), 78
beigen, blocked diagonalisation, 384
β-spin function, 29
block form of DODS matrix, 349, 382–384
Bohm, D., 678n
Born–Oppenheimer model, 7–8, 28, 348, 740
boundary conditions
   on orbitals, 53
braces ({ }) in computer language, 128
Brooks, F., 277n
brute-force, 22, 654, 668, 709

calibration
   of model by experiment, 740
canonical orbital, 92–94, 395
   delocalised, 111
CGUHF, 345
charge distribution, 100–102, 399
   basis partitioning, 404–406
   matrix representation, 401–404
   P matrix, 401
   two spin types, 400
charge-cloud notation, 34, 70, 74

charges and bond orders, 403
χ_i, defined, 29
CI, 29, 318, 543
  excitations, 546
    double (DCI), 547
    single, 546
  reference function, 572
  symmetry restricted, 550
  using determinants, 561
  closed shell, 115–121
Cohen, M. H., 679n
commuting operators, 509
completeness, 11, 19, 77, 98, 219n, 585, 781
conditional compilation, 201
configuration interaction, see CI
configuration space, 2
contraction, 251–253
  for core functions, 612
  of primitives, 32
Cook, D. B., 681n
Cooper, D. L., 585n
coordinate systems
  Cartesian (x, y, z), 224
  "diatomic" (ξ, η, ϕ), 224
  spherical polar (r, θ, ϕ), 224
coordinates
  space $(\vec{r}_i)$, 10, 28
  space and spin $(\vec{x}_i)$, 11, 28
  spin (s_i), 10, 11, 28
core
  effective potential, 639, 740
  frozen, 613
  penetration by valence electrons, 615
  precursor, 625, 628
  repulsion energy, 634
correlation, 542
  and double excitations, 547
  functional, 768
  in determinant, 17n, 51, 542
  in probability, 541
Coulomb
  law, 2, 3
  matrix, 82
    Kronecker form, 537

  symmetry contributions, 518
  operator, 44
cusp in electron density, 750

Dacre, P. D., 480n
damp factor, 319
Davidson, E. R., 651
Dawson function, 648
define statement (macro), 130
∇^2, the Laplacian, 3
δ
  Kronecker δ_{ij}, 85, 436
  notation for small change, 83
density function, 542
  and NOs, 589
  n representability, 754
density functional theory, see DFT
density matrix, 542
  perturbation corrections, 662
derivative
  of MCSCF energy, 734
  of non-variational energies, 735
  of SCF energy, 730–735
  of Schrödinger equation, 725
  of variational solutions, 727
  optimised parameters, 728
design
  of SCF code, 131
design criteria, 333
determinant
  expansion in minors, 66
  Slater (Φ_K), 15, 29
  vanishing, 56
determinantal method, 14
DFT, 750
  and DVM, 788
  existence theorem, 750–753
  ground-state density, 752
  Hohenberg and Kohn's proofs, 750
  Kohn–Sham equation, 757
  numerical integration in, 766, 773–776
  Voronoi cells, 774
  universal functional, 753
  variation theorem, 753–754

Different Orbitals for Different Spins, *see* DODS
Dirac, P. A. M.
  "bra" and "ket" notation, 70
  spin wavefunction, 3
direct
  CI, 553, 560–564
  SCF, 488
Discrete Variational Method, *see* DVM
do statement, 127
documentation
  importance, 138
  separate from program, forbidden, 277
DODS, 345
  and GUHF, 363, 382
  "doubly occupied" orbital in GUHF, 360
  doubly occupied orbital in ROHF, 442
$d\tau$, spin-space volume element, 32, 55
Dupuis, M., 533n
dV, spatial volume element, 32, 55
DVM, 786
  and DFT, 788
  basis functions in, 788
  eigenvalue problem in, 787

effective potential, 639
  criteria for model, 639
efficiency
  of coding, 124
Ehrenreich, H., 679n
eigen, implementation of Jacobi method, 179
eigenvalue, 11
  problem, 141–142
  sorting, 181
eigenvector, 11
  use as orthogonaliser, 207
Elder, M. E., 480n
electromagnetic field, 654, 657
electromagnetic radiation
  and RPA, 671

"electron density"
  as fundamental quantity, 749
  for single determinant wavefunction, 101
electron gas
  exchange potential, 764
  α in, 764
emacs, 334n
entanglement
  of space and spin, 575
η_k defined, 28
exchange
  functional, 765
  matrix, 83
    symmetry contributions, 518
  operator, 45
    local form, 46
exchange-correlation
  constraints on, 768–771
  hole, 768
expectation value, *see* mean value

Fermi hole, 620
$F_\nu(t)$, basic GTF integral, 244
WEB, 278–287
Fock matrix, 90
for statement, 127
Fourier component, 672
Fourier transform (of $1/r_{12}$), 247
Free Software Foundation, 334
frozen core, 613
ftp, of additional materials, 288, 652
Fukutome, H., 689n
functional, 749
  exchange-correlation, 758

G(R) matrix, (*see also* scfGR), 136
  formation, 136
  improved formation, 212
g77, 334n
γ_i (NO), 590
Gaussian product rule, 235
GAUSSIAN suite, 200, 250
Gaussian-type function, *see* GTF
gcc, 334n
Generalised Valence Bond, *see* GVB

generi, repulsion integrals, 266
genint, integral generator, 263
genoei, one electron integrals, 266
gensym modified **genint**, 489
Gerratt, J., 585n
getint, 152, 491
  direct version, 167
  full implementation, 307–308
  special for testbench, 190
  testbench model, 159
  typical usage, 153–154
Givens method, 176
gmprd subroutine, 137
gtprd subroutine, 140
GofR
  in GUHF, 353
GTF, 233
  integrals, 234–251
  primitive (η_k), 28, 30
    Cartesian, 31
    type of, 31
  properties, 233
GUHF, 345
  implementation, 350–356
GVB
  Heitler–London pairs, 597

Hamiltonian
  molecular, 2
    in atomic units, 4
  one-electron, 28
Handy and Schaefer's method for derivatives, 735
Handy, N. C., 561n, 735n
Hartree–Fock equation
  canonical, 48
  compared to Schrödinger equation, 316
  differential, 47, 76
  for ROHF, 444
  many shells (McWeenyan), 418
  matrix
    equivalent forms, 657
  matrix, derivation, 76–88
  matrix, full form, 94
  matrix, in orthonormal basis, 91
  non-variational methods, 779
  stabilty of solutions, 688
  total energy, 95
Hartree–Fock matrix ($\mathbf{h}^F$), 88
Hartree–Fock model
  constraints on, 344–347
  existence of solutions, 39
Hartree–Fock–Slater, *see* HFS
Hehre, W. J., 250
Hermitian
  conguate of matrix, 86
  matrix, 46, 414, 510
  operator, 11
HFS, 764
hole
  exchange and correlation, 769–771
  Fermi, 620
HOMO, 317
Householder method, 176
hybrid AO, 572

idempotency of **R** matrix, 435
ifdef statement, 204
inner shell, 612
integrals
  asymptotic forms, 742
  implementation
    data structures, 256
    overview, 263
    storage, 298–300
  over GTFs, 234–251
    efficiency considerations, 250
    electron repulsion, 245–249
    kinetic energy, 238–239
    nuclear attraction, 239–245
    overlap, 235–238
  over STOs, 223–232
    electron repulsion, 231
    kinetic energy, 229
    nuclear attraction, 229–231
    overlap, 228
  repulsion, notation, 34
  spin, 35

use of physical interpretation, 740
interface
of **subroutine**, 135
interpolation, during SCF cycles, 211
irreducible representation, 512
iteration
single, in SCF code, 142–144
termination criteria, 145

Jacobi method, 176

Kernighan, B. W., 335n
kinetic energy
density in DFT, 771
gradients, 772
operator, 3
King, H. F., 533n
Knuth, D. E., author of "Literate Programming", 270
Kohn–Sham, *see* KS
Kohn–Sham equation, 757
Kolar, M., 651
Koopmans' theorem, 50, 112
Krommes, J., author of FWEB, 270
KS, 766
matrix equation, 766–767

Lagrangian and normalisation, 87
Lagrangian multipliers, 41, 413
interpretation, 49
"lapidary" method in mathematics, 55
LaTeX, 270
LDA, 765
level shifter, 319
Levy, S., author of CWEB, 270
linear dependence
of orbitals, 56
linear expansion method, 14
for molecular oribitals, 78
for orbitals, 30
for wavefunctions, 23
shortcomings, 473
lister, symmetry-equivalent labels, 489, 493

Literate Programming, 270
Local Density Approximation, *see* LDA
Local Spin Density, *see* LSD
localised orbital, 113
looping construct, 128
Løwdin's rules, 71
LSD, 765
exchange-correlation functional, 765
LUMO, 317

macro, 130, 334
comparison with program segment, 155
for subscripting, 214
in a WEB, 271
USE, 155
use to extend language, 130
"magic numbers", forbidden, 130
maintenance, of codes, 124
manual page, Unix, 138
matrix
and array, 125
Coulomb, 82, 103–107
physical interpretation, 106
exchange, 83, 103–107
physical interpretation, 106
function of, 186–187
of basis functions, 79
of MO coefficients, 79
of MOs, 79
one-electron integrals, 80
overlap integrals, 79
partitioned, in CI, 549
product, code for, 137
response, 688
Matsen, F. A., 584n
McMurchie, L. E., 651
MCSCF, 446
GVB model, 596
paired excitation, 590
McWeeny, R., 418n, 512n, 583n
mean value
of energy, 20, 54, 74, 80, 395
minimal basis, 159

minors, of determinant, 63
MO, 77
    expansion coefficients (C_{ri}), 79
    spatial (ψ_i), 78
    spin (χ_i), 77
model
    general, 29
    Hartree–Fock, 29
model potential
    for atomic core, 626–629
modules, in a WEB, 270
molecular orbital, *see* MO
Møller–Plesset method, 29, 709–722
    derivative, 735
MP2, *see* Moller–Plesset method
Multi-Configuration SCF, *see* MC-SCF
multiple copies of codes, forbidden, 200

n representability
    condition for, 755
natural orbital, *see* NO
Nesbet's method, 554–556
Nesbet, R. K., 554n
next_label, generation of repulsion integral labels, 196
NO, 588
    and minimal expansions, 590
    for pairs in GVB, 598
    single determinant, 588
non-local operator, 45
normalisation
    and exchange-correlation, 768
    condition on MOs, 84–85
    of GTF, 261
    of single determinant, 56
    wavefunction, 20
Nozieres, P., 678n
nuclear motions, 7

"occupation numbers", 101
operator
    antisymmetrisation, 16
    Coulomb, 44
    exchange, 45
    kinetic energy, 3
    non-local, 45, 617, 760–762
    projection, 415
orbital, 222
    "defined", 97
    localised in VB, 583
    natural (NO), 588
    nodeless, 618
    "virtual", *see* virtual orbitals
orbital energies, 50
orbital model, 11–14
orthogonalisation, 184–189
    of HF matrix, 137–140
    constraints, 660, 673
    in Heitler–London VB, 566
    in VB and CI, 544
    of many shell HF matrix, 420
    reversing ("de-orthogonalisation"), 142
    strong, 599, 630
    to inner shells, 613
orthonormal, 19, 29
    basis, transformation to, 185
    orbitals in HF determinant, 40, 87
oscillations during SCF cycles, 210

pack/unpack, integer packing, 303
pair NOs, 598
pairing
    of spins in VB, 574
Pauli
    principle, 9, 10, 573, 613, 748
        effect in cores, 632
    repulsion, 614
    spin wavefunction, 3
permutation
    and spin eigenfunction, 576
    of electrons, 9
    of repulsion integral labels, 149–150
    operator
        product form, 574
    operator and spin operator, 579
    parity, 9
    symmetry in VB, 574

perturbation theory, 323–330, 654
  $2n+1$ rule, 726
  and derivatives, 725
  level shifter theory, 320
  Rayleigh–Schrödinger, 323
  self-consistent, 657
    energy, 667
    transition frequencies, 685
    vanishing perturbation, 684
Φ_K defined, 29
ϕ_k defined, 30
φ_k defined, 12
π-electron approximation, 165–167
Pines, D., 678n
Planck's constant ($\hbar$), 24
plane waves, 76
Plauger, P. J., 335n
Pople, J. A., 250
primitive function (GTF), 234, 252
primitive spatial product (in VB), 582
probability
  distribution function, 2, 100
  interpretation of wavefunction, 2
procrastination, 132n
projection operator, 415, 433–439
  core, 617, 631
  for symmetry orbital, 512
  in VB, 583
  R matrix as, 436
  spherical harmonic, 625
pseudo-orbital, 624
  arbitrariness, 621–624
pseudopotential
  atomic, 620
    dedendence on ℓ, 624
    model, 626
  definition, 619
  local form, 618
  non-local form, 618
pseudospectral method, 782
Ψ defined, 2
ψ_i defined, 29
putint, partner of **getint**, 298
  implementation, 300–303

quantity calculus, 26

R matrix
  code for generation, 134–135
  for each orbital, 411
  physical interpretation, 101
  representing determinant, 438
  singular, 438
Raimondi, M., 585n
Randić, M, 233n
Random Phase Approximation, see RPA
ratfor, 335
RCS version control, 201, 268, 276, 334, 337–338
  details, 339–342
repeat statement, 128
repulsion integral
  processing, 154–158
response matrix, 688
Restricted Hartree–Fock, see RHF
Restricted open-shell Hartree–Fock, see ROHF
reverse dictionary order, 299, 487
(R)GUHF, see GUHF
RHF, 344
Ring, P., 679
ROHF, 442
RPA, 669–677
  history of name, 678
Rumer, G., 583n
Rys polynomials, 245n, 251

SCF
  conventional, 165
  convergence aids, 321
  defined, 131
  direct, 165
  implementation
    Version 1, 364
  initial version of code, 147
  testbench for, 160
scfGR, **G(R)** implementation
  first model, 155
  for GUHF and DODS, 351
  improved, 213

with symmetry, 524
Schaefer, H. F., 735n
Schrödinger equation
  first derivative, 726
  time-dependent, 2
  time-independent, 5–7, 28
Schrödinger, E., 2 and *passim*
Schuck, P., 679
scount, counts equivalences, 520
second quantisation, 553
self interaction, 33, 45, 106
  correction (SIC), 106, 757, 761
self-adjoint, 19
Self-Consistent-Field, *see* SCF
self-interaction
  correction (SIC), 769
semicolons
  terminating statements, 136
shalf,transformation to orthonormal basis, 188
shell
  empty, 416
  HF matrix, 413
  inner, of electrons, 612
  occupation number, 411
  of GTFs, 250
  of orbitals, 410
  of orbitals in ROHF, 442
  R matrix, 411
simulation of *ab initio*, 744
single determinant
  energy expression, 74
  physical interpretation, 33
  time dependent, 658
single-determinant model, *see* Hartree–Fock model
Slater's rules, 72–73, 80, 546, 552, 591
Slater, J. C., 29, 221
Slater-type orbital, *see* STO
"Software tools", 335n
spatial orbitals (ψ_i), 29
spherical approximation, 221
spherical constraint, 346
spin
  functions, 12, 29
  in VB, 573
  Hamiltonian in VB, 576
  pairing in VB, 574
  space, 12
spin density, 400
  Q matrix, 401
spin eigenfunction, 442, 548, 577–579
  and permutations, 576
  constraint, 346
  in VB, 574
spin polarisation in UHF, 363
spin-orbital (φ_k, χ_i), 12, 29
stacks of matrices, 422–423
  sample WEBs, 423
standard model, 3, 23, 28, 74, 654, 740
STF, *see* STO
STO, 222
  in DVM, 788
  integrals, 223–232
strong orthogonality, 599
subroutine, 135
supermatrix, 107, 539
Sutcliffe, B. T., 583n
switch/case statements, 354
symG, symmetrise partial **G(R)**, 528
symmetry
  blocking of matrix, 511
  breaking, 533
  Dacre-Elder method, 480
  numerical examples, 502
  orbitals, 512–514
  permutation in VB, 574
  permutations of basis, 483
  self-consistent, 482, 511, 689
  shells of functions, 533
symmetry constraint, 346

tangle (part of WEB), 271
testbench
  for GUHF, 357–359
  for SCF, 160–162
    direct version, 169–171
    replacing, 214
    results for H_2O and ions, 163

output for GUHF, 359–360
Thouless, D. J., 685n
Tichy, W. F., author of RCS, 339
Time-Dependent Hartree–Fock (T-DHF), *see* RPA
"top-down" design, 125, 214
trace
  of matrix, 81
transformation
  during SCF cycle, 133
  Kronecker product notation, 534
  of basis in SCF, 506–508
  of products, 535
  of repulsion integrals, 133, 550, 668, 691–708

UHF, 344
  synonym for DODS, 345
unimodular transformation, 89
unitary transformation, 102, 110–111
unpack, *see* pack
Unrestricted Hartree–Fock, *see* UHF
updating eigenvectors, 208

valence bond, *see* VB
variation principle
  applied to many shells, 412
  constraints on, 42, 108–109, 474, 630, 728
  global nature, 658
  lack of freedom, 165, 173
  linear, 22
  nature of turning point, 395
  parametric, 21
  practical, 19
  theoretical, 19
  time dependent, 670, 680–681
variational collapse, 641
VB, 544
  chemical structures, 575
  Heitler–London, 544, 566–567
  non-orthogonality, 572
  polar structures, 544
  primitive spatial product, 582
  spin-free, 581

version control, *see* RCS
vertical ionisation energy, 360
"virtual" orbitals, 311–322
  for H_2O, 312
  in level shifters, 319
  interpretation, 312
  variational use, 317

wavefunction
  many-electron (Ψ), 3
weave (part of WEB), 271
  cosmetic features, 287
WEB system, 269–288
  "book" paradigm, 270
  example, $F_\nu(t)$, 278
  for blocked diagonalisation, 384
  for four-index transformation, 696
  for many-shell HF matrix, 427
  for ROHF, 445
  for scount, 520
  for stacks of matrices, 423
  for symG, 528
  for lister, 493
  for ump2, 712
well-behaved function, 13n
while statement, 128

A CATALOG OF SELECTED
DOVER BOOKS
IN SCIENCE AND MATHEMATICS

Astronomy

BURNHAM'S CELESTIAL HANDBOOK, Robert Burnham, Jr. Thorough guide to the stars beyond our solar system. Exhaustive treatment. Alphabetical by constellation: Andromeda to Cetus in Vol. 1; Chamaeleon to Orion in Vol. 2; and Pavo to Vulpecula in Vol. 3. Hundreds of illustrations. Index in Vol. 3. 2,000pp. 6⅛ x 9¼.
Vol. I: 23567-X
Vol. II: 23568-8
Vol. III: 23673-0

EXPLORING THE MOON THROUGH BINOCULARS AND SMALL TELESCOPES, Ernest H. Cherrington, Jr. Informative, profusely illustrated guide to locating and identifying craters, rills, seas, mountains, other lunar features. Newly revised and updated with special section of new photos. Over 100 photos and diagrams. 240pp. 8¼ x 11. 24491-1

THE EXTRATERRESTRIAL LIFE DEBATE, 1750–1900, Michael J. Crowe. First detailed, scholarly study in English of the many ideas that developed from 1750 to 1900 regarding the existence of intelligent extraterrestrial life. Examines ideas of Kant, Herschel, Voltaire, Percival Lowell, many other scientists and thinkers. 16 illustrations. 704pp. 5⅜ x 8½. 40675-X

THEORIES OF THE WORLD FROM ANTIQUITY TO THE COPERNICAN REVOLUTION, Michael J. Crowe. Newly revised edition of an accessible, enlightening book recreates the change from an earth-centered to a sun-centered conception of the solar system. 242pp. 5⅜ x 8½. 41444-2

A HISTORY OF ASTRONOMY, A. Pannekoek. Well-balanced, carefully reasoned study covers such topics as Ptolemaic theory, work of Copernicus, Kepler, Newton, Eddington's work on stars, much more. Illustrated. References. 521pp. 5⅜ x 8½.
65994-1

A COMPLETE MANUAL OF AMATEUR ASTRONOMY: Tools and Techniques for Astronomical Observations, P. Clay Sherrod with Thomas L. Koed. Concise, highly readable book discusses: selecting, setting up and maintaining a telescope; amateur studies of the sun; lunar topography and occultations; observations of Mars, Jupiter, Saturn, the minor planets and the stars; an introduction to photoelectric photometry; more. 1981 ed. 124 figures. 26 halftones. 37 tables. 335pp. 6½ x 9¼.
42820-6

AMATEUR ASTRONOMER'S HANDBOOK, J. B. Sidgwick. Timeless, comprehensive coverage of telescopes, mirrors, lenses, mountings, telescope drives, micrometers, spectroscopes, more. 189 illustrations. 576pp. 5⅝ x 8¼. (Available in U.S. only.)
24034-7

STARS AND RELATIVITY, Ya. B. Zel'dovich and I. D. Novikov. Vol. 1 of *Relativistic Astrophysics* by famed Russian scientists. General relativity, properties of matter under astrophysical conditions, stars, and stellar systems. Deep physical insights, clear presentation. 1971 edition. References. 544pp. 5⅝ x 8¼. 69424-0

CATALOG OF DOVER BOOKS

Chemistry

THE SCEPTICAL CHYMIST: The Classic 1661 Text, Robert Boyle. Boyle defines the term "element," asserting that all natural phenomena can be explained by the motion and organization of primary particles. 1911 ed. viii+232pp. 5⅜ x 8½. 42825-7

RADIOACTIVE SUBSTANCES, Marie Curie. Here is the celebrated scientist's doctoral thesis, the prelude to her receipt of the 1903 Nobel Prize. Curie discusses establishing atomic character of radioactivity found in compounds of uranium and thorium; extraction from pitchblende of polonium and radium; isolation of pure radium chloride; determination of atomic weight of radium; plus electric, photographic, luminous, heat, color effects of radioactivity. ii+94pp. 5⅜ x 8½. 42550-9

CHEMICAL MAGIC, Leonard A. Ford. Second Edition, Revised by E. Winston Grundmeier. Over 100 unusual stunts demonstrating cold fire, dust explosions, much more. Text explains scientific principles and stresses safety precautions. 128pp. 5⅜ x 8½. 67628-5

THE DEVELOPMENT OF MODERN CHEMISTRY, Aaron J. Ihde. Authoritative history of chemistry from ancient Greek theory to 20th-century innovation. Covers major chemists and their discoveries. 209 illustrations. 14 tables. Bibliographies. Indices. Appendices. 851pp. 5⅜ x 8½. 64235-6

CATALYSIS IN CHEMISTRY AND ENZYMOLOGY, William P. Jencks. Exceptionally clear coverage of mechanisms for catalysis, forces in aqueous solution, carbonyl- and acyl-group reactions, practical kinetics, more. 864pp. 5⅜ x 8½. 65460-5

ELEMENTS OF CHEMISTRY, Antoine Lavoisier. Monumental classic by founder of modern chemistry in remarkable reprint of rare 1790 Kerr translation. A must for every student of chemistry or the history of science. 539pp. 5⅜ x 8½. 64624-6

THE HISTORICAL BACKGROUND OF CHEMISTRY, Henry M. Leicester. Evolution of ideas, not individual biography. Concentrates on formulation of a coherent set of chemical laws. 260pp. 5⅜ x 8½. 61053-5

A SHORT HISTORY OF CHEMISTRY, J. R. Partington. Classic exposition explores origins of chemistry, alchemy, early medical chemistry, nature of atmosphere, theory of valency, laws and structure of atomic theory, much more. 428pp. 5⅜ x 8½. (Available in U.S. only.) 65977-1

GENERAL CHEMISTRY, Linus Pauling. Revised 3rd edition of classic first-year text by Nobel laureate. Atomic and molecular structure, quantum mechanics, statistical mechanics, thermodynamics correlated with descriptive chemistry. Problems. 992pp. 5⅜ x 8½. 65622-5

FROM ALCHEMY TO CHEMISTRY, John Read. Broad, humanistic treatment focuses on great figures of chemistry and ideas that revolutionized the science. 50 illustrations. 240pp. 5⅜ x 8½. 28690-8

Engineering

DE RE METALLICA, Georgius Agricola. The famous Hoover translation of greatest treatise on technological chemistry, engineering, geology, mining of early modern times (1556). All 289 original woodcuts. 638pp. 6¾ x 11. 60006-8

FUNDAMENTALS OF ASTRODYNAMICS, Roger Bate et al. Modern approach developed by U.S. Air Force Academy. Designed as a first course. Problems, exercises. Numerous illustrations. 455pp. 5⅜ x 8½. 60061-0

DYNAMICS OF FLUIDS IN POROUS MEDIA, Jacob Bear. For advanced students of ground water hydrology, soil mechanics and physics, drainage and irrigation engineering, and more. 335 illustrations. Exercises, with answers. 784pp. 6⅛ x 9¼. 65675-6

THEORY OF VISCOELASTICITY (Second Edition), Richard M. Christensen. Complete, consistent description of the linear theory of the viscoelastic behavior of materials. Problem-solving techniques discussed. 1982 edition. 29 figures. xiv+364pp. 6⅛ x 9¼. 42880-X

MECHANICS, J. P. Den Hartog. A classic introductory text or refresher. Hundreds of applications and design problems illuminate fundamentals of trusses, loaded beams and cables, etc. 334 answered problems. 462pp. 5⅜ x 8½. 60754-2

MECHANICAL VIBRATIONS, J. P. Den Hartog. Classic textbook offers lucid explanations and illustrative models, applying theories of vibrations to a variety of practical industrial engineering problems. Numerous figures. 233 problems, solutions. Appendix. Index. Preface. 436pp. 5⅜ x 8½. 64785-4

STRENGTH OF MATERIALS, J. P. Den Hartog. Full, clear treatment of basic material (tension, torsion, bending, etc.) plus advanced material on engineering methods, applications. 350 answered problems. 323pp. 5⅜ x 8½. 60755-0

A HISTORY OF MECHANICS, René Dugas. Monumental study of mechanical principles from antiquity to quantum mechanics. Contributions of ancient Greeks, Galileo, Leonardo, Kepler, Lagrange, many others. 671pp. 5⅜ x 8½. 65632-2

STABILITY THEORY AND ITS APPLICATIONS TO STRUCTURAL MECHANICS, Clive L. Dym. Self-contained text focuses on Koiter postbuckling analyses, with mathematical notions of stability of motion. Basing minimum energy principles for static stability upon dynamic concepts of stability of motion, it develops asymptotic buckling and postbuckling analyses from potential energy considerations, with applications to columns, plates, and arches. 1974 ed. 208pp. 5⅜ x 8½. 42541-X

METAL FATIGUE, N. E. Frost, K. J. Marsh, and L. P. Pook. Definitive, clearly written, and well-illustrated volume addresses all aspects of the subject, from the historical development of understanding metal fatigue to vital concepts of the cyclic stress that causes a crack to grow. Includes 7 appendixes. 544pp. 5⅜ x 8½. 40927-9

CATALOG OF DOVER BOOKS

ROCKETS, Robert Goddard. Two of the most significant publications in the history of rocketry and jet propulsion: "A Method of Reaching Extreme Altitudes" (1919) and "Liquid Propellant Rocket Development" (1936). 128pp. 5⅜ x 8½. 42537-1

STATISTICAL MECHANICS: Principles and Applications, Terrell L. Hill. Standard text covers fundamentals of statistical mechanics, applications to fluctuation theory, imperfect gases, distribution functions, more. 448pp. 5⅜ x 8½. 65390-0

ENGINEERING AND TECHNOLOGY 1650–1750: Illustrations and Texts from Original Sources, Martin Jensen. Highly readable text with more than 200 contemporary drawings and detailed engravings of engineering projects dealing with surveying, leveling, materials, hand tools, lifting equipment, transport and erection, piling, bailing, water supply, hydraulic engineering, and more. Among the specific projects outlined–transporting a 50-ton stone to the Louvre, erecting an obelisk, building timber locks, and dredging canals. 207pp. 8⅜ x 11¼. 42232-1

THE VARIATIONAL PRINCIPLES OF MECHANICS, Cornelius Lanczos. Graduate level coverage of calculus of variations, equations of motion, relativistic mechanics, more. First inexpensive paperbound edition of classic treatise. Index. Bibliography. 418pp. 5⅜ x 8½. 65067-7

PROTECTION OF ELECTRONIC CIRCUITS FROM OVERVOLTAGES, Ronald B. Standler. Five-part treatment presents practical rules and strategies for circuits designed to protect electronic systems from damage by transient overvoltages. 1989 ed. xxiv+434pp. 6⅛ x 9¼. 42552-5

ROTARY WING AERODYNAMICS, W. Z. Stepniewski. Clear, concise text covers aerodynamic phenomena of the rotor and offers guidelines for helicopter performance evaluation. Originally prepared for NASA. 537 figures. 640pp. 6⅛ x 9¼. 64647-5

INTRODUCTION TO SPACE DYNAMICS, William Tyrrell Thomson. Comprehensive, classic introduction to space-flight engineering for advanced undergraduate and graduate students. Includes vector algebra, kinematics, transformation of coordinates. Bibliography. Index. 352pp. 5⅜ x 8½. 65113-4

HISTORY OF STRENGTH OF MATERIALS, Stephen P. Timoshenko. Excellent historical survey of the strength of materials with many references to the theories of elasticity and structure. 245 figures. 452pp. 5⅜ x 8½. 61187-6

ANALYTICAL FRACTURE MECHANICS, David J. Unger. Self-contained text supplements standard fracture mechanics texts by focusing on analytical methods for determining crack-tip stress and strain fields. 336pp. 6⅛ x 9¼. 41737-9

STATISTICAL MECHANICS OF ELASTICITY, J. H. Weiner. Advanced, self-contained treatment illustrates general principles and elastic behavior of solids. Part 1, based on classical mechanics, studies thermoelastic behavior of crystalline and polymeric solids. Part 2, based on quantum mechanics, focuses on interatomic force laws, behavior of solids, and thermally activated processes. For students of physics and chemistry and for polymer physicists. 1983 ed. 96 figures. 496pp. 5⅜ x 8½. 42260-7

Mathematics

FUNCTIONAL ANALYSIS (Second Corrected Edition), George Bachman and Lawrence Narici. Excellent treatment of subject geared toward students with background in linear algebra, advanced calculus, physics, and engineering. Text covers introduction to inner-product spaces, normed, metric spaces, and topological spaces; complete orthonormal sets, the Hahn-Banach Theorem and its consequences, and many other related subjects. 1966 ed. 544pp. 6⅛ x 9¼. 40251-7

ASYMPTOTIC EXPANSIONS OF INTEGRALS, Norman Bleistein & Richard A. Handelsman. Best introduction to important field with applications in a variety of scientific disciplines. New preface. Problems. Diagrams. Tables. Bibliography. Index. 448pp. 5⅜ x 8½. 65082-0

VECTOR AND TENSOR ANALYSIS WITH APPLICATIONS, A. I. Borisenko and I. E. Tarapov. Concise introduction. Worked-out problems, solutions, exercises. 257pp. 5⅝ x 8¼. 63833-2

THE ABSOLUTE DIFFERENTIAL CALCULUS (CALCULUS OF TENSORS), Tullio Levi-Civita. Great 20th-century mathematician's classic work on material necessary for mathematical grasp of theory of relativity. 452pp. 5⅝ x 8¼. 63401-9

AN INTRODUCTION TO ORDINARY DIFFERENTIAL EQUATIONS, Earl A. Coddington. A thorough and systematic first course in elementary differential equations for undergraduates in mathematics and science, with many exercises and problems (with answers). Index. 304pp. 5⅜ x 8½. 65942-9

FOURIER SERIES AND ORTHOGONAL FUNCTIONS, Harry F. Davis. An incisive text combining theory and practical example to introduce Fourier series, orthogonal functions and applications of the Fourier method to boundary-value problems. 570 exercises. Answers and notes. 416pp. 5⅜ x 8½. 65973-9

COMPUTABILITY AND UNSOLVABILITY, Martin Davis. Classic graduate-level introduction to theory of computability, usually referred to as theory of recurrent functions. New preface and appendix. 288pp. 5⅜ x 8½. 61471-9

ASYMPTOTIC METHODS IN ANALYSIS, N. G. de Bruijn. An inexpensive, comprehensive guide to asymptotic methods–the pioneering work that teaches by explaining worked examples in detail. Index. 224pp. 5⅜ x 8½ 64221-6

APPLIED COMPLEX VARIABLES, John W. Dettman. Step-by-step coverage of fundamentals of analytic function theory–plus lucid exposition of five important applications: Potential Theory; Ordinary Differential Equations; Fourier Transforms; Laplace Transforms; Asymptotic Expansions. 66 figures. Exercises at chapter ends. 512pp. 5⅜ x 8½. 64670-X

INTRODUCTION TO LINEAR ALGEBRA AND DIFFERENTIAL EQUATIONS, John W. Dettman. Excellent text covers complex numbers, determinants, orthonormal bases, Laplace transforms, much more. Exercises with solutions. Undergraduate level. 416pp. 5⅜ x 8½. 65191-6

CATALOG OF DOVER BOOKS

CALCULUS OF VARIATIONS WITH APPLICATIONS, George M. Ewing. Applications-oriented introduction to variational theory develops insight and promotes understanding of specialized books, research papers. Suitable for advanced undergraduate/graduate students as primary, supplementary text. 352pp. 5⅜ x 8½. 64856-7

COMPLEX VARIABLES, Francis J. Flanigan. Unusual approach, delaying complex algebra till harmonic functions have been analyzed from real variable viewpoint. Includes problems with answers. 364pp. 5⅜ x 8½. 61388-7

AN INTRODUCTION TO THE CALCULUS OF VARIATIONS, Charles Fox. Graduate-level text covers variations of an integral, isoperimetrical problems, least action, special relativity, approximations, more. References. 279pp. 5⅜ x 8½. 65499-0

COUNTEREXAMPLES IN ANALYSIS, Bernard R. Gelbaum and John M. H. Olmsted. These counterexamples deal mostly with the part of analysis known as "real variables." The first half covers the real number system, and the second half encompasses higher dimensions. 1962 edition. xxiv+198pp. 5⅜ x 8½. 42875-3

CATASTROPHE THEORY FOR SCIENTISTS AND ENGINEERS, Robert Gilmore. Advanced-level treatment describes mathematics of theory grounded in the work of Poincaré, R. Thom, other mathematicians. Also important applications to problems in mathematics, physics, chemistry, and engineering. 1981 edition. References. 28 tables. 397 black-and-white illustrations. xvii+666pp. 6⅛ x 9¼. 67539-4

INTRODUCTION TO DIFFERENCE EQUATIONS, Samuel Goldberg. Exceptionally clear exposition of important discipline with applications to sociology, psychology, economics. Many illustrative examples; over 250 problems. 260pp. 5⅜ x 8½. 65084-7

NUMERICAL METHODS FOR SCIENTISTS AND ENGINEERS, Richard Hamming. Classic text stresses frequency approach in coverage of algorithms, polynomial approximation, Fourier approximation, exponential approximation, other topics. Revised and enlarged 2nd edition. 721pp. 5⅜ x 8½. 65241-6

INTRODUCTION TO NUMERICAL ANALYSIS (2nd Edition), F. B. Hildebrand. Classic, fundamental treatment covers computation, approximation, interpolation, numerical differentiation and integration, other topics. 150 new problems. 669pp. 5⅜ x 8½. 65363-3

THREE PEARLS OF NUMBER THEORY, A. Y. Khinchin. Three compelling puzzles require proof of a basic law governing the world of numbers. Challenges concern van der Waerden's theorem, the Landau-Schnirelmann hypothesis and Mann's theorem, and a solution to Waring's problem. Solutions included. 64pp. 5⅜ x 8½. 40026-3

THE PHILOSOPHY OF MATHEMATICS: An Introductory Essay, Stephan Körner. Surveys the views of Plato, Aristotle, Leibniz & Kant concerning propositions and theories of applied and pure mathematics. Introduction. Two appendices. Index. 198pp. 5⅜ x 8½. 25048-2

CATALOG OF DOVER BOOKS

INTRODUCTORY REAL ANALYSIS, A.N. Kolmogorov, S. V. Fomin. Translated by Richard A. Silverman. Self-contained, evenly paced introduction to real and functional analysis. Some 350 problems. 403pp. 5⅜ x 8½. 61226-0

APPLIED ANALYSIS, Cornelius Lanczos. Classic work on analysis and design of finite processes for approximating solution of analytical problems. Algebraic equations, matrices, harmonic analysis, quadrature methods, more. 559pp. 5⅜ x 8½. 65656-X

AN INTRODUCTION TO ALGEBRAIC STRUCTURES, Joseph Landin. Superb self-contained text covers "abstract algebra": sets and numbers, theory of groups, theory of rings, much more. Numerous well-chosen examples, exercises. 247pp. 5⅜ x 8½. 65940-2

QUALITATIVE THEORY OF DIFFERENTIAL EQUATIONS, V. V. Nemytskii and V.V. Stepanov. Classic graduate-level text by two prominent Soviet mathematicians covers classical differential equations as well as topological dynamics and ergodic theory. Bibliographies. 523pp. 5⅜ x 8½. 65954-2

THEORY OF MATRICES, Sam Perlis. Outstanding text covering rank, nonsingularity and inverses in connection with the development of canonical matrices under the relation of equivalence, and without the intervention of determinants. Includes exercises. 237pp. 5⅜ x 8½. 66810-X

INTRODUCTION TO ANALYSIS, Maxwell Rosenlicht. Unusually clear, accessible coverage of set theory, real number system, metric spaces, continuous functions, Riemann integration, multiple integrals, more. Wide range of problems. Undergraduate level. Bibliography. 254pp. 5⅜ x 8½. 65038-3

MODERN NONLINEAR EQUATIONS, Thomas L. Saaty. Emphasizes practical solution of problems; covers seven types of equations. ". . . a welcome contribution to the existing literature. . . . "–*Math Reviews*. 490pp. 5⅜ x 8½. 64232-1

MATRICES AND LINEAR ALGEBRA, Hans Schneider and George Phillip Barker. Basic textbook covers theory of matrices and its applications to systems of linear equations and related topics such as determinants, eigenvalues, and differential equations. Numerous exercises. 432pp. 5⅜ x 8½. 66014-1

MATHEMATICS APPLIED TO CONTINUUM MECHANICS, Lee A. Segel. Analyzes models of fluid flow and solid deformation. For upper-level math, science, and engineering students. 608pp. 5⅜ x 8½. 65369-2

ELEMENTS OF REAL ANALYSIS, David A. Sprecher. Classic text covers fundamental concepts, real number system, point sets, functions of a real variable, Fourier series, much more. Over 500 exercises. 352pp. 5⅜ x 8½. 65385-4

SET THEORY AND LOGIC, Robert R. Stoll. Lucid introduction to unified theory of mathematical concepts. Set theory and logic seen as tools for conceptual understanding of real number system. 496pp. 5⅜ x 8¼. 63829-4

CATALOG OF DOVER BOOKS

TENSOR CALCULUS, J.L. Synge and A. Schild. Widely used introductory text covers spaces and tensors, basic operations in Riemannian space, non-Riemannian spaces, etc. 324pp. 5⅜ x 8¼. 63612-7

ORDINARY DIFFERENTIAL EQUATIONS, Morris Tenenbaum and Harry Pollard. Exhaustive survey of ordinary differential equations for undergraduates in mathematics, engineering, science. Thorough analysis of theorems. Diagrams. Bibliography. Index. 818pp. 5⅜ x 8½. 64940-7

INTEGRAL EQUATIONS, F. G. Tricomi. Authoritative, well-written treatment of extremely useful mathematical tool with wide applications. Volterra Equations, Fredholm Equations, much more. Advanced undergraduate to graduate level. Exercises. Bibliography. 238pp. 5⅜ x 8½. 64828-1

FOURIER SERIES, Georgi P. Tolstov. Translated by Richard A. Silverman. A valuable addition to the literature on the subject, moving clearly from subject to subject and theorem to theorem. 107 problems, answers. 336pp. 5⅜ x 8½. 63317-9

INTRODUCTION TO MATHEMATICAL THINKING, Friedrich Waismann. Examinations of arithmetic, geometry, and theory of integers; rational and natural numbers; complete induction; limit and point of accumulation; remarkable curves; complex and hypercomplex numbers, more. 1959 ed. 27 figures. xii+260pp. 5⅜ x 8½. 42804-4

POPULAR LECTURES ON MATHEMATICAL LOGIC, Hao Wang. Noted logician's lucid treatment of historical developments, set theory, model theory, recursion theory and constructivism, proof theory, more. 3 appendixes. Bibliography. 1981 ed. ix+283pp. 5⅜ x 8½. 67632-3

CALCULUS OF VARIATIONS, Robert Weinstock. Basic introduction covering isoperimetric problems, theory of elasticity, quantum mechanics, electrostatics, etc. Exercises throughout. 326pp. 5⅜ x 8½. 63069-2

THE CONTINUUM: A Critical Examination of the Foundation of Analysis, Hermann Weyl. Classic of 20th-century foundational research deals with the conceptual problem posed by the continuum. 156pp. 5⅜ x 8½. 67982-9

CHALLENGING MATHEMATICAL PROBLEMS WITH ELEMENTARY SOLUTIONS, A. M. Yaglom and I. M. Yaglom. Over 170 challenging problems on probability theory, combinatorial analysis, points and lines, topology, convex polygons, many other topics. Solutions. Total of 445pp. 5⅜ x 8½. Two-vol. set.
Vol. I: 65536-9 Vol. II: 65537-7

INTRODUCTION TO PARTIAL DIFFERENTIAL EQUATIONS WITH APPLICATIONS, E. C. Zachmanoglou and Dale W. Thoe. Essentials of partial differential equations applied to common problems in engineering and the physical sciences. Problems and answers. 416pp. 5⅜ x 8½. 65251-3

THE THEORY OF GROUPS, Hans J. Zassenhaus. Well-written graduate-level text acquaints reader with group-theoretic methods and demonstrates their usefulness in mathematics. Axioms, the calculus of complexes, homomorphic mapping, p-group theory, more. 276pp. 5⅜ x 8½. 40922-8

CATALOG OF DOVER BOOKS

Math–Decision Theory, Statistics, Probability

ELEMENTARY DECISION THEORY, Herman Chernoff and Lincoln E. Moses. Clear introduction to statistics and statistical theory covers data processing, probability and random variables, testing hypotheses, much more. Exercises. 364pp. 5⅜ x 8½. 65218-1

STATISTICS MANUAL, Edwin L. Crow et al. Comprehensive, practical collection of classical and modern methods prepared by U.S. Naval Ordnance Test Station. Stress on use. Basics of statistics assumed. 288pp. 5⅜ x 8½. 60599-X

SOME THEORY OF SAMPLING, William Edwards Deming. Analysis of the problems, theory, and design of sampling techniques for social scientists, industrial managers, and others who find statistics important at work. 61 tables. 90 figures. xvii +602pp. 5⅜ x 8½. 64684-X

LINEAR PROGRAMMING AND ECONOMIC ANALYSIS, Robert Dorfman, Paul A. Samuelson and Robert M. Solow. First comprehensive treatment of linear programming in standard economic analysis. Game theory, modern welfare economics, Leontief input-output, more. 525pp. 5⅜ x 8½. 65491-5

PROBABILITY: An Introduction, Samuel Goldberg. Excellent basic text covers set theory, probability theory for finite sample spaces, binomial theorem, much more. 360 problems. Bibliographies. 322pp. 5⅜ x 8½. 65252-1

GAMES AND DECISIONS: Introduction and Critical Survey, R. Duncan Luce and Howard Raiffa. Superb nontechnical introduction to game theory, primarily applied to social sciences. Utility theory, zero-sum games, n-person games, decision-making, much more. Bibliography. 509pp. 5⅜ x 8½. 65943-7

INTRODUCTION TO THE THEORY OF GAMES, J. C. C. McKinsey. This comprehensive overview of the mathematical theory of games illustrates applications to situations involving conflicts of interest, including economic, social, political, and military contexts. Appropriate for advanced undergraduate and graduate courses; advanced calculus a prerequisite. 1952 ed. x+372pp. 5⅜ x 8½. 42811-7

FIFTY CHALLENGING PROBLEMS IN PROBABILITY WITH SOLUTIONS, Frederick Mosteller. Remarkable puzzlers, graded in difficulty, illustrate elementary and advanced aspects of probability. Detailed solutions. 88pp. 5⅜ x 8½. 65355-2

PROBABILITY THEORY: A Concise Course, Y. A. Rozanov. Highly readable, self-contained introduction covers combination of events, dependent events, Bernoulli trials, etc. 148pp. 5⅜ x 8¼. 63544-9

STATISTICAL METHOD FROM THE VIEWPOINT OF QUALITY CONTROL, Walter A. Shewhart. Important text explains regulation of variables, uses of statistical control to achieve quality control in industry, agriculture, other areas. 192pp. 5⅜ x 8½. 65232-7

Math–Geometry and Topology

ELEMENTARY CONCEPTS OF TOPOLOGY, Paul Alexandroff. Elegant, intuitive approach to topology from set-theoretic topology to Betti groups; how concepts of topology are useful in math and physics. 25 figures. 57pp. 5⅜ x 8½. 60747-X

COMBINATORIAL TOPOLOGY, P. S. Alexandrov. Clearly written, well-organized, three-part text begins by dealing with certain classic problems without using the formal techniques of homology theory and advances to the central concept, the Betti groups. Numerous detailed examples. 654pp. 5⅜ x 8½. 40179-0

EXPERIMENTS IN TOPOLOGY, Stephen Barr. Classic, lively explanation of one of the byways of mathematics. Klein bottles, Moebius strips, projective planes, map coloring, problem of the Koenigsberg bridges, much more, described with clarity and wit. 43 figures. 210pp. 5⅜ x 8½. 25933-1

CONFORMAL MAPPING ON RIEMANN SURFACES, Harvey Cohn. Lucid, insightful book presents ideal coverage of subject. 334 exercises make book perfect for self-study. 55 figures. 352pp. 5⅝ x 8¼. 64025-6

THE GEOMETRY OF RENÉ DESCARTES, René Descartes. The great work founded analytical geometry. Original French text, Descartes's own diagrams, together with definitive Smith-Latham translation. 244pp. 5⅜ x 8½. 60068-8

PRACTICAL CONIC SECTIONS: The Geometric Properties of Ellipses, Parabolas and Hyperbolas, J. W. Downs. This text shows how to create ellipses, parabolas, and hyperbolas. It also presents historical background on their ancient origins and describes the reflective properties and roles of curves in design applications. 1993 ed. 98 figures. xii+100pp. 6½ x 9¼. 42876-1

THE THIRTEEN BOOKS OF EUCLID'S ELEMENTS, translated with introduction and commentary by Thomas L. Heath. Definitive edition. Textual and linguistic notes, mathematical analysis. 2,500 years of critical commentary. Unabridged. 1,414pp. 5⅜ x 8½.. Three-vol. set. Vol. I: 60088-2 Vol. II: 60089-0 Vol. III: 60090-4

GEOMETRY OF COMPLEX NUMBERS, Hans Schwerdtfeger. Illuminating, widely praised book on analytic geometry of circles, the Moebius transformation, and two-dimensional non-Euclidean geometries. 200pp. 5⅝ x 8¼. 63830-8

DIFFERENTIAL GEOMETRY, Heinrich W. Guggenheimer. Local differential geometry as an application of advanced calculus and linear algebra. Curvature, transformation groups, surfaces, more. Exercises. 62 figures. 378pp. 5⅜ x 8½. 63433-7

CURVATURE AND HOMOLOGY: Enlarged Edition, Samuel I. Goldberg. Revised edition examines topology of differentiable manifolds; curvature, homology of Riemannian manifolds; compact Lie groups; complex manifolds; curvature, homology of Kaehler manifolds. New Preface. Four new appendixes. 416pp. 5⅜ x 8½. 40207-X

History of Math

THE WORKS OF ARCHIMEDES, Archimedes (T. L. Heath, ed.). Topics include the famous problems of the ratio of the areas of a cylinder and an inscribed sphere; the measurement of a circle; the properties of conoids, spheroids, and spirals; and the quadrature of the parabola. Informative introduction. clxxxvi+326pp; supplement, 52pp. 5⅜ x 8½. 42084-1

A SHORT ACCOUNT OF THE HISTORY OF MATHEMATICS, W. W. Rouse Ball. One of clearest, most authoritative surveys from the Egyptians and Phoenicians through 19th-century figures such as Grassman, Galois, Riemann. Fourth edition. 522pp. 5⅜ x 8½. 20630-0

THE HISTORY OF THE CALCULUS AND ITS CONCEPTUAL DEVELOPMENT, Carl B. Boyer. Origins in antiquity, medieval contributions, work of Newton, Leibniz, rigorous formulation. Treatment is verbal. 346pp. 5⅜ x 8½. 60509-4

THE HISTORICAL ROOTS OF ELEMENTARY MATHEMATICS, Lucas N. H. Bunt, Phillip S. Jones, and Jack D. Bedient. Fundamental underpinnings of modern arithmetic, algebra, geometry, and number systems derived from ancient civilizations. 320pp. 5⅜ x 8½. 25563-8

A HISTORY OF MATHEMATICAL NOTATIONS, Florian Cajori. This classic study notes the first appearance of a mathematical symbol and its origin, the competition it encountered, its spread among writers in different countries, its rise to popularity, its eventual decline or ultimate survival. Original 1929 two-volume edition presented here in one volume. xxviii+820pp. 5⅜ x 8½. 67766-4

GAMES, GODS & GAMBLING: A History of Probability and Statistical Ideas, F. N. David. Episodes from the lives of Galileo, Fermat, Pascal, and others illustrate this fascinating account of the roots of mathematics. Features thought-provoking references to classics, archaeology, biography, poetry. 1962 edition. 304pp. 5⅜ x 8½. (Available in U.S. only.) 40023-9

OF MEN AND NUMBERS: The Story of the Great Mathematicians, Jane Muir. Fascinating accounts of the lives and accomplishments of history's greatest mathematical minds–Pythagoras, Descartes, Euler, Pascal, Cantor, many more. Anecdotal, illuminating. 30 diagrams. Bibliography. 256pp. 5⅜ x 8½. 28973-7

HISTORY OF MATHEMATICS, David E. Smith. Nontechnical survey from ancient Greece and Orient to late 19th century; evolution of arithmetic, geometry, trigonometry, calculating devices, algebra, the calculus. 362 illustrations. 1,355pp. 5⅜ x 8½. Two-vol. set. Vol. I: 20429-4 Vol. II: 20430-8

A CONCISE HISTORY OF MATHEMATICS, Dirk J. Struik. The best brief history of mathematics. Stresses origins and covers every major figure from ancient Near East to 19th century. 41 illustrations. 195pp. 5⅜ x 8½. 60255-9

CATALOG OF DOVER BOOKS

Physics

OPTICAL RESONANCE AND TWO-LEVEL ATOMS, L. Allen and J. H. Eberly. Clear, comprehensive introduction to basic principles behind all quantum optical resonance phenomena. 53 illustrations. Preface. Index. 256pp. 5⅜ x 8½. 65533-4

QUANTUM THEORY, David Bohm. This advanced undergraduate-level text presents the quantum theory in terms of qualitative and imaginative concepts, followed by specific applications worked out in mathematical detail. Preface. Index. 655pp. 5⅜ x 8½. 65969-0

ATOMIC PHYSICS: 8th edition, Max Born. Nobel laureate's lucid treatment of kinetic theory of gases, elementary particles, nuclear atom, wave-corpuscles, atomic structure and spectral lines, much more. Over 40 appendices, bibliography. 495pp. 5⅜ x 8½. 65984-4

A SOPHISTICATE'S PRIMER OF RELATIVITY, P. W. Bridgman. Geared toward readers already acquainted with special relativity, this book transcends the view of theory as a working tool to answer natural questions: What is a frame of reference? What is a "law of nature"? What is the role of the "observer"? Extensive treatment, written in terms accessible to those without a scientific background. 1983 ed. xlviii+172pp. 5⅜ x 8½. 42549-5

AN INTRODUCTION TO HAMILTONIAN OPTICS, H. A. Buchdahl. Detailed account of the Hamiltonian treatment of aberration theory in geometrical optics. Many classes of optical systems defined in terms of the symmetries they possess. Problems with detailed solutions. 1970 edition. xv+360pp. 5⅜ x 8½. 67597-1

PRIMER OF QUANTUM MECHANICS, Marvin Chester. Introductory text examines the classical quantum bead on a track: its state and representations; operator eigenvalues; harmonic oscillator and bound bead in a symmetric force field; and bead in a spherical shell. Other topics include spin, matrices, and the structure of quantum mechanics; the simplest atom; indistinguishable particles; and stationary-state perturbation theory. 1992 ed. xiv+314pp. 6⅛ x 9¼. 42878-8

LECTURES ON QUANTUM MECHANICS, Paul A. M. Dirac. Four concise, brilliant lectures on mathematical methods in quantum mechanics from Nobel Prize–winning quantum pioneer build on idea of visualizing quantum theory through the use of classical mechanics. 96pp. 5⅜ x 8½. 41713-1

THIRTY YEARS THAT SHOOK PHYSICS: The Story of Quantum Theory, George Gamow. Lucid, accessible introduction to influential theory of energy and matter. Careful explanations of Dirac's anti-particles, Bohr's model of the atom, much more. 12 plates. Numerous drawings. 240pp. 5⅜ x 8½. 24895-X

ELECTRONIC STRUCTURE AND THE PROPERTIES OF SOLIDS: The Physics of the Chemical Bond, Walter A. Harrison. Innovative text offers basic understanding of the electronic structure of covalent and ionic solids, simple metals, transition metals and their compounds. Problems. 1980 edition. 582pp. 6⅛ x 9¼. 66021-4

CATALOG OF DOVER BOOKS

HYDRODYNAMIC AND HYDROMAGNETIC STABILITY, S. Chandrasekhar. Lucid examination of the Rayleigh-Benard problem; clear coverage of the theory of instabilities causing convection. 704pp. 5⅜ x 8¼. 64071-X

INVESTIGATIONS ON THE THEORY OF THE BROWNIAN MOVEMENT, Albert Einstein. Five papers (1905–8) investigating dynamics of Brownian motion and evolving elementary theory. Notes by R. Fürth. 122pp. 5⅜ x 8½. 60304-0

THE PHYSICS OF WAVES, William C. Elmore and Mark A. Heald. Unique overview of classical wave theory. Acoustics, optics, electromagnetic radiation, more. Ideal as classroom text or for self-study. Problems. 477pp. 5⅜ x 8½. 64926-1

PHYSICAL PRINCIPLES OF THE QUANTUM THEORY, Werner Heisenberg. Nobel Laureate discusses quantum theory, uncertainty, wave mechanics, work of Dirac, Schroedinger, Compton, Wilson, Einstein, etc. 184pp. 5⅜ x 8½. 60113-7

ATOMIC SPECTRA AND ATOMIC STRUCTURE, Gerhard Herzberg. One of best introductions; especially for specialist in other fields. Treatment is physical rather than mathematical. 80 illustrations. 257pp. 5⅜ x 8½. 60115-3

AN INTRODUCTION TO STATISTICAL THERMODYNAMICS, Terrell L. Hill. Excellent basic text offers wide-ranging coverage of quantum statistical mechanics, systems of interacting molecules, quantum statistics, more. 523pp. 5⅜ x 8½. 65242-4

THEORETICAL PHYSICS, Georg Joos, with Ira M. Freeman. Classic overview covers essential math, mechanics, electromagnetic theory, thermodynamics, quantum mechanics, nuclear physics, other topics. xxiii+885pp. 5⅜ x 8½. 65227-0

PROBLEMS AND SOLUTIONS IN QUANTUM CHEMISTRY AND PHYSICS, Charles S. Johnson, Jr. and Lee G. Pedersen. Unusually varied problems, detailed solutions in coverage of quantum mechanics, wave mechanics, angular momentum, molecular spectroscopy, more. 280 problems, 139 supplementary exercises. 430pp. 6½ x 9¼. 65236-X

THEORETICAL SOLID STATE PHYSICS, Vol. I: Perfect Lattices in Equilibrium; Vol. II: Non-Equilibrium and Disorder, William Jones and Norman H. March. Monumental reference work covers fundamental theory of equilibrium properties of perfect crystalline solids, non-equilibrium properties, defects and disordered systems. Total of 1,301pp. 5⅜ x 8½. Vol. I: 65015-4 Vol. II: 65016-2

WHAT IS RELATIVITY? L. D. Landau and G. B. Rumer. Written by a Nobel Prize physicist and his distinguished colleague, this compelling book explains the special theory of relativity to readers with no scientific background, using such familiar objects as trains, rulers, and clocks. 1960 ed. vi+72pp. 23 b/w illustrations. 5⅜ x 8½.
42806-0 $6.95

A TREATISE ON ELECTRICITY AND MAGNETISM, James Clerk Maxwell. Important foundation work of modern physics. Brings to final form Maxwell's theory of electromagnetism and rigorously derives his general equations of field theory. 1,084pp. 5⅜ x 8½. Two-vol. set. Vol. I: 60636-8 Vol. II: 60637-6

CATALOG OF DOVER BOOKS

QUANTUM MECHANICS: Principles and Formalism, Roy McWeeny. Graduate student–oriented volume develops subject as fundamental discipline, opening with review of origins of Schrödinger's equations and vector spaces. Focusing on main principles of quantum mechanics and their immediate consequences, it concludes with final generalizations covering alternative "languages" or representations. 1972 ed. 15 figures. xi+155pp. 5⅜ x 8½. 42829-X

INTRODUCTION TO QUANTUM MECHANICS WITH APPLICATIONS TO CHEMISTRY, Linus Pauling & E. Bright Wilson, Jr. Classic undergraduate text by Nobel Prize winner applies quantum mechanics to chemical and physical problems. Numerous tables and figures enhance the text. Chapter bibliographies. Appendices. Index. 468pp. 5⅜ x 8½. 64871-0

METHODS OF THERMODYNAMICS, Howard Reiss. Outstanding text focuses on physical technique of thermodynamics, typical problem areas of understanding, and significance and use of thermodynamic potential. 1965 edition. 238pp. 5⅜ x 8½. 69445-3

TENSOR ANALYSIS FOR PHYSICISTS, J. A. Schouten. Concise exposition of the mathematical basis of tensor analysis, integrated with well-chosen physical examples of the theory. Exercises. Index. Bibliography. 289pp. 5⅜ x 8½. 65582-2

THE ELECTROMAGNETIC FIELD, Albert Shadowitz. Comprehensive undergraduate text covers basics of electric and magnetic fields, builds up to electromagnetic theory. Also related topics, including relativity. Over 900 problems. 768pp. 5⅝ x 8¼. 65660-8

GREAT EXPERIMENTS IN PHYSICS: Firsthand Accounts from Galileo to Einstein, Morris H. Shamos (ed.). 25 crucial discoveries: Newton's laws of motion, Chadwick's study of the neutron, Hertz on electromagnetic waves, more. Original accounts clearly annotated. 370pp. 5⅜ x 8½. 25346-5

RELATIVITY, THERMODYNAMICS AND COSMOLOGY, Richard C. Tolman. Landmark study extends thermodynamics to special, general relativity; also applications of relativistic mechanics, thermodynamics to cosmological models. 501pp. 5⅜ x 8½. 65383-8

STATISTICAL PHYSICS, Gregory H. Wannier. Classic text combines thermodynamics, statistical mechanics, and kinetic theory in one unified presentation of thermal physics. Problems with solutions. Bibliography. 532pp. 5⅜ x 8½. 65401-X

Paperbound unless otherwise indicated. Available at your book dealer, online at **www.doverpublications.com**, or by writing to Dept. GI, Dover Publications, Inc., 31 East 2nd Street, Mineola, NY 11501. For current price information or for free catalogs (please indicate field of interest), write to Dover Publications or log on to **www.doverpublications.com** and see every Dover book in print. Dover publishes more than 500 books each year on science, elementary and advanced mathematics, biology, music, art, literary history, social sciences, and other areas.